Astrophysics and Space Science Library

Volume 459

The Astrophysics and Space Science Library is a series of high-level monographs and edited volumes covering a broad range of subjects in Astrophysics, Astronomy, Cosmology, and Space Science. The authors are distinguished specialists with international reputations in their fields of expertise. Each title is carefully supervised and aims to provide an in-depth understanding by offering detailed background and the results of state-of-the-art research. The subjects are placed in the broader context of related disciplines such as Engineering, Computer Science, Environmental Science, and Nuclear and Particle Physics. The ASSL series offers a reliable resource for scientific professional researchers and advanced graduate students.

More information about this series at http://www.springer.com/series/5664

Pieter C. van der Kruit

Jan Hendrik Oort

Master of the Galactic System

Springer

Pieter C. van der Kruit
Kapteyn Astronomical Institute
University of Groningen
Groningen, The Netherlands

ISSN 0067-0057 ISSN 2214-7985 (electronic)
Astrophysics and Space Science Library
ISBN 978-3-030-17800-0 ISBN 978-3-030-17801-7 (eBook)
https://doi.org/10.1007/978-3-030-17801-7

Cover illustration: Oort in 1955, From the Oort Archives

This Springer imprint is published by the registered company Springer Nature Switzerland AG
The registered company address is: Gewerbestrasse 11, 6330 Cham, Switzerland

Jan Hendrik Oort in 1973. From the Oort Archives

Jan Hendrik Oort in 1975. From the Oort Archives

This book is dedicated to

the memory of
***Jacobus Cornelius Kapteyn** (1851–1922),*
whom Oort described as
'mijn inspireerenden leermeester',[1]

and to

my wife Corry
for her love, understanding and support.

[1] My inspiring mentor, Oort (1936b, p. 4).

From a Letter from British Astronomer Roger John Taylor (1929–1997) to Jet Katgert-Merkelijn:

'... this arose through a 'Balloon debate'. These astronomers were supposed to be in a balloon—Hoyle, Lovell and Oort—but there was only enough gas to carry one, so two had to be thrown out. Speakers on behalf of each argued their case and then the audience (the University astronomy society) voted. I took the part of Oort. I am glad to say that he stayed in the balloon!

This happened in 1984.'

1 My name sir is Jan Hendrik Oort.
I'm well known where astronomy's taught.
I've a special notation
For Galactic rotation.
Those A and B constants of Oort.

2 The changed motion of stars with disc height
Has shown that there's mass without light.
Whether faint star or planet
Or black hole or granite
The Oort limit is certainly right.

3 A balloon requires plenty of gas
Much of which used to be hidden mass,
Until van de Hulst showed
That the Galaxy glowed
With a radio line from the cold gas.

4 His researches on the radio line
Were inspired by an idea of mine.
Now when studied with care
In galaxies far, their
Rotation it serves to define.

5 It is generally counted a riddle
What occurs in our Galaxy's middle.
But my Ann. Rev. review
Is top reading for you,
If you hope to decipher the riddle.

6 Superclusters are formed at high z(ed).
Before quasars appear they look dead.
Although I suggested
H may be detected,
In a paper that you've not yet read.

7 If a comet goes close to the Sun,
That orbit may be its last one.
But do not despair,
There is plenty more there
In the Oort cloud so far from the Sun.

8 The night skies in Europe are grey,
But radio waves come all day.
So I urged the creation
In our sea level nation
Of Westerbork's mighty array.

9 At the same time for working with light,
It is clear that a mountain is right.
So Frenchmen and Dutchmen
And Germans and such men
Put our telescopes on Chilean height.

10 Though my decades now number eight,
My knowledge is not out of date.
I still work with relish
For it's publish or perish.
In your hands dear sir is my fate.

Preface

I tried to imagine the situation of a historian of natural sciences, working some time in the twenty-first century on a monograph describing Jan Oort's scientific importance. If successful, this historian would have covered in his monograph a very substantial portion of the history of astronomy in our century.

Bengt Georg Daniel Strömgren (1908–1987)[2]

During 2015, when my biography of Jacobus C. Kapteyn had appeared in the Astrophysics and Space Science Library, colleagues asked me what my next project was going to be. Often the suggestion was made that this would have to be a similar biography of Jan Hendrik Oort. I was hesitant since I had known Oort well. But the idea gradually became more and more appealing and late in 2015 I started to form an idea of what would be involved; the task seemed overwhelming.

One of the first things I did was to have a talk with Jet Katgert-Merkelijn, who had made an extensive inventory of Oort's archives and next I had some initial email exchanges with Marc Oort, who is a grandson of Oort and holds a Ph.D. in astronomy. I decided to proceed; I estimated the project would take four years and that 90 days were required at the University Library Leiden, where the Oort Archives are kept. I went on to secure some funding that would be required to make the many visits and quickly obtained approval or positive response. The Kapteyn Astronomical Institute in Groningen, which is my home base, also supported the project. Marc Oort consulted his uncle Abraham Hans Oort, younger son of Jan Hendrik Oort, and arranged permission for me on behalf of the Oort family to consult the confidential parts of the Oort Archives.

I was at that time six years into retirement and had only few time-consuming duties or commitments left. I wrote the first draft of the first sentences of the first chapter at home on 5 January 2016, when the northern Netherlands were hit by a severe case of freezing rain and glazed frost that disrupted all traffic and forced the University to completely shut down. People skated in the streets. I was unable to

[2]In the Liber Amicorum written for Oort's eightieth birthday (*LibAm80* [2]).

leave my house for three days and that made for a good start. When the weather had improved, I made my first visit to the Oort Archives in Leiden. In the end I spent about 75 days in the University Library in Leiden. This was facilitated by having support to spend a similar number of nights in the van de Valk hotel Leiden, where I—and my wife who accompanied me on about two-thirds of the visits—became a regular customer. To my surprise it took me only three years to complete the manuscript. A major factor is the excellent inventory of the Archives that Jet Katgert has provided.

I was fortunate to be able to visit the Carnegie Observatories in Pasadena, California, in 2018, for a period of more than five weeks, during which I examined the archives of various astronomers Oort corresponded with at the Huntington Library and at the California Institute of Technology, and had a enlightening interview with Maarten Schmidt. I also visited the Oort Archives at the European Southern Observatory in Garching, Germany.

Sources. Publications on Oort frequently referred to or matters closely related to him have been given a special designation to be used throughout this book. For example, Jet Katgert's inventory of the Oort Archives is designated as the *JKM-Inventory* [1], and my Kapteyn biography as the *JCKbiog*. In Appendix C.1., these have been listed with reference to the complete bibliographical information.

There are two types of **diaries** in Oort's private archives. One type has the size of notebooks (on the order of 20 by 15 cm). These are catalogue number 9.3.1 in the *JKM-Inventory* [1], running from 1915 to 1960, with 1946 missing (but sometimes there are very long intervals between entries). I will refer to these as the *diary notebooks*. The other type consists of small appointment books that in addition to appointments often contain detailed notes on activities and impressions. These run from 1923 to 1991, again with some years missing, and are catalogue number 9.3.2 in the *JKM-Inventory* [1]. I will designate these as *pocket diaries*.

Like for the Kapteyn volume, this book is also accompanied by a **dedicated Website**: www.astro.rug.nl/JHOort. This not only provides links to publications by Oort, but also to publications about him. In addition, I give access to much of the public material I used. My modus operandi at the Oort Archives has been to make photographs on my tablet (an iPad) of all things I judged to be useful to have available. These are high-resolution pictures, and there are of order 30,000 of them. I collected these in separate directories for each item number in the *JKM-Inventory* [1]. I compressed all pictures by a factor 2 by 2, ending up with four times fewer pixels, and then turned the result into pdf files, again by item number. These are also made available on the website, with permission of the Leiden University Library. Of course, this does *not* include the private archives.

Jet Katgert-Merkelijn has provided a web page at the site of Leiden Observatory, which shows a number of high-quality scans of (mostly group-) photographs in which Oort appears: www.strw.leidenuniv.nl/oortfotos/.

The **audience** I had in mind while writing this book broadly speaking includes those interested in the history of astronomy that have a background in natural science (so not necessarily astronomers) or historians specialized in astronomical history.

I therefore have added **explanatory material** about fundamental astronomy and on Kapteyn; most of that has been adapted from my biography of Kapteyn *JCKbiog* [7]. More technical explanations that often require some mathematics have been put in separate boxes that can be ignored by those not interested in such details.

I will refrain from using **nicknames**. As a student of Oort, I addressed him as 'Professor' and he me as 'van der Kruit', but immediately after I had obtained my Ph.D. as 'Piet'. In the eighties, when I had become a senior staff member in Groningen, he and his wife proposed that I use their first names, but I have hardly done so. In this book, they will be 'Oort' and 'Mrs. Oort'. I will use first names when they are commonly used. So, for example, Oort's older brother Henricus and younger brother Arend Joan Petrus I will call 'Hein' and 'John', respectively, but I will not use 'Jo' for the latter as he was known by his family members and intimate friends. Oort's youngest son, Abraham Hans, who has been a great help, will be designated as 'Bram Oort'.

Genealogy: Sites with genealogical information and official records of certificates of births, marriages and deaths, have been listed in references [10–14].

Currency calculators: The sites used can be found in references [15–17].

Copyrights paintings of Oort: There are four paintings of Oort, all in the possession of Leiden University. Descendants of the painter may exercise rights on reproduction of these. In order to find these, I have contacted 'Pictoright' [18]; these only administered the rights of D. H. Nijland (Fig. 8) and arranged the fee and the permission. I also contacted 'Catawiki' [19], 'Schilderijen-site' [20], 'Historische Vereniging Oud Leiden' [21], and a number of galleries and auction houses that had recently been involved in the sales of other paintings from the same artists. Usually, they did not have this information and were not allowed to provide this anyway because of privacy regulations. I gave up in the cases P. Herfst (Fig. 12) and R. Martinez (Fig. 14). For G. de Wit (Fig. 15), Mrs. Catherine Lenssinck-Laimböck of Galerie Laimböck was able to direct me to Mrs. Asja de Wit, daughter of G. de Wit, who gave her consent to include a reproduction in this book.

It was a special privilege to be in the position to write a biography of Jan Hendrik Oort, since I am fortunate to have known him over a long period of time. I will document this briefly. As a student at Leiden University I attended two of his lecture courses, one on the solar system in the academic year 1964–65 and one on stellar dynamics in 1966–67. Since my Candidaats exam (roughly Bachelors) in 1966 I worked daily at the Sterrewacht, while Oort was director. My research projects as requirements for my Doctoral exam (roughly Masters) in 1968 were both done under his supervision. These concerned the distribution of high-velocity clouds on the sky in the model in which these resulted from an inflow of intergalactic gas (small project) and a survey of neutral hydrogen in a region on the sky around the Galactic Center, using the Dwingeloo Radio Telescope (large project). The latter grew eventually into my Ph.D. thesis which I defended in 1971, with Oort as 'promotor' (supervisor). From 1970 onward, I worked with the newly completed Westerbork Synthesis Telescope on nearby galaxies and Oort took an active interest in this and became in one case a co-author. He wrote an extremely positive recommendation for me to be awarded a Carnegie Fellowship at the Mount

Fig. 1 Jan Hendrik Oort and myself after my inaugural lecture as Professor of astronomy at the University of Groningen in January 1988. From the author's collection

Wilson and Palomar Observatories, Pasadena, California, to which I moved in August 1972. After my return to the Netherlands in January 1975, I worked at the Kapteyn Astronomical Institute at the University of Groningen, where I was appointed full professor in 1987. Throughout all this period, Oort kept a warm interest in my work. In January 1988, he attended my inaugural lecture (see Fig. 1). Oort was a source of personal inspiration and a warm human being, inspiring professor and colleague; I have tried to write this book in the most objective manner that this relationship allows.

I hope both this book and the accompanying website will enhance the appreciation for the work and person of Jan Hendrik Oort. I have dedicated this book in the first place to the memory of Jacobus Cornelius Kapteyn; after all Jan Hendrik Oort is a major part of Kapteyn's legacy.[3]

Groningen, Assen, Leiden, The Netherlands — Pieter C. van der Kruit
Pasadena, CA, USA
January 2016–February 2019

[3]This book is accompanied by the website www.astro.rug.nl/JHOort.

Acknowledgements

I am grateful to many individuals for help, advice and assistance and institutions for support. In no special order, except that it is more or less chronological, I like to acknowledge the following.

Jet Katgert-Merkelijn, for her advice, for providing a large directory with scanned photographs, and for answering many questions. She also provided me with recordings of interviews she had held in the nineties with Adriaan Blaauw and Henk van de Hulst. Without all the work Jet did organizing the Oort Archives and providing a detailed description of it, the writing of this book would have taken at least twice as long, probably even more. Marc Oort, grandson of Oort, who holds a Ph.D. in astronomy, arranged my access to the private parts of the Oort Archives and answered many questions. Bram Oort, son of Oort, gave the permission and provided me with photographs and copies of letters and speeches in the family and always promptly answered many questions; I am grateful for the trust he put in me.

The department Special Collections of the University Library Leiden, where I worked for about 75 days examining the Oort Archives; the support of the staff was superb! Per Olof Lindblad sent some correspondence between his father Bertil Lindblad and Oort. Jornaroos (Jetske) Hamaker-Nauta, niece of Oort, provided important information on Oort's parents. Henk-Jan Kooke, audiovisual department of the University of Groningen, for copying Jet Katgert's interviews from audio cassette tapes to mp3 files. Gineke Alberts transcribed these files. The library of the University of Groningen (both the central location and the branch at the Zernike Campus) provided access to publications. Jacqueline Strijker-de Vries of this library located old volumes of Hemel and Dampkring. Willem Otterspeer of Leiden University provided information on Leiden University during World War II. Uta Grothkopf, librarian and custodian of the historical archives at ESO Garching, helped me accessing the Oort Archives there. Mrs. Joke Wilkens-van Herk, for information on her father Gijsbert van Herk. Gail Bieger-Smith agreed to a telephone interview on her life as a Leiden student and on her experiences with Oort. Bruce Elmegreen for the photograph of Slipher's spectrum in Fig. 7.10 and Deidre Hunter of Lowell Observatory for arranging permission to use it. Ard Hartsuijker of

CAMRAS, the A.C. Muller Radio Astronomy Station, operating the Dwingeloo Telescope for amateur use, for providing old photographs of the Kootwijk and Dwingeloo dishes. Miller Goss (National Radio Astronomy Observatory, Socorro, New Mexico) and Ron Ekers, Dick Manchester, John Whiteoak and Diet Ostry, radio astronomers/technicians from Australia (Radiophysics, Commonwealth Scientific and Industrial Research Organisation, Sydney) for help with information on Eric R. Hill. Mrs. Catherine Lenssinck-Laimbock and Mrs. Asja de Wit arranged permission to include Fig. 14.

I thank John C. Mulchaey, director of the Carnegie Observatories, Pasadena, California, for hospitality. Dan Lewis of the Huntington Library, Art Collections and Botanical Gardens, and Loma Karklins, CalTech Archives and Special Collections, for access to the archives of a number of astronomy correspondents of Oort. Maarten Schmidt, Hugo van Woerden, Butler Burton, Harry van der Laan, Jet Katgert and Whitley Shane for allowing interviews (in written form by Whitney) on Oort, and Petra Huizinga for making transcripts of these. Lo Woltjer's health problems prevented me from interviewing him.

And I especially thank my 'proofreaders' ('proof' as the Dutch 'proef' in 'wijnproeven', wine-tasting) for reading large parts or in most cases all of the manuscript and giving detailed criticisms and suggestions. They are Bram Oort, son of Jan Hendrik Oort, Marc Oort, grandson of Jan Hendrik Oort and astronomer, Jet Katgert-Merkelijn and Butler Burton, Ph.D. students of Oort's (both Jet and Butler carefully checked my English and made many important remarks), Jan Willem Pel, long-time colleague astronomer in Groningen and contemporary student in Leiden for insightful remarks on astronomical matters, Klaas van Berkel and David Baneke, both historians supplementing my limited historical knowledge, Albert Jan Scheffer, biochemist and virologist, Ton Schoot Uiterkamp, chemist and environmental scientist, and my wife Corry.

All responsibility for the text remains with me and I am the only one that should be blamed for any mistakes or incorrect statements in this book.

I am grateful for the support and help from my editor at Springer, Ramon Khanna.

I had financial support from the following organizations: the Netherlands Research School for Astronomy (NOVA), the Netherlands Organization for Scientific Research—Exact Sciences (NWO/EW), the Jan Hendrik Oort Fonds and the Leids Sterrewacht Fonds. When the University of Groningen appointed me to the distinguished Jacobus C. Kapteyn chair in Astronomy in 2003, the Faculty of Mathematics and Natural Sciences provided me with an annual grant until my formal retirement in 2009; what was still left over has been added to the budget of this project. The Kapteyn Astronomical Institute provided me with working space and support; I thank the directors, Reynier Peletier and Scott Trager, for their hospitality, and Lucia van der Voort, Christa Meijerink, Martine ter Wal, Ramona Thijssens, Martin Vogelaar, Wim Zwitser, Eite Tiesinga, Valentin Bajrami and Leon Doddema, for administrative, secretarial and computer support.

This book is dedicated to the memory of Jacobus Cornelius Kapteyn, whose fame made Oort come to study in Groningen and who inspired him to become an astronomer.

I thank my wife Corry for her love and support; this book is also dedicated to her.

Contents

Chapter 1
Growing Up in Oegstgeest

There is no elevator to success. You have to take the stairs.
Anonymous.
There are no parents that do not secretly see their children as special.
And they are right, the possibilities are unlimited.
Godfried Bomans (1913–1971)

In 1981, Jan Hendrik Oort, then at age 81, published an invited autobiographical essay as introductory article in the 'Annual Review of Astronomy & Astrophysics', *AnnRev81* [3], with the title *Some notes on my life as an astronomer* (Oort 1981c). It ended as follows:

> 'A Dutch historian recently wrote a book on her life called *Omzien in Verwondering* (Looking Back in Astonishment). My own life has been, and still is, one of marveling about what lies ahead. If I were to write a book on it, I would rather call it *Looking Ahead in Wonder*.'

The Dutch historian and writer referred to is Annie Romein-Verschoor (1895–1978), who is also well-known for being an active protagonist of woman rights and for having definite Marxist sympathies; it is very unlikely though that her views had anything to do with Oort's choice for this quote. Indeed looking ahead was Oort's attitude in life and it does therefore not come as a surprise that an autobiography was never written. Although 'Looking ahead in wonder' would constitute a fitting subtitle to this book, it also would in my view not be appropriate.

In an 'In Memoriam' article published shortly after Oort's death, his longtime friend and colleague Hendrik Christoffel (Henk) van de Hulst [22] cited the reaction of American (and Indian born) astrophysicist Subrahmanyan Chandrasekhar (1910–

Godfried Bomans was a Dutch writer and television personality.

P. C. van der Kruit, *Jan Hendrik Oort*, Astrophysics and Space Science Library 459, https://doi.org/10.1007/978-3-030-17801-7_1

1995), who had received half of the Nobel Prize for physics in 1983 for his work on the theory of the structure and evolution of stars, but who had among other contributions to astrophysics also significantly extended the field of stellar dynamics initiated to a large extent by Oort. Chandrasekhar noted: 'The great oak of astronomy has been felled, and we are lost without its shadow'. This great tribute has been quoted too many times ti use it as a subtitle.

I chose *Master of the Galactic System* as the subtitle of this book, derived from the remark:

> 'In spite of the clouds of smoke, both literal and metaphorical, which obscure its form, the Galactic System seems at length to have met its master in the thoroughness of our Medalist, working quietly and thoughtfully in his room at Leiden',

made on the occasion of the presentation of the Gold Medal of the Royal Astronomical Society to Oort by Harry Hemley Plaskett (1893–1980), President, on May 10, 1946 [23]. After all, his fundamental work on the structure and dynamics of the Milky Way Galaxy is central to his many contributions to astronomy.

A biography of Jan Hendrik Oort will necessarily touch on almost every aspect of the development and organization of astronomy throughout the twentieth century, as Bengt Strömgren noted in the quote at the start of the Preface. After all, Oort's interests and contributions do branch out from his seminal work on the dynamics of the Galaxy into all aspects of modern astronomy and astrophysics, ranging from the physics of the interstellar medium and the study of the formation and evolution of stars, to the structure of external galaxies and of the Universe on large scales and in earlier times, and on smaller scales to the origin of comets and the evolution of the Solar System.

Keeping in mind we need to cover almost all developments and discoveries in astronomy during the twentieth century it is appropriate that this biography of Jan Hendrik Oort starts in the last year of the nineteenth century, the year of Oort's birth. For this we need to focus on a small town in the northern part of the Netherlands. The map in Fig. 1.1 shows the outline of the country in more recent times with pointers at the places where Oort lived and at other cities that play a role in this narrative.

1.1 Franeker

The northern province of Friesland, or Fryslân in the local language, is situated in the north of the Netherlands to the west of the province of Groningen. It includes the islands to the north except the most western and largest one, which is Texel, and Rottum, a group of three tiny, uninhabited islands in the east. The local population speaks the distinct language Fries, that relates somewhat to English. One of the best-known Frisian claims to fame is the 'Eleven Cities Tour', a skating tour (both a competitive race and 'recreational' tour) of almost 200 km, that is only held when the ice is good enough to hold the thousands of competitors and tourists [24]. It starts at the capital Leeuwarden in the center and leads past all the historical Frisian towns with medieval city rights (see [25]) before returning to Leeuwarden.

Fig. 1.1 Map of the Netherlands with places where Oort lived and worked (capitals) and other locations with two-letter codes for reference: F = Franeker, L = Leiden, G = Groningen, H = Haamstede, K = Katwijk aan zee, Ko = Kootwijk, N = Nunspeet/Hulshorst, O = Oegstgeest, W = Wassenaar, We = Westerbork. al = Alkmaar, am = Amsterdam, an = Antwerpen, ar = Arnhem, br = Breda, do = Dordrecht, dh = Den Haag, dv = Deventer, eh = Eindhoven, en = Enschede, ha = Haarlem, hd = Harderwijk, hr = Harlingen, le = Leeuwarden, ma = Maastricht, nij = Nijmegen, ro = Rotterdam, ut = Utrecht, vl = Vlissingen, ve = Velzen, wa = Wageningen, zu = Zutphen, zw = Zwolle. In 1900, when Oort was born, some features did not exist, notably the polders of Flevoland in the center; this was the *IJsselmeer*, which was the *Zuiderzee* until it was closed by the *Afsluitdijk* (Closure Dam) in 1932

Oort was not of Frisian descent even though both he and his father were born in Friesland, in two of the eleven towns, his father in Harlingen and Oort in Franeker, as the crow flies only 9 km apart (see Fig. 1.1). Oort may not have been Frisian, he was a avid skater and participated in the Eleven Cities tour more than once. He skated up to high age; in 1979, at age 78 he took a bad fall and finally gave up skating long distances (see later).

Franeker received city rights in 1374 and indeed does have a rich history, actually having been the seat of the second oldest university in the Netherlands. The University was founded in 1585, only ten years after the one in Leiden. In the revolt against Habsburg rule, Franeker very early on chose the side of William I, Prince of Orange (1533–1584), also known as William the Silent, who led the revolt and in the Netherlands is seen as the 'Father of the Fatherland'. Each of the seven provinces of the Republic of the Seven United Netherlands had the right to open a University and the reason that a University in Friesland was established in Franeker rather than any of the other cities, derives from this early commitment to the cause of the Netherlands. It appears that there were four main reasons for the province to wish to host a Frisian Academy. It would be cheaper than studying in Leiden, the parents were relatively nearby and could more easily keep an eye on their children, it would be profitable for the Frisian population and the money spent on education by local families that could afford it, remained within Friesland.

At first the University prospered. It offered university education in theology, law, medicine, the humanities, philosophy and mathematics, and natural science. It boasted some very famous professors, such as Petrus Camper (1722–1789), described as physician, anatomist, physiologist, zoologist, anthropologist, paleontologist, and naturalist. Camper occupied the chair in philosophy between 1749 and 1755. After a short stay in Amsterdam he established himself as professor in Groningen. Well-known students of the University of Franeker include René Descartes who was a student between 1628 and 1630, and Peter Stuyvesant (c.1612–1672), Director-General of colony of the New Netherlands, where he expanded the settlement of New Amsterdam that later became New York. Professors that taught astronomy and were of some prominence in their days are Adriaan Metius (1571–1635), Johannes Phocylides Holwarda (1618–1651) and, Jean Henri van Swinden (1746–1823). Mathematician Johann Samuel König (1712–1757), who had studied with Johann Bernoulli (1667–1748) and Daniel Bernoulli (1700–1782), was Professor of Philosophy and Mathematics at the University of Franeker from 1744 to 1749; he is remembered for a theorem in kinetics and was the subject of a plagiarism charge related to the principle of least action. König actually is part of Oort's academic genealogy (see Appendix A.6).

The University of Franeker began to decline in the eighteenth century, suffering from low student numbers and the competition with the University of Groningen. The latter had been founded in 1614 as the third university in the Republic of the Seven United Netherlands. A number of important professors left Franeker to accept an appointment in Groningen, such as student of König and Professor of Philosophy Antonius Brugmans (1732–1789), who also figures in Oort's academic genealogy. The University of Franeker was closed in 1811 by the Napoleonic occupation, together with those of Utrecht and Harderwijk, while the Universities of Leiden and Groningen were continued. After the Kingdom of the Netherlands was established in 1815, the University in Franeker was not re-opened (unlike the one of Utrecht), but became a 'atheneum', which again suffered from low student numbers and was finally closed in 1843.

During the period of decline following the heydays of the University, a wool carder by the name of Eise Jelteszn Eisinga (1744–1828) built an orrery, which now is the oldest surviving and operating one worldwide. This orrery was a dynamical scale model of the Solar System, which he built in his own house. It is on a scale of one to one billion (10^{12}) and shows the then known planets moving along their orbits hanging from the ceiling of his living room. It was built between 1774 and 1781 and still is accurate. Eisinga was a self-educated amateur astronomer, having profited from the presence of a university in his home town. He actually held an appointment of Adjunct Professor at the University of Franeker until it was closed in the Napoleonic era. The Royal Eise Eisinga planetarium, as it is called now, was Eisinga's response to the prediction that the conjunction in 1774 of the Moon and planets out to Jupiter (configuration of all of these objects more or less in the same direction from the Sun) would pull the Earth from its orbit.

At the time of the fin-de-siècle, Franeker was a rather moderate town, having about six or seven thousand inhabitants. With the University closed in 1811 and the remaining Atheneum in 1847, much activity had disappeared. The governing board of the province of Friesland decided in 1848 to use the building previously occupied by the University/atheneum to house a 'Kankzinnigengesticht', an asylum for lunatics, in Franeker. Up till then Frisian lunatics were nursed in Deventer, over 100 km to the south. The first patients—27 in number—arrived in Franeker in 1851 and the asylum grew steadily. Most inhabitants worked during the day on the farms surrounding Franeker. In 1897 a new second physician was appointed to the asylum by the name of Abraham Hermanus Oort (1869–1941).

1.2 Abraham Oort and Ruth Faber

Abraham Hermanus Oort had been born in nearby Harlingen. His father, Henricus Oort (1836–1927), was a descendant of a long line of theologians and clergymen [26]. Henricus had studied theology in Leiden and obtained a doctorate in 1860. In the same year he married Elisabeth Wilhelmina de Goeje (1838–1907) and became a vicar[1] who preached in Santpoort, a small village to the west of Amsterdam near the coast. They had eleven children. In 1866, Henricus had been appointed in Harlingen, where Abraham was born as their seventh child. After six years there Henricus Oort moved to Amsterdam, where he briefly worked as a Professor at the Athenaeum Illustre, the precursor of the University of Amsterdam, before being appointed in Leiden as a Professor in the Faculty of Arts and Philosophy (not Theology) in 1875. His specializations or 'teaching assignments' were Hebrew, ancient Israel and the Old Testament. He belonged to the movement of Liberal Protestants, of which the Leiden Professor in Theology Abraham Kuenen (1828–1891) was a prominent member.

In addition to his many scientific and theological publications, Henricus Oort was involved in providing biblical translation together with other clergymen or vicars.

[1] In the Dutch language there is a distinction between a 'pastoor' and other clergymen. The former is part of the clergy of the Catholic or Anglican churches. Preachers in protestant or reformed churches—as the case in the Oort ancestors—are often designated as 'dominee'. To avoid confusion I do not use the word pastor, but vicar or parson for protestant clergymen. Minister would have been possible too, but that can be confusing as a minister could also be a politician.

Fig. 1.2 Postcard showing the house where Oort was born. It is seen here at the extreme right. In the center the 'Koorndragershuisje' (Corn Bearer's House). The postcard is undated, but this view would be quite similar to what this part of Franeker would have looked like when Oort was born

He published a 'Bible for Young Children' together with Isaäc Hooykaas (1837–1894), who preached in Schiedam near Rotterdam and later in Rotterdam itself. Hooykaas was married to Henricus Oort's sister Petronella Everharda (1839–1916). This work, to which his brother Abraham Johannes Oort (1838–1917), a parson preaching in Zutphen in the eastern part of the country, contributed as well, has been translated into French and English. Together with Kuenen, Hooykaas, and others, Henricus Oort was involved in a translation of the Old Testament, based on Greek sources and 'according to strict scientific principles' with the aim to be more readily comprehensible than the official one, the 'States Bible'. This new translation became know as the 'Leiden Translation', but was not widely read outside Protestant circles. Eventually, Henricus Oort provided also a new translation of the New Testament.

When Henricus Oort moved to Leiden in 1875, his son Abraham Hermanus, who had been born in Harlingen on March 28, 1869, was only six years of age. After elementary school, Abraham went to the Leiden Gymnasium (grammar school) and after completing the final exam in 1887, he was ready to enroll in the University of Leiden. The family tradition was to take up theology and two older brothers had actually decided to do that. Instead, Abraham chose medicine. After obtaining his full qualifications to practice medicine in 1896, he spent some time in Heidelberg at the psychiatric clinic there as a volunteer, which changed his life and turned him to psychiatry [27]. The following year, on September 14, 1897, he was married in Leiden to Ruth Hannah[2] Faber (1869–1957), and took up a position as second physician of the asylum in Franeker.

[2]Aficionado's of language curiosities (like myself) will have noted that 'Hannah' is a palindrome. It actually is the longest palindrome among Dutch girl's first names. For boys that is Reinier [28].

Fig. 1.3 Plaque at the facade of house where Oort was born in Franeker. It resides to the right of the windows on the ground floor (compare Fig. 1.2). Photograph by the author

Ruth Faber, had been born on October 6, 1869 in the village of Ossendrecht, from Antwerp to the north just across the Belgian-Dutch border. She was a daughter of Jan Faber (1838–1889) and Henriëtte Sophia Susanna Schaaij (1840–1908). Jan Faber also came from a lineage of clergymen and was himself a preacher as well. He died relatively young, days before his fifty-first birthday and well before the marriage of his daughter Ruth with Abraham Oort. At the time of his death he preached in the Reformed Church of Heemskerk, near the coast to the northwest of Amsterdam, where he had been appointed only the year before.

The young couple Oort-Faber settled in Franeker at the address Zilverstraat 16 (see Figs. 1.2 and 1.3). This is actually very close to the Eise Eisinga Planetarium. Their first child, a boy that they named Henricus, was born on June 30, 1898 and the second, the subject of this book Jan Hendrik, followed on April 28, 1900 (Fig. 1.4). The latter's birth certificate is reproduced in Fig 1.5 and shows that the time of birth was recorded as 8:30 in the morning. The witnesses required for the certificate were Willem Ferwerda and Adriaan van Voorthuijsen. Ferwerda is described on the document as a 'boekhouder' (bookkeeper or accountant) and was 41 years of age. The only person of that name and born around 1859 that I can trace indeed lived in Franeker (at least in 1888) and had the profession candidate notary (again in 1888). Adriaan van Voorthuijsen (1872–1952) is easier to find. He was a physician and general practitioner in Franeker at the time, who earned his MD in Leiden. He and Abraham Oort differed only three years in age and would have known each other in Leiden during their student days. Van Voorthuijsen later moved to Groningen, where Oort saw him regularly when he was a student there.

Fig. 1.4 Abraham Hermanus Oort and Ruth Hannah Oort-Faber, father and mother of Jan Oort, in 1902. From the Oort Archives

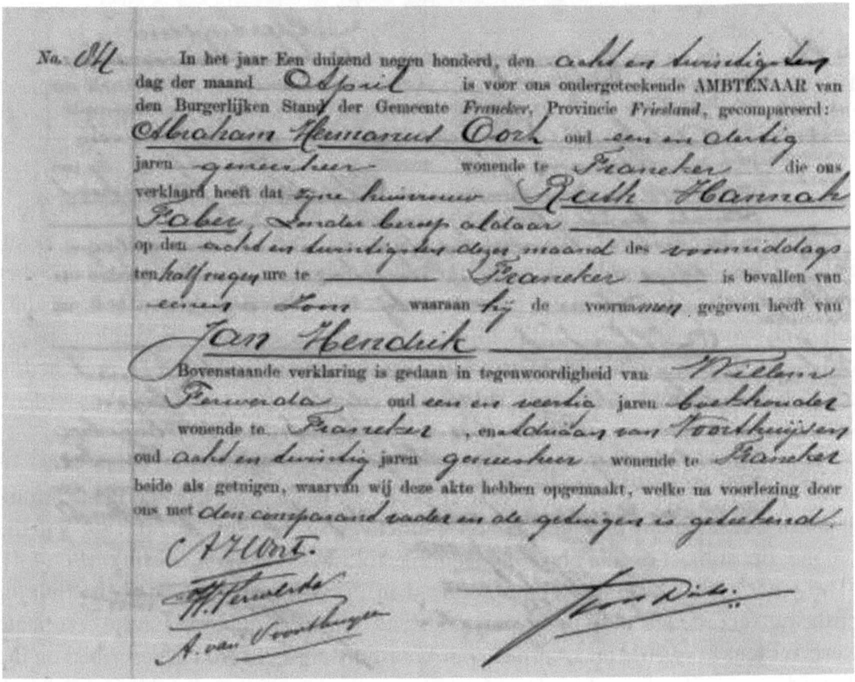

No. 84 In het jaar Een duizend negen honderd, den acht en twintigsten dag der maand April is voor ons ondergeteekende AMBTENAAR van den Burgerlijken Stand der Gemeente *Franeker*, Provincie *Friesland*, gecompareerd: Abraham Hermanus Oort oud een en dertig jaren geneesheer wonende te Franeker die ons verklaard heeft dat zijne huisvrouw Ruth Hannah Faber, zonder beroep aldaar op den acht en twintigsten dezer maand des voormiddags ten half negen ure te Franeker is bevallen van eenen zoon waaraan hij de voornamen gegeven heeft van Jan Hendrik

Bovenstaande verklaring is gedaan in tegenwoordigheid van Willem Ferwerda oud een en veertig jaren boekhouder wonende te Franeker, en Adriaan van Voorthuijsen oud acht en twintig jaren geneesheer wonende te Franeker beide als getuigen, waarvan wij deze akte hebben opgemaakt, welke na voorlezing door ons met den comparant vader en de getuigen is geteekend.

A H Oort.
W. Ferwerda
A. van Voorthuijsen

Fig. 1.5 Birth certificate of Jan Hendrik Oort [29]

Three more children would follow, but only after the Oort-Faber household had moved from Franeker to Oegstgeest near Leiden.

1.3 Oegstgeest

In 1899, the first Professor of Psychiatry in the Netherlands was appointed at the University of Leiden. Gerbrandus Jelgersma (1859–1942), who occupied this chair until 1930, had studied medicine in Amsterdam and had a major interest in the brain and neural system. At the time of the appointment he was Director-Physician in a sanatorium for neuropaths at Arnhem. Abraham Oort, although working at that time already in Franeker, became his first PhD student. As we saw above, Abraham Oort had spent some time in Heidelberg; he had worked there with Emil Kraepelin (1856–1926). Kraepelin experimented with persons performing concentrated mental work for periods of time up to a quarter of an hour or so and then investigated the influence of coffee, tea, alcohol, etc. on their performance. He had also collected data on children of various ages and adults performing mental arithmetic.

Abraham Oort was allowed to use Kraepelin's results to study the effects of age, training, fatigue, etc. on the speed of adding numbers. He analyzed these data, supposedly in his spare time, in Franeker and presented the results in a PhD thesis *Proeven met doorlopenden hersenarbeid genomen op schoolkinderen* (Experiments with continuous mental work performed on elementary school children), which he defended on November 26, 1900 at the University of Leiden. He obtained the Doctor's degree *cum laude*. Jelgersma was his promotor (supervisor) and Abraham Oort therefore became the first student who obtained a Doctor's degree under his supervision.

According to the agreements at the time of Jelgersma's appointment, the City of Leiden opened a new sanatorium 'Rhijngeest' in the village of Oegstgeest, only about 4 km to the northwest of Leiden [30, 31]. It was set up to be a clinic for 'temporary treatment of patients with neurotic disorders'. The estate Rhijngeest had been acquired by the City of Leiden in 1895 together with nearby Endegeest, a medieval castle surrounded by gardens and forest, where actually French philosopher and scientist René Descartes (1596–1650) had lived between 1641 and 1643. The City established at Endegeest an 'asylum for lunatics' in 1897. This in contrast with Rhijngeest, more or less on the other side of the street, which as we saw was a clinic for temporary treatment of patients with neurotic disorders, and run quite independently. There has been some confusion between the two. Rhijngeest was eventually (in 1950) renamed the Jelgersma-clinic, but has been discontinued in the 1990s. Since 1999 the town-hall of Oegstgeest is located there. Endegeest is still operated as a psychiatric hospital.

Rhijngeest was funded by the City of Leiden and was not part of the University. It was therefore the City Council of Leiden that decided on December 11, 1902 to appoint Dr. Abraham Oort to be the First Physician (Director) of the sanatorium [32]. Undoubtedly this was following recommendation from Jelgersma, who must have had trust in his first student. Abraham Oort accepted and moved to Oegstgeest in 1903. He remained in this position until his retirement.

The Oort family lived in the Director's house on the estate (Fig. 1.6). It was (and still is) surrounded by gardens and trees. The estate is very close to the current location of the offices of Leiden Observatory in the Oort Building and the Huygens Laboratory and can actually easily be seen from the higher floors of these buildings,

Fig. 1.6 Directors house on the estate Rhijngeest, where Oort grew up. From Wikimedia Commons [33]

including the 'Oort Room', where some of his books and other materials are kept. The first part of the name Rhijngeest refers to the river Rhine, of which the oldest branch ended in the North Sea near Leiden. It is now not more than a canal. The second part, 'geest', is an old name for a 'strandwal' or beach ridge, a bank formed by the tides on the sea by depositing sand parallel to the shores. Rhijngeest must have been a very comfortable place to grow up.

In addition to running the sanatorium, Abraham Oort was much involved in other matters related to psychiatry, such as developing tests for admission exams for schools for nurses specializing in psychiatry (designated as qualification 'black cross'). Oort describes his father in the *AIP-Interview* [4] as 'much interested in science. He made some simple brain experiments and was one of the first in the Netherlands to use psychological tests.' For a long time he edited new editions of a textbook on psychiatry; this book *Lectures about neurotics and lunatics and their nursing*, which had been written by Dr. J.C.Th. (Jan Christiaan Theodoor) Scheffer (1855–1905), Director-Physician of Endegeest, and published in 1906. Scheffer had died in 1905 at a young age of 49 and Abraham Oort took it upon him to update the book roughly every five years. He was a very visible and active member of an organization of physicians against the use of alcohol. The members promoted a medical approach to alcoholism, treating it as a disease similarly to the then also very common tuberculosis and syphilis. Abraham Oort was very much concerned with the effects of alcohol on the performance of the nervous system. The organization promoted complete abstinence of drinking as the only way towards a cure. The last of his propositions—a set of statements attached to a Ph.D. thesis, the last one often of general societal significance—posed that from then on in new asylums placement of alcoholics, even involuntarily, should be possible. The Oorts senior remained teetotalers all their lives.

In Oegstgeest, Abraham and Ruth Oort-Faber had three more children. Arend Joan Petrus was born on May 27, 1903, Jetske Sophia Susanna on June 24, 1904 and Emilie Annette on March 6, 1908 (Figs. 1.7, 1.8, 1.9 and 1.10). The first of these, Arend Joan Petrus was usually called 'John', derived from the middle one of his first names Joan (pronounced 'Yo-an'). He was in the family given the nickname 'Jo' (pronounced 'Yo'), and later intimates used this as well. Jetske was in family circles referred to as 'Jeppie' and Emilie as 'Emy', but I will not use these nicknames. I will however refer to the younger brother with John, as he was commonly known. The older brother Henricus was generally known as 'Hein' and I will adopt that as well.

Abraham Oort was very much interested in botany, which he had no difficulties pursuing in the large garden around the house and on the Rhijngeest premises. It was fenced off and closed to the public and the Oort children had a lot of freedom to play in the fields. It is reported that the Oort family kept chickens and generally enjoyed the outdoors, a quality that Oort kept for the rest of his life. Summer vacations were usually spent in nearby Katwijk aan Zee, where the Oort parents owned a vacation home that was called Sandy-Hook. Katwijk is less than 10 km from Oegstgeest, almost straight to the west on the coast. The Oort family kept and used the house until World War II, during which is was demolished by the German occupation authorities in the framework of the 'Atlantikwall' (Atlantic Wall) structures that were build by the Germans to defend the European coastline.

Oort went to elementary school in Leiden. It was designated as a 'public school for extended elementary education of boys'. In the Netherlands the grades ran (and still do) from 1 ('very poor') to 10 ('excellent'). Oort's grades almost all ran between 7 and 9, with the 9 very common for all subjects including reading and writing, arithmetic, history, geography, French, and drawing. Also for behavior, presence, homework he scored similarly high marks. Gymnastics averaged 7, but for 'diligence' his mark regularly was 10.

Oort's youth must have been happy in general and his school results were excellent and promising. The only troublesome aspects might have been the health of his mother. She had very serious surgery when Oort was eleven and spent about a year in Italy to recover. An aunt (a sister of Oort's mother) ran the household during that period. For respiratory problems, Oort's mother also spent some time in Davos in Switzerland. None of the other children, including Jan Oort himself, seemed to have suffered from this or any other serious disease, except that the older son Hein, who developed a serious case of diabetes and would die in his twenties.

Jetske (Jornaroos) Hamaker-Nauta (b. 1936), daughter of Oort's younger sister Emilie Annette, has provided me with a description of her grandparents and the Oort household at Rhijngeest.

'.. and there also was another villa where Prof. Suringar, the botanist, had lived who had planted many interesting plants. Now some patients lived there on their own expense. The house was tall, old, and small for a family of 5 children and the resident maid, who had to address the small daughters as Miss Jetske and Miss Emy. There were two washbasins in the house, in the bedrooms one used a bowl with water that was refreshed every day and which sometimes in winter was partly frozen. Only on Saturdays the geyser was used so everybody could take a warm shower.

At the front of the house the father had a study and there was a veranda that blocked much of the light in the living room. This veranda was used by the repair seamstress. [... The daughters played often] under the table, because the older brothers had used most of the space in the living room with their buildings and trains. [...]

The park was ideal for wandering around and making discoveries, in the holes under the rhododendrons and in the ponds. Ponds with water-boatmen (Notonecta glauca), beetles (Gyrinus natator), sticklebacks and salamanders. And sometimes there was a kingfisher in the 'Hudsonbay'! In the back were berry bushes and strawberry plants, which never had any ripe strawberries (blackbirds?) There was also a large tree with gold rennets. In early fall, the apples were carefully picked with a kind of net on a stick and then taken to the attic and turned every week; only when they had bad spots could they be eaten, ones in perfect shape only at Christmas.

[Then there were] chickens. Sometimes they had to pick hands full of addersmeat or young nettle for them, which was a chore, and sometimes to collect the eggs, wash and put a date on them. They never bought eggs. Young cockerels were exchanged with head nurse Dunnebier: you don't eat your own chickens if you can avoid it. That kindest nun, still dressed in a habit, also delivered broody hens so that there could be new chicks. Sometimes they were hatched under a lamp. [...] The mother was an animal friend, she showed how the mother hen taught her chicks pecking by pecking with her beak on a shelf, and she mimicked that with her nail, and the chicks understood what she meant. In the family she was called 'Grandma Chickens'. The mother also kept bees, she put a hood on her head and a lit pipe in her mouth. The honey was hurled by herself.[...]

The mother had socialist sympathies and she approved of the idea of Cooperations: jointly stocking merchandise, no pursuit of profit. Once a month a big box of groceries was delivered. [...] There was a great fear of tuberculosis. Understandably since 8 out of 16 of the father's siblings had died, mostly through the complication meningitis. His mother suffered herself from this disease. And so the father greatly feared all infectious diseases and infections, and therefore [my mother] was not allowed to play with other children and did not attend school [until age eight].'

And about her grandfather and Oort's father:

'He was Director-Physician of a psychiatric hospital in Oegstgeest, but you wouldn't tell if you met him: a little shy, a very modest man, quiet and perhaps almost boring, small, very friendly eyes, wearing a beret and having a small white beard. He was not a man of many words, telling stories, no, I did not learn that from him.

He had to take a rest once for a long period because of an upset stomach and he spent that time woodcarving. For all his grandchildren he cut with a small knife and much precision a wooden box with their name on it.'

And Jan Oort's younger son Abraham Hans (Bram) contributed the following:

Jan's parents both came from a long family tradition of preachers and theologians. Grandma Chickens and their older daughter Jetske went regularly to the 'Remonstrants' church [Dutch protestants, followers of Jacobus Arminius (1560–1609)], but the other children were not religious, actually more antagonistic ['in de contramine'] towards this trend in the family. Grandmother tells that especially Jan was antagonistic all the time, not just about religion but in general – 'Jantje contramine'. Jan was a real scientist without a belief in God although he had a feeling of there being some higher order in nature. He loved going into the wilderness; signs 'no entry' were a red flag for him to find out what it was that was forbidden behind them. I don't think he was ever fined – was always able to talk himself out of it.

Grandma chickens was a precious, gentle grandmother [...]. She often told about her long trip by (sail?)boat to Indonesia. She remembered the great impression to see Cape of Good Hope for the first time. The Faber family went to Indonesia for a new job as parson that her father had taken up. Her younger brother Paul Faber, was still a baby then and there was a cow on the ship to provide milk for the babies on board. Later uncle Paul became my favorite great-uncle – he retired at age 50 as a member of the High Court of the Dutch Indies; and lived to far in his nineties. [...]

Grandma Oort was a social worker in the slums of Leiden, often accompanied by Jetske and Emy. There were a lot of poverty and alcohol problems in Leiden. Jetske and Emy both went on in social work. Jetske remained single and worked as a nurse; she was the one that took care of Grandma Oort when she grew older. Emy married Jelle Nauta, a capable radiologist in Rotterdam.'

Indeed Paulus Frederik Karel Faber (1874–1973) lived to age 98; he far outlived his wife Clara Cornelia Steynis (1880–1937). Jetske went on to become Director of the Student Sanatorium in Laren. The radiologist is Jelle Haring Nauta (1907–1987).

In the *AIP-Interview* [4] with David deVorkin in 1977 Oort was asked about his religious background. After all, both his parents were descendant from generations of clergymen and theologians. Oort replied when asked whether he had received any religious training as part of his upbringing, while still living with his parents:

'Yes. The common religious training, lessons about Bible interpretation. Both my mother and father came from families of clergymen. My grandfather, the father of my father was a Protestant clergyman with quite liberal ideas. He was one of the founders of the more liberal Church in Holland (In colloquial English, the Netherlands is sometimes referred to as Holland, which actually is no more than the set of provinces on the west coast that were centers of trade in the seventeenth century. Especially cities in it like Amsterdam. The usage of Holland is just as incorrect as referring to the United Kingdom as England.). He later became Professor of Hebrew in Leiden. He was one of the three people who made a new translation of the Bible into Dutch, with extensive comments. It was quite a large undertaking. Several of my uncles (his sons) were pastors. The father of my mother was also a clergyman. [...]

My mother kept up her interests in that, at least in the early years of her marriage. But my father was less interested in Church matters.'

[DeVorkin: Did you go to Church regularly or do you go to Church regularly now?]

'No. But it was not so customary in Holland, I think, as it was in the United States to go to Church. There were many highly esteemed families who thought going to Church was not at all essential.'

1.4 The 'Hoogere Burgerschool'

After finishing elementary school in 1912, Oort entered a form of secondary education called Hoogere Burgerschool (literally Higher Civic School) or HBS. This choice needs some explanation, as it would seem obvious that like his father he would enter the Leiden Gymnasium. His ancestors as clergymen would also have had such secondary education. The HBS gave no formal right to enter university afterwards. So, what made him (and his parents) choose this type of school?

The HBS was the new type of secondary education introduced in 1863 as an alternative to the gymnasium. Responsible for it was Johan Rudolph Thorbecke (1798–1872), a politician and statesman from the Liberal movement, which promoted optimum personal development of individuals, a parliamentary democracy and more extended suffrage and civil rights. Thorbecke was responsible for the new constitution of October 1848, which limited the power of the King. He was Prime Minster in three periods between 1849 and 1872 and among his accomplishments was a revision of the school system in the Netherlands. This started with elementary schools and was followed by the establishing of the HBS in 1863.

At the HBS no Greek and Latin were taught—contrary to the situation at the Gymnasium, where these languages formed a major part of the curriculum—, the aim of the HBS being to educate young men (at that time not women) for leading positions in industry and commerce, rather than prepare them for academic professions. There was much emphasis on the natural sciences and the modern languages English, French and German. The level of education was very high; teachers in mathematics, physics and chemistry usually had a Ph.D. in their disciplines. In fact the relatively large number of Nobel Prizes in physics and chemistry for the Netherlands in the first years it was awarded (from 1901 onward) is often quoted as a direct result of the introduction of the HBS [34, 35]. Indeed Jacobus Henricus van 't Hoff (1852–1911) (chemistry in 1901), Hendrik Antoon Lorentz (1853–1928) and Pieter Zeeman (1865–1943) (physics in 1902), en Heike Kamerlingh Onnes (1853–1926) (physics in 1913), had all gone through the HBS and not the gymnasium that up until then had been the only route to University. Johannes Diderik van der Waals (1837–1923) (physics in 1910) had also not gone through the gymnasium or even HBS, eventually being awarded special permission to go to university by Thorbecke [34, 35]. Actually, special permission by the Prime Minister had been a way for some time for persons not qualified to enroll at universities. Aletta Henriëtta Jacobs (1854–1929), the first female student in the Netherlands (in medicine), was given special permission by Thorbecke himself to take exams in Groningen. It was one of the last acts of Thorbecke before his death.

So, the HBS did not qualify automatically for enrollment in university, not even for disciplines that did not strictly require an extensive knowledge of Greek an Latin. It was, however, possible to enter academia by passing an admission exam. An early example of a person following that route is Kamerlingh Onnes, who was born and grew up in Groningen [36]. It was here that the first HBS was opened; in fact the law was not even in effect when the City Council of Groningen applied with Thorbecke for support to open such an institution, which resulted in the first students being admitted in 1864. Kamerlingh Onnes entered a year later. The special admission exam that was required to enroll for academic studies was, for the University of Groningen at least, not prohibitively difficult. The early Nobel Laureates usually chose this route, although there were other ways of circumventing it. Medical students who came from the HBS could become physicians after completing academic studies, but could not defend a thesis and earn a PhD as a medical researcher. Some defended a thesis at a foreign University; others, like Jelgersma, were awarded a doctorate *honoris causa* (in Jergersma's case from the University of Utrecht).

A law making it possible to directly enter university after the HBS as a matter of routine was only adopted in 1917, the year Oort passed his final HBS exam, and was not in effect when Oort had to choose which type of secondary education to follow. By that time it had become an accepted course of action when young persons showed special talent for mathematics and exact sciences to go to the HBS and it did not take long before the majority of scientists in these disciplines took that route. In addition, the growing number of teachers required offered excellent job opportunities for those who obtained a PhD. So, Oort's choice must have been the natural thing to do.

Oort's older brother Hein went on later to study physics at the University of Leiden, so it is not unreasonable to suppose he went to the HBS as well. But the younger brother John was different and inherited the interest Abraham Oort had for plants. In the *AIP-Interview* [4], Oort says, referring to his father:

> 'He was also very much interested in nature, especially botany and plants; those interests were inherited by my younger brother, who became a naturalist already from his school days. I think in a way he was much more pronounced in his direction than I was. My mother had undergone a severe operation and had to go to Italy for a recovery period after that. She got letters from the children. In letters from me arithmetical problems

Fig. 1.7 Picture supposedly taken in 1902 and has Oort on the left with his older brother Hein. It was quite common at the time for boys of age two to be dressed as Oort is here. From the Oort Archives

were sometimes scribbled in the margin. When she got a letter from my youngest brother it was always about flowers, birds and mushrooms he had found. In fact, when he was still in the lower grades he discovered a mushroom in the woods which had never been found before in Holland. He was at that age already quite an expert. He later became a Professor of Plant Diseases at the University of Wageningen.'

Indeed, John studied biology and botany and eventually became a Professor of Phytopathology in 1949 and was Director of the quite well-known Laboratory of Mycology (the study of fungi) and potato research, which was renamed and became the Laboratory of Phytopathology at the Wageningen University & Research as it is now known. Consistent with what we saw above, John actually *did* go to the gymnasium in Leiden [38]. The younger sisters Jetske and Emilie went to the HBS for girls that had been instituted by that time (Fig. 1.11).

Oort at age fifteen is very likely among the boys in Fig. 1.12. I did send this picture to the Artificial Intelligence Department of Groningen University, where work is done in the area of face recognition, together with Oort taken from Figs. 1.8 (age eight) and 2.13 (age twenty). However, I was informed that more and better resolution pictures are required for an identification with any confidence. I then organized a poll in the Kapteyn Astronomical Institute among staff and students sending them eighteen

Fig. 1.8 Oort and his brothers and the older of his sisters around 1908. Brother Hein is at the top left, John bottom right and sister Jetske in the middle of the picture. The younger sister Emilie is not on this photograph; she was born that same year if indeed this date is correct. From the Oort Archives

Fig. 1.9 Oort (right) together with his older brother Hein. This picture was noted to date from around 1908, but could well be from somewhat later. From the Oort Archives

selected faces from the group photograph together with the same comparison ones. The two most frequent choices for identifications of Oort have been illustrated with the figure together with the comparison photograph at age 20. The two pictures on the right were the top two of the poll. My preference definitely is the lower of these two. Furthermore, the person in the top panel definitely appears of more than average height.

Fig. 1.10 Oort's sisters Emilie Annette (left) and Jetske. Photograph provided by J. Hamaker-Nauta

Fig. 1.11 Front of the building of the Hoogere Burgerschool Leiden. In the back we see part of the Pieterskerk, the main church building of Leiden, and now the location of many formal ceremonies of the University of Leiden. From Erfgoedleiden.nl [37]

Oort's private dairies go back to his teenage years. These are also kept in the Oort Archives in Leiden, but in parts not open to a general public. The oldest entry is dated December 24, 1915. Unfortunately, there is very little in these concerning his choice for Groningen as a place to study and his developing interest in astronomy. The

Fig. 1.12 Teachers and students of the Hoogere Burgerschool Leiden on the occasion of the commissioning of the new building in 1915. The older boys wear hats or caps, while the youngest have no headgear. The top two pictures on the right are the first (top) and second choice in the identification poll (see text) at the Kapteyn Astronomical Institute. My own preference is the lower one (see text). For comparison at the bottom Oort at age twenty from Fig. 2.13. From Erfgoedleiden.nl [39]

stories during his secondary school years mostly concern private matters and need no detailed discussion here. One point of interest is in connection to a girl named Kitty with whom he fell in love not long before the first pages were written, and who returned in the notes for some time, only losing prominence after he left Leiden and Oegstgeest for university studies in Groningen. It refers to the issue of pillarisation or pillar system ('verzuiling' in Dutch). This characteristic of society, which was very prominent in the Netherlands, refers to the segregation along political or religious lines. The girl concerned was from roman catholic parentage and a major problem was that she apparently was forbidden by her parents and her pastor to mix with boys from the protestant 'pillar'. Oort does see and meet her regularly, at parties, sports and other social events, but she mostly treats him somewhat coolly.

In the Netherlands, the pillarisation originated towards the end of the nineteenth century. The reform of the Constitution of 1848 by Thorbecke (see above) had improved the societal position of the Catholics, who eventually promoted the institution of catholic schools. In the more strict protestant circles opposition developed against the liberal views of Thorbecke and others, first under the leadership of Guillaume (Willem) Groen van Prinsterer (1801–1876) and later of Abraham Kuyper (1837–1920), who founded the first political party, representing the more orthodox protestants. In addition to the resulting protestant and catholic pillars, there was a third one which is usually described as social-democratic. This left the people not belonging to any of these, like atheists or Protestants or Catholics that had no strong ties to their churches.

So, where does Oort fit in all this? Whenever appropriate I quote from Oort's 'diary notebooks' (see Preface for explanation). Oort seems not too concerned with the backgrounds of others, so his upbringing was presumably more in this 'liberal' or 'general' category. This is in agreement with religion being an unimportant part of life in his parents' household.

Oort described in his diary notebooks his love for walking; he mentioned sometimes long walks along the beach, alone or with others, often in nearby Katwijk, the place where the Oorts had a second home for weekends or vacations. Maybe even more profound is Oort's love for skating. From the entry of February 6, 1917:

'What a fantastic and healthy sport is skating. This afternoon I skated with great speed together with Paul Reinshoed to Katwijk. where he wanted to show his school records. The weather was magnificent and the beach was covered with large blocks of ice, stacked on top of each other. The water in between had frozen as it was flowing.

I like these days more and more the simple, ordinary things. I believe that only there in the long run the deepest beauty can be found.'

These passions, especially skating have become legendary among friends and colleagues. He continued with some reading preferences:

'As I wrote before I rejoice in all beautiful things I see; I only read Dickens and some fairy tales, since these give a good feeling of cosiness and this love for tales by Dickens, and these tales by themselves are so beautiful and free of roughness and evil.'

In 1917 Oort easily passed the final exam of the HBS. His marks were very high for mathematics and exact sciences: '10' for mathematics and for mechanics, '9' for physics and '7' for chemistry. For languages he scored at just satisfactory levels, '7' for Dutch and '6's for French, English and German. Greek and Latin at a gymnasium might have been somewhat problematic. For drawing he had only a '5', maybe to no surprise for those like me who have seen his scribbles on the blackboard when he was lecturing. Geometrical drawing was better with a '7', but that involved the use of compass and ruler.

During his teens Oort developed an interest in astronomy. He has on a few occasions attributed that to reading books by Jules Gabriel Verne (1828–1905). Novels like Verne's were available in Dutch translations and these sparked his interests, as they have sparked the interests in astronomy and science in general before and after that of many teenage boys and in the course of time increasingly girls. In fact his developing interest induced friends of his parents to lend him a small telescope, with which he studied the skies at night. To quote from the *AIP-Interview* [4]:

'But I certainly got interested in science and astronomy already in my high school years. I don't remember exactly where my first interest in the subject came from. I suppose it was by reading books by Jules Verne who was very popular in those days. *Around the World in 80 days, the Voyage to the Moon* and *The Excursion of a Comet*. [...]

That was the sort of science fiction of those days, but in a way perhaps of a more literary nature than the present science fiction. Jules Verne was quite a witty French author. But he is still being read today, my grandchildren enjoy his books as much as I did. [DeVorkin: That's marvelous. Did you read anything like Flammarion?] I suppose very soon after I got into reading Jules Verne.'

The last book of Verne is not very well known these days. The original French title of this book is *Hector Servadac*, and is known in English as *Off on a Comet*. This shows that Oort read Dutch translations and apparently translated the Dutch title directly into English. His grandson Marc Oort tells me Oort read this book to him when he was a child and that he inherited some of the Verne books (but of the two-part *Hector Servadac* the first volume is missing). 'Flammarion' is the book *Astronomie populaire* by the French author Camille Flammarion (1842–1915), which was widely read as a popular book on astronomy. In the Netherlands it first appeared in translation in 1884 with the title *De wonderen des hemels* (The wonders of the skies).

> 'Already in my high school days I got a small telescope from friends of my parents and so I became interested in the stars but not in such a very pronounced way. I liked physics in school just as much and it was perhaps, a more exciting subject matter than astronomy in those days. When I went to the University of Groningen I went there partly because a well known famous astronomer, J.C. Kapteyn was teaching there. [...]
> Yes that was the main reason.'

The choice to take up studies in natural science had not come completely without any doubts. In the same entry on skating (see above) of February 6, 1917, when the final exams were getting close, he also noted:

> 'They want me to take up academic studies: ship-building, chemistry, mining engineer, I don't know what. But remember that the human souls rich in elves are so infinitely more beautiful than ship-building, chemistry, mining industry, never forget, that for you there is nothing else but to live a beautiful life only in your inner self. So live!'

And in July 1917 there is even mention in the diary notebooks of possible studies in mining egineering in Sweden and Norway and a visit with his father to Professor Knol in Den Haag. Wopke Aurelius Knol (1880–1932) [40] was a Professor of Mining Engineering at the Delft Institute of Technology (the precursor of the Technical University). Oort wrote: 'He was not able to provide much important information'.

Rather than enter the University of Leiden, as his older brother Hein had done to study physics, Oort in the end opted to go to Groningen. Although he was undecided whether to study physics or astronomy, he quoted the fame of Jacobus Cornelius Kapteyn (1851–1922) in Groningen as the main reason for him to choose to go to the north of the country for his academic studies. It seems that the choice for Groningen was also a result of wanting to avoid competition with brother Hein in Leiden [41]. He does note in his diary notebook on September 3, 1917: 'The only right choice left is to go to Groningen or to go to Utrecht for my studies', but without any background or further explanation. It probably refers to this resolve not to study physics in Leiden. No further clue is to be found in the diary notebooks on this issue. And about his parents he added in the *AIP-Interview* [4]:

> 'They were very liberal. They thought that if you really liked the subject very deeply you must follow your calling and try to study that subject. But when at Groningen I wasn't quite decided yet whether I would specialize more in physics or astronomy. It was the personality of Professor Kapteyn which decided me entirely. He was quite an inspiring teacher and especially his elementary astronomy lectures were fascinating.'

Fig. 2.0 Oort as a student in Groningen in 1921. From the Oort Archives

Chapter 2
Kapteyn and Groningen

Those that thoroughly study the lives of those, that on this Earth did accomplish something significant, are always struck by the phenomenon that the real strength of these people has been a shortcoming turned into a quality.
Godfried Bomans.

It is the supreme art of the teacher to awaken joy in creative expression and knowledge.
Albert Einstein (1879–1955)

Oort's scientific work covers almost all aspects of astronomy. This book is written with a general audience in mind, not necessarily with an astronomical background. The large scope of Oort's work demands knowledge of the general astronomical and astrophysical setting. I have therefore added a few sections on basic astronomical matters. Mathematical details have been put in separate boxes, but the text should suffice without these for those not so proficient in mathematics. For more background many good quality articles in Wikipedia are available; for Dutch-speaking readers I also refer to my lecture notes on 'Introductory Astronomy' that I taught for many years to the first-year students in astronomy, physics and mathematics at the University of Groningen [42].

For a good understanding it is also necessary to have some understanding of the person and the works of Oort's mentor, Jacobus C. Kapteyn. It was Kapteyn that drew Oort to Groningen for his academic studies and who definitely turned him towards astronomy. The first quote above actually refers to Kapteyn. For details I refer to my biography of Kapteyn *JCKbiog* [7], published in this same series. In the paragraphs below on Kapteyn or his contemporaries I have often taken pieces of text from this earlier book. This has also been the case for sections on the astronomical background. Rather than simply referring the present reader to my *JCKbiog*, I have taken such material often from that. This is certainly the case for almost all boxes in this chapter, but also for some paragraphs of text. However, these texts are not

P. C. van der Kruit, *Jan Hendrik Oort*, Astrophysics and Space Science Library 459, https://doi.org/10.1007/978-3-030-17801-7_2

identical; some editing has been done, some additions and clarifications have been added or sometimes sentences have been deleted.

2.1 Positional and Sidereal Astronomy

Box 2.1. Celestial coordinate systems

Positions of stars on the sky are designated in coordinate systems on a sphere, much like geographical longitude and latitude. The sky is viewed as a dome (the celestial sphere) and one could then use the angles *azimuth*, measured along the horizon, usually from the south, and *altitude*, the angle above the horizon, to define an object's position (Fig. 2.1, left). However, these angles change with time because of the Earth's motion and with the observers position on Earth.

The fundamental coordinates for an object are based on the rotation axis of the Earth, which gives celestial poles (seen in the zenith when standing at the corresponding Earth pole) and the equator. The coordinates are *right ascension (R.A.* or α), measured along the equator, and *declination* (Dec. or δ) as the angular distance from the celestial equator. The zero point of R.A. is the *vernal equinox* (♈ in Fig. 2.1, right), which is the position on the sky where the Sun crosses the equator at the time of the northern hemisphere spring (roughly 21 March). Figure 2.2 depicts the situation when the equinox ♈ happens to be on the meridian. R.A. is measured anti-clockwise and is expressed in hours (the full circle of 360° corresponding to 24 h), minutes and seconds.

The Sun moves along the celestial sphere in a year, along a great circle that is called the *ecliptic*. The equinoxes are the points of intersection of the (planes of the) equator and the ecliptic. The angle between ecliptic and equator (the *obliquity*) is currently 23°.5, but varies somewhat over time because of *nutation* (see below).

For studies of the Galaxy, the coordinate system is based on the Milky Way. The corresponding coordinates are *Galactic longitude* and *latitude* (l and b). Nowadays, the longitude zero point is the direction of the Galaxy's center as defined by the International Astronomical Union IAU; in Oort's early years the zero point was defined to be at the intersection of the Galactic and celestial equators (and the center was at longitude 327°). The Galactic equator makes an angle of 62°.9 with the celestial one.

First I turn to coordinate systems, which are necessary to define coordinates for positions of celestial objects on the sky. Oort always had a keen interest in and saw the importance of astrometry, the science of stellar positions, and he did fundamental work defining the coordinate system to study the Galaxy. Details are collected in Box 2.1. Suffice it to remember that in astronomy we use in practice three types of coordinates that we will encounter in this book. These are viewed by imagining the sky to be a sphere, which rotates around the Earth, with the objects attached to this sphere. The rotation is the reflection of the daily rotation of the Earth around its north-south axis. The first type of coordinate system is defined with respect to the horizon and the coordinates are called *azimuth* measured along the horizon and *altitude* or height above the horizon, all in degrees. These are dependent on both the position of the observer on the Earth and on the start of the observation. Obviously this is not suitable for catalogs where positions should be independent of place and time. The system used for catalogs rotates along with the celestial sphere and is therefore independent of the position of the observer and (unless the objects themselves move on the sky) the time of the observation. The universal system used is based on the poles and equator of the celestial sphere. The celestial poles are straight above the terrestrial poles and the celestial equator above the terrestrial equator. These coordinates are *right ascension* and *declination*, the first measured along the equator and the second perpendicular to it (see Box 2.1).

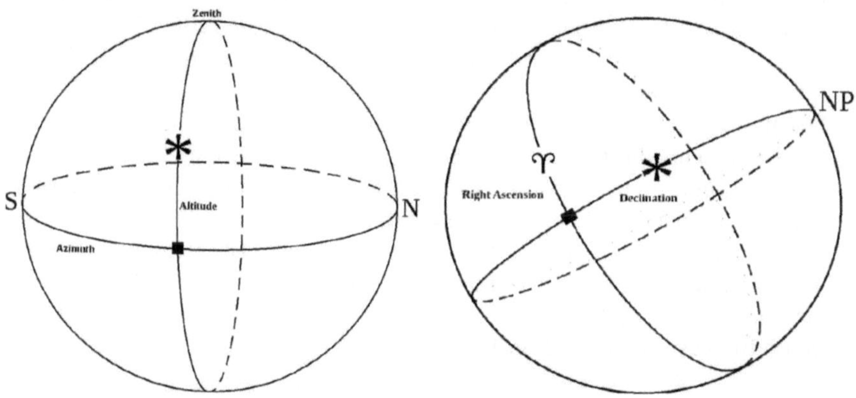

Fig. 2.1 Coordinates on the sky for a star in azimuth and altitude (left), and in right ascension and declination (right). The symbol ♈ indicates the position of the vernal equinox. Figures by the author

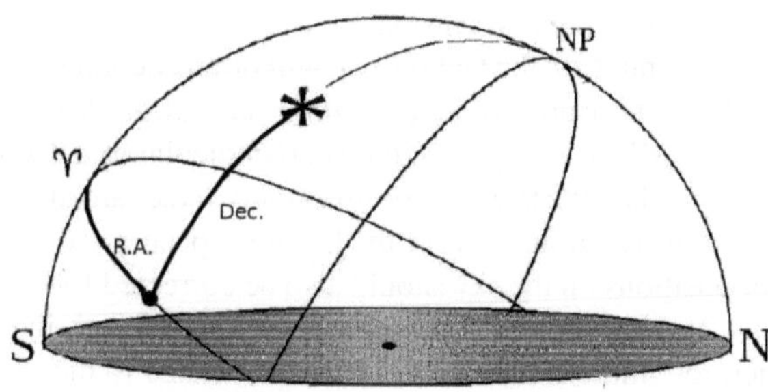

Fig. 2.2 Coordinate system right ascension (R.A.) and declination (Dec.) for a northern hemisphere observer. R.A. is the angle between the vernal equinox ♈ and the projection of the star's position on the equator ●, and Dec. the angle between the star ∗ and ●. Here the vernal equinox happens to be on the meridian. Figure by the author

Box 2.2. Precession, atmospheric refraction and aberration

The Earths rotation axis makes a slow gyrating motion similar to the motion of the axis of a spinning top on a flat surface. This is called precession. Due to this *precession* the position of the vernal equinox, and thus R.A. and Dec. of any object, changes slowly with time. This leaves the obliquity unchanged and the equinoxes move along the equator and complete a full circle in about 26,000 years; a shift (the *precession constant*) of about 50 arcsec per year. Precession was discovered around 130BCE by Hipparchus of Nicaea (c.190–c.120BCE). In addition, there are very small variations in the obliquity as a result of perturbations by the Moon and the planets. This is called *nutation*.

The Earths atmosphere is curved and its density changes with height. The light from a star through the atmosphere therefore follows a curved path which results in the apparent position being slightly higher in the sky. This effect is called *atmospheric refraction*. The effect changes with altitude, being maximal at the horizon to zero at the zenith. The effect is larger at bluer wavelengths; at optical wavelengths it amounts to about a minute of arc at 45° altitude.

Aberration of light results from the fact that the speed of light is finite (300,000 km/s), while the Earth goes around the Sun with approximately 30 km/s. Usually the effect is likened to the case of raindrops falling on a window of a moving train; even if the drops fall vertically they come down at a slanted angle on the window. The effect results in a shift of the star's position because the telescope has moved with the Earth during the finite time it takes the light to travel the length of the telescope tube. The effect is maximal in the direction perpendicular to the Earth's motion and of the order of a ten thousandth (30/300,000) of a radian or about 20 arcsec. It was discovered by James Bradley (1693–1762) in 1725.

The annual path of the Sun on the sky—the reflection of the orbit of the Earth around the Sun—is called the *ecliptic* and the position of the Sun at the northern spring, when the Sun stands directly above the equator, the *vernal equinox*; it is by definition the zero-point of right ascension.

Since Oort worked much on the Galaxy, we will often encounter coordinates relative to the Milky Way. These are *Galactic longitude* and *Galactic latitude*. Longitude is measured along the Milky Way and latitude perpendicular to it. Coordinates of objects on the sky, even when these would be absolutely fixed, are affected by changes in the orientation of the rotational axis of the Earth in space (*precession*). The actual measurements of positions on the sky should also be corrected for *refraction* of light which is deflected from a straight path while traveling through the atmosphere of the Earth. Furthermore there is *aberration*, which is related to the fact that we move through space along with the Earth at a velocity that is very small but not negligible with respect to the speed of light. The background and details are presented in the Box 2.2.

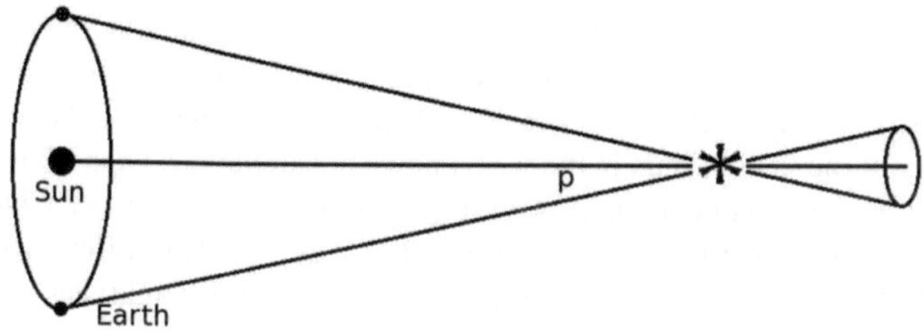

Fig. 2.3 Annual motion of the Earth around the Sun (left) is reflected on the sky by a small ellipse that is traced out by the star in the course of a year. The small angle 'p' is the parallax. Figure by the author

Box 2.3. Distances of stars

In the course of a year a star describes an ellipse on the sky as a reflection of the motion of the Earth around the Sun (see Fig. 2.3). The semi-major axis of this ellipse is its *parallax*. The first parallaxes were measured in the 1830s in Germany by Friedrich Wilhelm Bessel (1784–1846) of 61 Cygni from Königsberg Observatory (now Kaliningrad, Russia), from Dorpat Observatory (now Tartu) in Estonia, by Friedrich Georg Wilhelm von Struve (1793–1864) of Vega, and from Cape Observatory by Thomas James Henderson (1798–1844) of α Centauri.

The distance of a star is expressed in *parsec*, which is the distance at which the parallax is one parsec, so that one Astronomical Unit would subtend an angle of one second of arc at one parsec distance. Herbert Hall Turner (1861–1930), a British astronomer, is quoted as the one who introduced the parsec. A parsec is 3.1×10^{16} m or about 206,000 (the same as the number of arcsec in a radian) Astronomical Units, our mean distance to the Sun. Alternatively the *lightyear* is used; this is the distance traveled by light through vacuum in one year. One parsec is equal to 3.26 light-years.

Box 2.4. Brightness of stars

For historical reasons, astronomers express the *apparent brightness* of a star in *magnitudes*. This concept has come down to us from antiquity, which is why the scale was not rigorously defined at first. Eventually the scale became accurately established as a logarithmic one, such that 5 magnitudes correspond to a factor 100 (and one magnitude to a factor of about 2.512). This definition was established by Norman Robert Pogson (1829–1891), who proposed this scale in 1856.

To the naked eye the brightest star is Sirius at a magnitude of -1.5; the faintest that can be seen are about magnitude 6. This represents a factor of 1000 or so in brightness. The Pole-star (Polaris) is of magnitude 2.0, and the majority of the visible stars in the well-known constellations Ursa Major (Big Dipper), Cassiopeia or the belt of Orion (accent on the second syllable 'ri') have magnitudes between roughly 1.7 and 2.4 e.g., so they differ by only a factor two or so. But their distances range from 65 to 1300 lightyears, or by a factor 20.

In order to correct the apparent magnitude of a star for its distance from us the *absolute magnitude* is defined by calculating the magnitude of a star if its distance were 10 parsecs or about 32.6 lightyears. This definition of absolute magnitude was first used by J. C. Kapteyn in the 1890s (see my *JCKbiog* [7] for details).

If m is the apparent of a star and M the absolute magnitude, r the distance in parsecs, π the parallax in arcsec and L the luminosity in solar units,

$$M = m - 5\log r = m + 5 + 5\log \pi.$$

For the Sun the absolute magnitude $M = 4.8$ at visual wavelengths.

Box 2.5. Motions of stars

In the course of time stars change their positions with respect to each other due to their motions in space. This is called a star's *proper motion*. It was first noted by Edmond Halley (1656–1742) in 1718.

In addition to the proper motion due to the random velocity of a star through space, there is also a systematic pattern of streaming across the sky as a result of the Sun's motion with respect to the whole of the stars in its neighborhood (the *Local Standard of Rest (LSR)*.). This pattern consists of a tendency of stars to be streaming away on average from the point on the sky of the direction of the Sun's motion through space and this is called the *Solar Apex*. It lies in the constellation 'Hercules' and the velocity of the Sun with respect to the LSR is of order 20 km/s. This streaming pattern can be used to derive statistical distances for well-defined groups of stars, the so-called *secular parallax*.

Radial velocities are measured on spectra using the *Doppler effect*, named after Christian Andreas Doppler (1803–1853). The Doppler effect is the well-known property that the sound from a vehicle moving towards us has a higher pitch than when standing still and vice versa. This also holds for light, where a velocity away from us results in a shift towards redder, i.e. longer wavelengths. The lines in a star's spectrum can be used to determine its radial velocity.

The change in wavelength $\Delta\lambda$ compared to the wavelength λ of the line at rest is the same as the radial velocity $V_{\rm rad}$ compared to the speed of light c, or

$$\frac{\Delta\lambda}{\lambda} = \frac{V_{rad}}{c}.$$

This formula hols only for radial velocities much smaller than c. For a medium optical wavelength—say 5000Å or 500 nm— a 1Å shift corresponds to a radial velocity of 60 km/s, since $c \approx 300{,}000$ km/s.

Coordinates are determined by measuring the altitude and time at meridian passage of a star (or other object). The meridian is the line running from south on the horizon thought the zenith straight above us to the north back on the horizon (or vice versa). With geographic latitude of the Observatory this can be turned into a declination and the time into a right ascension when the time is related to the position of the vernal equinox at that moment. For this the sidereal time is used, which is zero when the vernal equinox is on the meridian and reads 24 h or zero again when this equinox returns to the meridian. Since our clocks are set with respect to the Sun, this is different from the usual time and lasts 23 h, 56 min and 4 s on our clocks. The telescope traditionally used for this is a so-called meridian circle, which can move only in the plane of the meridian about a horizontal axis.

Properties that I also need to introduce for a non-astronomical audience concern the brightness of stars, their distances and motions in space. I start with distances. These are notoriously difficult to measure, but often of fundamental importance to understanding of the cosmos. The fundamental manner is to look at the motion of the star on the sky with respect to the average of faint stars nearby on the sky, which are presumable much more distant. The almost circular motion of the Earth in its orbit translates in an annual elliptical motion of the star, the dimension of the major axis being directly related to its distance. Half the major axis is defined as the *parallax*. Even for the nearest stars this is less than a second of arc. By definition the distance of a star is 1 parsec (pc), when the parallax is 1 arcsec. It amounts to 3.1×10^{16} m. A lightyear is the distance light travels in one year and this is 9.5×10^{15} m; one parsec equals 3.26 lightyears.

Brightnesses of stars are expressed in magnitudes, dating back to the ancient Greeks. The brightest stars in the sky were designated magnitude one and the faintest the eye can see magnitude six. The definition has been made more precise in the course of time (see Box 2.4). In the modern definition a difference of 5 magnitudes corresponds to a factor 100 in brightness.

The magnitude of a star depends on its distance (and intrinsic luminosity). To refer to the intrinsic brightness of a star the *absolute magnitude* has been defined as the magnitude that the star would have if its distance were 10 parsec. It has been Kapteyn in Groningen who is responsible for this definition of absolute magnitude.

Finally we have to consider the motions of stars in space (see Box 2.5), which has been a major concern in Oort's work. Astronomers distinguish radial and tangential velocities. The tangential one is the component perpendicular to the line of sight. It manifests itself by a slow change of position on the sky, called the proper motion. It is measured in arcsec per year (sometimes per century) and depends not only on the velocity of a star in space, but also on its distance. The radial velocity is along the line of sight (in km/s and positive when away from us) and is measured by the shift of spectral lines in a stellar spectrum and does not depend on distance. These are shifted towards redder, i.e. longer wavelengths when the star is moving away from us.

2.2 University of Groningen and Kapteyn

Jacobus Cornelius Kapteyn was appointed Professor of Astronomy and Theoretical Mechanics at the University of Groningen in 1878. The background to this appointment has been described in much detail in my biography of Kapteyn *JCKbiog* [7] and I will only give a summary here with the most important points.

We have seen how the liberal prime minister Thorbecke had reformed the secondary school system. The reform of higher education took more effort and was successful only after Thorbecke's death under more conservative leadership. It went into effect in 1877. Contrary to the expectation that only the Universities of Leiden and Utrecht would survive, the University of Groningen was continued and the city of Amsterdam was allowed its (Municipal) University. So in 1877 there were not two or three, but even four universities in the Netherlands. In addition, the law provided for another very important change, which involved the expansion of opportunities for scientific research. Academic studies were to be seen as training and preparation not only for performing societal service in positions for which a scientific background was required, but also for independent practice of scientific scholarship. The emphasis was put on science rather than a broad general education, and this in turn led to increasing specialization in the curriculum. As a result, students received a more thorough but inevitably more narrowly-focussed training.

The number of subjects taught increased greatly and the new law resulted in a sharp increase in budgets for the universities, starting with a doubling in 1877 and another one in 1878. In part this was due to an increase in the number of professors, but their salaries went up as well. Their number went up to almost 200 nationwide, also twice the previous number. Not only were they required to offer more courses, research was now an increasingly important part of their duties. Gradually publications came to be a major criterion in the selection for appointments.

Soon after the introduction of the Higher Education Act of 1876, the Faculty of Mathematics and Natural Sciences in Groningen requested an increase in the number of professors and stated that the appointment of a Professor of Astronomy was considered 'urgent and most necessary', insisting strongly that the appointment of a Professor of Mathematics and Theoretical Mechanics was top priority. The outcome of all this was that Jacobus C. Kapteyn was appointed Professor in Astronomy and Theoretical Mechanics. The Royal Decree was published in December 1877 and Kapteyn took up the appointment in February 1878.

Kapteyn was born and raised in Barneveld, where his parents ran a boarding school for boys. He also had received his education there and after that had studied mathematics, physics and astronomy in Utrecht, where he defended his Ph.D. thesis in 1875. This thesis produced under the supervision of Cornelis Hubertus Carolus Grinwis (1831–1899), Professor of Mathematics, concerned the theory of vibrating flat membranes and would be designated nowadays as research in applied mathematics. Kapteyn had wanted to specialize in astronomy, but when he was to choose a subject for and perform his Ph.D. research there was no Professor of Astronomy. He had tried to transfer to Leiden, but financial reasons (presumably his father did not want to pay for it) had prevented that. However, at the time he obtained his Doctor's degree there happened to be a vacancy in Leiden at the Observatory for the position of 'Observator'. An Observator belongs to the scientific staff, working as an observer at the Observatory under the direction and responsibility of the Director and Professor. A formal translation of Observator into English would be 'observer'; however, I prefer to use the term Observator when it refers to a formal appointment at an astronomical Observatory. The Observator had no teaching duties, his primary task being to perform observations and reduce the data into publishable form. The Director, Hendricus Gerardus van de Sande Bakhuyzen (1838–1923), had known Kapteyn from the latter's efforts to try and come to Leiden for his Ph.D. and, after a trial period, had arranged for his appointment, also in 1875. As Observator in Leiden Kapteyn gained much practical experience in observational astronomy and astrometry, which resulted in him being selected for the new position in Groningen out of five candidates.

Unlike the two other State-funded Universities in Leiden and Utrecht, the University of Groningen, however, did not have an Observatory. Kapteyn unsuccessfully tried to raise funds for his own Observatory. A major argument for the Minister to refuse to allocate money for an Observatory in Groningen, was the negative advise he obtained from the Directors in Leiden and Utrecht, who clearly were not enthusiastic to share the available funds for astronomy with a third party. This did not discourage Kapteyn, who started to do significant work in astronomy, some of it facilitated by the use of the telescopes in Leiden during his academic vacations. First, he proposed a new way of solving Kepler's equation for orbits in the Solar System. These orbits, being either of planets, comets or asteroids (a swarm small bodies moving mostly between the orbits of Mars and Jupiter) were described by the three laws that Johannes Kepler (1571–1630) defined, and who derived an equation, named after him, that made it possible to calculate the position of an object in its orbit at any specified time when the parameters of the orbits were known. However, solving this equation had to be done by numerical techniques and that was a major problem at the time. Existing methods were very time-consuming and manpower-intensive, while the rapid discovery of large numbers of asteroids required quick and accurate determinations of their orbits before they would be lost again. This was a real threat until in the 1890s the Astronomisches Rechen-Institut in Berlin under Friedrich Tietjes (1834–1895) and Julius Bauschinger (1860–1934) took the responsibility of keeping track of asteroids and started the publication of circulars with positions, etc. Kapteyn's work was very much in line with his academic training as a mathematician. His method, based on a new series expansion, did not, however, constitute a significant improvement.

Kapteyn's aim in research was focused on the determination of the distribution of stars in space, a branch of astronomy building on the work of Frederick William Herschel (1738–1822). The pioneering research of William Herschel and his son John Frederick William (1792–1871) in providing star counts over restricted areas of the sky, had only led to a rudimentary solution to the 'Sidereal Problem', also referred to as the 'Construction of the Heavens'. William Herschel had provided a crosscut through the Sidereal System from 'star gauges' in a great circle on the sky almost perpendicular to the Milky Way. These counts were only down to magnitude 15 (visual) or so, as I found when I compared them to modern data [43]. For his program to improve on that, Kapteyn needed much deeper star counts, accurate positions, proper motions, and distances to stars. He concentrated first on the old problem of absolute declinations, since accurate declinations at two times are necessary to derive a proper motion. The determination of declinations depended on measurements of height above the horizon (altitude) and these were affected by effects of flexure in the telescope tube and structure, and of atmospheric refraction, both of which were known to exist, but not determined with sufficient accuracy. He designed a method to establish a set of absolute declinations for stars in selected parts of the sky to be used as a framework against which other declinations could be measured, at least when the geographical latitude of the Observatory was known independently. But this was also measured using altitudes and therefore equally affected by the effects of telescope flexure and atmospheric refraction (Fig. 2.4).

Kapteyn next devised a method to derive latitude independent of these effects, which only involved determinations of azimuths, *differences* of altitudes of meridian passages of two stars occurring at about at the same time with similar heights but on opposite sides of the zenith, or timing of passages through the prime vertical (the line east-zenith-west). He tested this in Leiden during academic vacations, using a so-called Universal Instrument (see Sect. 5.1 for an explanation) that can be pointed to every position in the sky (Fig. 5.7. in my *JCKbiog* [7]), which proved that the method would work. He published a paper on this in German with the title (in translation) *On a method to determine the altitude of the pole as much as possible free from systematic errors* [44]. This brought him in contact with David Gill (1843–1914), Her Majesty's Astronomer at the Cape of Good Hope and Director of the Royal Observatory in Cape Town (see Fig. 2.5), whom he had sent a reprint of this paper and who subsequently started a program to set up such a framework in the south.

Kapteyn also tried to measure distances of stars. He realized that the annual motion on the sky due to the orbit of the Earth around the Sun, the parallax, was predominantly east-west and therefore would result in small differences in the time of passage of a star through the meridian. This should then be determined with respect to fainter stars in the same area of sky that are assumed farther away. The problem is that the effect is extremely small and the timing has to be done to an accuracy of the order a few hundredths of a second of time. This can only be done by

Fig. 2.4 Jacobus Cornelius Kapteyn on an undated photograph. From Kapteyn Astronomical Institute

repeated measurements provided it is done with extreme consistency. Still, Kapteyn measured distances of a set of about a dozen stars (selected to have large proper motions and therefore presumably at relatively small distances) with the Leiden meridian circle (see Sect. 2.1) to an amazing degree of precision. He must have been an excellent observer. This was a major success, but also a failure, since this also showed that measuring distances of stars in wholesale manners was impossible using trigonometric techniques.

In the second half of the nineteenth century, star catalogs had been produced to great accuracy both in position and in apparent brightness, but only in the northern hemisphere. Friedrich Wilhelm August Argelander (1799–1875) had founded the Bonner Sternwarte in 1844 and between 1849 and 1863 he and his assistants had measured positions and magnitudes of stars in the northern hemisphere (north of declination −2°) down to magnitude 10, although complete only down to apparent magnitude 9. The resulting star catalog, known as the *Bonner Durchmusterung*, contained 324,198 stars. Although work was in progress on extensions to the southern skies, both from Bonn as far as possible—the *Südliche Durchmusterung*—and further from Argentina—the *Córdoba Durchmusterung*—, not much progress had been achieved in the decades after that. In the mean time photography entered the field of astronomy. David Gill had noted that on photographs of the Great Comet of 1882 many stars were visible. He reasoned that it should be possible to photograph the whole southern skies and measure positions of stars from these plates and estimate their magnitudes, and in this way produce a southern Durchmusterung.

He proceeded to build a special telescope and started taking the necessary plates, but failed to secure the manpower to do the actual measuring of the plates. In his correspondence with Kapteyn he slowly lured the latter into offering his services to perform that task. Kapteyn secured funds to do the measurements (and hire an assistant) and eventually, after twelve years of work, the resulting *Cape Photographic*

Fig. 2.5 David Gill, Director of the Cape Observatory in South Africa and Simon Newcomb (1835–1909), Director of the Nautical Almanac Office at the US Naval Observatory. Gill: From the *Album Amicorum* presented at the retirement of H.G. van de Sande Bakhuyzen as Professor of Astronomy and Director of Leiden Observatory in 1908. (Archives Leiden Observatory; see caption Fig. 3.1) Newcomb: from Henriette Hertzsprung–Kapteyn's biography of her father [45]

Durchmusterung was published by Gill and Kapteyn, the last of the three volumes in 1900. The *CPD*, as it is often referred to, contained positions and magnitudes of 454,875 stars south of declination $-18°$. It made Kapteyn a well-known and respected astronomer.

Kapteyn had not neglected his real ambition. He had decided that, now that measurement of distances of stars was not possible on a large scale, he had to try and determine stellar distances statistically. For this he used the fact that stars, including the Sun, do move in space with respect to each other. The motion of the Sun relative to the mean of stars around us, reflects in a systematic pattern in the proper motions of all stars on the sky, was first found by William Herschel in 1783. This gave the direction of the solar motion in space and when complemented with radial velocities also the actual speed of the Sun. Since this is about 20 km/s, or about four times the radius of the earth's orbit around the Sun per year, this provides a much larger baseline than classical parallax measurements, and that baseline furthermore grows with time. If all stars were stationary in space their proper motion would be a direct measure of their distance, but since stars do move themselves, this method can only be used in a statistical manner.

Kapteyn, with the help of his brother Willem Kapteyn, who was a Professor of Mathematics in Utrecht, set out to develop the mathematical procedures to determine the distribution of stars in space, once large quantities of star numbers as a function of their apparent brightness were known and large data-sets of proper motions would be available. He had to make three assumptions to make this possible. The first was that in every volume of space the distribution of intrinsic brightnesses of stars would be the

Fig. 2.6 Edward Charles Pickering, Director of Harvard College Observatory, and George Ellery Hale, Director of Mount Wilson Observatory. Pickering: Courtesy of Harvard College Observatory [46]. Hale: taken from Henriette Hertzsprung–Kapteyn's biography of her father

same as what it is near the Sun. The second assumption was that there is no material in space that absorbs or scatters starlight so that a distant star would appear fainter than what it would be had there been no such interstellar material. This assumption would haunt Kapteyn all his life. The third assumption was that everywhere stars would move randomly without any preferred direction and on average with the same speed.

The latter seemed reasonable, but not long after 1900 Kapteyn discovered from a careful study of available material that this was in fact not the case. He found that there were two preferred directions, when corrected for the motion of the Sun through space, opposite on the sky and lying in the Milky Way. This seemed to indicate that there were two collections of stars moving through each other in opposite directions, and Kapteyn referred to this phenomenon as the 'Two Star Streams'. He announced this in 1904 at a meeting of scientists in St. Louis, Missouri, that was held on the occasion of the 'Louisiana Purchase Exposition', the World Fair celebrating the centenary (actually a year too late) of the acquisition of the USA of a large piece of land from the French. Prominent in the organization of this meeting had been Simon Newcomb (1835–1909), Director of the Nautical Almanac Office at the US Naval Observatory (see Fig. 2.5), who knew Kapteyn well and had visited him several times in Groningen. Kapteyn was the first Dutch astronomer to attend such a scientific meeting in the United States and with this discovery he instantly became famous and a leading astronomer. Actually, his interpretation in the end turned out to be incorrect. It results from asymmetries in the distribution of space velocities of the stars, but that will be discussed later on in this book. Only after Kapteyn's death this was established definitely.

Fig. 2.7 Jacobus Kapteyn and his wife at Mount Wilson during their visit in 1909. They are sitting in front of the tent that was provided for them. A year later the Kapteyn Cottage had been built, where they stayed during their further visits until after 1914 the First World War prevented them from traveling to the California. From University Museum Groningen

The existence of the star streams made the determination of the distribution of stars in space not impossible, but certainly more complicated. Kapteyn realized that more information was necessary for a proper solution of that problem. What was needed was information on the distribution of stars and their properties over the whole sky down to much fainter limits than existed at that time. Since it was impractical to do this for all stars, Kapteyn proposed that Observatories should collaborate to provide as much information as possible on stars in a set of well-defined small areas of the sky. This proposal, his Plan of Selected Areas, that he published in 1906, was eventually adopted by various Observatories. In particular, Edward Pickering (1846–1919), Director of Harvard College Observatory, (see Fig. 2.6) took it upon himself to photograph all the over 200 area's from Harvard and its Southern Station in Peru. At a price though, since Pickering got the Plan extended with a set of special area's in the Milky Way, which meant much more work for Kapteyn.

Even more spectacular was the decision of George Ellery Hale (1868–1938), Director of Mount Wilson Observatory (see Fig. 2.6), to make Kapteyn's Plan the prime program for his new 60 in. Telescope—then the largest in the world—, followed by his offer to Kapteyn to be appointed as part-time research associate of the Carnegie Institution (that funded Mount Wilson Observatory). With this giant telescope star counts could be extended for the northern Selected Area's to magnitude 19, three

Fig. 2.8 Painting of Kapteyn by Jan Pieter Veth. It is a preliminary version of a painting given to Kapteyn and his wife on the occasion of his fortieth anniversary as a University Professor in 1918 (reproduced in Fig 3.7). See text for the full story. This is what Kapteyn looked like when Oort came to study with him in Groningen. The painting itself is property of Kapteyn's great-grandson and namesake

magnitudes fainter than for Harvard. Kapteyn first visited Mount Wilson in 1908 when the 60 in. became operational and returned each year with his wife until 1914, after which the First World War prevented them from sailing across the Atlantic. Figure 2.7 show them in front of the tent at Mount Wilson during the 1909 visit. After this a special cottage was built for them, which became (and still is) known as the Kapteyn Cottage.

The plates from Harvard and Mount Wilson were measured chiefly by Kapteyn in Groningen and resulted in catalogs that were mostly published after Kapteyn had died, but of which he must have seen some preliminary results. The parts on the 'Systematic Plan' were published in three installments in the Annals of the Astronomical Observatory of Harvard College between 1918 and 1924. The coordination after the deaths of Pickering (1919) and Kapteyn (1922) was taken over by Kapteyn's successor in Groningen Pieter Johannes van Rhijn (1886–1960). The corresponding part on the 'Special Plan', on which Pickering had been insisting so much, was not published until 1952. The Mount Wilson Catalog was finally published in 1930 as a special publication of the Carnegie Institution, eight years after Kapteyn had died. See [47] for the references to these publications. Frederick Hanley Seares (1873–1964) from Mount Wilson had done fundamental work on the determination and calibration of stellar magnitudes from photographic plates taken with the 60 in. Telescope.

This was the state of affairs in the University of Groningen when Oort enrolled in 1917 to study physics and astronomy, attracted to a large extent by the presence there of Kapteyn. Figure 2.8 shows a painting of Kapteyn around the time of his fortieth anniversary as a professor in Groningen in 1918. This is a trial version of a portrait presented to him and his wife by friends and colleagues. The final version was rejected by Mrs. Kapteyn and the painter, Jan Pieter Veth (1864–1925), replaced it with the

Fig. 2.9 Kapteyn Astronomical Laboratory in Groningen. This is where Oort worked as a student and assistant. From University Museum Groningen

well-known version of Kapteyn sitting behind his desk, which hangs in the Kapteyn Room in the Kapteyn Astronomical Institute (see Fig. 3.7). The rejected version was later painted over by Veth himself with an academic gown, cap, and attire and is now hanging in the Senate Room of the University Academy Building. The building, in which the Astronomical Laboratory was housed and where Oort must have worked as a student and later, as an assistant, is illustrated in Fig. 2.9. This picture was taken after it was formally named after Kapteyn by the Dutch government, at the proposal of the University of Groningen, at the time of Kapteyn's retirement in 1921. The building no longer exists; after a fire on the first floor in 1988, the governing board of the University seized the opportunity to tear it down.

2.3 A Student in Groningen

Oort started his studies in Groningen in 1917. He officially enrolled on October 3, as mentioned in the 'Groningensche Studentenalmanak'. This annual book served as the yearbook of the Student Association *Vindicat atque Polit*, that we will encounter in more detail later. An impression of Oort's life in Groningen can be gained from a few letters he wrote to his older brother Hein (Fig. 2.10). Hein studied physics in Leiden. The first letter was written on Saturday November 10, 1917, so not long after Oort arrived in Groningen. The address was Poststraat 2, which is at the corner with the Oude Boteringestraat, very near the central Academy Building of the University,

Fig. 2.10 Henricus (Hein) Oort in 1917. From the Oort Archives

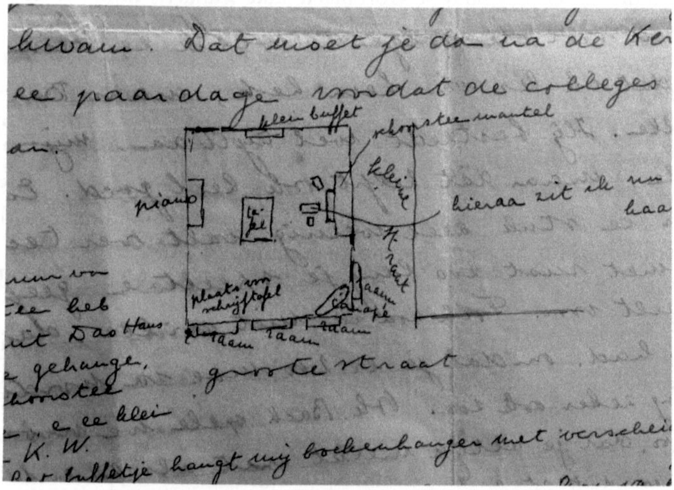

Fig. 2.11 Part of Oort's letter of November 10, 1917 to his older brother Hein. The sketch is of his living room. From the Oort Archives

Kapteyn's Astronomical Laboratory and the central square of Groningen, the Grote Markt. A part of the first page has been reproduced in Fig. 2.11.

> 'Dear Hein, I will now finally let you hear from me. Something that is the least pleasant of all things is not having your own home, and that was my situation when I occupied that room in the Gelkingestraat. It will be much better here and it already is. Before this I never was in my room and always was with Bart or the Voorthuysens, but I spend more time here now and slowly it also is becoming more and more my room, since I have my own books and put my own pictures on the walls. I will make a drawing so that you can imagine how nice it would be to sit together if you would come and visit.'

The names 'Bart' and 'van Voorthuijsen' appear regularly. It is not clear from the letter who Bart was; his last name is never mentioned. The best option is Barteld Roelof Damsté (1899–1920). His father, Rudolf Bartheld Damsté, was a physician in Leiden. Damsté enrolled in Groningen in October 1917, some two weeks later than Oort, and studied theology. He lived at an address different from Oort (but nearby: in the 1918 Almanak his address is Turftorenstraat 29a, also very much in the center of Groningen). He died on May 6, 1920 according to the municipal records of Leiden as an inhabitant of that city without a profession. He probably quit his studies and fell ill (or the other way around) in 1919 or 1920, since for those years he is no longer listed as a student in Groningen.

In an undated letter, Oort sent in a postscript greetings from Bart to Hein and mentioned that Bart is very busy so Oort has to take walks by himself. He added that the Voorthuysens are in excellent shape. This refers to Adriaan van Voorthuysen, whose name appeared on Oort's birth certificate (see Fig. 1.5). He had moved to Groningen in 1910 to become a school physician or medical officer, where he became much concerned with the fate of mentally disabled children, for which no appropriate schooling was provided. His pleas were heard by the Minister and in 1921 he was appointed as chief inspector at the Ministry for special elementary schools for such children. Van Voorthuijsen is known as the driving force behind the establishing of a special education for mentally handicapped children. At least for some time van Voorthuijsen and his wife would have provided a local home for Oort as old friends of his father.

The drawing in Fig. 2.11 shows at the bottom three windows overlooking a 'big street', the Oude Boteringestraat, and one to the left into the Poststraat. In the middle is a table and at the right the fireplace, in front of which Oort mentioned he was writing the letter. To the left was the piano and at the top a 'small cupboard', on which there was 'a carafe with water and 2 glasses and further the copper candlesticks with a candle in it, that picture from Dürer of the hermit that carries Jesus across a stream'. At the bottom-right is a sofa. This must be a living room and Oort apparently had a separate bedroom and washing facilities. The picture referred to is probably the 1511 woodcut by Albrecht Dürer (1471–1528) showing St. Christopher carrying the infant Jesus Christ across the river. Oort later had a copy of Dürer's well-known portrait of a rabbit in his living room.

'I like Dürer a lot these days. His pictures are só beautiful, and they bestow a lot of power upon you when you have the courage to look at them. You should try to find one also to put on the walls of your room. Bart and I put five cents in a piggy bank every day; we can then buy a book about Dürer hopefully by the time of the Christmas holidays.'

Oort described how enchanted he was at that time of Mozart and in particular one sonate (unspecified) that he was practising for his piano lessons. Also playing Bach was a pleasure. He continued:

'Do you read much? Or are you so busy with your work that you have no time for it? In case you do read, write me what it is. In the vacation you should see my astronomy lecture notes. Kapteyn and his astronomy slowly begin to attract me greatly. [...]

I wish you could hear this Mozart sonata, but when I finish this letter I will play it, maybe you can hear it then in the letter, and also the 2nd Invention. I could play those otherwise in the Christmas vacation.

Well, I have to stop now, since I also still need to have coffee and after that I will attend more lectures by Haga, always about heat and work. This morning we had capita [selected lectures] about grating spectra and Rowland in connection with that. His lecturing is very pleasant and not very difficult and his experiments are magnificent.'

Hermanus Haga (1852–1936) (see Fig. 2.18) was a Professor of Physics and Meteorology in Groningen between 1886 and 1922. In a letter of October 22, 1918, a year later roughly, Oort again wrote to Hein (there were very likely more letters that did not survive).

'Very nice that you like your work; I am busy myself also with the mathematics lectures, especially those of Barrau which require much homework. You probably did not study analytic geometry. Or did you? One set of lectures of Barrau just continue with Analyt. Math.; the other with algebraic curves, that will be covered in Leiden as well. Wolff lectures about integral calculus extensively in one course and about series in another. Further I follow Kapteyn's lectures, some about determination of orbits that promises to be interesting, but for the moment is about 'making calculations' and the 'Method of Least Squares' that is more than full of mathematics and especially integrals. I have found, as I wrote this morning, a very nice spot for my telescope, but had no clear night and now it is clouded also. That star chart is very nice, isn't it? At least it contains many stars.

In the time I have left I am busy working on chemistry, namely the capita by Jaeger and that takes a lot of time. Further I often play piano in the morning and work especially hard at finger exercises, That sonata looks difficult to me, but I will give it a try. We have a nice dance club, so I can meet some more Gron. girls and there are very nice ones among them. But since we have had only one lesson yet and I knew none of them before, I will tell you much more about it later.

Do you have time to read? I do not read all that much, or rather I don't get to it at all.

At the Harmonie I recently heard a very beautiful suite by Bach. You know, that suite in which there is a violin part that you played so often.'

Johan Antony Barrau, (1873–1953) was Professor of Analytic, Descriptive and Higher Geometry between 1913 and 1928. Before that he was a professor in Delft and after it in Utrecht. Julius Wolff (1882–1945) taught higher algebra, differential, and integral calculus and the theory of functions from 1917 to 1922, after which he became a professor in Utrecht. Frans Maurits Jaeger (1877–1945) was a Professor of Inorganic and Physical Chemistry (1909–1943). The *Harmonie* was a clubhouse that was housed in a major building in the center of Groningen and contained a well-known concert hall. The orchestra suite probably was the third (BWV 1068) in D-major and the part for practicing violin its 'Air'. The letter ends with a postscript:

'Will you please tell mother there is no need to send that dictionary, but she should send those lecture notes of Kapteyn that I asked for! I have plenty of butter and cheese for the moment, but still I am hungry all day, like in fact most people here are.'

There is another undated letter that I wish to quote from. It refers to an armistice, so it must have been not too long after the Armistice of Compiègne of November 11, 1918, the end of World War I. Oort wrote how the bells of the big tower of Groningen, the Martinitoren, started to chime and mentioned that these had not been rung since the start of the War in 1914.

'How terrific life is! It is as if music has the power to make you understand the deepest secrets and connect the greatest to each other. It holds something mighty that gives you life and character! [...]

You learn to see that all great things are a single part of this world, that everything is connected. Dimitri Fyodorowitch Karamazow and Grouchenka with Alischa and these then with Tschaikowsky and Bach [...]. You should read Dostojefsky (that is the Brothers Karamazow). It is very beautiful.'

The names Oort quoted (I have kept his spellings) are from the famous book *The brothers Karamazow* by Fyodor Mikhailovich Dostoyevsky (1821–1881). And following this introduction Oort gives a long discourse on hyperbolic tangents, that he has discovered lately, and fills a paragraph full with equations. Then he returns to Dostoyevsky and asks Hein to remind Bart of the fact that he has the latter's copy of the book.

The final letter to Hein is dated October 1, 1921. He tries to console his brother, who apparently has to choose between going to Delft for another four years, maybe a different, but then technical, second study or becoming a teacher (of physics one would think) and Oort refers to a discussion they had on this before the summer. Delft would give more opportunities: teacher in mathematics and physics, engineering, etc. Oort considers himself as a lucky man and an optimist.

'I am not really able to put my thoughts to words or to think clearly. I would like to sit somewhere by myself to come to a clear view of the important things in life. These thoughts now only concern trivia and special issues and that leaves behind a perpetual feeling of dissatisfaction and discontent.'

There is no further correspondence with Hein Oort. He died of diabetes in June 1922. There are letters from his younger brother John (mostly about biology matters), but no letters from Oort that shed light on his early Groningen years.

Oort did not stay very long in this room in the Poststraat. The 'Groningensche Studentenalmanak' lists him at that address only in the issue for 1918 (published late 1917). In the 1919 issue the address is Martinikerkhof 17a. This refers to the old graveyard around the major church in Groningen, the Martini church near the Grote Markt. It is the oldest church in town, named after Saint Martin of Tours (316–397), who is the patron saint of the Bishopric of Utrecht. Groningen belonged to that for a while. It dates back to the thirteenth century and became a protestant church in 1594. The associated tower is almost 97 m high and an important landmark. In all later volumes of the 'Groningensche Studentenalmanak', Oort's address is Brugstraat 7a. He needed some time to find a room to his liking. The Brugstraat is also not too far from the center of Groningen, very close to the second major church, the Der-AA kerk (Der Aa church). It was first built in the fifteenth century, so is not as old as the Martini church; the tower is 76 m high, also second to the Martini tower.

It is remarkable that except for the mentioning of the chiming of the bells of the Martinitower, no reference is made in the letters or notes to World War I, while this was raging when Oort came to Groningen. It apparently went by without much direct effect on student and other daily life in Groningen.

Not only did Oort have a lively and active interest in music and literature and did his interests in mathematics, physics and especially astronomy deepen, he was also enjoying other parts of a student's life. We have already seen that he was involved with dancing. He also became a member of the Groninger Studenten Corps *Vindicat atque Polit* (Maintain and Civilize), the fraternity of which is the oldest student associations in the country (founded in 1815). It maintained, and still does so, a society building located on the Grote Markt, which is called *Mutua Fides* (Mutual Trust). It is shown in Fig. 2.12 as it appeared in Oort's time. *Vindicat*, as the association was and is usually called, organised dance clubs such as Oort referred to in his letter to Hein. In Fig. 3.0 Oort is shown wearing the ceremonial cap 'Corpsmuts' that one was entitled to wear after completing the initiation on official occasions. One wore the cap for the first time in a procession or serenade, but—unfortunately for Oort—during WWI such ceremonies were suspended and the picture was probably taken in *Mutua Fides*. *Vindicat* was strictly for men. The female students in Groningen had their own association, the Groninger Vrouwelijke Studenten Club *Magna Pete* (Strive for the Great), which had been founded in 1898. The two societies merged in 1970.

Oort has been listed as member of the 'central faculties board' for 1921 in the Groningensche Studentenalmanak, on behalf of the Philosophical Faculty (which later would become Mathematics and Natural Sciences), of which he then was secretary. This probably is the umbrella body of the faculty associations that were concerned with the student interests. This was a function that involved work and time, and shows that he felt responsibility for representing students in such bodies

Fig. 2.12 *Mutua Fides*, the Society of the Groninger Student Corps *Vindicat atque Polit*, of which Oort was a member during his student days in Groningen. This pictures dates from around 1890. From Wikimedia Commons [48]

and did not shy away from such activities. He appeared to have had no other formal function in 'Vindicat'.

The cost of university studies was not insignificant. The Oort parents had three sons, each of whom pursued university education: Hein physics in Leiden, Jan astronomy in Groningen, and John biology in Utrecht. For a number of years at least two of them had to be supported at the same time. At that time the two younger daughters were at the HBS for girls. As an indication, the annual cost of a university student living away from his parents' home amounted around 1900 to something like Dfl. 1200, which corresponds to a current € 16,000 or so. Some students had some additional income from tutorships, etc., but there is no indication Oort had any such extras. The academic studies of their sons and the education of their daughters was a major financial burden for the Oort parents, even though as Director of a medical institution the salary of Abraham Oort would have been relatively high.

Oort's father wrote on April 8, 1919:

'First congratulations with the 'tentamen' that you passed [A tentamen is an examination at the end of a lecture course; the final exams are taken when the required 'tentamens' have all been completed successfully]; the number of these is large when you do these separately over the various curriculum years. I decline to send you each time some money as a token of congratulations. But this time I do send you f. 80. [This is € 480 in current purchasing power.]. I hope this will relieve some of your urgent needs. Mother advised me to put some pressure on you, so that you learn to match the expenses with your income, that otherwise you should come to Oegstgeest for your studies. But, as you will understand, mother was also the first to say: send him some money, and said as a sigh of hope and trust, that you did so well and have no debts. She found out later that among the bills to pay there were some from last year, but anyway... enough of this; you will understand and we are both convinced that you are not a squanderer. Mother would urge you to send your laundry before Saturday; she waited for it in vain this week.'

The company that was responsible for transporting goods, van Gend & Loos, had special cardboard boxes for sending laundry. The van Gend & Loos company originated in Antwerp by a merging in 1809 of two stagecoast companies owned by the brothers-in-law Jan-Baptist van Gend and Petrus Josephus Loos. The Netherlands and Belgium were a single nation between 1815 and 1839, and after the separation the company continued in both nations. Due to competition by the railways they gave up passenger transport in the 1880s and continued as a transport company. It was taken over by Dutch Railways in the 1920s and merged with the DHL (Adrian Dalsey, Larry Hillblom and Robert Lynn) parcel service in the 1980s. It was very common for students in those days to send their laundry home and their parents to send it back using this service.

Once a year there was a major activity of *Vindicat*, the gala ball. This was, as the indication 'gala' indicates, a formal event in evening dress. The male students brought their dates, which were of course often recruited from *Magna Pete*. It involved a ball as the main event, followed by a supper. It was often preceded by a concert by a student choir ('Bragi' was and is the name of a prominent one) or a play performance by a student theater company. It took place in the *Harmonie* and participants wore formal dress. Figures 2.13 and 2.17 have been taken during the 1920 ball. Oort could have put the experience from the dancing lessons here to good use.

In a letter to his parents of February 2, 1922 he related the following on his skating and dancing activities:

'Today two weeks ago I skated for the first time on the Paterswoldse Meer [a lake near Groningen] with Schilt, as I wrote you. The subsequent week was so cold that every step in the street was a misery. Still Piet and I undertook on Wednesday a skating tour on the Reitdiep protected by high dikes. The ice was enormously thick but poor with floes and on

Fig. 2.13 Oort, other students and their dates during the gala ball of Student Association *Vindicat atque Polit* in 1920. From the Oort Archives

account of the strong wind there were holes in the ice. There was nobody else on the ice, which gave a lonesome and desolate feeling with that weather.

On Friday the wind ceased somewhat and the temperature rose and I could not resist the temptation to skate down the Damsterdiep, with very good ice but still some wind. When Piet and I left the ice we suddenly felt like going for a dance and it could not have been more fortunate that the same evening there was a masked ball in Frigge organized by members of the Clubhouse that we had not planned to go to. We called up and were still able to secure tickets and we rounded up four more house mates and went. There we hired domino's [From the French: wide cloaks with a hood that priests used to wear in winter, usually including a mask]. [...] It was all very pleasant. After the unmasking it became even more unusual. I had supper with Annie Mossel, a beautiful girl, but engaged, and Piet with Toet Mees. August Mees, one of our former house mates, invited us to his table where there were some girls too many. The supper was very animated, but I did not eat much as there was lots of serpentines, covering all tables with thick layers of paper. The dancing went on until 4 o'clock, but even so I was present at the laboratory at 9 o'clock.'

Fig. 2.14 'Old Four' team of the Groningen student rowing club *Aegir*, after they won the Varsity race of 1918 with Oort as the cox. From the Oort Archives

'Piet' was apparently living at the same address. According to the Studentenalmanak his full name probably was Pieter Hendrik de Waard, a student in the Faculty of Law, who registered with the University in 1918. Schilt is Jan Schilt, a somewhat more senior astronomy student, whom we will encounter again later.

Another activity that Oort took up actively and extensively is rowing. He actually enjoyed rowing (and sailing) throughout much of his life. In Groningen he became a member of the Groninger Studenten Roeivereniging (rowing club) *Aegir*. In Norse mythology, Aegir, which means 'sea', is a giant living in the ocean, but also the God of sea fare and of beer. The rowing club had been founded in 1878 as an official association under the auspices of *Vindicat*. The club was and is involved in recreational rowing, but also took part in competitions and still does.

The most important student's rowing competition is the Dutch 'Varsity' race, which is a rowing regatta (series of contests) between students from the various Dutch universities. In Oort's days, it was held on the North Sea Canal, which joins Amsterdam straight to the west to the North Sea, in the second half of May or early June. Its origin is a contest or challenge organized in 1878 between *Njord* and *Laga*, the equivalents of *Aegir* in respectively Leiden and Delft. From 1882, the Utrecht rowing club *Triton* also joined, but for *Aegir* it was reported to be too far from Groningen to compete. This must refer to the difficulty and expense of moving boats over large distances. The number of boat classes— and competitors—increased. The first formal Varsity (organized on 'neutral ground'; or water really) was held in 1883.

The main race is the 'Old Four' (coxed four, sometimes in English referred to as Varsity Four), which means that there are four rowers with one oar each, two on the starboard side (right-hand) and two on the port side. The coxswain (usually shortened to cox) has a rudder to steer the boat and usually sits in the stern. The 'Old' refers to the fact that the best and most experienced rowers formed the team. The cox was chosen for being relatively lightweight. The winner of this race was and is considered the winner of the Varsity. The distance covered was 3 km. *Aegir* participated in the Varsity for the first time in 1918 and Oort was cox of the Old Four. *Nereus* (Amsterdam) was also competing by then. *Aegir* won the race by six boat-lengths and thus also the Varsity (Fig. 2.14). It would take until 1956 before they would win again.

A year later Oort once more participated in the race as cox of *Aegir*, but that time they ended third. He also participated as cox of the 'Young Four', consisting of not so senior students, but that combination did not win either. In 1920 the old four did not compete due to illness of one of the rowers, but it is not recorded who the cox of that team was. For the 1921 race it has only been recorded that the *Aegir* team did not do well, but the members of the team have not been listed. But it is recorded that in the singles (skiff) *Aegir* member E. A. Kreiken ended second and saved the honor of the Groningen rowing club. This was Egbert Adriaan Kreiken (1896–1964), another astronomy student. Kreiken was born in the same house in Barneveld as J. C. Kapteyn (for the full story see my *JCKbiog* [7]); he was some four years ahead of Oort and obtained his Ph.D. from the University of Groningen in 1923.

On January 2, 1920 Oort sent his mother, then in Davos, a postcard from Berlin (see Fig. 2.15), where apparently he and three other students had spent Christmas and New Year. The text on the postcard is (Translation by Jet Katgert-Merkelijn in *Cat2000* [5] with minor changes.):

> 'Dear Mother, You cannot half imagine how I am enjoying my stay here. Everything is cheap, or rather almost everything. I have been to the large Opera House twice. I have never seen anything so magnificent and beautiful, and that for *f* 1, 20 [*f* 1 in 1919 corresponds to about € 6 now], extraordinarily beautiful seats. You can buy everything here, also foodstuffs, and mostly cheaper than in Holland. Yet, the situation in the streets is terrible. Everything needs painting, trams are horse-drawn, humping and bumping up and down Friedrichstrasse. We are staying here in the center of town (Gendarmen Platz) in a quiet and very pleasantly decorated hotel for DM 15, breakfast not included. We traveled a full day comfortably in the D-train from Cologne to Berlin, which we boarded in Hannover (2nd Class for *f* 4,00, almost 500 km). It was not crowded. For the Germans who are not so rich as we are now, the food situation is really difficult. Tonight we will see the *Oresteia* of Aeschylus in the Grosse [should be Grosses] Schauspielhaus, which has been built recently. If I could I would stay here for a full month. The large streets and everything else! Many greetings to all in Davos. This is us in front of the Reichstag building.
>
> Yours, Jan.'

Fig. 2.15 Oort (second from right) and three companions visiting Berlin presumably over Christmas 1919 and New Year 1920. The Reichstag is in the back. From the Oort Archives

In the catalog for the 2000 Oort exposition, *Cat2000* [5], the date written on the postcard is given as January 2, 1919. However the date of the (re-)opening of the Großes Schauspielhaus, designed by architect Hans Poelzig (1869–1936), is November 28, 1919 [49] with a performance of the *Oresteia* of Aeschylus (c.525–c.456BCE) directed by Austrian-American Max Reinhardt (1873–1943). Presumably performances continued for some weeks. Oort might erroneously have written down the old year; after all the new year was only two days old.

2.4 Courses

Oort has stated repeatedly that he came to Groningen to study with Kapteyn, but was undecided whether he would major in astronomy or physics. We see from his letters to his brother Hein, that mathematics did interest and attract him also. The lectures by Kapteyn turned him definitely to astronomy. There is very little in his letters or his diary notebooks of those days that refers to that process directly.

One interesting piece that shows his deep interest in astronomy concerns a sky viewing session with his younger brother John. This must have been in Oegstgeest at the end of 1918.

> 'New Year's Eve. Jo and I have used Mister Knappert's telescope to make a walk through the Universe, so that I can hardly write as a result of the cold. You can make really long walks in it, before you have had enough of it, there is such a diversity of wonderful things in the Universe. We have spent most of our time admiring Saturn, that can be seen very well through this telescope. And the nice thing about Saturn is because of the ring you can better envisage the spherical nature of the whole body. You observe exactly where the ring disappears behind the main body and at two points you can look in between the ring and the disk, even though the ring is almost in the plane of view. You also see the shadow it throws onto the disk as a thin dark line, and the motion of the planet with respect to a faint star nearby was very nicely noticeable after an hour. Jupiter with its four moons, that was extraordinarily bright, and with its equatorial bands is still to be found in Gemini. And the second thing we studied were star clusters. I found three of them and all were easily resolved into stars.
>
> I had planned to walk by myself a bit through Endegeest since it was new year's eve. Things went differently; anyway I am very sad really that it already will be a new year again. It is always sad, at least that is what I feel, when something old disappears and then I always start to have sympathies for it. If I could still support the old 1918, that has now been pushed away, no shot away, from the Earth and every other body in the Universe, I would have done so. How ungrateful mankind is!'

The Oort Archives contain lectures notes from courses that Kapteyn taught. These encompass three small booklets. The first has the title *popular astronomy*, which should be interpreted as general or elementary astronomy in current parlance. The second page specifies that the course by Kapteyn was given on Thursday between 1 and 4 in the afternoon. The 'Groningensche StudentenAlmanak' indicates that Kapteyn gave a lecture of that title in both Oort's first and second year between 1 and 3 pm, followed by a course on 'spherical astronomy' between 3 and 4 pm. The notes start with the following (see Fig. 2.16):

> 'Historic development of the freshness and the impartiality of the observing.
>
> The history of the explanation of the universal system.
>
> Until recently almost all we knew in the area of astronomy was based on the observations of the motions of the stars. These motions have been observed from the earliest times onward. With these motions the relative positions of the fixed stars did not change. Each star in itself seems to make a circular motion in ± 24 hours around the same point (the pole) on the sky. Observations have led automatically to the assumption of

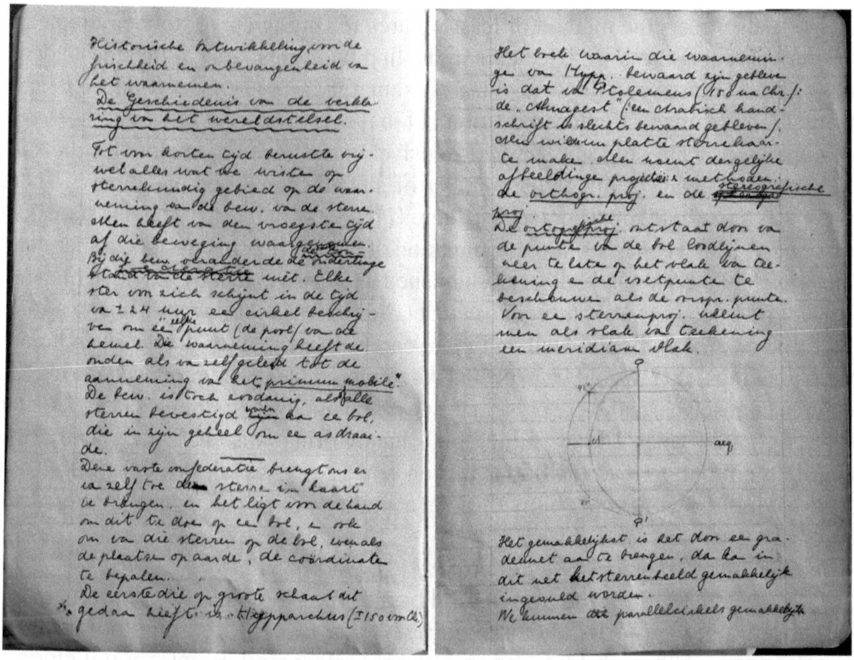

Fig. 2.16 First two pages of Oort's notes of the lecture course 'popular (=general) astronomy', given by Kapteyn in Oort's first year as a student in Groningen. See text for partial translation. From the Oort Archives

> the 'primum mobile', i.e. the motion is such that all stars appear to be attached to a sphere that as a whole rotates about an axis.
>
> This fixed configuration leads us automatically to chart the stars, and it seems obvious to use a sphere for that, just like locations on earth.
>
> The first person who did this on a larger scale is Hipparchus (~150 before Chr.). The book in which these observations have been preserved is that of Ptolemeus (150 after Chr.) [Claudius Ptolemy (87–>170)]: the 'Almagest'(: an Arabic handwritten version is all that has been preserved). People wanted to produce flat star-maps. Such pictures were called projections: 2 methods: the orthogr. proj. and the stereographic proj.'

The lectures then develop in great detail the theory of spherical astronomy. But it does put everything in a historical context by explaining the observational procedures and by indicating how the description and eventually the understanding of the properties and phenomena emerged. The course is restricted to planetary motions on the sky and coordinate systems, including precession. It also treats in detail solar and lunar eclipses and corresponding regularities like the Saros period.

The second notebook concerns *spherical astronomy* and starts with spherical geometry and astronomical coordinates. The third notebook is again on *popular astronomy* and concentrates on planetary orbits and Kepler's laws that describe these orbits. It also treats the method of least squares for fitting curves to measurements. It is probable that these were lectures in the first two years, after which the students were supposed to obtain the 'Candidaats' exam. One would expect Kapteyn to treat his method of solving Kepler's equation but that is not the case.

Kapteyn probably did not lecture extensively for astronomy majors for the subsequent 'Doctoraal' exam, as the number of students would be one or two. There likely was more individual contact and there Kapteyn would teach students statistical astronomy, which concerns the determination of the distribution of stars in space and this branch of astronomy had been established to a large extent by Kapteyn himself.

Kapteyn's lectures were thorough and detailed, always starting with observations. Oort's notes are detailed as well, but more carefully written (or drawn when it concerns drawings) then would be expected when these notes were produced during the lectures themselves, and were presumably worked out from draft notes later. At later times, certainly when I knew him, Oort's handwriting was characteristically tiny and often difficult to read (as Butler Burton remarked: ' small, but very neat and precise; just like the man himself'). As a student it was just as neat and precise, but of more common size.

In his autobiographical essay *AnnRev81* [3], Oort noted:

> 'When in 1917, at the age of 17, I began my studies in Groningen I became almost immediately inspired by Kapteyn's lectures on elementary astronomy. Although I had been strongly interested in astronomy since my high school years in Leiden, and this had influenced my choice of the University of Groningen because Kapteyn was there, I was, in 1917, still undecided between physics and astronomy as my major direction; I remember that I was so impressed by the way he taught elementary celestial mechanics that I tried to convey my new insight to friends who had likewise just entered the University, but were studying humanities. But I do not believe that I succeeded in conferring to them the full appreciation of the fascination of celestial dynamics.
>
> Perhaps the most significant thing I learned – mainly, I believe from Kapteyn's discussion of Kepler's method of studying nature – was to tie interpretations directly to observations, and to be extremely wary of hypotheses and speculations. In the first part of his course, Kapteyn refrained, for instance, from introducing the notion of 'force' to replace the measurable quantity 'acceleration'. He disliked intricate mathematical formulations which prevented one from 'seeing through' a theory; he feared the danger that the formulae might make one lose sight of the essentials. This was, of course, before quantum mechanics brought home the fact that one's sight is insufficiently developed to 'look through' the deeper domains of physical science without the aid of mathematics.'

Oort also talked about his Groningen studies in the *AIP-Interview* [4], from which I take a few parts (Fig. 2.17):

> 'But when at Groningen I wasn't quite decided yet whether I would specialize more in physics or astronomy. It was the personality of Professor

Fig. 2.17 Oort, in a detail cut from a photograph taken during the 1920 gala ball of Student Fraternity *Vindicat atque Polit*. From the Oort Archives

Kapteyn which decided me entirely. He was quite an inspiring teacher and especially his elementary astronomy lectures were fascinating.

He was a man who insisted very much in his teaching on seeing very clearly through a problem to avoid going into mathematical arguments at an early stage of the problem. He had the feeling that he first had to try to grasp the problem more or less in its entirety before going to mathematics. He considered mathematics as an indispensable tool but not an ideal tool for approaching physical and astronomical problems. And it seems this spirit was taken over by many of his pupils. In this connection I remember the attitude taken by his pupil W. de Sitter, who worked in the very early days on the expanding Universe, and was one of the first to actually apply the general theory of relativity to astronomical problems. He was very much more of a mathematical mind than Kapteyn or I, but he had the same feeling that mathematics was indispensable but contained a danger that it might sometimes lead you away from the actual insight into the problems.'

Fig. 2.18 Hermanus Haga and Frits Zernike, whose physics courses Oort followed. From University Museum Groningen

In the interview Oort was also asked about his contacts with other professors (in addition to Kapteyn and van Rhijn), while a student in Groningen (Fig. 2.18).

> 'Yes. I was interested in mathematics. I even went to lectures given by a famous French Professor Denjoy in Utrecht on point ensembles. But that was an interest of a more aesthetic nature and not an interest that would ever lead me to take up mathematics as a profession. I was too interested in nature itself. There was one professor in Groningen who certainly had considerable influence on my education. This was Professor Zernike, a theoretical physicist and also experimental physicist. He received the Nobel Prize for the phase-contrast microscope that he invented, but nowadays best known for his theoretical work in optics. In a way he was the worst teacher imaginable, he never prepared his lectures and often repeated the same stuff over again the next week because he had not prepared anything new. Many people found him difficult to follow in lectures but I learned more from him than from almost anybody else.'

French mathematician Arnaud Denjoy (1884–1974) was Professor in Utrecht between 1917 and 1922; he was famous for his work in calculus and in particular integral theory of real functions. Frits Zernike (1888–1966) was a 'Lector' (an academic level, just under that of 'professor, this rank is no longer used in Dutch universities) from 1915 and from 1920 onward a professor until 1958. He won the Nobel Prize in physics in 1953.

Oort sometimes wrote down some thoughts on loose pieces of paper, which got inserted in his diary notebooks. There are two that I think are worthwhile quoting here, since they illustrate well Oort's fascination with science and his approach to scientific inquiry. The first was found in the just inside the back cover of the notebook covering the period '$\frac{1}{2}$ 1917–1919'.

'I had thought in studying mathematics and natural science I would come closer to nature, that I would understand how everything moves and is connected with all that is useful; and now I find that one has to envisage things that do not exist, that also have no form, and have absolutely no use. Because he that has clear insights in natural science, will penetrate these without knowing a space that doesn't exist, and in which nothing exists, and then will not waste his time to study questions like whether a curve exists that cuts an axis in 2 points will have its largest maximum distance from that axis! [incomplete sentence] Does one really have to start that way to see the beauty in the laws of nature? Then I much more appreciate Kepler; he looked and looked and then saw in front of him all that we know now, he started unknowingly, as purely and as beautifully as he knew how to act.'

In this the reference to Kepler points at the influence Kapteyn had on Oort. The other is a loose envelop in the diary notebook entitled '1917', but the postmark on the envelop is from December 1918. It reads:

That miracles exist, I don't doubt. Whoever reflects on the Universe even for a single moment, will no longer go about his ways self-assuredly and say, 'there are no miracles; mankind will explain everything'. Why is everything natural that we humans call natural; in the substance of the matter I think it is all as curious as can be.'

Oort's studies went along successfully. His curriculum included—in addition to Kapteyn's lectures—mathematics lectures and practical instructions from Professors Barrau and Wolff (who we met already above), physics from Professors Haga and Zernike. Chemistry was taught by Professors Jeager and Hilmar Johannes Backer (1882–1959). The curriculum also included some mineralogy and geology, and the Professor was Jan Haitzes Bonnema (1864–1941). Oort's academic degrees were all obtained *cum laude*: the Candidaats (roughly bachelors) on July 2, 1919 and the Doctoraal (Masters) on November 5, 1921. His *cum laude* for the Candidaats was one of three out of 62 in total that academic year for the Faculty of Philosophy (would be Mathematics and Natural Science today) and for the Doctoraal one of two out of 40. Eventually he would obtain his Ph.D. also *cum laude* in 1926, the only one in his Faculty and one of three out of 20 for the whole University that academic year. Within the uncertainties of small number statistics this is the same as today (5–10%). Oort had definitely established himself as one of the top students. The Doctoraal diploma (now Masters) concludes course work; it entitles the holder to proceed towards the Ph.D. without requiring additional course work. Someone at that level was addressed 'Drs.', abbreviating 'Doctorandus', meaning 'to become a Doctor'. This title is now replaced by the Masters. Oort's Doctoraal exam occurred a few months after Kapteyn's retirement at age seventy at the end of the academic year. Kapteyn actually left Groningen immediately upon retiring. Pieter van Rhijn was the new Professor of Astronomy and Director of the Kapteyn Astronomical Laboratory.

Kapteyn was very active in an organization that was called the 'Natuurkundig Genootschap' or Natural Sciences Society of Groningen. This is the current Royal Natural Sciences Society or KNG. It is a learned society that was founded in 1801 and still is very active.[1] Kapteyn had become a member of the Board in 1886 and in 1904 was elected chairman of the scientific chapter, which he remained until his retirement in 1921. This department organized scientific presentations and Oort undoubtedly must have attended at least some of these, although he had not been recorded as a member. For more on the KNG see my *JCKbiog* [7].

2.5 First Research

After he had obtained his 'Candidaats' in February 1919, Oort became involved in the research at the Astronomical Laboratory. He wrote regularly to his mother on his activities, such as in the same letter as on the skating and dancing quoted above:

> 'I slowly look for plans for a possible dissertation. That is not easy, since there are two options. One is the safe side, in which case you select a subject like a better determination of the speed of the Sun for various classes of stars, that will almost certainly yield a useful result, something that is needed in astronomy. The disadvantage is that not much entirely new will come out of that, although I will have at the same time to search for stars with high velocity (that I told you about) that almost certainly form an interesting and little studied group. The advantage of that is that I can get probably support from the laboratory and will have computers and eventually a publication. Van Rhijn, whom I visited Tuesday evening, thinks this is a good subject, but wants to consult Kapteyn about it first.'

But before he got involved in research projects at Kapteyn's Astronomical Laboratory, he must have spent quite some time (judging from the extensive notes) on a piece of research that did not get published, but is present in the form of undated notes in the Oort papers, most likely written when Oort was an undergraduate. It concerns the tides on Earth. It refers to a—as it turns out—completely incorrect explanation of the tides on Earth by Kapteyn, and it compares lunar positions with actual tides in 1919. So, it has probably been produced around that time, when he was in his second or third year.

Kapteyn had published in 1909 a paper on the origin of the tides on Earth [50]. It was based on a lecture he held before the 'Mijnbouwkundige Vereeniging', the Association for Mining Engineering, in Delft. This was undoubtedly associated with what is now the Technical University of Delft, where mining engineering was being taught. Kapteyn's explanation is completely wrong. He explains the influence of the Moon by a pull on the oceans that speeds up the rotation where the Moon is seen at the horizon and leaving it unaffected directly below it. This, he argued, results in the conclusion: 'Under the Moon we have low tide and 90° away from it high tide'. The correct explanation is that on the side of the Earth towards the Moon the attractive force at the surface of the oceans is larger than at the bottom. This correct explanation actually stems from work long before Kapteyn by Isaac Newton (1643–

[1] I have been chairman of the Board since 1998, so what else can I say.

1727), Pierre-Simon Marquis de Laplace (1749–1827), and others. It has been known for a long time that the Sun exerts a similar influence on the Earth, although less in magnitude. On the side away from the Moon an additional effect is the centrifugal force resulting from the Earth moving around the center of gravity of the Earth-Moon system; the two effects together produce a more or less symmetrical result. It is a bit of a mystery how Kapteyn came up with this theory, while the tidal force explanation was widely known and accepted. A more detailed discussion of the lecture has been given in my *JCKbiog* [7].

Oort's notes started with: 'The tidal wave that we observe is the one that originates from streams at the places where the Moon is on the horizon'. So Oort adopts Kapteyn's erroneous explanation for the tides. Probably Kapteyn had mentioned it in one of his lectures (but see footnote 2 in Sect. 3.2 on books with the correct explanation that Kapteyn gave Oort to read!) The style of the notes is quite typical for Oort's way of working throughout his life. To illustrate that I reproduce a small part of his reasoning. He first deduces that there should also be a wave directly under the Moon as a result of 'the decrease in the weight of 1 cm^3 of water' at these positions. In a footnote he explained that his is so because 'the pressure of the water on the ocean floor should be the same as at the floor 45° away from it (otherwise there would be no equilibrium) and since the water is lighter under the Moon the height has to be larger'. This actually is *not* fundamentally different from the correct explanation. The height of the wave he estimated as about one centimeter, depending on the depth; it will increase substantially when the wave hits a continent. After concluding that the most pronounced wave originates in the Pacific Ocean, he estimated the speed of the wave, which he figured depended on the depth, at least as long as the size in the horizontal direction was larger. In the Pacific Ocean this then came out as '±1200 km per hour'. He next investigated the relative influence of the waves resulting from the Moon and the Sun, estimating how the solar wave delays or speeds up the lunar one, and he then compared spring tide (when the two enhance each other) with lunar phases (full moon, etc.), all with little success.

This is very much Kapteyn's way of doing research by first exploring the generalities and consequences of a phenomenon by making order of magnitude estimates to arrive at a general understanding and a possible way of explanation before embarking on a comprehensive, exhaustive, theoretically complete and exact investigation.

It is of interest to ask to what extent Kapteyn's way of lecturing affected Oort in his teaching. In a course on the planetary system that Oort gave in 1964–1965, which I followed and preserved the notes from, he in any case started out very much like Kapteyn, discussing in depth the historical development of the understanding of the planetary orbits going back to Aristarchus of Samos (c.310–230BCE), Aristotle (384–322BCE) and Hipparchus. And indeed he provided quite some detail when discussing Kepler and his planetary laws. In particular he treated in detail the way Kepler determined that Mars' orbit is elliptical, explaining how Kepler threw out all previous notions that every motion should be a combination of circular motions at constant angular speed and set out instead to determine Mars' orbit directly from observations by Tycho Brahe (1546–1601) without making any preordained assumptions.

Oort's approach is very reminiscent of Kapteyn's predilection of Kepler's methods.[2] In the lectures Oort was a bit short on tides, providing the correct explanation for the phenomenon without mentioning Kapteyn and noting it takes three days for the tidal wave to reach us from the Pacific Ocean, so that spring tide is delayed by that much from conjunction of Sun and Moon. He spent more time on the slowing down of the Earth's rotation by the tidal wave and the consequently slowly increasing distance to the Moon from conservation of angular momentum. All of this is not very different from the structure of Kapteyn's lectures according to Oort's notes. Oort lectured with a rather soft voice, looking at the blackboard and with his back towards the students, who all always clustered in the first rows of the lecture room. Even though he used lecture notes, these were sometimes from different years in different weeks, using different mathematical notations. He did not use textbooks. Yet, he was generally considered a very inspiring professor.

The student society *Vindicat atque Polit* published in their yearbooks in addition to information for students also a section on issues of general interest (articles, poems, etc.). In the 1921 issue, Oort published in that section an account of the work of Kapteyn (Oort, 1921). That same year Kapteyn had retired and he and his wife then lived with their daughter and son-in-law in Amsterdam, already suffering from what would prove to be his fatal illness. On December 29, 1921, Kapteyn wrote his only letter to Oort that is present in the Oort Archives (see Figs. 2.19 and 2.20 and Appendix B.1 for a full translation): 'The sympathetic manner in which the article has been written gives me much joy, but it has given me special pleasure that it was You who wrote it.' He invited him to visit him in Hilversum some time 'and I hope we have the opportunity to discuss and review many things from our beloved science.' Kapteyn never moved to the house he and his wife had bought in Hilversum, but died in Amsterdam in June 1922.

Kapteyn's satisfaction that it was Oort who wrote this tribute, shows Kapteyn's high esteem for Oort and his view that he was his most promising student. This is confirmed by the following story. In the Strömgren archives (held in the archives of Lund University Library) there is a letter from Mrs. Kapteyn, dated December 28, 1936, to Mrs Strömgren, wife of Svante Elis Strömgren, Director of Copenhagen Observatory (Petra van der Heijden's contribution to the *Legacy* [6], p. 48). It was a reply to a card the Strömgrens had sent Mrs. Kapteyn and in it they apparently mentioned that their son Bengt Georg Daniel Strömgren (1908–1987), later a famous astronomer, had been invited by Otto von Struve (1897–1963) to spend some time at Chicago. Mrs. Kapteyn wrote:

> 'I do hope that Bengt is not remaining for good in Chicago; much that I love our American friends but I have the feeling that each country has the right to keep their big man to herself. Prof. Oort is now in Leiden at the Observatory, I know he was my Darling's best beloved student and

[2]This in turn I inherited as a student from Oort. It has resulted in me presenting all my students after defending their Ph.D. theses, as a personal gift a copy of the fascinating book *The Sleepwalkers: A history of man's changing vision of the Universe* by Arthur Koestler (1905–1983) [51], which presents the same point of view and contains a comprehensive biography of Kepler.

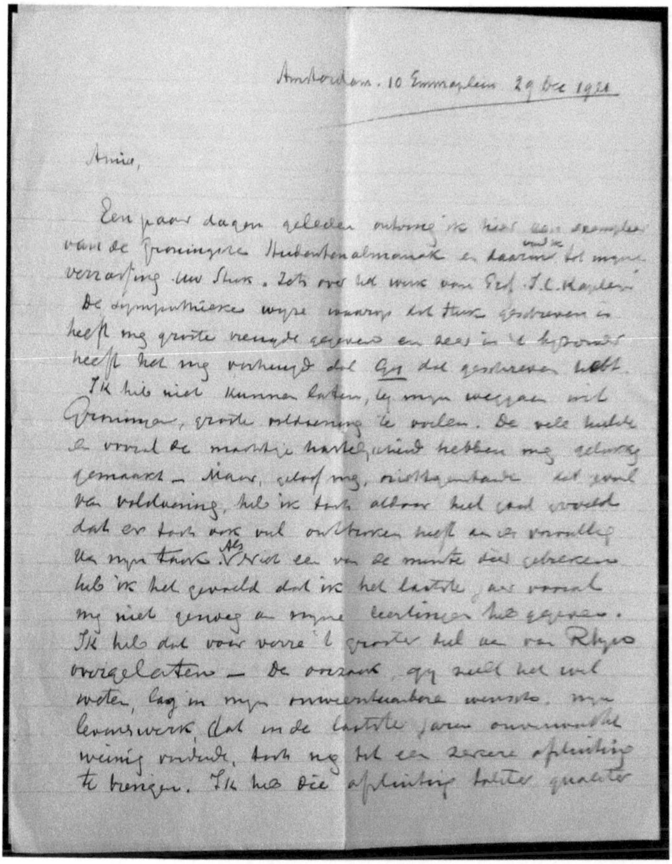

Amsterdam. 10 Emmaplein 29 Dec 1921

Amice,

Een paar dagen geleden ontving ik hier een exemplaar van de Groningsche Studentenalmanak en daarin tot mijne verrassing uw Stuk: Iets over het werk van Prof. J.C. Kapteyn. De sympathieke wijze waarop dat stuk geschreven is heeft mij groote vreugde gegeven en zeer in 't bijzonder heeft het mij verheugd dat Gij dat geschreven hebt.
Ik heb niet kunnen laten, bij mijn weggaan uit Groningen, groote voldoening te voelen. De vele hulde en vooral de machtige hartelijkheid hebben mij gelukkig gemaakt – Maar, geloof mij, niettegenstaande dat gevoel van voldoening, heb ik toch altoos heel sterk gevoeld dat er toch ook veel ontbroken heeft aan de vervulling van mijn taak. Als niet een van de minste der gebreken heb ik het gevoeld dat ik het laatste jaar vooral mij niet genoeg aan mijne leerlingen heb gegeven. Ik heb dat voor verre 't grootste deel aan van Rhijn overgelaten – De oorzaak, gij zult het wel weten, lag in mijn onweerstaanbare wensch, mijn levenswerk, dat in de laatste jaren onrustbarend weinig vorderde, toch nog tot een zekere afsluiting te brengen. Ik heb die afsluiting [illegible]

Fig. 2.19 First page of the only letter of Kapteyn to Oort that is present in the Oort Archives. Second page in Fig. 2.20. See text for details and a complete English translation of the text Appendix B.1

assistant and I felt glad that he did not accept the positions they offered him in America, but came back to Holland. And o? let him choose his wife from your country. We have found the international marriage a great drawback.'

The last remark refers to the unsuccessful marriage of her second daughter Henriette Kapteyn to the Danish astronomer Ejnar Hertzsprung. Actually, the final piece of advise was unnecessary, as Bengt Strömgren had married Danish Sigrid Caja Hartz already in 1931. Otto Struve (1897–1963) was Director of Yerkes Observatory.

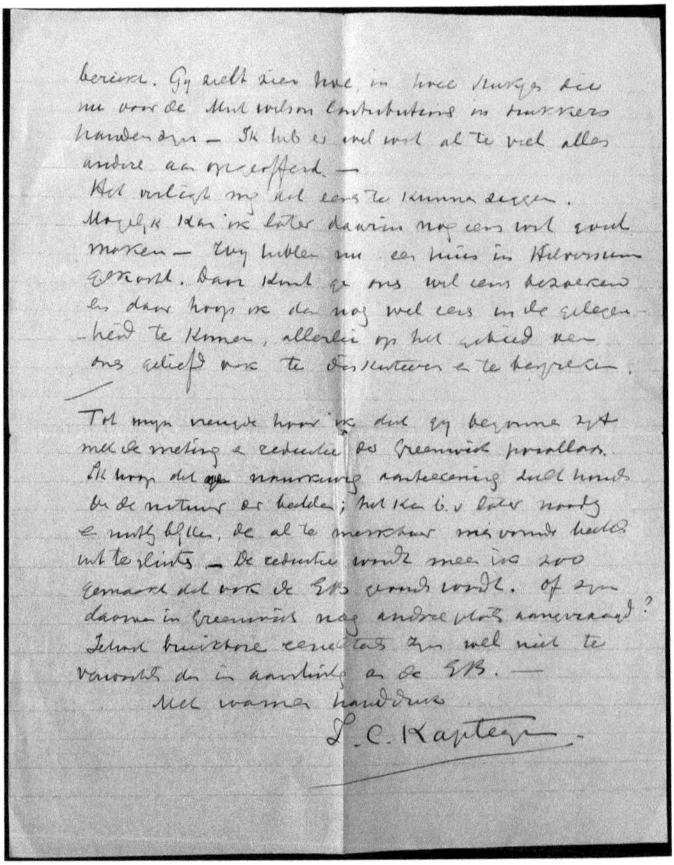
bereiken. Gij zult zien hoe, in hoeveel stukjes die
nu voor de Mount Wilson Contributions in drukkers
handen zijn — Ik heb er wel wat al te veel alles
andere aan opgeofferd. —
Het verlangt mij dat eens te kunnen zeggen.
Mogelijk kan ik later daarin nog eens wat goed
maken — Wij hebben nu een huis in Hilversum
gekocht. Daar komt gij ons wel eens bezoeken
en daar hoop ik dan nog wel eens in de gelegen-
heid te komen, allerlei op het gebied van
ons geliefd vak te discuteeren en te bespreken.

Tot mijn vreugde hoor ik dat gij begonnen zijt
met de meting en reductie der Greenwich parallax.
Ik hoop dat ge nauwkeurig aanteekening zult houden
bij de metingen der hoeken; het kan b.v. later nooit
een nuttig blijken, de al te [illegible] hadden
uit te sluiten — De reductie wordt mee in zoo
gemaakt dat ook de E.B. gevonden wordt. Of zijn
daarom in Greenwich nog andere plots aanwezig?
Verder bruikbare resultaten zijn wel niet te
verwachten dan in aansluiting aan de E.B. —
Met warmen handdruk
J.C. Kapteyn.

Fig. 2.20 Second page of the only letter of Kapteyn to Oort that is present in the Oort Archives. For first page and further information see Fig. 2.19

2.6 Stars and Galaxies

Before continuing with Oort, we need to put things in current perspective and for this I will briefly go into the properties of stars and galaxies.

Spectral Classification

The earliest systematic work on stellar spectra was performed by Father Pietro Angelo Secchi SJ (1818–1878), a Jesuit astronomer. He was Director of what is now the Vatican Observatory, originally established in 1774 as the Observatory of the Roman College. This Observatory was confiscated by the Italians after Secchi's death, but was reactivated at the Vatican around 1890 by the Holy See and still exists today. Although commonly known as the Vatican Observatory, the official name is Specola Vaticana. Secchi did pioneering work in stellar spectroscopy and in the 1860s de-

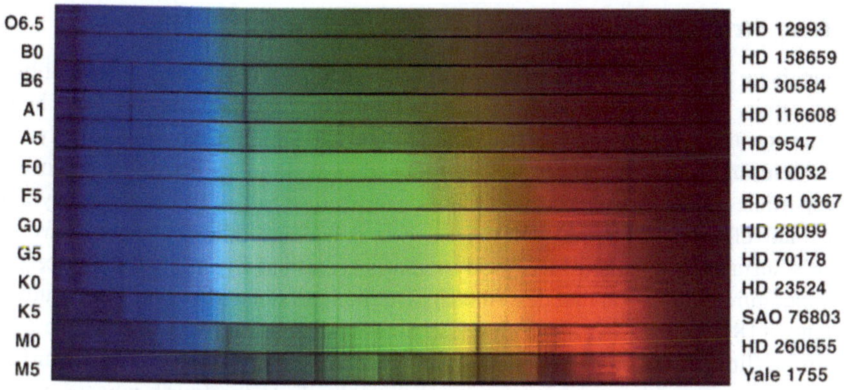

Fig. 2.21 Spectra along the Harvard/Draper sequence. From Wikimedia Commons [52]

signed the first classification scheme of stellar spectra. His earliest set of 'Secchi classes' encompassed three types, which also correlate with the colors of the stars (and with modern classes).

For the modern spectral classification we are indebted to Edward Pickering of Harvard College Observatory (see Fig. 2.6) and some of his female staff (known as Pickering's harem). It is named after Henry Draper (1837–1882), who was a physician as well as an accomplished amateur astronomer. Draper was one of the first to photographically record the spectrum of a star at his private Observatory and to recognize the similarities of the many stellar spectra he obtained to that of the Sun. After his death his widow supported the spectroscopic work of Pickering through a fund called the Henry Draper Memorial. Pickering obtained spectra of hundreds of stars and these were classified in a new system, mainly by Williamina Fleming (1857–1911). The result was published in 1890 as *The Draper Catalogue of stellar spectra photographed with the 8 in. Bache Telescope as a part of the Henry Draper Memorial* [53]. Pickering, in conjunction with Antonia Caetana de Paiva Pereira Maury (1866–1952) and Annie Jump Cannon (1863–1941), among others, refined and revised it and applied it to about 400,000 stars. These female astronomers, together with others belonging to Pickering's 'harem', such as the discoverer of the period-luminosity relation among pulsating stars Henrietta Swan Leavitt that we will meet later, deserve enormous credit for their work and the term 'harem' does not really express their importance. This resulted in the *Henry Draper Catalogue* and its extension, published between 1918 and 1936. The number of spectral types was restricted and the labels chosen earlier rearranged, leaving the sequence OBAFGKM that we still use today. This final sequence has been illustrated in Fig. 2.21; note the systematic change from blue to red (going from O to M or hotter to cooler), and the systematic variation in appearance and strength of spectral lines. For reasons having to do with previous incorrect understanding of stellar evolution through the spectral type range, the blue (O-star) end of the range is often called 'early type' spectra and, moving through the sequence towards M-stars, 'late' types.

Stellar Structure and Evolution

Stars have a very wide range of properties. For example, look at the amount of light they emit. They can be intrinsically very bright or very faint, the brightest being some ten thousand times more luminous than the Sun and the faintest only one ten thousandth. The Sun is in this respect rather average, as is true for almost all properties. The light from the stars has a general color, which can be relatively blue or relatively red. In this respect, too, the Sun is average. The light emitted from a star's surface is so-called black-body radiation, emitted by any body of a certain temperature. It has been first described by German physicist Max Karl Ernst Ludwig Planck (1858–1947) and has a characteristic dependence on wavelength, the peak shifting (linearly) to bluer wavelengths and the total amount emitted per unit surface area increasing with increasing temperature (actually to the fourth power). So the color of the light from a star (the relative amounts of red and blue light) is a measure for the temperature on the stellar surfaces, and ranges from 30,000 K for the hottest to less than 3,000 K for the coolest. For the Sun the surface temperature is about 5,800 K and the peak is in the corresponding black-body emission is in the middle of the optical spectrum, the wavelength range to which our eyes are sensitive. The surface of a star is not a surface in the usual sense but the layer at which light that is emitted can escape from the star into space.

This range of luminosities and temperatures is most conveniently displayed in the so-called Hertzsprung–Russell (or HR) diagram (see Fig. 2.22), named after Ejnar Hertzsprung (1873–1967) and Henry Norris Russell (1877–1957). It was independently found by these astronomers. Hertzsprung worked at the Potsdam Astrophysikalisches Observatorium at the time. He was Henriette Kapteyn's husband, meeting to her in Groningen before Herzsprung was being introduced to George Hale at Mount Wilson in California by Kapteyn.

In the Hertzsprung–Russell diagram, most stars are on the Main Sequence, running from top-left where the hot luminous stars reside, to the cool, faint stars in the bottom-right. Stars spend most of their lives on this sequence. While there, they 'burn' hydrogen in their extremely hot (of order ten million Kelvin) interiors; in these nuclear reactions, hydrogen is converted into helium and this releases energy, which is eventually radiated away into space—after slowly drifting up to the surface by repeated absorption and re-emission; in the Sun this takes on the order of a hundred thousand years. The part of the star where this happens is relatively thin and is referred to as its surface. The temperature inferred from the color of the light corresponds with that at the surface of the star.

Stars are made up of almost three quarters hydrogen and about one quarter helium, the other chemical elements accounting for almost nothing to a few per cent. But, interestingly, these elements heavier than helium always occur in roughly the same relative proportion. In the Sun the chemical elements heavier than helium account for 2% of the mass.

There also is a large range in stellar masses and lifetimes; the bright, hot stars at the top-left of the Main Sequence (MS) turn out to be the most massive stars (tens of solar masses) and the least luminous, red ones have masses of one tenth of that of the

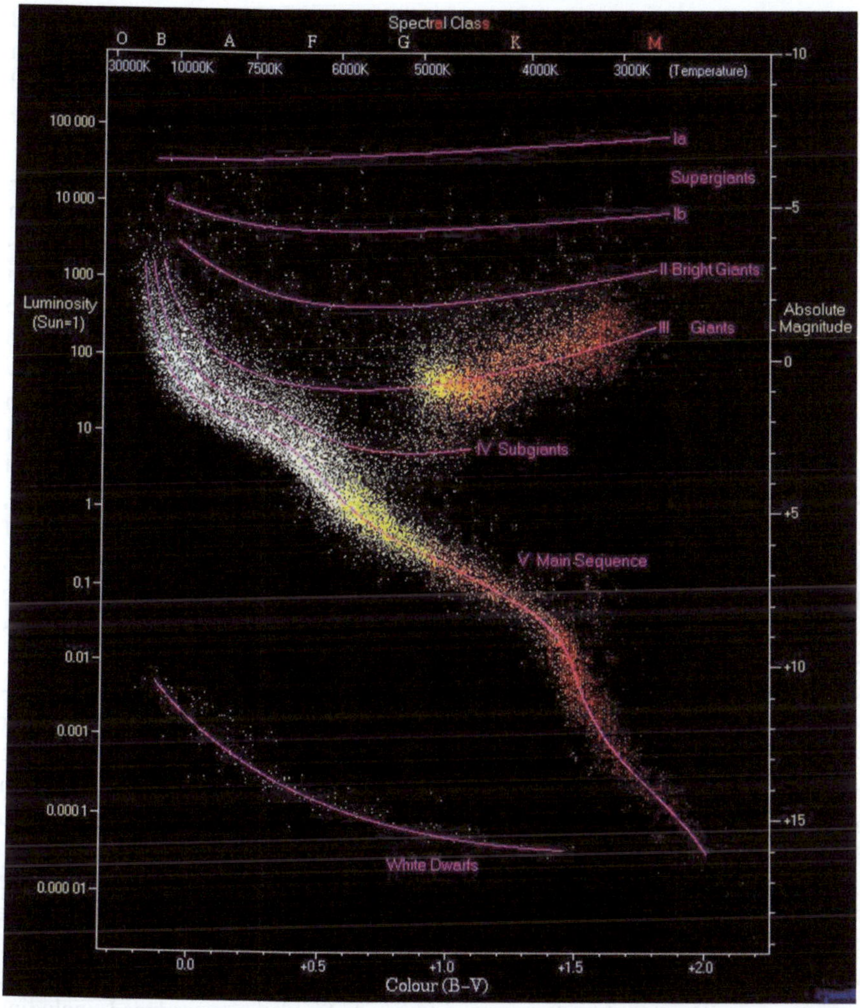

Fig. 2.22 Hertzsprung–Russell, or HR-diagram. The horizontal axis is an indication of the temperature at the surface of the star. At the bottom it is expressed as the color index $(B - V)$, which is a measure of the amount of blue light compared to visual light. At the top the axis is labeled with spectral types and actual temperatures in Kelvins. The vertical axis is a measure for the production of energy by the star in the form of light; on the left as luminosity expressed in that of the Sun, and on the right the absolute magnitude. Source: The Hertzsprung Russell Diagram by Richard Powell [54]

Sun. Stars have very different lifetimes on the Main Sequence, the massive, bright hot stars 'burn' their hydrogen into helium at a much faster rate then the light, faint, and cool ones. The life of a star on the MS ends when the hydrogen in the central part, by which is meant about the inner ten percent of the mass of the star, has been consumed. The Sun again is average; its life on the Main Sequence lasts about 10 billion (1×10^{10}) years. Its current age is about 4.6 billion (4.6×10^9) years and is about halfway through its Main Sequence stage. The most massive stars live on the Main Sequence for one million years or less and the lightest ones one hundred billion (1×10^{12}) years, which is about ten times longer than the current age of the Universe.

Stars form from collapsing gas clouds that heat up as a result of release of gravitational energy when they contract, and which they cannot radiate away rapidly enough when the protostar gets denser and denser. On the Main Sequence, stellar interiors have become sufficiently hot to 'burn' hydrogen into helium. What happens when the hydrogen is exhausted? Stars comparable to the Sun in terms of mass become red giants as the outer envelopes expand and follow a 'giant branch' running up to the top-right of the HR-diagram. The central parts contract and heat and eventually they will burn helium into carbon and oxygen. By that time the star is situated in the HR-diagram in the clump of stars halfway along the giant branch, and then it moves up again more or less along the giant branch (the so-called 'asymptotic giant branch'). After a relatively short period of time they throw off their outer envelopes, which then form an expanding shell around the star. This may happen in the form of a nova, when the star suddenly brightens for periods of months before fading away. This shell phase is called planetary nebula, a historic name that was given because of the appearance of these shells in small telescopes. The naked cores of the stars eventually end as faint, very compact white dwarfs (bottom of the HR-diagram). Stars that are initially more massive than about 5 times the Sun will become neutron stars rather than white dwarfs. In neutron stars gravity is so strong that matter becomes even more compact as the electrons are, so to speak, compressed into the protons to form neutrons. These heavy stars will after exhausting their hydrogen and helium expand enormously and become supergiants. They then go through a number of short phases of nuclear fusion where they burn their atomic nuclei into heavier and heavier chemical elements up to iron. The fusion stops here because from iron onward it *takes* energy to merge atomic nuclei into heavier ones. When the nuclear reactions stop and the production of energy in the center ceases, the star will contract violently and the energy released in that collapse will cause it to explode as a supernova. During this phase more chemical elements are produced, including the heaviest ones that exist in nature. Since supernovae throw out out most of their matter into space, they enrich existing gas between the stars with the products of the nuclear reactions. This results in subsequent stars having more elements than just hydrogen and helium. This process is called (stellar) nucleosynthesis. What remains after a supernova is a neutron star when the original star is not too massive (5–10 solar masses $M_{\odot}$), but if it is very heavy, a black hole is formed that is so compact that not even light cannot escape from it.

The Milky Way and Other Galaxies

The stars we see at night are all part of the Milky Way galaxy. Its main structure is revealed as a faint band of light around the sky from which it derives its name. Since the Galaxy is spatially predominantly a flattened disk, in which we are immersed, the Milky Way stretches as a great band over the full sky. In Fig. 2.23 we see a composite

Fig. 2.23 A panoramic view of the Milky Way. This view shows the full 360° with the direction towards the center of the Galaxy in the middle. It is made up of many separate photographs and has been produced by the European Southern Observatory ESO. In broad terms the bottom part can be seen best from the northern hemisphere, the top part from the south. The two blobs to the right of the Milky Way in the upper part are the Magellanic Clouds, companions to our Galaxy. The two small dots also somewhat to the right of the Milky Way but in the lower part are the star cluster the Pleiades (extreme bottom) and the nearest spiral galaxy, the Andromeda Nebula. Other bright objects are planets that happened to be in the field when the pictures were taken. (European Southern Observatory)

photograph of the full 360°. The most prominent part in the middle can best be seen from the southern hemisphere; this corresponds to the direction towards the center of the Galaxy. In the northern hemisphere it can only rise above the horizon by a small angle (for the Netherlands slightly less than 10°). In the south also the two companions to the Milky Way Galaxy, the Large and Small Magellanic Clouds are visible; in the north the nearest large galaxy, the Andromeda Nebula (see Fig. 7.9) can be seen with the naked eye, but only under the best of circumstances.

By and large, stars can be divided into two Stellar Populations. The stars that we see in the sky almost all belong to the flat structure that we see as the Milky Way, and this is called the disk of the Galaxy or Population I. Population I stars can be relatively young, i.e. younger than a few billion years. Around the disk there is a much sparser populated extended structure, the halo. The halo stars, Population II, stars are distributed over a more or less spherical volume, but much concentrated towards the center of the Galaxy. These stars are less prominent, except when they form dense conglomerations of some some hundred thousand stars that are called globular clusters. These Populations II stars are roughly as old as the Galaxy itself (i.e. up to 12 billion years) and contain less (and often much less) than 0.1% in terms of heavier elements. The Hertzsprung–Russell diagrams of globular clusters miss the upper part of the Main Sequence, where the younger stars reside. These diagrams thus look like the diagram in Fig. 2.22, except that they miss the (roughly) white upper-left part. Actually for any cluster of stars, the turn-off where the Main Sequence stops is a good measure for the age of that cluster.

The Universe is full of galaxies, some similar to our own, some very different. Figure 2.24 is intended to indicate what the Galaxy would look like if we could view it from the outside. Here I have chosen two other galaxies that resemble our own fairly well. The picture on the left shows the view if we would be situated such that we would see the disk more or less from above (or below), and on the right if it would be seen edge-on. The face-on view shows a so-called spiral structure, which is closely related to where star formation is currently taking place. This dominant part is the disk. On photographs like the one on the right of Fig. 2.24, this flat disk also dominates the light. The halo or Population II is not very prominent; its central parts can be seen as the bright concentration near the center. Astronomers nowadays observe frequently from Observatories in the Andes in northern Chile. Anyone who has been working at one of these places in about June, when the center of the Galaxy is almost overhead in the zenith in the middle of the night, will testify that the Milky Way looks very much like NGC 891 from there (compare Figs. 2.23 and 2.24). Even in actual properties NGC 891 is almost a twin to our Galaxy (see my contribution to the *Legacy* [6]).

So the picture is the following: The disk is Population I; its stars are very young to rather old (up to about 10 billion years) and form a flat disk, which rotates with a velocity of a few hundred km/s, while the relative velocities of stars are of the order of tens of km/s. The Sun is part of Population I. The halo stars of Population II are all very old and since they occupy a more or less spherical volume, this does not rotate rapidly as a whole, but the stars have individual velocities of 100–150 km/s with respect to each other and a large collective speed with respect to the stars in the disk. Very early on, the Universe consisted almost entirely of hydrogen and helium (with only a tiny trace of lithium and beryllium), so these first stars of Population II have very little or sometimes virtually no chemical elements heavier than helium.

The canonical picture of galaxy formation is that Population II formed early, from a collapsing gas cloud, supplemented by in-falling and merging of smaller companion systems. Originally this gas consisted only of hydrogen and helium, which was formed about three minutes after the Big Bang. The gas was enriched by stellar nucleosynthesis in the very first generation of stars that would have no heavy elements at all at birth (sometimes referred to as the Population III) and subsequent later stars of Population II, so that eventually it contained small amounts of chemical elements heavier than helium. This gas dissipated its random motions by collisions

Fig. 2.24 Two galaxies like our own Milky Way Galaxy. On the left M81, which is seen at a moderate angle; on the right NGC 891, which has the disk edge-on (compare to Fig. 2.23). National Optical Astronomy Observatories (NOAO), [55]

of gas clouds, so that rotation became dominant, and settled in a flat disk, where stars have formed up to the present time as Population I stars.

There is a wide range of morphologies among galaxies that are classified as the so-called Hubble morphological types, proposed in the 1920s by Edwin Powell Hubble (1889–1953), although to a significant extent based on earlier work by John Henry Reynolds (1874–1949), which Hubble failed to acknowledge.[3] The various morphological types in this classification scheme can be interpreted as a range in the relative proportion of the two Stellar Populations, ranging from pure Population II in elliptical (sometimes called early type) galaxies to exclusively Population I in 'pure-disk' (late type) spiral galaxies. The spirals are divided furthermore in a range from Sa to Sc, where Sa systems have thinner, tightly wound spiral arms, while in Sc galaxies they are thicker and more open. The picture of galaxy formation and evolution is evolving rapidly at present as a result of the enormous improvement in modern observations and numerical simulations. The interaction between all physical processes that play a role in galaxy formation is still far from understood, but there is much evidence that merging of smaller fragments and cannibalizing of small satellite systems must be an important factor.

[3]For more background and discussion of the issue of Hubble's discoveries and his apparent lack of giving credit to others when appropriate, see the proceedings of the conference *Origins of the expanding Universe: 1912–1932* [56].

Fig.3.0 Oort as a student with the 'corpsmuts', probably after completing initialization to the Student Association *Vindicat atque Polit* (see section 2.3). From the Oort Archives

Chapter 3
Stars of High Velocity

The most exciting phrase to hear in science, the one that heralds new discoveries, is not 'Eureka!' but rather 'hmm....that's funny...'
Isaac Asimov (1919–1992)

Talent alone won't make you a success. Neither will being in the right place at the right time, unless you are ready. The most important question is: 'Are your ready?'
John William (Johnny) Carson (1925–2005)

3.1 The Kapteyn Universe

After having obtained his Candidaats (Bachelor) in 1919, Oort started to do some research projects in astronomy under Kapteyn. But before going into that I need to discuss the progress Kapteyn had made on his program to study the distribution and motions of stars in space since 1917 when Oort had entered the University of Groningen.

When Oort studied in Groningen, Kapteyn was involved in his final attack on the problem of the 'Construction of the Heavens' (see Sect. 2.2). For a more elaborate treatment of all this I refer to my Kapteyn biography *JCKbiog* [7]. More background can be found in the excellent review by Robert William Smith [57]. This attack was 'final' in the sense that it was the last thing he could do before his retirement in 1921, but in the publication it was described as 'a first attempt'. Anything more than that had to await the progress of the Plan of Selected Areas, which was in full swing, certainly as far as the star counts in the Selected Areas from photographic exposures provided by Harvard and Mount Wilson were concerned, that were being measured in Groningen.

P. C. van der Kruit, *Jan Hendrik Oort*, Astrophysics and Space Science Library 459, https://doi.org/10.1007/978-3-030-17801-7_3

Box 3.1. Secular parallax

This is for a group of stars the statistical distance indicated by their proper motions on the sky as a reflection of the Sun's motion in space. If the Sun moves in space in the direction of the Antapex with a velocity $V_{\odot}$ and if a star's position on the sky is an angle λ away from the Apex, the component perpendicular to the line of sight is $T_{\odot} = V_{\odot} \sin \lambda$.

The expected proper motion $\nu_{\odot}$ along an arc on the sky between the star and the A(nta)pex is then is $\nu_{\odot} = \pi T_{\odot}/4.74$, where π is the parallax in arcsec (and $r = 1/\pi$ the distance is in parsecs); the value 4.74 serves to change the units of km/s of T into arcsec/year of ν.

For a group of stars the secular parallax is the mean π estimated from their proper motions using this relation, giving each star a weight corresponding to what fraction of the Sun's speed is reflected in the proper motion. This weight is $\sin \lambda$. The secular parallax of a group of stars is then

$$\langle \pi \rangle = \frac{4.74 \langle \nu \sin \lambda \rangle}{V_{\odot} \langle \sin^2 \lambda \rangle},$$

where the triangular brackets indicate mean values.

The attempt was based on the three assumptions. The assumption of no preferred direction of motion in space to derive stellar distances in a statistical manner was replaced by allowing for Kapteyn's Star Steams, the properties of which were studied by his students, such as Herman Albertus Weersma (1877–1961) (see Fig. 3.1). This and other information could be used to derive the 'luminosity curve' (now known as the luminosity function, which I will use from here on) that describes the relative number of stars in a volume of space of various luminosities or absolute magnitudes. This curve or function then would be assumed the same everywhere in the Stellar System. Pieter van Rhijn (see also Fig. 3.1) in his Ph.D. thesis of 1915, had done extensive studies of existing material to estimate the mean distance or parallax of stars as a function of their apparent magnitude and their proper motion. For this he used Kapteyn's method of *secular parallax*, in which the distance of a star can be estimated assuming that its proper motion on the sky reflects the motion of the Sun through space. Since stars have their own individual motions this method only works statistically (see Box 3.1).

The final assumption that had to be made was that interstellar space was fully transparent, so that the apparent brightness of a star was not affected by loss of light during the journey of light towards us. This had been a great worry for Kapteyn and he had published extensively on the subject in American journals [58–60]. In 1909 he had proposed that if there were general scattering of light by dark, unseen matter it should be more effective at bluer than at redder wavelengths and that therefore more distant stars with the same intrinsic properties would appear redder. He

Fig. 3.1 Herman Albertus Weersma and Pieter Johannes van Rhijn, Kapteyn's assistants during respectively 1908–1912 and 1914–1921 with Frits Zernike in between. Weersma: From the *Album Amicorum* presented at the retirement of H.G. van de Sande Bakhuyzen as Professor of Astronomy and Director of Leiden Observatory in 1908 [61]. Van Rhijn: From Wikimedia Commons [62]

assumed it to be due to 'Rayleigh scattering', named after Lord Rayleigh, John William Strutt, 3rd Baron Rayleigh (1842–1919), and then the scattering should be inversely proportional to the fourth power of the wavelength. Rayleigh scattering is due to individual molecules, that are much smaller than the wavelength. In reality the scattering is off dust particles, the distribution of sizes of which (more comparable to the wavelength) have to be taken into account and interstellar extinction is in first approximation inversely proportional (linearly) to the wavelength. Therefore Kapteyn had assumed much too steep a wavelength dependence. From studies of stars at visual (eye-sight) and photographic (bluer sensitivity in early emulsions) wavelengths he deduced a very sensible value according to modern standards for the reddening of 0.31 ± 0.06 magnitudes per kiloparsec (kpc), later improved by van Rhijn in his Ph.D. thesis to 0.195 ± 0.03. Reddening here is the change of the color index, the difference of the magnitudes of a star in the two wavelength bands. Kapteyn inferred from this a loss of visual light of order a full magnitude per kpc (using a wavelength dependence according to Rayleigh scattering) and therefore this would have a very significant impact on his analysis. But he worried about possible systematic effects such as correlations of the mean color of stars with absolute magnitude, so that at larger distances he would systematically study stars with different intrinsic colors than nearby.

The problem seemed solved with the observations of the globular cluster Messier 13 in Hercules by Harlow Shapley (1885–1972) of Mount Wilson Observatory (he studied 1300 stars!). He used a distance of 10,000 parsecs or 10 kpc. The currently accepted distance is 6.8 kpc. Shapley noted that if Kapteyn's value for the reddening were correct, the smallest (bluest) color index he should have observed would be at least two magnitudes redder than he had actually observed. Shapley concluded from this that the reddening of starlight in interstellar space could not exceed 0.0001 magnitudes per parsec, which is 'an amount', he concluded, 'completely negligible in dealing with the ordinary isolated stars. In the light of this result we are probably justified in assuming that the non-selective absorption in space (obstruction) is also negligible' [64]. Like all astronomers of the time, Kapteyn accepted this result as convincing evidence for the absence of interstellar extinction.

The preparatory studies for Kapteyn's final attack were published in three extensive papers in the series of 'Publications of the Astronomical Laboratory at Groningen' between 1916 and 1920. What is involved is sufficiently clear from the titles: *On the number of stars of each photographic magnitude in different Galactic latitudes* by van Rhijn, *The secular parallax of the stars of different magnitude, Galactic latitude and spectrum* by Kapteyn, van Rhijn, and Weersma, and *Numbers of stars between definite limits of magnitude, proper motion and Galactic latitude for each spectral class, together with some other investigations* by Kapteyn and van Rhijn [65–67]. With these preparations finished, Kapteyn and van Rhijn published a paper in 1920 in 'The Astrophysical Journal' with the title *On the distribution of the stars in space especially in the high Galactic latitudes*, [68]. It starts out as follows:

> 'The investigation contained in G.P. [Groningen Publications] 27, 29 and 30 were carried out for the purpose of making possible an elaborate treatment of the arrangement of the stars in space. At least two more publications will be necessary to complete this investigation. Now that, after so many years of preparation, our data seem at last to be sufficient for the purpose, we have been unable to restrain our curiosity and have resolved to carry through completely a small part of the work, even though, by doing so, the rules for strict economy of labor cannot be altogether adhered to.
>
> The present paper is the outcome of this more or less provisional work. In the main it relates to the stars as a whole between Galactic latitudes ±40° and ±90°. It is provisional in that the extremely valuable parallaxes, now placed at our disposal by Mitchell, have not yet been used, and moreover because the thorough discussion which will have to be made of the measured parallaxes in all Galactic latitudes is not yet complete. We expect little to be changed, however, in the definitive treatment of the subject, which will form a part of the more comprehensive work.
>
> A somewhat less refined investigation of all stars (irrespective of spectrum) in Galactic latitudes 0°, 30°, 60° and 90° has also been added.'

Box 3.2: The Principal Laws of Statistical Astronomy

In the notation as in the Ph.D. thesis of Willem Johannes Adriaan Schouten (1893–1971) under Kapteyn's guidance in 1918 [69]:

h_m, apparent brightness of a star of magnitude m; $h_m = 1$ for $m = 0$.
i, absolute brightness corresponding to absolute magnitude M.
$N(h_m)$, observed number of stars of magnitude m.
$\pi(h_m)$, mean parallax of stars of magnitude m.
$D(r)$, density (number per unit volume) of stars at distance r.
$\varphi(i)di$, percentage of stars between i and $i + di$.

For any star $i = h_m r^2$ with r and $\pi(h_m)$ determined from proper motions, secular and other parallaxes.

If $N(h_m)$ and $\pi(h_m)$ are known from observations, $D(r)$ and $\varphi(i)$ follow from:

$$N(h_m) = 4\pi \int_0^\infty D(r)\varphi(h_m r^2) r^4 dr,$$

$$N(h_m)\pi(h_m) = 4\pi \int_0^\infty D(r)\varphi(h_m r^2) r^3 dr,$$

but this involves simultaneously solving two integral equations.

The procedure that Kapteyn and van Rhijn followed was to first use as much information as possible, in particular secular parallaxes, to derive analytic expressions (mathematical formulae) for the star counts as number of stars per interval apparent magnitude and per interval parallax. They combined this information for all stars more than 40° from the Milky Way. With some not very critical assumptions, such as the spread in parallax of stars at a certain apparent magnitude and parallax, they derived expressions for the luminosity function, the relative number of stars in each interval of absolute magnitude, and for the total density of stars as a function of distance from the Sun. In technical terms (see Box 3.2 for the mathematical expressions), the star counts and the luminosity function can be used (by inversion of integrals) to derive the density distribution of stars as a function of distance from the Sun. The procedure they used was developed in 1910 by Karl Schwarzschild (1873–1916), Director of Sternwarte Göttingen at that time.

The luminosity function they found was a Gaussian, the number of stars per volume element increasing as the mass decreases, with a maximum at stars of absolute magnitude 8 (about 20 times fainter than the Sun) and then decreasing again. This is not too bad according to present information, except that there is no such decrease for faint stars and the maximum rather is a point faintwards of which the function remains more or less constant.

Then they could use more extended star counts as a function of apparent magnitude together with this luminosity function to derive the run of star density as a function

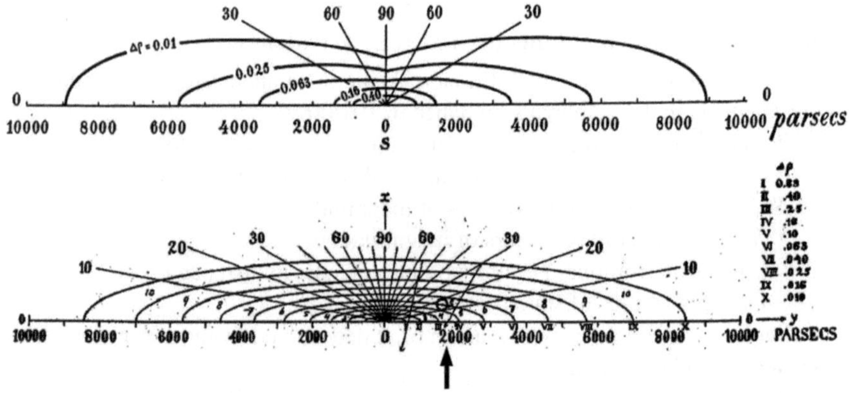

Fig. 3.2 Kapteyn Universe. At the top the model derived from star counts from Kapteyn and van Rhijn [68] and below after fitting ellipsoids from Kapteyn [70]. In the lower illustration the small circle (indicated by the arrow) denotes the position of the Sun, but has been incorrectly drawn too far from the center (see text)

of distance. This comes down to assuming the Sun is in the center, as the density distribution varies only with distance and therefore would be symmetric with respect to the Sun in every direction. They did all this first from counts averaged around the Milky Way and then repeated the procedure by averaging counts in bands on the sky around +30° and −30° out of the Milky Way plane, the same for ±60°, and finally in areas around the poles at ±90°.

This resulted in the figure illustrated at the top in Fig. 3.2. James Hopwood Jeans (1877–1946), who also is one of the founders of the theory of stellar dynamics, coined the term *Kapteyn Universe* for this model of the Stellar System. It was a major accomplishment. It should be said in all fairness that a similar, though more mathematical and less widely appreciated version of this had been derived independently and at about the same time by German astronomer Ritter (Knight) Hugo Hans von Seeliger (1849–1924).

But Kapteyn did not stop here, but took an important and fundamental next step. He asked how the system could be in equilibrium considering the distribution of matter (stars) and the stellar velocities. This is the first application of galactic dynamics, the field that Oort advanced significantly later. Kapteyn published another article, also in 'The Astrophysical Journal', entitled *First attempt at a theory of the arrangement and motion of the Sidereal System* [70]. He first fitted the derived density distribution to ellipses (Fig. 3.2 bottom), so that he could calculate forces relatively easily. Next he looked at the vertical distribution, treating it as stratified atmospheric layers. The average height stars can attain from the symmetry plane then depends in a known manner on the average speed *and* the distribution of all gravitating matter in a column. The mean velocity of stars was known, so Kapteyn could calculate from his density distribution the total mass and deduce from that the average mass of stars had to be 1.4–2.2 times that of the Sun. This value seemed very reasonable in view of the fact that the mean total mass of the two components of binary stars had been found to be 1.6 that of the Sun. Kapteyn's distribution of stars in the vertical direction was actually very good even to modern standards and his analysis rather close to more recent ones.[1] There was no need for additional 'dark matter'. Oort was to improve on this significantly in an important paper in the early 1930s.

For the horizontal direction in the plane of the system, Kapteyn proceeded as follows. He calculated the forces again from the observed distribution, using the average stellar mass inferred from the analysis of the vertical distribution, and found that in the regions near the Sun a centrifugal force was required corresponding to a rotation speed of about 20 km/s over a significant range of radii. Now his two star streams moved at about twice that value with respect to each other and therefore he inferred that the stars rotated around the center of the system at 20 km/s in two opposite directions. From some more detailed considerations Kapteyn decided that the Sun was some 650 parsecs from the center of the system. From the distribution of Cepheids (a kind of periodically variable star; see Sect. 3.1), Hertzsprung had estimated that the Sun is 38 parsec out of the symmetry plane. Unfortunately Kapteyn drew the position incorrectly in Fig. 3.2, three times too far from the center, maybe by confusing lightyears and parsecs.

However, while the vertical analysis was essentially correct, there were two flaws with the Kapteyn Universe in the direction of the plane of the Milky Way, designated from now on as 'Galactic plane', which were soon found—to a large extent actually from work by Oort, as we will see. The first was the fact that the universally accepted notion of no interstellar absorption was incorrect. It did exist and was significant, but only near the plane of the system, so that Shapley's inference of transparent space was correct, except for a thin layer around the plane of the Milky Way. The second was Kapteyn's interpretation of the presence of preferred directions among the stellar motions in space as two opposite streams; the observations could just as well be interpreted as an asymmetry in the distribution of these motions, the largest mean motion being the directions of the apparent streams. The phenomenon really is a reflection of the distribution of velocities of stars being non-uniform, as was pointed out within a few years by German astronomer Karl Schwarzschild (see Sect. 6.2). Such an asymmetry, in which the mean motion in one direction is larger than in others, cannot be distinguished from two opposite streams on the basis of proper motions on the sky alone. Kapteyn rejected Schwarzschild's interpretation on the basis of the information available that the kind of stars that make up both streams were fundamentally different and the fact that one stream contained 1.5 times as many stars than the other. Astronomers generally were divided on this issue; for example, Arthur Stanley Eddington's (1882–1944) leading 1914-book on the theory

[1]The vertical distribution of the Kapteyn model can be fitted well with the parametrization $\mathrm{sech}^2(z/z_o)$, proposed in 1981 on the basis of surface brightness studies in external, edge-on galaxies by myself and Leonard Searle (1930–2010) [71]. Using Kapteyn's density distribution one gets a very good fit with $z_0 = 650$ parsec, while the modern value is more like 700 pc. Kapteyn used a mean vertical velocity of 10 km/s, while the now accepted value is slightly larger at 13 km/s or so. Modern astronomers actually use the velocity dispersion (see Sect. 4.5); this then is 17 km/s.

Fig. 3.3 Arthur Stanley Eddington and Jacobus Cornelius Kapteyn. It is not known when, where and by whom this was taken. Oort was presented with a copy of it by S. Chandrasekhar (letter of July 7, 1954), who claimed it had been taken during the 1922 IAU meeting in Rome. This is impossible; see Sect. 3.4. Chandrasekhar wrote that John Louis Emil Dreyer (1852–1926) of Armagh Observatory, Northern Ireland, had it in the form of a postcard from the photographer and gave a print to Harold William Newton (1893–1985) of Royal Greenwich Observatory, who presented another print to him. From the Oort Archives

of stellar dynamics, *Stellar movements and the structure of the Universe* [72], had not resolved the matter (Fig. 3.3).

The situation at the time Oort started his first research can be illustrated well with Fig. 3.4. These drawings have been published by Willem de Sitter (1872–1934) in his book *Kosmos* of 1934. This is a Dutch translation of his *Kosmos: A course of six lectures on the development of our insight into the structure of the Universe, delivered for the Lowell Institute in Boston, in November 1931* [73], but the figures are only included in the Dutch version. The figures are credited to Oort (by de Sitter). The top panel shows the confusing situation at the time Oort was a student in

Groningen. Shapley had determined distances to many globular clusters. For this he used stars called Cepheids (after the prototype δ Cephei), which are stars in a late stage of evolution that pulsate periodically due to some internal instability and the frequency of pulsation is closely related to the star's luminosity or absolute magnitude, as Henrietta Swan Leavitt (1868–1921) at Harvard had discovered between 1908 and 1912 in the Magellanic Clouds (companions to the Milky Way Galaxy). Determining the period is then a means of determining the distance, provided that the relation between the period and the absolute magnitude has been calibrated. Shapley found that the globular clusters were distributed in a more or less spherical volume concentrated towards a center in the Milky Way at a large distance from us. The confusing situation was that the Kapteyn Universe on the left in Fig. 3.4 is displaced from Shapley's globular cluster system. The Kapteyn model was rather good in the vertical direction, but strongly affected by absorption in the radial (horizontal) one. The system of globular clusters was too large; Shapley's distance to its center was 20 kpc [74], while the currently accepted value is about 8 kpc. This was partly due to an incorrect calibration of the luminosity–period relation of the Cepheids used for the distance scale of the globular clusters. Furthermore globular clusters at low Galactic latitude near the Milky Way suffered from significant interstellar absorption and seemed fainter and more distant, while Shapley was convinced that space was transparent. The Kapteyn system is two or three times too small and Shapley's two times too large. In many American college textbooks (and presumably wider in the scientific and general community) the presentation of the differences in their respective pictures 'places Kapteyn and Shapley as intellectual rivals and their systems as totally antithetical [, while] the historical reality is much different' (Owen Gingerich in the *Legacy* [6], p. 191; see also Sect. 3.1).

The bottom panel of Fig. 3.4 shows the understanding in the thirties. This is very close to the present situation, except that the size is somewhat too large. Oort plotted the Sun at 10 kpc from the center of both the disk and the globular cluster system, while the currently accepted values is more like 8 or 8.5 kpc. The disk is now assumed to have a radius of 18–20 kpc (Fig. 3.5).

While Oort still was a student, on April 26, 1920, the famous 'Great Debate' took place in Washington at the National Academy of Sciences between Harlow Shapley and Heber Doust Curtis (1872–1942) on the subject *The scale of the Universe'*. Shapley worked at the time at Mount Wilson, but was considered for the directorship of Harvard College Observatory, which at long last had become vacant after the death of Edward Pickering the year before. Curtis worked at Lick Observatory, where he had studied galaxies, photographing them with the 36 in. (91 cm) Crossley Telescope. This telescope had been given to Lick Observatory by Edward Crossley (1841–1905), British politician and astronomer, where it originally had been installed at his Bermerside Observatory, but the British weather was so poor that he decided to give it away so that it was installed on Mount Hamilton in California in 1895. The telescope is well-known for many early photographs of galaxies (and other nebulae). For more details I refer to papers by Michael Hoskin (b. 1930) and Virginia Louise Trimble (b. 1943) [76, 77].

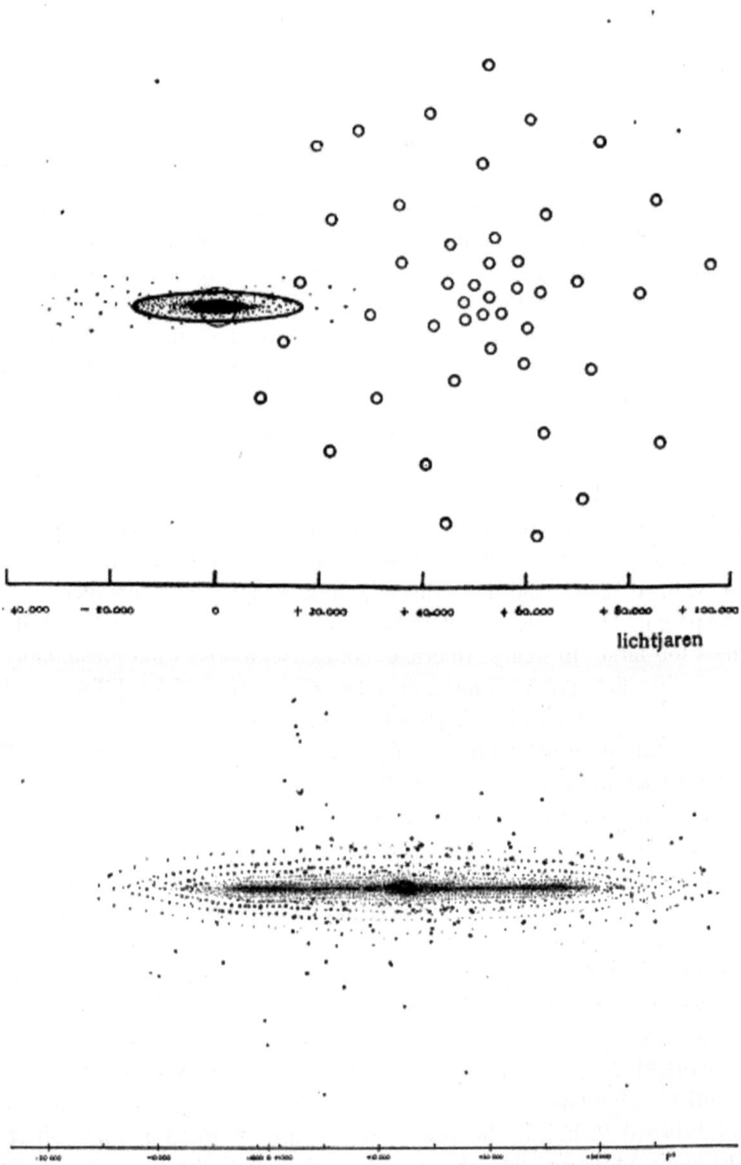

Fig. 3.4 Drawings that were published by Willem de Sitter in his book *Kosmos* of 1934 and were attributed to Dr. Oort. The lower one used here was published by Oort himself in his inaugural lecture as Extraordinary Professor, reprinted as Oort (1936b). At the top the situation according to the general understanding when Oort did his first research, with the Kapteyn Universe on the left displaced from Shapley's system of globular clusters. The bottom figure the more correct structure of the Milky Way galaxy. The top panel has a scale centered on the Sun in lightyears, but runs from -13 to $+33$ kpc, and the bottom scale is in parsecs running from -15 to $+40$ kpc

Fig. 3.5 Globular cluster Messier 53 (M53) as observed by Hubble Space Telescope. It is a rather outlying cluster in the Milky Way at a distance op about 20 kpc from both us and the Galactic center. ESA/Hubble and NASA [75]

In the debate Shapley took the point of view that the globular cluster system was large and that the spiral nebulae and the Kapteyn Universe were part of it. Curtis defended the opinion that spiral galaxies were structures similar to the Milky Way Galaxy and more distant than the globular clusters. He referred to them as 'Island Universes', a term going back to Immanuel Kant (1724–1804). In 1920, the distance to e.g. the Andromeda Nebula was not known, but this was not so for the companion galaxies to the Milky Way, the Large and Small Magellanic Clouds. Already in 1913 Hertzsprung had used Cepheids to determine the distance to the Small Cloud, calibrating the relation between the period and luminosity of the Cepheids by local examples of these stars [78]. He made an arithmetical error that curiously enough was not corrected in the editing, proofreading, etc. He found a parallax of 0.″0001, or a distance of 10 kpc, but had this corresponding to 3000 lightyears. That should be 30,000! He possibly divided by about 3 rather than multiplying. It is not a misprint since he concluded that the Clouds maybe part of a plane parallel to and detached from the Milky Way at 2000 lightyears. The true distance to the Small Cloud is more like 60 kpc (~200, 000 lightyears).

Unfortunately, Shapley does not comment in any detail on the Kapteyn Universe in his autobiography *Through rugged ways to the stars* [79], written in 1969. Kapteyn is mentioned only once, being 'of the older school, but one of the most famous European astronomers', when he met him on Mount Wilson in 1914.

> 'I took Kapteyn my first measurements of the distance of Globular clusters and said, 'The method that I've been telling you about—the method of getting the distances of objects with the variable stars in Messier 3—gives this sort of result. I'd like to have you see it.' He looked and suggested that I check my observations again. In other words, he would

not accept the result. But he was kind about it, because I was a nice young man and he was a nice old man.

There was no question about my having hit the jackpot, but Kapteyn was attacking the distance problem in a very different way, as a part of proper-motion studies. He did not want to give up his own methods, which were good but very rough. It was eight years after I announced my distance to the Milky Way center before Kapteyn halfway accepted it.'

Indeed Kapteyn—and van Rhijn—seriously questioned Shapley's distances to globular clusters (see Sect. 16.6 in my *JCKbiog* [7]), but it seemed that Oort had not been so dismissive. Shapley does not mention Oort at all in his autobiography, so we cannot deduce anything from that.

There is in this context an interesting remark in the *American Institute of physics* interview with Bart Bok [80]:

'[DeVorkin: Kapteyn spent his whole career looking for general absorption. And was always convinced that it was there.]

'In some ways, but van Rhijn said,'Kapteyn has proved there is none.' Yes. And then de Sitter [this is wrong, it was van Rhijn] did some early work in his thesis—for he was Kapteyn's student too—on color effects, and found no color effects with very poor color indices. And therefore, van Rhijn says, 'There's no such thing as absorption.' [...] In the end, van Rhijn was very good friends with Shapley, and he always told me, he didn't dare to tell Kapteyn how he thought Shapley was right, but that he thought Shapley had been right about the general structure of the Milky Way system. And then van Rhijn kept on saying, 'But there is a local system as well, in which we live, and that local system can only be investigated properly by the techniques of Kapteyn'.'

3.2 High-Velocity Stars

Oort's first astronomical work was performed in collaboration with Jan Schilt (1894–1982), who was a student six years Oort's senior in Groningen. The paper (Schilt and Oort, 1923) was a response to a call by the 'Bachiene Foundation' in 1921 to compete for an essay prize. This foundation had been founded in 1918 on the basis of the testament of the daughter of Dutch politician Philippus Johannes Bachiene (1814–1881), who was a member of parliament for the liberal party and strong supporter of Thorbecke. This daughter, Marianna Carolina Francisca Baart de la Faille-Bachiene (1847–1912)—her name is not even mentioned on the Foundation's Website—stipulated that the funds she made available were designated for 'either prizes about subjects in natural sciences, namely human physiology, zoology or astronomy, or making grants available to promising but impecunious students in those areas'. Except for the first years, the Foundation has mostly done the latter, more recently in the form of prizes for excellent research in the areas indicated.

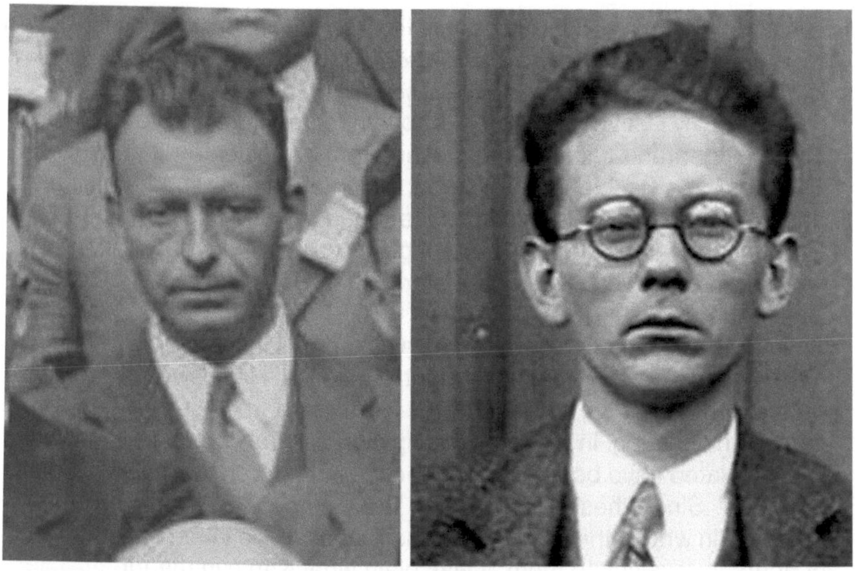

Fig. 3.6 Jan Schilt (left) and Willem Jacob Luyten. The picture of Schilt is taken from the group photograph of the General Assembly of the International Astronomical Union at Harvard in 1932. Luyten's picture is from 1928. Both from the Oort Archives

The Bachiene foundation resides at the University of Leiden in the Faculty of Law together with the Thorbecke Foundation; both foundations have the same board.

The prize Schilt and Oort competed for (and won) had been defined by Kapteyn, probably through de Sitter, and the call was published in the British journal 'the Observatory' [81]. It was aimed at deriving 'from the [...] proper motions of the stars belonging to the [...] spectra types [...] F, G, K, M the frequency of a component of the linear peculiar velocity as a function of its amount'. The explanation specified that 'it is supposed that the theory of the star-streams, in the form given to that theory by Eddington, is accepted. Peculiar velocity then means the velocity freed from both the Sun's motion and from the stream-motion.'. Kapteyn still had Eddington on his side in the explanation of the star streams. The deadline for submission of manuscripts was March 1, 1921. The announcement ended with: 'The author to whom the prize is awarded receives a calligraphed certificate on parchment, together with a sum of money amounting to at least five hundred guilders.' The certificate would have gone to first author Jan Schilt. One would think the prize money was divided between Schilt and Oort; in current purchasing power it corresponds to some €3100 (Fig. 3.6).

The current chairman and secretary of the Bachiene Foundation were not aware of the whereabouts of the archives. The library of the University of Leiden also had no idea where more information about the Foundations, the Prize and this particular competition would be located. There is some information however in the autobiography of Willem Jacob Luyten (1899–1995), who was a student in Leiden at that time and obtained his Ph.D. under Hertzsprung in 1921. I quote from this book, *My first 72 years of astronomical research: Reminiscences of an astronomical curmudgeon, revealing the presence of human nature in science* [82]:

'During my last year in Leiden, Kapteyn retired from his professorship at Groningen and came to the Observatory at Leiden as a consultant. That year, the Bachiene Foundation, which annually issued a call for a prize essay in one of the sciences, announced one in astronomy. The Foundation was asking for a thorough discussion of solar motion and individual stellar velocities of the stars in the Boss Preliminary General Catalogue.

As my interest in stellar motions had awakened, I thought of writing in for it and asked Kapteyn what he thought of the idea. He answered, 'You are perfectly free to do so, but I should tell you that at Groningen we have one copy of the catalog, in which all the Upsilon and Tau Components which we calculated have been entered, and we have given this copy to Schilt and Oort. Since these Upsilon and Tau Components were absolutely essential in writing the prize essay and it would have taken me hundreds of hours to recalculate them, it was plain what Kapteyn meant.

This was my first introduction to high level scientific manipulation. I never forgot it. In due course, Schilt and Oort got the prize.

Many years later a similar essay prize was called for on the variable star RZ Cassiopeia. The only person who had all the available observations was Aernout de Sitter, son of the Director of the Observatory, You can guess: Aernout de Sitter got the prize.'

Luyten was a special character. Oort described him, when he visited Harvard, where Luyten worked at the time, in his pocket diary for 1924 as follows:

'Continuation Friday March 21. [...]

I talk a lot with Luyten, although he has the flaw that he is always looking for a quarrel with anybody and always wants to talk about lesser qualities of others. On top of that he often brags a little and is not very refined in his expressions, but still is pleasant company and has a good view of many things and people, and furthermore has a good taste for music and many other things.'

The Schilt-Oort paper was entitled *The frequency of a component of the linear velocity for stars brighter than $3^{m}_{.}8$ of spectral types F, G, K and M, derived from the proper motions of Boss' Catalogue*. This catalog is the *'Preliminary General Catalogue of 6188 stars for the epoch 1900, including those visible to the naked eye and other well-determined stars'* [83] and had been prepared by Lewis Boss (1846–1912) and published in 1910. Boss was Director of Dudley Observatory in Albany, New York. Schilt and Oort derived mean motions of stars of different spectral classes (from about 13 km/s for A and F stars increasing to 18 km/s for M stars) and they used it to construct 'density laws' for these classes. The paper was published as a separate publication (Fig. 3.7).

The first research paper that Oort completed by himself was published in the first volume of the 'Bulletin of the Astronomical Institutes of the Netherlands' (BAN),

Fig. 3.7 Kapteyn at his desk. Painting by Jan Pieter Veth. It resides in the Kapteyn Room in the Kapteyn Astronomical Institute. The person in the top-right is David Gill. Kapteyn Astronomical Institute

dated September 1922. The title was *Some peculiarities in the motions of stars of high velocity*, Oort (1922). In the *AIP-Interview* [4], Oort responded as follows to a question on when he actually first worked with Kapteyn on research as opposed to taking classes:

'Rather early in my third year, I believe. [That would mean late 1919.] The first two or three years usually are taken up by a kind of general physical mathematical education at universities and it is only after the first examination, Candidates examination, that research is taken up by the students. [...] My own earliest research was on the phenomenon of stars of high velocity which certainly fell outside the regular domain I would say of research of the Observatory at that time. But it was certainly close to Kapteyn's interests.

Yes, I knew fairly well what they were doing at the time at the laboratory and had some discussions with him [Kapteyn] on this subject, but only at the end when it was finished and he could give an opinion on the possible interpretation which was valuable to me at the time. But I certainly don't want to imply that there was any difficulty in approaching him. It was just perhaps, ordinary shyness of a young student. [...]

The stars of high velocity [...] turned out to be a subject of rather fundamental importance which I didn't foresee when I started to work on them at that time. I was just puzzled by it. But in essence they foreshadowed the rotation of the Galaxy which at that time had not yet been conceived. The thing we grew up with in Groningen was the so-called Kapteyn system which was supposed to be more or less symmetrical around the Sun. This later turned out to be quite erroneous because Kapteyn and his co-workers had not realized that the absorption in the Galactic plane was as bad as it turned out to be. They had good reasons for their underestimate of the absorption. They had investigated it for objects too far out of the plane such as globular clusters, where the absorption would generally be small. It led to wrong conclusions when extrapolated to directions in the plane where the absorption is indeed very large. But these high velocity stars did not fit at all into the Kapteyn system and he thought at the time they might be just outsiders moving through the Stellar System such as we knew it at the time and would be more or less unrelated to it. But later on it turned out that they were members of the large Galactic system as it had been outlined by Shapley in the globular clusters. And so they were outsiders in a way to our region of space but not to the Galactic system as a whole. It was Bertil Lindblad who first suggested this. [...]

[...] before one realized sufficiently how serious the interstellar absorption was there was no sufficient reason to have great doubts about the general size of the Galactic system. Although at that time we thought that the Galactic system would be like the spiral galaxies we didn't know either where the large spirals were and how large they were because we didn't know the distances. But from the earliest days of which I can remember my own thoughts, I think I was convinced that they were of a similar nature. There were many different galactic systems in the world so to say and the fact that the Kapteyn system did not show distinct spiral structure was not so difficult to understand because you see it exactly edge on and everything would lie on top of each other so it would be very difficult, still is difficult, at the present time, to unravel the spiral structure. So that was no matter of great concern at this time. But what was a matter of great concern was the system of globular clusters which Shapley first investigated. But nobody at that time, Shapley included, had anything like a clear idea of what the connection was between the stellar system and the system of globular clusters. I think the idea more or less, was that the Kapteyn system would be a big cloud of stars drifting around in a much larger system that was outlined by the globular clusters. Now I believe that Kapteyn himself and van Rhijn probably also did not worry sufficiently about this. For one thing at that time it was still doubtful whether the distances that Shapley had derived were correct and that they might be much closer than Shapley thought. I have no clear recollection now of what I thought myself about that.'

The general notion was that the spiral nebulae were systems quite similar to the Kapteyn Universe, all floating around in a larger system, maybe related to Shapley's system of globular clusters. But the Shapley system would also contain individual stars in addition to globular clusters and spiral nebulae. Such stars would then have velocities with respect to us that would have to be larger than the escape velocity from the Kapteyn system. It were these stars that Oort turned to and he asked the question how they could fit into all of this.

Now Walter Sydney Adams (1876–1956) and Alfred Harrison Joy (1882–1973) of the Mount Wilson Observatory had already noted that the velocities of stars of high velocity were all directed towards one hemisphere on the sky [84], and this is what motivated Oort to look into this in more detail. It had a direct relation with dynamical issues, which had attracted him from the beginning. It is instructive to quote from his 1981 autobiographical paper *AnnRev81* [3]:

> 'Practically from the beginning of my study, I had been fascinated by the possibility that *dynamics*, which had so long been practically a monopoly of astronomers investigating the Solar System, might one day be applied to the so vastly greater system of stars. An impressive foreshadowing was Kapteyn's 'star streams' which Karl Schwarzschild subsequently interpreted as an ellipsoidal distribution of stellar motions.
>
> A strong stimulation had come to me from reading a semi-popular book by Newcomb which Kapteyn had lent me. In later years there were the inspiring books by Eddington, *Stellar Movements and the Structure of the Universe*, and by Jeans *Problems of Cosmogony and Stellar Dynamics*, both of which fundamentally influenced my education as an astronomer.'

The book by Simon Newcomb might very well have been *Popular Astronomy* first published in 1878, of which Kapteyn's copy is still present in the Kapteyn Room of the Kapteyn Astronomical Institute in Groningen.[2] The books by Eddington and Jeans are available on the Web [85, 86].

The measurement of radial velocities of stars had a major stimulus by the completion of the 60 in. Telescope at Mount Wilson in 1908 and Adams had been a major provider of such data. For his study Oort compiled a new set of stars with known radial velocities. He first used a then recent compilation of stars with radial velocities by Joan George Erardus Gijsbertus Voûte (1879–1963) (pronounce as 'vowtuh'), then working in the Dutch East Indies, where he was involved in founding the Bosscha Observatory in Lembang (Java). This Observatory was made possible by rich tea-planter and amateur astronomer Karel Albert Rudolf Bosscha (1865–1928) from Bandung. The *Catalogue* was compiled under the auspices of the International Astronomical Union and was circulated privately (and published in a Dutch physics periodical in the East Indies). Also new data were provided to Oort privately by John Stanley Plaskett (1865–1941), at that time Director of the Dominion Astrophysical Observatory in Victoria at Saanich, British Columbia.

[2]This book, that both Kapteyn and Oort then would have read, contains a quite correct explanation of the origin of the tides on Earth, in quite puzzling contrast with what I discussed in the previous chapter.

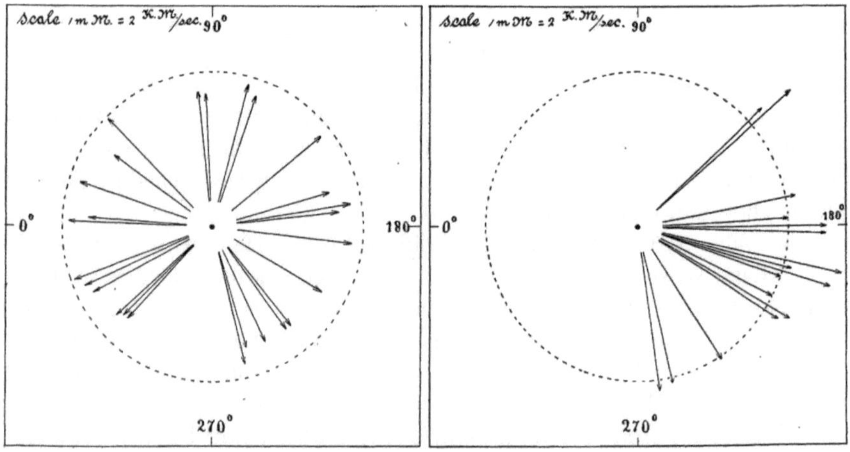

Fig. 3.8 Figures from Oort (1922) on stars of high velocity. See text for explanation

Oort's notes on this study in his Archives contain many pages of lists of stars and further reductions. It is important to correct the velocities for the motion of the Sun in space, although the precise value of that correction is not too critical, since the Sun's speed is about 20 km/s, much smaller than those of the stars considered. Actually, in a letter to his mother, Oort complained at the little progress that he could make because of the tedious and extensive work of making these calculations.

First he looked at the stars with proper motions $>0''.003$ per year; Oort argued that this sample was more or less complete. The transverse velocity (the velocity perpendicular to our line of sight, nowadays usually referred to as tangential velocity), could be calculated if the parallax was available (23 out of 42 stars). The mean transverse velocity was 35 km/s and Oort assumed that the mean for the whole sample was 25–30 km/s. When he divided that sample into those stars with *radial* velocities between 50 and 61 km/s and those with 62 km/s or larger (up to 90 km/s), he found the result reproduced in Fig. 3.8. Here each radial velocity has been plotted as a vector in which the *direction* is that of the projection of the radial velocity onto the Galactic plane. That means the Galactic longitude of the star when the radial velocity was positive and 180° from it when it was negative. The degrees around the border are Galactic longitude, the angle measured along the Milky Way at that time with respect to one of the points where the Milky Way and celestial equator cross. In modern terms the direction towards the center of the Galaxy is 327° in those units (following Oort and early radio astronomers in the Netherlands). The *length* of the vector is the magnitude of the *total* radial velocity (*not* the projected one). The dotted circle indicates 62 km/s. At about this value a dramatic change can be seen, from essentially random in all directions to restricted to less than half the plane.

Next he looked at the Galactic latitudes of the radial velocities. Here he found a similar break at 63 km/s. The stars with larger velocities all had latitudes between −40° and +40°, but for those with smaller ones, 9 out of 40 were at higher latitudes.

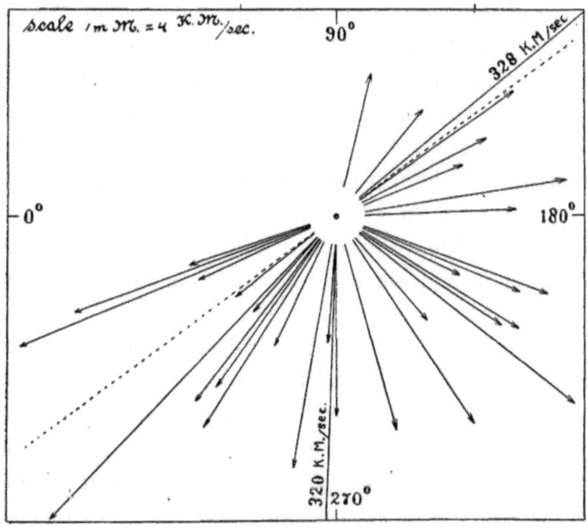

Fig. 3.9 Another figure from Oort (1922). See text for details

This confirmed that a distinct and abrupt change occurred at 61–63 km/s. If the mean transverse velocity was 25 km/s this corresponded to 66 km/s full three-dimensional space velocity.

In Fig. 3.9, Oort showed the same plot for all stars that have radial velocities larger than 90 km/s. This was not a statistically complete sample. The scale here was twice as large, but still two stars with velocities over 300 km/s do not fit. The effect that the velocities point to roughly one hemisphere persists.

Oort then looked at the transverse velocities. He had 61 of those from a catalog compiled by Pieter van Rhijn. Most of these had no measured radial velocities, but their proper motions and parallaxes could be used to infer their transverse velocity. Here Oort then found a similar effect, namely for all stars with transverse velocities in excess of 80 km/s and sufficiently accurate parallaxes, the longitudes to which these were directed in space were restricted to between 145° and 325°. This is the dotted line in Fig. 3.9 and the stars were restricted to the area to the lower-right of that line. This was a very fundamental discovery that would eventually have profound effects on the emerging understanding of the structure of the Galaxy.

Finally Oort took the stars (56 in number) for which the total space velocity could be determined sufficiently accurately and was larger than 65 km/s. For that collection of stars the centroid of their motion in space came out as 74 km/s in the direction of longitude 236°, to the lower right corner in the figures, and only some 7° out of the Milky Way plane.

Before examining the significance of this longitude a few words about this phenomenon in relation to modern understanding is in order. Kapteyn, in his dynamical model for the Sidereal System, had deduced that the two star streams showed that the system of stars was rotating in two separate swarms with 20 km/s or so in two opposite directions around a center 650 pc from the Sun. This meant that the center of the Kapteyn Universe had to be 90° away in longitude from the directions of the streams. This put the center at longitudes 77° or 257° and Kapteyn had chosen for the latter since that is in the southern constellation Carina where the Milky Way is very bright. This is more or less in the middle of Oort's sector of high-velocity stars. In reality the streams are manifestations of anisotropy in the velocity distributions (see Sect. 6.2) as proposed by Schwarzschild, and we will see later that stellar dynamics demands the largest mean velocities to point towards and away from the Galactic center, that is the center has to be *in* the directions of the streams, 347° or 167° longitude. The asymmetry studied by Oort in the stars of high velocity shows the rotation of the Galactic disk with respect to the much more slowly rotating globular cluster system and the individual stars therein (all of which have large random space velocities). Remember this is distinct from Kapteyn's star streams which are related to the random motions of the stars in the rotating local part of the Galactic disk.

Some high-velocity stars are rather nearby and therefore they are not uncommon in the solar neighborhood; but they are not bright. One of the nearest actually is Kapteyn's Star, which has an apparent magnitude of 8.9 and is at a distance of only 3.9 pc (only 28 stars or multiple star-systems are nearer than this). It has the second highest proper motion, namely 8.7 arcsec per year! Its tangential velocity is about 160 km/s and the radial one 245 km/s. The total space velocity of over 290 km/s is higher than the rotation of the Galaxy, which is due to the fact that Kapteyn's star moves in the retrograde direction around the Galactic center. It has been proposed that it once was part of the largest globular cluster in our Galaxy (ω Centauri), and recently it has been found to have at least two planets in spite of its low content of the necessary heavy chemical elements. For more details, also for some credit issues for the discovery, see Sect. 7.9 in my *JCKbiog* [7]. Kapteyn discovered it in 1897 while preparing the *Cape Photographic Durchmusterung* and it then was the highest proper motion known. It was surpassed by Barnard's Star of magnitude 9, 5, discovered by Edward Emerson Barnard (1857–1923) in 1916. Its proper motion is over 10 arcsec per year, but it is at only 1.8 pc. After Proxima- and α Centauri it is the third nearest star from the Solar System. Both Kapteyn's and Barnard's Stars are now known to be high-velocity stars and part of the halo of our Galaxy, crossing the disk of the Galaxy near the Sun at this time.

In his 1922 paper, Oort then gave some attention to the necessity to exclude high-velocity stars in determining the Sun's apex and space velocity. But the most telling part of the paper comes only at the end in a few paragraphs under the title *Remark. The total attraction of the Milky Way system*. This is stellar dynamics and the real interest of Oort. The section started with stating that Kapteyn had estimated that the total attraction of the Galactic System corresponded to a star falling from infinity having a space velocity here of 81 km/s. Actually, as Oort noted in a footnote, Kapteyn had not published this value, but had written 'about 100 km/s'. Now we have seen that Kapteyn had inferred from the vertical dynamics that the mass of the average star was of order 1.5 times that of the Sun, and Oort simply reduced that to a mean mass equal to that of the Sun. He assumed that indeed the stars he studied were 'foreigners' (in his notes he used the Dutch word 'vreemdelingen', strangers) and came from 'immense distances outside the Galactic system'. To illustrate that he noted that a star moving at 100 km/s near the Sun must have been at 240,000 parsecs 3×10^9 years ago, the lower limit to the age of the Earth from 'researches on radioactive minerals by Rutherford and others'. New Zealand physicist Ernest Rutherford (1871–1937), 1st Baron Rutherford of Nelson, pioneered dating the age of the Earth through techniques using radioactive decay of atoms. Oort did not specify where he got this value from, but there certainly seemed to have been a consensus that the Earth's age was at least a few billion years. The sharp limit of 66 km/s then must be the escape velocity of the Stellar System, or as Oort wrote in *AnnRev81* [3]: 'Following a suggestion by Kapteyn I hypothesized that the limit of 65 km/s was related to the velocity of escape from the Kapteyn System, and that the high-velocity stars might be interlopers from outside the System.' But this would mean, as Oort deduced, that the mean mass of the stars is 0.65 times that of the Sun. This—he noted—fitted well with the observation that the average luminosity of stars is 16 times smaller than the Sun.

This study was part of Oort's qualification for the 'Doctoraal' exam, which entitled one to start Ph.D. thesis research. The only requirement left for a Doctor's degree than was the successful writing and defense of a Ph.D. thesis.

3.3 Leiden Observatory

Before continuing with Oort's future I will first have a look at developments at Leiden Observatory when Oort was completing his Doctoraal (Masters) program.

Leiden Observatory was founded in 1633 when a quadrant, originally constructed by Willem Janszoon Blaeu (1571–1638), was bought for Willibrord Snel van Royen or Snellius (1580–1626) by Jacobus Golius (1596–1667) and installed on the roof of the Academy Building, then as now the central focus of the University of Leiden. A quadrant is an apparatus that can measure angles accurately; it had been used by Snellius to perform a new measurement of the circumference and the radius of the Earth with the method first applied by Eratosthenes of Cyrene (c.275–c.195BCE). This involved measuring the (north-south) distance between two cities which Snellius did by triangulation of church towers. For a network of cities one then has to measure the angles of as many triangles between them as possible; if then the distance between one pair of relatively nearby cities is known, one can in this way find the distance of two cities much farther apart, in his case Alkmaar and Breda. The difference in geographical latitude can be measured with respect to the altitudes of the Sun or stars at any moment and that gives the fraction of the circumference of the Earth. Quadrants can of course also be used to measure angles between stars.

It took more than two centuries before the famous 'Sterrewacht' was built under the leadership of Frederik Kaiser (1808–1872); it was officially opened in 1861 (see Fig. 3.11). Kaiser is seen as the modernizer of astronomy in the Netherlands, which was still relatively insignificant when he was appointed. For an extensive descriptions of the history of Leiden Observatory, see [88–90], and for an excellent introduction to Kaiser, [91].

Fig. 3.10 Frederik Kaiser, director of Leiden Observatory under whose leadership the building of Fig. 3.11 had been accomplished. From the Archives of Leiden Observatory

Kaiser (Fig. 3.10) was born in Amsterdam in an immigrant family from Germany. Frederik's father died when he was eight years of age and his upbringing was taken over by his uncle Jan Frederik Keyser (1766–1823), who also lived in Amsterdam but had changed his name into a Dutch spelling. He was a private teacher of mathematics, astronomy and the like. He gave Kaiser a thorough education, including much astronomy which was one of his passions. Gerard Moll (1795–1838), who figures in Oort's academic genealogy (Appendix A.7), also owed some of his education to Keyser. Moll eventually became Director of Utrecht Observatory and Professor of Astronomy. He put Kaiser forward for the position of 'Observator' (assistant observing with the instruments) at Leiden Observatory. In spite of his lack of formal education, Kaiser was appointed in 1826. He took university courses and obtained the 'Candidaats' degree in 1831. He made quite a name for himself when Halley's comet was due to return in 1835. He revised calculations made earlier by renowned scientists and predicted the time of passage of the perihelion to an astounding accuracy of 1.5 h, where many celebrated astronomers had been wrong by much larger time intervals, sometimes even 9 days. [The perihelion is the point in the orbit of a Solar System object where it is closest to the Sun.] In the same year University of Leiden promoted him to Doctor *honoris causa*.

In 1837, the Director of Leiden Observatory, Pieter Uylenbroek (1797–1844), resigned to become full-time Director of the physics department. Kaiser was appointed in his place. In 1840 followed an appointment as Extraordinary Professor of Astronomy and in 1845 as full Professor. But it was not until 1853 that a new Observator was appointed, Jean Abraham Chrétien Oudemans (1827–1906). However, Oudemans left in 1856 to take up a professorship in Utrecht and Martinus Hoek (1834–1873) succeeded him as Observator in Leiden. Oudemans moved to the East

Indies as a geographer in 1859 and Hoek also left Leiden to become professor in Utrecht in Oudemans' place. Hoek was the Professor of Astronomy in the years that Kapteyn studied in Utrecht. Hoek died in 1873, leaving Kapteyn without a professor under whom to write an astronomical Ph.D. thesis, and Kapteyn had done most or all of his thesis research by the time Oudemans returned to Utrecht. The position of Observator in Leiden, vacant when Hoek left, was taken up by Nicolaas Mattheus Kam (1836–1896), who remained active as an astronomer, albeit in his spare time after he took up a position as teacher at the 'Hoogere Burgerschool' (HBS) (see Sect. 1.4) in Schiedam in 1869. Schiedam lies just to the west of Rotterdam; it is my hometown and the HBS there was my secondary school.

Kaiser felt very restricted in his work by the location of the Observatory on an observing platform on top of the roof of the central Academy Building. The position was far from ideal when it came to the stability of his instruments, and he appealed regularly to the Curators, the governing body of the University, and to the Government, to provide funds for the construction of a dedicated Observatory building on solid ground. Curator can also be translated as 'trustee'; however, I prefer a translation as close as possible to the original Dutch term. Kaiser often referred to the building of the Pulkovo Observatory near St. Petersburg as a model for what he envisaged for Leiden. In the end his efforts paid off and funds became available from a whole range of sources. But the major share had to come from the Government and in 1857 the Minister of Internal Affairs, who was responsible for universities, finally reserved sufficient funds in his budget to enable Kaiser to realize his dreams. This minister was Gerrit Simons (1802–1868), another student of Gerard Moll in Utrecht. However, this budget was voted down in Parliament for other reasons and further frustrating delays followed. But then another minister, Anthony Gerhard Alexander Ridder [Knight] van Rappard (1799–1869), who succeeded Simons in 1856 when the latter's health deteriorated, happened to be a strong supporter of Kaiser, and the costs for the construction of a Leiden Observatory were finally securely included in the government's budget. Then things moved fast and the Observatory was in place by 1861 (see Fig. 3.11).

The total costs of the construction of the Leiden Observatory building had originally been estimated at well over Dfl. 100,000. (this is equivalent to a purchasing power of about €2.2 million today), as Kaiser wrote in his communication *Nachrichten über die neue Sternwarte in Leiden* to the German journal 'Astronomische Nachrichten', where he announced the completion of his Observatory to the astronomical world [92]. He did not state the final costs.

Kaiser's main activity in astronomy concerned positional astronomy, i.e. astrometry. This indeed was central in the research of the cosmos at the time, astronomers being very much occupied with measurements of positions of stars, planets, comets and increasingly of asteroids in order to find their orbits. His major instrument was the meridian circle (Fig. 3.12), which had arrived in 1861; it had been constructed by the German firm of *Pistor and Martins* from Berlin. The diameter of the aperture lens was 15 cm and the focal length about 2 m, which means it was a state-of-the-art instrument when it was delivered. It was located at the ground-floor level under a roof that could be opened and served as the main instrument of Leiden Observatory

Fig. 3.11 Leiden Observatory as it appeared on the frontispiece of the Annals of Leiden Observatory, volumes 1 through 8 (1868–1902). From the Archives of Leiden Observatory

Fig. 3.12 Meridian circle of Leiden Observatory (left) and Hendricus Gerardus van de Sande Bakhuyzen at its eyepiece. From the Archives of Leiden Observatory

until it was finally decommissioned and taken apart in the fifties of the twentieth century; it has been in the Dutch National Museum for the History of Science and Medicine Boerhaave in Leiden since.

There were two telescopes in the domes of the Observatory, which could be used for measuring relative positions of celestial object, the largest being a new 18 cm telescope (from the well-known firm of Merz and Sons in München) in addition to the older 15 cm telescope from the same company that had been acquired in 1838. Essential for positional astronomy is the availability of excellent clocks. The main clock was one that gave sidereal time, which is measured relative to the stars rather than to the Sun, so that a sidereal day of 24 sidereal hours lasts 23 h, 56 min and 4 s on our clocks. The Observatory clock had been constructed by the famous Andreas Hohwü (1803–1885), a Danish clock-maker who had settled in Amsterdam. Furthermore there was a large suite of smaller instruments for astronomy and geodesy. It was a modern and well-equipped Observatory.

After the new Observatory had been completed, Kaiser requested the appointment of a second Observator. He had a young student, Hendricus Gerardus van de Sande Bakhuyzen (Fig. 3.12) in mind, but the request was denied. Van de Sande Bakhuyzen had already done some voluntary work for Kaiser between 1861 and 1862, but had left and become a teacher at a secondary school. The position was allowed towards the end of 1863 and Kaiser chose for Andreas van Hennekeler, who eventually (in 1868) left as well to go into teaching. Kaiser complained much about the problem of finding suitable candidates for his Observator positions; students were much more interested to finish their studies as quickly as possible so as to be able to work in secondary education. In the end he had to resort to German astronomers and in 1868 Karl Wilhelm Valentiner (1845–1931) was appointed.

Valentiner took over as temporary Director when Kaiser died in 1872 until van de Sande Bakhuyzen was appointed as Kaiser's successor a few months later. Under Kaiser, Leiden Observatory had established itself as a major center for positional astronomy, not in the least as a result of Kaiser's extraordinary ability to determine extremely accurate positions of celestial objects. This prominent international reputation was reflected in the decision by the *Astronomische Gesellschaft*, the leading astronomical society of continental Europe, in 1875 to hold its bi-annual meeting in Leiden, the first time it convened outside of Germany and Austria. This brought some 30 leading astronomers, mainly from Germany and Austria, to Leiden [93].

Van de Sande Bakhuyzen appointed his younger brother Ernst Frederik (1848–1918) as Observator. By that time the latter had not yet completed his Ph.D. thesis, which he finally did in 1879 on the subject *Bepaling van de helling der ecliptica, uit waarnemingen verricht aan de sterrenwacht te Leiden* (Determination of the obliquity of the ecliptic from observations obtained at Leiden Observatory). Prior to this—in 1875—Valentiner left to take over the directorship of the Mannheim Observatory and the resulting vacant position had been filled by Jacobus C. Kapteyn. However, when Kapteyn left to Groningen to take up a position as Professor in 1875, the position remained vacant. For the sake of completeness I mention that the Curators had allocated funds to appoint an 'assistant' on a permanent basis to make all the necessary calculations for the reduction of observations and other matters to alleviate the workload of the Observators. In March 1876, Adrianus Eecen (1846–1918) was appointed in this position. He was a mathematician, who wrote a thesis *Over de torsie van een elliptischen cylinder* (On the torsion of an elliptical cylinder). He left in December 1877 to become a teacher at the gymnasium in the city of Assen (my current residency, some 25 km south of Groningen), and was replaced by Thomas Joannes Stieltjes (1856–1894). According to van de Sande Bakhuyzen in his Annual Report, Stieltjes was 'a very promising mathematician'. He was to become a *famous* mathematician later, his name surviving as one of the eponyms of the 'Riemann-Stieltjes integral' in mathematics.

De Sitter's Appointment in Leiden

In the summer of 1908 H.G. van de Sande Bakhuyzen reached the age of seventy and had to retire from his professorship in astronomy and as Director of the Observatory. The position was, not unexpectedly, first offered to Kapteyn, already in late 1907. This offer was generous and must have been very gratifying to Kapteyn. It would have given him direct access to the Leiden telescopes, including the photographic one that he so disappointingly had seen being allocated to Leiden. And the workforce in Leiden and the support from the University were far from insignificant. But still he declined. The professional reasons for this are not too difficult to imagine. At that time the work on the Plan of Selected Areas had just been started and much effort on Kapteyn's part would have been required to coordinate things not only on a global basis, but also in relation to the measurements of the plates and the reduction of the data in Groningen. It would have been the wrong time to leave Groningen. The telescopes in Leiden, including the photographic one, were not much suited to work related to the Plan, which required photography and spectroscopy from large telescopes at dark, excellent sites. The fact that Kapteyn had at that time only just been appointed as part-time Research Associate at Mount Wilson undoubtedly played an important role as well. His annual absence for four or five months would not be compatible by the position of Director of the Observatory and would have made an effective leadership difficult (Fig. 3.13).

Also astronomy had changed and Leiden Observatory had been lagging behind developments elsewhere. On an international scale, the focus had changed from positional astronomy to astrophysics, from measurements of positions to studies of physical properties and structure. Measuring the cosmos was replaced wit understanding the cosmos. Much of this had to do with the enormous rise of astronomy in the United States, where observational studies of astrophysical issues were facilitated with new telescopes, techniques, methods, approaches and ideas. Contrary to this situation, in Europe much of the astronomical research effort remained focused on astrometry, on traditional work with existing methods and techniques.

This is sometimes ascribed to the Carte du Ciel effort, which took up a considerable fraction of the resources, both in terms of manpower as well as in telescope time and budgets. The work on it was almost entirely concentrated to European Observatories. The Carte du Ciel was a consorted effort, coordinated and directed by the Observatoire de Paris to photograph the whole sky for two purposes: (1) an *Astrographic Catalogue* with accurate positions and magnitudes of all stars down to magnitude 11,

Fig. 3.13 Willem de Sitter around 1898, as he appears in the *Album Amicorum* presented at the retirement of H.G. van de Sande Bakhuyzen as Professor of Astronomy and Director of Leiden Observatory in 1908. (From the Archives of Leiden Observatory; see caption Fig. 3.1)

and (2) a *Carte du Ciel* proper, consisting of deeper exposures and distributed on photogravure plates. The taking of the plates for the first part was only finished around 1950, but although all plates were eventually measured and the results published, the reduction from rectangular coordinates to celestial ones was often never completed. Eventually the results were incorporated as first epoch exposures for proper motions in the framework of the European Space Agency's astrometric satellite Hipparcos. The second part proved prohibitively expensive and was eventually superseded by the *National Geographic Society—Palomar Observatory Sky Survey*, taken with the wide-angle 48 in. Schmidt Telescope at Palomar Observatory in the 1950s and its subsequent southern extension. A Schmidt telescope, first designed by Bernhard Schmidt (1879–1935) from Germany, is a reflecting telescope, where a glass corrector plate at the front of the telescope tube modifies the light such that a large field of view results in the prime focus of the mirror. An extensive exposition of the influence of the Carte du Ciel on European astronomy is given in *American astronomy: Community, careers and power, 1859–1940* by John Lankford [94], especially his final chapter 'Astronomy compared'. Lankford concluded, that the cause of Europe lagging behind American astronomy indeed has to a large extent been the enormous efforts required for the Carte du Ciel and the fact that American Observatories did not participate in this, which gave them the option to develop observational astrophysics next to astrometry. But an equally important reason has been the insistence on a very uniform approach to the Carte du Ciel's execution by the French leadership and the resulting rigid definition of the observational procedures, which left little room for innovation and progress. In addition, other factors contributed as well, as pointed out in Jones' *Was the Carte du Ciel an obstruction to the development of astrophysics in Europe?* [95], concluding 'that there have been not enough people with necessary interests, skills and training' in Europe anyway.

Leiden was not taking active part in the Carte du Ciel, but still shared in the malaise. In his autobiography, Antonie Pannekoek (1873–1960), who had done a Ph.D. thesis in Leiden in 1902, compared the atmosphere in Leiden around 1905 to the 'smell like in catacombs, of deadly rigidity and boredom'. Volume 8 of the Annals of Leiden Observatory had appeared in 1902, but volume 9 appeared only in 1915, thirteen years later(!). It contained only 5 papers. The number of Leiden articles in the Astronomische Nachrichten similarly was minimal. Between 1900 and 1908, H.G. van de Sande Bakhuyzen published two contributions in that journal; he published typically twice per year in the proceedings of the Royal Academy and annually produced his report on the proceedings of the Observatory. Obviously, the production of significant scientific results had not completely vanished, but had fallen to a minimum.

In any case, Kapteyn declined the offer to come to Leiden. But he put forward and strongly promoted Willem de Sitter, his first Ph.D. student. De Sitter had been born in the Frisian city of Sneek, in a family that had produced many lawyers, and had come to Groningen to study mathematics. During his undergraduate studies he attended Kapteyn's lectures and measured photographic plates in his laboratory. There he met David Gill, during the latter's third visit to Groningen in 1896. Gill invited de Sitter to come to the Cape to prepare a Ph.D. thesis. De Sitter went to the Cape in August 1897 and left in December 1899. The intention was to make photographic parallax observations; but the telescope he was to use was not completed in time, so in the end he began under Gill's auspices to work on the satellites of Jupiter.

While at the Cape, de Sitter married Eleonora Suermondt (1870–1952) in 1898. After returning to Groningen he was appointed Assistant to Kapteyn and defended his Ph.D. thesis *Discussion of heliometer observations of Jupiter's satellites made by Sir David Gill, K.C.B. and W.H. Finlay* in 1901, published in a different form as [96]. The satellite system referred to here consists of only the four major ones, Io, Europa, Ganymede and Callisto, which are similar in mass to our Moon and had been discovered by Galileo Galileo in 1610.[3] The Galilean satellite system of Jupiter remained of major interest to de Sitter throughout his life. He eventually published a comprehensive study of its structure. An interesting account of this is in his George Darwin Lecture *Jupiter's Galilean satellites* [98]. He ascribed the stability to the peculiar characteristics of the inner three satellites that dominate it. These are very close to a system of three bodies in circular orbits and in one orbital plane in exact resonance. The periods of these satellites around Jupiter are in a ratio of 1 : 2 : 4 (Io, Europa and Ganymede respectively), so that the mutual attractions and distortions of the orbits are far from random but return systematically in time and result in orbits locked in resonances. This provides a long-term stability to the whole system. The

[3] Just beating Simon Mayr (or Maris) (1573–1624) by one day [97].

Fig. 3.14 Ernst F. van de Sande Bakhuyzen as he appears in the *Album Amicorum* presented at the retirement of H.G. van de Sande Bakhuyzen as Professor of Astronomy and Director of Leiden Observatory in 1908. From the Archives of Leiden Observatory; see caption Fig. 3.1

actual situation is complicated by the effects of the somewhat different planes and shapes of the orbits of the satellites and the influence of the fourth satellite, Callisto, which is more or less in a 3 : 8 resonance with Ganymede (the outer one of the major three), and understanding these effects was de Sitter's long term subject of study.

In 1908, de Sitter was only 36 years of age, but had been and was very productive scientifically. Taking over Kapteyn's position would have to wait until 1921, when Kapteyn would turn seventy, but by that time de Sitter would be approaching fifty himself. With the limited number of professors in the University, it was essentially impossible to be appointed to a second chair in astronomy in Groningen. The position in Leiden was his chance, and undoubtedly Kapteyn realized this as well. There were two problems though. The first was that de Sitter was interested in theoretical problems, not only related to the satellites of Jupiter but increasingly to relativity theory. His first paper on this was only published in 1911 (on the incorporation of relativity in the theory of planetary orbits), but he must have been thinking about relativity well before that. In the paper he quoted as a major stimulus a publication by Jules Henri Poincaré (1854–1912) in 1905 (Einstein does not even get mentioned). De Sitter of course had done observational work during his period at the Cape and had been contributing heavily to Kapteyn's program to determine the distribution of stars in space, but his experience was very different from running an Observatory and directing an active research program involving telescopes. Secondly, Leiden still had van de Sande Bakhuyzen's younger brother Ernst Frederik (see Fig. 3.14) on the staff as the senior Observator, who at age 60 was much more senior than de Sitter. In the end, de Sitter was appointed as Professor of Astronomy and E.F. van de Sande Bakhuyzen as Director of the Observatory. This left the Observatory in the hands of an observer (although the younger van de Sande Bakhuyzen proved to be an uninspiring leader) and added a very talented theoretician to the University to set up a strong research and educational program.

Kapteyn lost de Sitter as the senior assistant in his Laboratory and important collaborator, just at the time he went for his first visit to Mount Wilson. But he must have been happy for de Sitter. On August 10, 1908, when it still seemed that de Sitter would be appointed both Professor and Director, Kapteyn wrote to Gill: 'You will probably have heard already from de Sitter, that he has been appointed in Leiden as a successor to Bakhuyzen. I am exceedingly happy, for if he had lost this chance, he might have had no other for an indefinite time. Only I would have liked that there had been appointed two successors, one as a Director to the Observatory, one as a Professor. Only Government would not hear of such an arrangement.' In the end the politicians did listen though.

Reorganization of 1918

On March 3, 1918 (a Sunday) Ernst Frederik van de Sande Bakhuyzen, died unexpectedly. On Monday morning, de Sitter was asked to take over the tasks of the Director on a temporary basis until a new Director could be appointed. The resulting reorganization of Leiden Observatory is an important chapter in the development of astronomy in the Netherlands in the twentieth century. It has been described in some detail in the *Legacy* [6] by Wolter R. de Sitter (1936–2009), grandson of Willem de Sitter. Except for David Maarten Baneke's (b. 1979) paper on Dutch astronomy in the interwar years [100] and until an English version of his book (quoted in [90]) will be published, most other publications on the subject are in Dutch, such as two entertaining articles also by Baneke (titles translated): *As if by magic* on the reorganization itself and *You would not expect him to make a mess of the stars, would you?* on the blocking of the appointment of Pannekoek as Deputy Director by the Prime Minister because of his socialist views [101, 102]. The issue is addressed only briefly in English by de Sitter in his 1933 short history of the Observatory (see below).

There is no question about the state of astronomy at the Observatory (so excluding the work that de Sitter did) when the younger (E.F.) van de Sande Bakhuyzen died; the elder (H.G.) van de Sande Bakhuyzen was still alive. Antonie Pannekoek (1873–1960) had obtained his Ph.D. in Leiden in 1902. His thesis was on *Lichtwechsel Algols* (brightness variations in Algol), a well-known variable star also called β Persei, that is an eclipsing binary where the two components occult each other as seen from Earth. He still worked in Leiden, teaching as 'privaat docent' (unpaid by the University) on the history of astronomy. In his memoirs [103], he described the atmosphere at the Observatory under the brothers van de Sande Bakhuyzen as extremely conservative.

> 'In this environment, where everything happened in the traditions of twenty or thirty years earlier, where there was only endless computation and without anything ever being finished, where the new ways of astronomy were hardly appreciated, all enthusiasm must eventually disappear. Later Kapteyn once remarked to me: I never understood how

you kept up with it so long. [...] I dreaded every Monday morning, when I had to attend the weekly conference in the Director's office, where there would be some chatter, and every one mentioned what they had done that week—or made something up—and I realized that every week was in large measure the same, everything just having trickled along a bit. I then always felt a smell around me like in catacombs, of deadly rigidity and boredom'.

The focus of Dutch astronomy had changed to Groningen, but there the Astronomical Laboratory was minimally staffed. There were Computers, but Kapteyn had to obtain external grants for their continued employment. The yearbooks of the University of Groningen lists the employees. In addition to Kapteyn himself there was one paid assistant, who had a Ph.D. or was preparing for it. This position had been occupied by de Sitter (1901–1908), Weersma (1908–1912), Zernike (1912–1914) and van Rhijn (since 1914). Starting in 1918/19, Egbert Kreiken became a second assistant, albeit without a salary from the University. Until 1910 the only further employee was T.W. de Vries with the job description 'klerk' (clerk). In 1910 de Vries' job description changed to 'amanuensis', best described as senior laboratory assistant, and the staff was extended with another 'klerk', J.L. Jansen. This was the situation in 1918, when the reorganization in Leiden started. In all fairness, it should be mentioned that from 1919 onward three Computers ('rekenaars'), doing various numerical tasks, were added at the Laboratory (J.M de Zoute, D. van der Laan and G.H. Muller) and that it is likely that these were already present, but not on permanent or long-term contracts. They might have been funded through outside sources (Kapteyn had such support from the Carnegie Institution for many years in the framework of the Plan of Selected Areas). Computers usually had some form of secondary education after elementary school, such as 'Meer Uitgebreid Lager Onderwijs' (MULO), comparable to junior highschool in the USA.

Leiden was by far the largest astronomical institution in the Netherlands, not only with two professors, but also a fully equipped Observatory. Its personnel had not changed much since 1908, the ten years of the directorship of E.F. van de Sande Bakhuyzen. The 1908 situation is summed up by de Sitter in a *Short history of the Observatory of the University at Leiden 1633–1933* [104], as follows: 'In 1908 [...] there were three Observators and one 'assistant for the computations' (whose title was changed to that of 'Conservator' in 1911), besides a staff of 5 full-time and 3 half-time Computers, and three mechanics and instrumentmakers.' Most importantly, the University of Leiden seemed determined to continue to support its Observatory and astronomy program. After all, in 1908 they had hired de Sitter, but left the Observatory in the hands of the younger van de Sande Bakhuyzen. The salaries were very low (Kapteyn pointed this out in a letter, noting that some had taken other part-time jobs to support themselves and their families), but there was a clear indication that in 1918 the Curators were ready to support the continuation of the Observatory again.

De Sitter kept a detailed diary of the reorganization, in large part reproduced in the chapter by his grandson Wolter de Sitter in the *Legacy* [6]. He noted he was not feeling well the Sunday van de Sande Bakhuijzen died, but his wife wrote to

Fig. 3.15 Left : Ejnar Hertzsprung as a young man. Right: Antonie Pannekoek. Courtesy of the University of Aarhus [107] for the Hertzsprung photograph. Pannekoek's picture is taken from the *Album Amicorum* presented at the retirement of H.G. van de Sande Bakhuyzen as Professor of Astronomy and Director of Leiden Observatory in 1908. (From the Archives of Leiden Observatory; see caption Fig. 3.1)

Kapteyn immediately. As it happened, Kapteyn had been in Leiden the Sunday before (February 24) and had discussed the future of the Observatory with de Sitter, probably in anticipation of van de Sande Bakhuyzen's retirement expected that summer (he had turned 70 in January). Apparently they both felt that the Observatory was in poor shape, but decided not to intervene then but wait 'until I [de Sitter] had fully regained my health'. Kapteyn wrote back immediately and urged him to accept the temporary appointment, so that he could remain in charge and have a maximum influence. Of possible candidates to come to Leiden, Kapteyn strongly supported Pannekoek. Next Sunday (March 10), Kapteyn again visited de Sitter in Leiden. Their conclusion was that the most desirable situation would be for de Sitter to assume the directorship and for Hertzsprung (still at Potsdam) and Pannekoek to become assistant Directors (see Fig. 3.15).

Ejnar Hertzsprung

Hertzsprung, who had been born in Denmark in 1873, was trained as a chemical engineer in Copenhagen, and obtained his degree in 1898. A biography of him by Dieter Bernard Hermann (b. 1939) was published in German under the title *Ejnar Hertzsprung, Pionier der Sternforschung* [105], and I have taken some details from this below. Hertzsprung worked in St. Petersburg, but in 1901 he went to the University of Leipzig to study photo-chemistry. He returned to Denmark in 1902 and worked at the University of Copenhagen Observatory and at the (private) Urania Observatory. Here he had learned astronomy and observing techniques, although not without problems, as he has been reported to have shattered the objective lens of the 12 in. telescope of Copenhagen Observatory, generously forgiven by the Director Svante Elis Strömgren, see [106]. In 1909 he was appointed by Karl Schwarzschild at the Astrophysikalisches Observatorium in Potsdam. Before that, Schwarzschild had been Director of the Sternwarte Göttingen and had just moved to Potsdam to manage the Observatorium in 1909. Hertzsprung's greatest contribution is his work on the systematics of stars according to spectral type, luminosity and evolution. This resulted in the *Hertzsprung–Russell diagram*, independently developed by Henry Norris Russell, which has been instrumental in understanding stellar evolution.

While Hertzsprung worked in Potsdam, Schwarzschild recommended him to Kapteyn, who decided to introduce him to Hale and bring him along on one of his visits to Pasadena. In 1911 Hertzsprung paid a lengthy visit to Groningen and Kapteyn. Klaas van Berkel noted in the *Legacy* [6] (page 161): 'In Groningen, Hertzsprung however not only got to know Kapteyn much better, but also his daughter Henrietta [should be Henriette], and when he left for the US in June 1912 the two of them were engaged.'

Henriette Mariette Augustine Albertine Kapteyn (1881–1956) and Ejnar Hertzsprung married in 1913. They had one child, a daughter Rigel Hertzsprung (1916–1993), named after the brightest star in the constellation Orion. The marriage was not a happy one and they divorced formally on January 19, 1937. They had separated long before that date. In his biography of Hertzsprung, Herrmann quoted a letter of Hertzsprung to his sister of 'May 23, 1923, a few months before his fiftieth birthday', that 'he and Hetty have decided to separate and to dissolve their ten year old marriage. Daughter Rigel was 7 years of age at that time.' Many persons, including Adriaan Blaauw (1916–2010), who had studied with Hertzsprung, have confirmed the widespread opinion that Hertzsprung was a workaholic with little interest in social and other non-professional matters. Klaas van Berkel wrote in the *Legacy* [6]:

> 'It is unclear however how quickly Kapteyn came to realize just how un-social Hertzsprung was. If we have to believe Luyten, he should have known this before the couple was married. 'For three or four months [after his departure for the US] she [Henriette Kapteyn] never heard a word from Hertzsprung, and finally Kapteyn wrote to the Director of Mount Wilson Observatory about it. Eventually Hertzsprung's fiancee received a postcard from him on which he wrote: 'So sorry not to have written before, but the skies here are so beautiful, and, you see, where the stars are, there I'm happy.' (Luyten, *My First 72 Years* [see [82]].) Since Kapteyn had traveled to the US for his yearly stay at Mount Wilson together with Hertzsprung and must have returned in September or early October, this story, if true (and Luyten says he had heard it from Henriette Kapteyn herself), implies that only after returning from the US, Kapteyn was informed about Hertzsprung's lack of attention.'

Leiden Observatory After 1918

The solution de Sitter arrived at—surely to a large extent following the advise of Kapteyn—was to have a structure with three departments, each with their own line of research and led by the Director and the two Deputy Directors with delegated authority. He himself would be the Director and lead the theoretical department. He had moved into the area of relativity, introducing the work of the physicist Albert Einstein, who was still a German national at the time, to the Anglo-Saxon world. After having met Einstein in Leiden (see also Fig. 6.3), they later developed the theory of the expanding Universe together, eventually leading to the famous Einstein-de Sitter Universe model, that expands forever, but stops precisely at infinity [99]. De Sitter and Kapteyn envisaged a situation where Pannekoek would head the Fundamental Department, under which the astrometry would reside, and Hertzsprung a new astrophysics department.

There is no need here to describe the reorganization in detail, so I restrict myself to mentioning that de Sitter, undoubtedly with Kapteyn's help, gained support for this plan. However, financial allocations, in particular with a view to appointments, took a long time to be approved. The Prime Minister of the liberal party, Pieter Willem Adriaan Cort van der Linden (1846–1935), who had been in function since 1913, had doubts about Pannekoek, but would probably not block the appointment. Pannekoek was a prominent and very active member of the socialist movement. After the general elections of July 1918, however, a new government was formed under the conservative catholic Charles Joseph Marie Ruijs de Beerenbrouck (1873–1936). He proved to be a very strong opponent of an appointment of Pannekoek and refused to allow him to be put forward. Eventually, Pannekoek was appointed at the University of Amsterdam, for which the city of Amsterdam was responsible, and the Municipal Council, unlike the national Parliament, had no problem with his socialist background, or at any rate did not let it stand in the way of an appointment. De Sitter and Hertzsprung were only appointed Director and Deputy Director in the course of 1919. No alternative for Pannekoek was identified to lead the Fundamental department.

The reorganization of Leiden Observatory was extremely successful, although not all the funds and extra positions for Computers etc. were realized (see Baneke's article *As if by magic* [101]). Dutch astronomy, in particular that in Leiden, prospered under the leadership of Kapteyn's protégés de Sitter, Hertzsprung and somewhat later Oort. A letter by de Sitter to Kapteyn of May 24, 1920, provides entertaining reading. By that time de Sitter was in Arosa, Switzerland on a cure for his tuberculosis, brought about by an overdose of ether for anesthesia,. The letter referred to a disagreement about the managerial organization of the Lembang (Bosscha) Observatory in the Dutch East Indies, which in Kapteyn's view should be supervised by a national committee, whereas de Sitter was reluctant to give up authority that he felt was part of his directorship.

'The Minister is preparing his plans for specialization and concentration. In Leiden this has been interpreted in such a manner, that each university specializes in a certain area of each discipline. As concerns astronomy, you know my proposal. In short, Leiden specializes in everything except the special items of Groningen and Utrecht. Nijland wrote to tell me that Utrecht for astronomy does not think about specialization, but about concentration, meaning in Leiden. The advise from Utrecht ends: 'The equipment in Leiden must be unstinting, while Utrecht and Groningen, each according to their nature can be furnished on a more modest scale.' Anyhow, to me it seems probable Leiden will be able to obtain equipment and staff on a more liberal scale than Utrecht and Groningen.'

Kapteyn maintained that he did not really intervene in the case of the appointment of Hertzsprung, who after all was his son-in-law. But it is not difficult to imagine that he would be pleased to see his daughter Henriette and his granddaughter Rigel leave Potsdam and settle in the Netherlands. But it is also very well possible that Kapteyn realized that for the future of *Dutch* astronomy there had to be a strong institution in the country, and that Leiden Observatory, if continued to be supported by its university as it was then, would be in a much better position to become this leading institute than his less generously supported Laboratory in Groningen. It does not mean he would give up his Laboratory, van Rhijn and Groningen astronomy, but that it would be the best way forward for Dutch astronomy to establish a modern, strong and extended Observatory in Leiden.

The correspondence between Hertzsprung and Kapteyn, kept in the Archives of the University of Aarhus, reveals a consistent picture of Kapteyn advising Hertzsprung about the progress of the reorganization and what his best course of action should be. There is no sign of any interference by Kapteyn with a view to gaining any improper advantage for Hertzsprung. Their tone is interesting. Initially (1906) they corresponded in English and address each other as 'Dear sir'. By 1912, Hertzsprung wrote to 'My Dear Professor Kapteyn', signing with 'Ejnar Hertzsprung', while Kapteyn answered in German to 'Lieber Hertzsprung' (as always signing with 'J.C. Kapteyn'). After the marriage of Ejnar and Henriette, Kapteyn continued for a while in the same fashion, but by 1917 Hertzsprung started to correspond in Dutch (although still based at Potsdam), addressing Kapteyn as 'Lieve Vader' (Beloved father), while Kapteyn addressed him as 'Beste Ejnar' (dear Ejnar), sometimes 'Amice', and signed with 'Vader'. The letters after he had settled in Leiden concerned to a large extent Hertzsprung's efforts to obtain financial support for instrumentation that had been promised to him when he accepted his appointment as (extraordinary, i.e. in addition to the structural chairs) professor at the University of Leiden (Fig. 3.16).

What happened after Pannekoek's appointment had been refused? Who was to lead the fundamental department? I quote from the Kapteyn biography by Henriette Hertzsprung–Kapteyn:

'One time when Kapteyn visited the Hertzsprungs in Leiden the matter was discussed over dinner. The daughter had a sudden thought and said: 'Father, would you not take this position temporarily? You know this kind of work intimately. And because of your upcoming retirement you will have sufficient time available.' Hertzsprung immediately supported this unusual idea. Kapteyn laughed and said that the idea had not even

Fig. 3.16 Willem de Sitter. From the Archives of Adriaan Blaauw

occurred to him. After some thought he said: 'It is not such a strange idea, however, and I will give it some serious thought.' He felt that it was urgently necessary that order had been brought to the chaos in the material collected and its old fashioned state. Leiden Observatory had a special place in his heart, he felt at home there, was acquainted with the staff and was highly regarded by all. Someone like that could only succeed. He wrote to Hertzsprung on 18 April 1920:

'Now that you are serious about this, I have given it some thought. I am prepared to accept a position as 'adviser' for one day a week to work on this... This is inspired only because of the extensive plan of yours and only to help the Observatory. Personally there is in this new obligation little that is really attractive to me...' It did take away a major part of his freedom to do what he wanted, which he was looking forward to. 'Es ist gerade diese völlige Freiheit, welche für mich das Emeritat anziehend macht [It is exactly this complete freedom that makes the status of emeritus for me so attractive]', he wrote to Prof. Eberhard in Potsdam, with whom he had developed a good friendship over the last years.'

Gustav Edward Eberhard (1867–1940) received his doctorate under Hugo von Seeliger at München, and from 1918 until his retirement he was a senior observer at the Astrophysikalisches Observatorium at Potsdam.

'The conclusion was that de Sitter was extraordinarily pleased with the proposal that Kapteyn would temporarily take upon him the Deputy directorship, for which he would spend a week in Leiden every two months. He would stay with his children, which was a great joy on both sides. 'Now you will be the assistant of your former assistant', they teased him, but they were also proud for the greatness of his mind which, when it came to matters that he felt were important, did not care about convention. There were some who felt that a special title had to be found for him, as they felt this modest position was not doing justice to his dignity, but such considerations were foreign to him and his unselfish wish to serve was more to him than whatever title. So this way he came back to Leiden Observatory, the place where his astronomical career had started half a century earlier. He made new research plans, which would scrutinize the adopted fundamental systems of star positions, and at the same time oversaw the reduction of older observations at Leiden to final results.'

Indeed, after his retirement in the summer of 1921, Kapteyn took up a part-time, equivalent to one day per week, appointment in Leiden as Deputy Director and head of the fundamental (astrometric) department.

3.4 Yale and Schlesinger

In 1921 Oort was appointed assistant at the Kapteyn Astronomical Laboratory. In a letter of April 30 (year not given, but clear from the context that it is 1921) he wrote to his mother:

'Yesterday morning Professor van Rhijn asked me to see him and informed me that he wants me to become his assistant. I told him my concerns. The concern 'Schilt', that has been enhanced with a similar concern 'Kreiken' (as he has been rejected for the Indies), he felt was not relevant because it would be for a longer period anyway. He wanted, if at all possible, to turn it into a well-paid job (so no assistantship), so that I could stay on it longer and would not for financial reasons have to take a job as teacher. I accepted the very nice offer that I also regard as a real honor, in principle, but want to discuss this extensively with you, which he felt was understandable and I am therefore absolutely free to accept or not to accept.

See you Monday night,
Many affectionate greetings and a kiss from
Jan'

In a postscriptum he adds that 'Van Rhijn would prefer to keep this secret for the moment.'

Egbert Adriaan Kreiken was still working on his thesis which he defended early in 1923, a little more than six months after Kapteyn's death. Kreiken's thesis, *On the colour of the faint stars in the Milky Way and the distance of the Scutum Group*, had Pieter van Rhijn as thesis supervisor. Kreiken first became a teacher, but then went on to pursue an active and successful career in astronomy, for many years at the Lembang Observatory, later in Liberia and Turkey. He is remembered for his founding of the Ankara Observatory, which in 2013 was named the University of Ankara Kreiken Observatory. For more details see my *JCKbiog* [7, 108–110].

But obviously Oort took the offer and became assistant to van Rhijn. Actually the date of the offer was before Kapteyn's formal retirement, so it was made by van Rhijn in anticipation of his new role as Professor of astronomy and Director of the Astronomical Laboratory after the summer. The concerns Oort mentioned would have been that he felt that Schilt and Kreiken were more senior than he was and would be entitled to be offered the position first. Jan Schilt, as we have seen when I discussed the Bachiene Prize paper (see Sect. 3.2), was six years Oort's senior and might also have been a possible candidate for the position. Schilt moved to Leiden in 1922 as assistant and obtained his Ph.D. in Groningen under van Rhijn in 1924 with a thesis *On a thermo-electric method of measuring photographic magnitudes*. He spent most of his career in the USA, where he eventually became chair of the Astronomy Department of Columbia University from 1933 to his retirement in 1962.

As soon as he had reorganized Leiden Observatory after his appointment as Director, Willem de Sitter had been concerned with the future. The appointment of Hertzsprung took some time to complete, and the one-day-per-week appointment of Kapteyn as Adjunct-Director to lead that department obviously was a temporary measure. After the failure to appoint Pannekoek there was no other candidate in the Netherlands suitable to assume the position to lead the fundamental department. At the same time, Kapteyn was concerned that his interim appointment would not last too long. Finding suitable successors had been a problem for some time. There was also no obvious alternative candidate in the Netherlands in 1908 than de Sitter to appoint as Professor in Leiden and in 1918 as Director. Kapteyn had these problems before.[4] He must have considered his assistant Herman Weersma as a possible successor, but in 1912 Weersma left to take up a position elsewhere and his next assistant Frits Zernike already left in 1914. Eventually he did find Pieter van Rhijn. But on November 3, 1912, nine years before his retirement, he wrote to his longtime friend and collaborator Anders Severin Donner (1854–1938) of Helsingfors Observatory (Helsingfors is Swedish for Helsinki): 'At present my greatest worry is to find a successor that will keep the laboratory from going into a decline. My assistant [Weersma] leaves me. Do you know anyone—assiduous, practical, reliable – who without too much pay [...] could develop in the next 9 years to become my successor?' Donner had subsequently approached his student Yrjö Väisälä (1891–1971), who considered, but declined to take the option. He remained in Finland and was one of the founders and the first Director of the Turku Observatory.

On June 13, 1920, Kapteyn wrote de Sitter on the issue of his temporary position as Adjunct-Director and the prospects of identifying possible successors [111]:

> 'Now the great task that concerns me. I am prepared to accept a possible temporary appointment as 'interim Adjunct-Director' with the duties as described in your letter of 2 June. I wish to stay not a single instant longer than the moment a good man has been found to accept

[4] I became aware of the following after completion of my *JCKbiog* [7]. I am grateful to Kalevi Mattila of Helsinki Observatory for providing this information.

> the appointment permanently, a man who can devote his full youthful energy to it. A salary of Dfl. 3000 is totally satisfactory.
>
> As to the question whether someone can be found in the near future as permanent adjunct? I have rather good hope for Oort and Schilt. Hintze I do not know well enough. But the first two will have to have more training in Leiden in addition to their Groningen studies. I see this matter as follows. A man like Oort should finish his Ph.D. first here in Groningen. If he succeeds in writing a thesis that attracts attention by astronomers, he should be assistant in Leiden for some time to develop his good qualities. Then he could be the man you are looking for. I think there is no reason to think that within a year there will be another suitable man in the Netherlands.
>
> Would the time come that Pannekoek will have more chance not to be refused by the government, then I will immediately retire.'

The amount of Dfl. 3000 corresponds to almost 4 times the annual wage of an unskilled worker. I do not know who Hintze was; very likely Coert Hendrik Hins (1890–1951) was meant. Hins was a student of de Sitter at the time, and completed a Ph.D. thesis in 1925 on *Inleiding tot een catalogus van plaatsen en eigenbewegingen van 1533 roode sterren* (Introduction to a catalogue of positions and proper motion of 1533 red stars). In his thesis he thanked Kapteyn, 'whose inspiring example I only had the privilege to work with for a short time'. Kapteyn possibly might have come to know him well, but only after he became Adjunct-Director in Leiden and had learned to write his name correctly. Hins has been at Leiden Observatory until he moved to Lembang Observatory in 1946.

Kapteyn wrote again on June 26, 1920, apparently on a draft proposal for the Minister. De Sitter had mentioned the possibility of appointing a foreigner at the Adjunct-Director's position and mentioned the names of Leopold Courvoisier (1873–1955), a Swiss astronomer working in Babelsberg-Berlin Observatory, and Toivo Ilmari Bonsdorff (1879–1950), a Finnish astronomer at Helsingfors. Kapteyn did not think they were up to it and would prefer to find a Dutch astronomer. He again put forward Oort and Schilt, saying, that 'I would certainly support your desire to hire a man like Oort or Schilt to Leiden, either as assistant or as volunteer. I think that Schilt would, if necessary, work as volunteer. His father seems to be rather wealthy and he (the son) has told me more than once that he has no deadline but wants to be a student as long as he can do useful work.' De Sitter did not formally propose any of these foreigners for appointment. The appointment of Kapteyn took some time to be formalized. He took up his Leiden duties after his retirement, in the fall of 1921, but was stricken by his fatal illness after a few months and died in June 1922.

Oort in the mean time, had become assistant in Groningen. Some time before the summer of 1922, de Sitter would have discussed with him the possibility to come to Leiden and offered him a position, although with the condition that Oort first work some time abroad to gather experience with astrometry, so that he might eventually lead the fundamental department. All of this must have happened later than March, 1922, because there is no mention of it in correspondence of Oort with his parents.

In a letter to his parents of March 1, 1922, Oort complained about the slow progress of his work on high-velocity stars, but did not mention anything related to his future. Nor on March 12, where he mentioned that high-velocity stars take all his time. But also that 'The girls of our dance club will probably offer us an extra evening next Wednesday with 'schrikkeldansen' ['leap dance', where the women invite the men to dance], which could be fun.' Then there is a lapse of letters until later that year when Oort actually had moved to Yale Observatory.

In early May 1922 the first General Assembly of the International Astronomical Union took place in Rome. The official National Committee for Holland, as the Netherlands was designated there, consisted of de Sitter, Hertzsprung, Nijland and Baron Jacob Evert de Vos van Steenwijk (1889–1978) as secretary. The latter had a Ph.D. in astronomy, but coming from a family of politicians, he became, after a period as teacher, mayor of the city of Zwolle and later Haarlem, and subsequently Queen's Commissioner in the province of North-Holland. He remained actively involved in the astronomical community in the Netherlands. Albertus Antonie Nijland (1868–1936) was Professor of astronomy in Utrecht and Director of Utrecht Observatory. It is known that Oort attended too. He was, of course, much too junior for that but it must have been de Sitter or van Rhijn that suggested this and helped to make it happen. In the newspaper of the 1970 General Assembly in Brighton, Oort mentioned that he 'happened to be passing the Easter vacation [should be Whitsun/Pentecost], during which the meeting was held, with a student friend who lived in Rome for a year. Not surprisingly, he has not been listed as a formal participant and he is not mentioned in the formal proceedings (see note in Appendix C.1 on the IAU Transactions). He does appear, however on the group photograph (see Fig. 3.17). The remark of Schlesinger in Sect. 3.4 confirms Oort's presence in Rome.

As a sideline I remark that Kapteyn wanted nothing to do with the Union. During the meeting he was fatally ill and would die within weeks. He had vehemently opposed the founding of the IAU and the decision of the Royal Academy of Sciences (the precursor of the KNAW) to join. The reason was that strong sentiments had prevailed to ban the 'Central Powers', particularly Germany, Austria, from membership of international organizations, even scientific ones. Kapteyn had been indignant about this and with his Groningen friend Gerardus Heymans (1857–1930), Professor of Philosophy and Psychology, had tried to keep the KNAW from agreeing to join. When they failed, Heymans gave up his membership, and Kapteyn never went to any of its meetings from then on. Kapteyn's protégé's de Sitter and van Rhijn did support the Union and went to Rome. Although German astronomers started participating in General Assemblies on a personal basis (e.g. in Leiden in 1928 under the presidency of de Sitter), membership of Germany came only in 1951.

Fig. 3.17 Detail from the photograph with participants at the first IAU General Assembly in 1922. Oort is in the lower-right hiding partly behind Henry Norris Russell. Directly above Oort is Ejnar Hertzsprung and at the extreme left Frank Schlesinger. For the full picture with many more identifications, see [112]. From the Oort Archives

In the *AIP-Interview* [4], Oort told the following:

'Professor de Sitter, who was Director of the Observatory in Leiden, wanted me to take a position in Leiden. He thought it would be good for my general education to go for a certain period to a place where they did fundamental work. He, at that time, was rather befriended with Professor Schlesinger at Yale and so I got a research assistantship at Yale Observatory for two years during which I learned a lot about various subjects which I would never have entered upon without that and which were quite useful to me in my later life but were not directly connected with the sort of work I had started out at Groningen and that I really wanted to continue in. At Yale I worked on the problem of latitude variation, for instance, which was quite far way from the subjects I had so far been studying. But I consider these years as having been very useful. They gave an interest of problems of fundamental astronomy that I felt was capitalized in later, and which certainly influenced my future lectures.'

To the question of whether he knew he was going to go back to Leiden after two years, Oort replied: 'Yes, that had been the idea from the start. The job problem was easier in those days.'

Box 3.3 The International Astronomical Union

The International Astronomical Union (IAU) was founded in 1919. Its mission is to promote and safeguard the science of astronomy in all its aspects through international cooperation. Its individual members are professional astronomers from all over the world, active in professional research, education or instrument development in astronomy. Currently, the IAU has over 10,000 individual members in 90 countries worldwide. Of those countries, 70 are national members.

The scientific and educational activities of the IAU are organized by its Scientific Divisions, Commissions and Working Groups. The IAU Secretariat is hosted by the Institut d'Astrophysique de Paris, France.

The key activity of the IAU is the organization of scientific meetings. Every year the IAU sponsors a number of international IAU Symposia and every three years the IAU holds a General Assembly, which offers six IAU Symposia and a range of Joint Discussions, Focus Meetings, Special Sessions, and business and scientific meetings of Divisions, Commissions, and Working Groups.

Among the other tasks of the IAU are the definition of fundamental astronomical and physical constants, unambiguous astronomical nomenclature, promotion of educational activities in astronomy and informal discussions on the possibilities for future international large-scale facilities. Furthermore, the IAU serves as the internationally recognized authority for assigning names to celestial bodies and surface features on them.

The IAU works to promote astronomical education and research in developing countries in collaboration with organizations such as UNESCO.

The IAU Website is at www.iau.org. A comprehensive history of the first 50 years of the IAU has been written by Adriaan Blaauw, *History of the IAU, Birth and first half-century of the International Astronomical Union* [113].

Although we do not know when exactly de Sitter offered Oort the possibility of joining Leiden Observatory and gaining foreign experience in astrometry, it *is* known that de Sitter made contact with Frank Schlesinger (1871–1943) about the matter during the IAU meeting in Rome in May 1922. Schlesinger was Director of the Yale University Observatory since 1920 and he and Willem de Sitter were good friends. Schlesinger furthermore was one of the leading American astronomers and an expert in astrometry, especially using photographic techniques (Fig. 3.18).

Frank Schlesinger had worked earlier at the Yerkes Observatory of the University of Chicago at Williams Bay, Wisconsin. George Ellery Hale had founded the Observatory in 1897 with funding provided by business man Charles Tyson Yerkes (1837–1905) and Hale had succeeded in erecting there the largest refractor telescope in the world. This was a telescope with a lens as objective that in this case measured 40 in. (about 1 m) diameter. Schlesinger had been employed at Yerkes,

Fig. 3.18 Frank Schlesinger in 1914. From Wikimedia Commons [114]

where he had developed techniques for measuring trigonometric parallaxes using photographic methods on such 'long-focus' telescopes. This is important for large plate-scales that is essential to astrometry. He married Eva Hirsch (18??–1928) in 1900 and their only child, Frank Wagner Schlesinger, was born in 1901. In 1905, he moved to the Allegheny Observatory of the University of Pittsburgh, where he became Director. In 1914, the Thaw Telescope was taken into service there, a refractor telescope with an 30 in. (77 cm) objective lens and a focal length of 47 ft (14.4 m), especially designed for long-focus photographic studies of stellar parallaxes. The telescope is named after William Thaw (1818–1889), a businessman in transportation and banking, and funded by his son William Thaw Jr. in memory of his father. But in 1920 Schlesinger had assumed the directorship of Yale University Observatory. To understand his incentive for this move we need to go back to the developments of astrometry.

One of the most important projects undertaken in the nineteenth century, was the *Astronomische Gesellschaft Katalog* or AGK. We have seen that between 1849 and 1863 Friedrich Argelander at the Bonner Sternwarte had measured positions of all stars in the northern hemisphere, complete down to apparent magnitude nine and an incomplete sample of stars of magnitude ten. The resulting star catalog, known as the *Bonner Durchmusterung*, contained 324,198 stars. The AGK was the follow-up with much more accurate positions using meridian circles (see Sect. 2.1) at various Observatories, which was started in 1861, encompassing about 200,000 stars, complete down to magnitude 9. Together with eleven other Observatories, Leiden Observatory had been involved in its execution. Schlesinger set out to redetermine the positions of the AGK stars and find their proper motions, using wide-field photographic camera's. As we will see he also made plans for a parallax and proper motion telescope in the south, complementing the effort for the *Cape Photographic Durchmusterung* of Gill and Kapteyn and it was Yale that was prepared to fund such a project. Schlesinger was the world expert in astrometry, publishing the first *General Catalogue of Stellar Parallaxes* in 1924. His national and international standing was very high; tributes to this are his elections to the office of President of the American Astronomical Society (1919–1922), and of the International Astronomical Union (1932–1935). He was the ideal man for de Sitter to send Oort to before he would assume the leadership of the Leiden fundamental (astrometric) department.

The two men spoke to each other during the 1922 Rome IAU General Assembly about Oort. De Sitter wrote Oort from Rome (actually on paper with the letterhead of the General Assembly) on May 10, 1922. This was the last day of the meeting and apparently Oort had already left Rome.

'Amice,

I have to write you immediately on an important matter. I will come straight to the point. Prof. Schlesinger of Yale University offers you a position at his Observatory for one year. The salary is $ 900 + a room in the Observatory. You will have to devote all your time to the Observatory for 10 months, while 2 months are vacations. You will need to start on September 1, or at the latest October 1. I propose you come to talk about this as soon as I am back in Leiden.

There is a bit more text on details, such as that the offer still needed to be finalised when Schlesinger would be back in the USA. In a follow-up letter of June 7, 1922, de Sitter replied to some questions by Oort, suggesting the 'all your time' that Oort was to dedicate to the Observatory would mean 'a certain number of hours during the day + a few hours during clear nights. I would think that some time will be left for your own work'. Also de Sitter mentioned salaries of 'Conservators' and so in the Netherlands for comparison (Dfl. 4000–5000 at that time, while Schlesinger's 900 dollars corresponded to Dfl. 2500–3000), which meant that what Schlesinger offered was not unreasonable, especially for one that had not yet obtained a Ph.D. Oort had apparently inquired about Schlesinger's southern hemisphere telescope, but de Sitter wrote he did not know anything about this. The letter is followed by a telegram from Oort to de Sitter on June 8, which said: 'Decided for America. Preferably October. Oort'. Things then moved fast. On June 12, 1922, de Sitter wrote that he had just received a telegram from Schlesinger, saying 'Confirmed'.

All of this coincided with the death of Oort's brother Hein of diabetes on June 8, the same day as Oort's telegram to de Sitter. It should be realized that diabetes at that time was often lethal, treatment by insulin had not yet been discovered. The first treatment of a diabetes patient with insulin was only performed on a human by Frederick Grant Banting (1891–1941) in January of 1922, and although public use spread very rapidly—Banting won the Nobel Prize for the discovery of insulin already in 1923—it came too late for Hein Oort. He apparently had decided that life was not worthwhile anymore and had traveled to Vienna, where he lived a life as if there were no diabetes. His parents went to look for him and they were in Vienna when he died; Mrs. Oort-Faber sent a letter to the other children in the Netherlands to notify them of his death [115]. This left a deep scar in the Oort family; his parents, especially his mother, never really recovered from this. Oort himself has told his son Abraham that he accompanied his father to Vienna to look for Hein, but that then must have been somewhat earlier (maybe after leaving Rome) to be consistent with the letter of his mother and his telegram to de Sitter from Groningen. Unfortunately, Oort had also told his son that he had no recollections of this traumatic episode. His diary notebooks are of no help; except for a short, unrelated remark on March 8, his pocket diaries have no entries between December 6, 1921 and September 22, 1922 when he had arrived in New York. The collection of his pocket diaries starts only with 1923.

De Sitter's son Aernout (1905–1944) probably knew Oort's younger brother John well. De Sitter continued the letter of June 12:

> 'We heard about the death of Hein late in the evening from Aernout. He went to see Jo briefly. It is very sad; we were really sorry for Jo and the girls, but they seemed to be fine. It is a major loss for your parents—but I believe that both for them as well as for Hein himself now is better than extending life a few more years with the certainty of an approaching end.'

Schlesinger had written to de Sitter again, also on June 8.

> 'My dear de Sitter, I was very glad to get your cable of the 8th, which according to our code signified that Mr. Oort was ready to accept our assistantship and could begin his duties on October 1. I have sent the recommendation for his appointment to our Corporation and it will be acted upon tomorrow. There is no doubt that the action will be favorable, in which case I shall cable you 'confirmed', which will mean that the business is now complete and that we shall look forward to Mr. Oort coming here about October 1. Perhaps it may be well to state the conditions. The salary is $ 900 a year, beginning October 1, Mr. Oort being entitled to two months' vacation beginning August 1, 1923. It is understood that the appointment is for one year, but if both parties are willing it may be renewed. Mr. Oort's duties will consist of assisting with the automatic zenith tube for determining the latitude. I will see to it that he gets as large a variety of experience as will be feasible at our Observatory in one year.
>
> The room at the Observatory to which he will be assigned is a large one and we will install a comfortable sleeping cot. If he wishes to occupy the room as an office and sleeping quarters, he will be expected to provide his own bed clothes including blankets, and will be expected to keep the room looking like an office in the daytime.

I was well impressed with the little I saw of Mr. Oort at the meeting; and especially in view of your high recommendation of him. I am looking forward to becoming better acquainted with him and to have him work with us next year.

I have not forgotten that I left you in a hurry at lunch on the last day of the meeting without having paid for my share. I propose to keep this account open for three years and to repay you in kind in Cambridge in 1925.

My best regards to Mrs. De Sitter and yourself. Remember me also to the Kapteyns when you write or see them, and let me know how he is getting along.

Very truly yours,
Frank Schlesinger'

The 1925 IAU General Assembly was to take place indeed in Cambridge, UK.

This letter was written only nine days before Kapteyn's death. In a letter on different matters of July 26, 1922, Schlesinger wrote:

'Although you and Hertzsprung had prepared us for the worst with regard to Kapteyn's health, the news of his death came as a shock. The world has lost its greatest astronomer. Personally we feel we have lost one of our best friends.'

Before that, Schlesinger had written to Oort (through de Sitter) on June 17, the day before Kapteyn died, sending him his letter of appointment.

Fig. 3.19 Oort family probably in 1922 after Hein's death, but before Oort's departure for Yale. From the left: Jetske, father Oort, John, Emilie, mother Oort and Jan Hendrik Oort. From the Oort Archives

'I have already stated in my letters to Professor De Sitter how great a satisfaction it is to me to know that you are to be with us for a year. I am looking forward with pleasure to seeing you early October and to your working with us during the year.'

So, Oort was to interrupt his work in Groningen on stars of high velocity, but had the prospect of a new experience in the USA and afterwards a position at Leiden Observatory. It was a difficult time, coinciding closely with the loss of his older brother Hein and Kapteyn, his 'groote leermeester' (great mentor) in astronomy. The family picture in Fig. 3.19 must have been taken after Hein's death, but before Oort's departure for New Haven. They do not look very cheerful.

Fig. 4.0 Oort in 1924 on the Cranberry Isles, off the coast of Maine. From the Oort Archives

Chapter 4
Via Yale to Leiden

Make sure you work at a place which gives you opportunities to learn, travel, explore, interact with intellectuals and develop new work skills.
Abhishek Ratna

'I think of Princeton as being lazy and good-looking and aristocratic –
you know, like a spring day. Harvard seems sort of indoors —'
...... 'And Yale is November, crisp and energetic.'
Francis Scott Key Fitzgerald (1896–1940)

4.1 Yale University Observatory

So, in September 1922 Oort left the Netherlands for a stay of at least one year at Yale Observatory. Yale University, located in New Haven, Connecticut, is the third oldest institution of higher education in the United States of America. It was founded in 1701 (Harvard University is the oldest and dates back to 1636) as a collegiate school and has been a university since 1887. The name 'Yale' comes from Elihu Yale (1649–1721), who was a merchant. He had been born in America (Boston), but grew up in England and became President of the British East India Company in Madras (India). Acting upon a request for help to provide a new building for the Collegiate School of Connecticut, Yale in 1817 sent a large gift, consisting among other things of books and bales of goods. The latter were sold for an appreciable amount of money, from which the building was funded and after it had been finished it was named after Yale. The school became Yale College and eventually, when it grew into a university, the name Yale became attached to that also. Yale University is one of the members of the Ivy League, really a collegiate sports conference of

Indian marketing manager and management coach [116].
From *This side of Paradise*.

P. C. van der Kruit, *Jan Hendrik Oort*, Astrophysics and Space Science Library 459, https://doi.org/10.1007/978-3-030-17801-7_4

Fig. 4.1 Yale Observatory (undated, but likely at the time Oort worked there). On the right is the heliometer dome. Manuscripts & Archives, Yale University [117]

eight institutions of higher education, among which Harvard and Princeton, that are viewed as among the most prestigious universities in the USA (and are indeed high up on current lists of university rankings).[1] Although sometimes confused with the lock manufacturing company of the same name, Yale University in fact has nothing to do with it. That company was named after Linus Yale Jr. (1821–1868), the inventor of the pin-tumbler lock, who established it in Stamford (as it happens also Connecticut) in 1868 together with a partner. Both Linus and Elihu Yale are from the same family of Welsh descent though.

Yale College (the graduate school) very early on started an educational program, that in 1861 led to the first Ph.D's that were awarded in the USA. One of these was on an astronomical subject, on how to determine the orbit around the Sun of a meteorite when its velocity and its direction at entry into the Earth's atmosphere were known [118]. Yale Observatory (until 1920 also known as Winchester Observatory) had been founded in 1880. William Louis Elkin (1855–1933) started astrometry at Yale with a 6 in. heliometer. A heliometer is a telescope, of which the objective lens is cut in two and in which the two halves can be moved separately. When the images of two stars are made to coincide the relative positions of the two halves can be used to measure the angular separation of those two stars on the sky. It is called a heliometer, because the first examples were built to measure the diameter of the Sun. Other means of splitting the image in two with some optical element are possible too, and have been applied in later heliometers. The heliometer at Yale had been installed in 1882 and Elkin joined Yale in 1884 (Fig. 4.1).

Elkin had studied in Strasbourg, and worked for some time on astrometric problems with David Gill at the Cape Observatory. Gill and Elkin wrote an important paper on the subject, *Heliometer determinations of stellar parallax in the southern*

[1] Scott Fitzgerald, whom I quoted at the beginning of the chapter characterizing these three, entered Princeton University, but dropped out before finishing his studies.

hemisphere [119]. In the introduction Gill described that he met Elkin, while visiting European Observatories before he went to the Cape. Elkin was preparing an 'Inaugural Dissertation' (equivalent to a Ph.D. thesis) in Strasbourg on the parallax and orbit of α Centauri. The distance is a little over 4 lightyears; α Centauri is the nearest star to the Sun, but it also is a multiple star system consisting of a binary (A and B) and a third component at a larger separation, which is at present closer to the Sun (called Proxima Centauri). The binary system (α Centauri A and B) and Proxima are separated at the moment by about 2° on the sky as seen from Earth and Proxima is just over 0.1 lightyears closer to us than the α Centauri AB system. Now, Gill had just purchased a heliometer and Elkin was invited to come to the Cape to use it. After he had arrived by the end of 1880, a month was spent on preparations. But then (Gill described) 'I was suddenly recalled to England on urgent private matters. [...] Dr. Elkin occupied my house in my absence, and remained as my guest, and as a member of the family circle, till the completion of our programme. He sailed from the Cape on May 16, 1883. His work from first to last has been a labour of love.' This project was aimed at determining parallaxes of nine stars, among which α Centauri and Sirius.

At Yale, Elkin also made various fundamental contributions to the study of meteors. He constructed batteries of cameras with rotating shutters to measure trails of meteors and the speed with which these moved along them. With two of such devices he was also able to find the distance of the trails. He measured 238 parallaxes of stars with the Yale heliometer, which was a major contribution to the field. He was Director of Yale Observatory from 1896 to 1910, when he resigned from the directorship for health reasons. No successor was appointed, but instead there was first an acting Director and later an administrator looking after affairs. A Professor of Mathematics was appointed Director in 1918 with the understanding that the Observatory remained closed until the end of World War I and until an astronomer was found for the position. This mathematician, British-born Ernest William Brown (1866–1938), had an active interest in astronomy and worked on lunar tables and orbits of planets and asteroids. Mostly routine work continued at the Observatory.

During that period plans were developed to build a 'polar telescope', designed by Charles Sheldon Hastings (1848–1932), Professor of Physics and interested in optics. He had been involved in the construction of major telescopes at Allegheny Observatory near Pittsburgh—named after the range of that name, which is part of the Appalachian ranges—, but also at Swarthmore's Sproul Observatory, both in Pennsylvania. The new design was called a polar telescope, because the major component was a steel shaft parallel to the rotation axis of the Earth, i.e. pointing at the celestial pole. It had a 30 in. (76 cm) coelostat, which is a flat mirror that reflects the starlight up the shaft. At the bottom of this was a 15 in. (38 cm), 15 m focal length objective lens, which formed an image at the end in a room on top of an 18 m high tower. In the shaft a smaller, parallel guiding telescope resided with the same focal length. The observer could then conveniently work from a heated room. The coelostat tracked the star by moving at the diurnal rate and since the image in the focus rotated, the plate holder also rotated at the diurnal rate. The telescope was named the Loomis Memorial Telescope after Elias Loomis (1811–1889), for many years Professor of

Natural Philosophy and Astronomy at Yale. It had been built for photographic work on parallaxes, but after completion in 1915 it remained unused until the new Director would be appointed. After that it ironically was used mainly for stellar photometry. The disadvantage of the construction is that the reflection on the coelostat gives rise to astigmatism ('blurring' as a result of loosing symmetry about the optical axis of the system), which in addition to being a nuisance also varies with temperature.

In 1920, Frank Schlesinger was appointed Director. A detailed description and discussion of the history of Yale Observatory has been published by Ellen Dorrit Hoffleit (1907–2007) [120], who rose to become as a senior research astronomer at Yale. In this book she described Schlesinger as 'the father of modern astrometry' and as 'wizard of parallaxes'. I have already described (see Sect. 3.4) that he took the initiative of re-measuring the positions of the *Astronomische Gesellschaft Katalog* or AGK, which gave rise to the *Yale Zone Catalogue*, with major contributions by Ida Barney (1886–1982). In addition he started work on the *Bright Star Catalogue* to which Ida Barney also contributed. This catalog, which contains all (a little over 9,000) stars brighter than magnitude 6.5 (so slightly fainter that the naked eye can see), has now had its fifth edition; in the course of time Dorrit Hoffleit also was deeply involved with this. Schlesinger seemed to have been of the opinion that women were unsuitable for theoretical work, but excellent observers and reducers of data. Schlesinger came to Yale because the University was willing to support his plan for a large southern telescope for astrometric work. But he also was interested in problems of geographical latitude and variations of the orientation of the Earth's rotation axis, not only in space but also with respect to the surface of the Earth. This was the problem Jan Hendrik Oort was supposed to work on with a special telescope, the zenith telescope.

4.2 To America

Oort left for the the United States of America in September 1922. In his diary notebooks there are some entries related to his voyage to New Haven. There is a loose set of nine sheets, preceded by a small note, saying that he composed a set of impressions to be sent to a number of people, and followed by a list of ten addresses. The idea is apparently that after reading it one should sent it on 'after not too long a period' to the next address on the list. It starts with four addresses in Groningen (the first is Brugstraat 7^{a}, where he used to live, then some family and good friends (such as the van Voorthuysen family, who now lived in Utrecht after Adriaan had become chief inspector at the Ministry) and ends with his parents. Except for a Miss J. S. S. Faber—most likely his aunt Jetske Sophia Susanna Faber (1862–1945)—and his parents, all these addresses have a small cross, indicating that it ended up with the Oorts in Oegstgeest, possibly skipping the aunt. The first page was dated September 15, 1922. Here Oort wrote:

> 'It is so beautiful outside now. You cannot easily imagine yourself being on this big passenger steamer with bright lights all over it now sailing at

full speed on the Atlantic Ocean. The sky is dark, sometimes with a small opening between the clouds and then you can see a few stars and then again it is dark. The sea forms a big disk around us and we have left already quite some horizons behind us since we saw land for the last time yesterday afternoon, which consisted of a tall lighthouse standing on a small island off the coast of south-west England.

We arrived at [Plymouth] yesterday morning and stayed there for two hours. It was foggy but still there was much to see, a few departing destroyers, a number of freight ships and later the coast covered with forests and the largest hotels and factories of Plymouth in the distance, further down the bay. After departing we sailed along the coast and you could beautifully see all those caves and gorges in the high rocks penetrated by sparkling waves. That was a fantastic morning!'

On a different loose sheet, dated September 17:

'The boat still sails on and it looks like something strange that cannot be completed. There is a circle of water around us, that is black now, with a black border separating it from the sky, where stars can be seen here and there. That circle remains the same and does not move away from the center. Everything always keeps going, with an amazing perseverance without ever slowing down. One should admire Columbus, since he did not know how many hours he still would need to sail and how many days with always the same continuous circle around him.

[...] You know what is so strange? That the little plane on which we sail looks so small. In Katwijk it always seemed that the horizon was so far away, here it looks as if it is only a few kilometers from us!'

The second sheet of the set of notes was written after arrival in New Haven. It was dated Monday September 25, 1922.

'New Haven. It is a strange idea that I now am on the other side of the big pond and that all those I am writing to now are unreachably far away. [...] The voyage over the ocean was in most respects extraordinarily pleasant, even though the weather was not really good. It started the afternoon of the first day when we entered Boulogne. The boat of the pilot that came to our side, sometimes disappeared completely under the waves. Dinner was served with care and I did not eat much. To bed early and I fell asleep soon and did not notice the storm that we had that night in the Channel. Next morning I found to my horror that the water almost ran over my bath, because the ship tilted so much and I did not feel very well until we reached Plymouth but it did not get worse. The 'Rotterdam' was very quiet there so we had a beautiful view and not very many people were seasick. My taste for sea voyages got much better.

We had dancing parties two evenings on the deck, which were not so nice since few people danced and there were quite a few annoying persons. The presence of unpleasant people was one of the less nice points of the trip. especially later when I knew them by sight. You cannot escape people, in particular when the weather is poor and you cannot sit

Fig. 4.2 SS Rotterdam (IV) of the 'Holland-Amerika Lijn', on which Oort traveled to the USA in 1922. From Wikimedia Commons [121]

outside. I have sometimes looked with jealousy at people in third class. Among them there were quite a few interesting persons, at least at first sight. One of them was an old lady, who every morning when the weather was fine sat on a beam with a fat book with Russian characters, probably a Bible. And there were a few very pretty Slavic young ladies with colorful scarfs and also a fair number of expensively dressed people that seemed out of place and appeared to belong really to the upper class. They did all sorts of games there like hide and seek or tag and the poshest and the poorest participated with great enthusiasm. I think I will take third class on the way back, but maybe the people will be different then. First class is beautiful with many lounges, dining rooms, music chambers, reading rooms, etc. and many promenade decks. There is plenty of room for everyone. I have been there only twice. They were very strict in that respect and the get-togethers with the Hamburgers and Liesje Storm van Leeuwen had to take place at a staircase or across a fence. They were not even allowed in second class.'

Oort traveled with the S. S. Rotterdam of the 'Holland-Amerika Lijn' (HAL), in full the NV Nederlandsche-Amerikaansche Stoomvaart Maatschappij. The HAL had been founded in 1871 and had operated out of Rotterdam when the 'Nieuwe Waterweg', the ship canal that provided a short connection of Rotterdam harbor with the North Sea, was opened in 1875. The S. S. ('Screw Steamer', not 'Steam Ship' as often supposed), to distinguish from a P. S., or Paddle Steamer), had been in service between 1909 and 1939 after which it has been demolished. It was just over 200 m long and displaced more than 24,000 tons. It could accommodate 520 passengers in first class, 555 in second and 2500 in third, and had a crew of 470. The records of the HAL are kept at the Archives of the city of Rotterdam. For outbound journeys there even is a separate alphabetic list (on microfiche) of passengers, so one can search under a particular letter, such as 'O', and then find all ships in chronological order and for each ship all passengers with names starting with 'O'. A visit I made to these Archives confirmed that Oort traveled on the S. S. Rotterdam, leaving Rotterdam on September 13, 1922. According to the 'State of Passenger Moneys' he indeed traveled in second class, booked from Groningen, for a total fare of Dfl. 350.50, which would correspond now to a purchasing power of about €2,500 (Fig. 4.2).

'The Hamburgers' refers to Hartog Jacob Hamburger (1859–1924), who was a Professor of Physiology and Histology in Groningen, and his wife. Hamburger has been chairman of the Natural Sciences Society (an organization dedicated to bringing science to a broad audience; see Sect. 2.5) between 1907 and 1924 and it is not unlikely that Oort knew him from his Groningen student days. The report of the KNG for the years 1921 and 1922 commented for September 1922 on an extended absence of their chairman Hamburger:

> 'Prof. Hamburger has been invited by Johns Hopkins University in Baltimore to deliver a number of lectures. When he arrived in America, he received invitations from quite a few other places. He accepted these and gave lectures in Baltimore, New-York, Toronto, Cleveland, Washington, Philadelphia, Boston and Newhaven. In these lectures he addressed the chemical side of Physiology.'

The Hamburgers would be on their way to Baltimore on the same trip of the S. S. Rotterdam in September 1922 that Oort had been on. I will come to Liesje Storm van Leeuwen later.

> 'On Friday morning we arrived in Quarantines at Sandy Hook still before New York and had a terribly long doctor's examination so that we only disembarked around noon and then still had to wait for luggage inspection. I did not sleep very well that night and was very tired. On top of that I was suddenly all by myself in the large barrack of the H.A.L.; the Hamburgers had left already and Liesje S.v.L. with whom I was going to spend a day in N.Y. and then see her off, found out that her brother had met the ship and they had to leave that same afternoon. In spite of this I very much enjoyed my first moments in N.Y.! The entrance to the harbor was so beautiful. It was foggy when we entered the bay and suddenly you saw coming out of the fog the Statue of Liberty that even from the 'Rotterdam' emerges as something majestic and kind, even before you see the new land, where so many of the people that have entered the harbor have left their home countries for with some trepidation. This then is an excellent statue to provide encouragement and new energy. The decision [by immigrants] to leave their familiar land for a completely unknown future had been very hard. Among my fellow passengers there were nice and courageous people, but the worst were the returning German-Americans.

New York is like alcohol, certainly the first time you visit. Especially the pace and the noise add to the excitement. All traffic moves very fast and then stops suddenly when signs changes, because Broadway and Fifth Avenue have sign posts such as our trains do. They do use red and green lights or big signs at almost every intersection. It is a lot of fun to watch all that action from a bus. I traveled a fair distance with that the first afternoon and at the same time had a good exploration of Grand Central Station, where the trains for New Haven depart. I wandered around a long time before I started to be somewhat familiar with the lay-out. It has various floors of enormous size, especially the lower ones, where one can find the connections to the subways. It is quite some fun to sit in such an 'express train', where you cover the distance of 40 streets or so in a few minutes (the subway has 4 tracks in the busiest places, 2 for express trains and 2 for local trains). It is very easy to find your way around in N.Y., since all streets are straight and numbered, but to find the quickest way to get somewhere remains difficult.

The morning of the second day that I spent there, I saw among others Brooklyn Bridge and the Woolworth Building, although I had no time to climb the tower. I returned to the hotel, sitting in the back of a streetcar and from there saw Broadway all the way to the beginning. It really looked like a gorge in the rocks, much like the Aareschlucht [a gorge in the river Aare in Berner Oberland, Switzerland], where you also see only a small stripe of sky and no more than dark street below. The most interesting thing in N.Y. seems to me however the view from the boat before you see the Statue of Liberty. You then look at Manhattan from the other side (that is the island, on which the central part of N.Y. is located), you see an enormous mass of skyrocketing buildings, that from that vantage point are projected against one another.

In the afternoon, I left for New Haven, which is 130 km N.E. of N.Y., and is connected by electric trains. You go uninterrupted through suburbs [Oort called them villa villages] along the coast. New Haven itself looks like a beautiful town and in that respect it is an exception to most American cities, although I have no real experience with that. A large part of the center is occupied by university and student buildings, of which a very large number exist. There is for example a beautiful, large building, in old Gothic style, that has just been completed, with a large number of small stained glass windows and even artificial sagging roofs and broken windowsills! In spite of that I think it is a beautiful building; I will later send a photograph. It consists of various departments and low gates provide entrances to even smaller, cosy courtyards separate from the larger campus in the middle. I heard that some lady had donated 6 million dollars for it (to compare with the 1 million dollar that had been provided for the Peace Palace [in Den Haag]). I saw a room in it where the son of the Director of the Observatory lives, also decorated in old style, open fireplace, wide windowsills where you can sit, and wooden

Fig. 4.3 Lanman-Wright Hall on Yale Campus as it appears today. From Wikimedia Commons [122]

walls, which all are very beautiful. The luxury and excessiveness of it all, especially as far as the students are concerned, is really amazing. But then Yale is one of the most expensive universities. I will have my dinners in a large student-house and expect to learn more about all this and will some time write about it. But for the moment this: the majority of the students work for some of their income and there are only a few that get all their study money from their parents. There is a special office in the University, where you can go to look for a job.'

So, Oort arrived in New Haven on Sunday September 24. The building he described must be the Wright Hall (see Fig. 4.3), which was completed in 1912 and at the time the most recent addition to Yale campus. It was originally named for Henry Parks Wright (1839–1918), the first Dean of Yale College, but renamed Lanman-Wright Hall after a major renovation in 1993, funded by Yale benefactor William Kelsey Lanman (1904–2001).

The notes end with a description of a visit Oort made a few days later (September 27, he did not waste much time) to Woodstock in the State of New York. He continued the notes at about midnight from a place called 'The Maverick'. The original plan was that he would travel there with Liesje Storm van Leeuwen before moving on to New Haven, but when her brother had picked her up in New York, they might have decided that he should come and spend a few days at Maverick before taking up his job in New Haven. This would appeal to his curiosity and longing for adventure, and since his appointment at Yale was to start October 1, he had to do this as quickly as possible. Woodstock began as a village settled by the Dutch, located in the Catskill Mountains. Early in the twentieth century a colony called The Maverick was founded in nearby West Hurley. It was populated by writers, musicians and artists, who occupied primitive housing. It featured theater and concert facilities and since 1915 festivals were held, which included music and theater performances. This actually is the same activity as the famous 1969 Woodstock festival. By the time of Oort's visit there were primarily classical concerts at the festival, which took place in August. Oort could still meet many musician and other artists, that were in residence there.

'Around midnight. You will never guess where I am writing this. My bed is bordered by 4 still unpainted poles that have recently been sawed off. In the glow of a paraffin lamp I can only see with difficulty what I wrote and also what the room looks like: rough wooden walls, hardly any furniture, only a few beautiful carpets. Outside there is forest all around and the squirrels (a sort of small 'eekhoorn') make a lot of noise. In the little moonshine there is, I have seen little of the surroundings, but here in this area, in the Catskill Mountains, on the slopes of which this house is situated, Rip Van Winkle had his one hundred year's sleep. [Washington Irving's (1783–1859) short story is rather popular in English literature lessons in Dutch high schools.] Maybe you now know where I am, close to the Hudson, some 175 km from New York.

Liesje Storm van Leeuwen had invited me to come and stay here for a few days before I start my work in New Haven. My trip here was curious. I was not informed about the situation here and the way to reach it, so that, when I stepped off the train in Kingston at 7.15, it turned out that there was no connection to Woodstock, so I had either to walk 18 km or find a car. The latter turned out possible for $ 3.-, since there happened to be another passenger who also had to go that way. So, at high speed we drove through the moonlit landscape (all major roads are asphalted here, so people can go much faster. In New Haven all streets, avenues and roads in the neighborhood are asphalted; I have seen no streets of gravel yet). Suddenly the driver turned into a narrow, rocky path, where he made the car jump over stones and moguls without any scruple. We went along like that for a while through a forest, until in the end we saw a wooden house, out of which a Jewess appeared who informed us where Miss Storm lived. She seemed delighted she had somebody to talk to, so the instructions were extensive. We drove on and at the place indicated there was a wooden house not too far from the path; when I entered I immediately heard Liesje's voice and to my surprise I found myself in the middle of 4 Dutchmen (It is not very easy to talk English all the time and in New Haven nobody speaks Dutch). The room that I entered, like the one I am in now, was made of rough wood, in the middle there was a bright, lit fireplace, that looked self-made.

When I woke up the next morning I found out that there was beautiful nature all around, large forests with various trees mixed together and many flowers and nice valleys. Down in the valley there was a shower, also self-made and very pleasant, just in the middle of the forest, somewhat higher up was a well and its water was channeled through troughs to the shower. Since I now have to end this letter I will briefly tell about some more curiosities; the Maverick, where I am now, is on a lonely mountain slope, where they have built a number of small wooden houses, occupied in summer, mostly by artists that I am not too impressed of. The people with whom I stay, built these houses themselves. They easily make a living by that. The boy has a Ford, that he uses to get all kinds of things for us and takes us to all sorts of curiosities such as to a beautiful 'Fancy Ball' somewhere in the middle of the forest, where I danced with them and that was very nice. I met a beautiful Canadian woman, Mrs. Adams, who earns her living in New York designing clothes and who has an excellent taste for that, judging from what she was wearing herself. I talked with here and heard all sorts of interesting things about New York and life here at the Maverick. They drank a lot during this ball (which seems very much the custom in America, in spite of abolition), not just hoe-brewed, but also hard liquor. But the life these people lead is really nice; I think Piet would immediately come to America if he had seen it, everything being primitive but pleasant at the same time. The Cramers went to other places often in their Ford, like all the way to Maine with cots and a tent and they then stayed away for a week or more. The Ford they bought for f. 175.-. I would like to travel to California that way.

Well, all the best. I hope you are all well.

Jan.

P.S. I returned on a boat along the Hudson river. Very pretty, although not as pretty as the Rhine. They never lock doors in Maverick. At one of the houses I saw a complete tea-set on a shelf standing outside. That will stay there until the occupants return next summer. It is unbelievable that they leave the doors wide open all the time when they leave for the evening, with undoubtedly money and precious things inside.'

It becomes clear who Liesje Storm van Leeuwen was and how Oort knew her when one studies the genealogy of the family Storm an Leeuwen [123] in the context of 'Maverick'. Actually 'Storm' is a degeneration of the word 'Stam', which means a family or tribe. She was Alide Marie Catherine Storm van Leeuwen (1889–??), who according to this genealogy lived in Woodstock. With the name Alide it would be not unreasonable to suppose that her family and friends called her Liesje. She was 33 years of age by that time and had settled, or went to settle, in Woodstock at the Maverick. She might have been a musician or artist, but we cannot be sure. It also follows from that genealogy that it was her brother Arnold Storm van Leeuwen (1885–1948), a military officer and government employee, that must have met her at the boat in New York. Arnold must have resided in Woodstock, since his first child (also called Alide) is recorded as having been born at Woodstock in 1922. But how did Oort know her? She was 11 years his senior, so she could probably not be someone he knew from school or university. The answer is that the oldest of her brothers was Willem Storm van Leeuwen (1882–1933), a physician, who in 1920 had been appointed Professor of Pharmacology in Leiden, where his specialties were allergic disorders, particularly asthma. Oort's father must have known him professionally and it must have become clear that his younger sister would travel with the same ship as Oort. Oort himself had a keen interest in art and music and therefore would have liked to visit Maverick, but the original plan that he would travel there with her after arrival in New York was superseded by her brother Arnold coming to New York to pick her up. Professor Willem Storm van Leeuwen died at a rather young age, and his wife might either shortly after that or maybe at some later time have moved to Woodstock as well, since she died there in 1956. Painters and artists Konrad Cramer (1888–1963) and his wife Florence Ballin (1884–1962) were well-known inhabitants of Maverick (see [124]).

4.3 The Zenith Telescope

When Frank Schlesinger (see Sect. 3.4 and Fig. 4.4) accepted the directorship of Yale Observatory, his highest priority was to establish a large telescope in the southern hemisphere for parallax work. He was also interested in latitude variations. In positional work in astronomy high accuracy is required to determine proper motions and parallaxes of stars. Part of the problem is an exact knowledge of the position of the celestial pole and therefore the geographical latitude of the telescope used. This is related to the matter of 'absolute declinations' that will come back later in this book in connection with the Leiden Expeditions to Kenya.

The position of the celestial pole changes as a result of variations in the orientation of the rotation axis of the Earth in space (precession and nutation) and because Earth moves with respects to the stars (and the speed of light is finite; aberration). These effects were explained in Box 2.2. But there is also an effect of the Earth's rotation axis moving with respect to its mantle, so to the surface of the Earth itself. This polar motion gives rise to a variation of latitude. It results from motions in the core and mantle of the Earth, and in recent times also from melting of ice sheets on Greenland (and to a lesser extent Antarctica, since it is near the rotation axis) as a result of global warming. There are three components, an annual term, a periodic one and a slow variation. The periodic term (dubbed Chandler wobble) has a period of about 435 days and is related to coupled atmosphere-ocean circulation. The effects of polar motion are at a level of order 15 m on the surface of the Earth, which does not sound like much, but 10 m is one four-millionth of the Earth's circumference and corresponds to a change of the axis of 0.3 arcsec.

The effect of latitude variation was first discovered by Karl Friedrich Küstner (1856–1936) in the 1880s. Küstner later became Director of the Bonner Sternwarte, but he at the time still worked at the Berlin Observatory (before it moved to Babelsberg, Potsdam). The historical record was actually addressed in detail by Schlesinger in a note in 1936 [125], setting the record straight that not Seth Carlo Chandler, Jr. (1846–1913)—whose name is attached to the Chandler wobble—was the discoverer, but rather the first one to confirm it. The effects of the polar motion have to be disentangled from aberration and it was actually Küstner's objective to measure the aberration constant when he discovered latitude variation.

The straightforward method to derive one's latitude is to take a circumpolar star (so close to the pole that it never sets) and measure its zenith angle (the angle from the zenith) or the latitude above the horizon, when it crosses the meridian above

Fig. 4.4 Frank Schlesinger at the 23rd Meeting of the American Astronomical Society (AAS), of which he was President at the time, held at the University of Michigan, Ann Arbor, in September 1919. University of Chicago [126]

and below the pole and take the mean. But that presents the problem that in both measurements the altitudes and the times are different and so are the systematic effects such as flexure of the telescope and atmospheric refraction (see Box 2.2). To calculate the mean you have to add these zenith angles and even if these systematic affects were the same, they do not cancel. What was used extensively is the so-called Horrebow–Talcott method, first proposed by the Danish astronomer Peder Horrebow (1679–1764) and rediscovered and extended by the American civil engineer Andrew Talcott (1797–1883), who used it for geodesy (see Box 4.1). This method uses two stars with known declinations that cross the meridian at about the same time at opposite sides of the zenith at about the same altitude or zenith angle. The altitude of the pole then follows from the *difference* of these two zenith angles. Since telescope flexure and refraction are essentially identical (approximate simultaneity assures equal atmospheric conditions such as temperature, pressure and humidity, etc.) these are eliminated by the subtraction. Of course one needs to know the declinations of these two stars from other measurements. The method is explained in more detail in Box 4.1.

Box 4.1 Determining one's geographical latitude

The **Horrebow–Talcott method** to determine a geographical latitude, which is the same as the determination of the pole's altitude, works as follows.

Take two stars that have declinations such that they pass the meridian at very nearly the same altitude (or zenith distance), one south of the zenith and the other north of it. We need stars for which the declinations are known.

Let these be $\delta_{\rm south}$ and $\delta_{\rm north}$, and measure their zenith distance (which is 90° - altitude) $z_{\rm south}$ and $z_{\rm north}$. Then the geographical latitude ϕ is

$$\phi = \tfrac{1}{2}(\delta_{\rm south} + \delta_{\rm north}) + \tfrac{1}{2}(z_{\rm south} - z_{\rm north}).$$

Flexure and refraction are the same because of the similar altitudes and times, so they cancel when the zenith distances z are subtracted. However, the problem is that these systematic errors change with time, because they depend upon temperature, density (atmospheric pressure) and maybe even humidity. Therefore this only works when observations are obtained very closely spaced in time. This method requires *absolute* declinations of the stars used. Furthermore, since it is necessary to have the meridian passages of the two stars closely spaced in time their right ascensions have to be closely the same (or not far from 12 h (180°) apart if one passes below the pole).

Schlesinger, as a leading astrometrist, had a keen interest in latitude determinations and in February 1922 (not long after his arrival at Yale) published a paper on this subject, *On progressive changes in latitude* [127], summarizing evidence that indeed the latitude of any place on Earth did change significantly. The International Polar Motion Service operated six stations (the International Latitude Observatories, all at the same approximate latitude of 39°09′) around the world, but three of these had been closed because of World War I and other causes and Schlesinger argued that the remaining three at least should be kept operating. This should not come as a surprise when it is realized that after obtaining his Ph.D. from Columbia University in 1899, Schlesinger was employed to set up the station at Ukiah, California, which formed part of this network. He must have decided to return to this matter again at Yale using a photographic 'zenith telescope'.

The history of the technique used by Schlesinger and Oort goes back to Kapteyn, who had been interested in these matters also and had actually proposed a method to measure absolute declinations in the 1880s. I will first briefly describe Kapteyn's contribution. In 1890, a short note appeared in the 'Astronomische Nachrichten', written by Friedrich Robert Helmert (1843–1917), the Director of the Geodetic Institute and a Professor of Geodesy at the University of Berlin [128]. Helmert reported that with regard to the orientation of the Earth's axis (from the geographical latitude), no perceptible change had been observed in Berlin and Potsdam during the first half of 1889, but a strong change had occurred in the second half of that year, amounting to no less than 0.″5 to 0.″6. This was confirmed by observations from Prague and Strasbourg. Although he had already reported this during a geophysical conference, he also wanted to draw attention to his findings by alerting astronomers asking them whether their observations were consistent with his. Kapteyn in Groningen read this note and sent a letter to Helmert, who offered it to the 'Astronomische Nachrichten' for publication.

In this note, [129], Kapteyn did not discuss the effect itself, but pointed out, that there was a very accurate method to measure one's geographical latitude using photography of stars that pass through the zenith, the point in the sky directly above the observer, through an instrument that used the principle of the 'reflex zenith-tube'. It had been designed around 1850 by Sir George Biddell Airy (1801–1892), Astronomer Royal and Director of the Royal Greenwich Observatory. It had been used extensively to monitor the star γ Draconis to determine the constant of aberration (see Box 2.2) and the star's parallax. Take an objective lens and point it to the zenith; light from a star going through the zenith is then reflected below the lens by a basin filled with mercury. The mercury surface is located at a little more than half the focal length from the objective, so that an image of the star is formed just below it. Then mount a photographic plate at this position. It will partly block the lens so the plate should have a rather small size. This can be used to measure the zenith distance of a star with great accuracy by measuring the angle between the star and the position of the optical axis on the plate.

Now Kapteyn proposed not to measure absolute zenith distances, but to use the instrument to measure *differences* in zenith distance. A star that passes close to the zenith, will leave a trail on the photographic plate; this trail will have a slight curvature, since the daily path of a star on the sky is a small circle centered on the pole. Then take a second star that also goes through the zenith, but on the other side from it with respect to the first star; then rotate the objective and plate by exactly 180° and observe that second star. Then there are two trails; the one from the second star will also be curved but in the other direction, so that the trails can be distinguished. Now if the polar height changes, for example such that the first star crosses at a larger zenith angle, the second star—being on the other side of the zenith—will cross at a smaller zenith angle. But since the plate has been rotated between the measurements, the distance between the two trails has changed in the second measurement compared to the first one and the method can be used for accurate measurement of *changes* in the positions of the poles. Since observations are done at the zenith, the mercury 'mirror' ensures exact orientation towards the zenith. Also note that this method does not suffer from effects of light deflection in the atmosphere and flexure in a telescope tube.

Kapteyn went on to discuss the possible sources of error and evaluated their seriousness and how this could be corrected. He contacted the Henry brothers in Paris, Paul Pierre (1848–1905) and Mathieu-Prosper (1849–1903), on the performance of their objectives, to judge down to what magnitude stars could be recorded. Based on a Henry objective of 33 cm diameter, 343 cm focal length and allowing for blocking, loss of light in reflection, etc., Kapteyn estimated that stars of magnitude 7.6 would still leave a useful trail on the photographic plate, and he therefore concluded that the method was feasible. As far as I am aware, there was no published response from Helmert.

Schlesinger also adopted this approach and it was Oort's task to operate the instrument, obtain and measure up the plates and determine the latitude variation. I do not know whether Schlesinger took the idea directly from Kapteyn. The US Naval Observatory was also operating a photographic zenith tube for latitude variations, so its use was possibly developed independent of Kapteyn's publication.

Not much has been recorded on the project. The American Astronomical Society decided in 1923 to include annual reports of American Observatories as part of its publications. The first series covered the twelve months ending June 1923 [130]. The report for Yale Observatory by Schlesinger for that period contained the following:

> '*The Zenith Camera*, a ten-inch photographic instrument intended for the determination of latitude variations. Experiments with this instrument have been continued throughout the year by Mr. Schlesinger and Mr. Oort. Devices for making it completely automatic have been installed. Some preliminary plates measured by Mr. Oort indicate a satisfactory degree of accuracy, but it was unfortunately found that the construction of the lens unnecessarily limits the field. As this defect was due to a misunderstanding, Mr. McDowell had generously offered to re-figure the curves gratis and the objective is now in his hands.
> The lens was returned to the Observatory in August, 1923. Its field is much improved.'

And a year later:

> '*The Zenith Camera*, a ten-inch photographic instrument intended primarily for observing variations in latitude. The lens for this instrument was sent on April 30, 1923, to the maker for correction. It was returned to the Observatory in September, from which date experiments were resumed and were continued throughout the year. On 113 nights Mr. Oort and Mr. Stearns secured 144 latitude plates and 35 for miscellaneous purposes; they have measured 107 of these plates.'

And for the period ending June 1925:

> '*The Zenith Camera*, a ten-inch photographic instrument intended for observing variations in latitude. Mr. Stearns continued to obtain plates throughout the year. The observations necessary for this experiment are now complete; it remains to finish the reductions and to discuss the material.'

Carl Leo Stearns (1892–1972) had come to Yale as a graduate student, while working as an instructor at Wesleyan University in Middletown, Connecticut. He wrote a Ph.D. thesis on efforts to perform stellar photometry with the Loomis Memorial Telescope. He spent his career mostly as an astrometrist, Professor of Astronomy and eventually Director at Van Vleck Observatory of Wesleyan University.

The staff of Yale Observatory was small; there was the Director, the Director emeritus (Elkin) and a Professor of Mathematics (Brown, who had been interim director), but these were not active in any observational program. There were three assistants, one temporary assistant (Oort), one Computer (a person performing calculations), one secretary and three technicians. With limited staff and Schlesinger's priority being a telescope in the south it is no surprise things were left unreduced.

In a biographical memoir of Schlesinger, his successor Dirk Brouwer (1902–1966) wrote [131]:

> '... several experiments that were taken up at various times, and that were crowded out by the larger projects. Among these were attempts to devise a method that would permit the accurate measurement of the moon's position by photography as well as experiments with a photographic zenith telescope.'

This is not really the full story. In letters between Oort and Schlesinger the subject came up often in the first years after Oort's time at Yale. For example in a congratulatory letter to Oort on the successful defense of his thesis and his obtaining the Ph.D. degree in May 1926, Schlesinger wrote that he intended to measure up some of the plates.

> 'Mr. Stearns, as you know, continued observing with our zenith camera until July of last year. I have been looking a little into the results and I think I may have to do some measuring. I intend to get at this work just as soon as my current duties permit. I will of course send you the manuscript when it is prepared'

But not much happened. In May, 1928 Schlesinger announced he would visit Leiden and proposed to discuss the matter. Apparently they decided that all the material should be send to Oort in Leiden and that the further measurements and the analysis would be done there. In a letter of August 22, 1928, Schlesinger wrote that he had sent by express all the material to Leiden, 'the latitude plates together with the two notebooks and the manuscript that is related to them'. These notebooks still are in the Oort Archives. On December 17, 1928, Schlesinger wrote:

> 'Professor de Sitter has just written to me that he cannot buy the Flagstaff machine [probably a measuring machine developed at Flagstaff Observatory], so that I think it an excellent idea to borrow the one at Amsterdam if you can.'

There also is an extensive draft of a letter in which Oort reports in great detail on the results. It turned out that he had one of the Computers of the Observatory, Gerrit Pels (1893–1966), measure up the plates and together they analyzed the results. These were disappointing, since there were many systematic errors that could not be corrected for. The draft letter, which is undated, concluded as follows:

> 'I needn't say that I hate to send you this letter. The results were a very great disappointment to me, but perhaps they are a still greater disappointment to you; for you get them at once. I was introduced to them by small bits. It will be tried again, I hope, with a better attached plate holder.'

So, there apparently was a structural fault in the instrument. There is no letter that could be a direct response from Schlesinger to this, but these measurements in Leiden must have been completed before 1930. In any case, on September 5, 1930, Schlesinger wrote:

> 'You will understand that I have not been able to do anything about the latitude work, except to think about it at odd moments. It seems to me that, after all I have not taken sufficient precautions against a shifting of the lenses, and when I set the instrument up again I must reconsider the mechanical arrangement. My guess is that, in spite of the very gentle way in which the objective is rotated, the two components have not maintained the same relative position throughout the group interval. I am naturally very anxious to resume this work as soon as circumstances permit.'

On December 16, 1934, Schlesinger acknowledged the failure of the project and said he would look into the mechanical structure and the foundations of the instrument. He did indeed do this and reported on May 23, 1935, in a letter prior to a visit to Leiden, that tests of the optical properties of the objective lens showed that there were faults in its figuring. After that there is no more mention of the project.

4.4 Life in New Haven

Oort's life in New Haven and at Yale Observatory was full. There are various entries in his diary notebooks that show that work took up a large fraction of his time. But he got to know people as well, also outside astronomical circles. In particular the Torey family. I quote from the *AIP-Interview* [4], where Oort was asked about who he did primarily meet in astronomy:

> Principally Professor Schlesinger and his wife. But the friends I was just now thinking of were a Professor of Oriental Languages and his wife and daughter. I'd been introduced to them by a cousin of mine who was a Professor of Oriental Languages in Leiden. They became like second parents to me when I was in Yale. His name was Charles Torrey.
>
> [Did you correspond with him after you went back to Leiden?]
>
> Especially with his wife. His wife was an author and we did a good deal of reading over there in the evenings together. We read aloud to each other and I learned a lot from that.
>
> [Do you recall where you lived? Did you live with them?]
>
> No, I lived at the Observatory. Between their house and the Observatory there was a street lined with beautiful elms; the elms are no longer there, I suppose.'

DeVorkin remarked 'No, that's Dutch Elm disease.' and Oort replied 'Which we can't help.' This concerned Charles Cutler Torrey (1863–1956), who taught Semitic languages at Yale since 1900, and studied Christian and Islamic religious sources and their origins and the relations between the two religions. One of his books (actually a set of lectures) is available in full on the Web and is called *The*

Jewish foundation of Islam [132]. He also translated the Gospels into Aramaic. In addition he worked as an archaeologist, co-founding in 1900 the American School of Oriental Research, later named the Albright Institute, in Jerusalem, of which he was the first Director (1900–1901). He was actively involved in excavations himself. His wife was Marian Edwards Richards (1877–1947). She indeed was a writer, using the name Marian Richards Torrey, and her apparently most successful book was *My outrageous cousin*, published in 1929. They had a little daughter at the time Oort was in New Haven, Anne (Nancy) Torrey (1920–2009), born when her parents were already at a relatively advanced age.

It is a bit of a puzzle who that cousin of his was, that Oort mentioned had introduced him to the Torrey's. At that time (1922) the relevant professor was Arent Jan Wensinck (1882–1939), who is described as an islamologist, historian of Syrian mystic theology and Professor of Semitic Languages. He was Professor at the University of Leiden between 1912 and 1939. However, he is, as far as I have been able to ascertain, not related to Oort, certainly not as first cousin. However, he studied with Michael Jan de Goeje (1836–1909), Professor of Syriac, Chaldean and Arabic languages, who was a brother of Oort's paternal grandmother. But he died too early to have introduced Oort to Torrey. The remaining possibility is Karel Hendrik Roessingh (1886–1925), who had been a Professor of Theology since 1916. Now Oort's uncle Henricus Lucas Oort (1861–1925), a minister in Utrecht and other places, had in 1919 married Rolina Gijsberta Roessingh (1865–1965). Her father, Petrus Hendrik Roessingh (1840–1916), and Karel's, Petrus Marius Eliza Roessingh (1858–1930), were brothers, so Karel and Rolinda were cousins. What Oort should have said is that the person who brought him in contact with the Torrey's was a cousin of his aunt.

There is in the Oort Archives an extensive letter to his parents, which gives an excellent impression of Oort's time in New Haven. The year is not in the heading, but 23 February was on a Saturday in 1924. I include most of this letter in three separate parts, starting with the top sentences and continuing with the middle part.

'Saturday evening, 23 Febr., 12 o'clock.

Dear father and mother,

There is so much I should write about that I don't know where to start, the more since I hardly have time left to write.'

[The letter first adressed his astronomical work (see below) and from the context there the time is very likely midnight and not noon.]

'– So you now <u>start</u> to get a good impression in this letter of me, when I am at work. That [sic!] is only partly true though, and I am to the contrary going out very much, visiting people, have dinners at their homes, skating, going to the movies with others, dancing, etc.! Actually, now I exaggerate again, but I enjoy myself here next to my work, as much as one can possibly enjoy oneself in a city.

Sunday morning 12 o'clock.

As I wrote you I have become friends with Doris Thompson. You would be startled [later inserted 'a bit'] by her American manners and many of her American character traits, but she is basically a lovely and sensitive girl, surprisingly little affected by her good looks and popularity. She also

Fig. 4.5 Oort during his two years in the USA, probably in New Haven. It has not been possible to find out who the two ladies are, since there are no notes or names on the back of the photograph in the Oort Archives. They might be two of the Welch daughters though

differs from other girls in that she seems to have a good intellect. She is very good at directing my attention away from my work and it is very pleasant to see her or go out with her. Monday night we went skating on a pond in Woodbridge in the moonshine; the evening before last there was a big party in the Lawn Club, where she is member of a committee and I saw her in the settings where she is most happy and probably is best at home, every half minute a different dancing partner, so that it was impossible to dance with her for more than ten steps at a time.'

So, Oort still enjoyed skating and dancing and found ways of doing that with others. He had some observations of the country he now lived in (Fig. 4.5).

'I start to like this country more and more, especially the cheerfulness and ease here (that are maybe much influenced by the large amounts of sunshine), and the sense of humor of people and the strength to put unnecessary worries aside, because of which they can better enjoy the beauty and magnificence of life and it often occurs to me that in spite of the greater superficiality in general, the society and people are more advanced here than in Holland. Sometimes I really hate the idea to move to Leiden, especially this year when you will be away [The Oort parents spent a year in the East Indies]. I dread to be again among gloomy, sluggish, not-open-to-anything-new people in Leiden, it is reminiscent of confinement, as if there is no way out from there once I will have arrived, and maybe that actually is true. In that respect I always had a kind of aversion to Leiden when I came home from Groningen, and in addition I don't know any of the astronomers there (I am glad to be with Schilt

again); don't misunderstand and think that I always exaggerate my inner feelings; it is just a feeling based on things not very real and maybe influenced by a few gloomy memories of Leiden.
And when you are back in Oegstgeest things are completely different and part of the time I will be away from Leiden.
Still if I had the choice I would rather move myself and all my things to Groningen, and maybe eventually even here, although I am not certain about that.'

Oort mentioned some people that had written him from the Netherlands and he described the Welch family in New Haven that he had visited. Mrs. Welch gave him cakes and marmalade for his lunches at the Observatory and he had gone to the movies with the middle daughter Francis ('Emily is in France, and Betty in college'). Obviously Oort had an active social life. He also shared some observations of American life and society with his parents.

'Everybody I meet has read 'Hans Brinker and his silver skates', which is almost the only thing they know about Holland. They are extremely disappointed when I tell them I never read that book.'

Hans Brinker; or, the Silver Skates: A Story of Life in Holland is a novel by American author Mary Elisabeth Mapes Dodge (1831–1905), first published in 1865. The novel is written for children and has been set in the Netherlands about the early 19th-century. The book's title refers to the beautiful silver skates to be awarded to the winner of an ice-skating race. Hans Brinker and his sister will participate and hope to win so they can pay for a doctor to cure their sick father. About halfway through the book these is a story read in a reading-class about the 'Hero of Haarlem', who saved the country from a flood by plugging a leak in a dyke with his finger (Fig. 4.6).

'People make many mistakes in the pronunciation of their own language, even preachers, public speakers and professors, etc.

Fig. 4.6 Oort's *ex libris*. It is present in diary notebooks and pocket diaries during his stay at Yale, in one case with a letter to his parents from Pasadena in July 1924. He probably designed it while residing in New Haven. From the Oort Archives

There is a complete absence of hypersensitivity in many areas and of hidden unhappiness and sorrow. People are more practical here and more natural, happier and healthier (the doctors, churches and medical services work only with one thing: 'don't worry!', [...]
Nobody here knows Ibsen.
Business spirit here: When a boy asks a beautiful girl for a ball it is a kind of trade and one negotiates the number of dances.'

4.5 Oort's Further Research at Yale

In addition to his work on the latitude variations with the zenith telescope, Oort got involved in other work as well at Yale Observatory.

A first piece of work shows Oort's interest in dynamics once more. The matter addressed is the equipartition of energy. The motion of an object represents a certain amount of energy, the so-called kinetic energy which is half the product of the object's mass and the square of its velocity. If energy were equally shared between stars, those with large masses should have smaller velocities than less massive ones. The study also related to Oort's ongoing work on stars of high velocity. Indeed such an effect seemed to be present among stars with known distances that are intrinsically bright and also for intrinsically faint ones. In Oort (1924c), he showed that this was not a genuine result; among the intrinsically fainter stars there is a proportion of high-velocity stars (as we have seen with radial velocities larger than 65 km/s), while these are not abundant among intrinsically bright stars. Allowing for this effect removed all evidence for a general increase of velocity with decreasing mass. He used for this a sample of stars with spectroscopic parallaxes obtained at Mount Wilson. Spectroscopic parallaxes are estimates of a stars distance on the basis of their spectra, which show both their spectral types and whether they are dwarfs or giants in the Hertzsprung-Russell diagram (see Sect. 2.6) and in this way one can estimate their absolute magnitudes and thus their distances. The errors in these properties at that time were of order one magnitude (see e.g. [133]), or a factor 1.5 in distance. Especially for M-stars, where the differences in absolute magnitude of dwarfs and giants are largest, and for stars with velocities less than 65 km/s, the ratio of the mean velocities of dwarfs and giants is close to unity, while an M-giant has a mass of one or a few solar masses $M_{\odot}$ and an M-dwarf of 0.5 $M_{\odot}$ or less. For spectral types from F to M the combined value for this ratio in velocities was found to be 1.04 ± 0.05.

The same result was obtained in Oort (1924b), in which he compared single stars and binaries. Oort estimated the mass ratio of components in binaries as 0.7 (so on average binaries are 1.7 times as heavy as single stars), and equipartition of energy should give rise to a ratio of mean velocities of 1.27. Oort found that the ratio actually is 1.06 ± 0.05, but classification of whether or not a star is single or binary (on the basis of spectra) remained uncertain. This result was presented by Oort (1924e) at the 31st Meeting of the American Astronomical Society (AAS) that was held on December 27 and 28, 1923 at Vassar College, Poughkeepsie, New

Fig. 4.7 Group photograph of attendants to the 31st meeting of the American Astronomical Society in 1923 at Vassar College, Poughkeepsie, New York. Oort is near the top, just to the right of the middle (see also Fig. 4.8). The person in the front row straight below Oort is Frank Schlesinger (fifth from the right) and to the left of Schlesinger stands Henry Norris Russell. See [134]

York. Vassar College had been founded in 1861 by Matthew Vassar (1792–1868) and is particularly known as being the first American higher education institution that granted academic degrees to women. Matthew Vassar had made his fortune as a brewer. In the group photograph (Fig. 4.7) Oort stands in the middle near the top. There are two more young Dutch astronomers in this picture (but unlike Oort these spent their careers further in the USA); directly below Oort is Peter van de Kamp (1901–1995) and two persons to the left Willem Jacob Luyten (see Fig. 4.8).

The most interesting study (in relation with what will follow in terms of Galactic structure) is in Oort (1924d), where he investigated the possible relation between stars of high velocity and globular clusters. To put things in perspective: As we have seen, the situation as known now is that both are members of the halo or Population II components of the Milky Way Galaxy and *are* intimately related. Harlow Shapley had determined the distances of globular clusters from Cepheids. These are stars in a late stage of their evolution, where they pulsate periodically due to some internal instability and the frequency of pulsation is closely related to the star's luminosity or absolute magnitude. Henrietta Swan Leavitt at Harvard had discovered between 1908 and 1912 that in the Magellanic Clouds (companions to the Milky Way Galaxy) such stars, which all should have very closely the same distance, showed a tight period-luminosity relation. This presents a means of determining the distance, provided that the relation has been calibrated. Actually, the most common type of Cepheid (called short-period Cepheids or cluster variables at that time) in globular clusters are RR Lyrae stars, a late (helium burning) evolution stage typical for old stars with small abundances of heavy elements in the halo. Oort noted first that such short period

Fig. 4.8 Three young Dutch astronomers at the 31st meeting of the American Astronomical Society in 1923 at Vassar College, Poughkeepsie, New York. Detail from Fig. 4.7. On the left Willem Jacob Luyten, to the lower-left of Oort Peter van de Kamp and to the right Oort. [134]

Cepheids outside globular clusters had high velocities with characteristics similar to high-velocity stars.

Oort looked into the possible relation between stars of high velocity and the system of globular clusters (see below). Globular clusters had a very peculiar characteristic, namely that they avoided the Milky Way on the sky. Only one of them had a Galactic latitude less than 5°. It was possible that this 'Zone of Avoidance', which also holds for spiral nebulae for which this name was first used, would be the result of the presence of 'dark absorbing matter', for which Oort referred to a wide-angle picture of the Milky Way. But Oort quoted Shapley's authority that 'there are several indications which favor the hypothesis that globular clusters actually do not exits near the central plane of the galaxy'. Eventually, of course, Shapley was proven wrong in this and interstellar absorption really is the explanation for the Zone of Avoidance. The top panel of Fig. 4.9 (indicated as 'Fig. 1') shows the effect. The horizontal axis is the sine of Galactic latitude (zero for latitude zero, plus or minus unity for latitude ±90°) and vertically the number of stars. Clearly, low Galactic latitudes are being avoided in the distribution of globular clusters.

Now, if there were indeed a relation with high-velocity stars then these should avoid having velocities pointing away from (or towards) positions in the Galactic equator. Oort noted at the outset that globular clusters do have radial velocities of on average 140 km/s or so, well comparable to that of high-velocity stars. He used a sample of high-velocity stars provided by Gustav Benjamin Strömberg (1882–1962) of Mount Wilson Observatory. In addition he used stars with lower velocities from a variety of sources. For all these stars he calculated the antapices, that is the directions (and the corresponding positions on the sky) from which the three-dimensional velocities of these stars were pointing. In the second panel from the top the distribution of Galactic latitudes of all his high-velocity stars are given. There is a marked minimum around latitude zero.

However a 20% error in the parallaxes would correspond roughly to an error of 0.04 in the sine of the latitude, which has significant consequences for the figure. It

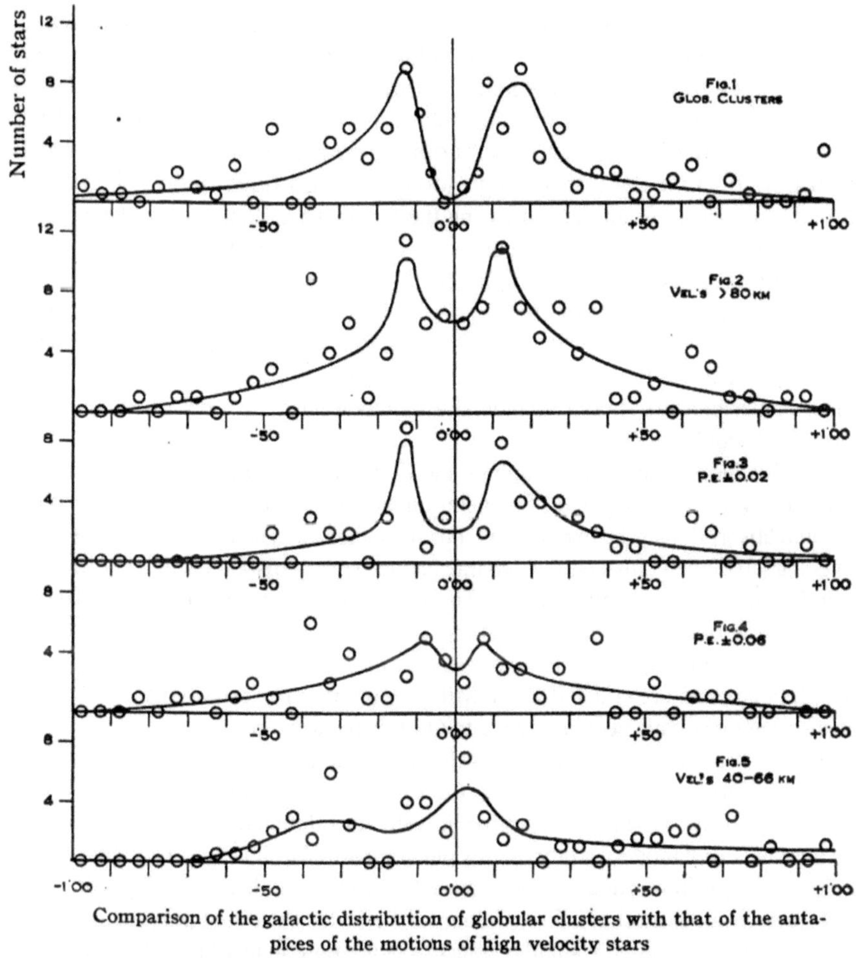

Fig. 4.9 Figure taken from Oort (1924d) comparing various samples of high-velocity stars (lower four panels) with globular clusters (top panel). In each panel the circles are the data and the lines are drawn to guide the eye. For explanation see text

would fill in a minimum near the Galactic plane at 0.0 on the horizontal axis. Oort therefore separated the sample according to accuracy of the parallax and the resulting distributions are the third and fourth panels from the top he designated 'Fig. 3' and 'Fig. 4'. There is a strong minimum accompanied by strong peaks in the first one of these, where accuracies are best. The stars of high velocity show the same avoidance of the Milky Way in the directions of their velocities as the distribution of positions of globular clusters. To complete the case, Oort also did the same exercise for a sample of stars with lower velocities, for which he found earlier that the asymmetry in the longitudes found for high-velocity stars disappeared. That is the lowest panel and indeed no minimum in the Galactic plane is evident.

What to make of this? Oort suggested that stars of high velocity have actually escaped from the (system of) globular clusters. But he also concluded that 'Our knowledge about the motions and the arrangement of the clusters, and especially about the velocities of these stars, is, however, so preliminary that it does not seem of much use to work out the consequences of such a hypothesis before more data about high total velocities have been collected.' He added a surprising footnote, addressing the fact that the Zone of Avoidance for the clusters may be the result of an outer ring of absorbing material which by gravity would affect the motions of the high-velocity stars. That would mean that the mass of this ring had to be that of 10^{20} $M_{\odot}$ or 10 billion times the mass of the Kapteyn system. The then rather new theory of General Relativity of Einstein, for which Oort cited a publication by de Sitter for more details, would predict then that the light of the globular clusters should be blue-shifted much more than observed, so that the hypothesis could be discarded.

Finally, Oort added a paragraph on the Zone of Avoidance in the distribution of spiral nebulae on the sky and of planetary nebulae (see Sect. 2.6). He seems to have favored absorption as the explanation, but since there was no evidence for absorption in *c*-stars and open clusters, the nebulae had to be very distant (which is true for the spirals but not for the planetaries). The designation *c*-star is an old one for stars with narrow line spectra and such stars can have various spectral types and can be relatively distant. Open clusters are clusters of stars with less regular shapes than the globulars and are situated in the Milky Way disk. The possibility of interstellar absorption must have been attractive to Oort from early on.

Indeed the Zone of Avoidance results from dust extinction in interstellar space, explaining in a natural way the result Oort found for the globular clusters. But why would high-velocity stars have few or no velocities directed away from this zone? I do not know of a modern study on this phenomenon, but it is very well possible that it not at all spurious, but a real fact. We are talking about high-velocity stars whose three-dimensional motion is in the plane of the Galaxy and that pass the position of the Sun within a relatively small distance, say no more than 1 kpc. These must have had a long pathway through the disk of the Galaxy. Now first look at 'common stars' in the solar neighborhood, the ones in which Oort earlier had looked for effects of equipartition. This is not the case, as Oort indeed found, but the characteristics of their motions are as follows. The property of interest is the velocity dispersion. It is in many cases possible to describe the distribution of stellar velocities in a certain volume by a Gaussian (see Box 4.2 and Fig. 4.10). The property used to denote a typical mean value is called the dispersion, defined as σ in Box 4.2. For any distribution it is the r.m.s. value (for root-mean-square): calculate the deviations from the mean for every value, square that, add them up, and take the square-root of this sum. In statistics this is also called the standard deviation.

For common stars near the Sun the velocity dispersion is in round numbers 50 km/s in the direction towards and away from the center of the Galaxy, 35 km/s along or opposite to rotation and 20 km/s perpendicular to the plane. As we saw there is no equipartition nor dependence on stellar mass, but more recently it has been found that there is a relation of the velocity dispersions with age of the stars. The velocity dispersion in the neutral hydrogen gas, from which stars form, is about 10 km/s and

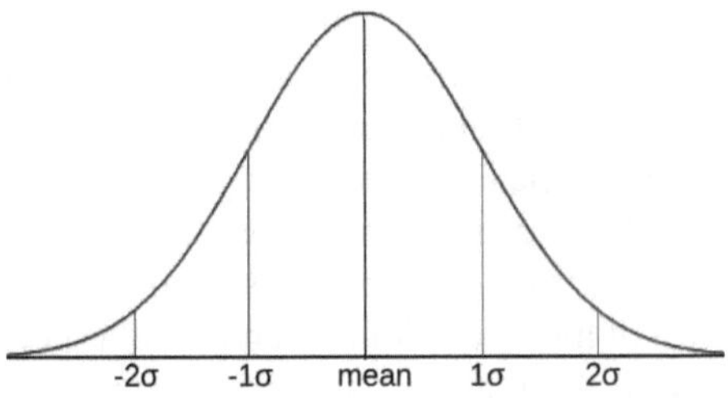

Fig. 4.10 Gaussian distribution. For details see Box 4.2

among young stars, say O- and B-stars, it is similar. But for larger ages it grows until it reaches the values just quoted for stars a few billion years old and then levels off with age. The reason for that is not fully understood, but it appears that there are at least two effects that contribute to this 'secular evolution'. In both cases stars are deflected in their orbits by variations in the gravitational field. In the model first suggested by Lyman Strong Spitzer, Jr. (1914–1997) and Martin Schwarzschild (1912–1997) from Princeton, the fluctuations result from condensations in the interstellar medium, later identified as 'Giant Molecular Clouds' (with masses up to 100 million Suns) in which stars form. In a model proposed by Lodewijk (Lo) Woltjer (b. 1930)—one of Oort's later students and son of Leiden staff member Jan Woltjer—and Basil Barbanis, then from Columbia University, New York, it is the effect of spiral structure. Currently it is often assumed that both effects contribute. In any case, stars are being scattered and in particular move to larger vertical distances where the disturbing effects are less so that the secular evolution levels off. It has become clear in recent times that merging of small satellite systems must have occurred frequently during galaxy evolution, especially at early times, which would also dynamically heat up (increase the random motions) stellar disks.

Box 4.2 Normal or Gaussian distribution

Many distributions in nature follow this distribution. Its equation is

$$F(x) = \frac{1}{\sqrt{2\pi}\sigma} \exp - \frac{(x-\mu)^2}{2\sigma^2},$$

where μ is the mean value and σ the standard deviation. The first term ensures that the integral, the total area under the curve, equals unity. It has been illustrated in Fig. 4.10.

In the distribution 68% lies less than one standard deviation σ away from the mean μ, 95% within 2σ and 98.7% within 3σ

Now these scattering effects must also work on high-velocity stars that are part of the halo, when they cross the disk. But the stars that move *in* the plane of the Galaxy will have long path-lengths through the layers where the scattering takes place and many will therefore have been scattered out of the plane in the course of time. Detailed study is required to see whether the relevant timescales work out, but this may be a possible explanation for the effect that Oort found.

Finally, Oort and Marsh (1924a) studied the secular parallax of stars of thirteenth magnitude, based on proper motions determined by Adriaan van Maanen in 38 fields. Hannah M. Marsh was an assistant that worked for some years at Yale during the 1920s. Adriaan van Maanen (1884–1946) had obtained a degree in astronomy in 1911 in Utrecht, but had worked mostly under Kapteyn; upon recommendation by the latter to George Hale he went to Mount Wilson Observatory, where he worked for most of his career. Van Maanen was a specialist in measurements of proper motions and the observations referred to here were obtained with the 60 in. Telescope at Mount Wilson. Proper motions are used to obtain secular parallaxes, that is the statistically derived distance for a group of stars (see Box 3.1). These have to be corrected for the motion of the Sun in space and crucial in that is the direction of this velocity, the Apex. Systematic effects arise easily from incorrect assumption on the position of the Apex on the sky. Van Maanen's plates had been measured by Mrs. Marsh. There had been some suspicion for quite some time that systematic errors in the proper motions were present in the Boss Catalogue prepared in 1910 by Lewis Boss; we already encountered this catalog in relation to the Schilt-Oort prize-winning essay for the Bachiene Foundation (see Sect. 3.2). The result of analysis (along lines proposed by Kapteyn) was that indeed there were systematic errors in the Boss Catalogue, unless the Apex changed significantly with spectrum or apparent brightness of the stars, which was unlikely. The results agreed very well with work in Groningen by Peter van de Kamp, a student who obtained a Ph.D. with van Rhijn in 1926 (actually everyone called both of them Piet). The Oort and Marsh work was presented at the 32nd Meeting of the American Astronomical Society, held at Dartmouth College, Hanover, New Hampshire on August 3–6, 1924. Oort did not attend this meeting (he was traveling from California to Leiden at that time, as we will see) and this paper must have been presented by someone else, one would think Schlesinger, who was present according to the group photograph [135].

I continue with the letter Oort wrote to his parents on February 23, 1924, where he tried to explain his astronomical research.

> 'As you see it is quite late; the zenith telescope program now is such that I have to stay up late to change the plate and then again before the morning [meaning before sunrise] I have to get up briefly to take the plate out. That is rather tiring, but sometimes I sleep very well in the morning. This week however we had mediocre weather and therefore I had time, as I wrote before, to finish a short paper about the motions of bright and faint stars that I started so long ago. I was almost done typing it when last Thursday morning I received a letter from Strömberg from Mount Wilson, an answer to a second letter I had written and in which I had asked him for some velocities that they had finished calculating at Mt. Wilson that I cannot do here since I don't have time for that. Furthermore the plan I had with these was rather silly and I would not have come to do it myself in case he would not send them. The Director of Mt. Wilson is reluctant to give away material, but Str. wrote me he had not been able to restrain the temptation to write me anyway; I could use them in the end as control check on my own calculations.

The reason why I wanted them were:
On the sky there are star clusters, small round areas where stars are concentrated enormously, so-called 'globular clusters'. Of these clusters we know a large number thanks to Shapley, for example their distances (that are enormous compared to those of ordinary stars) and these show a strange phenomenon, namely that the are completely absent in a belt of 10° width in the Milky Way, while outside that, especially not far from the Milky Way, there are a lot of them. The velocities of these clusters are large and I had already noted when I wrote my B.A.N. article that they seemed to be moving to the same hemisphere as the high velocity stars, and already then I was looking in vain for a direct connection between the two (e.g. one can imagine that the high-vel. st's have escaped from these clusters or something like that). When that is the case there is a real chance that the motions, the full velocities, of these high-velocity stars would also avoid this narrow zone in the Milky Way and it now appears from these velocities of Strömberg that this rather convincingly is the case, in spite of the fact that all those motions are almost || [perpendicular] to the Milky Way, that is to say between 6° and 30° from the Milky Way, but almost none between 0° and 6°.

I hope you can follow this quick story a little bit. When it is really true, and that is what it looks like, then it is I think an important thing, even though the explanation may be very different from the idea I started with.

Of course I was very excited about this, since I had very little hope it would work and I am only afraid that Strömberg will find the same thing one of these days, although that thought really makes me feel rather childish. Still I will not have much rest before I have finished a small 'note'. And that takes time since I will have to re-calculate and check all these velocities.

Schlesinger is not here this week, but Monday he will advise me what to do with this.

Maybe there is more in it, and then it could be combined with a few other things, that I have done with the high-velocity stars and also the note on the weak and bright stars, and it may be turned into a [Ph.D.] thesis.

Do not mention it to Schilt, I want to work it out better and with more confidence to see if it is real.'

4.6 Oort's Future in Leiden

It had been the idea from the start that Oort would move to Leiden after his stay at Yale. Naturally, Oort kept in contact with de Sitter and letters about his future were exchanged with many details. Oort wrote on December 15, 1922 a long letter to de Sitter, mostly about his worries that there were systematic errors in proper motions in declination, as actually Kapteyn had suspected. The last paragraph had an interesting turn.

> 'There is not much to say yet about the results of the zenith telescope, but there is another matter about which I would like to write to you: Professor Schlesinger asked me whether I would like to stay here for a few more years. Very probably it would be possible for him to offer me a 'substantially better' position and the work would be concerned with spectra with a view towards experiments about the determination of radial velocities with an objective prism (something I would like to spend some time on anyway – even if it would not happen). The experiments will be done here first with a small telescope, specially built for that and when successful it seems likely that funds for a larger telescope would be possible. I did not ask for details of this 'better' position, but it would be possible for me to come to Holland every summer. He would also be able to keep the position open until I have fulfilled my military service in Holland and maybe even completed my Ph.D. thesis (although I have little hope for that).

An objective prism is (in the case of a refractor) a thin prism in front of the objective lens at the front of the telescope, so that every star image is a small spectrum rather than a point.

Oort mentioned that the decision had to be taken before the budget request for the coming academic year would have to be filed. But he was also concerned with his future, noting that the longer he would stay in America, the smaller his chances would become for obtaining a position in the Netherlands. Now, it had been Schlesinger's main goal when he came to Yale, to build a telescope for parallax measurements in the southern hemisphere. It appears that he had offered Oort a position in relation to this effort. Eventually the place where this should be erected was chosen to be Milner Park (elevation 1730 m) at the campus of the University of Witwatersrand near Johannesburg in South Africa, 4 km west of the Union Observatory (see below). The construction started after Oort's departure from Yale and the dedication took place in June 1925. Actually, this 26 in. Yale refractor was moved to Mount Stromlo near Canberra, Australia and made operational there in 1958, where it was completely destroyed in the firestorm of January 2003.

Not much more has been recorded on the prospects to come to Leiden until March of 1923, when both de Sitter and Hertzsprung wrote Oort almost simultaneously about his future in Leiden, which is unlikely to be a coincidence. The extensive letter by Hertzsprung is dated March 6. He started by saying he should have replied much earlier to Oort's letter of November 4 (1922). The letter first concerned radial velocities from objective prism observations and how these can be used to see which stars belong to the Magellanic Clouds, followed by a longer part on the relation between Oort's high-velocity stars (Hertzsprung calls them 'eenzijdige hardlopers', biased runners) and globular clusters and RR Lyrae variables. His remarks on Oort's future contain some observations on his own experience and I will quote that part here. In the mean time there seemed to have developed a possibility that Oort would become strongly involved with the Yale Telescope, to the extent that Schlesinger had offered him to go to South Africa for a number of years to take charge of the project.

'The French say: Nothing is as definitive as a temporary measure. Be careful that such a temporary situation (as your involvement with Johannesburg) does not turn into a position for life – if you do not want that.

When I left Copenhagen to go to Germany, I had no intention whatsoever to stay there forever. I had good prospects to return to Denmark when the new Observatory [...] would be completed. The new Observatory never materialized. [...]

As far as I am concerned I myself have taken as a starting point that I would want to be there where – everything considered – I would be most useful for astronomy. How important the national environment is, is a very personal matter. My own experience is that when one is young, one tends to underestimate the circumstances, especially when one has had an intelligent upbringing. One would then, ceteris paribus [all other things being equal], achieve most in the atmosphere in which one grew up, at least when one felt at home in that atmosphere.

My opinion about you is – you will have to decide yourself whether I am right – that you belong best in a Dutch environment, that is to say that you will contribute most to astronomy under these circumstances. I cannot imagine that you would be happy to be in America forever – or even less so in South Africa. Someone who, like you, has a theoretical inclination needs persons with a similar interest to talk to. You will have more difficulty finding that in America than here. It is even worse, since you will encounter in America – owing to their own incapability – often a rather hostile attitude towards theoreticians. It is often not appreciated in America as it should be, since they are unable to judge their value. Remarks like: 'Every German student can do that calculation' are typical in that respect. I should add though to this somewhat nasty remark, that on the other side Europeans equally poorly judge the American effective look upon things and sometimes even regard that with disdain, which is not justified by the result achieved in America. Anyway, you will have thought about this yourself.'

There may be some truth in these remarks. Europe harbored important theoreticians in the first decades if the twentieth centrury, like Albert Einstein, Arthur Eddington, James Jeans, Karl Schwarzschild, Willem de Sitter, while important American astronomers, such as Henry Russell, George Hale, Harlow Shapley, Edward Pickering, Edwin Hubble were more observationally inclined. For example, in the *Legacy* [6] David Hyam DeVorkin (b. 1944) stated 'American astronomers who performed some sort of research were typically observers', and Klaas van Berkel pointed out that in the Netherlands at least 'the institutional circumstances were – or were considered to be – more favorable for astronomers with a theoretical inclination' (David deVorkin in the *Legacy* [6], p. 142, and Klaas van Berkel, *Ibid.* p. 168–169). Whether the somewhat hostile tone of Hertzsprung was justified, is another matter. Hertzsprung added to these outspoken remarks that it was not at all sure that Schlesinger's telescope would go to Johannesburg. But he said, he hoped it would, since then there would be the option of a good collaboration with the southern telescope Leiden Observatory intended to build there (a project that was promised to Hertzsprung when he joined Leiden). And he noted that 'a temporary stay in Johannesburg would be profitable for you, but I hope it would not be extended so much that there will be no chance to have you join Leiden.'

De Sitter wrote a day earlier. I quote it in full.

> 'Dear Jan,
>
> The other day your parents came here to speak to me about your plans and the question of your going to the southern hemisphere for Professor Schlesinger. I promised them to write you myself. I am doing this in English, so that you can, if you wish, show my letter to Prof. Schlesinger and talk it over with him.
>
> It may be that you are now to make a decision that will to a great extent be decisive for your future scientific career. I do not wish to influence you in making your decision, I only wish to aid you in putting the question to yourself as sharply as possible.
>
> I think, it comes to this in the end: do you wish to become a theoretical astronomer or an observer? In the first case you should try ultimately to come to Leiden, in the second you should stay in America, or in any other good climate. In *both*[2] cases a trip to South Africa will do you a lot of good.
>
> An observing astronomer must be in a good climate, a theoretical astronomer must be in a scientific atmosphere. I know of no instances of great theoretical astronomers who have worked successfully away from general scientific surroundings. Of course Leiden has not a monopoly of the latter, but the position as Adjunct-Director at this Observatory offers exceptional opportunities to a theoretical astronomer. By a theoretical astronomer I do not mean a mathematician: Kapteyn and Hertzsprung I consider theoretical astronomers.
>
> For a theoretical astronomer in the wider sense I think the position as Adjunct-Director would be nearly ideal. As you know the Observatory comprises of three departments: the fundamental, the astrophysical and the 'theoretical' (i.e. astromechanical) departments. The Fundamental Department includes, of course, the meridian-circle, but it by no means consists of this instrument exclusively. The fact that the Pole and the vertical are the only directions that can be fixed fundamentally lends to the plane through these directions a paramount importance, but there is a great many fundamental problems that can, and must, be solved by other methods. The Leiden Observatory must not be fettered by tradition, if it thinks that these other methods are preferable over the old ones.
>
> The chiefs of the three departments, the two Adjunct-Directors and the Conservator (Dr. [Jan] Woltjer) have a large degree of freedom. They have under them a staff of scientific assistants to whom the details of the execution of their plans can be entrusted. They themselves must be

---

[2]De Sitter's typewriter must have had an option to produce text in italics.

leaders, capable of choosing for themselves, though of course in consultation with the Director, the problems for their department and the means of solving those problems.

As regards the Fundamental Department especially, I hope to be able to find an Adjunct-Director to whom I can leave the direction of this department almost entirely. I do not wish to exercise control over his work except in very broad lines: it must be really important fundamental work, not casual investigations, that can be done anywhere and by anybody.

The post is at present unoccupied, as you know, by the death of Kapteyn. I cannot appoint you to it at once for two reasons. Firstly you are still a little too young to put you at the head of men who are much older than you. Secondly I do not know how you will develop, whether you will be a mathematician, or an astrophysicist, or a general theoretical astronomer. If you were a duplicate of either Hertzsprung or Woltjer I might have to try and find another man who more fulfills the demands of the case. I want a man who, to continue in the same stile [style], is a duplicate of Kapteyn or of Newcomb. I am willing to wait some time until I can get such a man, but I do not like to wait longer than is absolutely necessary, for at present all the work of the Fundamental Department falls on me, and makes a too heavy demand on my time.

If your ambition lies in the direction of what I have called general theoretical astronomy, I honestly think you could not find anywhere a better position than the one I could offer you here, as soon as you are qualified for it. The best you can do in that case is to get as much and as varied experience as you can. You cannot have a better teacher as Schlesinger. As to your going south in charge of the big telescope, this would certainly be very valuable for your astronomical education, but I should say from this point of view one year of it would be sufficient, and you should also try to gain experience of other instruments and methods, including those of fundamental astronomy. Of course I do not know whether Professor Schlesinger would send you out for one year only. I cannot promise to wait much longer than the beginning of 1925, i.e. until our present meridian programme will be completed.

I say *if* your ambition lies that way. Whether it does no one but yourself can decide, and it would be foolish of me to wish to influence you. No good can come of constraint in such matters. [sic!] Only I may be allowed to say that I hope it does, for in my opinion many important problems of the near future lie in the direction of fundamental astronomy.

I have in this letter spoken out my mind rather freely, and I hope you will consider it as confidential, and show it to nobody else but Professor Schlesinger.

Yours most sincerely,'

[signed W. de Sitter]

Oort wrote back on March 25. He first said that he had decided 'not to go for 5 years to the photographic telescope in the southern hemisphere, but to stand for the adjunct-directorship in Leiden', although he still needed to become convinced himself that indeed he was really qualified for it. He noted de Sitter's apparent high opinion of him and he also confirmed that his ambition was in the area of general theoretical astronomy, and not to become an observer pur sang. He noted that he had shown the letter to Schlesinger, who 'immediately advised to forget about the southern plan'. Oort also announced that he would come half July to the Netherlands to fulfill his military duties. Schlesinger had helped to arrange this already some months earlier.

In the Netherlands, military service (or conscription) had been instituted by the French occupation in 1810. Dutch men served in the French army; they were selected by drawing lots. After the Kingdom of the Netherlands had been proclaimed in 1815, conscription did not cease to exist, however. The selection of those required to serve in the army continued to take place on the basis of drawing lots. The drawing system remained in use until 1938. Oort had been been tested as fit for service and on this occasion his height had been recorded as 1.723 m (5′7″), about 1 cm *taller* than the average man that reported for the draft in the 1920s [136]. His compact, trim, and neat personal presentation did not give the impression he was even of average height. Oort had been formally drafted into the Dutch army on January 17, 1920 in the fourth regiment infantry, but his actual having to report for boot camp ('eerste oefening' or initial training) was postponed because of his academic studies. Basic training would be for five and a half months, of which two initial, general training. The rest was (in case of academicians at least) devoted to training as petty or non-commissioned officer. This postponement was for one year at a time and in Oort's case had been renewed every year until the summer of 1924. But upon his request the final postponement was withdrawn and he reported for basic training in July of 1923. How had this come about?

Before this decision not to go to South Africa was taken, Schlesinger had already written a letter to the Minister of War of the Netherlands, dated March 6, 1923.

'Dear Sir,

Mr. J.H. Oort, Phil. Doct^{S} of the University of Groningen, has been research assistant at this Observatory during the past year. I have been extremely well impressed with his abilities and I am sure he is going to be one of the band of scientists for which your country has won the admiration of the world. I have made him an offer to take charge of a large telescope that this Observatory is about to erect somewhere in the southern hemisphere, very possibly South Africa. He informs me that he has not yet completed all his military obligations, and that to do so will require from two to six months, depending upon whether he is asked to go through the training necessary to attain under-officer's rank. If the latter, then I greatly fear that it will not be possible for us to wait for him. I should therefore strongly urge that if he can with propriety be excused from more than two months service that we should consider this a great courtesy on your part, and that, in my opinion, it would as well be a scientific service.

Very respectfully yours,
Frank Schlesinger
Director'

The title Phil. Doct^{S}, that Schlesinger used, is the Dutch form of a Masters Degree, which until recently was called doctorandus or Drs. After the Drs. degree students worked solely on their thesis research. To what extent Oort himself seriously considered the possibility of playing a major part in Schlesinger's southern telescope project is not clear from his letters, pocket diaries or dairy notebooks. We do know that Schlesinger's request to the Minister was actually granted, and that he came to the Netherlands in the summer of 1923 for his military obligations. His military passport mentions he entered basic training on July 23, 1923 and was sent on 'groot verlof' (long furlough) on August 30 of that year. So he seemed to have gotten away with no more than six weeks. He went on 'herhalingsoefening' (repeat training) for periods of a few weeks in 1927 and 1928. I have not been able to find him registered as passenger on any of the ships of the Holland Amerika Lijn back from thc USA to Rotterdam in May to July of 1923 (there are no name-indices in the Rotterdam Archives for inbound trips, and I have gone only through second class passenger list for the relevant trips), but he is registered as a second class passenger on the S. S. Veendam, leaving Rotterdam on October 10, 1923 for New York (Fig. 4.11).

The correspondence Oort had with de Sitter after his return from military service shows that it had been decided (probably in conversations during the period that Oort was in the Netherlands) that he should look into matters of absolute positions of stars and reliability of proper motions for the study of the structure of the Stellar System. Leiden Observatory had started a program in this area, most likely initiated by Kapteyn when he was Adjunct-Director. Towards the end of 1923, Oort wrote that his plans were to stay in New Haven until the end of April. He had earlier already taken

Fig. 4.11 Oort (front) during his basic training as part of his military service in the summer of 1923. From the Oort Archives

up plans to visit Mount Wilson Observatory for two or three months, but Schlesinger also suggested that he visit Harvard and Dudley Observatory in Albany (New York) for a few weeks. On February 1, 1924 Oort wrote de Sitter that he had received the telegram formally offering him an appointment in Leiden as Conservator and that he had sent a telegram back confirming his acceptance. He also expressed understanding that de Sitter did not yet find him qualified to take up the leadership of the department of fundamental astronomy. And on February 7 he confirmed that he had no plans to stay in America, reported that Schlesinger had also advised him to take the Leiden position, and stated that 'as long as I am not married – and I have no plans for that at this moment –, the matter of the salary is not important for me.'

4.7 Back to the Netherlands

From the notes Oort made in his diary notebooks, pocket diaries and some letters it is obvious that he enjoyed his time in the United States and would not have objected to, or maybe even have welcomed, extending his stay there. In his diary notebook he wrote not very long after his arrival in New Haven (January 3, 1923):

> 'Already now I start to apprehend a bit the idea of going to live in Leiden for the rest of my life, and at the same time I start to believe I would long back to America at the same time like Co Kramer and Storm van Leeuwen [Konrad Cramer and (probably) Arnold Storm van Leeuwen from Maverick] and feel cramped and tied there, especially [in] ancient and old-fashioned Leiden. I do not feel at all happy to come back to the old things, those old things with which (in Leiden *much* more than in Groningen) something dark and black is attached.'

And in a note of May 8 (when he was visiting Pasadena, California) he entered the remarks:

> 'I receive a letter from de Sitter, that makes me even more pessimistic, because there in Leiden I do not get one of the senior positions but one of the most junior ones, working under Hins and V.d. Bosch, for a salary of f. 3600,–. One good thing is that probably in November I will go to Greenwich. De Sitter's letter is short and he does not even take the time to say anything about my draft papers, which I find depressing.'

An annual income of Dfl. 3600 would correspond today to about twice the income of an unskilled worker, and must have been disappointingly low. It should be said though that Oort had not even completed his academic studies with the defense of a Ph.D. thesis. Coert Hendrik Hins and Willem Hendrik van den Bos (1896–1974) were assistants/astronomers at Leiden Observatory. Not much later (July 8), actually de Sitter wrote some remarks on the drafts in a 'pleasant letter'. On some loose pages in his pocket diary for 1924, Oort wrote (dated July 21 from California) 'I promised de Sitter to come back and start with him Kapteyn's plan for fundamental declinations. This promise is more important than any temporary feeling of unhappiness.' But he also privately raised the option of leaving Leiden again after one or two years.

Then there is the matter that he still had to obtain his Ph.D. degree. This is the subject of the final part of the letter to his parents of February 23, 1924. He started that with a sudden change of subject.

'And then now a point on which I want to ask you for advise. If there indeed is something of substance in this phenomenon, and if it can be worked out into a dissertation, with whom shall I then obtain my doctorate?, that is Hertzsprung versus Van Rhijn. When I think of a thesis defense then I should first have to decide on that and preferably very soon, because then I could arrange the publication before I leave. When Leiden or Groningen would agree to print this, everything is fine, but I fear that neither of these places do print dissertations anymore and then I could maybe persuade Schlesinger or Boss to print it.

We have discussed this problem before. In favor of Hertzsprung is, that he will without doubt help me to make it significantly better, that he will give all sorts of useful suggestions and provide good criticism, while Van Rhijn seems not so much interested in this subject and in the second place will not put forward any new points of view. Also, in the future I will have to deal much more with Hertzsprung than with Van Rhijn. On the other side of all this there are these points:

1° I promised Van Rhijn half-and-half to submit my thesis with him (at my own initiative, stupid enough).

2° I would prefer one hundred times more to obtain my degree in Groningen than in Leiden, because that is where I did my studies and where all my friends still are, and this last argument is probably what influences me most.

3° is Zernike in Groningen, and I would rather defend my propositions to him than to Ehrenfest.

Now, I should go back to work, which is a shame with this magnificent weather. Going skating will cost me all afternoon and tonight I am reading with the Torrey's and will have no time to work also.

At the moment there is much to do with the zenith telescope and that does not come at the right time, especially since I wanted to try to go to Boston at the end of the week.

I will write more soon, and also what Schlesinger's opinion is about these high velocities. Don't think too much of it for now.

Many cordial greetings for Jo, Jetske and Emy and especially for you,

from your you loving

Jan'

Propositions go along with a Ph.D. thesis and are at least for a large part points of view or theorems unrelated to the subject of the thesis. Paul Ehrenfest (1880–1933) (see Fig. 6.3) was an Austrian-born theoretical physicist in Leiden. Boss is not Lewis Boss of the famous catalog Oort used, but his son Benjamin Boss (1880–1970), who also served as the Director of both the Dudley Observatory in Albany, New York, and the associated Department of Meridian Astrometry of the Carnegie Institution of Washington.

Fig. 4.12 Oort visiting Lick Observatory in 1924. The two pipe-smoking men are on the left Robert Julius Trumpler and on the right Charles Donald Shane. From the Oort Archives

Already in May 1924, Oort started to prepare for leaving Yale and New Haven as he planned to travel extensively across the USA before returning to Europe. In his pocket diaries we see that he made last visits to many people in New Haven, such as: May 6 – Packing and making visits to Mrs. Schlesinger, Mrs. Brook, Mrs. Mildred Stearns, Mrs. Vivier for lunch, Mrs. Browne calls up. Last supper with Torreys, sit outside, see Mrs. Welch on the way home. He said farewell to the Observatory on May 7 and traveled to New York and from there to Dudley Observatory in Albany, New York, where he stayed for a few days with Benjamin Boss and his wife, on to Buffalo and then to Chicago and Kansas City, eventually arriving in Flagstaff, Arizona on May 23. After a trip down Grand Canyon on mule's back he traveled on to Pasadena, California, near Los Angeles, admiring on the way the yucca's and desert scenes. He reached Pasadena on May 28, and already the next day, May 29, he traveled up to Mount Wilson Observatory.

> 'With Brown in the Observatory bus up the mountain (2000 m), where I make acquaintance with van M. [van Maanen] Beautiful wide view with steep chasms. Everything is very beautiful and fine so as to become fond of the stars immediately, especially when seen from the Monastery, and of the view from its balcony. Very nice meals, all free, and everyone together, workshop men, bus drivers, janitors, night assistants, etc. Saw the 100″ mirror that is worked on to remove astigmatism, went to the Monastery to finally write miss W. and later back to the 100″ and midnight lunch. Beautiful clouds seen below us.'

The aim of this trip was to spend some time at Mount Wilson Observatory and talk to the staff there. Oort stayed in Pasadena until July 22, so altogether almost two months, when he moved on to Lick Observatory south of San Francisco. In Pasadena he talked a lot with Adriaan van Maanen and other staff members: George Hale, Walter Adams, Gustav Strömberg, Frederick Seares, Harold Delos Babcock (1882–1968) and others. He visited 'the Torrey's', which probably refers to Charles Torrey's brother Joseph (1862–1952), since Oort mentioned that he talked extensively with a Helen Torrey, who loved art and whom he liked very much and invited to go to an 'Egyptian Theater'. Joseph indeed had a daughter Helen Maria (1901–1990). There was sight-seeing (with the Babcock's to Balbao Island near Los Angeles) and in his notes the words rattle snakes and tarantula spiders show he at least looked out for these animals and maybe saw them.

Oort traveled on to the Bay area, stopping at Pebble Beach near Monterey and ending up visiting Lick Observatory, which he reached July 28. Lick Observatory (see also Sect. 3.1), is located on Mount Hamilton near San Jose, some 80 km south of San Franciso. It had been built around 1880, being funded by James Lick (1796–1876), who has been described as a carpenter, piano builder and land baron. Oort met various astronomers there, in particular Charles Donald Shane (1895–1983) and Swiss-borne Robert Julius Trumpler (originally Trümpler, 1886–1956). Figure 4.12 shows Oort at Lick during this visit with various staff and the two astronomers just mentioned.

From San Jose he traveled on via San Francisco, Berkeley, Denver, Detroit, Chicago to New York, where he arrived on August 10. The next day he went on to Boston and visited Harvard, unfortunately finding Harlow Shapley absent, but where he talked extensively with Willem Luyten.

Before departing for Europe, Oort spent a few days at the Cranberry Isles with the Torreys. Mrs. Torrey has written Oort various times from this location through the years (for example in August of 1923, when Oort was in the Netherlands for his military service), always signing with "Aunt Marian'. The Torrey's very likely owned a second house there. The Cranberry Islands are located just off the coast of Maine, some 50 km mostly south and a bit east of Bangor, Maine. He apparently got there by boat from Boston. Great Cranberry Island, where the Torrey's resided, was where the photograph in Fig. 4.0 must have been taken.

From Maine Oort traveled to Quebec in Canada. His dairies show that he booked his trip back to the Netherlands from Pasadena at the end of June. He left from Quebec 'on Mount Royal' (I assume this refers to the ship, named after a hill in the city Montreal, which name also derives from Mont Royal) on August 22, 1924.

Oort's trip back to the Netherlands was uneventful according to his pocket diaries; he read, played games (shuffle board) and danced, sometimes with pleasure, sometimes with disappointment (one Scottish girl was not very good at it, another danced 'like a frog'). He traveled via England, probably landing at Liverpool. At least he went to London by train, where he almost ran out of money and could not buy breakfast. On the boat across the North Sea he did travel first class but could not pay for a cabin. He arrived in Leiden on August 31, 1924.

Fig. 5.0 Oort in 1925. From the Oort Archives

Chapter 5
Rotation of the Galaxy; Marriage

Look across the Universe,
and you'll see that almost everything is rotating.
The Earth rotates on its axis as it orbits the Sun.
And the Sun itself is rotating. As you can probably guess,
we even have galaxy rotation with our Milky Way Galaxy.
Universe Today [137]

When you realize you want to spend the rest of your life
with somebody, you want the rest of your life
to start as soon as possible

When Oort arrived back in Leiden, his parents were still on a long stay in the East Indies: they wrote letters from Buitenzorg, the current Bogor, some 60 km south of Jakarta. This was the administrative center of the government, where also the palace of the Dutch Gouverneur-Generaal, the highest colonial authority, was located. On September 1, 1924, Oort was introduced at the 'Sterrewacht' and received his formal appointment as 'Conservator' with an annual salary of Dfl. 3600. This was as de Sitter had offered him earlier (see Sect. 4.6) and I repeat that this would correspond to about twice the annual income of an unskilled worker. He spent much time talking to Jan Schilt, his old study mate from Groningen. Schilt in his Ph.D. thesis in Groningen, written under the supervision of van Rhijn, and defended in 1924, had developed a photometer that could be used to measure stellar magnitudes on photographic plates by illuminating a stellar image with a pencil beam of light and measuring the light transmitted through the plate with a galvanometer designed by Frits Zernike (and of course comparing it to the unblocked beam). In Leiden Schilt had just completed an an improved version; the so-called Schilt photometer remained in use for many years [139]. Schilt married Johanna (Jo) Timmer (1898–> 1982) in Groningen in February 1925, before moving to the Mount Wilson Observatory later that year, from where he went to Yale Observatory at the invitation of Schlesinger in 1926 (Fig. 5.1).

From the movie 'When Harry met Sally'.

P. C. van der Kruit, *Jan Hendrik Oort*, Astrophysics and Space Science Library 459, https://doi.org/10.1007/978-3-030-17801-7_5

Fig. 5.1 Leiden Observatory photographed from the air with a military airplane in 1924. The picture is taken from the south. The canal is part of the original Leiden town defenses and is called 'Witte Singel'. The main telescopes are the 10 and 6 in. refractors on the main building, and the 13 in. photographic telescope on the right on top of the (detached) photographic department. The (also detached) west tower on the left has housed various instruments. The west wing (also on the left) contained the Director's residence; between it and the main building was the meridian room containing the 6 in. meridian circle. Above the Observatory the botanical gardens and at the very top the main auditorium and academy building of the University. From Wikimedia Commons [138]

I note in passing that Oort's notes in his pocket diary were in Dutch when he lived in New Haven, but when he arrived in Leiden he started writing in his diary notebooks and pocket diaries almost always in English, at least for a few years after which he resorted back to Dutch.

On the evening of his first day in Leiden, Oort had dinner with the de Sitters, who lived in Oegstgeest. And already on the next day, September 2 he started in the Fundamental Department. His sisters had come along with their parents to the East Indies, as Oort's mother sent the best wishes from 'the children and father' in a letter from Buitenzorg. His brother John had moved to Utrecht to study biology and botany. On September 3, Oort met van Rhijn at the railway station who came for a presentation to the Netherlands Astronomers Club or NAC, the professional society for astronomers in the Netherlands. It had been founded in 1918 and still flourishes; it is also open to Belgian (Flemish) astronomers. Already on September 12 he took the train to Groningen (in the evening), where he stayed for a few days, and met with old friends and his Professors Zernike and van Rhijn. There seemed to be hardly any time to adapt and settle down (but jetlag was still a thing of the future). Indeed almost from the minute Oort arrived in Leiden, de Sitter set him to work; the subject was the matter of absolute declinations.

5.1 Absolute Declinations

On Sect. 2.2, I discussed Kapteyn's work on the problem of absolute declinations. The determination of declinations depended on measurements of height above the horizon (altitude) and these were affected by effects of flexure in the telescope and by atmospheric refraction, which were quantities not known or determined with sufficient accuracy. Kapteyn had devised a method to reduce the problem to the determination of the altitude of the pole and next designed a method to establish this altitude 'free from systematic errors'. It involved measurements of azimuth differences of altitudes of meridian passages of two stars occurring at about at the same time with similar heights but on opposite sides of the zenith, and timing of passages of the prime vertical (see below). All of these were unaffected by the flexure and refraction. The matter of absolute declinations was a major concern of the Fundamental Department of Leiden Observatory, eventually—with de Sitter and later Oort as the driving forces – resulting in two Expeditions to Kenya (see also *The Kenya Expeditions of Leiden Observatory*, by Jet Katgert-Merkelijn [140]).

The advantage of going to the equator is that in order to measure the declination of a star from that position, one in principle only needs to measure its azimuth at rising or setting and that is not affected by (vertical) atmospheric refraction or telescope tube flexure. Consider Fig. 5.2. The shaded plane is the Earth's surface with the half-sphere the sky above it. The celestial poles are on the horizon, respectively at the north-point 'N' and the south-point 'S'. The point 'Z' is the zenith. The diurnal paths of three stars are indicated; they rise in the eastern sky perpendicular to the horizon and set in the west. Those positions can be measured and the direction to the north is

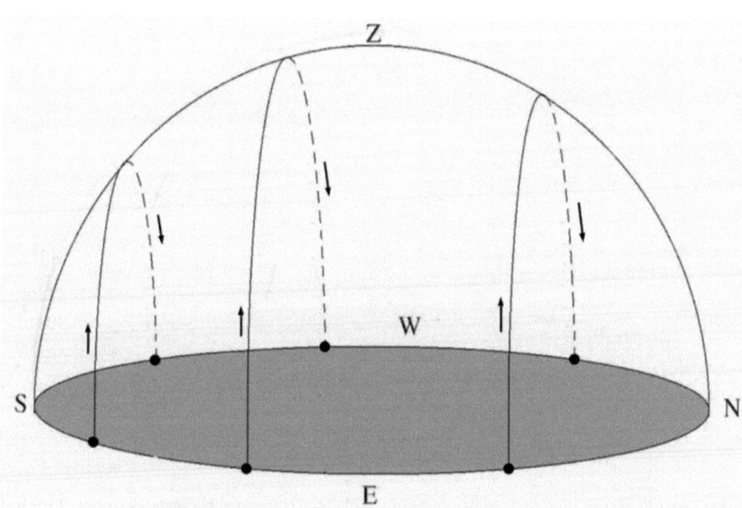

Fig. 5.2 Diurnal paths of three stars from a location on the Earth's equator. For explanation see text. Figure by the author

by definition halfway in between these points. In the usual convention the azimuth is the angle between the position on the horizon of the south (S), and that where the star rises or sets. On the equator that then is equal to 90° plus the declination and it is therefore a direct measurement of declination without any effects from refraction or flexure. In practice one would not be able to measure the azimuth when the star is exactly on the horizon, but say 10° above it, and then apply a very small correction.

Carl Heinrich Ludwig Sanders (1864–1937), a Dutchman (so pronounced with the 'a' in Sanders as in 'father'), who had worked as an agent for a British coffee company in Portuguese Congo, had eventually become a coffee plantation owner in Matuba (in the province Cabinda there, in present-day Angola), at a geographical latitude of $-5°17'$. He had been interested in measuring geographical positions, and had published since 1899 various reports in the astronomical journal 'the Observatory', and for this work had been in regular contact with E.F. van de Sande Bakhuyzen at Leiden Observatory. His problem was that he had no ephemeris and instrumentation to determine his latitude with high accuracy. In 1917 and 1919 he published two notes, also in 'the Observatory', on *Determination of latitude and declinations* [141]. He expressed some reservations that the proposal he was making might not be original. He proposed a method that can be illustrated with Fig. 5.3, where we see the diurnal path of a star from a position not too far from the equator (for clarity the distance PM has been exaggerated here). The pole 'P' now is at a relatively small angle φ above the horizon. The azimuths at rising and setting now are not directly related anymore to only the declination. But one can also measure the zenith angle at the meridian z_m and with these two measurements it is possible to solve for both the latitude of the observer (or the angle of the pole above the horizon) and the declination of the star.

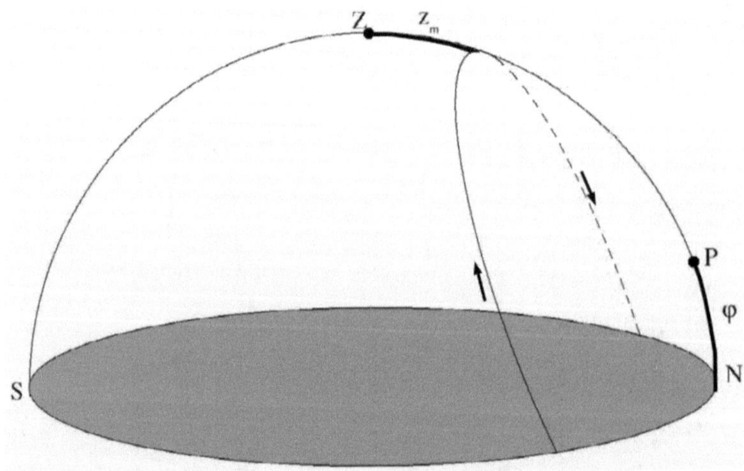

Fig. 5.3 Diurnal path of a star from a location not too far from the Earth's equator. The half-circle $S - Z - N$ is the meridian and the celestial pole is at P. The thick arcs are the geographical latitude φ and the zenith angle at meridian passage z_m, used in Box 5.1.The azimuth A (not indicated) is the angle along the horizon from the point S to that where the star rises or sets. Figure by the author

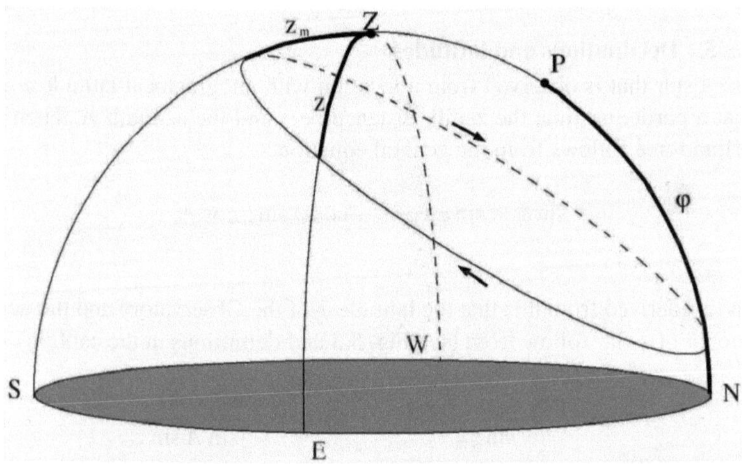

Fig. 5.4 Diurnal path of a stars from a location at relatively high latitude. The half-circle $E - Z - W$ is the prime vertical. The angle z_m is again the zenith angle at meridian crossing and z the zenith angle at the crossing of the prime vertical. Otherwise as in Fig. 5.3. Figure by the author

Of course this involves the measurement of a zenith angle (or altitude) and this is liable to refraction and flexure, but this star can be chosen such that this is performed at rather a high altitude, where these effects are small (and the error in the correction even much smaller that the effect itself). This method works in principle from any place on Earth, but Sanders' point was that it would be very accurate when performed near the equator.

He made at the same time make another proposal about the determination of one's latitude (see Fig. 5.4). Here the pole 'P' is higher up (φ is now much larger). We see a star's diurnal path (it is a circumpolar star that does not set, but it could equally well have been a star that does set and rise). Now Sanders not only proposed to measure the zenith-angle on the meridian (z_m) but also at the prime vertical z. The meridian traces a great circle on the sky through the zenith and the north and south points, and similarly the prime vertical traces such a great circle but now through the east and west points on the horizon. As one can appreciate when mentally rotating the pole up or down, the relative values of z and z_m vary with latitude and this can be used to determine the latter. Sanders also showed that the errors in these two zenith–distances result in a relatively small error in the latitude. Sanders concluded that the problem of the declinations could be attacked using three stations, one on the equator and two at moderate northern and southern latitudes.

Box 5.1 Declinations and latitudes.

Take a star that is observed from a location with geographical latitude φ and let at a particular time the zenith distance be z and the azimuth A. Then the declination δ follows from the general equation

$$\sin\delta = \sin\varphi\cos z - \cos\varphi\sin z\cos A.$$

Sanders derived from this that the latitude φ of the Observatory and the declination δ of a star follow from (see Fig. 5.3 and definitions in the caption)

$$\tan\varphi = \frac{\sin A - \cos z_{\rm m}}{\sin z_{\rm m}} \qquad \tan\delta = \frac{\sin A\cos z_{\rm m} - 1}{\sin A\sin z_{\rm m}}$$

This works well at low latitudes (or large zenith angles), where A can be well determined and $z_{\rm m}$ should not be too close to zero.
For the case in Fig. 5.4 (again see definition in the caption) he found

$$\cot\varphi = \frac{\cos z_{\rm m} - \cos z}{\sin z_{\rm m}} \qquad \cot\delta = \frac{1 - \cos z\cos z_{\rm m}}{\cos z\sin z_{\rm m}}.$$

Here small $z_{\rm m}$ (when the sine is small) should be avoided.

All relevant equations are collected in Box. 5.1. Sanders had not only made these suggestions, but actually in 1920 published a study based on his observations on *Latitude variation observations at Matuba, Cabinda* with Harold Spencer Jones (1890–1960), who at that time was assistant at the Royal Greenwich Observatory and would later become Astronomer Royal [142]. Now Sanders' work had been brought to the attention of Willem de Sitter by Ernst van de Sande Bakhuyzen. Sanders fortunes turned and in 1922 his plantation almost went bankrupt and he sold it. He remained in Portuguese Congo, doing surveyor work for the Portuguese government, but de Sitter also gave him an annual stipend to make measurements for Leiden Observatory. Eventually, in August of 1926, he joined Leiden Observatory as 'chief assistant', but that was after Oort had been appointed in Leiden.

Maybe it is useful to briefly summarize what the staff of the Sterrewacht looked like around 1925/1926. In addition to the Director and his adjuncts (de Sitter and Hertzsprung) and Conservator Oort, there was one more Conservator. This was Jan Woltjer Jr. (1891–1946) who had done his Ph.D. under de Sitter on a thesis with the title *Investigations in the theory of Hyperion*, which he defended in 1918. Hyperion is a moon of the planet Saturn, relatively close to Saturn, a bit more distant from it than the largest moon Titan and is locked in a 4:3 resonance with Titan. The study is reminiscent of de Sitter's work on the moons of Jupiter. Woltjer had been appointed Lecturer of Theoretical Astronomy (one rank below full Professor) at

Fig. 5.5 Carl Sanders and Coert Hins, the latter in the foreground, taken from the staff photograph of Leiden Observatory in 1931 (Fig. 7.7). The person to the left of and partly covered by Sanders is Pieter Oosterhoff. From the Oort Archives

the University of Leiden and Conservator at the Observatory in 1920. Then there were two Observators, one was Willem Hendrik van den Bos (1896–1974), who was working on a Ph.D. thesis under de Sitter on micrometer measurements of binary stars (defended in 1925), and the other was Coert Hendrik Hins. Hins worked in astrometry; his thesis was in Dutch, the title being in translation *Introduction to a catalog of positions and proper motions of 1533 red stars*, defended in December 1925 with de Sitter as supervisor. He worked with the meridian circle of Leiden Observatory and would play an important role in the first Kenya Expedition that came out of the work of Sanders. Hins and Sanders are depicted in Fig. 5.5, taken in 1931. In addition to this the scientific staff had assistants; in 1924 these were Jan Schilt, who was about to leave, and the other Dirk Brouwer. The latter would write a thesis under de Sitter, like that of his mentor on the satellites of Jupiter, which he defended in 1927. After this he went to the USA, spending his career at Yale Observatory of which he became Director in 1941. Two more young astronomers that would rise to fame, Gerrit Pieter (later Gerard Peter) Kuiper (1905–1973) and

Pieter Theodorus Oosterhoff (1904–1978) were still to come; they were appointed assistants in 1927.

De Sitter had assigned Oort to work on the subject of absolute declinations immediately after the latter had joined Leiden Observatory. There is some indication from Oort's private notes that this had been agreed upon between de Sitter and Oort before he came to Leiden. Actually, in his history of Leiden Observatory, written in 1933 [104], de Sitter mentioned Oort's arrival as 'the addition to the sphere of activities of statistical astronomy', and does not mention astrometry in connection with him. Oort set to work and quickly published two papers, one on the actual observations of Sanders (Sanders and Oort 1925b) and one with de Sitter on a *Provisional scheme for the determination of fundamental declinations from azimuth observations* (de Sitter and Oort 1925d).

In the first paper Sanders' approach was used to test to what extent these observations could help to resolve the issue of determining declinations 'independent of errors in the vertical refraction, and with an instrument in all respects different from a meridian circle, especially now that Kapteyn and others have drawn attention to the serious consequences of our uncertainty about the systematic errors of the declination in existing fundamental catalogues'. Sanders and Oort used the observations that Sanders had done from Matuba. The instrument used for the measurements was a 'universal instrument', which is a small telescope on an alt-az mounting. It was originally designed for geography and therefore also designated as a theodolite. The aperture (primary lens) had a diameter of 54 mm and the focal length of the telescope was 6 cm. So it was a relatively small instrument (Fig. 5.6). The telescope was of the 'broken' type, which means that a prism or small mirror deflected the light so as to ease observing through the eyepiece.

The alt-az mounting means that the telescope can move around two axes, one towards the zenith (when properly aligned) and one in the plane of the horizon. This means that the azimuth and altitude can be read off immediately from the dials on the axes, so that it could be used to accurately measure azimuths (angles along the horizon) of objects or differences of azimuths. The circles from which the positions were read had a diameter of 35 cm. The alt-az mounting is much easier to build than the parallactic mounting of telescopes, in which one axis points at the celestial pole and one is in the plane of the equator. This more complicated structure has the advantage (of course with proper guiding) that the image stands still in the focal plane, which is convenient for long exposures. Modern telescopes employ the alt-az mounting, where now the rotation of the image in the focal plane is corrected under computer control by rotating the instrument and detector.

The universal instrument was mounted on a pier of cement and masoned bricks, well founded in the ground and separated from the observing floor, and was enclosed in a shed with a sliding roof. For the azimuth reference a mark was painted on a wall almost one kilometer away, which was used in the afternoons and early mornings to calibrate the zero-point of the azimuth and the collimation (the alignment of the optics and optical axis with the telescope and its physical axis), which depended on temperature. The latitude of Matuba had been determined by the (Horrebow-)Talcott procedure (see Box 4.1) in 1919; it was based on declinations of stars in the Boss

Fig. 5.6 Sanders' universal instrument. It is looking straight up in this picture. The eye-piece is at the middle-right in between two microscopes for reading of the vertical altitude circle. Two further, slanted microscopes are present to read off the azimuth from the horizontal circle at the bottom. From the Archives of Leiden Observatory

Catalogue that we encountered before (e.g. the Schilt-Oort paper for the Bachiene Prize). It was this *'Preliminary General Catalogue of 6188 stars...'* for which doubts on the accuracy of the declinations had been raised by Kapteyn and others.

The latitude is required to make corrections for the calculation of the declinations from the measurements of the azimuths (see Fig. 5.4) and for the fact that stars were not observed on the horizon. However, the corrections stemming from the latitude are hardly affected by errors in the latitude itself. The observations had been made during the period January 1923 to November 1924, so up to the time Oort came to Leiden. The observing records confirm (as Sanders had mentioned before in publications) that the weather conditions at Matuba are far from ideal and better observing sites should be available elsewhere near the equator. It appears that only a fifth or so of the available nights were suitable for observing. In total 42 stars were observed between declinations $-25°$ and $+5°$. In the end the declinations determined were between 0 and 1.5 arcsec larger than in the Boss Catalogue with a significant variation with declination itself. Sanders and Oort argued that the fact that the declination corrections determined from observations when stars were in the east or in the west, were very similar supported the correctness of their results. The final conclusion was that indeed with such a small instrument near the equator independent checks of declinations in meridian circle catalogs could be obtained.

The paper with de Sitter presented a plan to determine a fundamental system of declinations from azimuth observations in a systematic manner, 'free from sources affecting the observation of zenith distances' and covering the whole sky. For this they proposed that a special instrument be built, consisting of a horizontal telescope of 90 mm aperture and focal length about 1 m. It would be a 'broken' telescope so that the observer could more easily access the eyepiece. The azimuth could then be read off from a horizontal circle of at least half a meter diameter by a set of two or four microscopes 'firmly connected to the movable frame' on which the telescope would be mounted. The central part of the proposal was to use a station near the equator, but in order to cover the full sky they proposed to also employ two station, one north and one south, about 35° from the equator.

For the equatorial station they proposed to use the methods developed by Sanders, but for the two other stations they gave preference to the method designed in the 1880s by Kapteyn, that I mentioned above (Sect. 2.2). This had brought Kapteyn into contact with David Gill, who had carried out an observing program from the Cape Observatory using this method. The results were, however, never published. However, Kapteyn's approach was used in the Ph.D. thesis of Leopold Courvoisier, in which he determined the latitude of Strasbourg Observatory in the Elsass in 1901 [143]. Courvousier, whom, as we saw, had been considered by de Sitter for a position in Leiden, had actually proposed a somewhat simplified scheme, which was judged by de Sitter and Oort to be easier to implement. Although intrinsically less accurate, they concluded that Courvoisier's scheme was to be preferred when many more observations were added to increase the measurement accuracy, while it also did not involve the assumption (as in Kapteyn's method) that the vertical refraction east and west of the meridian would be equal.

In the end the proposal by de Sitter and Oort led to the Kenya Expeditions in which Oort was to play a central role. The stations away from the equator were never realized.

Although Oort's attention had to be directed at least for a part of his time towards the completion of his Ph.D. thesis and subsequently his work on the rotation and dynamics of the Galaxy, he remained involved in fundamental work related directly or indirectly to astrometry. I will only mention here his astrometry-related publications in the years up to 1930 briefly without any detailed discussion. Together with student W.N. Doorn (employed 1925–1927 as assistant) he published a note on the results for the the solar motions when determined from radial velocities of stars of various absolute magnitudes and found no definite evidence for magnitude-dependent systematic differences (Oort and Doorn 1925f). He also conducted an *Investigation of the constant of refraction from observations at Leiden and at the Cape*, using old data collected in Leiden with the meridian circle between 1882 and 1897 by H.G. van de Sande Bakhuyzen and a similar and simultaneous program for the Cape Observatory (Oort 1928a). The observations were published in the Annals of Leiden Observatory (Oort 1928c).

5.2 Thesis on High-Velocity Stars

Oort still had to obtain his doctorate. It was decided that it would be on high velocity stars and that his formal supervisor would be Pieter van Rhijn, so that the defense and awarding of the degree would take place at the University of Groningen. The thesis would be printed as part of the 'Groningen Publications'.

The supervision of van Rhijn had to proceed mainly by mail and there are a fair number of letters in the Oort Archives from this period. The van Rhijn Archives in Groningen (in a repository of the University) are not very extensive and lack any correspondence before the 1930s. Most of van Rhijn's letters make reference to notes he had made on drafts of parts of the thesis, but the files do not contain the drafts themselves that apparently have not survived. At first, van Rhijn addressed Oort as 'dear Oort' in 1924, but in the course of 1925 it became 'dear Jan Oort', soon followed by simply 'dear Jan'. He signed the letters with either 'Pieter v. R.' and later simply 'Piet'. In a letter accompanying remarks on a draft of the first part of the thesis in May 1925, van Rhijn ended the letter with a rather old-fashioned and patronizing note: 'Keep going like this, my son!' ('Ga zo door, mijn zoon!'); or was that meant ironically? From then onward the tone became more like among equals.

The thesis was defended on Saturday May 1, 1926, and had the title *The Stars of High Velocity* (Oort 1926b). Defending means that he was interrogated about the thesis work and the propositions for 45 min by professors of the Faculty (Mathematics and Physics) in a public ceremony. In a preface to the thesis, Oort thanked all those that contributed to his education as an astronomer, first the Groningen professors (in other disciplines than astronomy), then Schlesinger, de Sitter and Hertzsprung. Then a special paragraph for Kapteyn: 'In particular I commemorate Professor Kapteyn, whose clear lectures and inspiring personality had the largest influence on my choice for astronomy as my 'major'.' And a special thanks to van Rhijn as thesis supervisor.

For the background of the study I refer back to Fig. 3.4, and the discussion there of the Kapteyn Universe and Shapley's globular cluster system. It should be kept in mind that Oort's picture of the Universe was still more like in the top panel of that figure. The important feature is that in reality (the lower panel) the globular cluster system has a small rotation velocity, while the disk of the Galaxy, of which the Sun and the Kapteyn Universe are part, has near the position of the Sun a rotation velocity of just over 200 km/s. As a result, the globular clusters and the high-velocity stars have a large, asymmetric motion with respect to us, and that is the reason why they all seem to come from one hemisphere of the sky around the direction of local Galactic rotation. This fundamental observation was illustrated by Oort in his thesis on the first pages in a figure that I reproduce in Fig. 5.7. In this figure the sphere representing the sky has been unfolded. The horizontal, central line is along the Milky Way and the sharp points at the top and bottom represent the Galactic poles. Galactic longitude is measured along the horizontal axis, running from 0° at the left to 360° at the right. The center of the Galaxy as we now know it (and which also is the center of the system of globular clusters) is at longitude 327° in the figure (in the current system

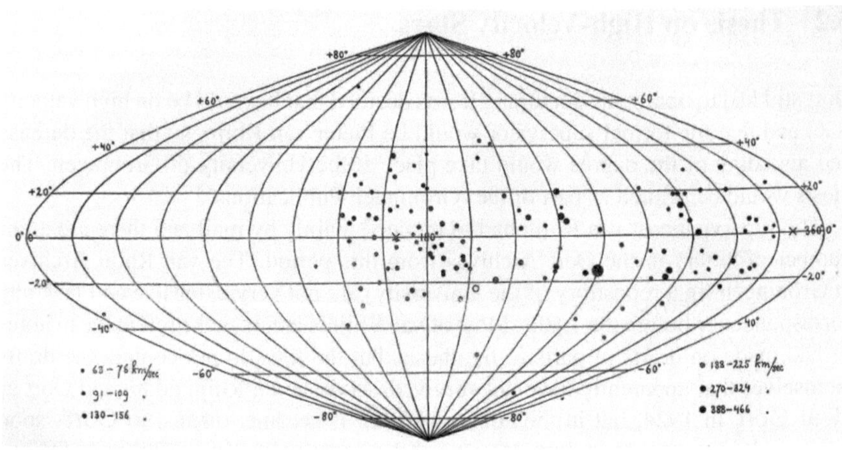

Fig. 5.7 First figure in Oort's thesis. It shows the distribution of apices of all his high-velocity stars. From Oort (1926b)

it is longitude 0° by definition) and the Galaxy rotates in the direction of longitude 57°. This is where the high-velocity stars should on average appear to be coming from, and that is indeed what we see, when we remember that what Oort plotted were the directions where the stellar velocity vectors were pointing *towards* and the distribution is indeed centered roughly on longitude 237°.

The thesis consisted of three chapters. The first concerned the general characteristics of the motions of the high-velocity stars and in this Oort first presented up-to-date figures confirming the onset of asymmetry at 63 km/s as he had found earlier (updated versions of Figs. 3.8 and 3.9). There were two major issues in the thesis and the first part contained a thorough discussion of these. The first major issue was the fact that Oort maintained that there was no asymmetry among the stars of low velocity and that this changed suddenly at a spatial velocity of 63 km/s. He found that when he plotted the frequency of occurrence of spatial velocities it followed a Gaussian curve with great accuracy (Fig. 5.8), except for a tail on the high velocity side, the stars he studied. To Oort, the excellent fit to a Gaussian for all stars except those of high velocity indicated a very random distribution. He added to this that a closer inspection showed no evidence whatsoever for any asymmetry in the motions of the stars below 63 km/s. Thus there a very abrupt and strong dichotomy between low and high velocities. However, this contradicted a recent, and what would prove to be important, discovery by Gustav Strömberg, who was involved in measuring stellar velocities at Mount Wilson. The discovery concerned the so-called (Strömberg) asymmetric drift.

To understand what this is consider the following. For a star in a circular orbit in de Milky Way around the Galactic center, there is always exact compensation of the attractive force by the centrifugal force. This may also be formulated as that the gravitation (the potential energy) and the motions (the kinetic energy) are in

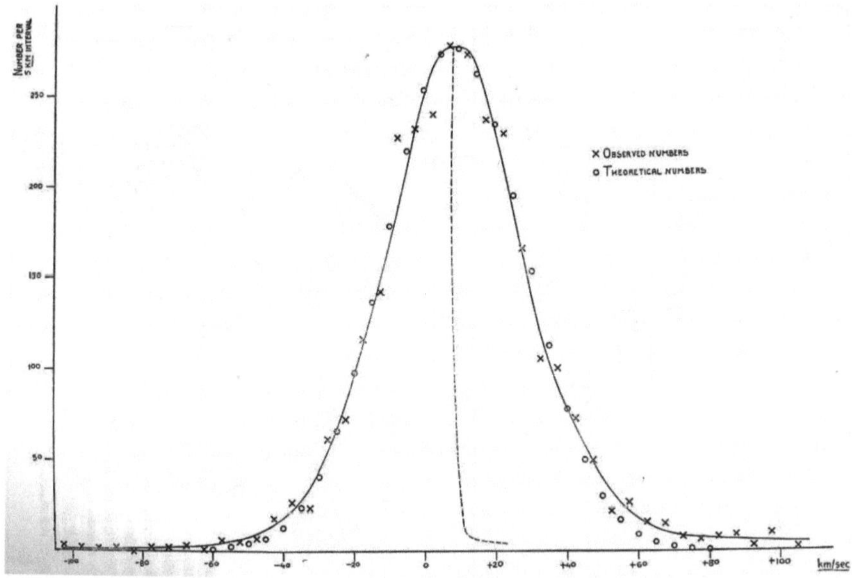

Fig. 5.8 Frequency of space velocities of stars. The crosses and the full-drawn line are the observations and the small circles indicate a theoretical Gaussian distribution. The fit is excellent, except for the excess due to the high-velocity stars on the right. From Oort (1926b)

balance. To be precise, this is often stated as the 'virial theorem', which says that for a system in equilibrium (averaged over time), twice the kinetic energy minus the potential energy equals zero. When all stars at a certain radius in the Galaxy would move in circular orbits they would have no relative motion; their velocity dispersion would be zero. But in reality the orbits are slightly different from circular. Then the stars do have random relative motions and there is therefore also kinetic energy in these random motions, the amount depending on the magnitude of the velocity dispersion. That means that the collective motion around the center must be less than that expected from circular orbits and that collections of stars with a certain average random motions or velocity dispersion therefore must systematically lag behind in Galactic rotation. The larger the velocity dispersion, the larger the lag. This gives rise to asymmetries in the distribution of the stellar velocities in space.

Strömberg had also studied the asymmetry in the velocities of the high-velocity stars (first noted by Walter Adams and Alfred Joy as mentioned above) and had published his results, which were quite similar to Oort's, in the same year, 1922. And Strömberg had followed this up with a paper in 1924 [144, 145], in which he showed that 'a very definite relation exists between group-motion and internal motions for different classes of stars'. This is what came to be called asymmetric drift. He believed that the 'asymmetries' were a fundamental property of space. The coordinate system of the spiral nebulae was probably that of absolute space and our Stellar System would move with a velocity of some 300 km/s with respect to that. The

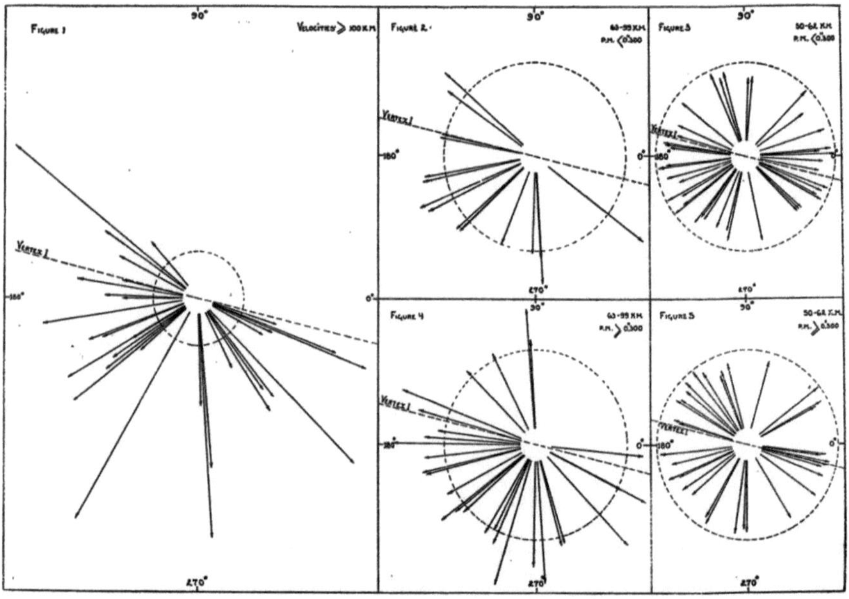

Fig. 5.9 Distribution of radial (not total) velocities of stars in Galactic longitude. On the left the highest velocities (in excess of 100 km/s) and on the right 63–99 and 52–62 km/s respectively (on top the nearer stars, parallax larger than 0.3 arcsec, and at the bottom the more distant ones). Below 63 km/s the distribution covers all directions. From Oort (1926b)

globular clusters and the high-velocity stars would move with much lower speeds with respect to that absolute system. And the asymmetry would persist to lower and lower velocities and therefore Strömberg maintained that there was an asymmetry in the motions of stars even of low velocity. Oort disagreed and felt that there was a clear and abrupt dichotomy, the high-velocity stars having a distinct asymmetry, but below the sharp cut at 63 km/s motions would be perfectly symmetrical. In his thesis Oort went through a very detailed analysis to show that the sudden change at 63 km/s was a distinct transition from symmetry to asymmetry. Much later Strömberg has turned out to be correct. Figure 5.9 shows Oort's analysis, using only measured radial velocities. At 63 km/s the distribution changes abruptly, from random below that value to very restricted above it. Oort felt that this was an important issue and—even though his thesis was also published in the series of the Kapteyn Astronomical Laboratory—published a separate paper about it as well (Oort 1926a).

There is an interesting loose note in the Oort Archives in the file with material related to his thesis research. It reads:

> 'I think that there is little or no significance in interpreting a phenomenon by means of a mathematical formula, as Strömberg does, and then beginning to speculate around this formula, without being able to give even a rough numerical test with any physical theory. It appears to me that a precise investigation of all details would be a great deal more

likely than a premature, and un-understandable generalization could be. In addition he has made mistakes: e.g. asymmetry in lower velocities and ignoring planetary nebulae with retrograde velocities >100 km/s.'

Oort is true here to Kapteyn's example 'to tie interpretations directly to observations, and to be extremely wary of hypotheses and speculations' (in *AnnRev81* [3]). The 'formula' Strömberg started out with concerns the frequency distribution of velocities (an assumed product of two Gaussians), representing a Schwarzschild ellipsoidal distribution of motions—rejecting Kapteyn's interpretation of two streams; I will come back to this issue in Sect. 6.2.

The second major issue, next to the asymmetries, was that of the velocity of escape from the Kapteyn system. As noted above, Oort in his 1922 paper had proposed that the sharp limit of (in that paper) 66 km/s corresponded to this escape velocity, and that this would mean that the average mass of stars in the system would be smaller than what Kapteyn had inferred (who had estimated the velocity of escape at 100 km/s). So, Oort in this thesis first proposed again that the high-velocity stars were 'foreigners' intruding into the Kapteyn system and set out to derive a better value for the escape velocity. For this he again looked at the vertical (perpendicular to the Galactic plane) direction near the Sun. Kapteyn had treated this as if the stars could be compared to a gas in isothermal equilibrium; isothermal means that the temperature, here in the form of the average motion or velocity dispersion, is everywhere the same. Arthur Eddington and in particular James Jeans had developed the fundamentals of stellar dynamics that could be used for a more formally correct treatment.

Indeed, Oort had carefully studied these developments. We saw that in *AnnRev81* [3], Oort mentioned 'the inspiring books by Eddington, *Stellar Movements and the Structure of the Universe* (1914), and by Jeans, *Problems of Cosmogony and Stellar Dynamics* (1919), both of which fundamentally influenced my education as an astronomer.' (Sect. 3.2). In particular James Jeans (see Fig. 5.11) had developed the fundamentals of stellar dynamics (his chapter 'X. The evolution of star clusters' has all the basic formulae) and he even had applied his approach in a paper, *The motions of stars in a Kapteyn Universe* in 1922, [146]. Oort used what came to be called one of the Jeans equations, relating in a flat system the velocity dispersion of the stars in the vertical direction to the gravitational field for a system in equilibrium.

Since both the stellar distribution and the velocity dispersion are known, it then is possible to calculate the forces (much like Kapteyn had done in his famous 1922 paper, but now through improved procedures). Using these, Oort estimated the escape velocity from the Kapteyn Universe in the vertical direction to be 167 km/s. For this he had to extrapolate Kapteyn's stellar distribution beyond Kapteyn's last shell (about 2 kpc; see Fig. 3.2). For illustrative purposes he repeated the calculation in case the density immediately dropped to zero beyond this last shell and found an escape velocity of 104 km/s. He was forced to conclude that 'the real escape velocity is larger than the 63 km-limit of the high-velocity stars'.

The second part of the thesis contained a detailed and careful collection and discussion of the available velocities of stars of various types (spiral nebulae turned out to give very conflicting results and he does not pursue this in any detail). In the end

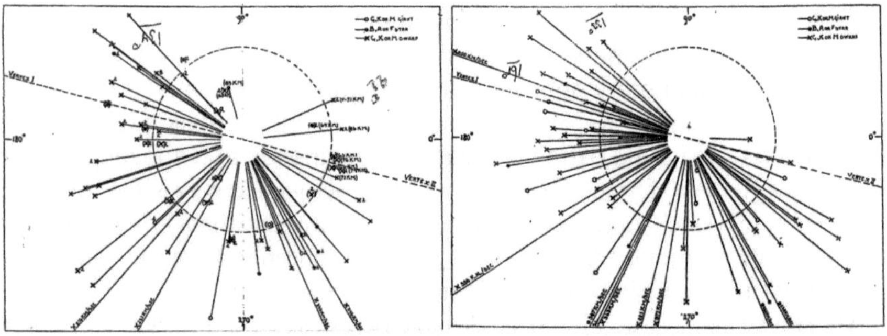

Fig. 5.10 Two examples of figures from Oort's thesis that display the distribution of high velocities. The left-hand panel has a 'representative sample of near-by stars' and the one on the right the other stars, except some brighter ones. Contrary to his earlier paper, here the length of the arrows corresponds to the *projection* of the total space velocity onto the Galactic plane. From Oort (1926b)

Oort presented a catalog of 233 high-velocity stars (and a supplementary 37 doubtful cases). The thesis has plots of the distribution of the velocities of these stars that are interesting to reproduce. Figure 5.10 shows velocities projected onto the Galactic plane. Oort believed that there is some evidence for two maxima in the distributions, as if 'we are penetrating two such [with high internal velocities] systems (moving towards 170° and 310°)'. In broad terms this would be towards and away from the Galactic center as we know it now. Oort did not perform a statistical analysis on the significance of this. He also plotted the Galactic latitudes of the apices (the directions the stars seem to come from) and interpreted this as 'possibly a preference to two narrow bands on each side of the Milky Way' (Fig. 5.11).

In a final discussion, Oort considered the possibility that the high-velocity stars are part of 'the larger Galactic system', outlined by the globular clusters, that penetrate our local space. Since they move with respect to the globular cluster system, the system of high-velocity stars must move with respect to this larger system. He expressed the view that 'maybe a velocity of escape of 63 km/s can still be maintained', which would not be the escape velocity from the 'entire Galactic system', but from the local system only. But how this could be accomplished was not made clear. He did, however, mention the possibility that the high-velocity stars are part of a larger system, but that would have to be rotating with respect to the globular clusters (and RR Lyrae stars). Finally, Oort noted that if the island Universe theory were correct, there must be a outer ring of absorbing matter to block the view of galaxies close to the Galactic equator. However, this all did not really lead to a consistent picture.

The final section of the thesis gives a discussion of what the best values are for the velocity of the Sun with respect to the common stars and what the average peculiar velocity is of those stars. These are 'constants' needed with high accuracy in statistical astronomy.

Fig. 5.11 Sir James Jeans as he appeared on the cover of the British (satirical) magazine Punch of February 27, 1935. From the Oughtred Society, which is dedicated to the preservation of the slide rule, see [147]

Like every Dutch Ph.D. thesis, it was accompanied by a set of propositions ('stellingen') not related to the topic of the thesis, that the candidate would be expected to be able to defend as well. Oort had fourteen of these. The full set can be seen in English translation in Appendix B.2. Examples are: 'VIII. Shapley's opinion that it should be possible to use apparent diameters of open clusters to deduce a system of relative distances can be contested', or 'XI. Eddington's claim that equipartition of energy is a third order effect compared to the general exchange of speed between two encountering stars, is incorrect'. It shows his broad understanding and thinking about the problems in the field. In his Archives there are detailed notes on how to defend these statements. For example, the background behind Proposition VIII on the diameter of open clusters (which *is* the way Trumpler actually proved that interstellar absorption existed) is his conclusion that the intrinsic variation of diameters among these clusters appeared to be too large. I will not discuss these in more detail.

To conclude this section, in Figs. 5.12 and 5.13 Oort and his guests are shown at the dinner after he obtained his doctor's degree. Note the formal dress that was common then (it is common now only for the person that defends the thesis). In the second of these I collect the members of his family and those of his future wife, who is sitting next to him. How he met her will be the subject of the next section.

Fig. 5.12 Oort, his future wife, family and guests at the dinner on the occasion of Oort's thesis defense on May 1, 1926. The location has not been recorded. Most guest must have been friends from Oort's student days. The family members have been collected in Fig. 5.13. On the extreme right are Willem de Sitter (sitting) and Pieter van Rhijn (standing). From the Oort Archives

5.3 Johanna Maria Graadt van Roggen

Oort kept up good friendships with quite a few persons, as his pocket diaries show. Two names of male friends that turn up frequently over a long period of time are those of 'Chris' and 'Willy' and I will explain first who these were. Chris is Christiaan Hooykaas (1902–1979), who was both a first and a second cousin. Oort's grandfather Henricus Oort had worked with Isaäc Hooykaas on Bible translations and they were brothers-in-law, since Hooykaas had married Henricus's sister Petronella (see Sect. 1.2). These had a son, also called Isaäc Hooykaas (1866–1944), who, following the family tradition, was a clergyman. He married Eva Maria Faber (1878–1949), sister of Ruth Faber, Oort's mother. Now Christiaan Hooykaas was a son of Isaäc and Eva and therefore had a double family line with Oort both through his father and his mother. Christiaan studied Indonesian Languages and Literature in Leiden and obtained a Ph.D. in 1929. He then moved to the Dutch East Indies as teacher and various other professions. His wife Jacoba Hindrika van Leeuwen Boomkamp (1901–1965), whose father again was a preacher, was a writer, publishing a children's book in 1941. She is best remembered for her diary notes from the Japanese

Fig. 5.13 Oort's family in the photograph at his thesis defense dinner in Fig. 5.12. At the top Oort and his future wife, and at the left his sister Emilie. At the bottom from left to right Oort's brother John, sister Jetske, father, future brother-in-law Coenraad, future father-in-law (sitting), mother and future mother-in-law

prison camp, where she was detained in World War II. Christiaan Hooykaas was set to work in the infamous Burma Railroad. They both survived and eventually moved to England in 1951 when Christiaan was appointed Reader of Old Javanese at the University of London. But that was far in the future; he and Oort were students at about the same time and their family relation strengthened their ties.

'Willy' is less straightforward to trace. Bram Oort, the younger son Oort, thinks he most likely was physicist Wilhelm (Willie) Gerard Burgers (1897–1988). He and his brother Johannes Martinus (Jan) Burgers (1895–1981) were students in Leiden under Ehrenfest and both ended up as Professors at the Technical University Delft. Usually the younger of the two is referred to as 'Willie', which Oort invariably wrote as 'Willy', at least when they were young. We will see soon though, that Willy indeed is the Wilhelm Burgers in question. It is quite possible that they knew each other from the Burger's student days in Leiden, although he was three years older than

Oort. He might have been a good friend with Oort's brother Hein, who was more of the same age, but then after Hein's death Burgers and Oort might have become good friends.

Although Oort had been good or even excellent friendly relations with females, and actually kept up some correspondence—at least for some time—with young ladies he knew from his days in the USA or Groningen, up to 1925 there seemed not have been any serious relationships that might have lead to a marriage. In the *AIP-Interview* [4], Oort was asked how he had met his wife and he responded;

> 'At a university celebration where she was. My brother knew her first because she came from Utrecht and he was studying botany or biology in Utrecht at the time. So he formed a link of introduction.'

Indeed John knew her first; Bram Oort has sent me a photograph taken in 1923, where she and John and at least one more person are on a sailing trip together. It is not entirely clear when this introduction happened. In Oort's pocket diary for 1925 the first entry that refers to her is on June 23: 'Meet Mieke Graadt van Roggen'. For the rest of that day he went to an 'Open air play about Alva and William the Silent. Very beautiful. I go with Emy and we have a fine time together in the rain and cold, absorbed in the play'. Emy is of course his sister Emilie. The play is about the Dutch revolt against the Spanish; Fernando Álvarez de Toledo y Pimentel (1507–1582), the 3rd Duke of Alba, headed the Spanish army on behalf of King Philip II of Spain (1527–1598) against the revolt lead by William of Orange. It is possible that this was performed in the framework of the 350th anniversary of Leiden University that year. For the next day it was his other sister's (Jetske) birthday and at night he went to an 'Ehrenfest colloquium with Babcock and van Maanen'. Apparently both were visiting Leiden. On Thursday, June 25th he had a

> '... busy day with Babcock and with preparations for the Sempre Ball to which I go with Mieke who is very lovely. The concert is beautiful, Birgitt giving some of her best songs and all people in evening dress and with flowers. We dance till 5.30 and I do not go to bed.'

And on Saturday June 27:

> 'In the evening Mieke, Netty Gorter, Jep and Emy, Cor v.d. Stadt, Jo and I go out on a bicycle trip to Katwijk. Dance a few minutes in Het Zwaantje'

And on Sunday June 28:

> 'Netty, Jetske, Emy, Mieke, Jo and I go sailing. Bring Mieke to the train.'

Jep, Emy and Jo are the nicknames of Oort's sisters and brother. Het Zwaantje (the little Swan) is the local name for bar/restaurant De Zwaan overlooking the sea from the esplanade; it goes back to an inn in 1602. The Sempre Ball (or Gala) was (and still is) the annual major dancing party after the concert of the Sempre Crescendo student orchestra of the of the Leiden Student Association Minerva. This was not really '*meeting* at a university celebration', but brother John may have introduced her to Oort for him to invite to the Sempre Ball. The pocket diary shows no indication of a celebration in Utrecht in the weeks before or at that time (Fig. 5.14).

Fig. 5.14 Johanna Maria (Mieke) Graadt van Roggen with her father in 1907 (left), and with her brother Coenraad Jan (Coen) in 1910 (right). From the Oort Archives

The pocket diary continues with notes on work, much horseback riding, a trip to London in July, on to Cambridge, where the General Assembly of the International Astronomical Union was held, a visit to the Royal Greenwich Observatory and back to Leiden (July 27). In August there was a trip to Hilversum, where he also visited Kapteyn's widow and back in Leiden talked to the Schlesingers. On Sunday 23, Oort mentioned that 'Mieke G.v.R.' and her cousin (or maybe nephew, as that is the same word in Dutch) have come from Den Haag by bike and with a large group to go on a picknick and a sailing trip. Oort noted: 'I like Mieke very much at some times, at others we seem to differ so widely that I am almost afraid of her.'

On August 26, Oort traveled to Utrecht to see his brother (but not Mieke) and on to Arnhem where the Burgers family lived. He was glad to see 'Willy' again. After that there is a trip with John, Willy and Chris to Switzerland and the Alps. On October 3, brother John and friends (among which 'Mieke') came to Leiden for the annual celebration of the 'Relief of Leiden', the breaking of the Spanish Siege of Leiden, on that date in 1574. The pocket diary has a small 'memo' section on each right-hand page and for the week ending October 17. Oort wrote there:

> 'I work much and the division of my interest and my doubts over two matters[1)] make me very tired.
> [1)]Mieke and my dissertation.'

In his diary notebooks, there is an entry for September 22, 1925, in which Oort expressed some longing to a longterm relationship and his fear that he had become less idealistic and more earthly-minded. The next entry is for October 26, when he visited the Graadt van Roggens in Utrecht and slept overnight in his brother's quarters. Three days later he wrote 'I really believe I am starting to fall in love with Mieke'.

Johanna Maria Graadt van Roggen had been born in Nijmegen on 7 January, 1906, so she was nineteen years of age when she and Oort (then twenty-five) met. Among family and friends she was known as 'Mieke', derived from her second name, but in this book I refer to her as Mrs. Oort. Her father was Willem Graadt van Roggen (1879–1945) and her mother Jeanne Laurense Hélène Henriëtte van de Water (1881–1974). She had an older brother Coenraad Jan, who had been born in 1904.

Willem Graadt van Roggen studied law in Groningen, but had to interrupt his studies when he fell ill and subsequently developed tuberculosis. During his long years of convalescence, when he had moved back to his birthplace Nijmegen, he wrote poems (as a teenager he already had published some poetry), which were published as a collection *Het schouwende leven* (the inspective life) in 1903 and a novel *'Tubeculeus'* (1904) about life of tuberculosis patients in a sanatorium. He did also get involved in editing work for literary journals. He recovered after a very long time and eventually obtained the Candidaats Degree (Bachelor) in 1911.

Graadt van Roggen and his wife, Hélène van de Water, got married in 1903. She was the daughter of a cigar producer and merchant, Coenraad Jan van de Water (1824–<1916) in Zaltbommel, about 30 km south of Utrecht on the river Waal, the main distributary branch of the Rhine. She had wanted to become a singer, but instead chose to be a housewife and mother (years later Oort's daughter Marijke would become a teacher of opera singers, especially in foreign languages, much to her grandmother's delight).

Graadt van Roggen discovered after a while that he could not really make a living from writing novels and poetry and moved into journalism, becoming without any training or formal qualifications editor-in-chief of a newspaper in Breda in the south of the country, and after some years he was asked to take that role in the 'Leeuwarder Courant', the major newspaper of Frisia, in 1910. He filled that position for a short time and the next year he became editor-in-chief of the 'Utrechtsch Provinciaal en Stedelijk Dagblad' (the provincial and municipal newspaper for both the province and the city of Utrecht). Under his leadership this turned into a successful newspaper with a liberal affiliation, and that is when he and his family moved to Utrecht. He became very active in local politics and social life. In 1916 he became involved in an initiative to organize the first *Jaarbeurs* (Annual Fair) in the Netherlands and of course his newspaper gave much attention to this. It was held in 1917 and became a major success with 690 companies participating. As a result Graadt van Roggen was asked to become Secretary General of the institution that organized these annual fairs and in 1918 he gave up his editorship of the newspaper. Instead he started a magazine 'De Jaarbeurs' and promoted commerce and industry, especially in Utrecht. But he remained interested in many other things as well. It has been recorded that in 1917 he made a trip along the Italian and Serbian battlefronts on invitation by the Austrian Government (the Netherlands was neutral during World War I), of which he published reports in the Utrecht newspaper. In 1921 he traveled by boat to Finland that had just gained independence from Russia. In 1923 he had become member of the Utrecht municipal council in which he was very much involved with matters of trade and industry. He regularly appeared on the radio trying to convince leaders in trade and industry to come to his Jaarbeurs.

After October 1925, things moved fast between Oort and his future wife. Johanna Maria Graadt van Roggen stayed with the Oorts in Oegstgeest, in late December 1925. She had been attending the HBS for girls, but did not continue with academic studies (as was still not custom for girls in these days, even when they were qualified). Instead she went to the 'huishoudschool', the school for housekeeping, which was a school to prepare young girls for the role of house wife or domestic servant. This school, also sometimes referred as 'cooking school', not only taught girls cooking, housekeeping and nutrition, but also sewing, domestics crafts and recording household expenses. She must have been much too intelligent for this.

In the *AIP-Interview* [4] Oort replied to the question whether she was interested in a career:

> 'No. She was much interested in poetry and art in general and still is. She never imagined that she would get a career in that line. Her ideal was more to build up a happy household and a harmonious life.'

Their younger son Abraham Oort informs me as follows (my translation):

> 'According to her she found my uncle Jo attractive, but she was more impressed with Jan. She instantly 'fell in love' with him. I think she felt more attracted to my father's more quiet and introvert character. This in contrast to the family in which she grew up with extravert, intelligent parents. In the 'Graadt van Roggen' family in Utrecht there was little piece and quiet and much argumentation among each other – she was not too good at that and therefore felt less 'at home'. But in spite of that she really loved her father and older brother Coen. But between her and her mother there was more tension.'

And from another set of remarks (original in English):

> 'She loved her poetic father but the relationship with her mother was not intimate. She must have been intrigued meeting my father, Jan Oort, a quiet and self-reliant man. The introvert Oort family was the opposite of the extravert Graadt van Roggen family. My parental grandparents were teetotalers while my maternal grandparents enjoyed a glass of wine.'

Oort himself was not a teetotaler, even though indeed his parents were (see Sect. 1.3). Some letters by Johanna Maria are kept in the private archives, dating back to November 5, 1925, when they had met not long before. But by January, 1926 they clearly had a relationship. On February 12, 1926 (according to the postmark on the envelop) she wrote that since he had left many more flowers have been delivered and she lists the persons these came from (among them the van Rhijns). And on the 16th she wanted to know if he had not lost his ring. It seems that they had formally become engaged to marry days before that. In the Netherlands, engagement rings were exchanged by the partners. These rings would eventually become wedding rings, but until that time were worn on the left hand and later as wedding ring

Fig. 5.15 Oort and his future wife, probably on the occasion of their engagement to marry in February 1926. From the Oort Archives

on the right one (however, Catholics did things the other way around). An engagement usually was accompanied by some reception or party, but since flowers were delivered later the engagement was probably celebrated privately and friends and acquaintances notified later. Figure 5.15 shows the couple when engaged (Fig. 5.16).

The wedding took place on Tuesday May 24, 1927 in Utrecht, followed by a reception. Another reception at the Oort's home Rhijngeest in Oegstgeest had taken place before the wedding on Saturday May 14. The wedding announcement mentioned that their future address in Leiden would be Plantsoen 35, about a 15 min walk from the Observatory. Figures. 5.17 and 5.18 show the official marriage certificate. The upper part has all the formalities, including a statement that their their birth certificates have been provided and another one that the parents had given their permission, which at that time was necessary for persons under age thirty. The lower

Fig. 5.16 Oort family in the summer of 1926. Sitting from left to right father Abraham Oort, Oort and his fiancée, mother Ruth Oort-Faber, brother John and his fiancée Titia (Pivy) van Mesdag. Standing sisters Emilie and Jetske Oort. From the collection of Abraham H. Oort

part has the signatures. On the right from top to bottom the signatures of Oort and the new Mrs. Oort-Graadt van Roggen, the parents Mr. Oort and Mrs. Oort-Faber and Mr. Graadt van Roggen. At the top left the witnesses, which were the groom's brother John Oort and the bride's brother Coen Graadt van Roggen, followed by that of the civil servant conducting the ceremony and Mrs. Graadt van Roggen-van de Water at the bottom. The newly wed couple is shown in Fig. 5.19.

5.4 Discovery of Galactic Rotation

Oort had been appointed Conservator to the Leiden Observatory. In 1926 he was also became 'privaat docent' at University of Leiden. This was an unpaid position at the University to teach courses in one's field. One therefore needed an other source of income, which in Oort's case of course was his position at the Observatory. He accepted this appointment with a public lecture ('openbare les') on November 17, 1926 (Oort 1927a). It had the title *Non-light-emitting matter in the Stellar System.* It was published in Dutch in 'Hemel and Dampkring' ('Sky and Atmosphere'), the periodical of the Dutch amateur astronomers and meteorologists. I have earlier provided an English translation as Appendix A. in the *Legacy* [6]. Here is a summary with some excerpts.

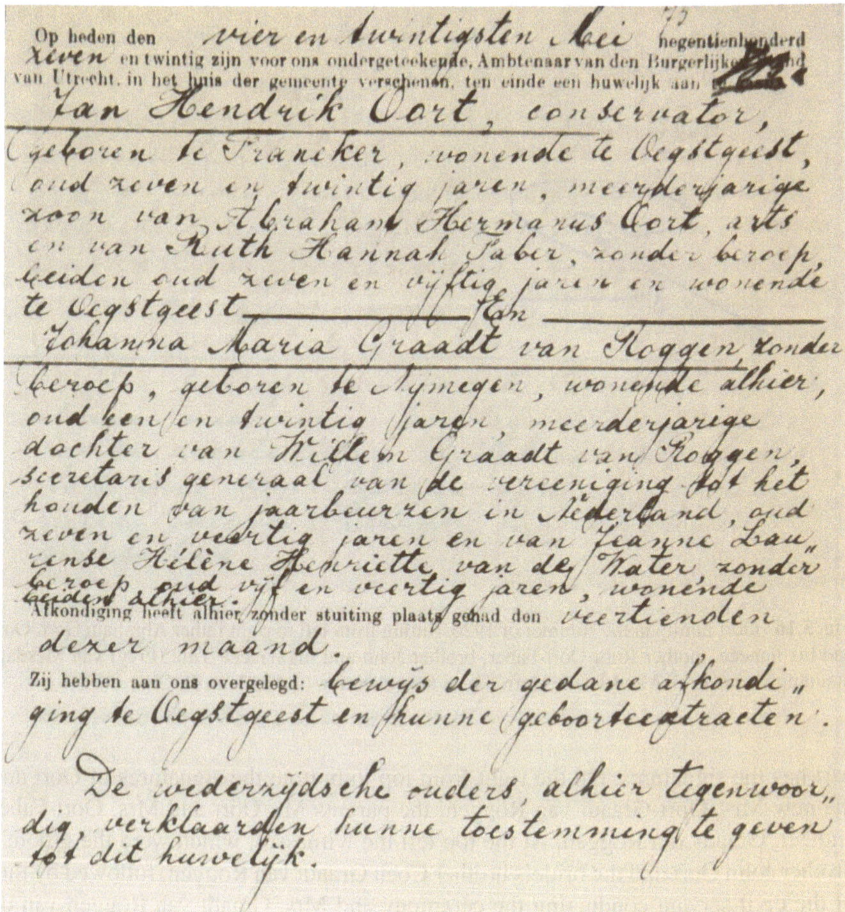

Op heden den vier en twintigsten Mei negentienhonderd zeven en twintig zijn voor ons ondergeteekende, Ambtenaar van den Burgerlijken Stand van Utrecht, in het huis der gemeente verschenen, ten einde een huwelijk aan te gaan:

Jan Hendrik Oort, conservator, geboren te Franeker, wonende te Oegstgeest, oud zeven en twintig jaren, meerderjarige zoon van Abraham Hermanus Oort, arts en van Ruth Hannah Faber, zonder beroep, beiden oud zeven en vijftig jaren en wonende te Oegstgeest — en

Johanna Maria Graadt van Roggen zonder beroep, geboren te Nijmegen, wonende alhier, oud een en twintig jaren, meerderjarige dochter van Willem Graadt van Roggen, secretaris generaal van de vereeniging tot het houden van jaarbeurzen in Nederland, oud zeven en veertig jaren en van Jeanne Laurense Hélène Henriette van de Water, zonder beroep, oud vijf en veertig jaren, wonende beiden alhier.

Afkondiging heeft alhier zonder stuiting plaats gehad den veertienden dezer maand.

Zij hebben aan ons overgelegd: bewijs der gedane afkondiging te Oegstgeest en hunne geboorteextracten.

De wederzijdsche ouders, alhier tegenwoordig, verklaarden hunne toestemming te geven tot dit huwelijk.

Fig. 5.17 Upper part of the marriage certificate of Jan Hendrik Oort and Johanna Maria Graadt van Roggen on May 24, 1927. For details see [148]

Oort started out by describing the researches of William Herschel up to the Kapteyn Universe and Kapteyn's early work on dynamics: 'In 1922, Kapteyn crowned his work with an investigation into the equilibrium of this Stellar System'. He then discussed how the calculation of forces can be used to estimate the amount of 'not directly observable mass in the Galactic System'. Next he turned to the work of Shapley and the displacement of his system of globular clusters from the Kapteyn System. It is possible, although not likely, that he showed lantern slides, which then might have been his drawings in Fig. 3.4. This led him to the matter of interstellar absorption, quoting Kapteyn's early determination of it and Shapley's research in globular clusters.

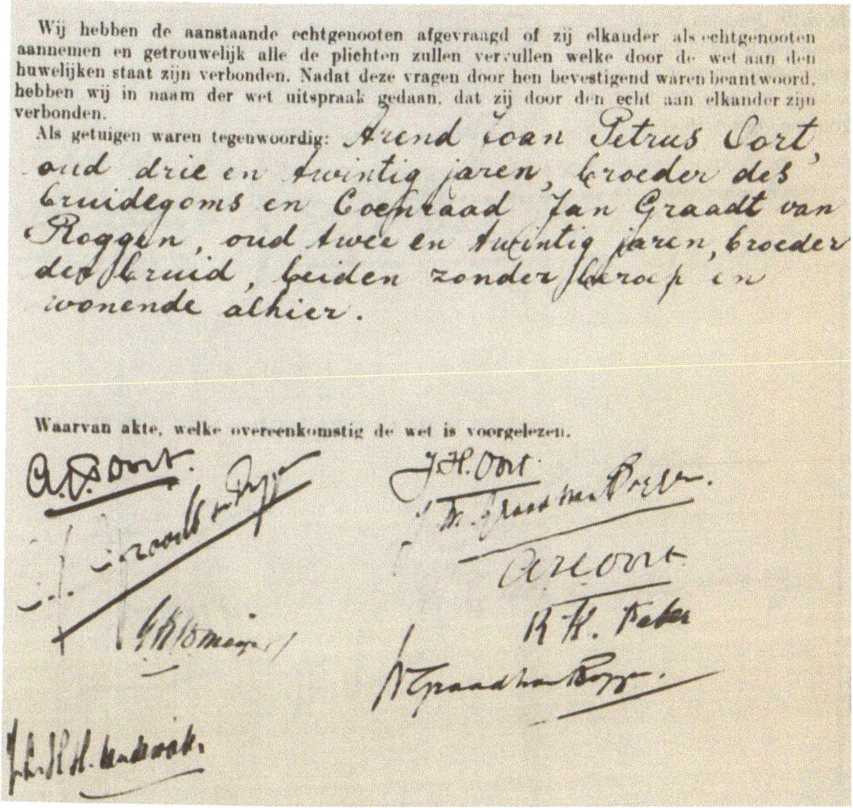

Wij hebben de aanstaande echtgenooten afgevraagd of zij elkander als echtgenooten aannemen en getrouwelijk alle de plichten zullen vervullen welke door de wet aan den huwelijken staat zijn verbonden. Nadat deze vragen door hen bevestigend waren beantwoord, hebben wij in naam der wet uitspraak gedaan, dat zij door den echt aan elkander zijn verbonden.

Als getuigen waren tegenwoordig: Arend Joan Petrus Oort, oud drie en twintig jaren, broeder des bruidegoms en Coenraad Jan Graadt van Roggen, oud twee en twintig jaren, broeder der bruid, beiden zonder beroep en wonende alhier.

Waarvan akte, welke overeenkomstig de wet is voorgelezen.

Fig. 5.18 Lower part of the marriage certificate

'In all studies about the extent of the Stellar System it has been assumed that no significant absorption of light in space takes place. Through the years the confidence in the global correctness of this assumption has become so widespread, that it may not be uninteresting to go back to the times when one worried seriously about this and to remind oneself of how fragile the foundations of our current belief are. When the first results about the finite dimensions of our Stellar System became known, it was natural to ask the question whether this finiteness of the Stellar System was not coming about because dark bodies were floating between the stars, or extended masses of dust or gas, that were obscuring the light of distant stars. These masses would then have such an absorbing capability that all stars beyond the Stellar System described above would be hidden from our view. The possibility was considered that the Stellar System would in reality extend significantly beyond this disk-like cloud and that the observed increasing scarcity of stars results from more and more of them being hidden by an absorbing

Fig. 5.19 Johanna Maria Graadt van Roggen and Jan Hendrik Oort on their wedding day, May 24, 1927. From the Oort Archives

medium. One may object that then the distribution of this material would necessarily have to be very improbable in order to explain the flattened form of the observed cloud of stars. There are good reasons to assume that the determined *thickness* of the disk, that is the size of the cloud of stars perpendicular to the plane of the Milky Way, is roughly correct. But one doesn't have very convincing arguments to defend the dimensions *in* the plane of the Milky Way. It was found that the sun is not very far from the centre. Now, throughout the history of astronomy a general fear has been ingrained against any claim for such a splendid position. In the course of time, the place where humans live has stepped down from a centre around which the whole world turned in seven crystal spheres to our current Earth that itself must circle around a Sun that has three-hundred thousand times more mass. And together with this Sun it must circle in an even more massive system. – When it was noted that the Sun, and therefore us, were given a more or less central position in the Galactic System, some astronomers became suspicious. I would like to quote *Kapteyn* on this:

I showed that the data used in this investigation are about as well represented if we assume that the star-density is the same at different distances from the sun, provided that we introduce an absorption of light of a determined small amount.'

Oort has stated (e.g. his *AnnRev81* [3]) that he did not recall ever to have believed anything else but that galaxies are systems outside and similar to the Sidereal System. Having explained that absorption (or rather scattering) would be selective and more effective in the blue than in the red and that stars therefore would appear redder with increasing distance, Oort noted that

'The results of studies of this, already started by Kapteyn in 1908, in the end proved negative. Shapley's remarks concerning such an effect in the very distant Globular Clusters, and those of Lundmark with respect to the spiral nebulae have shown that the selective scattering of light is so small, if it exists at all, that it can safely be neglected.'

Knut Emil Lundmark (1889–1958) was a Swedish astronomer, who worked at the time at Mount Wilson Observatory. Oort explained that the absence of selective scattering of light does not necessarily mean that starlight remains unobstructed and Oort quoted Eddington for further explanation of this. If gas is ionized,

'the free electrons in such an ionized gas have a much higher capability to scatter light than the molecules in a non-ionized gas and furthermore, these electron have no preference for blue light. All colours are being scattered equally, so that a priori we can expect no colour change for distant stars from these gasses, even if a substantial part of the light is being scattered.'

Oort next looked into the possibility that dark absorbing matter would exist, and that it is in the form of 'small solid particles (such as grains of sand, for example), distributed throughout the Galactic System'. He turned attention to the Milky Way and the dark areas that can be seen in it.

'The opinion that the dark regions are in reality holes in the Milky Way through which one views the empty space beyond, is now rejected by most astronomers. The coincidence of the bright nebulae with the dark areas has led to the conviction that most dark fields originate from an absorbing nebula, which hides the stars in the Galactic System behind it.'

The distribution of globular clusters has a minimum in the Milky Way plane, for which Oort felt the most likely explanation was that there is absorption in a 'relatively narrow ring around the Milky Way'. To support that view, he quoted the fact that for the small, faint nebulae that as a class of objects are known as spiral nebulae, there is a similar Zone of Avoidance. And only the year before Hubble had resolved stars in the Andromeda Nebula (see Fig. 7.9). If, on this basis (Fig. 5.20).

> 'the spiral nebulae can be considered as large stellar systems far outside ours, and if the hypothesis that all objects farther removed from us than 30,000 light-years and seen in the directions of the Milky Way are invisible to us, then it is to be expected that no spiral nebulae should be observed in the Milky Way. This has been found to be correct.'

The lecture concluded as follows (Fig. 5.21):

> 'Do we then have to imagine that the absorption of spiral nebulae is caused by an enormous, opaque Galactic ring outside the system of globular clusters, or should we look for the cause in a very weakly absorbing medium that fills space around us out to very large distances beyond the globular clusters? The possibility exists that a completely different explanation will be found, but for the moment that of an absorption of light is the least contrived.'

The picture remained unclear, unsatisfactory and confused. Strömbergs idea of the local collection of stars moving with respect to the greater system was not a bad idea, but still missed the point. But Oort is ready to challenge the Kapteyn Universe, suggesting that the fundamental starting point of no appreciable absorption has to be abandoned. Furthermore, this is an illustration of a characteristic that Oort showed throughout his career: a remarkable intuition in what direction has to be taken to make progress. His early choice to work on high-velocity stars without knowing what fundamental change in the view of the structure of the Sidereal System it would cause, constitutes another, first indication of this quality that served him throughout his career.

The breakthrough came with a set of papers in 1925 and 1926 by the Swedish astronomer Bertil Lindblad (1895–1965). Lindblad had obtained a Ph.D. at Uppsala and still worked at the time as a 'docent' at that university. In 1927 he was to become Director of Stockholm Observatory. Some of the crucial papers were published in a Swedish journal [149]. Lindblad first noted that the direction of Kapteyn's star streams was almost exactly perpendicular to that of Strömberg's asymmetric drift (see Sect. 5.2). I quote from the first paper in [149]:

> 'We assume that the Stellar System may be divided up into a series of sub-systems having rotational symmetry around one and the same axis, with different speed of rotation at the same distance from this axis and consequently having different degrees of flattening. The inner most flattened systems have a high star-density, though decreasing with decreasing speed of rotation; the systems with the highest speed of rotation are assumed to form the Milky Way clouds. In the extreme outer systems the space-density of the individuals, stars or globular clusters, is relatively low. The latter sub-systems show, on account of their low speed of rotation a strong asymmetric drift in velocity at right angles to the radius vector of the big system, when the velocities of their members are measured from a star, like our sun, moving as a member of the Milky Way cloud.'

Fig. 5.20 Oort and his wife at their wedding surrounded by their families. Sitting in the front are Jetske (left) and Emilie Oort. In the background from left to right: Engelina Graadt van Roggen-Kan and her husband Coen, Mr. and Mrs. Oort, Mrs. and Mr. Graadt van Roggen, John Oort and his fiancée Pivy van Mesdag. From the Oort Archives

Fig. 5.21 Two astronomers that played an important role in Oort's early work on the Galaxy. Left Harlow Shapley and right Bertil Lindblad at the IAU General Assembly at Harvard in 1932. From the Oort Archives

Lindblad noted the important point, that the direction of the center of the greater system, inferred from this, would coincide with that of Shapley's system of globular clusters. Lindblad proceeded to apply Jeans' theory of stellar dynamics to such a system and using the information available ended up with a greater system containing 180.10^9 ('180 billion') $M_\odot$. The density near the Sun is one solar mass per 7 cubic parsecs. The Milky Way cloud, of which the Sun is part, rotates at 350 km/s around the center at 10,800 parsecs in 190.10^6 years. These values are of the same order as the values of modern determinations.

But what about the star streams of Kapteyn? Lindblad wrote in the same paper:

'In the next approximation we take into account the evident fact that the inner sub-systems, the Milky Way clouds, show a very great richness in structure, suggesting considerable local accumulations of matter of the character of 'local systems'. In addition to the general field of force due to a 'smoothed' stellar system we must take into consideration the effects of local gravitational fields due to the accumulation of stars or nebulous material. In the paper referred to in the beginning [this is the second paper in [149]] I have shown that the effect of such local disturbing masses of great size would be primarily star-streaming in the Galactic plane at right angles to the line of the asymmetric drift of high velocities, and thus nearly along the radius vector of the big system. The two star-streams by Kapteyn fulfill very nearly this condition.'

So the star streams are the result of local disturbances of the gravitational field of the larger system by the local system, the Kapteyn Universe. The shortcoming of Lindblad's otherwise visionary proposal is that he still saw the Kapteyn Universe as an example of a local cloud, with its own gravitational equilibrium but otherwise separate from similar systems, that are shielded from view but are part of a grander rotating system.

This conception of a single large, rotating sidereal system of which the Kapteyn universe was only a part was the breakthrough that Oort required to see how all the pieces of the puzzle fitted in one complete picture. He has on many occasions expressed the point that he had failed to see in his thesis the truth while it was so obvious, but only came to the real insight after Lindblad's publications. He investigated the consequences and came to the insight how observations confirmed these ideas. The resulting paper, *Observational evidence confirming Lindblad's hypothesis of a rotation of the Galactic System* (Oort 1927c), was published on April 14, 1927. Only days before that, on April 8, Oort wrote the letter to Lindblad that has often been quoted (see Figs. 5.22 and 5.23). It has been put at my disposal by his son Per Olof Lindblad (b. 1927), like his father astronomer and (former) Director of Stockholm Observatory. I quote it in full.

'Dear Dr. Lindblad,

Enclosed I am sending you a paper in which you may be especially interested, as it gives an observational confirmation of your rotation hypothesis.

Not only from radial velocities discussed in this paper, but also from proper motions, of which a definitive discussion is being prepared for the printer, it seems quite probable that your explanation of the systematic motions of the stars of high velocity is the right explanation. I feel bound to express to you personally my admiration for this illuminating conception.

I must confess that I only got an insight into the real significance of your theory when reading some remarks in one of your latest papers on the subject; otherwise I would not have put forward the so little satisfactory suggestions I made in Groningen Publications N° 40 (section 15). [This is Oort's thesis and the discussion therein.] The main objection I had to your theory was the same as that which I always felt against Kapteyn's and Jeans' theory of the Galactic system, namely that it implied systems penetrating each other and rotating with respect to each other. It was considerably later that I saw that the different velocities of rotation required by your hypothesis could arise in a quite natural way. I then felt

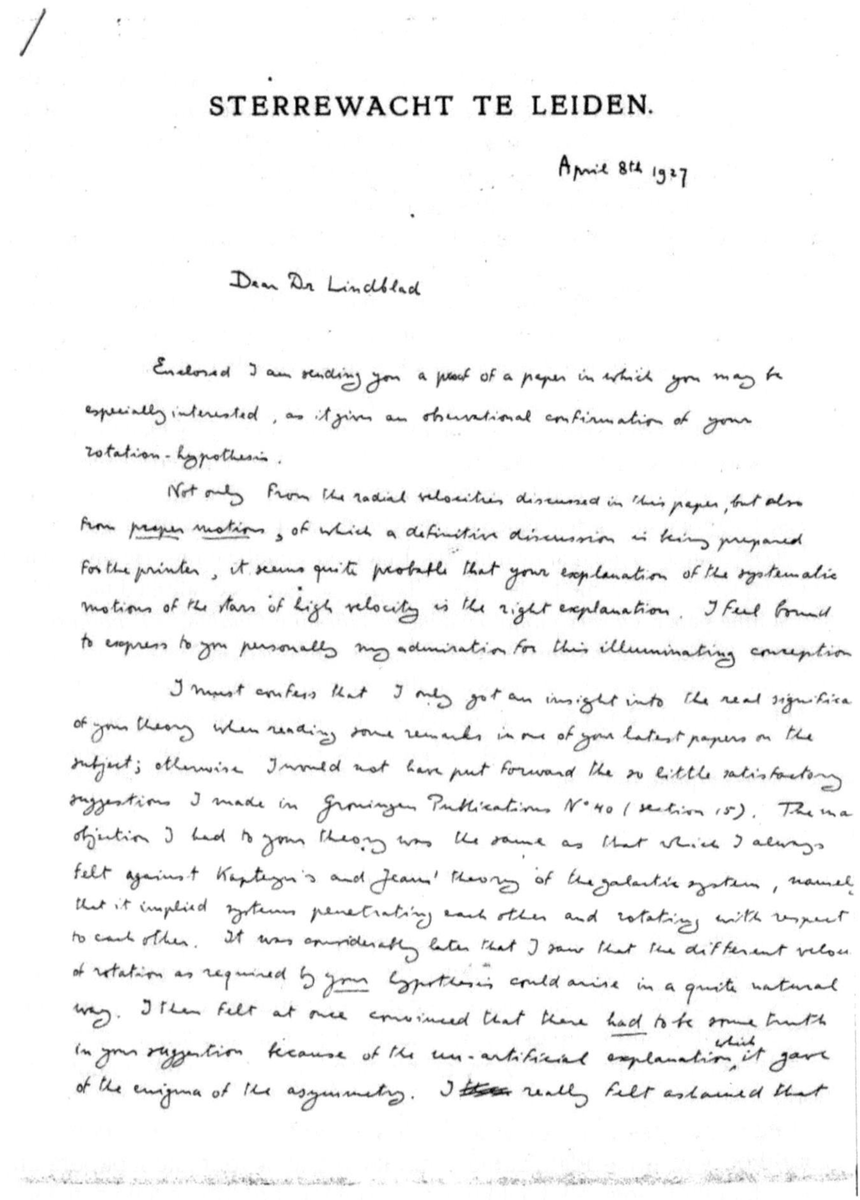

STERREWACHT TE LEIDEN.

April 8th 1927

Dear Dr Lindblad

Enclosed I am sending you a proof of a paper in which you may be especially interested, as it gives an observational confirmation of your rotation-hypothesis.

Not only from the radial velocities discussed in this paper, but also from proper motions, of which a definitive discussion is being prepared for the printer, it seems quite probable that your explanation of the systematic motions of the stars of high velocity is the right explanation. I feel bound to express to you personally my admiration for this illuminating conception

I must confess that I only got an insight into the real significa of your theory when reading some remarks in one of your latest papers on the subject; otherwise I would not have put forward the so little satisfactory suggestions I made in Groningen Publications N°40 (section 15). The ma objection I had to your theory was the same as that which I always felt against Kapteyn's and Jeans' theory of the galactic system, namel that it implied systems penetrating each other and rotating with respect to each other. It was considerably later that I saw that the different veloc of rotation as required by your hypothesis could arise in a quite natural way. I then felt at once convinced that there had to be some truth in your suggestion because of the un-artificial explanation which it gave of the enigma of the asymmetry. I ~~then~~ really felt ashamed that

Fig. 5.22 First page of a letter from Oort to Bertil Lindblad, dated April 8, 1927, on the discovery of Galactic rotation. For the second page see Fig. 5.23. Provided by Bertil Lindblad's son, Per Olof Lindblad

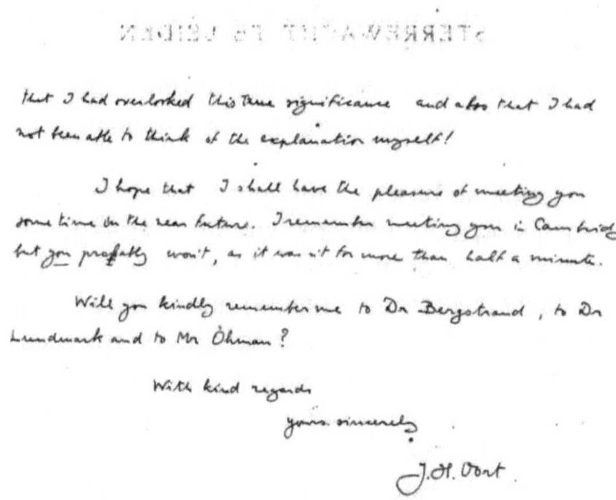
that I had overlooked this true significance and also that I had not been able to think of the explanation myself!

I hope that I shall have the pleasure of meeting you some time in the near future. I remember meeting you in Cambridge but you probably won't, as it was not for more than half a minute.

Will you kindly remember me to Dr Bergstrand, to Dr Lundmark and to Mr Öhman?

With kind regards
yours sincerely
J.H. Oort.

Fig. 5.23 Second page of a letter from Oort to Bertil Lindblad. For the first page see Fig. 5.22. Provided by Bertil Lindblad's son, Per Olof Lindblad

> at once convinced that there <u>had</u> to be some truth in your suggestion because of the non-artificial explanation which it gave of the enigma of the asymmetry. I really felt ashamed that I had overlooked this true significance and also that I had not been able to think of the explanation myself.
>
> I hope that I shall have the pleasure of meeting you some time in the near future. I remember meeting you in Cambridge but <u>you</u> probably won't, as it wasn't for more than half a minute.
>
> Will you kindly remember me to Dr Bergstrand, to Dr Lundmark and to Mr Öhman?
>
> With kind regards,
> yours sincerely,
> J.H. Oort'

Oort's handwriting was small and precise. It had been larger when he was still a student (e.g. Fig. 2.16), but by this time it had become what it remained the rest of his life. It took most people a bit of time to get used to this characteristic script, but once one had gained some experience with it, it was very well readable.

In Oort (1927c), he first described the situation with the Kapteyn Universe being much smaller than the Shapley system of globular clusters and then suggested that:

> 'As a possible way out of the difficulty we might suppose that the brighter stars around us are members of a local cloud which is moving at fairly high speed inside a larger Galactic system, of dimensions comparable to those of the globular cluster system. We must then postulate the existence of a number of similar clouds, in order to provide the gravitational potential which is sufficiently large to keep the globular clusters from dispersing into space too rapidly. The argument that we cannot observe these large masses outside the Kapteyn-system is not at all conclusive against the supposition. There are indications that

enough dark matter exists to blot out all Galactic starclouds beyond the limits of the Kapteyn-system.'

Oort refers here to his inaugural lecture as privaat docent (Oort 1927a), a publication in Dutch that few foreign readers would have at their disposal or would be able to understand.

'Lindblad has recently put forward an extremely suggestive hypothesis, giving a beautiful explanation of the general character of the systematic motions of the stars of high velocity. He supposes that the greater Galactic system as outlined above may be divided up into sub-systems, each of which is symmetrical around the axis of symmetry of the greater system and each of which is approximately in a state of dynamical equilibrium. The sub-systems rotate around their common axis, but each has a different speed.'

Of these sub-systems the one consisting of the globular clusters rotates then slowly, while that of the local, low-velocity stars rotates fast, presumably at about 300 km/s, the speed of the high-velocity stars with respect to us. Since their relative motions are of order one-tenth of that, their orbits must be nearly circular.

To consider the consequences, Oort took an important step: 'I shall altogether disregard the idea of a number of separate galactic clouds and take into consideration only the forces arising from the greater Galactic system as a whole.' So, he ignored the matter of the perturbations due to local condensations of matter, so that the greater system is primarily a smooth distribution with only large-scale motions. Then there is circular symmetry and the gravitational force is directed towards the center of the greater system and that varies only with distance from this center. This results in a situation where the period of revolution around the center varies in general with distance from this center, the inner parts usually going around in a shorter period of time than parts further out. This is called 'differential rotation' and all parts of the Galaxy, except those at the same distance from the center, move with respect to each other. If the rotation period were everywhere the same, the separation between two randomly chosen parts would not change; this is called solid-body rotation, but in general in galaxies this is not the case. For example, the Solar System is in differential rotation, the further out we go the longer planets take to complete a full orbit around the Sun. A gramophone record is an example of solid body rotation. Oort went on to investigate the effects of this in the form of systematic patterns in the motions of stars on the sky.

The pattern to be expected has been illustrated in Fig. 5.21. This figure applies to the case of a flat rotation curve, in other words the rotation velocity (in km/s, not angular velocity) is the same everywhere. That actually is what we now observe to be the case in the outer parts almost in all spiral galaxies, and also in our own. In the top-left panel the Sun is the point in the center and I consider points on a circle with a radius of 1 kpc. I have assumed the distance to the center to be 8 kpc towards the bottom of the page. The rotation speeds are indicated by the arrows; they are perpendicular to the direction to the center and this results in the arrows being somewhat slanted on the left and the right. The panel at the top-right shows the residual velocities relative to the Sun. These are small of course, so I have blown

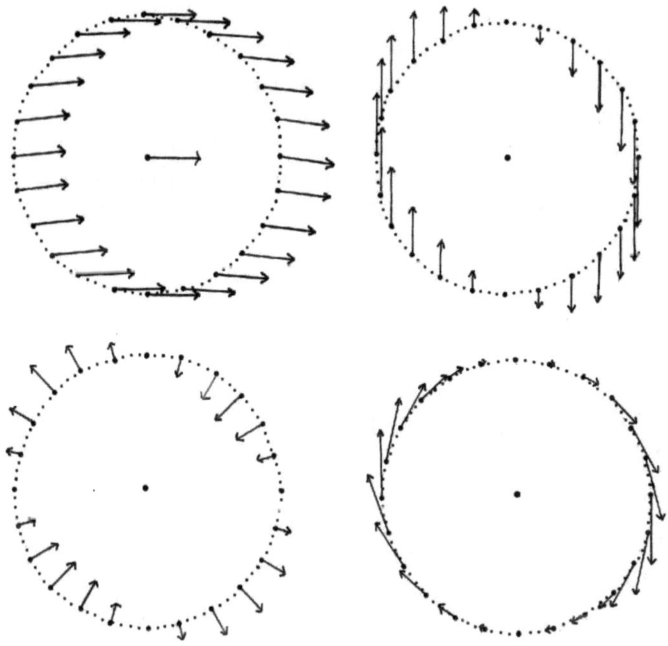

Fig. 5.24 Pattern of motions as a result of differential rotation. The top-left panel shows the rotation velocities for a set of positions 1 kpc from the Sun (which is the dot in the middle). All rotation velocities have the same magnitude and are perpendicular to the direction to the Galactic center, which is 8 kpc towards the bottom of the page. Galactic longitude 0° then is towards the bottom and 90°, the direction of rotation, towards the right. The panel in the top-right shows the relative velocities with respect to the Sun; the scale of the velocity vectors has been blown up by a factor of twenty. The bottom panels show the radial (left) and tangential (right) components of these relative velocities. Figure by the author

up the scale of the arrows by a factor of twenty. The residuals with respects to the point directly towards the Galactic center and the anticenter are zero, which is a result of the adoption of a flat rotation curve; in general these would correspond to the changes in the rotation velocity with galactocentric distance. The bottom panels show the observables, the radial and tangential velocities, the latter being related to the proper motions.

We can see qualitatively what kind of pattern results. The radial velocity of stars at the direction to the Galactic center, which is (new) longitude 0°, is zero, since the rotation velocities are perpendicular to the line of sight of the observer. Then they go to a maximum away from us at 45° and back to zero in the direction of Galactic rotation at longitude 90°, to a maximum towards us at 135° and again to zero at 180°. The whole thing repeats itself for the rest of the circle. So there is a systematic pattern with a double sine-wave; in technical terms a 'sine $2l$'.

Box 5.2 Differential rotation and the Oort constants.
Let the rotation velocity at a distance R from the Galactic center be $V(R)$ and the corresponding values at the position of the Sun $R_\circ$ and $V_\circ$. The tangential and radial velocities with respect to the Sun at longitude l (with the Galactic center at $l = 0°$) and relatively small distances r (compared to the distance to the Galactic center) are then

$$V_{\rm rad} = Ar \sin 2l \ ; \ \ V_{\rm tan} = Ar \cos 2l + B\,r.$$

with

$$A = \frac{1}{2}\left(\frac{V_\circ}{R_\circ} - \left.\frac{dV(R)}{dR}\right|_{R_\circ}\right) \ ; \ \ B = -\frac{1}{2}\left(\frac{V_\circ}{R_\circ} + \left.\frac{dV(R)}{dR}\right|_{R_\circ}\right)$$

$$A - B = \frac{V_\circ}{R_\circ} \ ; \ \ A + B = -\left.\frac{dV(R)}{dR}\right|_{R_\circ}$$

For a flat rotation curve we have

$$\frac{dV(R)}{dR} = 0 \ ; \ \ A = -B = \frac{1}{2}\frac{V_\circ}{R_\circ}$$

Oort first considered the case of a point mass in the center as for planetary orbits; in that case Kepler's third law relates the size of the orbit of a planet or other body to the period T (and hence its velocity V). For circular orbits of radius R, Kepler's law becomes

$$T^2 \propto \frac{R^2}{V^2} \propto R^3 \ \ \longrightarrow \ \ \frac{dV(R)}{dR} = -\frac{1}{2}\frac{V}{R} \ ; \ \ A = \frac{3}{4}\frac{V_\circ}{R_\circ}, \ \ B = -\frac{1}{4}\frac{V_\circ}{R_\circ}.$$

Oort also used the supposition of Lindblad that the principal part of the greater Galactic System is an ellipsoid of constant density. Then the radial force is proportional to R and

$$A = 0 \ ; \ \ B = -\frac{V}{R}.$$

In the proper motions (the tangential velocities) the pattern has the same shape, but now is a cosine (a sine function shifted by 90°) and it is offset from zero. In the case illustrated it starts at zero at longitude 0° (zero because of the flat rotation curve), goes to a maximum at 90° and then back to zero in the anticenter at 180°

and so forth. So it is a rotation with a pattern as in the radial velocities superimposed after shifting by 90° ('cos 2*l*').

The mathematical formulae have been collected in Box. 5.2. The amplitude of the patterns is governed by the amount of differential rotation (how much the rotation velocity varies with distance from the center) and the distance of the objects concerned. First look at the radial velocities and consider a set of stars at the same distance from us. The amplitude of the double sine-wave is then the product of that distance and a constant, that Oort called '*A*', and which can be calculated from the amount of local rotation and its change with distance from the center. '*A*' is the quotient of a velocity and a distance and is usually expressed in km/s per kpc (Oort originally used parsecs). For reference the currently accepted value is about 13 km/s.kpc. For tangential velocities the pattern with longitude is a double cosine-wave with the same amplitude (*A* times the distance from us) with an offset that is the product of a similar constant, that Oort called '*B*', and again the distance of the object from us. In practice, one does not measure the tangential velocity, but the proper motion, which is proportional to the tangential velocity divided by the distance, so that the amplitude and offset in proper motions become independent of distance, which is very convenient. *B* is then the angular velocity of the rotation in the pattern of proper motions onto which the double-cosine is superimposed. For a flat rotation curve the definitions of *A* and *B* are such that we get $A = -B$ and the currently accepted value for *B* is indeed about −13 km/s.kpc.

Figure 5.25 reproduces a sheet of paper from Oort's Archives on which he derived the relevant formulae (with a closed line around it), although here it is 'only for the case that all mass is in C'. The figure was used to define the geometry of the problem. He did not yet use the notations *A* and *B*, but he did so in Oort (1927c) in which he presented the results (see Fig. 5.26). In that paper he gave both the results for the Keplerian rotation, corresponding to a single central point mass like in the Solar System, as well as for the case of an ellipsoid of constant density as Lindblad had used in his papers (see Box 5.2 for the details).

The main point of the paper was to confirm the predicted pattern in observations and then to determine the value of the constants; it did not take long for these to become universally known as the 'Oort constants'. In his lectures Oort referred to them as 'two constants that I will call *A* and *B*', as if this had occurred to him then and there. He also needed to fix the direction to the center, and of course for this he took that of Shapley's globular cluster system at longitude 325°.

For the observational verification, Oort started with radial velocities. It is true that some astronomers had found systematic patters on the sky before, such as Strömberg, but there most of the effect had to do with the high-velocity stars. Oort put forward as a first good piece of evidence a table from Jan Schilt of so-called *c*-stars (giants or supergiants with narrow line spectra). Dividing the plane of the Milky Way in six sectors of 50° indeed produced a pattern as predicted. But Oort went further with the help of Computer Gerrit Pels to collect data on various sets of stars. In the end they produced a table with some 500 stars distributed over ten types. For each type they fitted the double-sine wave and with the average distance derived values for constant *A* and a probable error. For all but one category they found positive values

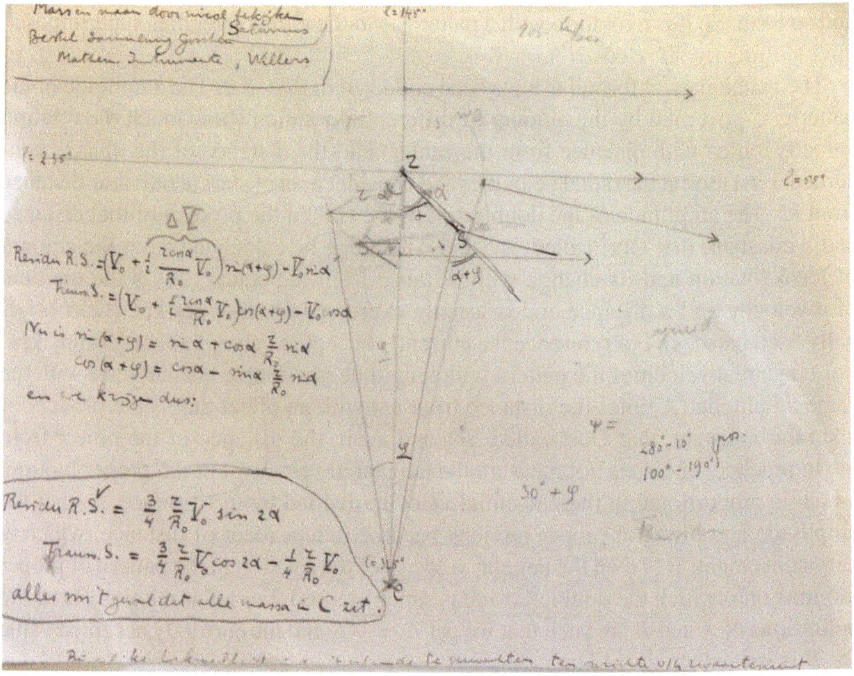

Fig. 5.25 Derivation of the streaming pattern for the case of a Keplerian rotation curve (i.e. like the planets in the Solar System). From the Oort Archives

Fig. 5.26 First publication of the Oort constants. The derivation, which is nowadays part of probably all lecture courses in Galactic and extragalactic astronomy is not part of the paper; all it says is that '... it is easily seen that ...'. From Oort (1927c)

Let us now consider a group of stars at a distance r from the sun and let us suppose that r/R is so small that all terms of second or higher order in r/R can be neglected, then it is easily seen that the residual velocity caused by the rotation is equal to

$$rA \sin 2(l-l_0)$$

in radial direction, and to

$$rA \cos 2(l-l_0) + rB$$

in transverse direction, if l_0 represents the galactic longitude of the centre (about 325°), l the longitude of the stars considered, R the distance of the sun from the centre,

$$V=\sqrt{RK}$$

the circular velocity near the sun,

$$A=\frac{V}{4R}\left(1-\frac{R}{K}\frac{\partial K}{\partial R}\right)$$

and

$$B=A-\frac{V}{R}$$

and a mean of 31.0 ± 3.7 km/s.kpc. That was excellent confirmation. In five cases they could also fit the longitude of the center as an additional free parameter and found 323° ± 2°.4, very close to the direction of Shapley's globular cluster system.

It seems that Oort had wished to present the evidence in a graphical form. In the notes reproduced in Fig. 5.27 he plotted the values for '$B1 - B2$ stars and weak c-stars', but found the results too unconvincing to publish. He wrote: 'After all not suitable for publication'. And as clarification 'Deviations still predominantly resulting from too low numbers, I think' and 'There are still too few points to represent the form of the curve even somewhat convincingly'. As it would not strengthen his case he decided to stick to the tabular material.

For the proper motions the situation is more complicated as the double cosine is offset from zero by the constant B. In order to proceed, Oort needed to make an assumption of the 'character of the gravitational force'. The system needed to be flat, but could be more spherical in the inner region. He assumed that it is a superposition of two forms, the Lindblad constant ellipsoid one in which the force is directly proportional to distance R from the center and the Keplerian case of a single, large central mass, inversely proportional to the square of R. He then took the values of A as determined from radial velocities (which would give the amplitude of the variations) and used proper motion data to fix the offset, which he then related to the ratio of these two forces near the Sun. The offset is constant B and with A this can be related to the gradient in the force field (see Box 5.2 for details). Oort found that the ratio of the 'Lindblad' and 'Kepler' forces near the Sun had to be 0.11.

With observed values for the Oort constants it is possible to make better estimates for the distance to the center and the local rotation velocity (Box 5.2). Strömberg's value for the motion with respect to the globular clusters is 286 km/s and with respect to the high-velocity stars it is 272 km/s. Adopting the latter value for the rotation speed near the Sun, Oort found that the distance to the center would be 5.9 kpc, much smaller than the 13–25 kpc found by Shapley. Even when the constant density ellipsoid were taken away completely, it only increased to 6.6 kpc. No conclusion was given; Oort just noted that 'the distance to the centre of the globular cluster system is very uncertain'. In a note added in proof he reported that a new determination of the precession constant increased the ratio between the two forces from 0.11 to 0.29 and decreased the distance to the center from 5.9 to 5.1 kpc.

It is remarkable that Oort did not feel uncomfortable with the necessary large amount of absorption that was required and accepted it without much discussion, only referring for support to his inaugural lecture as privaat docent (Oort 1927a), which was in Dutch and not readily available in the rest of the world.

I close this chapter with some remarks by Bartholomeus Jan Bok (1906–1983). Bart Bok has been a major figure in astronomy, especially in the early development of radio astronomy and in studies of the interstellar medium. He studied in Leiden and earned his Ph.D. in Groningen under van Rhijn in 1932. He next went to the USA, married astronomer Priscilla Fairfield (1896–1975) and for many years

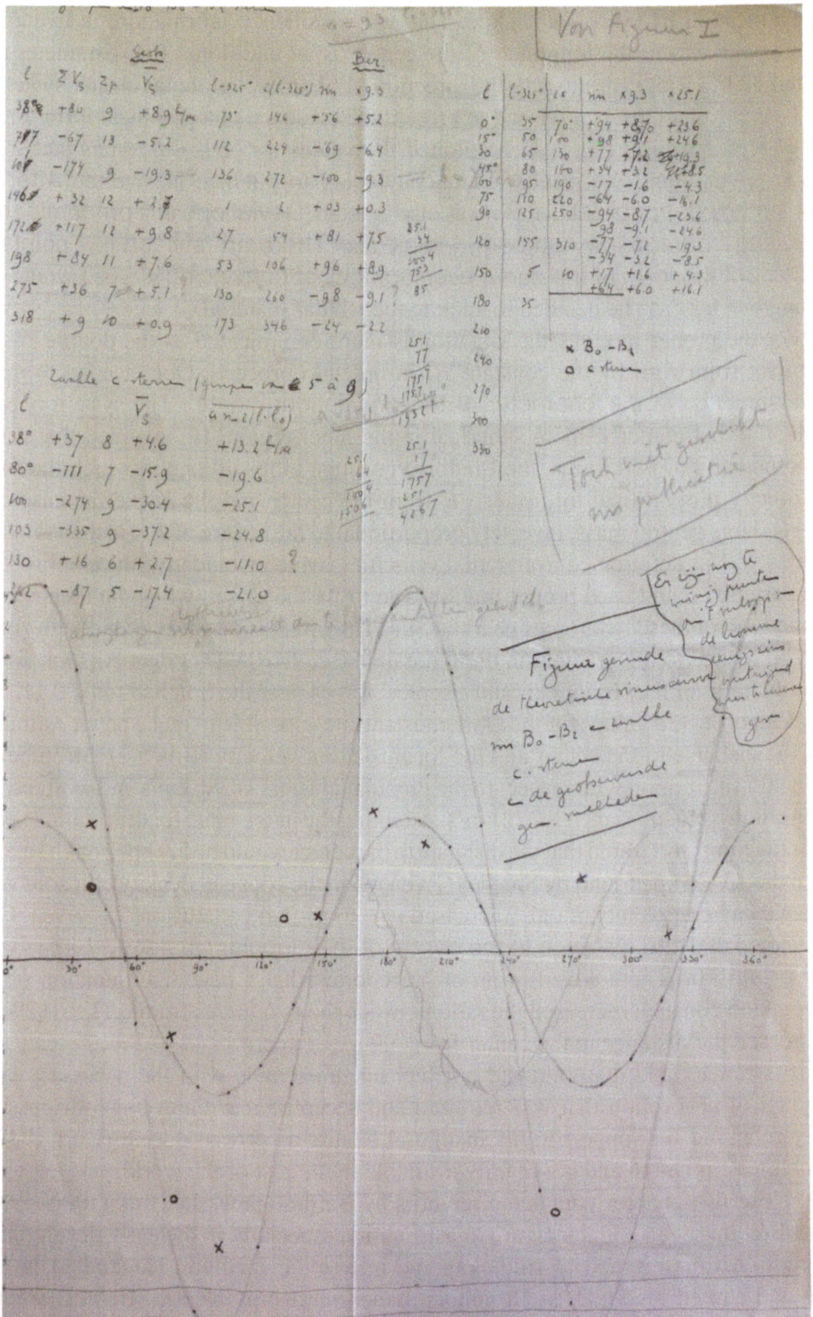

Fig. 5.27 Sheet of paper from the Oort Archives in which he plotted radial velocities of stars against longitude. The hand writing reveals that he decided against using this in a publication (see text for details). From the Oort Archives

worked at Harvard University, becoming eventually associate Director of the Harvard Observatory. In 1957 he was appointed Director of Mount Stromlo Observatory near Canberra, Australia and ten years later he became Chairman of the Astronomy Department and Director of Steward Observatory in Tucson, Arizona. In the *American Institute of Physics* interview with David DeVorkin, [80], Bok said:

> 'I had good mathematics teachers, but nothing great. But in astronomy of course, there was Oort and de Sitter and E. Hertzsprung. I had small courses from all of them. And the ones by Oort made by far the greatest impression, for Oort was then lecturing on the new theory of Galactic rotation of Bertil Lindblad, which was then considered the real great thing, because it explained the asymmetry of stellar motions. And that was really the thing that started it off. Lindblad himself had set up this rotational principle, but he had made it so complicated, full of differential geometry and oscillating [must be 'osculating'] planes, it was hellishly complicated to read. He never thought of the simple Oort formulae sort of thing.
>
> [DeVorkin: Relating it to observables?]
>
> Bok: The simple observables, you know. And then Oort was lecturing on it to us, to Kuiper, Oosterhoff [Oosterhoff] and me, and maybe one other person, maybe Duram [Doorn] was there, who's been dead for many years. Anyhow, he was lecturing to three or four of us, and Oort got bogged down in Lindblad's figures. Then he began to draw the simple diagrams for differential rotation. And got the sin (2 l) formula out of that, for the radial velocity.'

And in his contribution to the *LibAm80* [2], the Liber Amicorum for Oort's 80th birthday, Bok wrote:

> 'Anyhow, Jan Oort was presenting (on Monday afternoon at four) [this must have been in 1926] a series of seminars for Doorn, Kuiper, Oosterhoff and Bok on Lindblad's theories of Galactic rotation. As I remember it – others may have different recollections! – Jan told the four of us one Monday that he had got bogged down in Lindblad's complex mathematics and that there would be no lectures for two Monday afternoons to follow. And then there came the first Monday lecture after the crisis [sic], a lecture in which Jan Oort basically developed the simple formulae for the double sine-wave effect of Galactic rotation in radial velocities and the corresponding formulae for the effects in proper motions. The four of us realised that we were listening to an amazing new step that Bertil Lindblad had overlooked in his differential geometry oriented approach to the subject! Jan gave us the simple formulae simply derived, and we left the lecture room thoroughly aware of the fact that the effect of Galactic rotation on radial velocities is proportional to the average distance of the group of stars under consideration. [...]
>
> We learned one other thing on that first afternoon: the effects on proper motions in Galactic longitude were highly dependent on precessional corrections, and very much affected by possible systematic errors in

the fundamental system of proper motions—as they are today. Oort's constant A seemed determinable: B was a different matter.
[...] By 1927 all of us interested in understanding stellar motions were made to feel at ease, for the great mysteries of the early and middle twenties had been resolved.

Chapter 6
Galactic Structure and Dynamics

It is the harmony of the diverse parts,
their symmetry, their happy balance;
in a word it is all that introduces order, all that gives unity,
that permits us to see clearly and to comprehend at once
both the ensemble and the details
Jules Henri Poincaré (1854–1912)

Success is no accident.
It is hard work, perseverance, learning, studying, sacrifice
and most of all, love of what you are doing or learning to do.
Pelé (Edson Arantes do Nascimento) (b. 1940)

Oort's discovery of the rotation of the Galaxy was a breakthrough in sidereal (i.e. concerning the stellar distribution) astronomy. But before it would be recognized as such, it had to be accepted by the professional community. The first thing for him to do was to strengthen his case and arguments. But for wide acceptance there needed to be independent confirmation and support from one or more authorities in the field. The confirmation came quickly, notably by his former study mate in Groningen Jan Schilt, but particularly by Canadian astronomer John Stanley Plaskett of the Dominion Astrophysical Observatory. The authority that soon praised Oort's work in a very positive review was no lesser person than Arthur Eddington.

But the matter also raised some more questions, some of which would occupy much of Oort's time throughout in his career. These were mainly the following:

- What are the accurate values for the Galaxy's rotation speed, the constants A and B, and what is the distance from the Sun to the Galactic center?
- How do Kapteyn's Star Streams fit into all of this? And what is the dynamical structure of the stellar distribution in the solar neighborhood?
- What exactly causes the interstellar absorption? If it is dust in between the stars, what are the properties of the particles and how is it distributed in space?

P. C. van der Kruit, *Jan Hendrik Oort*, Astrophysics and Space Science Library 459, https://doi.org/10.1007/978-3-030-17801-7_6

- If the spiral galaxies are structures similar to the Milky Way, what can we say then about their structure and dynamics and can we identify the spiral arms in our Galaxy? And what about the nature of spiral structure?

6.1 Confirmation

The paper in which Oort announced his result had no detailed information on the data he used and was also mostly concerned with radial velocities and much less with proper motions. Oort set out to rectify these shortcomings. In quick succession he published three more papers on the subject. The first of these papers (Oort 1927d) concerned proper motions. In the original paper (Oort 1927c), the effects of differential Galactic rotation had been illustrated using observations of radial velocities of stars, while the effects of proper motions were presented only in a preliminary fashion and turned out to be not very convincing. Proper motions (which are essential in deriving the Oort constant B) have the problem that they are affected by precession and the corresponding motion of the equinoxes. Precession (see also Box 2.2) is the effect that the rotation axis of the Earth slowly precesses in space around the direction perpendicular to the Earth's orbit, and the vernal equinox, the point where the Sun crosses the equator on the sky and the zero point of right ascension, moves along the equator and the celestial poles trace small circles. The period of the precession cycle in which the equinox completes a full circle is roughly 26,000 years; this sounds like a long time but the resulting annual shift of the equinox is some 50 arcsec/year, so very significant. And since proper motions involve measurements of positions at two epochs, this is an effect that definitely needs to be taken into account. In particular, an incorrect correction for precession mimics a general rotation among the proper motions, and translates into an incorrect value for Oort constant B.

In fact, astronomers had been looking for a 'rotation of the galaxy', by which they meant a systematic component in the proper motions along the direction of the Milky Way. For example Carl Vilhelm Ludwig Charlier (1862–1934) of Lund Observatory in Sweden had claimed in 1913 that the effects was $0\overset{''}{.}35$ per century, and only the year before Oort's publication, British astronomer and historian of astronomy John Knight Fotheringham (1874–1936) of Oxford found a value of $0\overset{''}{.}15$ per century [151, 152]. Oort's rotation of the Galaxy was different, but the effect of his constant B was to a large extent the same.

Oort (1927d) addressed the effects of Galactic differential rotation on proper motions and the determination of the precession constant. Here his extensive knowledge of and experience with astrometry that he gained from Kapteyn and Schlesinger could be put to excellent use. But before that Oort used some more radial velocities to improve his determination of A and the longitude of the center of the Galaxy. The new value for A was 19 ± 3 km/s kpc (closer to the modern value) and for the center a Galactic longitude (in the system used at the time) of $324° \pm 2°$, only one degree from Shapley's center of the globular cluster system. He found also approximately the expected increase of the amplitude of the radial velocities with average distance for groups of stars; these distances were derived from secular parallaxes. There was one problem and that was with stars of spectral type *Md*. These are long-period variables, often of the type designated by prototype Mira, which are stars late in their evolution (after helium-burning), that are about to shed their outer layers and become planetary nebulae. These stars failed to show any sign of differential rotation in their radial velocities. Although Oort felt this to be 'somewhat alarming', he did not let this stand in the way of his theory. 'Personally I do not believe, however, that these apparently discordant results should be considered as an insurmountable difficulty in the way of a general galactic rotation' (Oort 1927d).

Proper motions have two components, one parallel to the Milky Way (proper motion in Galactic longitude) that has to be used to determine B, and one perpendicular in latitude. The sample Oort used consisted of some 800 stars of mostly spectral types O and B (divided into three groups). He noted that usually the precession constant was determined with the assumption that there is no systematic rotational motion around the Sun, since any such effect would be produced by precession. But constant B would have the same effect. He realized that there is another way, namely to use the condition that there should not be any systematic proper motion in latitude. Taking large numbers of stars from the Boss *Preliminary General Catalogue*, Oort fitted for the latter condition and found a significant correction to the universally accepted motion of the equinox as given by Newcomb.

In the end the results were presented in two diagrams, see Fig. 6.1. Although there is significant scatter, the results are unambiguous. The fit on the left, concerning proper motions parallel to the Milky Way, was based on the values for the constant A and for the longitude of the center as just determined. The vertical shift of the double cosine curve is constant B and this is found to be -24 ± 5 km/s kpc. The panel on the right shows his best fit to the uncorrected latitude proper motions determined

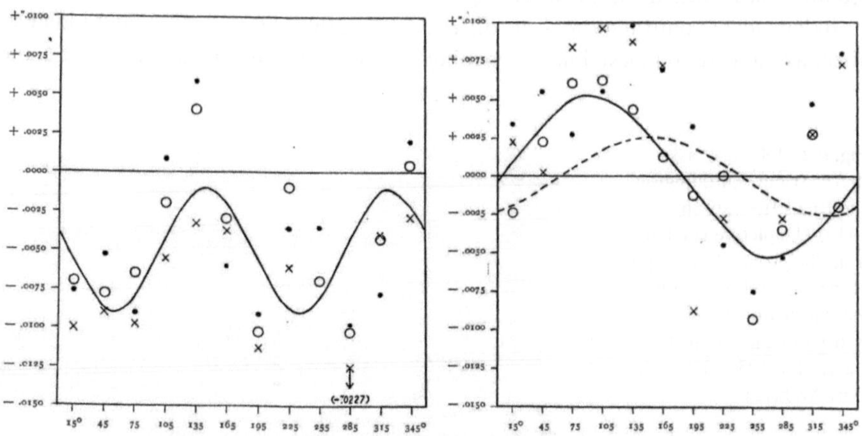

Fig. 6.1 Two figures from Oort (1927d). On the left the proper motion in longitude as a function of longitude with the best fit. On the right the proper motion in latitude, also as a function of longitude. The symbols designate three different samples of O and B stars. For the details see the text

with the new value for the precession constant (full-drawn line). Its use removed all systematic latitude proper motions that were in the Catalogue. The dashed line is the best fit in case the then universally accepted value of Simon Newcomb for the precession constant had been taken. Clearly Newcomb's value was in need of revision and the condition that there should be no systematic latitude proper motion worked well.

The values for A and B together with a local rotation velocity of 272 km/s implied a distance to the center of the Galactic System of 6.3 ± 2.0 kpc and a contribution of a point mass with an inverse square force of $50 \pm 12\%$ to the total force near the Sun. The mass concentrated near the center would be 6×10^{10} (60 billion) Suns. Oort ended his paper (1927d) with two remarks. The first concerned the Magellanic Clouds, which Oort suspected on the basis of their relatively low radial velocities were companions to the Galactic System. The second was a short discussion on the radial velocities of planetary nebulae, which are—as we now know—late stages of evolution of stars like the Sun. These are concentrated towards the center of the system and showed a larger average speed in directions close to the Galactic center. No explanation was offered.

The other papers were shorter. Oort (1927e) concerned Cepheids. They are also late stages of stellar evolution, but of stars more massive than the Sun. They seemed to partake in the rotation of the Galactic System, but that conclusion could have profited from more data. Oort stressed that his conclusion that an inverse square law could contribute significantly to the local force should not be taken as proof of a large central concentration of mass. And finally, Oort admitted he should have mentioned important work by his old fellow student from Groningen, Jan Schilt, on a possible rotational term in the distribution of proper motions. In Oort (1927f) his sources for radial velocities of stars and planetary nebulae were collected. Oort's four papers on Galactic rotation (1927c–f) were published in a period of half a year, viz. between April 14 and October 10, 1927. And he also got married in that same period. Success requires hard work and dedication (Fig. 6.2).

Independent confirmation of the differential rotation did not take long to materialize, as I already remarked in the opening sentences of this chapter. The first came

Fig. 6.2 John S. Plaskett (left), Arthur S. Eddington and Ejnar Hertzsprung in 1932. The picture is taken from the group photograph of the General Assembly of the International Astronomical Union at Harvard in 1932. From the Oort Archives

Fig. 6.3 Arthur S. Eddington (front left) in Leiden in 1923. In the front on the right Hendrik A. Lorentz and in the back from left to right Albert Einstein, Paul Ehrenfest and Willem de Sitter. From the Archives of Leiden Observatory

from Jan Schilt, who had worked on closely related subjects and now was at Yale. He published two papers related to Oort's suggestion [153, 154] in which he showed that proper motion studies of other samples of stars than Oort had used, confirmed his proposal of differential rotation. The second confirmation came from John Stanley Plaskett of the Dominion Observatory [155] and involved radial velocities of samples of fainter O- and B-stars. And a major authority on the subject, Arthur Eddington (see Fig. 6.3), praised Oort in a *Council note on the rotation of the Galaxy*, this being the Council of the Royal Astronomical Society [156], and subsequently devoted his 1930 Halley Lecture to it [157].

I quote from the first of Eddington's publications. After explaining Oort's ideas and methods, he continued:

> From the value of C [this is Oort's A] we can determine V if R is known, or vice versa. Thus, if we take Shapley's smallest estimate of 13,000 parsecs for our distance from the centre of the Galactic System, we find $V = 330$ km. per sec. There is independent confirmation for this. Strömberg has found a velocity of 272 km. per sec. (subject to a large probable error), for the mean velocity of the globular clusters. Since the latter extend over the whole system, their mean velocity represents roughly the relative velocity between the stars in our neighbourhood and the centre of the system; the velocity is, moreover, in the right direction transverse to R. The central mass which would be necessary to control this motion is 27.10^{11} times that of the Sun. This can scarcely be regarded as excessive considering the enormous extension now attributed to the Galactic System.
>
> These figures would be modified if we allowed for the fact that the mass is not entirely concentrated to the centre. Oort himself estimates R as

> 6000 parsecs, but it is not possible to expect much accuracy at present. What is especially needed is a more accurate determination of V by observation of radial velocities of globular clusters. Whilst the investigation fits in beautifully with much of our knowledge, it leaves certain difficulties outstanding. Star-streaming would presumably be an independent phenomenon since the stream-vertex is more than 20° from the longitude of the center of rotation $l_{\circ}$ – a difference too large to attribute to errors in the determinations. Oorts hypothesis scarcely seems to admit of the existence of separate star-clouds in the system. We are accustomed to think of the stars within say 1000 parsecs of the Sun as constituting a separate star-cloud; but evidently the cloud will be sheared as the rotational motion proceeds. In fact, without hypothetical interpretation, the direct evidence of the radial velocities is that the stars in the mean are falling in along the axis $l_{\circ} - 45°$ and expanding along the axis $l_{\circ} + 45°$, which makes it impossible for a limited cloud to keep its shape. Oort was inspired by a theory proposed by Lindblad which postulated a number of subsystems rotating within the Galactic System, but his results seem rather to minimise the role of subsystems. Finally we wonder how such a state of rapid revolution of the stars all in one direction round the centre of the galaxy could ever have arisen. But the cause of rotation of celestial objects always leaves us wondering.

Indeed the 20° 'vertex deviation' and Kapteyn's Star Streams were not understood yet in the context of the Galactic differential rotation picture.

Jan Schilt and Oort had been close friends, while students in Groningen and later colleagues in Leiden, but in 1925, Schilt and his wife Johanna (Jo) Timmer left for the USA. They kept up a regular correspondence and I will relate some of that here. The first letter dates from October 13, 1925 and has an amusing paragraph. It was written by Schilt at Mount Wilson. He addressed Oort as 'dear Oort' and signed with 'J. Schilt'.

> 'It is hailing, it snows and there's a thunderstorm. I am caught in the Laboratory and dare not cross to the Monastery. It has been clouded for four days and I have not taken a single plate [...].
>
> I believe the Kapteyn Cottage is unoccupied very often and I think we will make use of it regularly. It is a bit unpleasant to have to leave Jo by herself in Pasadena, although there is more necessity then to get to know others and speak English. – You might have expected that after one month in America I should give my opinion about America and Americans, but I will not do that now. A lot of what you told us we feel is 101% exaggerated. One thing (*sit venia verbo* [pardon the word]) I can judge already at present and that concerns your sublime (referring to their looks) American girls, their beauty and graceful dress. Oh my, what a poor judgment you have in that respect. I am convinced in America I would rather spend my whole life as a bachelor than etc.'

Schilt's appointment was apparently for one year, because already in January Schilt wondered what he would do next year; he regretted that had refused an offer Schlesinger had made and wondered if he could manage to stay legally in the USA.

He did not get along very well with Seares, whom he found authoritarian and high-handed and he felt it would not be good for the Observatory if Seares were to succeed Adams as Director. He also mentioned to Oort that he had insisted successfully that Strömberg tone down the rebuttal to Oort's position on the asymmetry in the stellar velocities. Strömberg reacted to Oort's (1926a) note in the Observatory by another note in the same journal and the tone indeed is not nasty [158]. In March there was another offer from Schlesinger to come to Yale, which Schilt accepted. And he became more positive about American girls. In May Schilt complained that thc immigration formalities were very slow; everything Schlesinger had tried failed. In the end they left the USA for Canada and re-entered, presumably on temporary visa.

Schilt had been working on very similar research as Oort and in a sense had been close to the right solution when he concluded that there are systematics changes of proper motions with Galactic longitude. In a letter of April 1, 1927, Schilt wrote from Yale and referred to Oort's rotation work.

> 'Schlesinger told me with a half word about your result that the galaxy rotates. The fact is much expected and I had not expected anything else. It really strikes me that you succeeded in showing this with radial velocities, if I have understood Schlesinger correctly. Although I am a priori critical about this, I will not desist from cordially congratulating you even before I have seen anything detailed. Schl. was very enthusiastic about it. I also miss our old discussions at lot. As you remember we had always $10\times$ as many things to talk about as we had time for.'

In an undated letter (but it must have been late May or June 1927, since he congratulated 'dear Oort and Mieke' with their marriage) he came back to this and wrote

> 'I must apologize for first writing about the stars – BAN 120 [this is (Oort 1927c), the discovery paper of Galactic rotation] provides a lot of food for thought. After the first reading I was almost completely in agreement and I have talked about your work during the Journal Club.'

Schilt had minor problems with the conclusion that there had to be a rather large central mass, but was not surprised about the distance to the center as he believed that Shapley's distances were 'according to me incorrect'. He showed himself a convinced and complete supporter of the Galactic rotation paper.

There is a bit of non-astronomy, but only at the very end, when Schilt revealed the great news that they had bought a car ('it is a Chevrolet, so not a Ford but a real car') and had been exploring Connecticut. The letter ended with:

> 'Last week van de Kamp and his wife were here. They had been married on the same day as you and he told us that van de Bilt also got married that same day. What is the chance of 3 Dutch astronomers getting married on the same day?'

Indeed Jan van de Bilt (1876–1962), Reader (roughly Associate Professor) of astronomy in Utrecht, also married (for the second time) on May 24, 1927. A cursory search has not revealed the date of Peter van de Kamp's marriage, but there is no reason to doubt Schilt's account and that he and his second wife Olga Rademacher also married on that date.

6.2 Dynamics

The model of the rotation of the galaxy was no more than an empirical one, lacking any fundamental explanation of the dynamics, the equilibrium—assuming a steady state—between the spatial distribution of matter (stars, gas, dust, etc.) and the gravitational field that results from that on the one hand and the motions of the stars in this field on the other. Oort had simply described how a large system with differential rotation would result in a systematic pattern of radial and tangential velocities. But he was ultimately interested in the workings of the whole system. There was still no explanation of how Kapteyn's star streams fitted in. The more fundamental problem of the dynamics was addressed and treated in detail in Oort (1928b), a paper with much algebra with the title *Dynamics of the Galactic System in the vicinity of the Sun*.

Oort started the paper with an introduction that had an important statement, namely: 'The stars are considered to form part of *one* rotating system, but with varying peculiar velocities' (Oort's italics). The idea of a cloud (as the Kapteyn system) for the local stars comparable to spiral galaxies well within the Shapley system of globular clusters had been abandoned. The issue of the nature of the spiral nebulae had been settled in the mean time by determination of their distances. Around 1917 distances were already estimated using novae (see Sect. 2.6) or 'brightest stars' of the Andromeda Nebula (M31; see Fig. 7.9) or the nearby spiral galaxy M33, which is comparable in many respects to the Large Magellanic Cloud (I pointed that out in [159]), and is close to M31 in space. For these, astronomers like Heber Curtis, Harlow Shapley and Knut Lundmark had claimed distances of a few hundred kpc. However, Adriaan van Maanen had confused the issue by claiming he had observed rotation in the spiral galaxy M101 (see Fig. 7.9) by measuring proper motions of features in it. This widely cited result turned out to be incorrect for reasons never made perfectly clear [160, 161]. By 1925 Hubble had settled the issue, using Cepheids and the period-luminosity relation discovered by Henrietta Leavitt, to determine the distance to M31 and M33 as being of order 300 kpc, not giving any references to these earlier studies! A list of all distance determinations up to 1929 has been provided by Way [56].

Oort addressed the matter of absorption as follows:

> Of the various objections against the tentative model of the Galactic System considered the most serious one is perhaps that a very eccentric position of the sun, as implied by the theory, is contradictory to observations of the density of the stars. It certainly would be so if we assumed that the light of all stars reaches us without obstruction. We know, however, that even in our immediate neighbourhood considerable obstruction does take place in extended regions covered by dark nebulae. The above difficulty may possibly be only apparent and caused by the observed density distribution in the direction of the galaxy being as much determined by the distribution of dark matter as by that of the stars themselves. Some evidence on this point may be found in the peculiar avoidance of Galactic

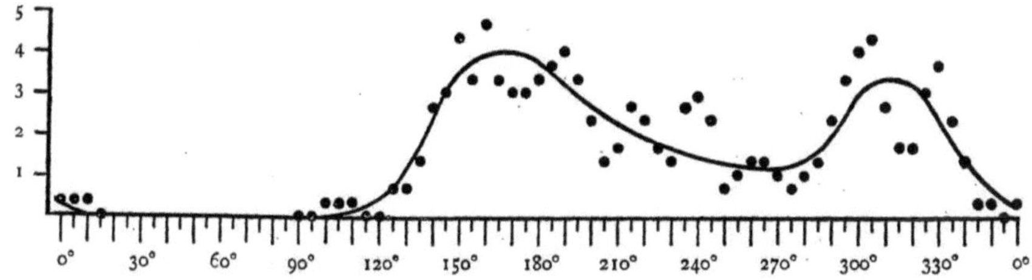

Fig. 6.4 Distribution of the antapices of high-velocity stars in Oort (1928b). The overlap of longitudes 300°–0° in the original figure on the left has been removed

regions by globular clusters, a phenomenon for which the hypothesis of light obstruction seems to give the most plausible explanation.'

Next Oort presented the then current state of affairs concerning the stars of high velocity. The distribution of their antapices (direction to which their space velocities are pointing) is shown in Fig. 6.4. The restriction to one hemisphere roughly centered on longitude 235°, the direction 90° away from the center of the globular cluster system, is still present. In the older work Oort had assumed that the velocity of 63 km/s corresponded to the escape velocity from the local (Kapteyn) system and this could not be maintained in the newer picture. So he replaced that with the view that this amount corresponded to the increase in the local rotation velocity necessary to attain the escape velocity from the greater Galactic System. To illustrate this new point of view Oort presented the figure reproduced here as Fig. 6.5. In this figure stars have been plotted for which the space velocity relative to the Sun had been determined reliably; with the velocity in the direction opposite to Galactic rotation vertically and horizontally the component directed towards or away from the center of the Galactic System. The Sun is the small circle in the middle of the empty, dashed circle on the right. This circle has a radius of 19.5 km/s. The dots are bright stars of a complete sample with velocities in excess of and complete to 19.5 km/s and this is why the small circle around the Sun is empty. The square to the upper-left of the Sun, just inside this small circle, is the mean velocity of these bright stars as seen from the Sun and the offset is the velocity of the Sun. The square has a full-drawn circle around it of 65 km/s; it encompasses most of the bright, presumable nearby stars. The small circles are a sample of high-velocity stars, selected to have a velocity with respect to the Sun in excess of 63 km/s. The rotation of the system near the Sun is assumed to be 300 km/s and the velocity of the center is indicated by the cross in the middle, which is also the center from which the longitudes around the border have been drawn. The large broken circle represents a velocity of 365 km/s with respect to the center (plus-sign), which Oort identified as the escape velocity from the Galactic System. No stars should be found outside this circle. This produces the asymmetry in the velocities of high-velocity stars, since these should be absent around the direction opposite to the local vector of Galactic rotation to the left. The star outside the graph on the left is Cincinnati 2018–2019, which had a space velocity of over 1000 km/s

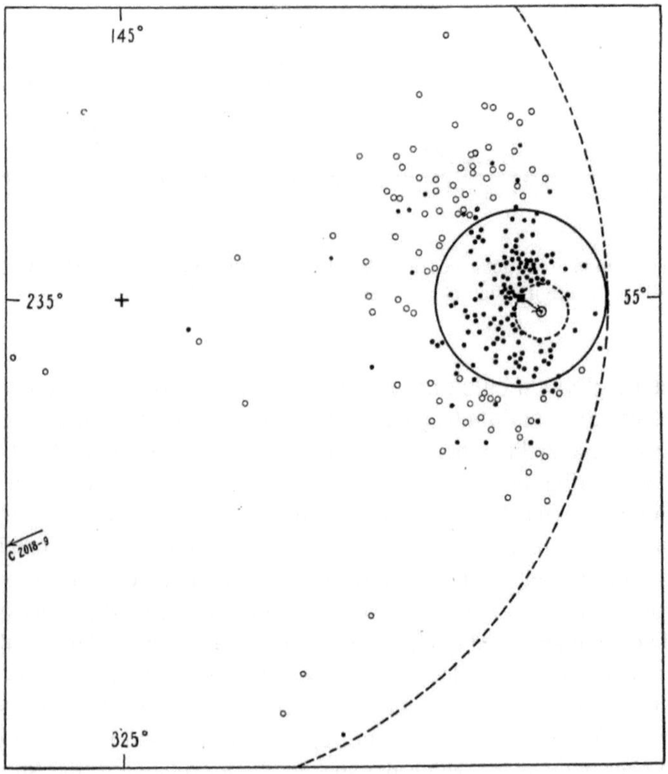

Fig. 6.5 Distribution of space velocities of bright stars (dots) and high-velocity stars (circles) in the new interpretation. The solid circle corresponds to 63 km/s with respect to the mean of stars near the Sun (solid square; the small circle is the Sun) and high-velocity stars are outside it. The large broken circle around the Galactic center (plus-sign) represents the escape velocity from the Galactic System and no stars should be seen to the right of this line. The tickmarks on the borders are Galactic longitude. From Oort (1928b)

using its spectroscopic parallax; however this was reduced to 600 km/s (but still outside the graph) if its trigonometric parallax were used.[1]

The next assumption Oort made was that the interpretation of Kapteyn's Star Streams is indeed that according to Karl Schwarzschild, which means that it is the result of the average random (in Oort's parlance 'peculiar') velocities being asymmetric and larger in the direction of the vertex of the streams.

In a later paper in the Dutch magazine for amateur astronomers and meteorologists, 'Hemel and Dampkring', Oort (1943a) explained how Kapteyn's Star Streams fitted into the notion of rotation of the Galactic System. Figure 6.6 comes from this publication. In the top part the velocities of stars are plotted with the base of each

---

[1]This pair of halo $K2$ stars (HD134439/40) is 28.6 pc from us and has a space velocity of 585 km/s. It is on an elliptical orbit that takes them out to a maximum radius of some 50 kpc [162, 163].

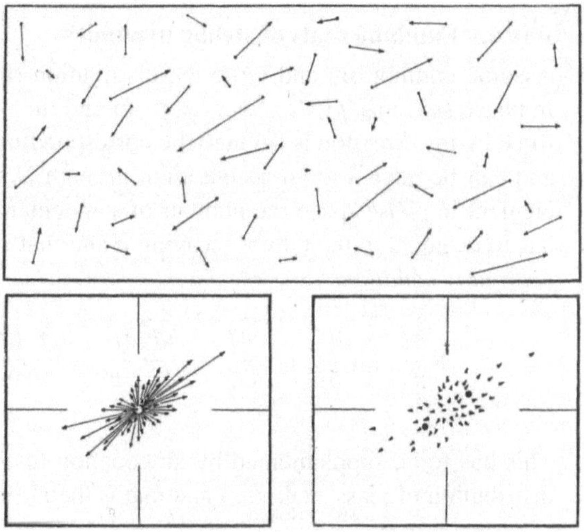

Fig. 6.6 Oort's explanation of the construction of a velocity diagram. For explanation see text. From Oort (1943a)

arrow the position in space of the star it belongs to. In the bottom left the vectors have all been shifted to a common origin and the distribution shows the systematics in the velocities in the volume of space referred to (that at the top). In the lower-right panel only the points of the vectors have been plotted and a velocity diagram usually has this representation with the arrow-points replaced by dots. From in particular the panel at bottom left it can be seen that such a distribution can be explained both as an asymmetric distribution or as a superposition of two opposite streams.

Stellar dynamics, or as it is nowadays often called galactic dynamics, starts with two fundamental equations. I have summarized this in mathematical formulae in Box 6.1; for a more background and details see Oort (1965a), [159] (particularly chapter 7 by I. King), or [164]. The first is the continuity equation, also sometimes referred to as the Liouville equation after the French mathematician and applied physicist Joseph Liouville (1809–1882) or the collisionless Boltzmann equation. after Austrian-Hungarian physicist and mathematician Ludwig Eduard Boltzmann (1844–1906). The designation 'collisionless' refers to the fact it is a simplified version of an equation to describe the distribution of particles in a gas, since in the application in stellar dynamics there are no effects of collisions between the stars. At any time at any position in the system (or rather a small volume around that) there is a number of stars. A small interval of time later, most stars in that small volume have been replaced by different ones. They come from neighboring volumes when they had the right velocity. But also their velocities have been altered by the acceleration due to the gravitational field. In the steady state that has been assumed, not only the number of stars should not have changed, but also the precise distribution of space velocities within that volume should have remained the same (which is a continuity statement). This is a much more powerful condition than would appear at a first glance.

Box 6.1 Fundamentals of stellar dynamics

Assume equilibrium and write the distribution function of stars of mass M in phase space as $f(M, x, y, z, u, v, w)$ and the potential as $\Phi(x, y, z)$. The force in any direction is (minus) the corresponding derivative of the potential and can be related to an acceleration through Newton's law. In any volume element in phase space the number of stars entering it in an interval of time has to be equal to the number leaving it and that condition translates into the *continuity equation*

$$u\frac{\partial f}{\partial x} + v\frac{\partial f}{\partial y} + w\frac{\partial f}{\partial z} - \frac{\partial \Phi}{\partial x}\frac{\partial f}{\partial u} - \frac{\partial \Phi}{\partial y}\frac{\partial f}{\partial v} - \frac{\partial \Phi}{\partial z}\frac{\partial f}{\partial w} = 0.$$

This has to be supplemented by an equation that relates the potential to the distribution of mass $\rho(x, y, z)$ and that is the *Poisson equation*

$$\frac{\partial^2 \Phi}{\partial x^2} + \frac{\partial^2 \Phi}{\partial y^2} + \frac{\partial^2 \Phi}{\partial z^2} \equiv \nabla^2 \Phi = 4\pi G\rho(x, y, z).$$

In the case of the Galaxy Oort assumed axial symmetry for which he used cylindrical coordinates with the lower case notation ϖ, θ, z and corresponding velocities upper case Π, Θ, Z (ϖ, a notation already used by Jeans, is pronounced 'curly pi' or 'pomega'). The two equations then become

$$\Pi\frac{\partial f}{\partial \varpi} + Z\frac{\partial f}{\partial z} - \left(\frac{\partial \Phi}{\partial \varpi} - \frac{\Theta^2}{R}\right)\frac{\partial f}{\partial \Pi} - \frac{\Pi\Theta}{\varpi}\frac{\partial f}{\partial \Theta} - \frac{\partial \Phi}{\partial z}\frac{\partial f}{\partial Z} = 0.$$

$$\frac{\partial K_\varpi}{\partial \varpi} + \frac{K_\varpi}{\varpi} + \frac{\partial K_z}{\partial z} = -4\pi G\rho(\varpi, z)$$

The second equation is the Poisson equation, named after French mathematician Siméon-Denis Poisson (1781–1840). It describes how strong and in what direction the gravitational force is resulting from any distribution of matter. If we know the distribution of matter, this equation allows us to calculate the gravity at any arbitrary position. These two equations combined define the problem of finding how the distribution and the kinematics in a stellar system are in equilibrium with the gravitational field. In the 1928 analysis, Oort only used the continuity equation, which means that he studied stars as moving in a gravitational field that might correspond to more gravitating matter than provided by these stars alone.

Oort assumed that the Galactic System is in equilibrium and supported that assumption with the fact that the Sun is 'certainly ten billion years old' (we now think it is half of that age) and rotates around the Galactic center in 200 million years (not too bad!) and that after 50 revolutions the system should be well mixed. For his analysis Oort assumed that the velocity distribution is exactly Gaussian in all directions, with in principle unequal dispersions (mean velocities) in the principal directions. So the velocity distribution is *ellipsoidal*, as proposed by Karl Schwarzschild. The observed vertices of the streams however, do not necessarily line up with the direction towards the Galactic center, but for the moment Oort left this to worry about later. This meant he had to leave the orientation of the ellipsoid free, that is to say it has three in principle unequal axes and the longest axis points in an arbitrary direction. Another way of saying this is that in the directions of the three perpendicular principal axes it has the shape of a Gaussian curve as often encountered in statistics, but the dispersion of all these three can be different. Oort then inserted this in the continuity equation and this gave him a complicated form of that equation.

It had already been assumed that the system is symmetrical about the rotation axis and Oort furthermore used the assumption that the plane of the system also is a symmetry plane. These significantly simplified his form of the continuity equations, so that a number of important consequences could be drawn. In the first place it followed that the long axis of the velocity ellipsoid had to point to the center of the system. This was however, not in agreement with observations. It is the same thing as the observation that directions of Kapteyn's Star Streams were not coincident with the direction to the center of the the globular cluster system. After all the direction of the streams really is the direction of the largest mean motions and that is the longest axis of the velocity ellipsoid. The difference is some 20°. This has become known at the *deviation of the vertex* and has persisted; it is now understood as due to local irregularities in the mass distribution in the disk of the Galaxy, such as spiral arms.

In the second place he found that for any group of stars a well-defined ratio exists, determined by the Oort constants A and B, between the velocity dispersion in the direction of the center of the Galaxy and that in the direction of rotation. Oort also found that the velocity dispersion perpendicular to the plane of the Galaxy should be equal to the radial one, in contradiction to what was observed. As a first remark on this he argued that this result would not be applicable to the Galaxy where the density distribution is very flat and therefore the force towards the center is independent of height above the plane for small vertical distances.

At the beginning of the paper Oort had noted that 'Thirteen years ago almost the same mathematical problem has been extensively treated by Eddington and Jeans', without further reference. One might think this must refer to the books mentioned in his *AnnRev81* [3] article, namely Arthur Eddington's *Stellar Movements and the Structure of the Universe*, and of James Jeans' *Problems of Cosmogony and Stellar Dynamics*.[2] And indeed they might be a basis for his work, except that these books were published in 1914 (not really thirteen years before 1928) and 1919. In the paper Oort wrote after presenting the material above (Fig. 6.7):

> 'After the present paper had been practically written my attention was drawn to two papers, one by Eddington and one by Jeans, deal-

[2]For details on these books and their availability on the Web see [85, 86].

Fig. 6.7 Karl Schwarzschild and Sir James Hopwood Jeans. From Wikimedia Commons [165]

> ing with very much the same problem as discussed in this and the preceding section. This is especially true of Eddington's article in which the most general case of a stellar system with ellipsoidal velocity law has been exhaustively treated. Instead of adopting a definite system of co-ordinates and introducing the orientation of the axes of the velocity ellipsoid [...] Eddington follows the more logical way of considering general curvilinear co-ordinates [...]
> The above equations are a special case of Eddington's general formulae and his results have been used in checking these equations.'

The papers by Eddington and Jeans that Oort referred to were published in the British journal 'Monthly Notices of the Royal Astronomical Society' [166, 167] in 1915, indeed thirteen years before Oort's paper. But this is very curious as this was (and is) one of the most read and authoritative journals. How could he possibly have missed them? They were published when he still was a teenager, but if Kapteyn drew his attention to the books by Eddington and Jeans just mentioned, he surely would have directed him also to these papers, had he not already found them himself. The section ends with Oort acknowledging 'helpful discussion on the subject of the present section' with Dr. Woltjer, certainly someone who would follow these aspects of theoretical astrophysics. So, Woltjer probably was the person that drew Oort's attention to the papers. But why only mention it at this place in the paper? Was Oort keen to show that he had made his analysis independently?

Box 6.2 Hydrodynamical or Jeans equations
James Jeans, [146], translated the continuity equations of Box 6.1 into a set of equations that contained observables. For any population of stars there is at a given point in space a density ν, a mean velocity $\overline{\Theta}$ in the θ direction—the other two mean velocities assumed to be zero—and three velocity dispersions (e.g. $\langle \Pi^2 \rangle^{1/2}$ in ϖ). Jeans assumed axial symmetry and multiplied the continuity equation by a velocity and then integrated over all velocities. This resulted in three equations. In the radial direction

$$\nu K_{\varpi} = \frac{\partial(\nu\langle \Pi^2 \rangle)}{\partial \varpi} + \frac{\nu}{\varpi}\left[\langle \Pi^2 \rangle - \langle (\Theta - \overline{\Theta})^2 \rangle\right] + \frac{\partial(\nu\langle \Pi Z \rangle)}{\partial z}.$$

In the tangential direction it has only cross terms and is not useful:

$$0 = \frac{2\nu}{\varpi}\langle \Pi(\Theta - \overline{\Theta})\rangle + \frac{\partial}{\partial \varpi}(\nu\langle \Pi(\Theta - \overline{\Theta})\rangle) + \frac{\partial}{\partial z}(\nu\langle (\Theta - \Theta)Z \rangle).$$

And in the vertical direction

$$\nu K_z = \frac{\partial}{\partial z}(\nu\langle Z^2 \rangle) + \frac{\nu\langle \Pi Z \rangle}{\varpi} + \frac{\partial}{\partial \varpi}(\nu\langle \Pi Z \rangle).$$

The cross-dispersions (as $\langle \Pi Z \rangle$) in these equations are likely to be small and are often assumed to be zero in first approximation.

One thing should be said. Both Eddington's and Jeans's analyses treated very general cases, while what Oort presented was focused on the Galaxy and the solar neighborhood. And he was on the whole less interested in general mathematical theory that could not be compared directly to data; Oort always preferred directly applicable analysis over fundamental approaches not directly tied to what he was studying. Could he have ignored those papers because they were too little focused on the problems in the Galaxy at hand?

Another approach Oort took to compare his results with was along the lines of Jeans in *The motions of stars in a Kapteyn Universe* [146]. In that paper (published in 1922) Jeans introduced what became known as the hydrodynamical or Jeans equations. This procedure is to rewrite the continuity equation in terms of density, mean velocity and velocity dispersion (a measure of the scatter around the average velocity, see Sect. 4.5), which are observable—at least in principle—quantities (see Box 6.2 for mathematical details). These equations can be rewritten for the case of the ellipsoidal velocity distribution, but by themselves they have a more general validity.

I already mentioned that the first result Oort had found was that the longest axis of the velocity ellipsoid should point towards the center of the Galaxy. We have seen that this is not fulfilled in the actual observations, which is a result of local perturbations of the pure axisymmetric case and actually also of the too strict assumptions of the exact Gaussian form of the velocity distribution. Next look at the ratio of the velocity dispersions in the plane of the Galaxy. Oort then found that this ratio was prescribed by the gravitational field, and in particular the gradient of that at the position considered. This gradient translates into the value and gradient of the circular velocity and that in turn implies it can be expressed in terms of Oort's constants A en B (for the formula see Box 6.3).

As Oort showed next, this important result can also be obtained from the Jeans equations and therefore is a much more general result than would follow from the use of the Schwarzschild ellipsoidal velocity distribution. Actually, as Oort also pointed out it can be recovered from the approach that Lindblad used. This involves the fundamental idea of epicycle theory that Lindblad had introduced in 1927 [168]. It concerns the orbits of stars in the Galaxy that have small deviations from a purely circular motions. Such orbits can be described as small deviation superimposed on a circular orbit. Lindblad described these in a frame rotating with the circular orbit, so that the *mean* position of a star remained the same; the small peculiar motion resulted then in the star sometimes being ahead of this center and sometimes behind, sometimes further from the center and sometimes closer in. It turned out that these motions can be described as a two-dimensional equivalent of the oscillation of a pendulum. The oscillations in the radial and tangential directions are correlated in the sense that they have the same period of a full oscillation and the deviation is maximum in one coordinate when it is zero in the other; this results in the motion in this co-rotating frame being a small elliptical orbit around the mean position. Figure 6.8 illustrates that; it has been taken from two much later publications by Oort. On the left we see the motion in the co-rotating frame and on the right in the inertial frame. In general the resulting orbit is of course not closed in the inertial frame; but in special cases it can be closed in a properly chosen reference frame, for example when the star goes through two epicycles in the same time as the center of the epicycle goes once around the Galaxy. This is an example of a so-called 'inner Lindblad resonance' and that concept will return later in this book.

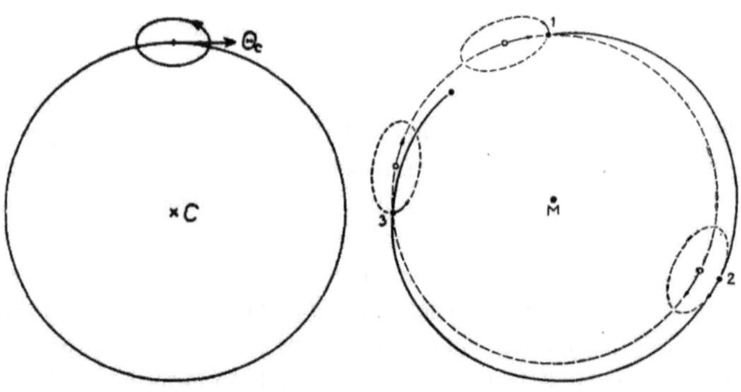

Fig. 6.8 Epicycles to describe orbits for stars with small peculiar velocities. From Oort (1965a) on the left in a co-rotating frame and (1943a) on the right in an inertial frame

Box 6.3 Oort constants and dynamical parameters of the Galaxy

The mean motion in the radial $\overline{\Pi}$ and vertical $\overline{Z}$ directions are zero, but in the tangential direction $\overline{\Theta}$ the motion is the rotation as a whole. The velocity dispersions are related by

$$\frac{\langle \Pi^2 \rangle^{1/2}}{\langle (\Theta - \overline{\Theta})^2 \rangle^{1/2}} = \sqrt{\frac{A-B}{-B}}.$$

Stars with small peculiar velocities move in small epicycles in a co-rotating frame (see text and Fig. 6.8). The axis ratio of the epicycle is $\sqrt{-B/(A-B)}$ and the period in the epicycle equals $2\pi/\kappa$, where the *epicyclic frequency* is

$$\kappa = 2\sqrt{-B(A-B)}.$$

The circular velocity Θ_c corresponds to a centrifugal force equal to K_ϖ. Then from the Jeans equation we get the asymmetric drift equation (see Sect. 5.2)

$$-K_\varpi = \frac{\Theta_c^2}{\varpi} = \frac{\overline{\Theta}^2}{\varpi} - \langle \Pi^2 \rangle \left[\frac{\partial}{\partial \varpi} (\ln \nu \langle \Pi^2 \rangle) + \frac{1}{\varpi} \left\{ 1 - \frac{\langle (\Theta - \overline{\Theta})^2 \rangle}{\langle \Pi^2 \rangle} \right\} \right]$$
$$+ \langle \Pi V_z \rangle \frac{\partial}{\partial z} (\ln \nu \langle \Pi V_z \rangle).$$

In practice (the second and third term) on the right are negative.
Oort noted that one may assume that the motions in the vertical direction are decoupled from the ones in the plane and that then the cross-dispersion on the right (the final term) would be zero.

It is of interest to put this in current context by looking at the motion of the Sun. For the solar neighborhood (at 8.2 kpc from the center), the best value for Galactic rotation is 220 km/s and the best values for the Oort constants are $A \sim 13$ and $B \sim -13$ km/s kpc. Stars near the Sun go through their epicycles in 170 million years, while the rotation around the center of the Galaxy takes 240 million years. The epicycle of the Sun (derived from its motion with respect to the mean of nearby stars) measures 0.48 kpc in the tangential (semi-axis) and 0.34 kpc in the radial direction. In the vertical direction the Sun oscillates around the plane (independently of the motion in the plane) and this oscillation has an amplitude of 85 pc and a full period of 80 million years.

The axis ratio of the epicycle and the average speed in the radial and tangential directions are determined by the Oort constants. It was easy to show, as Oort did in his paper, that the epicycle theory gave the same result as his analysis, namely that the velocity dispersion is the largest in the radial direction and that the ratio of the radial and tangential dispersions are determined by the local Oort constants. This is so, because these constants follow from the axis ratio of the epicycles. This might be confusing, since after all the epicycle is longer in the tangential direction and why then would the radial velocity dispersion be larger than the tangential one? This is because we see the epicycle in a frame rotating along at some Galactic radius. At any position, stars moving in the direction of rotation have their epicycle centers further out, where the rotation period is longer, so that these centers lag behind the rotation at that position. This reduces the tangential velocity in the frame considered, and vice versa.

An interesting consequence was that—since the ratio of the velocity dispersion depends only on the Oort constants—the observed mean motions of stars could be used to improve the estimated value of B (the determination of A was considered to be much more accurate than that of B from proper motions) and the new value of B then could then give an improved estimate of the distance to the center of the Galaxy. This turned out to be about 10 kpc, much closer to Shapley's distance to the center of the globular cluster system.

Oort next turned to the mean motion in the tangential direction (the ones in the radial and vertical directions were assumed to be zero because of symmetry) and he showed that the equation he had derived could also be recovered from Jeans' hydrodynamical equations (see Box 6.3). It showed that the average velocity in the tangential direction has to be less than the circular one following from the radial force. The latter corresponds to a circular orbit in which the attractive force from gravity is exactly balanced by the centrifugal force. Furthermore, this difference has to be larger when the population of stars considered had a higher velocity dispersion. 'Such a state of things seems actually to exist', Oort wrote. Indeed, it explained the asymmetric drift that Gustav Strömberg had discovered (see Sect. 5.2) and furthermore was also in agreement with the observations of the high-velocity stars. Stars in the solar neighborhood with ages like the Sun's have a radial velocity dispersion of about 50 km/s (so the Sun is rather slow compared to stars of its cohort) and this collection as a whole has an asymmetric drift of about 20 km/s and therefore takes 10% longer to go around the center than at the circular speed.

Except for the deviation of the vertex there was another problem that did not seem to fit the theory and that was the vertical motion. Oort found that the velocity dispersion in the vertical direction had to be equal to the radial one, while it was (and is) observed to be significantly smaller. Actually, Oort noted that his derived condition was a special case of a more general result derived by Jeans, who found that this had to apply in *any* steady system with axial symmetry. Oort's solution to this was that the Galaxy is highly flattened and he proposed that the motions in the vertical direction would have to be decoupled from and thus independent of those in the plane of the system. The mathematical formulation of this situation later became known as the 'third integral'.

Oort succeeded in explaining all major observations and features in a coherent picture of the structure and dynamics of the Sidereal System. It showed another feature of his scientific insight and intuition: the deviation of the vertex and the third integral did not fit the model; so they were simply left to be regarded as problems for later rather than objections to the synthesis.

6.3 Interstellar Medium and Absorption

The question of the presence of absorption by an interstellar medium and—if it existed—the amount of it, remained an important item, that needed solution for a good understanding of the stellar distribution and extent of the Milky Way. There are two ways of studying the effect, either through absorption lines in spectra of stars, produced by atoms or molecules along the line of sight towards the star, or by a general extinction and reddening of starlight. In 1929 Oort published a study in the first category, using the Calcium K-line (Oort 1929a).

When the first spectra were taken of the Sun, there were many dark spectral lines due to absorption of the light emitted from the solar surface in its outer layers ('atmosphere'). These Fraunhofer lines were first seen visually in the solar spectrum early in the nineteenth century. One of the early investigators was Joseph von Fraunhofer (1787–1826), who used letters to identify them running from 'A' to 'K'. It was later found that the H- and K-lines, respectively at 3968 and 3933 Å (at the violet part of the visual spectrum), were due to singly ionized calcium. In astronomy that is indicated as CaII (CaI then is neutral calcium). Later it was found that stars of spectral types similar to the Sun showed these lines as well. But in 1904, Johannes Franz Hartmann (1865–1936) of the Potsdam Astrophysikalisches Observatorium found that some binary stars showed a stationary K-line, while the wavelengths of the other lines in the spectra of the stars varied as the components moved around each other. The stationary line therefore had to come from an interstellar gas. The Fraunhofer D-line is also observed from the interstellar medium; it actually consists of two closely spaced lines at 5890 and 5896 Å and these correspond to neutral sodium (NaI). The calcium in the interstellar medium is ionized by starlight, especially from hot stars that emit much ultraviolet light. Near such stars though there the calcium is *doubly* ionized (and becomes CaIII) and no CaII H and K absorption lines are being produced in their vicinity. In spectra of later type the interstellar CaII is mixed with the lines of the same ion in the stellar spectrum. But this is not the case in early type stars and that is why the work mentioned here concerns mostly B- or A-stars.

Otto von Struve (1897–1963) at Yerkes Observatory, had estimated the intensity of calcium lines in thousands of stars. He was of Russian descent but spent almost his entire career in the USA. He came to Yerkes in 1921, where he eventually became Director. Later he served as Director of McDonald Observatory in Texas and of the National Radio Astronomy Observatory. Actually, von Struve was a descendant of a long line of astronomers, dating back to Friedrich Georg Wilhelm von Struve (1793–1864), who as we saw in Sect. 2.1, was one of the first to measure a parallax, and his

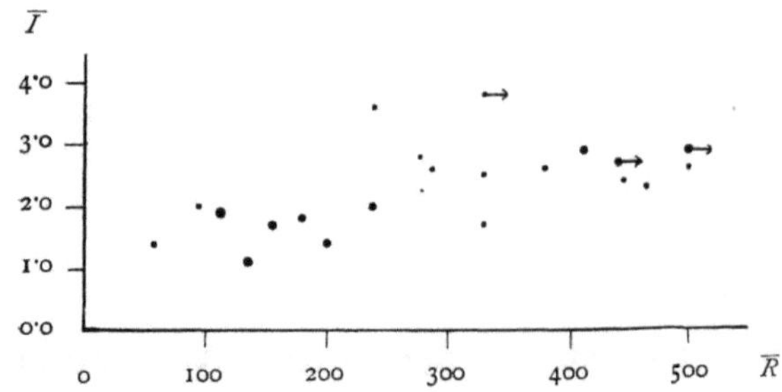

Fig. 6.9 Average intensity (or strength) $\bar{I}$ of Calcium K-lines in absorption in stellar spectra as a function of the average distance $\bar{R}$ of these stars. The intensities I are Struve's estimates; the unit is not important here. From Oort (1929a)

son Otto Wilhelm von Struve (1819–1905), both Directors of Pulkovo Observatory near St. Petersburg. Otto Struve's father was Gustav Wilhelm Ludwig von Struve (1858–1920) from Kharkov University, Ukraine, and his uncle Karl Hermann von Struve (1854–1920), Director of Königsberg Observatory and Berlin-Babelsberg Observatory. Otto Struve dropped the 'von' after he became a naturalized American.

Now, in his study Oort looked in detail at stars that Otto Struve had studied and used their proper motions as far as available to estimate secular parallaxes. In Fig. 6.9 the average intensity $\bar{I}$ of Calcium K-line absorption lines are plotted as a function of the average distance $\bar{R}$. There is a tendency for the intensity to increase with distance from the Sun, indicating evidence for a gaseous component to the interstellar medium. Oort also concluded that there was 'a slight tendency for larger intensities in the lower Galactic latitudes, indicating that the interstellar gas is concentrated towards the Galactic plane,'.

The interstellar medium can of course be used to study the properties of Galactic differential rotation as well, as was done extensively in a study by John S. Plaskett and Joseph Algernon Pearce (1893–1988) of the Dominion Astrophysical Observatory in Victoria at Saanich, British Columbia, [169]. Plaskett, as we have seen, had been one of the first to confirm Oort's differential rotation model. Over the years he and his colleague Pearce had collected spectra of all brighter (down to magnitude 7.5) B0 to B5-stars and had measured in all of these the radial velocity of the stars themselves and of the interstellar CaII H and K-lines. In the sample of 235 stars, they found about 160 'interstellar cloud velocities' suitable for analysis. Now they used the same method as had been used before to derive the solar motion from such measurements, but included the effects of differential rotation. They ended up with six free parameters, a value K for a possible systematic off-set of the velocities, the direction (coordinates of Apex and Antapex) and magnitude of the solar motion, but in addition a value for the product of mean distance of the stars and Oort constant A ($\bar{r}A$), and the longitude of the Galactic center. A second solution was made assuming the generally accepted best values for the solar motion. This showed that differential rotation was observable in the interstellar material as well as in the stars. The longitude of the center was 332°–335° rather than Oort's 323° or the 325° for the globular cluster system. Taking the best value for Oort constant A they found that the average distance of the clouds was some 450 pc. The systematic error K was only half a km/s or so and not significantly different from zero.

This induced Oort to redo the analysis, since he would not have accepted the difference in longitude of the rotation center with his value or that of the globular clusters. If interstellar gas showed differential rotation it had to be around the same center. This resulted in a short note (Oort 1930b). He rejected the systematic error K, confined the data to 20° or less from the Milky Way and took out some star clusters. Furthermore, he *assumed* the center was at longitude 327°, which he felt was the best value then. The result then was a solution for the solar motion and for the $\bar{r}A$, the mean of the product of r and A. Since he only used stars close to the Milky Way, he could not find a solution for the Galactic latitude of the Apex. The result could be compared directly to the best values available from stellar fits, which Oort took to be those determined by William Wallace Campbell (1862–1938) and Joseph Haines Moore (1878–1949) from Lick Observatory. This produced an excellent fit, the velocity of the Sun agreeing to better than one km/s (unlike Plaskett and Pearce) and the Galactic longitude of the Apex by 1° or less. The value of $\bar{r}A$ was now reduced (from 7.3 to 5.7 km/s). He made a separate solution using only stars with the strongest interstellar lines and found the longitude of the Galactic center derived from both the interstellar and the stellar velocities to agree within one degree (although with uncertainties of a few degrees). The analysis of the motions of the interstellar gas were in good agreement with other determinations after all. This is typical of Oort's approach; different solutions to the same problem should give consistent results, and if they would not agree then one should carefully re-analyze the data.

The presence of interstellar absorption by particles in the interstellar medium was finally universally accepted at about the same time, especially through the work of Robert Julius Trumpler at Lick Observatory. It is fortunate that the editors of 'Hemel and Dampkring' had asked Oort to review the situation for their readers, which he did in a fairly long paper (Oort 1931b). He started by recalling that in the Kapteyn Universe the vertical distribution of stars was rather good, but in the plane of the Galaxy it was very likely that it was affected by absorption, as he had suggested in his inaugural lecture as privaat docent (Oort 1927a). In the plane of the Milky Way, the apparent density fell off with distance from the Sun in all directions and Oort proceeded to estimate what amount of absorption was required to produced this fall-off in the case of a constant density with distance from the Sun. The answer was about one magnitude per kiloparsec, which is not a large amount. The observed rotation of the Galaxy suggested a larger extent than the Kapteyn Universe, probably as large as the Shapley globular cluster system. The amount of absorption required to produce the apparent disagreement was not unreasonably large.

Another option Oort put forward was to use the extent in the vertical z-direction for stars that appear concentrated to the Milky Way. This should not vary systematically with distance. Now it had been found for Cepheids by Kurt Felix Bottlinger (1888–1934) from Berlin and collaborators that the mean distance from the plane increased with distance from the Sun, unless there was an absorption of 1–2 magnitudes per kpc.

An important study that Oort next described was that by Trumpler (Oort still wrote Trümpler). He had obtained magnitudes and spectra of many stars in clusters and estimated the distances by comparing the apparent magnitude of stars with the absolute magnitudes of stars with very similar spectra. For the clusters these stars belonged to, Trumpler had also measured their diameters on the sky and classified them according to richness of stars and degree of central concentration. Within each class the diameters seemed to increase with cluster distance, unless these distances were corrected for an average interstellar absorption of 0.67 magnitudes per kpc.

The final piece of evidence came from following the original suggestion by Kapteyn that the absorption was selective in the sense that stars at larger distances would appear redder. Oort quoted studies by Stockholm astronomer Karl Yngve Öhman (1903–1988) and by Peter van de Kamp. The latter studied stars of spectral types between B0 and A5 of about apparent magnitude 10; their average colors were redder than those of brighter (more nearby) ones, and the effect was more pronounced for the early B-stars than for the A-stars, since the former should be more distant.

The case for absorption was overwhelming. Oort added that from the distribution of globular clusters on the sky he had estimated that the absorbing layer would have a thickness of order 100 pc. Kapteyn had originally suggested that the mechanism was 'Rayleigh scattering', which as we now know is scattering by individual molecules rather than by dust particles. However, Trumpler's data showed that the wavelength dependence was more like linearly proportional to the inverse of the wavelength, very different from the fourth power required by Rayleigh scattering. Oort did not mention dust or any other characteristic or property of the 'scattering medium' nor any candidate for it.

6.4 Family Life, Leiden and the IAU

After their marriage the young Oorts had settled in Leiden into a house on Plantsoen 35, just inside the canals around the medieval town. But already a year or so later they moved into a bigger house with the address at Cobetstraat 63. This was close to their first house but just outside the canals. Both were at walking distance from the Observatory and from the center of Leiden. The move to the bigger house might have been in anticipation of the birth of their first child, Coenraad Jan. He was born on December 5, 1928. In daily life 'Coen' (pronounced 'Koon') was his first name and since this was more than just a family nickname, I will adopt this also. Coen was a few months old in Fig. 6.10. Eventually the Oorts were to have three children. Marijke was born on April 25, 1931 and Abraham Hans (Bram) on September 2, 1934. Coen and Bram were named after close family; Coen after his uncle Coenraad Jan Graadt van Roggen and Bram after his grandparents (Oort's parents) Abraham and Ruth Hannah Oort. Marijke was not named to anyone in the family, according to Bram Oort.

Not too much can be learned about their private life from the letters in the private section of the Oort Archives. The Archives have only a few letters to and from Oort's parents (not surprising as they did not live very far apart). There are a few with the parents of Mrs. Oort, although these mostly contain congratulations on birthdays and so. Not long after their marriage, in October 1927, Oort wrote to his father- and

Fig. 6.10 This picture has been taken in 1929, when Coenraad Jan, the oldest child of the Oorts was only a few months old. Mrs. Oort (in the middle) is pictured with her parents-in-law and sister-in-law Jetske Oort. From the Oort Archives

mother-in-law how happy he was with their daughter as his wife. And some time in 1928 he thanked them for the financial support; the salary at the University must have been hardly adequate for a couple with child.

A major happening in Leiden must have been the General Assembly of the International Astronomical Union, which in 1928 was held there. In 1925, Oort had attended the second General Assembly in Cambridge, UK. This took place from 14 to 22 July. Willem de Sitter, Frank Schlesinger, and three others received honorary degrees as Sc.D. from Cambridge University and the meeting afterwards moved to London where the 250th anniversary of the Royal Greenwich Observatory was celebrated. President of the Union was William Wallace Campbell (1862–1938) of Lick Observatory and de Sitter was one of the five Vice-Presidents.

Oort's role was limited. After all, he had not yet defended his PhD thesis and had joined the staff of Leiden Observatory only recently. The meeting of Commission 8 on Meridian Astronomy, presided over by Frank Watson Dyson (1868–1939), Astronomer Royal and Director of the Royal Greenwich Observatory, had 'Dr. Oort' as secretary (not aware that his Ph.D. had not yet been awarded) [170]. At the 1922 Rome General Assembly, Stanley Samuel Hough (1870–1923), Director of the Cape Observatory had been appointed President of that Commission, but he had died in 1923. Dyson must have taken over his responsibilities; de Sitter was a member and must have had the secretarial responsibilities, which he would then have delegated to Oort. This Commission discussed extensively the matter of absolute declinations and

Fig. 6.11 Group photograph of participants of the third General Assembly of the International Astronomical Union at Leiden in 1928. The white bars on the top and the left sides indicate the location of Oort (see Fig. 6.12). From the Archives Leiden Observatory

possibility to measure these 'by the azimuths of stars at their greatest elongations'. Apparently de Sitter (and Oort) discussed their proposal to establish observing sites and send 'an observer with a special instrument to an equatorial, a northern, and a southern station'. The Union awarded Leiden financial support for this at 25,000 francs per annum over the next three years. This were French francs and the sum was of the order of Dfl. 4,000 at that time, or €25,000 today. At the end of the meeting (again Dr.) Oort was voted a member of Commission 8 on Meridian Astronomy and of a new Commission 33 on Stellar Statistics.

In 1928 the IAU held its third General Assembly in Leiden. Willem de Sitter was President. Figure 6.11 shows the attendants in session in the 'Stadsgehoorzaal', the Leiden Municipal Concert Hall. Oort in that audience and the full Executive Committee on the stairs of Leiden Observatory have been illustrated in Fig. 6.12. The meeting took place between July 5 and 13, 1928.

The meeting actually started in Den Haag, where the Minister of Education, Arts and Sciences, Marius Waszink (1881–1943) opened the congress in the 'Ridderzaal' (the Hall of Knights), the main auditorium of the seat of Parliament.[3] The number of participants in this third General Assembly of the International Astronomical Union was 243, up from 83 at the first General Assembly held in Rome in 1922 and from 189 attending the second in Cambridge in 1925. In addition, 91 'Ladies' accompanied the astronomers. The Dutch delegation was of course the largest, consisting of 46 astronomers (this must have included students) and 17 ladies. De Sitter had, using the President's prerogative, invited German and other astronomers from countries that

[3]The opening reception of the 1994 General Assembly of the IAU was also held there.

Fig. 6.12 On the left Oort in the group photograph of the General Assembly of the International Astronomical Union at Leiden in 1928. See the bars in Fig. 6.11 for his location there. At the right the Executive Committee of the IAU on the stairs of Leiden Observatory. In the front Henri-Alexandre Deslandres (1853–1948) from Paris, President Willem de Sitter, Shin Hirayama (1867–1945) from Japan, in the back Frank Schlesinger, General Secretary Frederick John Marrian Stratton (1881–1960) and Arthur Eddington. From Archives Leiden Observatory

were not IAU members. In his opening speech he said: 'Is is a great satisfaction for me to be able to say that very many of them – about one-eighth of our total number – have accepted this invitation. Nearly one-half of these belong to the great German nation, great, I mean, in the number and importance of its contribution to astronomy ...'. The chair of the Natural Science section of the Royal Academy of Sciences[4] spoke some conciliatory words about Kapteyn, who had strongly opposed the IAU as long as Germany and the central powers were forbidden to join, and who had never attended any meeting of the Academy since: 'Our two greatest Dutch astronomers, Huygens [Christiaan Huygens (1629–1695)] and Kapteyn, not even to mention the living, were no men of narrow nationalism; they took a strong stand for the universality of science. Especially Kapteyn has suffered much because of the alienation between nations that the war has caused.' During the General Assembly business meeting even the publication of the (Dutch) biography of Kapteyn by his daughter was noted. All of this must have felt very gratifying to Oort (if not to de Sitter and van Rhijn).

Oort's role again was not very prominent. The 'Transactions' do not mention any scientific presentations, but it can safely be assumed that Oort discussed outside the formal sessions his work on the rotation of the Galaxy, published the year before, and on stellar dynamics, which was to appear in November of 1928. Again, he acted as secretary of Commission 8 on Meridian Astronomy, but no reporting from himself has been recorded. Carl Sanders reported on a new project in Leiden on measure-

[4]The 'Academy' became the '*Netherlands* Academy' (KNAW) only in 1938.

ments of absolute declinations, together with Jean Jacques Raimond, Jr. (1903–1961) involving simultaneous observations with the meridian circle and a universal instrument. Raimond (which is pronounced as a French word) was a student in Leiden, who in 1934 would defend his Ph.D. thesis under Pieter van Rhijn in Groningen on *The coefficient of differential Galactic absorption*. Sanders and Raimond published a paper on their project later in 1928 [171]. The idea was to have two stars with approximately the same right ascension and observe then on the meridian (assuming known refraction and flexure); the meridian circle is used to measure the zenith distances, which gives the difference of the declinations and the universal telescope the angle between the two, which with the geographical latitude of the Observatory gives the sum of the declinations. Both declinations can then be computed. Oort will have had a clear interest in this, but probably was too busy with his Galactic work to be actively involved.

Oort must also have been participating in the meeting of Commission 33, which was presided over by Carl Charlier, Director of Lund Observatory in Sweden. The business was mostly concerned with defining the Galactic equator for the coordinate system and conversion of Galactic to equatorial coordinates and projected proper motions between these two systems. Tables of such conversions were to be provided and funds made available. Charlier also proposed a more 'natural distance unit', namely one million Astronomical Units (AU). Following von Seeliger, this was to be called the Sirius-Weite or *Siriometer*, since it was supposed to be roughly the distance of Sirius and that was the unit William Herschel had used. The absolute magnitude was then to be the apparent one at a distance of 1 Siriometer. The proposal was not accepted, which is all the better as the distance to Sirius is now known to be 8.6 pc, while one million AU amounts to only 4.8 pc, only a bit more than half of it.

At the end of the Assembly the membership of the Commissions was voted and Oort now appeared as a member of only Commission 33.

6.5 Leiden, Harvard or Columbia?

Now that Oort's star was rising and his reputation established, it should come as no surprise that he started to receive job offers, especially from the United States. There may have been more that never got beyond the stage of initial approach, but there are two that stand out as offers that he considered very seriously, namely from Harvard University in Cambridge, Mass., and from Columbia University in New York city (Upper-Manhattan). He eventually turned both down and it appears that after the second offer in 1930 had been rejected, he and his wife had made up their minds to henceforward make Leiden and the Netherlands their home-base for the rest of their lives. Figure 6.14 shows the letter from Harlow Shapley, Director of Harvard College Observatory, of December 22, 1928, offering him a professorship at Harvard University (Fig. 6.13).

Fig. 6.13 Cartoon in the newspaper Leidsch Dagblad during the IAU congress. The arrival of the participants. The Sun in front says *'What a terrible railway station this is, Manus'*; Manus is the Moon (Maan in Dutch) behind him. And Jupiter with the cigar says *'Can you get through this door with those rings, Saturn?'*. From the Oort Archives

The offer came months after the Leiden IAU General Assembly. The letter from Harlow Shapley was five typed pages long plus a sixth in handwriting. It started as follows:

'Dear Dr. Oort,

I was planning a few days ago to write you a note commenting on that recent excellent paper on Galactic motions [...]

But I have now decided to write you a more interesting communication, in fact, I should like to invite you to come to America.

Harvard University has naturally looked to me to recommend and arrange for the filling of a new professorship in astronomy. The position is in the University as distinct from the Observatory, and nominally it is a professorship for the instruction of university students. You may have heard of Robert W. Willson, and of the Robert W. Willson Professorship founded by his widow. The chair has just been founded, and we are concerned with its first occupant. Although the bequest expresses the desire of Mrs. Willson that the holder of the professorship should be primarily a teacher, and should follow the semi-laboratory methods of instruction that Professor Willson developed, the Trustees of the Fund are going to be very liberal in their interpretation, and we can frankly say that although a certain amount of undergraduate lecturing and responsibilities for the undergraduate general course in astronomy will be required of the Willson Professor, freedom for and expectation of active research in astron-

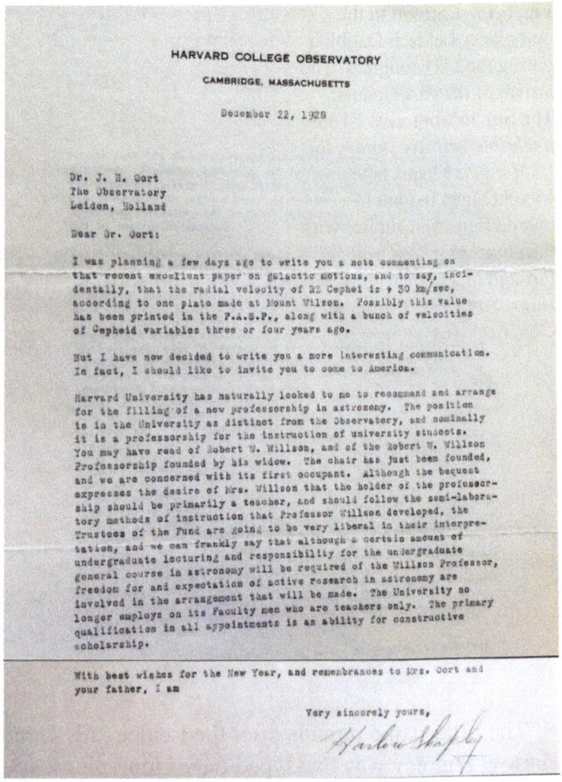
HARVARD COLLEGE OBSERVATORY
CAMBRIDGE, MASSACHUSETTS

December 22, 1928

Dr. J. H. Oort
The Observatory
Leiden, Holland

Dear Dr. Oort:

I was planning a few days ago to write you a note commenting on that recent excellent paper on galactic motions, and to say, incidentally, that the radial velocity of RZ Cephei is + 30 km/sec, according to one plate made at Mount Wilson. Possibly this value has been printed in the P.A.S.P., along with a bunch of velocities of Cepheid variables three or four years ago.

But I have now decided to write you a more interesting communication. In fact, I should like to invite you to come to America.

Harvard University has naturally looked to me to recommend and arrange for the filling of a new professorship in astronomy. The position is in the University as distinct from the Observatory, and nominally it is a professorship for the instruction of university students. You may have read of Robert W. Willson, and of the Robert W. Willson Professorship founded by his widow. The chair has just been founded, and we are concerned with its first occupant. Although the bequest expresses the desire of Mrs. Willson that the holder of the professorship should be primarily a teacher, and should follow the semi-laboratory methods of instruction that Professor Willson developed, the Trustees of the Fund are going to be very liberal in their interpre-
tation, and we can frankly say that although a certain amount of
undergraduate lecturing and responsibility for the undergraduate
general course in astronomy will be required of the Willson Professor,
freedom for and expectation of active research in astronomy are
involved in the arrangement that will be made. The University no
longer employs on its Faculty men who are teachers only. The primary
qualification in all appointments is an ability for constructive
scholarship.

With best wishes for the New Year, and remembrances to Mrs. Oort and your father, I am

Very sincerely yours,

Harlow Shapley

Fig. 6.14 Letter of Harlow Shapley of December 22, 1928, offering him a professorship at Harvard University. It is a collage of the first page and the bottom part of p. 5. Note the good wishes to Mrs. Oort and Oort's father, whom he might have met during the Leiden IAU General Assembly. Almost all of what is displayed is quoted in the text. From the Oort Archives

omy are involved in the arrangement to be made. The University no longer employs on its Faculty men who are teachers only. The primary qualification in all appointments is an ability for constructive scholarship.'

Robert Wheeler Willson (1853–1922) was the founder of the students' astronomical laboratory, where astronomy was taught using observing and demonstration apparatus. He had retired in 1919. The laboratory has been described by Harlan True Stetson (1885–1964) [172], who was in charge at the time the offer to Oort was made of the astronomy teaching at Harvard, except for the advanced students. The bequest (of $150,000) was 'to support a professorship of applied astronomy including navigation at the College itself as distinguished from the Astronomical Observatory'. Shapley noted in the letter that Pickering, his predecessor as Director, and Willson did not work closely together, since 'they had different views and saw little of each other'.

Shapley further estimated that the 'instruction schedule for the place will be aimed to consume less than half of a man's energy and time' and that 'Mr. Harry Plaskett has been called to assist in the direction of research students and in the instruction of graduate students' in astrophysics. This is the same Harry Hemley Plaskett that would deliver the Presidential address, when Oort received the

Gold Medal of the Royal Astronomical Society in 1946 and which inspired the subtitle of this book. He was the son of John S. Plaskett whom we met in connection with the confirmation of Oort's theory of Galactic rotation. With Plaskett's efforts and the growing involvement of Observatory staff, Shapley hoped to offer an educational program designed to attract the most competent students. The Willson Professor would, like Harry Plaskett, have a permanent office at the Observatory and have access to all its facilities. He even suggested a sabbatical year at the 60 in. reflector in South Africa. This Rockefeller Reflector was eventually installed in 1933 at Bloemfontein, Free State, South Africa. This was the Harvard Southern Station that had been earlier in Peru, most recently as Boyden Station at Arequipa. The 24 in. Bruce Telescope that was used for the southern part of the Plan of Selected Areas from Arequipa, had been moved to the Bloemfontein in 1929.

> 'To come to the more personal phase of this question, I should like to ask if you would be interested in the Willson Professorship. The Trustees of the Willson Estate naturally first hoped we would investigate American candidates, but after considering the matter fully, we have decided that there is no available American astronomer, of suitable youth, competence, and scientific standing. The Harvard standards are higher than for any other university in America, we believe. Though there are some American men who might be proposed for an instructorship or assistant professorship, we have gone into the situation and have convinced the Trustees that we should look towards Europe. For reasons that need not be specified, we think of you first of all. We feel that you would fit into the University community excellently; that your cooperation with the researches in physics and astronomy would be excellent; and I have assumed personal responsibility for your ideas, scientific capability, and ambitions. Your age, of course, is rather startling, but Harvard has heretofore appointed very young men to high positions.'

Shapley went on to explain that assistant professors and instructors have temporary contracts, but associate or full professors were permanent. A possibility would be for Oort to come on a temporary basis for one semester (February–June 1929). But if Oort would prefer a permanent arrangement from the beginning, he knows 'enough of your work and potentialities to recommend to the President of the University the making of a permanent appointment from the first. In such a case you could come to Harvard either on February 1, 1929 (a month delay would not be serious), or at the beginning of the next academic year, September 1929'.

Before concluding the letter, Shapley mentioned that he had discussed this with Schlesinger, who 'not only recommended that I invite you to come to America, but also expresses the belief that you would consider the matter favorably'. The last page is a PS in 'longhand. I would recommend to the University, in the event of your acceptance, an initial annual salary of $6500, if that is satisfactory to you. In addition here is 5% for an accumulating pension fund'. The $6500 correspond to about €76,000 now or about three times the annual wage of an unskilled worker. And he mentioned the months free from classes and examinations (June to beginning October) and undergraduate teaching (January and May).

Fig. 6.15 Oort in 1930. From the Oort Archives

The time scales were very short. The draft of Oort's reply was dated January 13, 1929. He was of course very honored by the offer, but remained dubious. 'I cannot but believe that, if I accepted, you would soon be disappointed considerably.' He noted he was at that time very happy in Leiden, and should not 'sacrifice so good a present for an always partly unknown future somewhere else, unless there would be very convincing grounds for such a change.' And he said: 'Very closely tied up with this is the personal factor, viz. whether Mrs. Oort and I would in the long run feel as happy in the States as we are here.' The access to the Harvard plate material and telescopes, for slitless or objective prism spectroscopy, photometry and radial velocity work must have made the offer very tempting. Oort liked the prospect of the Bloemfontein year very much too. But he remained noncommittal and had questions about the duties and responsibilities for the student's laboratory and the time available for research (Fig. 6.15).

Certainly Oort would have consulted de Sitter and Hertzsprung. There is no written record of that, but there are letters from van Rhijn, his former supervisor in Groningen, and from Schlesinger. Especially the latter seems to have been regarded by Oort as his most trusted adviser and mentor. Van Rhijn wrote a fairly long letter (in handwriting) on February 15. He first addressed the financial side.

> 'Indeed that would in the present circumstances weigh very heavily. In the long run it would be difficult to live on a salary of Dfl. 5000 [apparently the maximum Oort can expect unless there was some promotion]. It is true that there would be vacancies in something like thirteen years [de Sitter's retirement] and that you will have an excellent chance for promotion. But thirteen years is a long time for which any prediction is difficult. It would therefore not be good to stay here hoping that in thirteen years you will get a better astronomical appointment.'

Then van Rhijn commented on the research opportunities, which he judged to be quite good. On the other hand he could also work from Leiden in the south, using Schlesinger's Yale Telescope in South Africa, at least if de Sitter could afford sending him and his family there for a year. Then there was a comment on Shapley. 'I know that most American astronomers are more practical than theoretical. And they often publish too soon. They regularly make statistical blunders.' He reminded Oort of Kapteyn's admiration for American Observatories and quoted Ehrenfest as saying that Shapley is 'the only American astronomer that is able to think'.

Van Rhijn put forward six questions and indicated with the length of an arrow the weight of an affirmative or negative answer. Will de Sitter be able to arrange an increase in salary in Leiden? If not, go to Harvard. Will de Sitter have funds to let Oort go for a year to America or South Africa? Doubtful, but possible and if so, stay in Leiden. Will he, Mieke and the children be able to settle happily in America? Van Rhijn put a question mark, but if the answer is no, definitely stay in Leiden. Will he get major access to at least one telescope or be able to have people make observations for him regularly from South Africa? If no, definitely stay in Leiden. Would he have to spend no more than 2–3 days per week on teaching? If yes, maybe go to Harvard. Van Rhijn considered the matter of the salary (question 1) very important. But it was up to Oort to draw his conclusions once he has all the answers. But don't take too long: 'Think of Gill's words: The man who hesitates is lost'. Looking at all of this, it seems to me that van Rhijn is not very resolute and forceful, at least I have seen wiser and more helpful advise (and personally have had, for example from Oort).

Oort had great trust in Schlesinger and the latter's letter of advise on the Harvard offer was detailed and definitely much wiser than van Rhijn's, and certainly more helpful. Oort had written him almost immediately after the offer first came; the letter is in the Oort Archives only in a schematic draft form.

> 'A few days ago I received a most exciting letter from Dr. Shapley, inviting me to come to Harvard as a Professor at the University. My first thought was to obtain <u>your</u> advise. I wish you would spare a few moments of your time to write <u>me</u> your opinion. [...]
>
> The first question I would like to ask you is; would you think that at Harvard I could do perhaps a little more useful astronomical work than here? I feel that every step I take now might be entirely decisive for my whole life. I do not like to try the Harvard position for a year, unless I have pretty well decided to stay there, and I believe it might be very hard even to return to Europa if you get American children! [...]
>
> I needn't tell you that I could hardly believe my eyes when I read Prof. Shapley's letter.'

Schlesinger's letter was dated January 31, 1929 and started out by saying he received Oort's letter 'only a few hours ago'. Schlesinger started by pointing out the danger of a much more heavy teaching load than Oort was expecting. He mentioned the Stetson account just cited [172] and in particular the paragraph about the endowment for the chair, which stressed the aim of establishing a lasting tribute to Willson, and stipulated that the teaching 'shall be done for the most part by laboratory methods and not by lectures'. Schlesinger continued.

'I think you are minimizing the amount of energy that you would be putting into teaching. The number of students at the Laboratory has averaged more than 160 in the last twenty-five years and is now, I believe, nearly up to 200. More important than this, Dr. Shapley has interpreted very freely Mrs. Willson's wishes, but you can easily imagine circumstances in which this interpretation would not hold, or at least in which pressure would be exerted upon the Willson Professor to devote more and more of his energy to the primary object for which the professorship was founded. It is not at all inconceivable, for example, that Dr. Shapley may be called away from Harvard to fill a still higher post; or it may happen that a reorganization or change in administration in the University itself may tend to separate completely the professorship from the Observatory, as was the case when Willson was in charge of the Student's Laboratory. In any case, Jan, if I were you I should not accept this place unless I were prepared to give a good deal of thought and energy to teaching undergraduates, for any other course would seem to be outside of the wishes of the founder of the professorship.'

Schlesinger then summed up the advantages. Greater possibilities in America in general than in 'Holland'. Oort may expect to succeed de Sitter as Director in Leiden, but that would still be 13 years away, and returning from the USA to the Netherlands would be hard for him and his family, while the salary in Leiden would be less than what he would be used to by then (Schlesinger estimated Leiden would pay half as much as the Willson salary). He could of course accept the Willson chair and move on to other positions in a few years ('such as the new Observatory at the California Institute of Technology', which refers to the Palomar Observatory for which the Rockefeller Foundation had just provided sponsorship for a 200 in. Telescope). This Foundation had been established in 1913 by John Davison Rockefeller Sr. (1839–1937) and John Davison Rockefeller Jr. (1874–1960) to 'promote the well-being of humanity throughout the world'. Contrary to van Rhijn, Schlesinger did feel that 'you and Mieke are wise in not emphasizing too much the financial aspect'.

In reply to an apparent question by Oort, Schlesinger remarked on Harry Plaskett, saying he is 'an unusually good man', likely to succeed Shapley as Director in case that post became available. And he explained that Stetson was leaving *because* of the Willson Professorship, since 'the authorities at Harvard did not believe that Stetson could fill it acceptably', although Schlesinger estimated him as 'by no means inferior or even mediocre'.

The upshot of it all was that the position would involve much teaching and the situation concerning the professorship and the relation with Harvard University and the Observatory would be uncertain and potentially impossible. In the end, Oort and his wife decided to stay in Leiden. I quote from the letter Oort wrote on February 15, 1929 to his parents-in-law informing them first of their decision.

'For some time I have had the feeling that this [rejecting the offer] was the best solution, but was curious but also unsure what the advise would be in a letter that I expected from Schlesinger. It came a few days ago and was a very nice letter. Calm and well thought-out (as I expected) and at the same time very friendly. [...]

After what I had heard from Shapley and de Sitter, I had really expected to receive a very positive letter in the sense of advising me to go there, but the letter has hardly changed my thoughts, although possibly regarding other offers from America that I may get in the future.

I had a long conversation with Hertzsprung yesterday and decided in principle not to accept. In the evening Mieke and I visited Mr. and Mrs. de Sitter and discussed the matter again in detail, also concerning my growing need to obtain observational material that this offer has awakened in me, and we told them what our decision was with which they were very happy. [...]

In preparation for this 'important' visit de Sitter this afternoon had called up the secretary of the Curators to find out if there was anything definite from 'Den Haag' [the Hague as the place where the Ministry of Education resides] (the mayor is now outside the loop of course). The secretary called Den Haag and was informed that the papers were on their way. [...] Their content according to the message by phone is that the Minister is prepared to put an item on the budget for 1930 for a position as 'Lector' of Dfl. 2500 and to provide the same sum for 1929 as a special item for the lectures I have given as privaat docent. I therefore will get immediately Dfl. 1200 for last year and the same amount this year. This is very nice, much nicer than I had hoped for, and I will not be surprised if your letters gave an important push to turn it into such a positive outcome, father.

This will be added to my salary as Conservator and will make for a very nice income. Mieke and I at this moment feel like millionaires.'

The position of Lector or Reader is one grade below that of professor. The lectures that were given by a privaat docent were in general not covered by a salary. So, there had been some lobbying with the help of father-in-law Willem Graadt van Roggen, who had some contacts in government circles.

Oort wrote Shapley on February 17, 1929 to inform him. From his draft notes:

'As I have already written, the prospect of going to Harvard and starting on new work, with the possibility of trying out at least part of some vague ideals for which at Harvard there might well be a fine opportunity, attracts me more than I can say, especially as I feel an increasing want for more intimate contact with observations, or eventually, for making observations myself. It seems that a statistical astronomer could never get very far without this and in my opinion he would neglect his duty if he did not try his utmost to get the observations which he thinks he especially needs or will need in the future.

However, I am afraid that for me personally the Willson professorship would not be the right way to reach this. [...] And then the professorship is founded primarily for teaching of undergraduates. Two lectures a week is certainly very little, but with the responsibilities for between 100 and 200 students there would certainly be many additional duties. Unless I should give a very considerable amount of thought to these, and then minimize the already small quantity of astronomical work I am able to do, I should always have the unhappy feeling of failing in one of my principal duties.

Although I like to do some teaching neither my main ambition, nor I believe, my natural capability lie in the direction of teaching undergraduate students, of whom only a few would be more than casually interested in astronomy.'

The observational work that Oort was interested in concerned primarily proper motion and radial velocity data of stars for his work on Galactic structure and dynamics. But in his notes back to his early days in Leiden there is also a growing interest in studying external galaxies and he had been looking for a long time for suitable candidates to measure the brightness distribution from photographic plates. Access to a large reflector in the southern hemisphere will be a recurring issue.

The second serious job opportunity, this time for a professorship at Columbia University, came a year later. Schlesinger wrote him on January 25, 1930 to give (as it turned out in vain) an early warning (see Fig. 6.16).

'My dear Jan,

Here is some news which will at least interest you, and which I hope you will find important. It is that another American professorship is about to be offered to you. This is Columbia University, in New York City. The offer looks at first sight a good deal like the one you received from Harvard; but after thinking over the matter carefully, this professorship is, in my

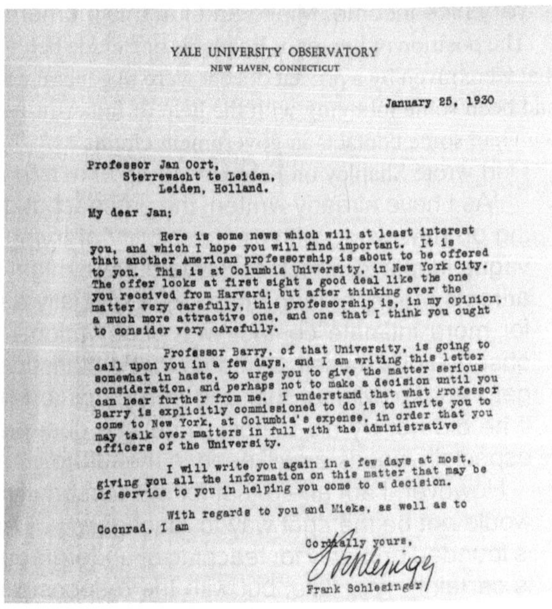

YALE UNIVERSITY OBSERVATORY
NEW HAVEN, CONNECTICUT

January 25, 1930

Professor Jan Oort,
Sterrewacht te Leiden,
Leiden, Holland.

My dear Jan:

Here is some news which will at least interest you, and which I hope you will find important. It is that another American professorship is about to be offered to you. This is at Columbia University, in New York City. The offer looks at first sight a good deal like the one you received from Harvard; but after thinking over the matter very carefully, this professorship is, in my opinion, a much more attractive one, and one that I think you ought to consider very carefully.

Professor Barry, of that University, is going to call upon you in a few days, and I am writing this letter somewhat in haste, to urge you to give the matter serious consideration, and perhaps not to make a decision until you can hear further from me. I understand that what Professor Barry is explicitly commissioned to do is to invite you to come to New York, at Columbia's expense, in order that you may talk over matters in full with the administrative officers of the University.

I will write you again in a few days at most, giving you all the information on this matter that may be of service to you in helping you come to a decision.

With regards to you and Mieke, as well as to Coonrad, I am

Cordially yours,

Frank Schlesinger

Fig. 6.16 Letter of Frank Schlesinger of January 25, 1930, announcing the impending offer for a professorship at Columbia University. Almost all of what is displayed is quoted in the text. From the Oort Archives

opinion, a much more attractive one, and one that I think you ought to consider very carefully.
Professor Barry, of that University, is going to call upon you in a few days, and I am writing this letter somewhat in haste, to urge you to give the matter serious consideration, and perhaps not to make a decision until you can hear further from me. I understand that what Professor Barry is explicitly commissioned to do is to invite you to come to New York, at Columbia's expense, in order that you may talk over matters in full with the administrative officers of the University.'

The correspondence about the offer went almost all through Schlesinger as an intermediary. There must have been letters to and from 'Dean Pegram', who was physicist George Braxton Pegram (1876–1958), Dean of the Faculty of Applied Sciences and chairman of the Physics Department, but there are no copies in the Oort Archives. Oort wrote Schlesinger, that 'I am not inclined to believe that it has been not without your influence that the Columbia authorities came to choose this strange way! This proposition to come to Columbia university mainly for the purpose of establishing an active research department I consider as a great honor. Your first letter arrived two days after Professor Barry's visit. It did not therefore fulfill its purpose in preparing me for the shock ...'. Prof. Barry had also expressed the view that what the authorities had in mind was a theoretical department, since New York was not a good place to erect a good telescope (Fig. 6.17).

But Oort then put forward, that 'Mieke and I put the question why try this at New York and not at Leiden?' Why indeed should they leave their native country? 'Especially Mieke is so much attached to Holland and many of its inhabitants that she definitely hates the idea of living in another country.' And what would Oort want scientifically?

> 'I should rather look for a <u>closer</u> contact with observational astronomy than I have here, than for the opposite. If a good fairy came along and told me that I could choose freely what I should most like to have I would choose one large telescope and the cooperation of another astronomer whose main ambition was observing and I believe that several astronomers of this kind could be found, so that together we could really get the things we most wanted to get for a study of the structure of the Universe.'

To be more specific, Oort described that what he meant by a large telescope was a 60 or 70 in. reflector in the southern hemisphere. So the situation was: Oort wanted a commitment to a large southern telescope now and there was a serious problem in that his wife was not very keen to move to America.

Hertzsprung felt he should get involved in the situation also and separately wrote a letter to Schlesinger explaining that Oort was happy in Leiden and that what could move him to go to Columbia was a large telescope in the South for measuring radial velocities. It is not at all clear that this intervention was helpful, but it made clear that a commitment to a southern telescope was required. On April 7, Schlesinger wrote:

> 'I have your letter relating to the Columbia situation, and I have also received one from Hertzsprung, to which I replied at once. Perhaps I

Fig. 6.17 Frank Schlesinger in 1928, very probably during the Leiden General Assembly of the International Astronomical Union. From the Oort Archives

had better make clear to you one of the things that I have already said to him; namely that it is a mistake to suppose that I am urging you to go to Columbia. I do think, however, that it is an extremely attractive offer and situation, and one in which you will be almost certain to achieve a notable success.

Columbia is, as I have already said, a very wealthy institution. [...] I believe that everything is possible at Columbia, and that, if a man were to achieve good success in any form of scholarship, the funds would eventually be forthcoming to give him all reasonable support and aid in his work. In this sense I think you might look forward to securing a great reflector in the southern hemisphere in ten years or so. But I feel sure that the authorities will be quite unwilling to make any promise now that such a telescope will be forthcoming.

It seems to me that the whole question hinges upon the conditions at Leiden and your estimate of your prospects there. [...] It is, of course, highly probable that you will succeed our good friend de Sitter as Director However, this promotion is at least ten years in the future, and any man of experience will tell that however likely this may be, much can happen, of a totally unforeseen character [...]. If I understand your circumstances properly, and if I understand your ambitions and make-up, it may very

> well be worth while for you to take this risk. On the other hand, you must realize that, after having refused two such places as those at Harvard and Columbia you are not likely to receive another (certainly not an equally attractive one) for perhaps the next eight or ten years.
>
> I did not know until your letter came that Mieke would be so averse to making the change. Of course, this is a great factor.'

Schlesinger made the point that he was aware of the many plans for southern telescopes, so that he would advise Columbia not to embark on another one. The resolution of the issue came quickly. In a letter of May 19, Oort informed Schlesinger that ' [...] they urged me to come for half a year, but remained vague about the future possibilities. Your letter put us in new doubts and serious worrying about what would be the right thing to do. [...] I really feel that I should not give up the surroundings and the work at Leiden, with which I am so happy, unless New York would hold prospects of important progress towards my ideals.'

On June 29, 1930, Oort wrote Schlesinger that 'at last we decided to stay in Holland'. Mrs. Oort wrote to her parents:

> 'Here is some pleasant and exciting news for you: We are not going to America! The answer came and was 'no Telescope', so it was immediately decided. Delightful, isn't it.
>
> It is a bit of a disappointment for Jan of course because of the great potential of a major piece of work, but there is also much telling against it, and we look forward to a future in Holland. Prof. de Sitter was happy also, as you can imagine. , [...]
>
> It must be a great relief to you!'

It appears that the reluctance of Mrs. Oort to move permanently to the USA has been an important factor in Oort's decision to remain in the Netherlands. Indeed, this was the end of a possible emigration the the USA. As Schlesinger had warned, no major offers came for some time after this one and Oort very happily remained in Leiden at the Observatory for the rest of his career and Mrs. Oort was very content with the outcome. Harvard and Columbia Universities may have been disappointed; there were no hard feelings about this though. Eventually, Harvard bestowed an honorary doctorate upon Oort, which is very special indeed, and Columbia awarded Oort a high honor, the Vetlesen Prize.

Fig. 7.0 Photograph of the Oorts in 1926. From the collection of Abraham H. Oort

Chapter 7
The Oort Limit and Galaxy Photometry

Then how do you know it's there?' He slows to a stop at a red light. 'Dark matter has a gravitational effect on other, objects. You can't see it you can't feel it, but you can watch something being pulled in its direction.
Jodi Lynn Picoult, (b. 1966) [173]

In a sense, the galaxy hardest for us to see is our own. For one thing, we are imprisoned within it, while the others can be viewed as a whole from outside. Furthermore, we are far out, from the center and to make matters worse, we lie in a spiral arm clogged with dust.
Isaac Asimov [174]

When he entered his thirties, Oort had established himself as a prominent astronomer and he seemed to have a clear view as to what he would undertake in the years to come, at least during the decade. He had decided to remain in the Netherlands for the foreseeable future on an appointment in Leiden where he had the freedom necessary to follow his scientific priorities and which would very likely lead eventually to the directorship of the Observatory. In private life he now was a married man and father of a son. In his pocket diary he wrote on July 30, 1930:

> Mieke and I are really very happy. It is a kind of quiet happiness that I never knew before. We are happy with little Coen, here on the house in the Plantsoen, with the rooms that are so 'familiar', and in Katwijk with all the space and the magnificent evening skies and the evenings which we spend together at home.'

Jodi Picoult is an American writer.

P. C. van der Kruit, *Jan Hendrik Oort*, Astrophysics and Space Science Library 459, https://doi.org/10.1007/978-3-030-17801-7_7

Box 7.1 Reduced proper motion

In statistical studies often the reduced proper motion H is used. It stems from an indicative absolute magnitude based on the proper motion that Hertzsprung used. When m is the apparent magnitude and μ the *total* proper motion (i.e. corrected for solar motion) in arcsec/year he defined

$$M_\mu = m + 5 + 5\log\mu,$$

reminiscent of the real absolute magnitude on the basis of the parallax π

$$M = m + 5 + 5\log\pi.$$

Hertzsprung's M_μ is the magnitude of a star when its proper motion is $0''\!.1$, and—apart from an additive term—would be the absolute magnitude if the proper motion were an accurate estimator of the distance. It was inspired by empirical equations that Kapteyn and van Rhijn derived to relate apparent magnitudes and proper motions to average parallaxes. Luyten seems to have introduced the notation H in 1923 and to have dropped the '+5', [175, 176]:

$$H = m + 5\log\mu.$$

In the 1920s many astronomers followed Luyten, among them Oort, and used this reduced proper motion. The '+5' came back later.

The only thing which Oort had failed to accomplish for his scientific work was regular access to a telescope, and in this respect he found himself in the same position as his great example Kapteyn had been. For his work on the dynamics of the Galactic System in the neighborhood of the Sun the major fundamentals had been laid, although improvements in the determination of the local parameters were necessary, while a more detailed and thorough study of the dynamics and structure in the vertical direction and the stellar composition of the local disk still remained to be made. In addition to this he must have felt that the main thing to do was to better understand the distribution and properties of the absorbing matter. And although he had no doubts that the Galactic System resembled the external spiral galaxies, the precise nature of the spiral structure in our system needed to be elucidated. For a better understanding of the dynamics of these spiral galaxies more data on these systems were required, in the first place the distribution of stars as indicated by the distribution of light. In bright systems or inner regions the rotation of some nebulae had been measured and possibly soon more would follow.

Before going into these matters in more detail, I will present a brief discussion of a paper (Oort 1930a), which was triggered by a question Hertzsprung put to Oort, namely what fraction of faint stars of considerable proper motion are indeed stars

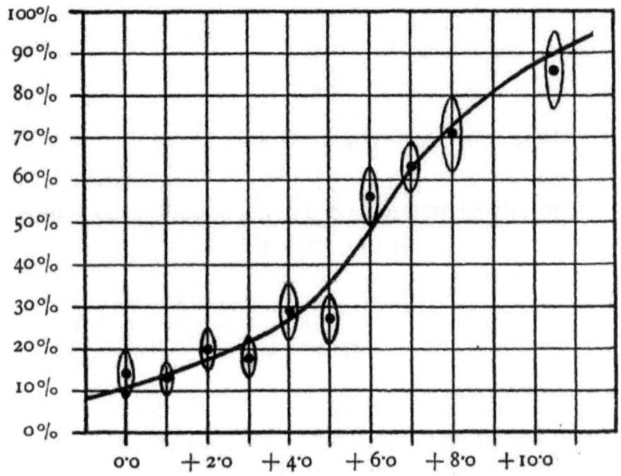

Fig. 7.1 Percentage of high-velocity stars (space velocity in excess of 63 km/s) as a function of the parameter H, the reduced proper motion (see Box 7.1). From Oort (1930a)

of high velocity. After all, care has to taken when deriving statistical properties of the local velocity distribution (or separately in the Kapteyn Streams), because the high-velocity stars need to be excluded. The division between high and low velocities would be of course the by now well-known 63 km/s. For his analysis and calculations, Oort chose Schlesinger's *General Catalogue of Stellar Parallaxes*, published in 1924 (see Sect. 3.4), and Oort's study concerned stars between apparent magnitude 3 and 9. He used an interesting concept, called 'reduced proper motion' (see Box 7.1). If all stars had the same tangential velocity, the measured proper motion could be used as a direct measurement of the distance and therefore its absolute magnitude. Hertzsprung had defined the property reduced proper motion, which is a kind of indicative absolute magnitude estimated from the proper motion. For high-velocity stars the value of H will of course be much too large for their apparent magnitudes.

Oort's results can be seen in Fig. 7.1. The percentage of high-velocity stars among stars with considerable proper motions is quite high. But even at relatively low H, the fraction is non-negligible.

7.1 Local Density

Box 7.2 The vertical force K_z.

Close to the Galactic plane the motion of a star in the direction perpendicular to it may be considered independent of that *in* the plane, the star oscillates in this z-direction only under the force (or acceleration per unit mass) in that direction $K_z(z)$ and the equation of motion has the simple form

$$Z = \sqrt{Z_\circ^2 + 2\int_0^z K_z(z)dz},$$

where Z is the vertical velocity at height z and $Z_\circ$ that in the plane at $z = 0$. Now assume a *tracer* population of stars with density distribution $\Delta(z)$ and velocity dispersion $\langle Z^2\rangle^{1/2}$. The continuity equation (cf. Box 6.1) can then be used to solve for the vertical force:

$$K_z(z) = \frac{1}{\Delta(z)}\frac{\partial}{\partial z}\left[\Delta(r)\langle Z^2\rangle\right].$$

Oort used the special case of a Gaussian velocity distribution (or the sum of two or three Gaussians), with velocity dispersion(s) $\langle Z^2\rangle^{1/2}$ independent of z. When the force $K_z(z)$ is known, the density distribution $\rho(z)$ of *gravitating* matter that produces such a force, can be recovered from the Poisson equation (cf. Box 6.1)

$$\frac{\partial K_z}{\partial z} = -4\pi G\rho(z).$$

In 1932 Oort published a long, comprehensive paper with the title *The force exerted by the Stellar System in the direction perpendicular to the Galactic plane and some related problems* (Oort 1932a). It ran 39 pages in the 'Bulletin of the Astronomical Institutes in the Netherlands'. The average length of BAN papers between the mid-twenties and mid-thirties was 4.5 pages, so Oort's paper was 8 times longer than average. It was an extensive study concerning in the first place the vertical (perpendicular to the Galactic plane) stellar distribution, kinematics and dynamics, but also addressed other aspects of Galactic structure up to a few kiloparsecs from the Sun, using K- and M-giants as tracers.

The method Oort employed was the following. He used compilations of stars with both known radial velocities and absolute magnitudes (and therefore distances) at Galactic latitudes above $\pm 40°$, so that a reasonable to large fraction of the space velocity was projected onto the perpendicular direction. He then determined the velocity Z for these stars from these projected velocities using a statistical correction for the component of velocity parallel to the Galactic plane. He found from this sample of almost 500 stars that for O-stars the dispersion was less than 10 km/s, but for A, F, G, K, and M-stars it increased from about 10 to about 15 km/s (for these classifications see Sect. 2.6). The distributions did not show any significant trend with distance from the plane, especially for K- and M-giants, on which most of his final analysis rested. He complemented all of this with the known space velocities of nearby stars. In the end this resulted in velocity dispersions as a function of spectral type and of absolute magnitude. The whole process involved, especially for the apparently fainter stars, omission of high-velocity stars, since often samples had been selected using proper motions.

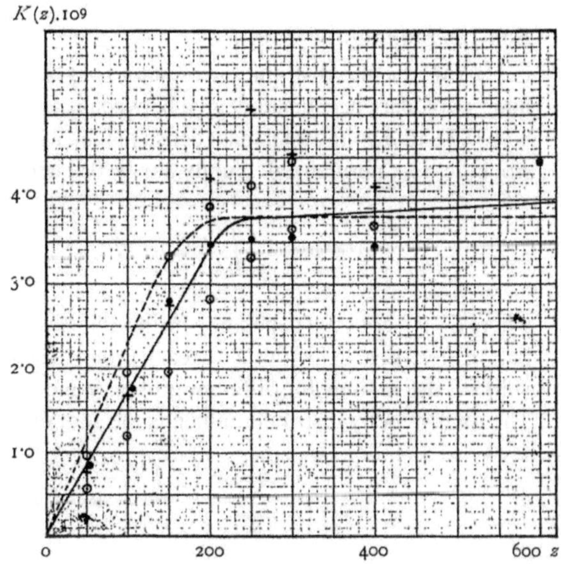

Fig. 7.2 Vertical force (or acceleration per unit mass) K_z with distance from the Galactic plane. The unit is cm/s^2. The symbols derive from different ranges of absolute magnitudes, crosses between −1.5 and −0.5 (roughly B-stars), dots −0.5 to +1.5, dots surrounded by circles +1.5 to +3.5 and circles +3.5 to +5.5. From Oort (1932a)

The second thing Oort needed was the space density distribution as a function of z. There were various sources in the literature, particularly papers by van Rhijn, Lindblad, and others. This gave density profiles as a function of absolute magnitude, corrected for the fact that the Sun was measured to be located some 39 parsecs south of the plane, as found by Hertzsprung and confirmed by others (e.g. Luyten). There is a markedly stronger concentration towards the Galactic plane for B and A-type stars, consistent with the fact that the velocity dispersions or mean velocities vary with spectral type.

The vertical force, following Oort usually written as K_z, can be determined when for samples of stars both the space density and velocity distribution are known (see Box 7.2 for mathematical details and expressions). Oort assumed that the velocity distributions were Gaussian (or sometimes the sum of two or three Gaussians) and that the velocity dispersions were independent of distance from the plane (which simplified the middle equation in Box 7.2) and arrived at the solution given in Fig. 7.2. Different symbols indicate results from stars in different ranges of absolute magnitude. The dotted line is an intermediate ('preliminary') result. At low distances from the plane the force (or the acceleration for a unit mass) rises linearly with distance from the plane. When he used data of Lindblad and from a 1927-thesis in Uppsala by John Harald Petersson (probably 1895–1979), the result was rather different, which he attributed to the studies having been made at lower Galactic latitudes where the studies.

Box 7.3. Star counts from K_z.
Let $\Phi(M)$ be the distribution of stars over absolute magnitude, $A(m)$ the number of stars per square degree between apparent magnitudes $m - 0.5$ and $m + 0.5$ and $\rho(r)$ the total mass density along a line of sight with r the distance from the Sun, normalized to unity at $r = 0$. Then using $M = m + 5 - 5\log r$

$$A(m) = C_1 \int_0^\infty r^2 \rho(r) \Phi(m + 5 - 5\log r) dr,$$

where C_1 is a constant depending on the units used. Now assume that the distribution of vertical velocities Z for each group of stars is Gaussian with different velocity dispersions for different groups (i indicating the group considered). The total velocity distribution then is the sum of these, say

$$\sum_{\text{all } i} \theta_i \, l_i \, \frac{l_i}{\sqrt{\pi}} \exp(-l_i^2 Z^2),$$

with θ_i and l_i constants. Using $z = r \sin b$, the mass density ρ follows from

$$\rho = \sum_{\text{all } i} \theta_i \,\mathrm{dexp}\left(C_2 \, l_i^2 \int_0^z K_z(z) dz \right),$$

where C_2 also depends on the units used and dexp denotes the decimal exponential function. The contributions of each group θ_i follow from $\Phi(M)$.

Having derived the vertical force up to about 500 pc, Oort turned the problem around. If for a group of stars (range of absolute magnitudes or of spectral types) one knows the velocity dispersion, the force field can be used to calculate what the vertical density distribution for that group of stars would be. And doing this for all types or absolute magnitudes of stars would give not only the number distributions of stars but also the counts in apparent magnitudes on the sky. Some mathematical details are collected in Box 7.3. Oort first calculated the counts for the Henry Draper Catalogue (see Sect. 2.6). The comparison came out very well for spectral types F and G. For the B- and A-stars the velocity law was very poorly known (there are few such stars at higher Galactic latitude) and the fits improved when the dispersions were chosen slightly differently within the uncertainties. A problem arose with the K- and M-stars. Oort attributed this to the uncertainties in the brighter part of the luminosity laws for these stars (that would then be giants) and he undertook a separate study of the calibration of the luminosity distribution of K- and M-giants in an accompanying paper (Oort 1932b). This concerned a new calibration of the absolute magnitudes of these stars that had been performed by Strömberg. Oort compared the distributions of radial velocities with those of proper motions, assuming that the average radial and tangential velocities were the same, and derived statistical distances from this. The luminosity function (distribution) that Oort deduced was significantly different from the one van Rhijn had published, but the corresponding revised absolute magnitudes improved the fit considerably for the K-stars, but less so for the M-stars, where Oort felt the luminosity law was rather uncertain still. Apart from this he arrived at consistent values from different types for K_z up to 500 pc.

There was an obvious need to determine the vertical force beyond 500 pc, since the asymptotical behavior could be used to estimate the total mass in a cylinder perpendicular to the plane. For this Oort used the counts as published by van Rhijn in the 'Groningen Publications' in 1917 in visual magnitudes, augmented by corrections to the magnitude scale by Pannekoek, and in 1929 in photographic magnitudes. Again much discussion was required to derive the best magnitude scales according to the latest data. These counts were restricted to high Galactic latitudes. The analysis proceeded in two steps. First the counts were compared to predicted counts for apparently faint stars that were also faint intrinsically and thus nearby, so that they provided a check on K_z within 500 pc. Having obtained good agreement here and thus having confirmed the full drawn curve in Fig. 7.2, Oort proceeded to estimate the force field at higher z.

He started with the best values for his constants A and B. These corresponded to a rotation velocity near the Sun of 300 km/s, and Oort then derived the corresponding value for the local radial force. As before he then assumed that this was made up of two components, one corresponding to a spherical mass around the center and the other a homogeneous ellipsoid. The force K_1 from the spherical mass was assumed to be 84% of the total (it is not explained how he arrived at this value, but it must have been afterwards after various tries). This corresponded to the analysis he had made before when he discovered differential rotation. Oort then considered two extremes, one called 'a' in which he took the vertical force at 5 kpc to be the same as that due to the central spherical mass and to increase linearly above 500 pc and a case 'b', in which at 5 kpc it was 42% of the value for 'a'. This is illustrated in Fig. 7.3. The

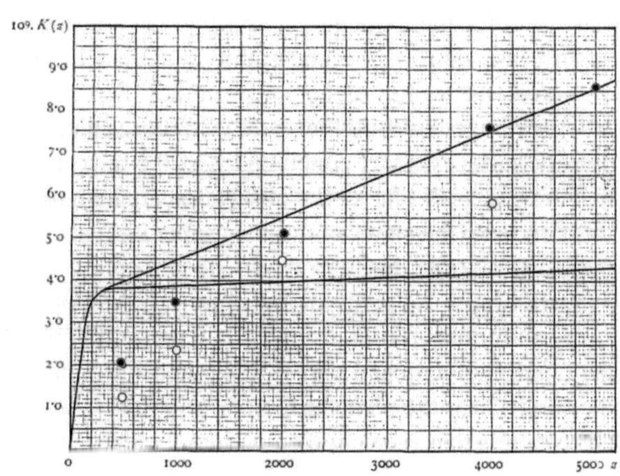

Fig. 7.3 Vertical force K_z extrapolated to 5 kpc from the plane of the Galaxy. The upper line is case a and the lower case b. See text for explanation. From Oort (1932a) with the symbols enlarged

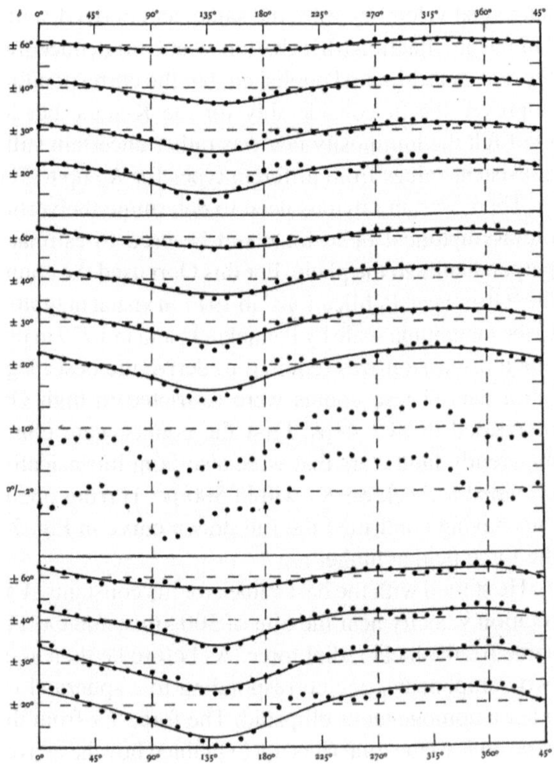

Fig. 7.4 Comparison of observed star counts with calculations as a function of Galactic longitude at selected latitudes. The longitudes on the horizontal axis run from $l = 0°$ through $l = 360°$ to $l = 45°$ on the left. In each panel the top curve is at $b = \pm 60°$, then $b = \pm 40°$, $b = \pm 30°$ and $b = \pm 20°$. In the middle panel data points are displayed also at $b = \pm 10°$ and $b = 0°/-2°$. The top panel corresponds to apparent magnitude $m = 14$, the central one to $m = 16$, and the bottom one to $m = 18$. The distance between two tick-marks on the vertical scale corresponds to a difference of 0.25 in the logarithm of the counts. For further explanation see the text. From Oort (1932a)

open circles are the forces derived from the spherical mass and the filled ones for the ellipsoid if it has an axis ratio of 1: 10.

With these forces the star counts at apparent magnitudes 8, 11, 14. 17 and 18 were computed and compared to observations. In fact this was done for four different assumptions for the velocity dispersions at $z = 0$ for the intrinsically fainter stars (and for the two cases a and b given above). These computations were taken as an illustration of the uncertainty since all the resulting counts represented the observed ones within acceptable accuracy. By comparing the counts at different latitudes (averaged over all longitudes) to those calculated for the case of plane-parallel density surfaces, Oort concluded that the agreement was good enough to deduce that the effects of interstellar absorption are insignificant above latitude 15°. He next went on to do the same analysis by separating according to longitude, which clearly indicated that the density of stars does increase towards the center of the Galactic System. Figure 7.4 gives an indication of the enormous amount of work involved. It shows the counts for three apparent magnitudes (14, 16 and 18) with longitude along the horizontal axis and curves separated for bands in latitude (in the middle panel the counts near the plane have not been fitted). Each point on these predicted curves corresponds to a very large computational burden.

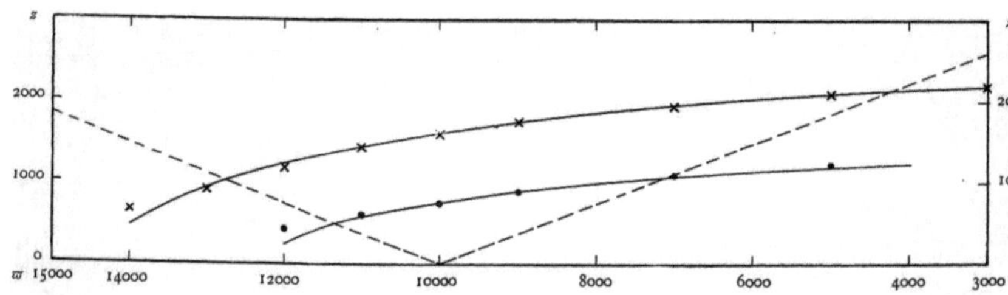

Fig. 7.5 Crosscut through the Galactic System. Symmetry is assumed with respect to the Galactic plane. The lines and symbols show surfaces of equal density of starlight, the one with the dots at 1/25th and the one associated with the crosses 1/100th of that near the Sun. The lines are ellipse fits centered on the Galactic center, which is to the right of the figure. The Sun is at the label 10000, which is the assumed distance to the center in parsec. The dashed lines indicate latitude 20°. From Oort (1932a)

A major result was Fig. 7.5, which translates the results of the computation in a cross section of the Galactic System. The curved lines indicate surfaces of equal 'density of photographic light' at values of 1/25th and 1/100th of that near the Sun. These lines are best fitting ellipses centered on the Galactic center (in this study assumed to be at 10 kpc from the Sun). The dotted lines (at latitude 20°) show the limits above which interstellar absorption is completely negligible.

The final section of Oort (1932a) concerned 'the amount of dark matter', which is not the absorbing medium but any matter that is known to contribute to the gravitational field, but cannot be accounted for by either known stars or gas and dust. Now Oort has to resort to Poisson's equation, which links the distribution of matter to the resulting gravitational forces (see also Box 7.2). To do that, he took his local force K_z and fitted it to the equations of the force field of a homogeneous ellipsoid and considered corrections to that due to further central spherical or ellipsoidal mass distributions until he found a good fit to K_z. The best estimate for the total density near the Sun in the Galactic System came out as 0.092 solar masses $M_\odot$ per pc^3 and the error was estimated as 20%. This is somewhat less than the 0.143 in Jeans' analysis of the Kapteyn Universe in 1922, in which he defined the Jeans equations. But it is close to what Kapteyn found in his 'first attempt', which Oort translated into 0.099 in the same units. This should be compared to known objects. First Oort analyzed the visible stars, for which he in the end adopted a value of 0.038 for stars brighter than absolute magnitude 13.5 or heavier than 0.22 $M_\odot$. But stars with fainter absolute magnitudes may make quite a significant contribution. In fact an extrapolation would give a

> 'total mass [...] dangerously near the value of 0.092 $M_\odot$ derived from K_z. We may conclude that the total mass of nebulous or meteoric matter near the Sun is less than .05 $M_\odot$ pc^{-3} or 3.10^{-24} g/cm^3; it is probably less than the total mass of visible stars, possibly much less.'

And in the summary (but not really treated in any detail in the paper):

> 'There is an indication that the invisible mass is more strongly concentrated to the Galactic plane than that of the visible stars.'

Fig. 7.6 Sombrero Nebula, also known as M104 or NGC 4594. This picture was taken with the Hubble Space Telescope, NASA/ESA and The Hubble Heritage Team STScI/AURA) [177]

The latter is still uncertain. A comparison to external galaxies was possible by calculating the mass in a column perpendicular to the plane of the Galaxy. Assuming that the same proportion of dark matter holds at all z, this came out as 80.4 $M_{\odot}$ pc^{-2} (per square parsec), and without any dark matter beyond 200 pc it would be 51.5 in the same units. This can be compared to the surface brightness seen from outside the system. This is known from star counts; in the pole of the Milky Way van Rhijn had found that the total light is equal to 0.0023 stars of magnitude zero per square degree, which would be half of the surface brightness of the Galactic disk as seen from outside. Oort estimated from an analysis of his data that 'one unit of light [that of a star of +6.0 photographic absolute magnitude] corresponds to 1.8 $M_{\odot}$. Had we considered the layer between $z = \pm 110$ pc only the result would have been 1.1 $M_{\odot}$.'

This compared well to an analysis Hubble had performed on the central part of the Andromeda Nebula (see Fig. 7.9), where a rotation velocity of 72 km/s at 200 pc from the center indicated a mass interior of that of 2.4×10^8 $M_{\odot}$ (maybe overestimated somewhat due to flattening). The ratio light-to-mass is then 1.3.[1] This is good agreement with the Galactic value, which would have been even better had Hubble known the current value of 780 kpc (he used 175 kpc). Hubble's was not the first measurement of rotation in an external galaxy. It had been measured earlier in the Sombrero Nebula (Fig. 7.6) by Vesto Melvin Slipher (1875–1969) at Lowell Observatory in Flagstaff, Arizona. At Mount Wilson, Francis Gladheim Pease (1881–1938) had confirmed the measurements. However, from this the ratio light-to-mass became 10. Oort used a distance of 3 Mpc (Megaparsec or one million parsec) for this calculation rather than the current almost 10 Mpc. He attributed the high value of the ratio to the dark dust-lane crossing the image, blocking a good fraction of

[1]In modern times we use the property M/L, the 'mass-to-light ratio', for this. The value for the solar neighborhood is of order 5 (within a factor 1.5 or so), so Hubble's 1/1.3 = 0.77 is too low.

the light. Finally, Oort estimated rotation velocities for the Magellanic Clouds and two globular clusters, ω Centauri, the largest in the Milky Way, and Messier 3. His distances were again much too small compared to what we know today, 26 and 29 kpc rather than about 165 and 200 kpc for the Magellanic Clouds, and 7 and 12 kpc rather than 16 and 34 kpc for the globular clusters. And these clusters are not supported by their rotation but rather by their random internal motions. But all of this was still unknown in the early thirties.

The value for the mass per cubic parsec near the Sun as derived from the vertical force became known as the 'Oort limit' and for decades there has been discussion to what extent known matter could make up for the total or whether dark, unseen, and unexplained matter was necessary. In 1932, Oort did not commit himself on whether there was a real problem in that the mass in known stars explained only about half the total mass indicated by the vertical force field, but would return to this issue over the years. In his famous chapter in 'Stars and Stellar Systems V' (Oort 1965a), he would report that 'roughly 40% of the local mass must be due to stars or gas of unidentified type, and, of which, therefore, we do not know the distribution in space.'

The numerical effort for the paper was enormous. Oort acknowledged the contribution of the Computing Department of Leiden Observatory at the end of the paper, and mentioned in particular Gerrit Pels and J.M. Kriest, presumably Johannes Marinus Kriest (1895–1981). The Computer Department was of substantial size, as required for efforts such as Oort's. In de Sitter's history of Leiden Observatory, written in 1933 [104], the Department was listed as consisting of 10 Computers. Figure 7.7 shows a group photograph of the staff of Leiden Observatory in June of 1931. The leadership is in the row in front of the steps, from left to right Hins, Oort, Hertzsprung, de Sitter and Woltjer. Computer Kriest is Nr. 16 and the person that through the years was most Oort's trusted Computer Gerrit Pels is Nr. 34.

This 1932 paper was a monumental one. Together with the discovery paper of differential rotation it constituted a major advance in the our understanding of the structure of the Galaxy both in terms of space distribution of the stars and the differentiation in various types of stars, the presence of dust and its confinement to a thin layer around the plane, as well as of an understanding of the kinematics and dynamics of the stellar distribution. The progress from Kapteyn's first attempt was probably more profound than that from Herschel to Kapteyn.

I close this section with a excerpt from the *AIP-Interview* [4], where Oort said:

> 'The conclusion arrived at from this investigation of the density distribution from motions perpendicular to the plane was that you could not account for the total force by the known stars. But it seemed very doubtful right away that this unknown matter that you were left with could be furnished by interstellar gas or interstellar dust. So right from the start it was fairly evident that it would have to be furnished by some unknown type of stars – fainter stars.'

[DeVorkin: Well in this particular paper in 1932, the value that you got for the density of the interstellar medium something on the order of $3x10^{-24}$ grams per cubic centimeter has been called by Struve and others the famous Oort limit. I was

1. G. van Herk
2. P. Th. Oosterhoff
3. C. H. L. Sanders
4. Ir. W. F. Kruyt-bosch
5. J. Uitterdijk
6. Dr. C. H. Hins
7. A. J. Wesselink
8. Miss H. A. Kluyver
9. G. P. Kuiper
10. Dr. J. H. Oort
11. A. de Sitter
12. Miss C. H. de Nie
13. H. Kleibrink
14. P. P. Bruna
15. Prof. Dr. E. Hertz-sprung
16. J. M. Kriest
17. M. D. Schepper
18. B. C. Mekking
19. L. Gaykema
20. J. C. Gaykema
21. P. de Haan
22. J. E. Prins
23. H. Zunderman
24. Prof. Dr. W. de Sitter
25. J. E. Stol
26. L. J. F. van Leeuwen
27. Miss W. van Heuzen
28. A. Luteyn
29. D. Gaykema
30. J. H. Kasten
31. H. M. Swaak
32. Dr. J. Woltjer
33. J. Nicaise
34. G. Pels
35. F. de Haas
36. E. W. de Rooy

(Mr. C. J. Kooreman was absent on military service)

Fig. 7.7 Staff of Leiden observatory in 1931 and their identification, as published in [104]. From the Oort Archives wondering if not only you were aware of that term but what the significance of it was?]

> 'It has been used far too much, because right from the start this was extremely uncertain and it still is at the present moment, I'm sorry to say. In a few years from now [1977] we'll have better data but it has taken a long time since 1928 to get those better data. But it was not so much the concern, as I said before, with the interstellar matter as with the unknown stellar population.'

7.2 External Galaxies

It is clear from his Archives that Oort had a long-time interest in extragalactic systems. Already as a student he drew up lists of galaxies apparently to select objects suitable for observations. He looked at the distribution of properties, such as that across the range of morphological types and had been much impressed by the photographic plates that Hubble showed him when he visited Pasadena and Mount Wilson in 1924. Now, an important uncertainty in the model for the rotation of the Galaxy was the actual value for the local rotation velocity. It had been estimated as the mean motion of the globular clusters and the high-velocity stars with respect to the Sun. But this was still very uncertain and it was of course not known whether the Shapley system had any rotation of itself. A better 'rest frame' against which to measure the magnitude of the rotation velocity would be external galaxies.

As it happened, Willem de Sitter had collected the necessary data that Oort needed for such an analysis. De Sitter's interests were more in the structure of the Universe. During the First World War, he had published three articles in the British journal 'Monthly Notices of the Royal Astronomical Society' explaining Einstein's General Relativity theory to the English-speaking community, all entitled *Einstein's theory of gravitation and its astronomical consequences*, in which he provided two solutions to the field equations [178–180]. In 1930, de Sitter had published a paper in the BAN *On the magnitudes, diameters and distances of the extragalactic nebulae and their apparent radial velocities* [181], in which he tested cosmological world models derived from Einstein's theory of general relativity with observations of galaxies. De Sitter collected data on diameters and apparent magnitudes of extragalactic nebulae to compare distances derived from either of these two properties, assuming average absolute values. He then compared these to radial velocities. Some of these were of course the radial velocities Hubble had used in his 1929 paper, in which he discovered the expansion of the Universe [182, 183]. Hubble had noted that almost all galaxies showed velocities away from us (their spectral line were shifted to the red) and the velocities or redshifts increased with distance from us. Actually of the 24 nebulae that Hubble used to make a distance versus radial velocity plot, all but four had been obtained by Slipher. Since then more radial velocities had been obtained by Slipher, Strömberg, Hubble and others. De Sitter did not specify where he had taken his values from, but used them to improve the determination of the expansion rate of the Universe; he also showed that his static solution obviously did not work. So he tried his second solution, in which there was an 'inherent expanding tendency of space'. However, this particular solution did not fit either, as in that model the 'path of a material point is a parabola'. At the end of the paper de Sitter mentioned that Eddington had drawn his attention to the work of Georges Henri Joseph Édouard Lemaître (1894–1966) on non-static, expanding solutions to Einstein's field equation, which eventually led to the Einstein-de Sitter Universe [184]. These were exciting developments and de Sitter and Oort must have discussed these issues frequently. Oort actually used the same data—de Sitter's—for external galaxies.

De Sitter used the unit 'A', the distance to the Andromeda Nebula taken to be 10^{24}cm (324 kpc), which did not receive wide following; fortunately, I would say, since that distance has been subject to substantial revision. Amusingly, de Sitter noted that 'No confusion with the Ångström unit, Å is to be feared. The ratio of the two units is 10^{32}.'

Oort (1930c) was aiming at using the extragalactic systems as an inertial frame against which the local rotation velocity might be determined. So, what he did (with the help of his trusted Computer Gerrit Pels) was to make a least-squares fit to the observed radial velocities, positions and de Sitter's distances of the nebulae on the sky, assuming that the radial velocities were the sum of an isotropic and linear expansion and a mean velocity, which would then be the velocity of the Sun with respect to the extragalactic systems and thus that of Galactic rotation. Linear expansion means that the radial velocity increases linearly with (is proportional to) distance and the isotropy that everything is the same in all directions. A least-squares solution is the one for which the sum of the squares of the differences between the observed values and those predicted by the fit is smallest. Unlike Hubble, Oort weighed the velocities according to their distances. Other earlier analyses like this, involving a mean velocity but no universal expansion had been performed by Lundmark at Lowell Observatory in 1924, and also at Mount Wilson by Strömberg in 1925 [185, 186].

Oort presented two solutions, one with and one without the Magellanic Clouds (he felt the one without the Clouds to be more reasonable as they would be companions to the Galaxy and in a sense part of it). The difference actually was very small; the resulting *direction* of the the linear velocity with respect to the nebulae was 'exactly opposite the rotational velocity of the Galactic System'—as expected—, close to earlier determinations by Lundmark, Strömberg and Hubble. The velocity was 380 ± 110 km/s, comparable to what Strömberg and Hubble had found (329 and 250–300 km/s), but different from that of Lundmark (650 km/s, using samples with distances estimated from stars or from mean luminosities in a cluster). Oort's solution for the expansion term, what we now would call the Hubble constant, was not very different from earlier determinations, which is not strange as the determinations had many velocities in common. Oort found (after translating back from his use of de Sitter's distance unit 'A') 432 ± 43 km/s per Mpc, where de Sitter had found 466 and Hubble 465 and 513 (and probable errors, like Oort, of order 50). The currently adopted value is about 220 km/s. And finally, Oort divided the sample into two parts, one to the north and one to the south of the Milky Way and found absolutely no evidence for deviations from isotropy (as is still the case now).

Fig. 7.8 Central parts of the Virgo Cluster. The giant elliptical galaxy M87 is at the lower left. Obtained from the Sloan Digital Sky Survey Skyserver [188]

Having obtained an independent estimate of the rotation velocity in the Galaxy, he continued this work with an investigation into the distribution of the intrinsic luminosities of galaxies and their peculiar motions (Oort 1931a). This had also been done for galaxies in large clusters by Hubble and Milton Lasell Humason (1891–1972). Humason (accent on the first syllable) was a staff member of Mount Wilson Observatory with an interesting background (see his biography [187]); he began as a 'muleskinner', driving mules up Mount Wilson with materials during the construction of the Observatory. In 1917 he became a janitor on the mountain top and not long after that a night assistant. This is a person that assists the observing astronomer by moving the telescope and performing other tasks. Humason turned out very skilled as an observer and in 1919 became a staff member, which he remained throughout his career. Big clusters of galaxies are large, sometimes very large conglomerations of galaxies. The best ones known is the Virgo Cluster (as the name suggests with its center in the constellation Virgo; see Fig. 7.8), which is only some 16 Mpc away and has a diameter of at least 2 Mpc. Within this radius it contains more than a thousand galaxies, the largest of which is a giant elliptical galaxy (actually on the sky it appears as almost spherical) known as Messier 87, which has a mass of over two trillion (2×10^{12}) Suns within a radius of 15 kpc or so. The Virgo Cluster contains also many spiral galaxies. Another important cluster is the Coma Cluster in the constellation Coma Berenices. The Coma cluster also contains over one thousand

Fig. 7.9 Andromeda Nebula M31 (left) is the nearest spiral galaxy to our own Milky Way Galaxy. It is number 31 in the list of Messier. The small galaxy bottom-center is M32 and the one higher up NGC 205, both companions to M31. The nebula Messier 101 (right) is among the nearest large spiral galaxies. M101 is also known as the Pinwheel galaxy. Its distance is about 6.5 Mpc. Unlike other pictures in this book, where north is up, this picture has north to the left; it has been taken using the Hubble Space Telescope. M31: ESA/Hubble and Digitized Sky Survey 2, M101: European Space Agency and NASA [190]

galaxies and is at a distance of almost 100 Mpc. A much larger fraction of the member galaxies is elliptical than in the case of Virgo.

Oort's study was disappointing because it indicated that either the distribution of luminosities of galaxies in clusters is very different from that of more isolated systems, or otherwise that the peculiar velocities of the latter, more local ones were on average much smaller. Oort derived absolute magnitudes of galaxies as a function of their type; he also found a new value for the expansion factor of the Universe that was about twice as large as he had found previously and this led him to conclude that the error of this value was at least a factor 1.5, maybe 2.

Oort assumed that the Galactic System, when seen from outside would resemble spiral galaxies, such as the Pinwheel galaxy M101 in Fig. 7.9. Because of our vantage point right within the system and the interstellar absorption, the spiral structure of our system was and remains very difficult to delineate. An attempt (although not a very convincing one) to derive the spiral structure from the appearance of the Milky Way on the sky had been published by Cornelis Easton (1864–1929) [189]. Easton, a Dutchman in spite of his last name, was a prominent journalist and newspaper editor. He had no formal training in astronomy but still did original research in that field (and in climatology) and published research papers in professional astronomical journals, including American ones. In recognition of his contributions, he received an honorary doctorate in 1903 from the University of Groningen with Kapteyn as his 'promotor'.

Another reason for Oort to be interested in nearby galaxies was their dynamics. We have seen above that he used data on the light distribution in the central parts of M31 and the observed rotation there to estimate the ratio of the mass in stars to the light they emitted. To study dynamics one needs the distribution of stars from the light distribution and their velocities from spectra. Since this will be a recurring subject in what follows, I will summarize here the history of galactic surface photometry and spectroscopy.

The first measurement of a light distribution (referred to as surface photometry as it involved the brightness of an area or surface brightness) in a galaxy was performed by John Reynolds in 1913 with his privately built 28 in. refractor near Birmingham, England, using exposures of up to 100 minutes [191]. He traced the surface brightness distribution of the Andromeda Nebula out to a radius of 7 arcmin, where it disappeared in the light of the background sky. For comparison, the distance between the Andromeda Nebula M31 and the companion M32 (straight below M31's nucleus in Fig. 7.9) is about 25 arcmin.

The procedure of photographic surface photometry is the following. The surface brightness corresponds to the degree of blackness of the photographic emulsion, which then needs to be related to the amount of light that had impacted upon it (the 'characteristic curve'). It is expressed as a 'density', which relates to the percentage of light transmitted through the emulsion when light is shone through it; density zero when all comes though, density 1 when this is only a tenth, density 2 at one hundredth, etc. This needs to be calibrated, which is done by exposing on the same plate an image of a set of spots of known relative brightness. This has to be done for a similar exposure time (because of so-called low-intensity reciprocity failure, which means that at low levels exposing a source of half brightness for twice the time no longer produces the same density). This and other idiosyncrasies of the photographic emulsion (under- and over-exposure, variations in fog and sensitivity across the plate) were difficult to control and often spurious results have been published.

The history of photographic surface photometry of galaxies has been reviewed by one of the founders of this work, Gérard Henri de Vaucouleurs (1918–1995), a French astronomer who worked most of his life in Austin at the University of Texas and McDonald Observatory [192]. De Vaucouleurs and his collaborators contributed heavily to the development of the techniques required. As we will see, both Oort and Bertil Lindblad went into this, Lindblad even attempting to measure the color distribution in a few spirals (e.g. [193]). A heroic effort to measure the diameters of a very large sample (300) of galaxies was performed by Erik Holmberg (1908–2000) of Lund Observatory in Sweden, on the basis of tracings along the principal axes of all these galaxies. His 'Holmberg radii' have been widely applied (and misused; see [194, 195]). It took him seven years! Of course measurements involved determining the photographic density point by point. Only rather late in the game, photographic surface photometry of full two-dimensional distributions became practical when done with computer controlled scanning machines. Even then the limitations of photographic emulsions compromised the reliability and much of this work has turned out to be incorrect. Probably the most successful such photographic surveys, for which the surface brightness distributions have been shown to be reliable by later digital measurements, has been performed in the early 1980s by myself and collaborators [196]. This 'Palomar-Westerbork Survey' of 16 spiral galaxies combined photographic surface photometry in blue, visual and red light using plates taken during 42 dark nights at the 48 in. Schmidt Telescope at Palomar Observatory with radio measurements of the distribution and velocities of the gas during 64 half-days at the Westerbork Synthesis Radio Telescope (WSRT) in the Netherlands. Starting with the 1990s two-dimensional digital detectors came into use, such as CCD's (charge-coupled devices as in electronic camera's) in which photons produce electrons that are collected in a two-dimensional grid in the device. After the exposure, the device can then be read out and the results stored so that a two-dimensional image can be produced. The linear response is a major advantage over photographic plates and CCD's quickly replaced photographic plates.

Measuring rotation is more tedious, because it involves spreading the light into a range of wavelengths. The rotation of spiral galaxies was discovered in 1914 by Vesto Slipher at Lowell Observatory in Flagstaff in the bright central parts of the Sombrero Nebula (Fig. 7.6) and independently in the same year in another bright spiral galaxy, M81 (see Fig. 2.4), by Maximilian Franz Joseph Cornelius Wolf (1863–1932) at the Landessternwarte Heidelberg-Königstuhl [197, 198]. Slipher exposed his spectrum for 80 hours, which meant that the plate had to be removed a number of times from the spectrograph and stored at low temperatures until the exposure could be continued on another occasion. I was in the fortunate circumstance that during a symposium at Lowell Observatory in October 2014, I could join a small group that was taken on a tour through their plate vault, which among others has the discovery plate of Pluto. We were shown this plate and also the one on which Slipher discovered nebular rotation. My colleague Bruce Gordon Elmegreen (b. 1950) took a picture of it through a magnification glass, which I reproduce in Fig. 7.10 after some contrast enhancement. The horizontal direction is that of wavelength, the vertical position along a line along the major axis through the galaxy. The inclined lines are barely visible, showing the enormous effort required to obtain stellar kinematical data at the time.

The problem is of course that to measure *stellar* motions one needs to be able to record the absorption lines in the spectrum and that means that one has to record a small dip in the continuum light. It is different for emission lines resulting from ionized atoms in so-called HII-regions ('HII' stands for ionized hydrogen) around young stars. The young, short-lived O- and B-stars are hot and emit much ultraviolet radiation which ionizes hydrogen and other atoms; when these recombine by encountering an electron, radiation is emitted at wavelengths corresponding to the transition in the atom or ion, and by observing the wavelength one can determine the radial velocity. For that reason rotation measurements were most successful and extensive in such applications to spiral and irregular galaxies, that have many young stars. Radio techniques have extended the determination of velocity fields in galaxies to those of neutral hydrogen gas and out to large radii.

Oort's wish to study the stellar dynamics of galaxies had to be restricted to bright inner parts of galaxies. Lindblad had suggested that spiral structure is related to instabilities invoked by the gravitational field of the central bulge. The measurements of

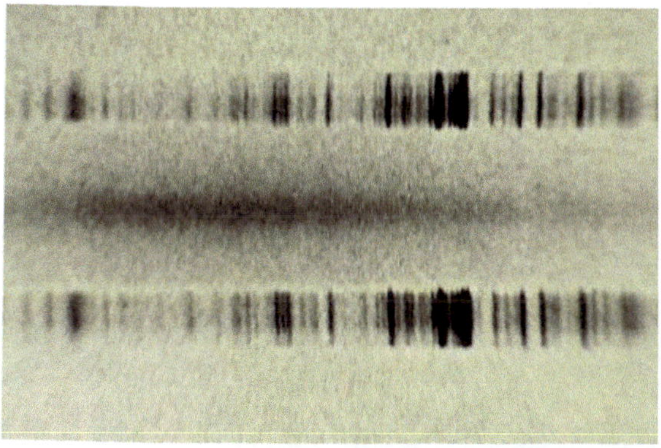

Fig. 7.10 Spectrum on which Vesto Slipher discovered the rotation of the Sombrero Nebula, obtained after an exposure of 80 h. Above and below is an exposure of an arc of a gas-discharge lamp with known spectral lines (usually from argon or neon) to calibrate the wavelength scale. Lowell Observatory Archives (with permission); picture made by Bruce Elmegreen

the rotation from the position of absorption lines in the spectrum were difficult enough; velocity dispersions from the width of these lines were well beyond observational possibilities. This has been feasible since the 1970s in the brightest parts of elliptical galaxies, but became reliable only when digital detectors became available. Measurements of stellar velocity dispersions in *disks* of spiral galaxies came much later still because the corresponding surface brightnesses are lower and the velocity dispersions much smaller. The first successful measurements of stellar kinematics in galaxy disks were performed in the 1983 by myself together with my Australian colleague Kenneth Charles Freeman (b. 1940) [199] using an image photon counting system (IPCS) developed by British astronomer Alexander (Alec) Boksenberg (b. 1936). This device uses an image intensifier tube which releases an electron for each incoming photon and subsequently multiplies it to a burst of 10^7 electrons which is then recorded by a TV-camera. Each individual photon event is located by fast image-centering software and displayed in real time on a monitor screen. Even with this or a similar device we had to expose for a full night per galaxy on the 3.9 m Anglo-Australian Telescope (AAT) at Siding Spring Observatory in Australia, and in our initial efforts multiple nights on a smaller telescope. Nowadays CCDs are used as detectors and galaxy spectroscopy is now routine procedure; modern integral-field spectrographs even measure the spectra over an extended piece of a galaxy.

Although determining reliably the distribution of stars or light in a galaxy was extremely difficult, Oort decided to become involved in it anyway. Probably his wish for a telescope on the southern hemisphere, that played a role in particular in his negotiations with Columbia University was to be able to do such studies. Not having been able to convince Columbia to support such an undertaking, he tried to get others to take plates for him. In 1932 Oort decided that he should try to obtain the necessary plates himself and he thought in the first place of the large reflector that was being erected at Perkins Observatory in Delaware, Ohio (Fig. 7.11).

Fig. 7.11 Perkins Observatory of Ohio Wesleyan University in Delaware, Ohio on an old postcard. From Observatories of Ohio [200]

7.3 Perkins Observatory

Perkins Observatory is associated with the Ohio Wesleyan University, which was founded in 1842 as a liberal arts college. Such institutions aimed at a broad, interdisciplinary education spanning the humanities, social and natural sciences. It was founded by Methodists, followers of John (1703–1791) and Charles (1707–1788) Wesley, originally aiming at a revival of the Church of England. The name of the Wesley brothers survives in that of the Ohio Wesleyan University, but also in many other institutions. It is located at Delaware, Ohio; this is located about 45 km or so north of the Ohio capital Columbus and it had a population in the early 1930s slightly less than 9000. Perkins Observatory has been named after Hiram Mills Perkins (1833–1924), who was Professor of Mathematics and Astronomy at Ohio Wesleyan University. In 1896 he donated the funds to establish an Observatory next to his residence. This Observatory was later turned into a student's Observatory for educating and training young people.

Perkins was a very religious man. He and his wife had no children, while his older sister never married, so—the story goes— there was no relative to inherit their extensive possessions (what about relatives of his wife?, one might ask), among which a fortune he had made as defense contractor during the Civil War, particularly selling hogs, raised at the pig farm of his family, to feed the soldiers. After his retirement in 1907 he decided to put his efforts (and money) into an Observatory of scientific importance with a telescope that 'should exceed in size all other telescopes in this country at that time, except the 100-inch at Mount Wilson.' The location finally selected is only a few kilometers south of Delaware.

After much delay, the building of the Observatory started in 1923, but there was no facility in the USA to cast a big enough mirror, while Perkins had stipulated it should be cast in the country. Initially, therefore a 60 in. mirror borrowed from Harvard was used in the telescope. The final mirror was cast by the National Bureau of Standards with a diameter of 69 in. (1.75 m) and installed in 1931, which made the telescope the third largest in the world (in addition to the 100 in. at Mount Wilson, the Dominion Astrophysical Observatory in British Columbia, Canada operated since 1918 a 72 in. (1.83 m) Telescope). Not much later operations were shared with Ohio State University in Columbus. Due to deteriorating light pollution conditions and the relatively poor Ohio weather (and low elevation) for astronomical observing, the telescope was moved in 1961 to Lowell Observatory in Flagstaff, Arizona.

In 1929 Harlan Stetson was appointed Director of Perkins Observatory. Before that he had been at Harvard University, where he was in charge of the astronomy teaching of undergraduates at the time Columbia had made its offer to Oort. Stetson was supervising the installation of the 69 in. mirror, when in April 1932, Oort contacted him about a possible visit he hoped to make to Perkins Observatory later that year. He would combine that with the General Assembly of the IAU, which was being hosted by Harvard University and held in Cambridge, Massachusetts.

'Recently, as our plans to go to the Union meeting in September have been taking a more definite step, I have been considering the possibility of staying a few months in the United States with the purpose of doing some observational work.

The work which I should especially like to undertake is a study of methods for determining surface brightnesses and diameters of extragalactic nebulae for statistical purposes. Such a work can hardly be done with any other instruments than large reflectors.

The only possible opportunity I see for trying out some of this work would be a stay at your Observatory, for in the present circumstances my wife and I cannot afford to pay the additional traveling and living expenses for a visit to Mt. Wilson. In some respects too I should prefer to come to the Perkins Observatory; I should indeed consider it a great privilege if you would consent to my visit and if, in case of a 1.5 to 2 months' stay, you would permit me to make use of its fine reflector.'

He also inquired into the possibilities 'of arranging some graduate lectures at the University (e.g. a brief course on statistical astronomy)' and he included a letter by de Sitter to support this request (Fig. 7.12).

Stetson replied within days (19 April), supporting the plan.

'I am very glad to receive your letter of April 11th, and to hear of the prospects of your coming to the Perkins Observatory. [...] October is likely to average excellent in Delaware. Our poorest months are December to March. [...] I think that something may be worked out in the way of a course of lectures on statistical astronomy in cooperation with our Graduate School at Ohio State University.

I am sure you would enjoy your stay here. We will do everything we can to see that our program is arranged to that you shall have a generous

Fig. 7.12 The 69 in. (1.75 m) Telescope of Perkins Observatory on an old postcard, proudly proclaiming (not shown) it to be 'the third largest in the world'. From Observatories of Ohio (see also Fig. 7.11)

share of the clear nights. The telescope is just about in working shape in the Cassegrain arrangement. I am expecting that we shall have the arrangement for photographing at the principal focus or at the Newtonian focus by the end of the summer. [...]

I recall with pleasure our occasional meeting at the Neighborhood Club in New Haven.'

The Neighborhood Club was a monthly meeting at Yale, instigated by Schlesinger, of a group of astronomers from New Haven's neighborhood, such as for example Shapley from Harvard, Russell from Princeton, Benjamin Boss from Albany, Frederick Slocum (1873–1944) of Van Vleck Observatory, John Anthony Miller (1859-1946) from Sproul Observatory, and of course others working at these and other institutions.

The Cassegrain focus is under the primary mirror, to which the light is directed through a central hole in the mirror after reflection by a small mirror at the top of the telescope near the prime (or 'principal' as Stetson called it) focus. This has a longer focus length and larger scale than at the prime focus. A Newtonian focus is also near the top of the telescope, but the beam of light is reflected just before reaching the prime focus by a small, flat mirror, tilted 45° with respect to the optical axis, so that a focus is formed just outside the telescope tube. For faint nebulae a focal length as small as possible would be required to provide a smaller scale to concentrate more light onto any spot in the focal plane, so the Newtonian of the prime focus would be preferred.

De Sitter had proposed that the visit should be in the context of a collaborative arrangement. On June 6, 1932, Stetson stated in a letter:

> 'We have the large mirror silvered and in operation with the Cassegrain, and I have now every reasonable assurance that the telescope will be available for short focus work in September. I have assurance from our President that you will be given an honorary appointment of Research Associate at the Perkins Observatory next fall in consideration of our proposed cooperation.
>
> I think I can also assure you of a source of some remuneration for a course of lectures on statistics, which we shall hope to have you give to graduate students. We are counting definitely on your being with us next fall.'

Oort's trip was partly financed through lecture courses, which was only possible through his rising fame. Mrs. Oort's travel of course had to be financed privately. Leiden Observatory might have supplied some contribution towards Oort's expenses, although the budget for research from the Minister was very limited. After all, this was in the middle of the Great Depression, which had started in 1929, and government budgets were extremely tight. For example, the Bachiene Foundation for many years, from the twenties through at least World War II, supplied an annual subsidy of order Dfl. 500 (about €2500 now), which was originally mainly justified as a contribution to the expenses for the 'Bulletin of the Astronomical Institutes of the Netherlands' (BAN); during the Great Depression sometimes also to pay for assistants, but not foreign travel. There is no specific request for a grant for Oort's 1932 trip.

7.4 To Magog and Harvard and on to Ohio

In Oort's diary notebooks there is a very extensive account of the journey to the United States, too long to even include in translation in an Appendix. Mrs. Oort also kept a journal to describe her experiences. It started on August 12, 1932, but was already ended on August 31. The booklet does contain a few notes on or recipes for American food and some assorted notes on American life. I will quote from both.

The Oorts were planning to leave their children, at that time Coen, who was three years old, and Marijke, who had just turned one in April, with Mrs. Oort's parents in Utrecht. In a letter of June 19, Oort wrote to his parents-in-law to say how grateful he was that they were willing to look after the children. He expressed the hope that this would also help in their upbringing, as 'Coentje can be difficult and irreverent and Mieke and I are often not very capable to coach him into the right direction'. In correspondence with Harlow Shapley, Oort also tried to arrange a visit to Harvard and to give some lectures (to raise funding for the trip, one might suppose), but this did not work out. In the end Shapley invited Oort to spend the second semester of the next year (February to May 1933) at Harvard to give a lecture course. This was still open when the Oorts started their journey.

They sailed on August 12, 1932 from Rotterdam, and like Oort's travel to New Haven in 1922, they made a stop at Boulogne and another one on the English coast (now Southampton rather than Plymouth). They actually were on the S.S. Rotterdam of the Holland-Amerika Lijn, the very same ship Oort traveled on to Yale in 1922. But they were in first class and had a very spacious hut. According to Oort's notes: '2 beds, large writing desk, 2 washbasins with warm and cold water, comfortable sofa and lazy chair, 2 long wardrobes and 2 portholes, size $3 \times 4\frac{1}{2}$ meter', the size of course referring to that of the cabin, not the portholes. It turned out that Jesuit priest Johannes Wilhelmus Jacobus Antonius Stein S.J. (1871-1951) traveled on the same ship. Stein had obtained a doctor's degree in astronomy in Leiden in 1901 under Hendricus van de Sande Bakhuyzen, and after a short period at the Vatican Observatory he had worked as a teacher in secondary education. In 1930 he had been appointed Director of the Vatican Observatory, where he was responsible for its move away from the Rome city lights to the papal summer residence and retreat Castel Gandolfo. While drinking tea ('Mieke serves with it delicious Dutch cake and flikken', a small round piece of dark chocolate), they discussed Stein's new Observatory and other issues.

Mrs. Oort commented a lot on Americans on the ship:

> 'There are quite a few American girls on board (the large majority of passengers is American and these are special characters.) All are noisy and dominant, slender and – painted. You get used to those faces with make-up and as Jan says: 'It even makes them look brisk', when they come down 'chalked up'. [...] American girls have a good dose of nonchalance and independence and for example during the dancing they sit down to play some solitaire while wearing the most beautiful evening dresses. The dancing is the strangest I have seen. The girls' posture is different since they put their heads on the boys' shoulders, eyes half-closed, so that the faces get the most sheepish expression. He does the same while his chin is resting on her forehead. [...]
>
> There is a lot of variety in clothing, but often it is most tasteful. There are no rules to observe; they choose clothes according to what they feel like, either the most exquisite dresses or just a sweater when coming to meals. Dresses are often low-cut in the back. [...] When selecting clothes one should pay attention to colors; but we Dutch women do not do that. Mrs. Frederick was surprised to note how badly women dress in Holland and that in general they do not give much attention to it. It is a remarkable fact that these days on the ship I have seen more beautiful or nice dresses than I have ever seen in a theater in Den Haag.'

After arrival in New York on August 21, they were met by Jan Schilt and his wife Jo with whom the Oorts stayed for a few days. They lived in a small apartment, of which Mrs. Oort noted: '3 rooms, bath room and kitchen, and all of that for 4500 guilders a year'. This was almost Oort's full salary. From New York they traveled on to Randolph, New Hampshire, where the Torreys now lived. They went by boat to Boston and by train from there. Charles Torrey had retired from Yale University earlier that year and they apparently had settled in Randolph (probably having given up their second home on the Cranberry Isles). Their little daughter Anne (Nancy), now eleven years of age, had greeted Oort with: 'I am sorry I did not write to you, Jannie!'. Her mother often referred to Oort by this affectionate nickname. The Oorts rested and walked in the beautiful mountains and countryside, up to 'Madison Huts (about 1200 meters above the valley).' This was the first time Mrs. Oort met the Torrey's: 'The Torrey's greeted us very warmly. and they seem to be very nice people to me. We had supper from paper plates, very practical but not very nice.'

From Randolph they traveled to Magog in the south of Quebec, not far from the city of Sherbrooke, just north of the Canadian border. They traveled through Newport, New Hampshire, from where they were to travel by train to Magog, but what was 'more or less the only train on the line to Sherbrooke had been canceled. We traveled by bus until, judging from the map, we were at a crossroad to Magog. After some negotiations we found someone prepared to take us to Magog for $ 3.–, where we arrived four hours earlier than we had planned.' The reason for going there was that they had promised to go and help Minnaert with observing a solar eclipse that would be visible there. Marcel Gilles Jozef Minnaert (1893–1970) (see Fig. 7.13) was Flemish by birth and first studied biology in Ghent. In 1914 he fled to Leiden having refused to report for military service. In Leiden he taught physics, but returned to Ghent to teach at the Flemish University there, which had been facilitated by the German occupation. He again fled in 1918, after which

Fig. 7.13 On the left Marcel Gilles Jozef Minnaert in 1935. On the right Helena Aletta (Heleen) Kluyver in 1931. To her right we see Gerrit Kuiper. Courtesy Leo Molenaar (left) and taken from the staff photograph of Leiden Observatory (right) in 1931 (Fig. 7.7)

he was sentenced in absentia to 15 years of forced labor because of 'connivance with the enemy'. He found a job in Utrecht, where he founded a school of solar research with the help of physicist Leonard Salomon Ornstein (1880–1941). The solar eclipse that Oort mentioned took place on August 31, 1932 and Minnaert had prepared to observe it from Magog. However, the effort was 'spoiled by clouds' [201]. In Oort's words:

> 'August 31. Eclipse day. The weather is better today early in the morning, but it gets worse and worse until it is completely cloudy at the time of the eclipse. Still we did some observations. The sudden darkness was impressive, just as the quick brightening afterwards.'

For the rest of the day they went sightseeing and talked to some local people. At lunch Oort tried to order some wine, but that turned out to be impossible. It could only be bought at State Commissioners. And they made a canoe trip in the evening, but they were caught by the infalling darkness. They had difficulty finding their landing jetty. Since it had cleared up they had had a beautiful view of the Milky Way.

From Magog the Oorts traveled to Boston by night train, arriving in Cambridge early in the morning of September 1. The IAU General Assembly took place there from 2 to 9 September, 1932. The Dutch delegation consisted of only four persons, Hertzsprung, Minnaert, Oort and 'Miss H.A. Kluyver'. Helena Aletta (Heleen) Kluyver (1909–2001) (see Fig. 7.13) studied astronomy in Leiden and worked at the Observatory. She was to obtain her doctoraal ('Masters') in 1934. She had come also to Harvard to work with Cecilia Helena Payne (1900–1979), who had been born in Britain, and moved to the USA because chances of an astronomy career for women were more favorable there. In her PhD thesis in 1925 she had shown for the first times that stars consisted mainly of hydrogen and helium (Shapley had disliked this conclusion, but did accept it a few years later). In 1934 she married Ukrainian-born astronomer Sergei Illarionovich Gaposchkin (1889–1984). Heleen Kluyver spent all of her career at Leiden Observatory, finally defending a PhD thesis in 1954 with Oosterhoff on RR Lyrae stars. Her formal title at Leiden Observatory was 'secretary'. She would later marry Computer Gerrit Pels and later still Gijsbert van Herk, when they had become a widow and widower.

Oort, together with Minnaert, was one of three 'Recorders' of the General Assembly and contributed to the final report of the meeting. He acted as secretary of Commission 8 on Meridian Astronomy, reported on 'the connection between the apparent distributions of faint stars and extra-galactic nebulae', and he was acting President of Commission 33 on Stellar Statistics. He was listed as member of that Commission for the coming three-year period. Figure 7.14 shows Oort and Mrs. Oort as they appear in the group photograph of the meeting, together with Harlan Stetson and his wife. Of this episode almost nothing is recorded in Oort's diary notebook.

Oort's notes continue on September 23 with a visit to the Oak Ridge Observatory, some 50 km out of Boston, operated by Harvard College Observatory.

> 'With Shapley to Oakridge in his old Buick sedan, of which I now am the owner. At 2 o'clock I received my 'license' and I feel very relieved and for the first time have the experience of owning a car.

Fig. 7.14 Oort and Mrs. Oort taken from the group photograph of the General Assembly of the IAU in Cambridge, Massachusetts in 1932. The person in front of Oort with glasses and mustache is Harlan True Stetson, Director of Perkins Observatory and the woman on the right Mrs. Stetson. From the Oort Archives

It is a nice autumn afternoon and while driving we have a pleasant talk about the beautifully colored trees and the windless, reflecting lakes, about Aernout and what future he would have, about Oosterhoff, Bok and Kuiper. We have a very pleasant conversation both on the way to there and back, maybe too pleasant to reach any definitive decisions.

The 1934 plan will be decided in the first months of 1933, and we will have to do some calculations about the cost of living here and how much we will need. Report on it in December.'

'Aernout', de Sitter's son (see Fig. 7.20), was working on a PhD thesis on stellar photometry with Hertzsprung, which he would defend in 1936. Oosterhoff (see Fig. 5.5), Bok (see Fig. 7.15) and Kuiper (see Fig. 7.13) were completing their theses as well (Bok in Groningen with van Rhijn in that same year, the other two in 1933 with Hertzsprung). The plan to lecture at Harvard had apparently shifted to 1934. Before he left Harvard, Oort had a chance to examine many plates taken with the Bruce Telescope at the Harvard Boyden Station in Peru.

'At my request Miss Mohr has selected a set of interesting plates and these are put one by one on a light-table. I feel like back in those years when I first looked at the sky with a telescope and searched for and found certain objects. I now look for the first time into the extragalactic world, Wolf's rich and nicely concentrated Coma cluster, contrasting with rich fields in the center of the Virgo group, and the Centaurus group; everywhere that great variety in form and concentration.

Fig. 7.15 Bart Bok, taken from the group photograph of the General Assembly of the IAU in Cambridge, Massachusetts in 1932. From the Oort Archives

Fig. 7.16 Picture of a 1925 Buick Sedan. The car the Oorts apparently bought from Shapley must have looked like this [202]

What strikes me are the nuclei, their smallness and high brightness. And then the fact that these are absent in the very long extended nebulae, even though there clearly is absorption.'

Jenka Mohr (1902–??) worked at Harvard as an assistant and astronomer.

The Oorts moved on to New Haven where they arrived on September 26. They now traveled with their own car, which they had bought from Shapley. It must have looked like the illustration in Fig. 7.16. It was the first time they used a car for an extended trip. They were very excited and in his pocket diary Oort recorded for every day of their travel how many kilometers they had covered. From these notes it appears they took the Lundmarks with them to New Haven, which Oort recorded was a '±170 miles' (about 275 km) trip. Of course there they stayed with the Schlesingers and relaxed for a few days. Heleen Kluyver had gone to Yale as well after the General Assembly and the Oorts and she and Jan Schilt (who would move to Columbia a year later) and his wife Jo Timmer also joined in some sight-seeing activities. The Oorts left New Haven on Saturday October 1.

7.5 Observing with the Perkins Telescope

Oort and his wife traveled via Pittsburgh, where they visited Allegheny Observatory and from there on to Delaware, Ohio. Car travel was still slow. Oort noted in his pocket diary that they left Pittsburgh at 2 o'clock (pm), 'drove on fast and only stopped at 8^h30^m' for the night. The distance covered Oort recorded as 146 miles (235 km) and this took them six-and-a-half hours, so they averaged 36 km per hour. They arrived in Delaware the next day (Tuesday October 4) (Fig. 7.17).

The weather Oort met was not very favorable. He wrote his parents, sitting across from his wife in the library of the Observatory on November 2, 1932:

> 'I have not had any relaxation yet. I had rather much to do with the preparations of my observations, my courses here and my lecture. Stetson is also very good at making you do various things; for example I have to give a radio talk, on November 19 I think, and write all sorts of reports on my courses and lecture. We get the impression he is using us as a means for publicity for his new Observatory and that can be annoying. I must admit though that he is very friendly and does his best to make

Fig. 7.17 Finding charts from Oort's observing program for the Perkins Telescope (top) and images from the Digital Sky Survey [203]. The inner rectangle (left and right) or circle (middle) have a size of one degree. The images are ten arcmin on a side. The galaxies are from left to right NGC 628, NGC 7217 and NGC 7331

things easy for us. Unfortunately he does not have much success; the weather has been extraordinarily bad for the kind of work I want to do and I have only obtained 2 or 3 useful plates in a single month! I would be happy with 3 or 4 clear nights, but I am doubting that I will get just one more. In addition, I can work only when there is no moon, but I had in no way expected it would be so unfavorable. It is a major disappointment, since I had looked forward to working on the plates in 1933.

Coming Friday (November 11) we go to Ann Arbor for a week and probably from there on to the Yerkes Observatory 100 km N.W. of Chicago. [...]

Life is nice here and although I am rather busy we see much of what is going on here. Mieke will have told you about our house and the village [This letter is missing]. Outside the moon periods we lead a complete night life; never go to bed before 2 o'clock. But when it is at least somewhat clear we are at the Observatory and drive back to Delaware around 2 or 3 [am]. Strange to get used so easily to such a different way of life; for example with a car and cheap gasoline. The distance to the Observatory is 7 km and does not amount to much. We make a roundtrip by car in 20 minutes to go and drink an ice-cream soda when we are thirsty. We drive for the most insignificant reasons.

It will be difficult for you to get an idea of our setting, but I will try to make a rough sketch. We live on a hill, about 100 meters from a major road, US 23, from Cleveland to Columbus and the south (major roads have all been numbered, and at every intersection or junction you can find these numbers, which makes it easy to follow the map, because you follow the same number for 1000 km to travel as direct and as fast as possible). The countryside is grassland, golf links, surrounded by trees along two small fences that have been set back a bit in the otherwise flat country, beyond this a rough, almost uncultivated land with grass and thistles where sheep and cows roam. On the other side of the road there is a small stream, the Olentangy River, it was an insignificant stream when we came, so that at a few shallow places we had difficulty to establish that it was in fact flowing and where we could easily cross. We were amazed to see the enormous bridges that had been built to cross this little ditch. Yesterday night we heard our ditch roar and when we repeated a part of our first walk this afternoon, we found a broad river full of mud and rapids. It was a beautiful afternoon. We had used the car to buy a big jug of apple cider somewhere in the countryside and went for another walk. A fine afternoon sun shone through the trees and there were beautiful colors of the yellow-brown corn sheaves and a few white horses that transported sand to dam a small tributary of the river. I was sorry I had forgotten my camera.

Mieke is doing well also; she has little work to do in the house and also not much at the Observatory, because it was arranged that I would have the necessary help from an assistant that knows the instrument, and that was really necessary. She is busy improving her English and I am happy to be able to bring her a stack of new books every now and then. She sometimes drives the car, but there is not very much opportunity to practice as I usually have the car to go to the Observatory. By the way, there are no driver's licenses in Ohio (this is farmers country!). You see children drive as well. Young Stetson, 14 years old, drives the car all the time, to go shopping or pick up or take his father or mother somewhere. In the middle of the night, as we go home, you only see trucks, usually forming trains of one truck with two trailers, with bright red and green lights on the sides to warn that it is something much larger than an ordinary car. When we arrive in Delaware we often stop at a lunchroom for truckers that is open all night, where you can buy a delicious hot sandwich with two sausages and a nice cup of chocolate with cream! Columbus seems far away most of the time. It is not a very pretty town and we do not have much time to go there. It is about 40 km from Delaware, which takes 30 to 40 minutes, but before you get somewhere in town it takes very much longer and there is not too much to do anyway.'

There was also publicity in the form of newspaper stories. Figure 7.18 shows an item from the Columbus Dispatch (date missing, but not long after their arrival):

'Her [Mrs. Oort's] duties as assistant to one of Holland's foremost astronomers will consist chiefly of operating the movable platform, upon which the huge telescope is placed, by means of a block which she is holding in the photograph; correcting the motion of the telescope by means of another block of three buttons; turning the telescope in any of the four directions, and rotating the huge dome with levers, while Dr. Oort fastens his eye on the visions in the telescope.

Wednesday was the first time that Mrs. Oort had visited the Observatory where she will spend much of the next three months. She examined

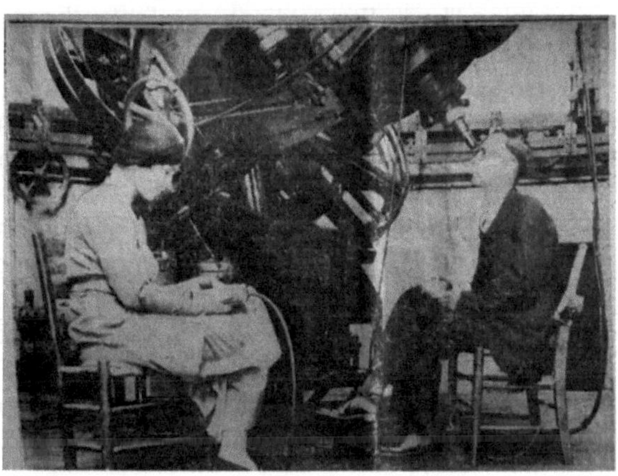

Fig. 7.18 The Oorts at the Perkins 69 in. Telescope in the Columbus Dispatch not long after they arrived in Delaware, Ohio. From the Oort Archives

Monday, December 5, 1932

Wife Of Dutch Astronomer Finds Americans Gregarious

Mrs. Jan H. Oort Likes Billboards, Thinks The United States Marvelous

By HELEN BEAL

"You Americans don't seem to like being alone. Always you group yourselves, herd together. Women have their clubs, the college all its little organizations—why, you even travel to Europe in groups!"

That's the most vivid impression of American life formed on her first trip to this country by Mrs. Jan H. Oort, wife of the Dutch astronomer who come to Perkins observatory last fall to use the world's third largest lens.

Sinclair Lewis and Fannie Hurst are the most popular American authors in The Netherlands, Mrs. Oort judged. "Everyone has read Lewis' 'Main Street' and 'Babbit.' They construct their idea of the United States from these books."

When the conversation veered to books, Mrs. Oort pulled a fast one and left the interviewer gaping with astonishment. "I have never read 'Hans Brinker and the Silver Skates,'" she said. "I never heard of the story 'till I came to this country.

Fig. 7.19 Another newspaper clipping, probably also from the Columbus Dispatch. From the Oort Archives

with interest the various levers and buttons which she will operate at her husband's command.'

Another clipping in Fig. 7.19 concerns an interview with Mrs. Oort. She used 'gregarious' probably as the best equivalent to the Dutch 'gezellig', with which she would have meant 'general togetherness that gives people a warm feeling' [204].

'You Americans don't seem to like being alone. Always you group yourselves, herd together. Women have their clubs, the college has all its little organizations – why, you even travel to Europe in groups. [...]

Mr. and Mrs. Oort are starting home to Holland Thursday after 'the most exciting experience of my life', to quote Mrs. Oort. 'Everyone has been so hospitable, and the United States – it is marvelous! Distances for instance. You Americans drive the length of my little Holland in one evening to hear a lecture. Your advertising, too – I have never seen anything like it. Everywhere one looks there is an advertisement. And such interesting ones. [...]

Holland is quite, quite different from Ohio', Mrs. Oort implied. 'Take newspapers, for example. Comic strips are almost unknown, but a whole page of them every night – that would be something', though Mrs. Oort did not say what.

[...] Scandal and personal items about famous people don't appear in European papers. Maybe we're not as curious as Americans. At any rate, if you were the wife of a Delaware astronomer in Leyden, no one would think of interviewing you. And they would ask your husband about astronomy, not about his personal life.'

On 24 November, 1932 Mrs. Oort also wrote an extensive letter to her parents after they had returned from Ann Arbor and Chicago.

'We have many nice moments together here with our gas stove and a cup of chocolate in the evening. It is so quiet here at night, I would not object to having no children. But the life of a woman is after all a permanent search for a balance between the noise of the children and the quiet of your husband. At home in Leiden I often feel no more than a mother and lose sight of my life with Jan. It is wonderful that this period has been presented to me and I am grateful for it. We feel so much stronger and refreshed to pick up our lives again at home.

Here everything feels new and young. Yes, they have much advertising, and they do spend their money, but is that really so bad, when you use your money to try new things? It results in people making plans; that young people have the opportunity to realize their ideals. In our country you only hear of people being fired, while here at all Observatories there are plans, big plans to build new telescopes. In Holland the rich people only investigate how to safely invest their money and here how they can best spend their money. In spite of all the unemployment and misery that must be present as well, it all looks luxurious.

[...] Today we celebrated 'Thanksgiving day' at the Stetsons with a large dinner with turkey and pumpkin-pie. Yesterday was a very rainy day so that we could go to bed at the normal time and this morning we went to the black church where we had planned to go to for some time. The service was fascinating! You have no idea how special it was to be there. The intimate, comfortable atmosphere and the enormously cute little church with a piano on a platform and the great mobility of the minister when he was preaching. The changes from wild enthusiasm to an ordinary story are very sudden. People show their approval by encouraging him all the time with 'that is very true', 'that is right' during his sermon. The singing also started suddenly and unexpectedly and the minister clapped his hands for encouragement. The singing was beautiful, especially since he had such a nice voice and fascinating personality. We talked to him afterwards and he said he enjoyed it and that had he known we were there he would have asked Jan to speak. What do you think of that? The attendance was among the poorest of the poorest blacks. Black children were most cuddly with their black curls and bright dark eyes. Children as small as Marijke were sitting quietly all the time, not being childlike for a single moment, they started to look around and hang over the backrest of the benches. I do not understand why all children behave themselves so much better than Coen and Marijke! [...]

About Jan's work; it is still without success, even though now he is gone all night and it looks really clear. You never know. Still Jan feels (despite eclipse and Delaware's bad weather) astronomically speaking it is not time lost. He has talked to many astronomers and formed new ideas and feels refreshed. It is a pity of course that he has been unable to start the work he had dreamed of for so long, but staying longer is not an option because the worst months weather-wise are coming now. [...]

> The trip to Ann Arbor and Chicago was extremely nice and it was fun to see another sky-scraper town, but it has a different atmosphere from New York. We stayed in the largest hotel in the world. So I have been at least for one night in an American hotel.'

The Oorts returned to Leiden just before the New Year. The trip back to New York went via Niagara Falls in snow and freezing cold. Oort mentioned that the odometer in the Buick told them that they had driven about 6800 km. What they did with the car upon departure has not been recorded. They also drove through Albany, where Benjamin Boss resided at Dudley Observatory, so maybe the latter took over the car or arranged for its sale.

They sailed from New York on December 20 on the S.S. Veendam. A letter from Stetson of February 11, 1933 records: 'You certainly had an exciting trip. I wish that I might have been with you to share that particular nautical experience. A roll of 25 is something to be remembered, and coming from a scientist I know that it is not exaggerated.' A roll is defined as the rotation about the bow-stern axis; an amplitude of 25° is considerable.

Oort had problems adapting to everyday life in Leiden again, which felt very bourgeois compared to the adventure of traveling in America. On January 14, 1933 he wrote in his diary notebook:

> 'The longing for adventure, for unknown territory, has never in my life been so strong just now that we have come back to the Cobetstraat and the Leiden of my bourgeois schooldays [this is a pun; he wrote 'Burgerschooltijd', 'Burger' meaning bourgeois, but also being the 'B' in his school type Hoogere Burgerschool] or HBS, a Leiden for which I have a feeling of hatred because of its colorlessness.' [...]
>
> 'The same with religion. I would like a new, different religion, an adventurous one. I sort of dislike the both misused and worn out existing religion. I think I would like a separate connection with some of my friends! A sectarianism of the worst kind! The strange aspect of the idea of a spirit is that it covers us all.'

Oort made only a few remarks in his diary notebooks on the observing. There is a paragraph about unspecified difficulties with the shutters and astigmatism in the Newtonian focus. Oort was able to secure some plates to take home to Leiden. The technique followed was to take an exposure of a galaxy and an out-of-focus image of stars with the same exposure time in one of Kapteyn's Selected Areas where the magnitudes of stars were known, preferably on the same plate. The reason for taking images out of focus was to spread the light of a star to avoid overexposing the photograpic emulsion. This could be used to calibrate the surface brightness measurements in the nebula. The Annual Report of Leiden Observatory for 1932 mentioned:

> 'Dr. Oort stayed some time at Harvard, where he studied faint extragalactic nebulae on Bruce plates, and visited some other Observatories in America. Most of his time was spent at Perkins Observatory, Delaware, Ohio, of which institution he was made a Research Associate, and where by the courtesy of Director Stetson he observed with the 69-

Fig. 7.20 Aernout de Sitter (left) and Herman Kleibrink, taken from the staff photograph of Leiden Observatory in 1931 (Fig. 7.7). From the Oort Archives

inch reflector, and gave lectures on statistical astronomy to the students of Ohio Wesleyan and Ohio State Universities. He returned to Leiden on December 30.'

In the following year some progress was made, according to the Annual Report:

'Dr. Oort has given much time to experiments and discussion of preliminary measures on plates taken by him at Delaware in the preceding year, in order to determine photometrically the distribution of light in extragalactic nebulae. Five plates containing 8 exposures of nebulae, together with extra-focal exposures of stars for the determination of the scale and the zero point, were measured by Mr. Kleibrink in the Hartmann photometer. One plate has been measured through with a registering photometer at Utrecht, by the kind permission of Professor Ornstein with the efficient cooperation of Dr. Minnaert.'

Herman Kleibrink (1910–1991; see Fig. 7.20) has been listed in the Annual Report for that year for the first time and as 'non-established Computer' (on a temporary contract). He remained with the Observatory his full career, particularly focusing on photography, in which he made quite a name, also outside astronomy.

The Hartmann photometer, designed to measure magnitudes of stars on photographic plates, compared the light transmitted through the photographic plate with a beam directly from the lamp. The latter was attenuated by passing through a photographic wedge of varying photographic density. The photometer contained a set of prisms such that the two images appeared in two eyepieces that could be viewed simultaneously each by one eye. The instrument had been first constructed by Johannes Franz Hartmann, (1865–1936), then working at Potsdam. The instrument in Utrecht employed a system in which the light was recorded with a photocell and the resulting current measured with a galvanometer. It was designed for a precise determination of the spectrum of the Sun over the full optical wavelength range [205, 206]. For 1934, the Annual Report of Leiden Observatory recorded:

> 'By courtesy of Professor Ornstein and Dr. Minnaert of Utrecht Dr. Oort and Mr. Kleibrink were enabled to continue measures with the registering photometer of the plates taken at Delaware in 1932. The measures made during the year include 11 exposures on extragalactic nebulae and 6 on standard areas with extra-focal images of stars.'

After that no further mention is made of the Perkins plates. Apparently there had been problems with the observations. Oort had arranged that further plates would be taken for him with the help of an assistant by the name of Marvin O. Cobb, who worked on a PhD thesis at Perkins, after which he must have left astronomy. Oort had a list (still in his papers), which he had also left with Cobb, of about twenty galaxies to observe, almost all spiral systems and with right ascensions suitable for the last and first months of the calendar year. Cobb wrote Oort in January that he had had very little chance for observing because of unusually bad weather, but also that there were guiding errors. Stellar images could shift on a plate by as much as 10 mm. 'Dr. Stetson has had some correspondence with the other Observatories where they have large reflectors. They say that it is impossible to keep a large mirror from shifting with change of position of the telescope. Consequently we are now working as rapidly as possible on a double slide plate-holder arrangement for guiding.' And Stetson mentioned on 11 February in a letter to Oort: 'I have become convinced that we had extraordinarily good luck on the 25-foot focus, and that we are probably right in surmising that the mirror shifts its position slightly with the changing hour angle of the telescope. I talked with Adams at Atlantic City [the New Jersey meeting of the American Astronomical Society, December 1932], and have also seen Curtis since you were here. They both feel that I am expecting too much in utilizing the long focus finder on prolonged exposures. Adam says the 100 in. will rather suddenly shift during an exposure adjusting itself to new lines of gravity.' Walter Adams was Director of Mount Wilson Observatory and Heber Curtis had worked at Lick Observatory, but had moved to the University of Michigan in 1930. The hour angle is the angle of the telescope away from the meridian. Some early results Oort apparently had sent were not unreasonable, since Stetson wrote on September 27, 1933, that 'you have done remarkably well in getting as good photometric data as you have.'

The proposed collaboration between Leiden and Perkins took time to take shape. Oort, with the consent of de Sitter who was also interested in the collaboration, even offered to send an assistant from Leiden to Ohio for half a year. But the letters by Stetson mentioned problems with the funding of the operations of the Observatory, and on January 15, 1934, his secretary wrote that Stetson 'had been appointed Research Associate at Harvard University'. The budget of Perkins Observatory had been cut and to 'meet the curtailed budget for this year's operations, I am taking a half-year leave of absence from Ohio Wesleyan and am spending the second semester here at Harvard' (March 26, 1934). This seemed to be the end of the collaboration. Stetson did not return to Perkins Observatory. None of the photometry Oort obtained has ever been published.

Oort next turned his attention to Mount Wilson. Pieter Oosterhoff in the mean time had completed a Ph.D thesis with Hertzsprung in April 1933 on *Effectieve golflengten en photographische magnituden van sterren in h en χ Persei* (Effective wavelengths and photographic magnitudes of stars in h and χ Persei). As so many of the young Dutch astronomers, after his PhD he had moved to the United States, where actually many of them stayed for the rest of their careers (see [207]). Oosterhoff had obtained a one-year fellowship of the Rockefeller Foundation first to Harvard and then to Mount Wilson, where he worked on variable stars. This was his field in which later would later make a name with his work on RR Lyrae variables, for which he found in 1939—back in Leiden—two distinct groups among the light curves that eventually have been named after him. Oort contacted him on November 7, 1933, with the request to obtain plates of a few galaxies, this time so-called early-type systems, which are galaxies that contain little or no dust. Oosterhoff was somewhat reluctant; he did not want to upset his Director by acting without his permission. Now Hertzsprung at some stage had also asked him to make exposures on some variables for him and when Oosterhoff had done so, Hertzsprung came back with a even more extensive request for plates of RR Lyrae variables. With some trepidation Oosterhoff had told Hertzsprung to discuss this first with Adams. It took Oosterhoff until May 20, 1935 to report to Oort that he had obtained permission from Walter Adams. He promised Oort to honor his request as best as possible with the limited clear time he had. Apparently Oort had written to Adams about it, proposing to use the 100 in., but observing time there was heavily oversubscribed. The 1935 Annual Report of Leiden Observatory mentioned that Oosterhoff indeed had taken 'some new plates' with the 60 in. at Mount Wilson, four of which had been forwarded to Leiden. In 1938, measurements and reductions of plates of three nebulae, were reported to have been completed, again using the microphotometer in Utrecht. These results in the end led to a study of the dynamics of two of these systems that we will return to later (Oort 1940a).

So, photometry of galaxies was still a bit too difficult. What did Oort have in mind? In a memorandum in Oort's papers he had made notes for himself to use in discussion on future work on galaxies. From the context is appears that he used it before he left Perkins Observatory to discuss with Stetson. Oort's aims were summarized as follows: (1) diameters to verify relative distances, (2) average surface brightnesses to do statistics, (3) detailed light and color distributions, (4) distribution on the sky of fainter galaxies to study absorption in the Milky Way (Zone of Avoidance), and typically for Oort: (5) 'Most importantly: unexpected new things that always come out of really good work.'

Fig. 8.0 Painting of Oort in 1938 by Dirk Hidde Nijland (1881–1955). It currently resides in the Oort Room of Leiden Observatory in the Oort Building on the campus north of Leiden. From the collection of the Faculty Mathematics and Natural Sciences, University of Leiden, see [208].

Chapter 8
Professor and Adjunct-Director

The road to success is always under construction.
Aloysius Paulus Maria (Louis) van Gaal (b.1951)

The spirals and the Galaxy bear the stamp of the past both in their regularity and in the irregularity of their structure.
Jan Hendrik Oort

It had been expected early on, possibly already at the time of Oort's appointment in Leiden, that he would succeed de Sitter as Director of the Observatory. That would have been in 1942, when de Sitter would reach the age of seventy. By that time Hertzsprung, the Adjunct-Director, would be one year from his retirement and it seemed not reasonable to assume that he would be appointed Director for such a short period. Oort would by then be 42. However, de Sitter's death and World War II would present a different outcome. Before turning to that I need to catch up first on a few other developments at Leiden Observatory. For more background on astronomy in the Netherlands in this period, see David Baneke's article on this issue [100], or (in Dutch) [90].

When Hertzsprung joined Leiden he was promised the provision of access to a telescope in the southern hemisphere. De Sitter had been very active realizing this and the first steps were taken, when in 1923 during a visit to Leiden by the Director of the Union Observatory near Johannesburg, Robert Thorburn Ayton Innes (1861–1933), an agreement was drawn up between the two observatories and approved by the respective governments. Under this agreement astronomers from Leiden 'shall have every facility to visit Johannesburg and to use the instrumental equipment of the Union Observatory for the purpose of extending to the southern

Louis van Gaal is a former Dutch soccer player and coach.
From the final paragraph of his inaugural lecture as Professor (Oort 1936b).

P. C. van der Kruit, *Jan Hendrik Oort*, Astrophysics and Space Science Library 459, https://doi.org/10.1007/978-3-030-17801-7_8

Fig. 8.1 Willem de Sitter and Ejnar Hertzsprung, around 1930 probably at the Observatory Leiden. From the Oort Archives

sky the researches that they have been making over the northern.' And 'in exchange the staff of the Union Observatory shall be allowed to enter the Leiden Observatory as research students for courses of instruction in theoretical and practical astronomy and have every facility in using the resources of the Leiden Observatory, subject to the approval of the Director.' De Sitter and Innes had known each other well since the days the former worked at the Cape Observatory in preparation of his Ph.D. thesis. Also there was cooperation with the Yale Southern Station, where Schlesinger had built a 26 in. (66 cm) southern reflector, also near Johannesburg. From the Leiden Annual Reports we know that Schlesinger quite regularity visited de Sitter and Leiden. Hertzsprung did not wait long and took off to South Africa later that same year (1923). He was to go for half a year, but in the end stayed for a year and a half. He enjoyed the opportunity to concentrate on research. He wrote to de Sitter (February 1924): 'I love working here, without any interference by other things, and—no wife who is jealous of my work.' Henriette Kapteyn and he had separated in the first half of 1923, to be formally divorced (much) later (Fig. 8.1).

At the same time Joan Voûte had established the Bosscha Observatory in Lembang near Bandung with the financial support of tea planters and amateur astronomers Karel Albert Rudolf Bosscha (1865–1928) and his nephew Rudolf Eduard Kerkhoven (1848–1918). This was formally opened in 1923. Although de Sitter might have hoped to direct its scientific program from Leiden (and provide in this way for Hertzsprung's telescope access in the south), it actually became an independent facility.

The proposal by de Sitter and Oort (1925d), to establish an observing station at a location close to the equator for fundamental declination determinations, resulted in an expedition to Kenya in the early 1930s. This was a full Leiden supported undertaking and the management of Leiden Observatory was solely responsible for it. It was to a large extent de Sitter himself who took care of the managerial aspects.

And then, in 1930, after years of lobbying and planning, the Rockefeller Foundation provided extensive funding for the erection of a major telescope in South Africa, which later would become the Leiden Southern Station. These preparations and the setting up of the facility concerned a major management effort, almost all of it coming down to the Director, de Sitter, and it was clear that a successor would have to take on all of this. It seems that de Sitter's style of directorship was not without a good share of authoritarianism and not too much delegation of work. Although in the de Sitter years Oort's actual active involvement was limited, eventually he would assume full managerial responsibility for these undertakings.

8.1 Succession of de Sitter

A very detailed account of the succession of de Sitter has been written (in Dutch) by Jet Katgert-Merkelijn [209], which has been used extensively in this section. Also material relevant to this discussion, in addition to what is in the Oort Archives, can be found in the Leiden Observatory Hertzsprung Archives [210].

Willem de Sitter died quite suddenly from pneumonia after a short illness on November 20, 1934, possibly related to the tuberculosis he had contracted earlier brought about by an overdose of ether for anesthesia, see Sect. 3.3. Oort (1935a) wrote a admiring obituary. Ejnar Hertzsprung was at that time Adjunct-Director of the Observatory and Extraordinary Professor of astronomy at the University. Extraordinary in this context means that the professorship was not part of the regular contingent of full, 'ordinary' professors. It was originally created for persons that taught at universities next to a their main employment or business. The fee was less than that of an ordinary professor, so during the Great Depression it was sometimes also used as a way of saving on professor salaries. Since 1930 Oort was Observator (the highest rank under the directorship) at the Observatory and Lector ('Lecturer' or 'Reader') at the University, the rank just below that of professor. Hertzsprung was 61 years of age and Oort 34. At first sight that would have meant that Hertzsprung would naturally become Professor and Director and Oort take over the positions Hertzsprung occupied. As it turned out, this was indeed what happened, but there is more to it.

Fig. 8.2 Willem de Sitter in 1933, painted by Johannes Carolus Bernardus Sluijters (1881–1957). This painting used to decorate a wall of the Leiden Observatory Director's office. From Leiden Observatory

Part of the staff did not look forward to a directorship by Hertzsprung, who was considered to be very strict and insensitive and to give little attention to the human sides of his subordinates. He was an workaholic and expected everyone working for him likewise to dedicate all his or her energy to astronomy. He was certainly keen to become Director. Actually, it appears that he had proposed to de Sitter some years before that (around 1930) that they would share the directorate. For de Sitter that was absolutely out of the question. Hertzsprung did not easily take no for an answer and wrote the President of Curators with the same proposal. However, this person wisely did not act upon this. When de Sitter was away for an extended period (September 1931–March 1932, when among others he gave his Lowell Lectures at the Lowell Institute in Boston, from which his book 'Kosmos' stemmed), he had left the administration to Oort rather than Hertzsprung, who after all as the Adjunct-Director would have been the natural choice. De Sitter might have felt it useful for Oort to gain some experience. But this had thoroughly upset Hertzsprung. Although there never was an open conflict on any serious matters, there had been friction before. De Sitter had been in a sanatorium in Arosa between 1919 and 1921 to recover from tuberculosis brought about by an overdose of ether for anesthesia, see Sect. 3.3, and Hertzsprung had acted as Director, but de Sitter even then had kept all financial authorities to himself rather than delegate these also to Hertzsprung (Fig. 8.2).

Oort, who left various personal (but usually undated) notes on the situation (unfortunately his diary notebook has no entries between 31 August 1934 and 20 June, 1935), wrote:

> 'Hertzsprung says: 'The most natural thing would be that I will be the next Director. There have to be very strong reasons to do otherwise.' And he does not see those.'

Obviously Hertzsprung wanted the position, since he was definitely more senior and must have felt it should come to him naturally. Oort did not want Herztsprung to be his superior and tell him what research he should be doing. From Oort's notes it appears that he was concerned that he would not have the freedom to pursue his own interests and to be able to claim a reasonable fraction of the Observatory's resources. On December 15, 1934, Oort noted:

> 'Hertzsprung wants me to see him at 9 in the morning. That makes me feel indignant. I lack the calmness Prof. de S. had to not let it affect him. I agree in principle with him, but I cannot accept to be told what to do...'

Was this an omen of the new situation where Hertzsprung as Director would treat Oort as a subordinate rather than an equal? Oort did have some support from Pannekoek. In a draft for a letter to Pannekoek, dated February 14, 1935, Oort noted:

> 'The committee that arranges the succession would need to conclude first, on the basis of advise it has received, that the easy solution: Prof. Hertzsprung the natural successor of de Sitter, is not desirable. Then ways can be considered to make the other solution acceptable to Hertzsprung, and if I was asked my opinion, I would be prepared, with the enormous respect that I feel for Prof. He., to do all I can to make things as easy as possible for him.'

Indeed, Oort expressed his respect for Hertzsprung in various notes. For example:

> 'What bothers me most in all this is my personal feeling of high regard and respect for Hertzsprung; [...]. I have the opinion that He. is getting less recognition than he deserves, that he is a greater man than many think, he is a greatness that I do recognize and therefore I look differently upon this and should do everything from my side to have him acknowledged.'

But he also noted:

> 'On the other hand He. is a person that overrates himself in areas that are outside his own. He says for example that he really had wanted to be a theoretician but for practical reasons used instruments. That is nonsens, to say the least. He followed his talents and those were certainly not in theory and it is a big overrating of him to think that he could have been a great astronomer also in that area. His 'practical' reasons are only to hide his inner inclinations, because he thinks that looks better. In that manner he will undoubtedly overrate his abilities as Director and feel, as Aernout [de Sitter's son, apparently in a conversation on the subject] says, that he can do anything better than others. Is that an inferiority complex? His attitude towards the human side of things in the world will be this total

lack of understanding of the feeling of others. That does not mean that I will be better suited to be Director. He has good qualities as well, and I, with my lack of self-confidence, in any case one bad one.'

He also had some observations about himself:

'After all, you always wanted this. You more than most other scientists have a longing for the mystical aspect of life, for the mysterious as Bremmer expressed it, and that for that reason you do not feel at ease among the usually more dispassionate scientists. There are only a few, like Ehrenfest in the past, with whom you felt this other connection. For this reason: You can do it as long as you are convinced that it is the right thing to do, when it is not a duty and a necessary evil, carried out without conviction.'

Bremmer was a friend who we will encounter later (see Sect. 11.1). There is a note, apparently a summary of points Oort wanted to raise in a conversation with some higher official in the University, probably at an advanced state of the proceedings to appoint a successor to de Sitter. In this the financial consequences for the University of various outcomes play a role. This was after all a time of economic crisis and government funding was minimal.

'My great respect for Hertzsprung. In case Prof. He. would get both appointments [Professor and Director], what would happen with me?
1. Relation to Hertzsprung, my own freedom. Also Rockefeller instrument and confidence that our collaboration will work.
2. My appointment as adj. dir. and Extraordinary Professor. I will certainly be very pleased with that. Salary now f. 6750,–. So no costs, almost.
3. Filling of the conservator position. There is room since a position of lector [reader] will become vacant. History of the Rockefeller Foundation: When this came about we had one observator for this plus myself. It was then expressed to be desirable to have two more. One has been filled, now another one? This [the case of no such position becoming available] is against the idea of the Rockefeller plan. Save by cutting budgets of institutes that are not so flourishing, you should not do that to what you regard to be a jewel.
Already save f 2250 + in the first 5 years on av. f. 2000,–. Rather make me Extraordinary Professor.
4. Is it absolutely impossible to raise a assistantship to a senior assistantship, so that at least one person can stay? The filling of temporary assistantships will be difficult enough. A. has a salary of f. 1410.-'

'A' must be Aernout de Sitter, who was an assistant. Oort's salary of Dfl. 6750 (annually) corresponded to about 4 times that of an unskilled worker; that of an assistant was comparable to the latter, so it was not a very well-paid job. And if a staff position had to be given up he was even prepared to accept an honorary professorship (without the corresponding salary).

The impression one gets from examining the complete set of Oort's notes is that he was prepared all along to accept that Hertzsprung would become Director and Professor of Astronomy. However, he was not going to give in before some important issues were resolved. First and above all, he had to be sure that he would be free to follow his own interests and choose his research topics. Oort was also interested to use the Rockefeller Telescope, and wanted assurance from Hertzsprung that he could have some share of the time. And he would need to get support for his work from the staff, such as the help of computers, but also a say in the hiring of the new staff like that to be allocated to fill the vacant position resulting from de Sitter's demise. There seemed to have been no problem for Oort and Hertzsprung to agree between them that Oort would have the same freedom as he had enjoyed under de Sitter, or as much as Hertzsprung had during de Sitter's directorship.

There was another important consideration that did not figure initially in Oort's notes and that only became apparent at a late stage. That was the Director's house. De Sitter lived there when he died, but it would become vacant as soon as his widow would have found another place to live. The Observatory housed the Director and Professor of Astronomy in a spacious residence, while Adjunct-Director Hertzsprung also lived at the Observatory in a somewhat smaller, but still comfortable house. The wing beyond the main building in Fig. 8.3 contains two smaller residences; Hertzsprung lived in the one furthest from the main entrance. At some stage Hins

Fig. 8.3 Leiden Observatory in 1933 from the northwest. The main building in the middle has the dome of the 10 in. (refracting) telescope over the entrance; the next largest telescope is the 6 in. refractor only just visible behind the smaller dome. Hertzsprung lived in the furthest part of the wing beyond the main building. In the distance we have the 13 in. photographic telescope and the 'photographic department'. From the Archives of Leiden Observatory

Fig. 8.4 Front of the Director's residence of the Observatory after the Oorts had moved in. The picture has been taken from the 'Sterrewachtlaan' in front of the main building. The person on the steps is the younger son Abraham. From the collection of Abraham H. Oort

Fig. 8.5 Director's residence from the garden side in the back, when the Oort family lived there. On the fence daughter Marijke Oort. From the collection of Abraham H. Oort

lived in the other part, but it is not clear if he did already at that time. There were also living quarters for the assistants and Observators ('bachelor' quarters in de Sitter's 1933 history of the Observatory), and residences for the chief instrument maker and the custodian (Figs. 8.4 and 8.5).

The residence of the Director was the westernmost part of the main wing. E.g. in Fig. 5.1 that means at the left of the main building. Hertzsprung lived on the eastern end of this main wing; his residence had been renovated as the final part of a program of enlargement and modernization of the premises that was completed in 1925 (Hertzsprung had been away on his 18 months stay at Johannesburg, when the works on his dwelling were executed). The Oorts still lived at their home in the Cobetstraat. Their third child, Abraham Hans (Bram) had been born on September 2, 1934, and apparently their housing with a family with three children had become smaller than desirable. Oort and especially his wife liked a garden and flowers and the Director's residence must have been extremely attractive to them. The residence was best used by someone with a family with children. If Hertzsprung were to become Director, he would of course have first choice to go living there, but the house was much too large for someone living by himself and he was probably not interested in the garden anyway (Fig. 8.6).

The formal procedure for the appointment was that the Faculty would make a recommendation to the Curators, who would make a proposal for the Minister. The Faculty consisted of the professors in the relevant Faculty, in this case of Mathematics and Natural Sciences, and that body consisted of those members of the Senate (the collection of all professors) in those sciences. Apparently, some felt one simply could not pass over Hertzsprung, others were concerned about his abilities to become an effective Director. The option that Hertzsprung would be Professor and Adjunct-Director and Oort Director and Extraordinary Professor was rejected by Hertzsprung. Oort seemed to have agreed to be Adjunct-Director and Extraordinary Professor and would take over a larger share of the administration, but only if he could be excused

Fig. 8.6 Director's residence and the main building from the back garden. Oort used to have his office behind the glass veranda in the middle. From the collection of Abraham H. Oort

from a fair part of his teaching. Hertzsprung then felt that this teaching should be taken over by Jan Woltjer and that Woltjer then should be Extraordinary Professor. But that opened the possibility that not Oort but Woltjer would take over the directorship when Hertzsprung would retire; after all Woltjer was only nine years Oort's senior and would be 52 when Hertzsprung retired in 1943. Oort was not ready to accept that.

In the end the Faculty voted on two options. The first was that Hertzsprung would be both Director and Professor (and presumable Oort Adjunct-Director and Extraordinary Professor). Of the thirteen present, six were for, five against and there were two abstentions. A majority but not an absolute one. So they voted also on the second option, which was to split the two positions, make Hertzsprung Professor and Oort Director. This was accepted with seven votes for, five against and one abstention. Obviously, the Faculty was divided on the issue, but the second proposal went as a recommendation to Curators. Hertzsprung, when he heard about it, said he would not accept that and argued that the first option should be implemented; after all that had been the first proposal voted on and had gained a majority. The solution was to let both men have an interview with the Minister. This was Hendrik Pieter (Henri) Marchand (1869–1956), who was about to resign (on May 18, 1935) for secretly having changed from being Dutch Reformed (and member of the corresponding political party) to Roman Catholic in 1934 (Fig. 8.7).

The Minster preferred a solution that apparently had been adopted in the Hospital of the University, in which Oort would be Director and Hertzsprung Professor and the adjunct-directorship abolished. Hertzsprung, according to his notes of the conversation, rejected that pointing to the situation in other natural sciences where the Professor and Director were one and the same person. Oort left no detailed notes, but appeared to have come to the interview ready to accept that Hertzsprung would be Professor-Director, if he were appointed Extraordinary Professor and his then vacant

Fig. 8.7 Entrance hall and staircase in the Director's residence of Leiden Observatory. Mrs. Oort always arranged for flowers. From the collection of Abraham H. Oort

Fig. 8.8 The Oort family at their vacation home Sandy-Hook in Katwijk. From the Oort Archives

Observatorship opened for recruitment. He wondered how to dress and seemed to have concluded that his Excellency could only be approached very formally and decided to wear a tail-coat.

In the end the Minster decided to have Hertzsprung appointed both as Professor and Director, and Oort Extraordinary Professor and Adjunct-Director. The first three appointments were proclaimed by Royal decree on May 27, 1935, followed by a Ministerial decision for the fourth. And indeed Pieter Oosterhoff was appointed Conservator per February 1, 1936. Hertzsprung asked the Curators to assign the Director's house (known as Sterrewacht no. 5) to Oort; it is not clear that this indeed was part of a pre-arranged deal, but that is not inconceivable. Hertzsprung probably disliked moving anyway (Fig. 8.8).

The upper floor of the Director's house was often occupied by another resident. For example between 1938 and at least 1942 this was Catharina Emilia te Lintum (1904–1969), who was working on a theological Doctor's thesis while teaching religion and Hebrew at the Leiden gymnasium (and probably also at a school in Wassenaar). She has taken some of the pictures of the house that have been reproduced in this section. As we will see Oort's oldest son later occupied this part of the house also for some time.

8.2 Inaugural Lecture

The family life easily adapted to the comfortable house at the Observatory. Not everything had been prosperous. A major blow to the family had occurred in August of 1933, when Mrs. Oort's brother Coen Graadt van Roggen died at the age of only 28. He had married Engelina Carolina Mary Kan (1906–1985) in 1928. It must have been a source of much grief for Mrs. Oort and for her parents.

Now that Oort had been appointed Professor (that it was an Extraordinary Professorship did not make much difference, except of course in the salary), he was entitled to present an inaugural lecture, accepting the appointment and being formally installed as a member of the Senate. This was delivered on Friday November 15, 1935 and had the title *The structure of galaxies* (De bouw der sterrenstelsels). It was held in Dutch and also published in Dutch in the amateur magazine 'Hemel and Dampkring' (Oort 1936b). It described Oort's view and thinking on the structure of our Galaxy and that of extragalactic systems at the time. I will mention the highlights and general thread of the narrative below; the full text, which is very illuminating in view of Oort's thinking in the middle of the (and his) thirties, appears in English translation in Appendix B.3.

In the first part of the lecture, Oort concentrated on and treated in much detail the distribution of galaxies on the sky. For this he used an illustration (that was his Figure 1; see Fig. 8.9) produced not long before by Shapley and his assistant Adelaide Ames (1900–1932). Ames tragically died in a boating accident at about the time that publication, *A survey of the external galaxies brighter than the thirteenth magnitude* [211], appeared. Oort pointed in the first place at the paucity of galaxies at low Galactic latitude, the Zone of Avoidance, resulting from absorption of light in the disk of the Galaxy. But a thing that also struck him was the irregularity of the unaffected distribution. Actually, what we see in the top-panel is a large conglomeration of bright galaxies quite close to the Galactic north pole. This is the Virgo Cluster at a distance of some 16 Mpc. But there is still more structure over the unaffected area. This already had fascinated Oort while working at Perkins Observatory, when he had some correspondence concerning this issue with Hubble. For example, the latter wrote to Oort on November 3, 1932:

> 'The most general result is that the nebulae approach random distribution as the limits of the counts are increased and that systemic deviations from this rule can probably be attributed to local obscuration in the Galactic system. The Zone of Avoidance follows pretty well the general distribution of known obscuring clouds as can be seen by tracing the zone on composite photographic maps of the Milky Way. The zone is irregular and is bordered by 'partial obscuration'; then comes the general field, where the number of nebulae per plate ranges from $\frac{1}{2}$ to twice the average with few exceptional areas and occasional clusters (about one cluster per 25-30 sq. degrees). The frequency curve for number of nebulae per plate approximated a normal error curve. [...]

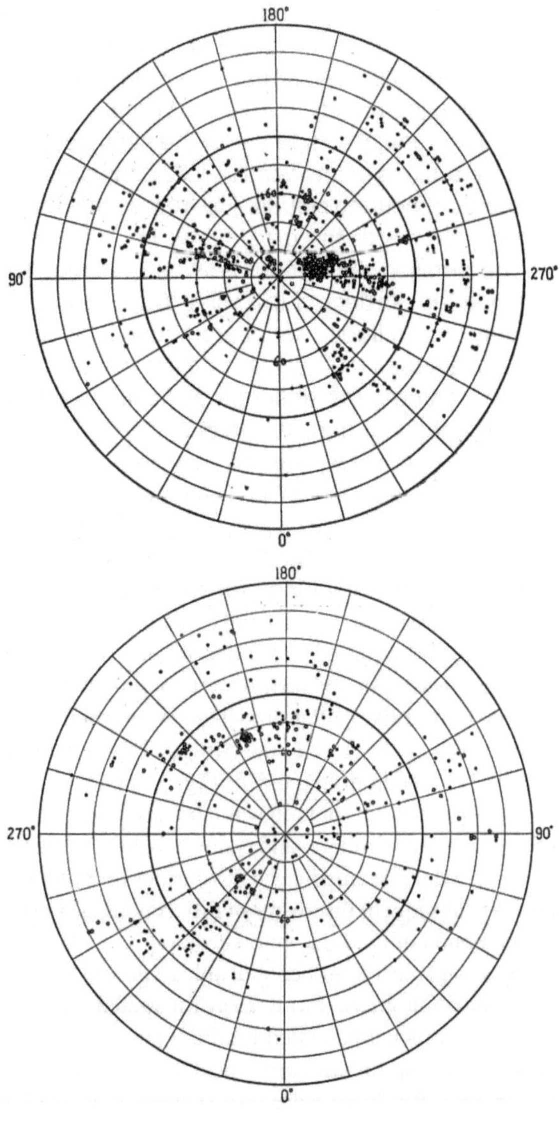

Fig. 8.9 Figure 1 of Oort's inaugural lecture. His caption: 'Distribution of the 1025 brightest galaxies across the sky; after Harlow Shapley and Adelaide Ames. Every panel shows half the celestial sphere bordered by the circle of the Milky Way. To study the irregularity of the distribution of galaxies one must restrict oneself to the areas inside the small thickly-drawn circles. In the areas outside that the distribution has been affected by the absorption of light in our Galaxy; close to the circle of the Milky Way it is almost opaque, so that there the galaxies are absent.'

The faint nebulae in the general field are not localized in transparent patches. They are distributed unevenly but the number per square degree to about 19.5 can be predicted with a p.e. [probable error] of 20 per cent. [...] In the direction of the Galactic center I suppose that absorption could be neglected as a first order effect down to about 45°. [...] I believe that the distribution of bright nebulae in the general field is little influenced by Galactic obscuration. When the Virgo Cluster is removed, the

> irregularity in the distribution is comparable with that commonly found among an equal number of faint nebulae.'

Hubble included a graph tracing out the Zone of Avoidance and per field a symbol representing the density of nebulae. And on December 9, 1932 he added:

> 'The zone of avoidance for the nebulae, together with the conspicuous large regions of low density, follow the general trend of the distribution of known obscuring clouds. This represents the rather narrow belt along the Milky Way with the great flares in Cassiopeia, Taurus and Ophiuchus. The detailed agreement between nebular and stellar densities is best along the borders of the Milky Way – obscuring clouds in these moderately high latitudes are near enough to affect both star[s] and nebulae. In high latitudes the nebular density is independent of star density. In very low latitudes, the absence of nebulae is more or less complete even where the star density is high.'

The case for absorption was very convincing. Although the distribution of the brightest galaxies was very irregular, the Universe appeared rather uniform on large scales.

Oort turned next to the structure of the Galaxy itself, going back to the work of Herschel and 'my inspiring preceptor' Kapteyn, in particular his 'Plan of Selected Areas'. He went into some detail explaining Kapteyn's search for absorption through reddening and returned to the distribution of nebulae to illustrate that this can help map the distribution of obscuring material, and arrive at a new picture of what the Galaxy looked like; a flat disk-like system rotating around a distant center with our position relatively close to the borders. It was here that he presented the picture (his Figure 2) in the bottom panel of Fig. 3.4, that de Sitter indeed attributed to him.

The discovery of rotation was next discussed, pointing first to Kapteyn's discovery of the two Star Streams and Schwarzschild's interpretation of anisotropic random motions, which seemed to be the correct interpretation. He then turned to Lindblad and the subsequent discovery of differential Galactic rotation and explanation of the phenomenon of the high-velocity stars.

A second reason for being unable to describe the structure of the Galactic System is our position internal to it. Oort illustrated three images of extragalactic nebulae, starting with what he called Figure 5 (see Fig. 8.12 here; Oort's order of the figure is odd). This spiral structure, most vividly seen in the whirlpool nebula M51, was fascinating. But there were also systems that had less pronounced spiral structure, or even ones in which it was absent (Oort's Figure 4; Fig. 8.11). Maybe in such systems the rotation velocity was smaller. Very fast rotating galaxies (supposedly) were extremely flat, such as in his Figure 3 (Fig. 8.10).

Here Oort came up with two issues that would have his attention and interest for a long time. The first was the question what our Galaxy would look like from a distance. It was very flattened so he supposed that it had to be a spiral galaxy like in his Figures 3 and 5. It seemed to agree with this in terms of size, mass and stellar composition. A major task would be to delineate the spiral structure in our Galaxy. But an important next problem was the nature of the spiral arms. Oort supposed they

Fig. 8.10 Figure 3 from Oort's inaugural lecture. His caption: 'The probably spiral galaxy N.G.C. 4565, that is seen edge-on.' Oort illustrated a photograph by George Willis Ritchey (1864–1945) with the 60 in. Telescope at Mount Wilson that was designed mostly by him. The picture here is from the the STScI Digitized Sky Survey [203] and compares well in quality to Oort's illustration. I have reproduced the image with the same orientation (south to the top) and sky dimensions as Oort used

Fig. 8.11 Figure 4 from Oort's inaugural lecture. His caption: 'The elliptical galaxy with more or less the maximal flattening N.G.C. 3115.' Oort used an earlier picture from the 36 in. Crossley Telescope at Lick Observatory, taken by James Edward Keeler (1857–1900). Further as in Fig. 8.10

could not be long-lived and that some influence from outside would have given rise to them. This issue would remain with him until beyond retirement.

The final paragraph of the lecture runs as follows:

'The spirals and the Galaxy bear the stamp of the past both in their regularity and in the irregularity of their structure. These are like stones

Fig. 8.12 Figure 5 from Oort's inaugural lecture. His caption; 'The spiral galaxy Messier 51'. Further as in Fig. 8.10

onto which fragments of inscriptions have been etched, of which one supposes that these relate if not to their origin, then at least to a special epoch in the past of the Universe. But their language cannot be read by us yet.'

8.3 Southern Skies

There are two initiatives that Oort was involved in—although de Sitter as Director had felt it to be his duty to take the lead—and both had to do with Observatories in Africa. The first is what would become the Leiden Southern Station in South Africa and the telescope that was erected there with the help of a major grant from the Rockefeller Foundation. Although this had started as a provision for Hertzsprung, when he accepted his position at Leiden Observatory, for a telescope in a good climate, de Sitter had taken the lead in overseeing the design, construction and erection of the telescope. Although Hertzsprung considered it as mainly his instrument, Oort had an interest in using at as well for his own research in statistical astronomy. In agreement with requirements of the Rockefeller Foundation, the funding and operations were administered by an especially established Fund, the Leiden Observatory Fund.

Fig. 8.13 The Rockefeller Astrograph of the Southern Station of Leiden Observatory on the grounds of the Union Observatory near Johannesburg. From Oort (1939b). From the Archives of Leiden Observatory

At the time of de Sitter's death, the telescope was still under construction. The site that was chosen was a natural one, namely the grounds of the Union Observatory near Johannesburg. Since 1923 there had been an agreement with the Director Robert Innes for an exchange of use of facilities between the staff members of the Union and Leiden Observatories. It was mainly Hertzsprung, who had used this opportunity, spending himself extended periods there. The work he did there was taking plates with the 10 in. Franklin-Adams Telescope, which had been donated by John Franklin-Adams (1843–1912), the maker of the first all-sky photographic survey. Hertzsprung and collaborators used it primarily to search for variable stars from repeated exposures. Since 1925, Willem Hendrik van den Bos had been posted there as a representative of Leiden Observatory, but in 1928 he had been appointed on the staff of the Union Observatory and his place had been taken by Hendrik van Gent (1900–1947). In 1941 van den Bos became Director of the Union Observatory. Van Gent worked on variable stars under Hertzsprung and would obtain a Ph.D. with him in 1932. With this major Leiden involvement with the Union Observatory, it was only natural to place the Rockefeller telescope there as well.

The Rockefeller Telescope was a twin 40 cm astrograph (see Fig. 8.13). An astrograph is a telescope designed especially for astrophotography; the Franklin-

Adams Telescope was also an astrograph. The instrument was described by Oort in some detail in a popular article in 'Hemel and Dampkring' (Oort 1939b). The telescope arrived in Johannesburg in September 1938. In 1937 Aernout de Sitter, had moved to Johannesburg as observer; he had obtained a Ph.D. degree in Leiden under Hertzsprung in 1936 on stellar photometry. In 1939 de Sitter moved on to the Bosscha Observatory near Lembang, Java and was replaced by Willem Christiaan Martin (1910–1945), who had defended a thesis in 1937 on photographic photometry, also under Hertzsprung. Although Oort had a keen interest in the telescope (as evident from his popular article just mentioned), not much research in his interest was carried out with the instrument; it was used extensively for variable stars and the Magellanic Clouds and in addition asteroids and comets, but not proper motions or fundamental astrometry. For completeness it should be mentioned that Leiden observers made use also of the nearby Yale Southern Station with its 26 in. telescope, also located in Johannesburg, about 4 km away. As an example, according to the Leiden Annual Report for 1939, the total exposure time that year was 329 h on the Rockefeller, 89 on the Franklin-Adams and 4 on the Yale telescopes all during 179 nights (Figs. 8.14, 8.15 and 8.16).

Closer to Oort's interest was the work on absolute declinations. The plan described in de Sitter and Oort (1925d) to establish three observing stations was pursued with vigor under the leadership of de Sitter. During the 1928 General Assembly, the IAU had awarded de Sitter a major grant for this project, which resulted in the first Expedition to Kenya to perform observations from the equator. Hins had effectively become the leader of the astrometric department and he went to Kenya in July of 1931 (see also Jet Katgert-Merkelijn [140]). Hins was accompanied by a young student, Gijsbert van Herk (1907–1999) (Fig. 8.17), who had studied astronomy in Amsterdam with Pannekoek. I quote from the Annual Report for 1931 by de Sitter:

> 'The purpose of the Expedition, which has been made possible by grants from the International Astronomical Union as well as from the Dutch government and several private persons and institutions in Holland, is to determine fundamental declinations at a station near the equator with a specially designed azimuth instrument, following the same method which was tried out some years ago by Mr. Sanders in the Portuguese Congo.
>
> The instrument was made by the firm of Cooke, Troughton and Simms, in accordance with the general plans of the Director, its principle features are a large and accurately divided horizontal circle and a reversible horizontal telescope of 10 cm aperture and 1 meter focal length.
>
> The instrument was shipped gratuitously to Mombasa by the Holland-Afrika Lijn; by the courtesy of the Director of this same company a reduction of 50% was made on the fares of the members of the Expedition.
>
> After many discussions Kenya (British East Africa) had been selected as the region holding the best promise for clear skies down to the horizon. From the reports so far available it seems that the expectations will be fulfilled. A definite site for the observing station could soon be selected,

Fig. 8.14 The 'horizontal' or 'azimuthal' instrument, designed by de Sitter for observations of absolute declinations from the equator. From [212], the Archives of Leiden Observatory

Fig. 8.15 The observing hut that had been erected at Equator, seen from the southwest. From [212]. From the Archives of Leiden Observatory

thanks, partly, to the help of Mr. A. Walter, superintendent of the British East African Meteorological Service.'

The scope of the expedition had grown. In the beginning it was envisaged that a single person would go to an equatorial site. But when it became evident that sites with at least some existing infrastructure in Ecuador (Quito Observatory) and Sumatra (Dutch East Indies) were unfavorable from the point of view of meteorological conditions, and that the best option was Kenya, where no significant inhabited center and electricity supply would be available, it was quickly realized that two persons would be required. The budget therefore had to be enlarged. Fortunately, the Pieter

Fig. 8.16 The observing station at Equator in Kenya. In the middle the zenith telescope and on the right the azimuth instrument. From [212]. From the Archives of Leiden Observatory

Fig. 8.17 Gijsbert van Herk taken from the staff photograph of Leiden Observatory in 1931 (Fig. 7.7). From the Oort Archives

Langerhuizen Lambertuszoon Fonds was prepared to pay for the telescope. This fund had been founded by the wealthy Pieter Langerhuizen (1839–1918), mayor of two cities (Bussum and Huizen) in 't Gooi, the part of the province of North Holland to the southeast of Amsterdam, who had inherited a fortune, but left this world unmarried and a large part of his wealth in the hands of this philanthropic institution. The Expedition examined three places where the major railroad crossed the equator, taking meteorological circumstances into account, but also topographical ones, such as Mount Kenya blocking the view of the horizon, or the proximity of Lake Victoria presenting the danger of fog. The final site was some 300 km from this lake, but fog turned out to be a problem still. Hins and van Herk sailed from Amsterdam on July 15, 1931.

> 'The site finally chosen by the Expedition is in the mountains northeast of Lake Victoria, about 6 km from the station Equator of the railway from Nakuru to Eldorit. [...] The observing house, which is large 5.2×3.7 m, has been built in such a way that the entire roof can be taken off, so that stars in all azimuths can be observed down to the horizon. [...] On a cement base two piers have been erected, one for the azimuth telescope, the other for the Leiden zenith telescope which has also been taken to Kenya.
>
> Regular observations were started about two months after the arrival at Equator. The program consists of observations of azimuth of bright stars of all declinations at a zenith distance of 82° to 83°. Even at these zenith distances the seeing appears excellent.'

'Seeing' is an astronomical term for the quality of the stellar images, both in terms of image size as well as regarding blurring and twinkling. The latitude of the observing station was determined with the zenith telescope to be only 43″.5 south, while the best maps gave 28″ south. The first value corresponds to 1.3 km. The work in Kenya was extraordinarily hard. The altitude was 2850 m; in spite of being in the tropics the temperature could easily fall to near freezing during clear nights. The site was located on the top of a small hill so it was not protected much from the wind. The observers had to point their telescope to low altitudes above the horizon, so there was also no shelter from the hut that held the telescope. The Sun has a diurnal orbit like all celestial objects (see Fig. 5.2), so all year long day and night each last for 12 h and nights were in principle long. Weather was unusually bad during almost all of 1932. From a letter of Hins to de Sitter on October 27, 1932 (in the Oort Archives; I have taken the translation of Jet Katgert-Merkelijn; see [140], also for more details):

> 'Those were curious observing nights; I can hardly call it observing, it is rather fighting the elements. At 7 o'clock generally a downpour, at 8 clearing up, but in such a way that the difference between wet and the dry sphere of our hygrometer is 0°.1, meaning a dew soaking everything. Then an icy north-easterly wind rises, bone chilling even though the thermometer does not fall very low, 7 or 8°. [According to van Herk to Katgert the temperatures were more like that of the freezing point] The writer, 2 meters away from the observer, can hardly hear the latter. In this way it

remains for a few hours. The floor is so wet that you have to watch your step in order not to slip; before each setting of the micrometer or reading of microscopes and level you have to rub everything dry; the dew-cap has to be removed to clean the objective a couple of times. Writing pad, observing lists, paper tape are all so soggy that the pencil hardly makes an impression. Then, at around midnight, sometimes earlier sometimes later, the upper atmosphere wins (I don't know enough of meteorology to put it better), and suddenly the sky is overcast.'

There are also stories of leopards and herds of wild boars. Hins and van Herk had a lot of help from a Carver family that lived on the nearby 'Canada Farm'. They had built in addition to the observing huts, two small sleeping quarters for themselves and a double hut for a motor-generator and battery cells. Astronomical observing in those days was very hard everywhere compared to that of the modern astronomer, who observes from a heated room through computer control. Observing at the telescope in the dome was hard; in Kenya it was almost inhuman. From the official publication of the expedition—where more details can be found [212]—: 'During our whole stay we never left the station together for longer than one day; with more or less regular intervals one of us took a short holiday to live at a lower altitude for a few days.' The observing campaign lasted from December 1, 1931 to February 25, 1933 (which means 453 nights). They observed for 123 nights, but only 55 of these were uninterrupted. It was a heroic effort indeed.

Part of the work was intended to form the Ph.D. thesis of van Herk. Oort acted as supervisor; it seems natural to assume de Sitter would have acted as such, had he lived, but now Oort indeed would be the most suitable choice. The expedition lasted until May 1933, after eighteen months of operation, when Hins and van Herk left Equator, after breaking up the station, shipping back the instruments and other material, except for things like the observing hut, batteries, a car, etc., that were sold (the hut was sold to the Meteorological Service). They had been away for almost two years. One may wonder what this did to the marriages of such persons. Hins had married Antoinetta Maria Cunera Hetterscheid (1889–1919) in 1917, but she died two years later at a very young age. Hins had remarried in 1921 with Else Auguste Scheiner (1890–1954) and their marriage survived the Kenya Expedition. Van Herk was unmarried at the time of the Expedition; he married Martha Doorn (1909–??) in 1934, but they divorced in 1941.

Van Herk's thesis defense took place on May 1, 1936; the title of the thesis was *Eenige uitkomsten van de waarnemingen in de jaren 1931–1933 te Equator gedaan voor de verbetering van het declinatiesysteem van Boss* (Some results of observations made in Equator from 1931 to 1933 to improve Boss' system of declinations). It contained the reduction of the data from both the zenith telescope and the azimuth instrument and a preliminary comparison with existing measurements. It became already clear that the reproducibility of the data was poorer than expected and the final accuracy likewise unsatisfactory.

Already during the period in Kenya, Hins and van Herk encountered difficulties from preliminary reductions of the data. One aspect of the data collection is to correct for collimation errors in the telescope, which means errors in the alignment of the optics. The classical method to do this is to 'reverse' the telescope (which is the reason why meridian circles could be lifted from their piers and 'turned around'), i.e. rotated along their optical axis. The same procedure had been followed with the azimuth telescope, but reproducibility remained a problem, even though they performed any test or took any precaution they could think of. In the end it was concluded that the instrument had suffered from 'hysteresis', meaning that the *direction* (clockwise or anticlockwise) in which the telescope was turned during reversal made a difference, as a result of small movements of internal parts. But that was found out only back in Leiden.

Each observation of a pair of stars, one in the east and one in the west, gave a measure for the zero-point on the horizontal circle. This was then calibrated by observing a mark (in standard jargon a 'mire') that had been put up at a distance; for accuracy they actually moved it farther out after a few months of observing. The error in the zero-point remained larger than expected. The final accuracy, after combining all error contributors and checking against all internal comparisons, turned out to be $\pm 0\rlap{.}''92$ in the declinations of their catalogue, much larger than they had expected (it should have been of order $0\rlap{.}''35$ or so). In the final publication [212] it is noted: 'So there must be some external sources of error, most probably connected with the unavoidable reversal of the instrument.' This was disappointing, but de Sitter had reacted when the first signs of the problems emerged by commenting that the expedition after all had an exploratory character. Useful results were obtained, but it could be done better. The sky coverage was also less than required for a good all-sky system of declinations, so the expedition had been too short. Oort with his keen interest in astrometry and appreciation of its fundamental importance, set out immediately to organize a second expedition. World War II intervened, however, and this second Kenya Expedition only took place around 1950.

8.4 Galactic Structure Allowing for Absorption

In the mean time Oort had concentrated his research on matters related to the structure of the Galaxy. In addition to obtaining improved values for the structural and kinematical parameters, his main aim was to understand the effects of absorption and ways to prove that the Galaxy is a spiral galaxy and delineate the spiral structure, at least in the neighborhood of the Sun. The way to proceed was through further analysis of the distribution of stellar properties. Such a comprehensive, new study was made possible by the availability of a substantial amount of new data in Kapteyn's Plan of Selected Areas, so I will discuss the progress of that project first.

A first part of the Plan had been to perform star counts, i.e. to determine the distribution of apparent magnitudes and this had been mostly the work of the American Observatories of Harvard and Mount Wilson. In the Harvard-Groningen Durchmusterung, the distribution of stellar apparent magnitudes in all Selected Areas down to a magnitude of 16 had been determined, and in the Mount Wilson Catalogue to magnitude 19 in the northern Areas (numbers 1 though 139; [47]). In this section all magnitudes quoted will be photographic ones. The next step would be to determine spectral types and most importantly, proper motions. The spectral types were taken up at two German Observatories, the Bergedorfer Sternwarte near Hamburg for the northern fields and the Postdam Observatorium near Berlin for the south. This has been described in more detail in Sect. 10.10 in my *JCKbiog* [7]. The first was executed in close collaboration with the Kapteyn Astronomical Laboratory under van Rhijn in Groningen. The northern work was eventually published as the *Bergedorfer Spektral-Durchmusterung der 115 nördlichen Kapteynschen Eichfelder*[1] by Friedrich Karl Arnold Schwassmann (1870–1964) and van Rhijn in five volumes between 1935 and 1953. The southern part, *Spektral-Durchmusterung der Kapteyn-Eichfelder des Südhimmels*, was published by Wilhelm Becker (1907–1996), partly in collaboration with Hermann Alexander Brück (1905–2000), also in five volumes between 1929 and 1938. The catalogues were based on objective prism plates (see Sect. 4.6). Oort must have had access to all or most unpublished material from these two Durchmusterungs as far as available at any particular time, which meant in Oort's case for stars brighter than magnitude 12 or so. For the northern Areas (number 1 to 115 again) Milton Humason at Mount Wilson used the 60 in. Telescope to determine the spectral type of on average 10 stars per Area down to magnitude 13.5 or so. This could of course not be done with an objective prism, but Humason used 'slitless spectroscopy' where the spectra were formed near the focal plane rather than near the objective of the telescope (a 60 in. mirror in this case).

Another major advance was the publication in 1934 of *The Radcliffe catalogue of proper motions in the Selected Areas 1 to 115* by Harold Knox-Shaw (1885–1970) and H.G Scott Barrett [213]. Radcliffe Observatory of Oxford University was built with funds from the trustees of Oxford benefactor John Radcliffe (1652–1714) in 1772 (an account of the Observatory's history has been written by Andrew David Thackeray (1910–1978) [214]). Knox-Shaw was 'Observer' (really Director) between 1924 and 1950; H.G. Scott Barrett had been assistant between 1908 and 1935 (interrupted). Radcliffe had a 24 in. Telescope; it was first used to try and measure parallaxes in the Selected Areas, but when that was unsuccessful, attention was re-directed towards proper motions. These were eventually measured in the 115 northern Selected Areas down to magnitude 14. This had been started in 1909 under Arthur Alcock Rambaut (1859–1923), taking photographs in duplicate, one being developed immediately, the other stored for re-exposure at least ten years later. The second stage was started in 1924 under Knox-Shaw. The catalogue, published in 1934, contained proper motions

[1] In particular Adriaan Blaauw, third Director of the Kapteyn Astronomical Laboratory, has been extremely annoyed by the absence of any reference to the Groningen Laboratory in this title, while a major share of the work had been performed there. Of course, van Rhijn, had he been a more forceful person, would not have let this happen.

Fig. 8.18 The Oorts at Leiden Observatory in 1935. From the Oort Archives

of some 32,000 stars. The Oxford Radcliffe site was actually vacated in 1935 and the Observatory moved to a site near Pretoria, South Africa (Fig. 8.18).

Another important source for Oort to use was a study by Peter van de Kamp and Alexander Nikolayevich Vyssotsky (1888–1973) at Leander McCormick Observatory near Charlottesville, Virginia. This Observatory had been founded with a gift from Leander James McCormick (1819–1900), who had together with his brothers Cyrus and William founded the McCormick Harvesting Machine Company, which produced a mechanical reaper invented by Cyrus. The Observatory was dedicated in 1885 and had a 26 in. refracting telescope. Van de Kamp and Vyssotsky had measured the proper motions of 18,000 stars with this telescope.

All these data together constituted a substantial fraction of what Kapteyn must have envisioned when he conceived the Plan of Selected Areas. But before embarking on an analysis of the available material with the aim to derive a new picture of Galactic structure, Oort needed to perform some preparatory studies.

In 1936, Oort published a thirty page long paper in the BAN with the title *Mean parallaxes of faint stars derived from the Radcliffe Catalogue of proper motions* (Oort 1936c), discussing in much detail mean parallaxes (or distances) for stars of a certain apparent magnitude derived from their proper motions in the Radcliffe Catalogue. This was of course vital information for any study on the stellar distribution in the Galaxy. Oort concentrated on components of the proper motions that were as close as possible to the direction of the major axis of the velocity ellipsoid, in other words in the direction in which on average random velocities in space were largest. This axis was assumed to point towards the center of the Galaxy. He next restricted himself to those Areas in which this was more than 66° from the motion of the Sun in space using the best value available for the coordinates of the Solar Apex. He compared that then to mean linear velocities derived from measurements of radial velocities,

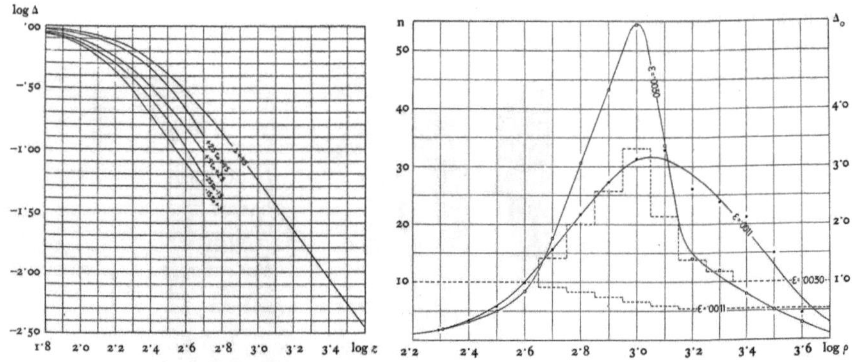

Fig. 8.19 Stellar density distributions derived from an analysis of proper motion and surveys in Kapteyn's northern Selected Areas. On the left perpendicular to the Galactic plane; on the right in the direction of Galactic latitude 7°. For details see the text. From Oort (1936c)

mostly restricted to a sample of 400 K-stars of about magnitude 7 or so, and a smaller sample (45) of fainter G- and K-stars. The ratio of the axes of the velocity ellipsoid was determined from nearby stars as 1 : 0.63 : 0.50 in respectively the direction of the Galactic center, of Galactic rotation and perpendicular to the Galactic plane, so for each Selected Area he could correct to the component towards the center.

There is quite a bit more detail in the analysis than need not to be considered here. The mean parallaxes vary smoothly with apparent magnitude and with Galactic latitude. And Oort found that when one compares stars with one one apparent magnitude fainter, this difference is on average for about two-thirds the result of the fainter stars having a larger mean distance and for one-third to them having a brighter absolute magnitude.

In the direction perpendicular to the Galactic plane the observations can be turned into a distribution of space density without having to worry about absorption. The result is reproduced as Fig. 8.19, panel on the left. The vertical axis is the (logarithm of the) density, the horizontal one the (logarithm of the) distance. Different lines indicate different absolute magnitudes. On the right in Fig. 8.19 the distribution is shown for latitude 7° for stars of apparent magnitude 14. Here the effects of absorption are significant and important to be taken into account. The two smooth curves show the space density of these stars (linearly on the vertical axis) as a function of distance (horizontal, logarithmic). The two curves are for two assumptions of the amount of absorption. The dust was supposed to be restricted to a layer in which the mean vertical distance is 56 parsecs and the curves refer to respectively 0.0030 magnitudes of absorption per parsec in the plane for the upper (strongly peaked) curve and 0.0011 for the other one. The paper also treated analyses for different spectral types (from the Bergedorfer and Mount Wilson data mentioned above). Oort finally found that the data gave a good constraint on the ratio of the velocity dispersion in the radial direction to that of rotation, namely 0.63 ± 0.02.

The other preparatory analysis Oort performed was presented in a shorter paper (7 pages only) with the title *A redetermination of the constant of precession, the motion of the equinox and the rotation of the Galaxy from faint stars observed at the McCormick Observatory* (Oort 1937a), based on the 18,000 faint stars observed by van de Kamp and Vyssotsky. These had also analyzed their data for these properties and had found the surprising result that there was a 30% difference in mean parallaxes north and south of the Galactic plane. Oort confirmed this conclusion from his own analysis of their data. 'However,' he wrote, 'in my opinion the reality of the result is extremely doubtful. Mean parallaxes derived from the distribution of the proper motions show no indication of it, ...' In his paper just discussed for magnitudes 12–14 the ratio was 0.95 ± 0.04. Also an independent analysis of the Radcliffe data with a somewhat different approach, that Armand van Hoof had performed under Oort's guidance, gave 0.99 ± 0.05. Van Hoof (1906–1989) had a Ph.D. in mining engineering and he taught topography at the University of Leuven in Belgium and would not long after this period be appointed Professor there to teach astronomy. He made a number of visits of a few months to Leiden.

Oort concluded in his paper:

> 'It is rather improbable that such a large difference in parallax would occur without a corresponding difference in number of stars per square degree. For these reasons I am inclined to consider the difference found by van de Kamp and Vyssotsky as spurious.'

This shows a typical characteristic of Oort; when a result did not fit his preconceptions, and when it was analysed, but the discrepancies were not going away or only partially, he would 'for the moment' stress the evidence that did support his point of view.

The paper concluded with a correction to the best value of the precession constant, which at that time was still dating back to Simon Newcomb, and improved values for the Oort constants: 19.0 km/s kpc for A and -13.7 for B (Oort gave the values in arcsec/year; for conversion 1 km/s kpc = 2.11×10^{-4} arcsec/year).

With these preparations Oort was ready to make an improved, and—at least for the moment—decisive attack on the question of the issue of *Absorption and density distribution in the Galactic System*. This work, Oort (1938c), was again published as a very long article in the BAN (32 pages this time) and can be regarded as the high point of statistical astronomy in the first half of the twentieth century. Its approach was presented by Oort in the first sentence of the summary as: 'The article is principally concerned with the following problem: What information can be obtained about the structure of the Galactic System if counts of faint stars in moderate and high latitudes are combined with absorption data from Hubble's counts of extra-galactic nebulae?' As we have seen, Oort had already corresponded with Hubble in his Perkins Observatory period and later on the distribution of nebulae and the extent of the Zone of Avoidance. Hubble had sent also extensive letters and advanced preliminary copies of the figures on the observed distribution of the nebulae that would eventually appear as figures 3 and 6 in Hubble's long publication on the issue, *The distribution of extra-galactic nebulae*, in 1934 [215].

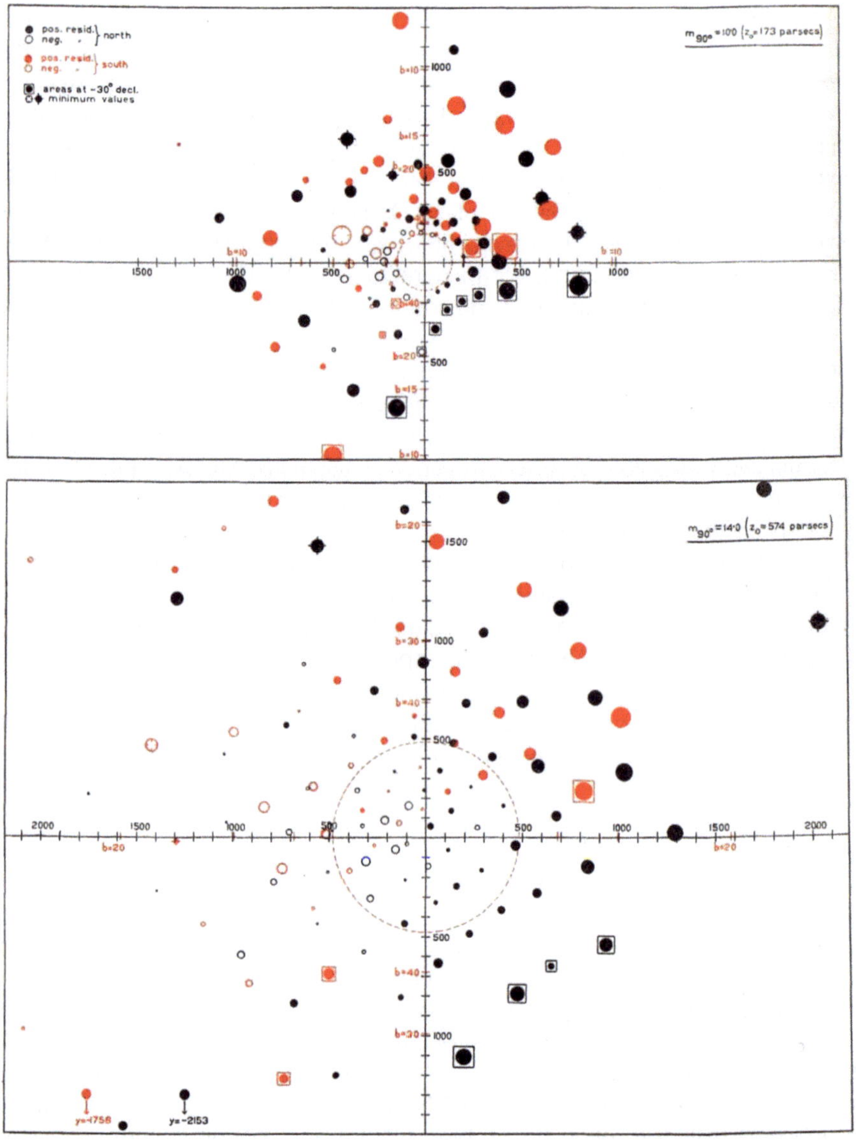

Fig. 8.20 Deviations of the actual number of stars from the case of a plane-parallel distribution. The three panels correspond to three planes parallel to the disk of the Galaxy, at mean distances of about 170, 400 and 1000 parsecs either side out of the symmetry plane. The horizontal extent is 2 kpc each way from the Sun. The center of the Galaxy is to the right. Each dot corresponds to a Kapteyn Selected Area, plotted at the position of the projected (logarithmic) mean distance of the stars. Filled circles are an excess, open circles a deficiency compared to a uniform plane-parallel distribution; black symbols mean north of the Milky Way, red circles to the south. The dashed circle indicates where latitude ±50° cuts through the planes. From Oort (1938c)

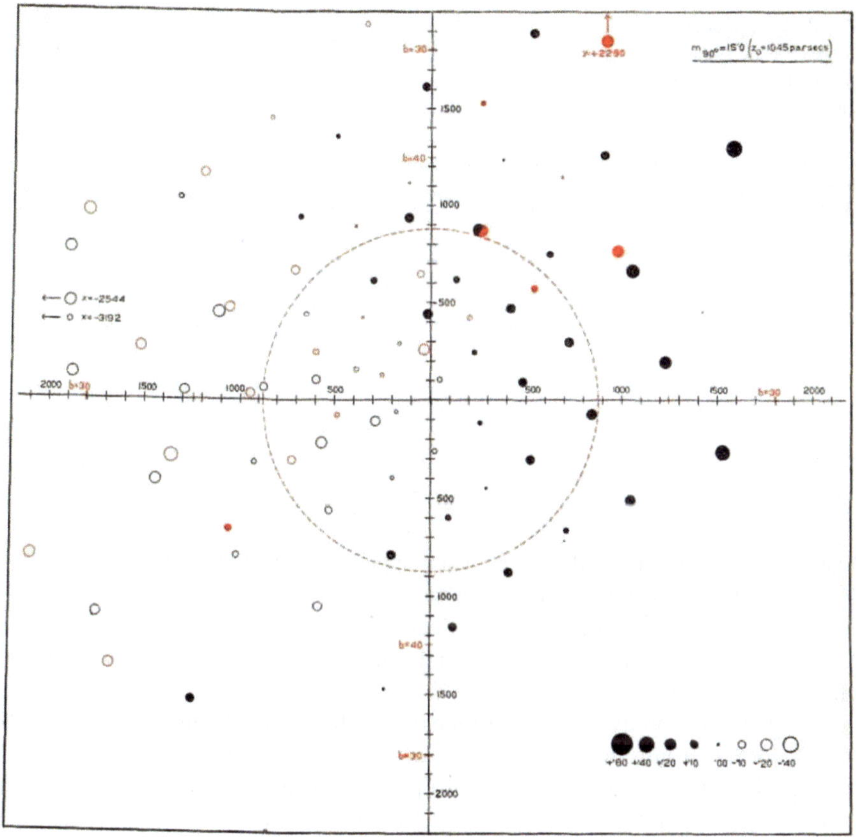

Fig. 8.20 (continued)

This distribution gave valuable information on the total absorption from us all the way to outside the Galaxy. But most stars are closer and suffer less extinction. So Oort first made an extensive study of the colors of faint stars in the Selected Areas. John Adelbert Parkhurst (1861–1925) of Yerkes Observatory had determined the colors of 1550 stars in Kapteyn's Selected Areas at declination +45° and these were published posthumously. Oort now determined the 'color excess' of these stars, using the spectral types of Humason and the Bergedorfer Durchmusterung. As we have seen, the color of a star is expressed as the difference in the magnitudes at two wavelength ranges; the color excess is then the difference with the color index expected from the spectral type (see Sect. 2.6). It was known that the colors of the Parkhurst stars (Parkhurst had noted it himself) were redder nearer to the Galactic plane, pointing to increasing absorption. There was also a study of 733 B-stars by Joel Stebbins (1878–1966) and Charles Morse Huffer (1894–1981) of the Washburn Observatory of the University of Wisconsin-Madison. They pioneered photo-electric techniques, using a photo cell at the telescope. Photoelectric photometry would replace photography

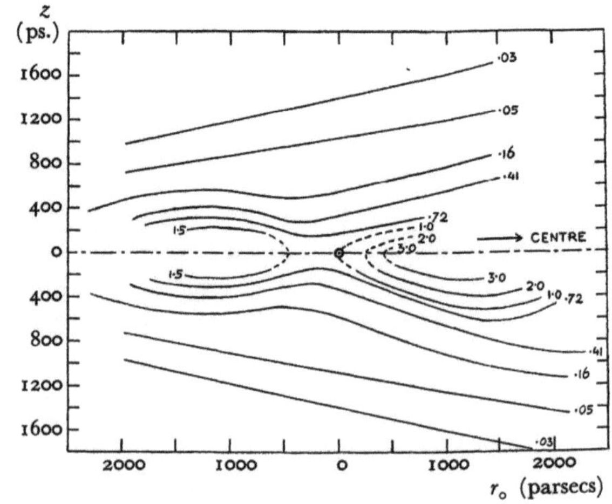

Fig. 8.21 Equidensity curves for the stellar distribution in a cross-cut through the Galactic disk. The Sun is in the center, the Galactic center to the right at 10,000 parsecs. From Oort (1938c)

to measure stellar magnitudes. They had mapped the reddening in the Galaxy using B-stars, that are bright and can be seen out to large distances and also have a rather uniform intrinsic color. Together with Albert Edward Whitford (1905–2002), also of Washburn, Stebbins had furthermore mapped the absorption using colors of globular clusters.

So, Oort had a pretty good idea of the distribution of reddening, but still needed to transform that into the actual amount of absorption. In other words, he needed to know how many magnitudes of absorption correspond to a particular amount of reddening. For this one could look at two stars of the same spectral type, one nearby and the other further away. Then one would measure the brightness as a function of wavelength for both stars and in this way the wavelength dependence of the absorption could be found. This can give clues to what the physical mechanism of the scattering is. Such a study had just been published by by Jesse Leonard Greenstein (1909–2002), who had obtained a Ph.D. at Harvard and then worked at Yerkes Observatory. Greenstein spent most of his career actually at the California Institute of Technology (CalTech) using the Mount Wilson and Palomar telescopes, where he became a leading expert in stellar spectroscopy and element abundances. Greenstein had calibrated photographic spectra of the Stebbins and Huffer stars in terms of the brightness as a function of wavelength using photo-electric techniques [216], and found the dependence of the absorption to be inversely proportional to the wavelength. Also Greenstein determined the actual amount of absorption by comparing stars still visible behind dark clouds to stars just next to it, which also gave him the factor to turn color excess into actual magnitudes of extinction.

The combined information gathered made it possible to conclude that 'in all probability a considerable fraction, if not all, of the absorption derived from the counts of nebulae must take place within a relatively small distance from the Galactic plane.' Oort then felt justified to approximate the situation by assuming that the *entire* absorption indicated by the nebulae took place in front of the stars considered. So with that assumption, for each Kapteyn Selected Area he proceeded to calculate the star density along the line of sight at three distances from the plane. He presented the results first in the form of deviations from a plane-parallel distribution, in which each plane parallel to that of the Galactic disk has a constant density. These results (which by the way represented an enormous amount of calculations and numerical work) were presented in the form of Fig. 8.20. From this he pointed out two important things. The first is that on the right-hand side the symbols are predominantly filled (i.e. showed an excess with respect to the plane-parallel distribution) and to the left predominantly open (a deficiency). This indicated a general and large-scale increase in stellar density towards the center of the Galaxy. Furthermore, the black and red symbols show, *grosso modo*, a similar distribution, which would mean that there is symmetry between both sides of the Galactic plane.

Oort continued by sketching from his data isodensity curves in a cross-cut through the Galaxy at the solar position in a plane at right angles to that of the Milky Way and along the direction towards the center. This has been reproduced in Fig. 8.21. It looks very much like what would be expected if the Galaxy had spiral structure like many external systems. The precise form of the 'arms' cannot be determined, but Oort is relatively certain about one thing and that is that 'the Sun should be between two spiral arms'. At larger distances from the plane the spiral structure is replaced with a general density gradient towards the center, amounting to a factor 1.39 in the stellar density per kpc.

This result in a rough, first approximation fits current understanding. In Oort's analysis, hot and bright O- and B-stars play a significant role and these are now known to be concentrated heavily to the spiral arms, where in a spiral galaxy star formation is preferentially taking place. So an analysis relying heavily on young-lived OB-stars enhances the real contrast between arm and interarm regions. Modern studies have revealed the spiral structure in our Galaxy in much more detail. The Sun then is located just inside a local arc or branch between the major spiral arms. Oort's analysis is along the lines that Kapteyn must have had in mind when he defined the Plan of Selected Areas and Oort's paper is therefore a major outcome of this Plan.

8.5 The IAU in Paris and Stockholm

The fifth General Assembly of the IAU took place in Paris between 10 and 19 July 1935, with Frank Schlesinger as President. The number of member states had grown to 24, China and the USSR having joined since the previous meeting. But neither Germany, nor Austria were part of the IAU. There were no guests from Germany and only 2 from Austria. These were difficult days with Hitler's Nazi party having taken

Fig. 8.22 Oort in the group photograph of the IAU General Assembly in Paris in 1935. From the Blaauw Archives

over control of Germany. In the opening session President Schlesinger mentioned that 'I have also received letters from German friends that make it clear that they would join the Union if it were not for difficulties of a financial order. I sincerely trust that they will have joined us before our next meeting.'

The delegation from the Netherlands was fifteen, according to the list in the Transactions, the proximity of Paris playing a role in that. However, this number must be incorrect. In addition to Oort, of the leading Dutch astronomers Hertzsprung and Minnaert were listed, but van Rhijn and Nijland were not, even though in the opening session both were recorded to have made statements. Also both presided over commission meetings, Nijland commission 27 on Variable Stars and van Rhijn commission 32 on Selected Areas. Hertzsprung was President of commission 26 on Double Stars. Hins acted as secretary of commissions 8 on Meridian Astronomy and 32 on Selected Areas, and Oort of commission 28 on Nebulae. Van den Bos (then at the Union Observatory in Johannesburg) was secretary of commission 26 on Double Stars. This is an indication of the legacy de Sitter and Kapteyn had left behind. Furthermore, Minnaert was secretary in commissions 12 on Solar Radiation and Solar Spectroscopy and 13 on Solar Eclipses.

A large fraction of the discussions concerned interstellar absorption, for which various desiderata for future research were suggested, such as 'colours of extragalactic nebulae at different Galactic latitudes, extension of systematic searches for Cepheid variables sufficiently bright to have their radial velocities determined' and 'studies of the surface brightnesses of extragalactic nebulae'.

In the final business session it was announced by the President, Frank Schlesinger, that Frederick Stratton would resign as General Secretary. 'The Union has good reason to regret that he feels he must no longer continue as secretary, and this feeling would be even more intense if we did not feel that we have found a worthy successor to him in Dr Oort.' It is not known of course to what extent Schlesinger was responsible for this, but he did have considerable influence. The next General Assembly would be held in Sweden. Carl Östen Emanuel Bergstrand (1873–1948) from Uppsala Observatory extended the invitation, to which Bertil Lindblad added, 'speaking as representative of the Stockholm Observatory [...]. We hope most sincerely that in due time the stars – your stars – will guide you all safely to our country in the North, which is not, however, as far away as some of you may perhaps be inclined to think.'

In the group picture of the General Assembly, Oort is difficult to locate, as he stands at a very moderate place (see Fig. 8.22), away from the leaders and officers of the Union. His role was to grow soon, now that he was General Secretary. As we will see, he was reconfirmed in that function in 1938 in Stockholm and actually stayed in office during World War II (Fig. 8.23).

The meeting in Stockholm gave Oort a good opportunity to visit Lindblad, and his wife Dagmar Bolin (1896–1986), in his own environment. Mrs. Oort came along; and they traveled by train, leaving Leiden early in the morning of Tuesday July 26, 1938 and stopping on the way at Copenhagen to see the Strömgrens, where they arrived after an overnight trip the next morning. On Thursday they traveled on to Stockholm. The General Assembly started on Wednesday August 3, preceded by a meeting of the IAU Executive Committee the day before.

The work as General Secretary in between General Assemblies was not negligible. It involved writing national delegations about all sorts of administrative and financial details, and preparing for the next General Assembly. Heleen Kluijver assisted Oort quite significantly, taking some of the burden from him. For the meeting in Stockholm an extensive report of the Executive Committee had been prepared for which Oort must have done a major part of the work. The meeting was presided over by the President of the IAU, Ernest Benjamin Esclangon (1876–1954) from France. The USSR had joined in 1935, but among the attendants there was no Russian astronomer (there were a small number from Eastern European countries). Germany, now under the control of the Nazi Party, was not a member, but eleven German astronomers attended as invited guests.

When in 1931 the IAU Statutes were up for renewal, new Statutes with By-Laws had been adopted at an extraordinary General Assembly in London in September of that year. An important feature had been the definition of individual membership. Such individuals could also be co-opted as members by the Commissions; if these did not come from adhering countries they were not listed as IAU members. In this way some leading German astronomers could participate in the work of the Commissions (and attend the General Assemblies), such as Otto Hermann Leopold Heckmann (1901–1983) from Göttingen (later Hamburg), Karl Küstner from Bonn, Albrecht Otto Johannes Unsöld (1905–1995) from Kiel, etc. Unsöld had a major contribution in the form of a paper on energy distributions in stellar spectra to the

Fig. 8.23 Mrs. Oort with Bertil Lindblad and his wife, Dagmar Bolin, in Sweden at the time of the Stockholm IAU General Assembly in 1938. This picture must have been taken by Oort. From the Oort Archives

meeting of Commission 36 on Spectrophotometry. Another person that was co-opted (by Commission 27 on Variable stars) was Richard Prager, who had been dismissed from the Babelsberg-Berlin Observatory in 1936 for being Jewish (as General Secretary Oort had some involvement with this affair; see below).

An interesting feature was the fact that two colloquia were organized as joint meetings of Commissions. One of these was on *The significance of the various branches of stellar astronomy for the problem of Galactic structure*. Lindblad presided over that and there were comprehensive review contributions by Bart Bok on star counts, absorption and extragalactic nebulae by Shapley and by Lindblad himself on the distribution and spectral types of special objects. In the latter Lindblad called the terms A and B 'Oort's coefficients of differential galactic rotation' (and should be considered the one that coined these terms) and Oort spoke on the evidence for spiral structure in the Galaxy.

An issue that came up was the matter of the functioning of the Union, triggered by a letter of Otto Struve, Director of Yerkes Observatory. In particular this concerned the role of individual astronomers in the Union and their participation in committees, etc., especially the role of those who were not members. The IAU in Stockholm decided that small sub-commissions could be nominated by the Presidents (as Oort wrote after the General Assembly to Struve). But this was not addressing the core of the dissatisfaction.

At the end of the meeting Eddington was appointed President for the next three years, Oort was re-appointed as General Secretary and it was decided that the next meeting would be held in 1941 in Zürich, Switzerland. Of course World War II intervened and Eddington and Oort had to stay in office (Eddington died in 1944; Oort's tasks were taken over as acting secretary by Walter Adams). Oort had to edit the proceedings of the meeting, which he sent to the press in December 1938.

Eddington, who had to leave Stockholm before the end of the Assembly, had a message read which included the following:

> 'We cannot foresee what may happen before we meet again in Zurich. On the astronomical side we can make some guess. The 200-inch reflector will be completed or approaching completion. The marvelous advance in the photography of the corona and prominences, which has been so memorable a feature of this meeting, will bear fruit in advancing knowledge. And our numerous commissions will patiently extend their researches. But on the international side no one dares prophesy. But, if in international politics the sky seems heavy with clouds, such a meeting as this at Stockholm is as when the sun comes forth from behind the clouds. Here we have formed and renewed bands of friendship which will resist the forces of disruption.'

There were at least two thorny issues that the IAU and Oort as General Secretary had to deal with, concerning the USSR and Germany. These have been described in Blaauw's History of the IAU (see Box 3.3, [113], and also [217]). The first concerned the Leningrad Astronomical Institute and the nearby Pulkovo Observatory during the Great Purge of 1936–38. The Directors of these two places, Boris Petrovich Gerasimovich (1889–1937) and Boris Vasilyevich Numerov (1891–1941) had been victims of that. Oort had tried various channels to get information about their whereabouts. Eventually he was told by Pulkovo Observatory that Gerasimov no longer worked there and that Numerov never had. It was an issue of discussions in Stockholm. Shapley and Struve wanted the IAU to make some statement. IAU Vice-President Tadeusz Banachiewicz (1882–1954) from Krakow, Poland, argued, referring to private sources, that it was not in the interests of the two astronomers to pursue the matter, so the IAU Executive in Stockholm decided not to do anything. It is believed both were executed.

The other issue was the dismissal of Richard Prager (1883–1945) from Babelsberg-Berlin Observatory by the Nazis because of his Jewish background. This occurred in 1936. Prager was an expert on variable stars and was in the process of publishing the second edition of an extensive source list of light curves of variables, the *Geschichte und Literatur des Lichtwechsels der veränderlichen Sternen*, an effort on which he had already published two volumes and was about to complete a third. When Prager was dismissed, the IAU could consider to take that under its own auspices, as Vice-President Harold Spencer Jones first suggested. In the end the IAU did not accomplish much until Prager himself managed to leave Germany and moved to Harvard, where Shapley had created some working arrangement for him. Oort proposed to the Executive that the IAU support this with two grants for 1939 and 1940, which made it possible for Prager to continue his work. His manuscript for the third volume was not released by the Germans however and he started the bibliography of variables discovered after 1930 to avoid duplication. He died in July 1945, after a serious illness [218].

8.6 McDonald Observatory Dedication

On May 5, 1939 the McDonald Observatory in Texas was dedicated and this was followed by a three-day symposium to which a selection of the leading members of the astronomical world were invited. Oort had been asked to give an astronomical presentation. The symposium was on the subject of *Galactic and extra-galactic structures*. The McDonald Observatory was built as a result of an endowment of William Johnson McDonald (1844–1926). He was a banker, who died unmarried with no direct heirs, and left most of his fortunes to the University of Texas to build a major Observatory in the state. This was completely unexpected and after some predictable hassles by contesting relatives and a subsequent settlement, the bequest of over $ 800,000 was received by the University in 1933. Since it did not have an astronomy department, the University of Texas decided to team up with the University of Chicago and Yerkes Observatory, and an agreement for joint operation was signed in which for thirty years McDonald would provide a telescope and Yerkes the astronomers to use it. When the contract expired in 1963, the Observatory became part of the Astronomy Department of the University of Texas at Austin. The site chosen was Mount Locke at an altitude of 2,070 m (6,790 ft) in the southwest of Texas near the Mexican border and the township of Fort Davis. The Observatory is some 800 km (500 miles) from Austin. The first Director was Otto Struve, Director of Yerkes.

The construction of the telescope was entrusted to the Warner and Swasey Company from Cleveland, Ohio, that had made quite a name in telescope construction in the USA, while being primarily a high-precision tool manufacturer. The specifications and concept were largely provided by Otto Struve. It resulted in an 82 in. (2.1 m) telescope at a very dry and clear site. When completed and dedicated in 1939 it was the second largest telescope in the world.

The invitation to the dedication came by letter of November 18, 1938 from Otto Struve. It included a three-day symposium as mentioned above. The meeting was sponsored by Warner and Swasey and participants were their guests. By separate letter of the same date, Struve invited Oort to a two months stay at Yerkes to lecture on statistical or dynamical astronomy. Oort replied on December 8, accepting the invitation to the dedication and symposium, but declining the one for the visit to Yerkes. He wrote:

> 'However attractive this plan appears to me, I am afraid I shall not be able to accept it this time. You may recall that I have been working on a programme for obtaining data on the distribution of light in extra-galactic nebulae. This work was started on a small and not entirely satisfactory scale at the Perkins Observatory (where we met in 1932). Later on, in 1935, a number of plates were taken for me by Dr. Oosterhoff at Mount Wilson. The results of these plates look satisfactory, but as the data obtained are still scarce I have been looking for an opportunity to obtain more Mount Wilson material. An unexpected opportunity appears now to offer itself in connection with your invitation to come to Texas in May,

Fig. 8.24 The Oort family in the back garden of the Director's residence of the Leiden Observatory. This picture has been taken around 1939–40. From the collection of Abraham H. Oort

and I am writing Dr. Adams, asking his permission to spend two months at Mount Wilson after the dedication. I am hoping that this stay will yield enough plates to continue with the work during some time.'

And he added that coming to Chicago after Mount Wilson would be too late for regular lectures, while he had already agreed to Shapley and Bok to come to a summer conference at Harvard between July 6 and August 17. And that coming before the dedication was also not possible as Hertzsprung was planning to go to South Africa and Oort would have to fulfill the duties of the Director (Fig. 8.24).

The trip of Oort in 1939 to the United States lasted almost five months. The children were at ages 10, 8 and 4, so leaving them behind in the care of others would be difficult. Therefore Mrs. Oort did not accompany him this time. But he wrote his wife regularly and these letters have been preserved in the private parts of the Oort Archives. It should be remembered that these were very difficult times; the threat of war with Nazi Germany was very real. The 'Anschluss' of Austria and Sudetenland had taken place in 1938, followed by the German occupation of Czechoslovakia; and the discrimination against Jews had culminated in the Kristallnacht in November 1938. Yet, hardly any remarks were made on the situation in either Oort's pocket diaries or the letters between him and his wife. There is also no recounting of discussions of the political situation among American astronomers.

Oort left on April 18, traveling by train via Paris to Cherbourg, where he boarded the 'Europa', one of the 'Turbine Ships' of the Norddeutscher Lloyd from Bremen. This and its sister ship 'Bremen' had set records to cross the Atlantic; the Norddeutscher Lloyd had held the Blue Riband, the unofficial record holder of speed in crossing the Atlantic, in the early thirties. They made the crossing in less than 5 days, so that with two ships offering a weekly service was possible for the company. From Oort's first letter, dated Thursday evening April 20:

> 'The vastness of the sea and air is magnificent and I have difficulty forcing myself to go downstairs to work. My cabin is also ideal. Much space, the porthole has the size of a window, so that the sunshine and sea air stream in pleasantly. All the staff are extremely helpful and friendly and one does not really notice that it is a German ship. Friday morning. There was no sign yesterday at all that would one remind of the fact that it was Hitler's birthday; had I not known this I would not have been aware. I had no opportunity to talk to anyone of the staff of the ship, but I would not be surprised if a large fraction of the personnel and crew on the ship would not be very positive about the new regime that takes away passengers and freight from them, so that their fine ship that would be the pride of any crew, is ignored and shunned. The newspaper on board does all it can to be neutral and not to provoke anybody. It indeed is very empty on board: 85 cabin passengers compared to 450 that can be accommodated. It gives a tragic feeling of enormous wastefulness. All those crew members, an orchestra of 20 musicians, four large elevators with personnel to operate them that are used only a few times a day, a restaurant upstairs with ten or fifteen waiters, manager, buffet chief, liftboys, etcetera, all for some seven passengers that eat regularly in that restaurant. [...]
>
> How are things at home? Has Marijke been very sad? I often think of you all and ask myself if it is good that I left you all to go on this trip. But I do believe I need to. And the time will pass.'

Upon arrival in New York on September 24, Oort saw the Schilts. Mrs. ('Jo') Schilt took him into New York to help him open a bank account and arrange his finances. He very much enjoyed 'both small Vermeers in the Metropolitan Museum'. 'Do you know what else I saw? That 3rd class wagon with two women and a child on the first bench by Daumier. I think Bremmer showed a reproduction of it recently. It is an extraordinarily beautiful painting'. The 'Met' has a few very famous paintings by Johannes Vermeer (1632–1675). Honoré Daumier's (1808–1879) famous painting depicts some of the impact of industrialization on society. Hendricus Bremmer was an connoisseur of arts. We will meet him later again as an adviser in matters of art to the Oorts (see Sect. 11.1).

Oort left by train from New York on the 26th and via St. Louis and San Antonio he finally got to Alpine, Texas on Saturday April 29. Travel was still exhausting and time consuming. But during lost hours waiting for trains, Oort visited some more musea and art galleries, according to his letters to Mrs. Oort. Alpine is the nearest railway station from McDonald Observatory. He was met by the Kuipers. Gerard Kuiper, who had met Sarah Parker Fuller (1913–??) at Harvard and married her in 1936, had done the bulk of the work in organizing the conference.

The Kuipers took Oort from Alpine to Fort Davis near the Observatory. Apparently Oort slept at the Observatory; he wrote he had had some problems with the altitude, but was adapting quickly. The Kuipers had entertained him the first days and on Monday May 1 he had been invited by the Struves in their home, which he described as beautifully positioned at the edge of an edifice.

> 'Struve asked me yesterday how I felt about becoming Director of Yerkes Observatory; he would himself then move to Chicago and become 'Head of the Department of Astronomy' at the University to revive the astronomical educational program and he would then also be able to keep directing the work at the McDonald Observatory, while the Yerkes people would keep the same share in that as they have now. I have not yet seriously considered this, since I have a strong feeling that I am so attached to Holland that there have to be exceptionally strong astronomical reasons for me to consider anything else in the first place. – But this is an active group and this is such a beautiful instrument (that will probably be second to the 100-inch at Mt. Wilson only, and even has some advantages over it), that I told Struve I will not reject the matter out of hand, without at least having looked at it from all sides. As far as I can see Struve's idea is that he and I would together lead 3 institutes, namely Yerkes, McDonald and Chicago. – But don't be worried, dearest. In spite of the great opportunities it presents for my work (for the direction my work is going many more opportunities than in Leiden) I have almost decided not to accept. I do want to think it over carefully. – Only the unexpectedly large, immense possibilities make me hesitate. They have the same allurement as this fantastic wide land here.'

Oort did not really come back to this except mentioning in a letter to his wife after he had left Texas that he decided to decline.

There are at least two very readable accounts of the dedication ceremony and the ensuing symposium, among which one by Otto Struve himself [219]. The dedication was held in the dome of the 82 in. Telescope, the speakers addressing the audience from the observer's platform. From the first paper in [219]:

> 'After the ceremonies the Warner & Swasey Company entertained a large company to a 'chuck-wagon dinner', cooked and served in the open below the summit [the program mentioned 'at the well'] by some of the leading ranch-owners of the county and their cowboys. On a later occasion the guests were entertained by the holding of a rodéo.'

The scientific program had been handled by Gerard Kuiper and Oort had had a quite detailed correspondence about this with him. The program consisted mostly of rather long talks. Most were for 45 min (a few 30 min), but the important contributions were assigned a full hour. There is no formal collected proceedings volume, but most papers have been published as separate articles in 'The Astrophysical Journal'. Already before the official inauguration, Harlow Shapley and Harry Plaskett among others had given 45 min general lectures for the wider audience at the ceremonies on topics of their own choice. Shapley had talked about the 82 in. Telescope, Plaskett described the process of the the figuring of the mirror and the results thereof. In his opinion 'the quality of the mirror of the 82 in. Telescope of the McDonald Observatory is unequaled by any mirror previously made and tested.'

The dedication had taken place on Friday May 5. On Saturday morning Oort started out with a 60 min lecture on 'Present Problems' as Kuiper summarized it in the program. Other major 60 min contributions were from Robert Trumpler on star clusters, German astronomer Wilhelm Heinrich Walter Baade (1893–1960), who worked at Mount Wilson, on photometric problems, Harlow Shapley on the distribution of extragalactic nebulae, Bertil Lindblad on the theoretical interpretation of spiral structure, Joel Stebbins on space reddening in the Galaxy, and Henry Russell on stellar masses. Another important contributor (45 min) was Indian astronomer Subrahmanyan Chandrasekhar, who held a Ph.D. from Cambridge and in 1937 had joined Yerkes Observatory, on star streaming and dynamics of Stellar Systems. Chandrasekhar's published version of his talk, which was very fundamental, ran for a record 151 pages in 'The Astrophysical Journal' [220]! Edwin Hubble gave an evening presentation on the subject of structural features of extragalactic nebulae.

Oort's contribution was entitled *Some problems concerning the structure and dynamics of the Galactic System and the elliptical nebulae NGC 3115 and 4494* (Oort 1940a). The problems he focused on were the matter of evidence for spiral structure in our Galaxy and some understanding of how elliptical galaxies, presumable less difficult to study than spiral galaxies, have been built. The first problem had to do with the vertex deviation. The distribution of stellar velocities in the directions in the plane of the Galaxy was indeed ellipsoidal, as Schwarzschild had proposed, but contrary to what was demanded by simple and fundamental considerations of galactic dynamics, the long axis of the ellipsoid was not pointing exactly towards the Galactic center. After briefly reviewing the history of the study of the Galaxy and of proposals that we live in a spiral galaxy (going back to Stephen Alexander (1806–1883), 'Professor of mathematics and astronomy in the College of New Jersey' [221]), and his recent own work that we just encountered, Oort turned to motions. Since in the spiral nebulae the spiral arms seemed to have colors corresponding to A- and B-stars, he proposed to examine such stars in the local parts of the Galaxy, and—since these were available in sufficient numbers—their velocities in space. He then was concerned not with the systematic motions but the peculiar ones, with which he meant their random velocities in space.

Oort reviewed the shape of stellar orbits and the velocity ellipsoid. The latter is the distribution of peculiar velocities at any point, e.g. the solar neighborhood. When the distribution of stars in space is uniform, the distribution of the number of velocities that occur then looks like the full-drawn curves in Fig. 8.25. At the origin the density of points is set to unity and the elliptical curves then connect the points of equal numbers. But if the space distribution is non-uniform (as in the case of spiral structure), the curves will be distorted. The deviation of the vertex could thus be the result of spiral structure. Oort considered a special case; his arguments are a bit involved, so readers not interested in the details may skip the next paragraph.

Each stellar orbit can be characterized by the two integrals (conserved properties), the total energy and the angular momentum. These two properties can be replaced by two others, e.g. in the epicycle description that Lindblad had first proposed, the

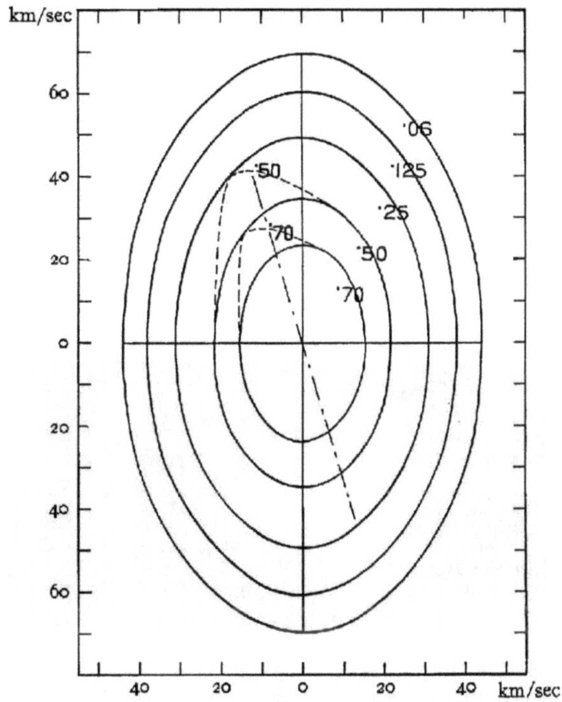

Fig. 8.25 Distortion of the velocity ellipsoid by a non-uniformity of stellar density. The values in the contours are densities relative to the center (which then is unity) and the dotted lines are distortions. Spiral structure may result in a deviation of the vertex as the dash-dotted line indicates. From Oort (1940a)

mean radius of the orbit (that is of the circle on which the center of the epicycle moves) and the velocity of the star in its epicycle at a certain point (e.g. when it is at the largest distance from the center). Now if the latter velocities have a frequency distribution (the relative numbers with which they occur) that is uniform over a region considered, Oort noted that 'the distribution of the centers of the epicycles will then evidently be directly related to the density distribution, while it also determines the velocity distribution at any given point.' He then goes through an approximate calculation of an example in which the density of centers of epicycles is non-uniform, namely that 'as seen from the Sun, the density of centers of epicycles between 248° and 271° Galactic longitude and from 1400 and 1900 parsecs distance were twice that in the corresponding longitudes at the other side of the Galactic center (between 19° and 42°).' This will distort the elliptical contours of the velocity ellipsoid from the purely elliptical shapes as illustrated in Fig. 8.25. These are old longitudes in which the Galactic center is at 327°.

This means that a deviation of the vertex (as observed for bright A- and B-stars of 15–20°) could be a consequence of changes in space densities of stars as in the case of spiral structure. But Oort warned that other causes are possible too, such as a sharp decline in star density not very far beyond the Sun in the outer regions of the Galaxy.

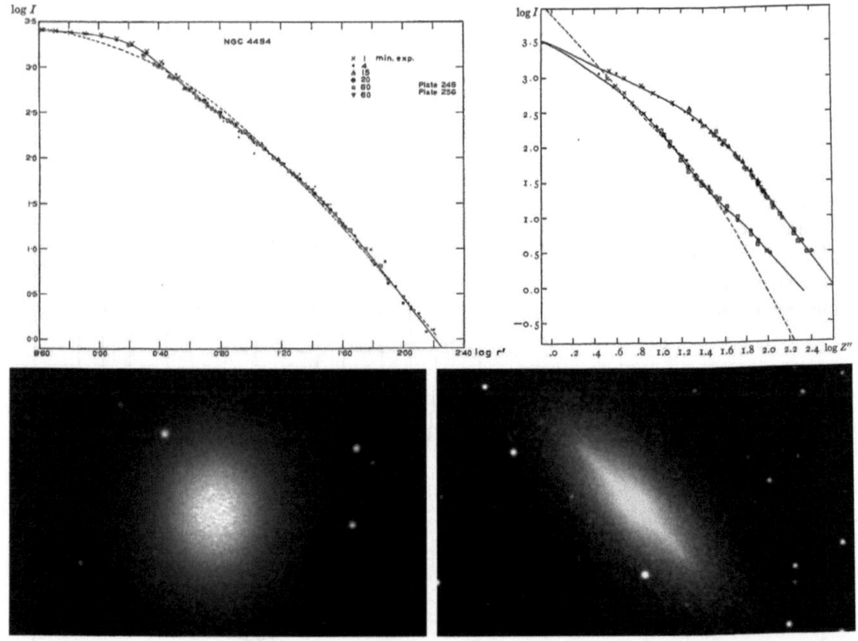

Fig. 8.26 Top: traces of the light profile of NGC 4494 (left) and NGC 3115. Different symbols denote different quadrants on the sky. In NGC 4494 the minor axis is corrected for the flattening. In NGC 3115 the lower curve is the minor axis and the upper one the major axis; the dashed curve is the de-projected luminosity density along the equatorial axis. From Oort (1940a). Bottom: NGC 4494 and NGC 3115 from the Digital Sky Survey [203]

The other part of the paper concerned what was foremost on Oort's mind at the time, and that is the light distributions in external galaxies. We have seen that his attempts to use the Perkins Telescope for this purpose failed, but the few plates that Oosterhoff had taken at Mount Wilson did pay off. It should be emphasized that very little detailed surface photometry was available still. At the McDonald dedication Oort presented data on two galaxies that had been photographed by Oosterhoff and on which extensive measuring had been done by Herman Kleibrink in Leiden. They have been illustrated in Fig 8.26 (bottom). NGC 4494 is an elliptical galaxy without a trace of a disk, but apparently free from dust and gas. NGC 3115 is a so-called S0-galaxy, which has a disk but no dust, gas or young stars. The absence of interstellar absorption helps the determination of the spatial distribution of stars, but the systems do not at all resemble our own Galaxy.

For the case of NGC 4495, Oort had used four plates taken by Oosterhoff on the Mount Wilson 60 in. Telescope. The way the measurements of the plates had proceeded was to make traces along lines through the center (usually twelve). The nebula was very regular, although the isophotes (lines of equal surface brightness) were not circular, but had an average axis ratio of 0.83. Oort presented the mean luminosity profiles averaged over all tracings after correcting for this flattening (see Fig 8.26, top-left). Here I need to say a few words about units. The surface brightnesses were expressed in units of the light of one star of apparent magnitude 25 per square second of arc. And that on a logarithmic scale. So the lowest point in Fig. 8.26 is at $\log I = 0$ or $I = 1$ and that is a level of one unit. In modern terms we would say this is 25 magnitudes per square arcsecond. It had been calibrated by taking out-of-focus (to avoid over-exposure) images of stars of known apparent magnitude. The highest value in the graph is at $\log I = 3.5$ or $I = 3163$ units. In modern notation this translates into 16.25 magnitudes per square arcsec. Now, at a dark site like Mount Wilson was in those days, the sky has a surface brightness of about 22 or 22.5 mag per square arcsec at photographic wavelengths and the faintest points constitute then an increase of the total observed sky brightness due to the galaxy by one tenth of the sky itself. Later photographic surface photometry (such as performed around 1980) went to a level of 1%, or 5 'magnitudes below sky'. In 1940, Oort's work was certainly state of the art.

Oort did, however, not go much beyond this presentation. The only thing he added was the dashed curve which is a parametrization that Edwin Hubble had proposed. Actually, just as in the case of the morphological classification scheme of Hubble, this was almost certainly also first proposed by John Reynolds of the UK, which Hubble again had failed to acknowledge (see Sect. 2.6 and [56]). Oort did however not go beyond presenting this profile and added no discussion on three-dimensional light distributions or dynamics. He did not comment on this in the paper, but it seems that his opinion on round or almost round nebulae (as NGC 4494) was that these are systems that do not rotate or very slowly.

For the case of NGC 3115 the paper is more elaborate. The minor axis profile is distinctly different from that along the major axis (see Fig. 8.26). Oort assumed that we see the galaxy essentially edge-on, which is still held to be true, but he did not argue the case for it. The isophote shapes are not elliptical (see Fig. 8.27), especially near the major axis. Oort commented on this but did not speculate on what this could mean (e.g. being the result of a distinct stellar disk). It is the same system he illustrated in his inaugural lecture (Fig. 8.10), where he described it as 'elliptical'. Oort used a distance of 1.5 million parsecs (Mpc); the currently accepted distance is almost ten times larger, 14 Mpc.

Oort then developed a method to de-project the observed distribution into a three-dimensional one, which can be done under the—probably very reasonable—assumption of axial symmetry. It does represent quite some computational work, which was done once more by his trusted Computer Gerrit Pels. Results of this are the dashed line in Fig. 8.26 and the isophotes in Fig. 8.27—lower part. The luminosity density was expressed in solar luminosities per cubic parsec and (if the distance is 1.5 Mpc) the central one comes out as 30,000 times that near the Sun, or about the same as in the center of the globular cluster M3, which is a quite normal one. The faintest contour is at 0.007 times that near the Sun.

Next Oort tried to study the dynamics, but that turned out to be a disappointment. He assumed that the density distribution would be reflected in the luminosity distribution. This assumption could be tested by calculating rotation velocities in the symmetry plane and comparing these to observations by Humason.

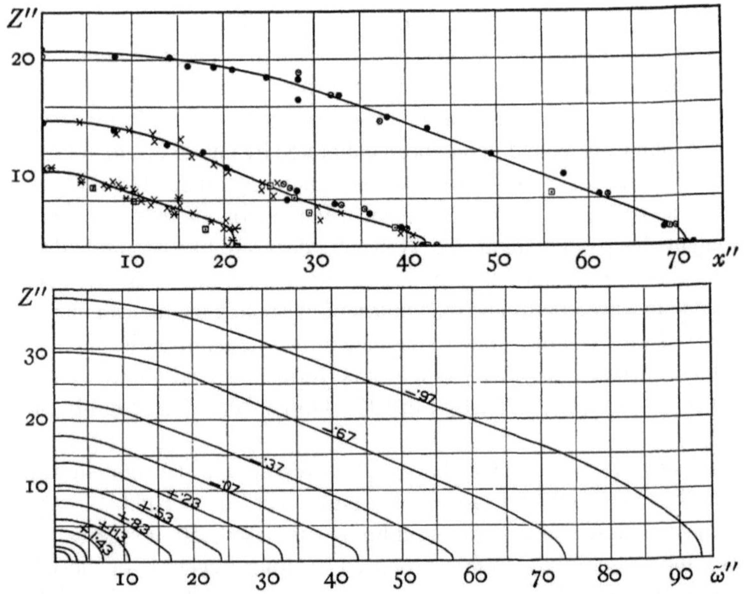

Fig. 8.27 Isophotes of the light distribution in NGC 3115. The top shows the observed ones and the bottom the surfaces of equal space density of light from a decomposition. Different symbols denote different quadrants on the sky. For further details see the text. The labels on the curves in the lower panel indicate luminosity densities on a logarithmic scale. From Oort (1940a)

Oort started the relevant section in the paper as follows:

'It was hoped, originally, that the distribution of the light in extragalactic nebulae would bear resemblance to the distribution of mass and that the investigation of the light-distribution might give some insight into the question of how far the most flattened elliptical nebulae and the regular cores of early-type spirals approach unstable configurations which could give rise to the formation of spiral arms. Humason's results on the rotation of NGC 3115 seem to indicate, however, that there is not much prospect in this direction; not, at least, for this nebula.'

Oort first calculated the potential (and gravitational forces) with the assumption that light traces mass. These are expressed up to a factor f, which is the mass-to-light ratio M/L (unity for the Sun; see Sect. 7.1), and a factor κ which is the real distance divided by the 1.5 Mpc he assumed (for the modern distance κ would be about 9). From this he calculated the circular velocity (the speed of a star in a circular orbit at that radius) at a number of galactocentric distances (which then had to be multiplied be a factor $\sqrt{\kappa f}$). He then deduced from this that M/L had to be at least 150. However, the measured rotation velocities could not be compared directly to these circular velocities. The difference between the two, after all, is the asymmetric drift. This was the effect (see Sect. 5.2) that due to kinetic energy in random motions, the collective rotation velocity is lower than the circular one necessary to compensate gravity and the amount depended on the magnitude of the random motions (the velocity dispersion). The mathematical formula associated with this is the lower one in Box 6.3.

So, what was the value for the velocity dispersion? Oort first looked at the vertical light distribution, and for the sake of the argument used an M/L of 200. Assuming that light traces mass, he could use the procedures that he developed to study the vertical force in the solar neighborhood (K_z in Box 7.2) to estimate the vertical velocity dispersion. He did this for two of the distances from the center where Humason had measured velocities, 16″.1 and 32″.2. The vertical velocity dispersions came out as about 250 and 190 km/s, and the circular velocities at these radii (for $\kappa = 1$) would be about 675 and 620 km/s. He then used a theorem of Jeans that said that in a system in exact equilibrium the radial velocity dispersion had to be equal to the vertical one (see Sect. 6.2). Although he had had doubts about this for the very flat disk of the Galaxy, he felt that in this case it was justified. He then knew all terms in the asymmetric drift equation (except for the last one, but that contained a cross-dispersion, and would be very small). This gave him circular velocities of 450 and 470 km/s. However, this was inconsistent with the 675 and 620 km/s derived above. If these were adopted, radial velocity dispersions of 400 to 500 km/s were required and such dispersions would result in very broad absorption lines and that was inconsistent the spectra of Humason. So, it was impossible to arrive at a consistent picture with a constant M/L.

So, he asked, 'How must the distribution of mass density be in order that both the observed distribution of light and the measured rotational velocities as well as the conditions for dynamical equilibrium can be satisfied?' This was approached with a very schematic model in which NGC 3115 consisted of three ellipsoidal shells of axis ratio 2/3. The last one extended from 20 to 70 arcsec from the center and beyond that the density was assumed zero. This could be fitted, but then M/L had to rise from about 20 in the inner parts to 250 in the third shell. This was very difficult to understand, as in the solar neighborhood Oort's value was about 2. Incorrect neglect of absorption in NGC 3115 would lower the M/L, but that 'would give impossibly low values in the central parts'.

Oort's final sentences were:

> 'There cannot be any doubt that an extension of the measures of rotation to greater distances from the nucleus would be of exceptional interest. Also, our knowledge of the system could certainly be greatly advanced if it were possible to obtain spectroscopic estimates of the peculiar velocities.'

Dynamical studies of external galaxies were not yet possible because of the difficulty of determining stellar velocities and dispersions at faint surface brightnesses. Milton Humason had measured four velocities at Mount Wilson and these showed a sharply increasing rotation velocity with galactocentric distance, rising almost linearly to 316 km/s at 32″.2 and then leveling off somewhat to 447 km/s at 45″.5. It is clear presently that Humason's data were far from correct. Not surprisingly, since it should be realized that it involved the recording of the weak stellar absorption lines in a spectrum which itself is already very faint. The best modern determination [222] by T.B. Williams (see below) shows a sharper increase to 200 km/s at 10 arcsec, but then a leveling off at 250 km/s at 20 arcsec and a constant rotation velocity at about that value out to 110 arcsec. Oort was much misled by the observations, for which in hindsight it had been much too early days. It is no surprise that no consistent picture emerged. However, Oort's paper showed how the analysis of such data would have to proceed, while his work also showed that photographic surface photometry was possible and how to determine three-dimensional light distributions from that. In spite of the failure to determine the dynamical structure, Oort (1940a) was a seminal paper.

8.7 Observing at Mount Wilson

After the McDonald dedication Oort traveled to Pasadena to obtain plates of nearby galaxies at Mount Wilson. From the train on May 10, 1939 he wrote to Mrs. Oort:

> 'In $1\frac{1}{2}$ hours we will be in Pasadena, where Adams has reserved a room for me in the so-called 'Atheneum', a sort of Faculty Club, that has been beautifully decorated, but not too expensive. I am curious how it will be in Pasadena. I have felt so unhappy there in 1924 that I return with some apprehension. But there is a large difference between then and now. The feeling that you are part of my life, even at a large distance, and that soon I hope we will continue our happy life, the realization that this is happiness, makes everything different. [...]
>
> [Continued from Pasadena] On the train from El Paso (where we left last night at seven-thirty) were also the Plasketts, Lindblad, Baade, Wright (the Director of the Lick Observatory), Trümpler (also from Lick) and Hubble. El Paso is in the middle of the desert, and when I woke up this morning there was nothing but desert to be seen from my window. [...] In the train, but also already in Fort Davis, I had many conversations with Baade and Hubble. The latter is a great man who wants to introduce me to all sorts of people, but I would much prefer to work hard. Baade is particularly nice. He likes Europe and Germany, but has preferred to stay here anyway, under the condition that he can return to Europe every two years. Both he and Hubble will help me. Baade will accompany me up the mountain coming Sunday and then will work with me for 3 nights, which is very nice and will prevent me from making many mistakes and loosing much time. I already learned a lot from Hubble and I can profit much from his plate collection to select objects for my program.'

On the back of an envelope there is a note that Oort must have written while in Pasadena:

> 'Hubble likes drinking a bit too much and loves society, Blakeslee and rich and genteel people. But on the other hand he is an extraordinarily serious worker and very diligent. He came to the Observatory directly after we had arrived and has a good sense of discriminating of who can do things and who cannot. Compared to Hubble, Lindblad has ways of

conducting himself that is enviable, as he calmly sits with his glass of whiskey on the terrace with the Hubbles, and I suddenly realize how different Americans are and how different I am from them. He fends off an assault by Zwicky in a beautiful way, smilingly remaining in perfect balance.'

Oort probably referred to science writer and Pulitzer Prize winner Howard Walter Blakeslee (1880–1952). Fritz Zwicky (1898–1974) was a Swiss-born astronomer, who came to the California Institute of Technology as a physicist to work with Robert Andrews Millikan (1868–1953), but turned to astronomy. He is credited as one of the major contributors to the discoveries of neutron stars and dark matter, for predicting gravitational lenses, but he also did work on rocket propulsion during World War II. He was known as a difficult and grumpy person, but also as a great humanitarian.

The observing went, as often, with ups and downs. Oort wrote to Mrs. Oort in detail about that; how he observed at night at the telescope and spent afternoons developing in the dark room, how he accidentally had a cover slip of a box of undeveloped plates that fogged two of them, how the weather was often poor, but how Baade gave him three of his nights on the 60 in. Telescope, etc. But he also wrote about concerts he was taken to and musea to see art. And the domestic things concerning the Oort children, such as Coen's 'moodily and impulsive character' and his choice for secondary school, Marijke teaching Bram to write, etc. But also his worries about her father, who had become seriously ill (but recovered).

The Annual Report of Leiden Observatory mentioned that he stayed at the Mount Wilson Observatory from May 10 to June 25. 'In total he obtained 122 exposures on elliptical and early type galaxies and 13 on globular clusters, using the 60 in. and 10 in. Telescopes'. The latter would be the 10 in. Cooke Telescope, a refractor that was used by Edward Barnard and others to produce a photographic atlas of the Milky Way [223]. However, it is a misprint and should be 100-inch, see Sect. 13.3. Oort also made a visit to Palomar Mountain, where the 200 in. Hale Telescope was under construction.

After leaving Pasadena, Oort first paid a visit to Lick Observatory, where Lindblad was spending some time. He next spent a few days in Lake Tahoe with Chester William Clark (1906–2000) and his wife Jessie May (1906–1989) and their two small children. The youngest of these children was a one-year-old daughter Marijke. He delivered presents from Mrs. Oort. Chester (Chet) Clark had graduated from the US Military Academy in 1931 and then had come to Leiden to write a Ph.D. thesis under physics Professor Wilhelmus Hendrikus (Willem) Keesom (1876–1956), which he defended in 1935. Clark became a career soldier in the US Army, ending up as a major general. The Oorts must have known them well from this Leiden period. Oort's youngest son Abraham remembered that Clark visited Leiden not long after the Oort family returned there after the war.

Next Oort went to visit Struve at Williams Bay, the location of Yerkes Observatory, where he arrived on July 4. He mentioned in a letter to Mrs. Oort: 'About the directorship we have not talked much. No new points of view have been brought forward, the threat of war is certainly a point of consideration, but I

Fig. 8.28 Oort and three Dutch astronomer-expats in 1939 at Harvard. From left to right: Bart Bok, Oort, Peter van de Kamp and Jan Schilt. From the Oort Archives

have the feeling that I then would let Holland down.' So that was the end of that. Actually, during all of Oort's trip this constitutes the only reference to the possibilities of war in Europe!

Oort next moved on to Harvard, where he first stayed for a week with Bart and Priscilla Bok and their children John and Joyce. 'They live in the countryside in Lexington, a hamlet at some distance from Cambridge in a beautiful area.' In Fig. 8.28 we see Oort with Bart Bok and two more Dutch friends, Jan Schilt and Peter van de Kamp. In the *LibAm80* [2], Bok presents this picture as having been taken at the Harvard Summerschool, 29 July 1937. The year must be wrong, Oort did not visit the USA in 1937; it would have been mentioned in that year's Annual Report of the Observatory (it mentions that Oort gave a lecture in Paris in July 1937).

It is clear that he now longed for home; after all, he had been away more than 3 months, Apparently, children of Marijke's age all seemed to develop scarlet fever, so he worried about his daughter contracting it too and wrote about that to Mrs. Oort (and she to him, one would say). But he had work to do, like observing for three nights at Harvard's Oak Ridge Observatory. He wrote to his wife on July 25 that he had observed with the 61 in. Telescope, but that the seeing had been poor and the results were 'not ideal'. He then took a room in Cambridge near Harvard, where he lectured for a few weeks. On the same July 25 letter, he wrote that the Torreys asked him to come by in Randolph. He decided to go for the weekend of August 13. He stayed in a hotel since Mrs. Torrey was very ill; Charles Torrey had written before that she had typhoid. But upon arrival in Randolph she appeared to be recovering. 'It appears that there is no danger as long as I do not eat there, which I will not of course.'

After visiting New Haven, Oort spent a few days in New York with the Jan and Jo Schilt and paid a brief visit to the van de Kamps at Swarthmore in Pennsylvania, before finally returning home, this time crossing the Atlantic with a Dutch ship, the S.S. Statendam (the name of an old dyke on Goeree-Overflakkee, one of the southern islands of the Netherlands, bordering the North Sea) of the Holland-Amerika Lijn. Oort was back in Leiden on September 2, 1939. World War II had just begun a day earlier with the invasion of Poland.

Fig. 9.0 Oort in 1939. From the Oort Archives

Chapter 9
World War II: The Watershed

War is not one disaster for millions.
War is millions of times a disaster for one.
Hans van Mierlo (1931–2010)

The Second World War made a deep cleavage,
not only in the lives of our generation, but also in the
development of astronomy. Looking back it appears as a wall
separating two different astronomies. The great change was
mainly brought about by the advent of radio astronomy.
Jan Hendrik Oort

9.1 Outbreak and First Years of the War

The secondWorld War started with the invasion of Poland, on September 1, 1939, one day before Oort's return from his extended trip to the USA. He was back just in time: from then on crossing the ocean became more and more hazardous, and eventually impossible for travel like Oort's. In the Netherlands the war started somewhat more than eight months later with the invasion and bombing of Rotterdam by the German military forces on May 10, 1940. According to his pocket diaries, Oort happened to be in Utrecht, where he had traveled to by bus the afternoon before, May 9. His pocket diary does not mention why he went there (it is noted that he did visit his dentist before leaving Leiden). Appointment notes are in pencil, but from May 10 on he recorded things that had happened with a pen across these notes. For May 10 the pencil notes in the background can still be discerned; these say 'Second Regiment Field Artillery' and 'draft 1923'. Now, Oort's military passport records that on November 25, 1938, he had been transferred to that regiment, so he probably had to report to it as part of the mobilization as a soldier that had been sent on long furlough. The barracks of

Henricus Antonius Franciscus Maria Oliva van Mierlo was a Dutch politician.
From *Reminiscences of Astronomy in the Twentieth Century*, Oort (1982e).

P. C. van der Kruit, *Jan Hendrik Oort*, Astrophysics and Space
Science Library 459, https://doi.org/10.1007/978-3-030-17801-7_9

Fig. 9.1 The Oort family in 1940 in Leiden. From the Oort Archives

the Second Regiment Field Artillery are in Ede, about 50 km east of Utrecht and that probably is where he went and spent the night. Oort had written across the pencil notes for May 10 (Fig. 9.1):

'Woken up at 4 o'clock by squadrons of aircraft and air defense artillery. At ±10 o'clock back to Leiden by car. Airplane shot down over the Merwedekanaal. Aerial combat over Leiden and Valkenburg. At night fire watch until ±5 o'clock. Three times air-raid alarm.'

The Merwedekanaal connects Amsterdam via Utrecht with the Rhine delta. It is precursor of the current Amsterdam-Rhine Canal that follows the same route between Amsterdam to the Rhine by way of Utrecht. Valkenburg was an air base—at the time still under construction—just southwest of Leiden. And on May 11:

'Warnings for air-raids on the radio. In the morning again aerial combat. During the day Leiden becomes unsafe because of paratroopers. In the evening had fire watch until 12 o'clock. We cannot go onto the Observatory lane [The street that leads onto the Observatory] after 8 o'clock.'

And for Sunday May 12:

'Dig a hole to dump garbage, while the guns of cruisers off the coast thunder and a thrush jumps into my hole to collect worms while I am digging. Wild stories about paratroopers. Until 12 o'clock fire watch with father.'

Monday May 13:

'On Monday also fire watch until 12 o'clock. There is a rumor Juliana [then Crown Princess] has left and also that the government and Queen [Wilhelmina] are leaving for England. Monday night rumors of Germans breaking through the Grebbe Line and having crossed the Moerdijk. Wesselink waits in the Astro-building. I stay up until 2 o'clock to wait for messages [on the radio, presumably], which finally do come at 2 o'clock.'

The Grebbe Line was a defense line against invasions, using inundation, that ran from the Grebbeberg, a hill near the Rhine to the southeast of Utrecht to the IJsselmeer, the central lake (inland sea) of the Netherlands. Oort's artillery regiment had fought there, but Oort himself apparently had been excused from that. The Moerdijk is a bridge south of Dordrecht that connects the southern part of the Netherlands to Holland, and eventually gives access to Rotterdam and Den Haag (See Fig. 1.1). Adriaan Jan Wesselink (1909–1995) was an astronomer in Leiden, who had defended his Ph.D. thesis under Hertzsprung in 1938. In 1946 he moved to South Africa, first as the observer at the Leiden Southern Station and in 1950 to the Radcliffe Observatory near Pretoria, and eventually in 1984 to Yale in the USA.

Wednesday May 15:

'Tuesday at ± 12 o'clock 16 meteorologists arrive. Bombing of Rotterdam, which we first become aware of by an enormous cumulus cloud and a fading cloud of smoke. Two times air-raid alarm this day. Airplanes throw out pamphlets. In the course of the afternoon the capitulation is announced.'

Finally on Thursday May 16:

'Meteorologist and father and mother leave. Transport of troops through Leiden.'

People certainly would have been scared of what would happen next. In the same pocket diary there is a note in the 'Memorandum' section at the end of the week ending May 11, which says: 'Bramke prepares a package for the'prison'.' The poor 5-year old son of the Oorts was bracing himself for the worst.

The history of the University of Leiden during the War has been documented in detail by Petrus Johannes Idenburg (1898–1989) [224, 225]. This has been an important source of information for this section. Idenburg was secretary of the Board of Curators of the University. The private papers of Oort contain a file on matters during the War (including interesting newspaper cuttings). There are a few sheets of paper that contain 'Stories of the Netherlands during the War'.

At first not too much changed and work at the University and at the Observatory proceeded to a large extent as before. The first signs of trouble came on October 21, 1940, when all professors and personnel were told they had to sign a so-called 'Ariër verklaring' (Aryan statement), which was a declaration about one's ancestors. It was in fact a first step towards the prosecution of Jewish citizens. One had to list parents and grandparents (of oneself and of one's spouse) and provide a declaration that none of the parents or grandparents of oneself or one's spouse were of Jewish descent. A meeting of the Senate to formulate a common stand had been forbidden by the Germans, who probably had heard of plans for this through a sympathizer with the Nazi regime, possibly Jacobus Johannes Schrieke (1884–1976), Extraordinary Professor in the Faculty of Law until 1941 and Secretary General of the Ministry of Law during WWII. Meetings of groups under 20 persons were allowed and from the members of the Senate a group was formed that secretly discussed this and other matters: this group came to be referred to as the 'Kleine Krans' (the small circle). Although initially there had been a move among professors to collectively refuse to sign this discriminatory statement, the large majority in the end decided to sign it, sometimes accompanied by a letter of protest. One would think that Oort would have signed as well, otherwise he would not have been able to continue his work, in fact could have been in serious trouble. This is at odds with a letter in his files, dated January 1944, from the 'Nationaal Spaarfonds', a kind of savings bank, pointing out that if he refused to sign the 'Ariër-verklaring', he would be considered a Jew and his savings would be transferred to Lippmann, Rosenthal & Co, a German controlled bank that 'administered' all belongings of Jews before they were deported. The Spaarfonds urged Oort 'in order to save yourself and us from unpleasantnesses we ask you to put your objections aside.' Probably the bank had been ordered to keep an independent record of these statements.

It is not mentioned in any of Oort's diary notebooks, pocket diaries, etc. whether or not he was an actual member of the 'Kleine Krans'. Willem Otterspeer (b. 1950), Professor of university history and long-time Curator of the Academic Historical Museum in Leiden, has provided me with information on a dinner celebration by a small group on February 8, 1941, the day of the Dies Natalis (birthday) of the University. It was held at the home of Berend George Escher (1885–1967), Professor of geology. The printed menu featured a woodcut made by his famous brother, artist Maurits Cornelis Escher (1898–1972), showing a fence closed with a padlock and the text: 'Van kwade trouw het beeld, versperrend naar vermogen; in onmin eens geteeld, in achterdocht getogen' ('An image of bad faith, obstructing all it can; once bred in quarrel, raised in suspicion'). Participants were advised to keep this document secret. Among these fifteen professors, Oort was indeed present, as were Jan Julius Lodewijk Duyvendak (1889–1954), Professor of oriental languages, Roeland Duco Kollewijn (1892–1972), Professor of law, historian Johan Huizinga (1872–1945) and physicist Hendrik Brugt Gerhard (Henk) Casimir (1909–2000). These fifteen could very well have been the majority of the Kleine Krans members. The Rector Magnificus at that time, Alexander Willem Byvanck (1884–1970), was not part of it. There is no entry referring to this dinner in Oort's pocket diaries. Saturday February 8, 1941 was two days after Oort had skated a large part of the Eleven Cities Tour. For that Saturday there is only a note saying 'storm and strong thaw'.

In the *WWIIStories* there is a remark 'Kleine Krans at Kollewijn's 28 April '42' (see below). This reinforces the supposition that Oort was a member of this group. However, for April 28, 1942 (actually also Oort's birthday) there are only some reminders that are unrelated to the meeting and a note 'Bios 8^h at Boschma'. Hilbrand Boschma (1893–1976) was a Professor of zoology and Director of the National Museum of Natural History in Leiden. The word 'Bios' has been clearly written (for a change). The secret meetings were apparently either not recorded in appointment books or in coded form.

On June 30, 1945 there is a referencd in*WWIIStories* to a meeting of the 'Kleine Krans': 'nice to be together again, although there are major changes compared to our original group.' But then, by that time it was safe to record this. There seems to be little room to doubt that Oort had been part of the 'Kleine Krans'.

Oort was also active in a group set up at about the same time, that was to look into re-structuring the University after the war. Obviously this had to be done with some secrecy. The Oort files contain a letter, dated October 25, 1940 by the secretary of the Senate, in which, on behalf of the Rector Magnificus, Oort was asked to become a member of a committee to investigate how the University should organized in the future. This is designated by the term 'Leidsche Universiteitsgemeenschap' (Leiden university community), which refers to all the workings of the University. I will refer to this as the 'Civitas committee'. There were sub-committees on Studium Generale, a general curriculum for all students, public outreach, a university house (as a general meeting place and the finances thereof), and a university periodical. Oort was asked to join the committee on general matters under the Rector's own chairmanship. Rector Alexander Byvanck, a Professor of Classical languages and archaeology, often failed to take a strong stand in the relation with German authorities. This committee is sometimes mentioned in connection with the Kleine Krans; it is not likely that they were actually the same since the members were appointed by the Rector, but there might have been considerable overlap (Fig. 9.2).

It was expected that the Aryan statement, which (by exclusion) identified everyone with a Jewish background, would be used against Jews. Indeed on November 21, 1940, the German occupying authorities ruled that all Jewish professors (and other personnel) had been relieved from their duties. One such professor was Eduard Maurits Meijers (1880–1954), who taught (Dutch and international) private law. His former student Rudolph Pabus Cleveringa (1894–1980), at that time Dean of the Faculty of Law, delivered an address at the regular time of Meijers' next course lecture five days later, on Tuesday November 26, 1940, in which he strongly condemned the dismissal of Meijers and other Jewish professors. In his lecture, he denounced the action of the Germans as a violation of international law. After the lecture the audience starting singing the 'Wilhelmus', the Dutch national anthem. Oort attended that lecture; his pocket diary mentioned for that date: '10 o'clock lecture Cleveringa', and the next day, November 27, he jotted down 'university closed'. Some of the students that attended Cleveringa's lecture had made typewritten copies of the text of the address and distributed these widely (for the full text in Dutch see [227]; Oort kept a copy in his Archives). Students went on strike. As a result the Germans closed the University, which stopped all teaching (but not the research). Most students went

Fig. 9.2 Gathering of students in front of the Academy building after the lecture of Professor Cleveringa on November 26, 1940, detesting the dismissal of Jewish professors and personnel. Oort attended this lecture. From Leiden University [226]

on to other universities, mostly Amsterdam. Cleveringa was arrested, but released in the summer of 1941. Meijers was arrested in August 1942, but survived various concentration camps. Cleveringa was taken hostage by the Germans in 1944, and also survived the war. Leiden University annually organizes various Cleveringa lectures all over the world to commemorate this courageous act (Fig. 9.3).

Oort's recollection as reported during the *AIP-Interview* [4] is somewhat inaccurate. Here are some parts:

> 'Yes, but the occupation came only in 1940. After that, of course, everything was closed but even the work continued to go on more or less normally at the university for some time [...] Among the professors who were dismissed was a very famous, (also internationally famous), Professor of law by the name of Meyers [Meijers] [...] the Dean of the Faculty of Law went into his class – it was a big class in the main university building – and delivered a speech in which he started by saying, 'I won't talk about his dismissal and I shall leave the people who did this, below us, but will concentrate on the greatness of the man dismissed by our aggressors.' This speech made such an impression on all his students

Fig. 9.3 Oort in 1941, from the Oort Archives

that on leaving the auditorium they sang the National Anthem and went on a strike. I happened to be present and was greatly impressed by the lecture. This occasion formed the beginning, in a way, of the active resistance in Holland. The speech by the Dean of the Faculty of Law was widely circulated during the rest of the War by the resistance groups.'

[DeVorkin: Did you partake actively in any of this resistance, anyone else at the Observatory?]

'No. We had a little group of professors in Leiden who came together regularly and discussed the problems the University faced in view of the German occupation. Most of the members of this group were put in hostage camps [...]. I wasn't and so we went down to live in the country for the rest of the war, not to take this risk unnecessarily, but I never did any real active work in the resistance.'

[DeVorkin: What was your contact during the occupation? What was your contact with the outside world, with the West? Was there any contact at all, was there indirect contact?]

'Practically none. There was no way, at least not in the second half of the war, of sending letters to foreign countries or to receive them. There was practically no contact.'

In order to obtain an impression of Oort's life in the first part of the War I quote from the *WWIIStories*, which would be his personal record of what he felt to be the most important developments. It starts with some of the resistance actions he seemed to have sympathized with.

'Already from the beginning the brothers Drion published'De Geus onder de Studenten' (The rebel among the students), the first clandestine publication.

Sept. '40. Preparation for actions against the Jews. Paul Scholten and Telders take the initiative for a protest.'

The brothers Drion are Jan (1915–1964) and Huib (1917–2004). They were students at the time; they both later became Professors of Law, Jan in Leiden and Huib at what now is the Economy Faculty at the Erasmus University in Rotterdam. Paul Scholten (1875–1946) was a Professor of law in Amsterdam and Benjamin Marinus Telders (Ben) (1903–1945) a Professor of law in Leiden.

'It is all done slowly; which is very clever. At first no hiring of Jewish civil servants.

26 Nov. '40. Meijers receives notice of dismissal. Cleveringa's lecture. *'I will not try to direct your thoughts to those that have produced this letter. The only thing I desire now is to keep them out of sight and below us and to direct your eyes to the heights where the bright figure resides of him who we are here for.'*

Owning a copy of this lecture is usually punished by a month in jail. There was no violence but it remained quiet and the students went on strike for a week. The next day Cleveringa was arrested and the University closed by the Germans. Also other univ. Protested against this measure. In Delft it also resulted in closure.

Charged with having roused the students to protest, Telders was imprisoned. A formidable opponent. In De Gids he had published an article in which he had condemned the betrayal by Seyss-Inquart (before the war) of Austria. Addressed its incompatibility with international law. With admirable mental power he kept himself alive through 4 years. In April '45 he died in concentration camp Bergen-Belsen.'

Austrian Arthur Seyss-Inquart (1892–1946) played a dubious role in the 1938 'Anschluss' of Austria, delivering his country to Nazi Germany during his two days as acting President. During the German occupation he was Reichskommissar in the Netherlands; after the War he was sentenced to death in the Nuremberg trials.

'The Germans wanted to re-open Leiden and Delft for the benefit of their own purposes. In April '41 Delft re-opened, after Easter exams were taken again in Leiden.

[...] In Aug. '41 it became known that some Leiden professors would be fired and replaced by Nazi sympathizers.

Kleine Krans advised to resign en masse. This was around October 2. [...] It made President-Curator abandon his plans for the moment.

Early 1942 serious troubles in A'dam. 22 Jan. A detonator bomb went off at the National Socialist Students Front. Punishment of the city, some prominent citizens, among which 5 professors [...] to the infamous camp in Amersfoort for half a year. [...]. Wife and two children of A'dam lecturer imprisoned when they could not find the man. [...]

Kleine Krans at Kollewijn's 28 April '42. Those present decide to resign and to have as many as possible follow their example. Beginning of May 80% resigned. Calmly accepted by the Germans.'

Kollewijn was taken hostage in 1941 and because of his health exiled to Hoogeveen in the north of the country. The meeting would have been at his residence without himself being present.

9.2 Personal Matters

For a while (until the summer of 1942) things went along more or less normally. The winters were rather cold; the Eleven Cities Tour (see Sect. 1.1) was held three years in succession between 1940 and 1942 (actually only 15 have been held between 1909 and 1997, which at the time of writing was the last one). Already in the winter of 1940, Oort noted in his pocket diaries various skating tours he made, among which the Eleven Cities Tour in Friesland.

> 'Sunday January 28, 1940. Skated near's Molenaarsbrug on the Old Rhine with Mieke, Coen, Marijke, Klaas and Anton. Thick layer of snow and took short walk in the snow back [...]'

The course of the main riverbed of the Rhine used to go through Utrecht and Leiden and is now referred to as the 'Old Rhine'. It caused many floods and was dammed off in the 12th century, opening the current course. 'Klaas' was Klaas Wiersma (1916–1993), a skating enthusiast who worked in the Faculty of Law and would become Professor of notarial law in 1946. Skating enthusiast 'Anton' was Anton Bicker Caarten (1902–1990), an expert on Dutch windmills who has written some authoritative books about this.

> 'Monday January 29. Trip though snow across the Afsluitdijk into a snow covered Frisia.'

The 32.5 km long Afsluitfijk, or Closure Dam, (see Fig. 1.1) was completed in 1932; it connects the province of North- Holland with Friesland and closed off the central 'Zuiderzee', which then became the IJsselmeer ('IJssel Lake, the IJssel being an important tributary of the river Rhine. There was a bus service across it.

> 'Tuesday January 30. Eleven Cities Tour. Sneek, IJlst, Sloten, Stavoren, Hindelopen, Workum, Bolsward, Harlingen with Anton Bicker Caarten. Back by train since there was too much snow on the Afsluitdijk.'

This probably was the first time Oort skated the Eleven Cities Tour; the previous ones had been in 1927 and 1933 and there is no pocket diary in the Archives for precisely those two year, nor is it mentioned in the diary notebooks. In 1940. Oort skated through nine cities (also counting the starting city Leeuwarden) and managed to complete 116 of the 199 km.

Oort's skates, probably the most recent ones he used, are shown in Fig. 9.4. These are of the type 'Friese doorloper'. The word 'doorloper' (from the verb 'continue') refers to the property that the 'iron' (the metal part made of a hard kind of steel) continues all the way to the end of the wooden frame under the heel.

The next year Oort again skated the tour with Bicker Caarten. They traveled by train, at least in the pocket diary there is a schedule via Utrecht and Zwolle to

Fig. 9.4 Oort's skates, donated by the Oort family to the University of Leiden. These skates have to be tied tightly under firm shoes or boots. They are now part of the collection of the Academisch Historisch Museum in Leiden. From the Archives of Leiden University

Leeuwarden, leaving Leiden on February 4 at 14:14 and arriving 20:02. For the next day a remark is made that they attended a concert.

> 'Thursday February 6: With Anton Bickers Caarten. Eleven Cities Tour. ± 6:30 started from Leeuwarden. It is still dark and it is 14° below zero until Dokkum and back to Barthlehiem in the morning dawn; from there via Vrouwbuurtstermolen and Berlikum to Franeker, Harlingen until Sloten, where at 6 o'clock I still have 68 km to go. For the sake of safety I choose to go back by train. Anton continues with two farmer sons called Schipma and arrives in Leeuwarden at 11 o'clock.'

This time the tour was skated anti-clockwise, contrary to the year before.

In 1942 again some skating tours have been listed in the corresponding pocket diary. Coen does a 80 km tour on 'his new doorlopers', However, Oort does not participate that year in the Eleven Cities Tour.

> 'Thursday January 22: Today Eleven Cities Tour. It is 15° below zero, in Friesland a few degrees more. 5200 participants. Anton and Jelle go by themselves. I give up my plan at the last moment as I do not feel fit enough to make it and my weight is so low now (56 kg) that I will need extra food. Both make it, Jelle completely exhausted and had to be towed for the last 35 km.'

'Jelle' was Oort's brother-in-law Jelle Nauta, the husband of his sister Emy; family lore has it that upon return Nauta burned his skates and never skated again. Oort regretted never having completed an Eleven Cities Tour; he does note (with some pride one would think) that his son Abraham completed the Tour in 1956 (finishing at 19:45!). There actually is in the Oort Archives a handwritten list of skating tours he made, but this covers the period after WWII.

Summer vacations were initially still possible, in addition to withdrawing to the family second home Sandy-Hook in Katwijk aan Zee. For example, for August 1940 there are some notes in the diary notebooks from 'Westerflier, Hoendelo', a place where they must have spent the summer. There are earlier notes from June 1935 from the same place, so it must have been a location they liked. Hoenderloo (with two o's) is situated some 40 km north of Arnhem in the forest-rich central part of the Netherlands called the Veluwe. On official maps a mark 'Westerflier' is present, about 1 km to the west of Hoenderloo. There is no house or other building there at present (on Google Maps), but there might very well have been a vacation home there or other esidence in the 1940s.

The closure of the University meant that there was no teaching. Research went on, although communication with the outside world had become difficult. Oort was able to keep in some contact through Bertil Lindblad in neutral Sweden. In the USA, Bart Bok headed the Committee for the Distribution of Astronomical Literature (CDAL), established in 1940 by the American Astronomical Society (AAS), to foster the exchange of literature between the USA and occupied Europe. The committee used intermediaries, like William Otto Brunner (1878–1958) in Zürich, Switzerland, Director of the Eidgenössische Sternwarte, and Bertil Lindblad, and others in the UK and USSR [228]. Oort wrote a few letters to Brunner in Zürich, who also acted as an intermediary to forward letters or write for inquiries on Oort's behalf to American astronomers. Oort must through this channel, though with considerable delay, must have heard of the death of Frank Schlesinger in July of 1943. He wrote an obituary in Dutch in September of 1944, which was not published until 1945 (Oort, 1945a).

> 'For him it was a calling to advance classical, positional astronomy and to improve its foundations, and he always resisted the temptation to use large instruments to explore new areas and discover unknown treasures. Those working with him had to share this fate. He once told me, smiling, that he had entertained the idea when he came to Yale Observatory, to have painted over the entrance: 'Those that enter here, should give up any hope of discoveries'.'

On May 12, 1941, Oort's father, Abraham Hermanus Oort, died at the age of 72. The announcement in the newspaper 'Leidsch Dagblad' mentioned only the date of the funeral [229]. There are no entries in Oort's pocket diary or diary notebooks, nor any notes on a speech he might have delivered at the funeral. The private Oort Archives contain some formal documents concerning the winding up of affairs such as the execution of his testament. Father Oort had retired as Director and head-physician of the Rhijngeest Sanatorium in 1934 at the age of 65. He was a modest man; I have not been able to locate any 'in memoriam' or other article about him at the time of his death. There is only the short article on the occasion of the 40th anniversary of his obtaining qualifications as physician [27].

9.3 Astronomers Conferences

In 1941, the first Netherlands Astronomers Conference had been organized by Marcel Minnaert from Utrecht. These conferences were meant to bring together the Dutch astronomical community, including students, for more than the single day of meeting of the Netherlands Astronomers Club.[1] More information (including photographs) is available on the Website of the NAC (the Club that is) [230]. The meeting took place June 3 to 5, 1941 in Doorn, which is not far from Utrecht. Oort was present and gave a presentation on international collaboration and another one on elliptical nebulae (see Fig. 9.7).

The next year, 1942, the second Astronomers Conference took place at the same location, July 1 to 3. Again Oort attended, this time speaking about fundamental reference systems. Photographs of this meeting can also be found on the NAC site [230]. Oort's Archives contain the program of this meeting. The participants left on Wednesday July 1 from Utrecht by bike or tram (and on foot in Doorn) at about lunchtime. After arrival and tea, a number of astronomers presented short contributions on their work and in the evening there was a 'causerie' (a very informal discussion) by Théodore (Fjeda) Walraven, at that time a student in Amsterdam, on the 'social importance of astronomy', followed by a walk, soccer, recitation and music, apparently all in one evening. On Thursday the senior astronomers from the four institutes (Oort for Leiden) presented an overview of the ongoing research and after lunch there was again a walk in the woods with a stop for two more presentations, one by van Herk on the role of amateurs in astronomy and one by Oort on the fundamental reference frame of proper motions. In the evening Henk van de Hulst addressed the issue of astronomy in antiquity and after that again there was music, this time by the IAAT (the inter-academic astronomers trio; members unknown) who among others played music by Georg Philipp Telemann (1681–1767). The last morning there were again presentations by participants on their own work. After table-tennis finals (the competition must have been going on between presentations and meals) and lunch people left.

This is a good opportunity to introduce some Dutch astronomers that have played a role in Oort's life and that we will meet again later on these pages. They have been indicated in the captions to Figs. 9.5 and 9.8. I will take them in alphabetic order (Fig. 9.6).

Adriaan Blaauw has been mentioned before, but I will briefly describe his background (see also [231]). He studied in Leiden and in 1935, already before obtaining his Candidaats (bachelor) exam, he came to work at the Observatory. In his interview with Jet Katgert-Merkelijn, the *JKBlaauw-interview*, he mentioned that Hertzsprung was Director and young students were given variable stars for which they had to estimate magnitudes (from photographic plates) and then publish a so-called 'bouquet' of 25 or so stars as a contribution to the BAN. After about 10 stars, Blaauw had enough of that and found a way to go and work for Oort without upsetting

[1] Both the Club and the Conference are often referred to as the NAC. To my delight there are plans to change the 'Club' finally into the more professional 'Society' soon (2019).

Fig. 9.5 Participants in the first Netherlands Astronomers Conference in 1941. Oort is at the extreme left. Next to him are van Herk and Minnaert. The person in the back with beard and pipe to the left of the sundial is Jacob Houtgast, immediately to the right of the sundial stands Adriaan Blaauw and in front of him (looking down) Lukas Plaut. To the right of Plaut we have Leendert Binnendijk. In the front on the left Fjeda Walraven, next to him Henk van de Hulst and with his arm on the pedestal of the sundial Kees de Jager. From the Netherlands Astronomers Club [230]

Hertzsprung too much. In 1938, after he had obtained his doctoraal (master) degree, he moved to Groningen, where he became assistant, and eventually after the war, in 1946, obtained his Ph.D. with van Rhijn.

Leendert Binnendijk (1913–1984) was a student in Leiden, who worked for Hertzsprung on a project to study the Pleiades, the star cluster of which one can see the brightest six (or so) members with the naked eye. At a distance of about 135 pc it is one of the nearest clusters. Binnendijk determined photographic magnitudes and used other properties of these stars (colors, proper motions) from other sources, mostly Hertzsprung, to study the cluster and the field stars around it. The project was not finished until after the War had ended (when Hertzsprung had retired and left Leiden), so he finished the analysis under Oort and became Oort's second Ph.D. student when he defended his thesis in January of 1947 (see Appendix A.4). Binnendijk went to Sproul Observatory at Swarthmore College in Pennsylvania, where he worked with Peter van de Kamp. He spent most of his later career at the University of Pennsylvania in Philadelphia, working primarily on binary stars (Fig. 9.7).

Hendrik Christoffel (Henk) van de Hulst (1918–2000) also has been mentioned before. He studied in Utrecht with Minnaert, under whose supervision he would defend a Ph.D. thesis on *Optics of spherical particles* in Utrecht in 1946. At the time of the Astronomers Conferences he was mostly working on the theory of lines in the

Fig. 9.6 Oort and Henk van de Hulst during the first Netherlands Astronomers Conference in 1941. From the Netherlands Astronomers Club [230]

solar spectrum. We will soon meet him again in relation to Oort's interests in dust absorption in interstellar space and a possible emission line in the radio spectrum.

Jacob (Jaap) Houtgast (1908–1982) obtained his Ph.D. in Utrecht working on the solar spectrum. Although he collaborated extensively with Minnaert, his actual supervisor was (Belgian) physics Professor Léon Rosenfeld (1904–1974) and he wrote a dissertation on *The variations in the profiles of strong Fraunhofer lines along a radius of the solar disc*. He made a major contribution to the new field of radio astronomy, especially in relation to radio emission from the Sun (Fig. 9.8).

Cornelis (Kees) de Jager (b. 1921) had come to Utrecht to study physics, mathematics and astronomy in 1939 and was a junior student at the time of the Astronomers Conferences. When in 1943 the German occupation authorities demanded a declaration of loyalty of all students (that they would refrain from any action against the Germans), he refused and went into hiding, actually in the underground parts of the Sonnenborgh Observatory in Utrecht.

Lukas Plaut (1910–1984) was born in Japan, grew up in Germany, and studied astronomy in Berlin until 1933, when his Jewish background forced him to leave Germany. He went to Leiden and studied with de Sitter and Hertzsprung under whom

Fig. 9.7 Oort giving his presentation during the first Netherlands Astronomers Conference in 1941. The number 4594 is the NGC number of the Sombrero Nebula illustrated in Fig. 7.6. From the Netherlands Astronomers Club [230]

he obtained his Ph.D. in 1939 defending a thesis on *Photografische photometrie der veranderlijke sterren CV Carinae en WW Draconis* (Photographic Photometry of the variable stars CV Carinae and WW Draconis). The circumstances of the German occupation forced him to go to Groningen, but he was arrested and put in labor camp, later transported to the concentration camp Fürstenau in Germany. He survived and returned to Groningen after the War was over.

Théodore (Fjeda) Walraven (1916–2008) studied in Amsterdam with Anton Pannekoek. He was to come Leiden after the War in 1946 and obtained his Ph.D. in Amsterdam in 1948 on a thesis on the *Line Spectrum of Delta Cephei* (Fig. 9.9).

During 1942 the situation became progressively worse. The Germans started arresting leading citizens such as professors, politicians, clergymen, artists, writers, and others, and sending them to a special internment camp at Sint-Michielsgestel, 20 km north-west of Eindhoven. They were hostages to be killed in retaliation of sabotage and other resistance activities. The number executed was not very large, but life remained uncertain. The first group was one of 460 persons that were confined there on May 4, 1942, but at its peak there were some 700 persons living in this center.

Fig. 9.8 Participants in the second Netherlands Astronomers Conference in 1942. In the second row on the steps, we see from the left Blaauw, van Herk and Oort. In the back the second sitting person from the left is Adriaan Wesselink, next to him Johannes van Tulder and two persons further to the right Adriaan van Woerkom. The person sitting in the front in short sleeves is Henk van de Hulst and the one with sunglasses Jaap Houtgast. From the Netherlands Astronomers Club [230]

Astronomer Marcel Minnaert was interned there for almost two years between May 1942 until April 1944 (so he was not at the second Astronomers Conference). He taught physics to fellow hostages. In Leiden, some close friends of Oort were also taken hostage; some from the Kleine Krans or from the Civitas committee. Since Oort had probably been active in both committees he feared he would eventually be taken hostage as well and after the summer of 1942 he and his family went into hiding.

For the record the following: Adriaan Blaauw has also been jailed, but this was for allegedly listening to clandestine broadcasts of the Dutch radio transmitted from the UK and for spreading news he had heard. This did not happen until February of 1945, so it is a different story, not to be confused with the hostage-taking in 1942. He was released after friends and family bribed some minor German officials. Van Rhijn [232] mentioned that a computer by the name of Müller had been arrested at the same time and that they employed 'an apparatus used for registering time signals' for this.

Fig. 9.9 Pieter van Rhijn (left) and Oort during the second Netherlands Astronomers Conference in 1942. From the Netherlands Astronomers Club [230]

9.4 Beyond Statistical Astronomy

The field of statistical astronomy had been founded around the turn of the century mostly through the efforts of Kapteyn in Groningen and von Seeliger in München. It was designed to solve the 'Sidereal Problem'—the structure of the Stellar System—by determination of the distributions of stars in space and over luminosity (later also spectral type), and the distribution of their velocities and had the ultimate aim to understand the dynamics, the equilibrium between the space distribution of gravitating matter and motions. Kapteyn's Plan of Selected Areas was conceived to address these matters. Oort, being a student of Kapteyn, continued this; his 1938c paper on the density distribution and absorption in the Galaxy was a culmination of this program, in which he used star counts, proper motions, radial velocities, and spectral types obtained in the context of Kapteyn's Plan.

This involved going beyond the traditional methods of statistical astronomy. For one thing he wanted to understand spiral structure and had turned his eyes to external galaxies; after all in his study of NGC 3115 his primary aim had been to identify dynamically unstable situations that Lindblad had proposed might be associated with spiral structure. And he wanted to understand the distribution of interstellar dust and its properties to guide interpretations in view of interstellar absorption. For all this he needed observations. Oort was not a pure theorist; the fundamentals of stellar dynamics were not defined by him, but rather by Jeans and Eddington. Although not a pure observer either, for his work he was in need of observations—if necessary obtained by others— and his rejection of American job offers had been caused for a significant part by a lack of access to telescopes. His publications in the first years of the War were almost the last ones related to statistical astronomy and astrometry. After 1945 not more than a few papers were concerned with these disciplines, although he corresponded extensively with American astronomers on the progress of surveys and sometimes requested data, such as in the 1950s from Nancy Grace Roman (b. 1925) at Yerkes (she had been a student of Peter van de Kamp at Swarthmore), Milton Humason at Mount Wilson and Gerald Edward Kron (1913–2012) at Lick.

I will not go into detail on the papers Oort published just before and in the early years of WWII, but restrict myself to summing up the questions that were addressed and, briefly, the results. The papers were only a few pages in the 'Bulletin of the Astronomical Institutes of the Netherlands'. With limited contact with the world outside of occupied Netherlands, there was not much hope to obtain new data sets or copies of recent publications from American Observatories after 1940. In Oort (1939a) he used new radial velocities of RR Lyrae stars obtained by Alfred Joy at Mount Wilson to define the rotation velocity of the Galaxy. This came out much lower than for globular clusters (he found 130 km/s for the RR Lyrae stars compared to 274 km/s for the clusters), which was surprising. Globular clusters also contain RR Lyrae stars, which are late stages of stars of about the solar mass and he had to conclude that—contrary to what was then believed—that the RR Lyrae stars in the Halo could not have escaped from these clusters. The space distribution of RR Lyrae stars at high Galactic latitudes as determined by Harlow Shapley could be used together with these same velocities to determine the vertical force field at larger distances from the plane of the Galaxy using the continuity equation. At larger distances the force appeared to stay level with height above the plane of the disk and this could be used to estimate the total mass of the Galaxy and the local rotation velocity. This was done in collaboration with Adrianus Jan Jasper van Woerkom (1915–1991), who was a student at the time. The total mass came out as $1 \times 10^{11}\ M_{\odot}$ and the rotation velocity near the Sun at 240 km/s (Oort and van Woerkom 1941b), not much different from what we use today (Fig. 9.10).

Jules Jacques Marie Adriaan de Kort (1910–1994) had tried at Oort's suggestion to estimate the amount of rotation in the system of globular clusters that was allowed in view of its essentially spherical shape and this was published in a paper in the BAN in 1941 [233]. It turned out that this restriction was not very sensitive and no useful limits were obtained. He also tried to estimate the number of unseen, obscured clusters in the central parts of the Galaxy, but this produced unrealistic numbers of a few thousand. Now Oort (1941c) tried to reconcile the number of clusters in the inner regions with dynamical constraints. Oort found that there could not be a very large central concentration of mass and that the rotation velocity of the Galaxy near the Sun would probably be less than the value of 270 km/s that was used up till then. De Kort was a Jesuit priest, who studied astronomy in Leiden. After the War (in

Fig. 9.10 Oort during the second Netherlands Astronomers Conference in 1942. From the Netherlands Astronomers Club [230]

1949) he would obtain a Ph.D. under the supervision of Oosterhoff. After working for many years at the Vatican Observatory and having taught at the Jesuit-related Georgetown University in Washington, D.C., he came back to the Netherlands to lead the Astronomical Institute at the Catholic University Nijmegen (now Radboud University).

In 1942 five papers appeared consecutively in the BAN. Oort (1942b) used stars at large distances from the plane of the Galaxy to determine the position of the Galactic pole, and thus the orientation of the plane. It was preceded by a longer paper by Johannes Jacobus Maria van Tulder (1917–1992), who also worked as a student for Oort. Van Tulder dropped out of astronomy, but eventually obtained a Ph.D. in the social sciences on a statistical study of job mobility in 1962. He was appointed extraordinary (honorary) Professor on statistical methods in sociology at the University of Leiden in 1966. At Oort's suggestion, van Tulder had used objects concentrated to the Galactic plane, such as planetary nebulae, and the zone of avoidance to determine the position of the pole and from the curvature on the sky the distance of the Sun from the Galactic symmetry plane [234]. The pole was $\sim 1^\circ$ away from previous determinations and the Sun 13.5 ± 1.7 pc from the plane (consistent with the currently accepted value of about 12 pc). This is less than the roughly 30 pc which was found in a number of earlier studies, probably for a large part due to neglect of absorption. Oort's analysis used the star counts in the Selected Areas as given in the Mount Wilson Catalogue (see Sect. 2.2) and essentially confirmed van Tulder's position of the pole.

At the same time Oort published two more papers with van Tulder and one with Oosterhoff. The ones with van Tulder concerned long-period variables. These are cool stars in their late stages of evolution, when they have left the Main Sequence and are pulsating, sometimes semi-regularly, with periods from about a hundred days to a few years. Often amateur astronomers conduct observational monitoring programs for such stars. Since they are cool giants or supergiants, they are situated generally in the Hertzsprung-Russell diagram on the upper-right (see Fig. 2.22). They are moderately old to quite-old stars in a Galactic context. Since they are bright, they can be seen out to relatively large distances.

Oort and van Tulder (1942d) used surveys of long-period variables to determine their space distribution both from the distribution on the sky and from their radial velocities. They found that their space density in the Galaxy falls off rapidly with distance from the center, as fast as the fifth power of that distance. The axes of the velocity distribution or ellipsoid were determined also, as well as the value for the Oort constant A. The latter came out somewhat low, but the velocity distribution was rather asymmetric. In the next paper (Oort and van Tulder 1942e), the latitude distribution of long-period variables was used together with Oort's vertical force K_z to estimate their distances. This resulted in improved absolute magnitudes, which came out brighter than thought previously.

Oort (1942f) addressed the matter of the dynamics of stars with large velocity dispersions, such as the long-period variables. Such stars make larger excursions in distance from the center and therefore are sensitive to differential rotation and density gradients. Oort's analysis showed that for stars with peculiar velocities in excess of 25 km/s the density gradient in the outer Galaxy is less steep than near the Sun.

The Oort and Oosterhoff (1942c) paper concerned some Cepheids (see Sect. 3.1) that seemed to have an unusually large peculiar velocity with respect to Galactic rotation. They came to the conclusion that these stars are probably relatively distant in spite of being rather bright on the sky, because they would be situated behind an area of low absorption, in agreement with low color excesses (interstellar reddening).

In 1943, Oort published three consecutive papers in the BAN, all concerned with the accuracy of fundamental catalogs, which are vital for proper motions studies. He had started writing a textbook on Galactic Structure and Dynamics, which I will discuss in more detail below, and while preparing a chapter on astrometry and proper motions, he felt obliged to consider the accuracies of fundamental catalogues. The first two of these three papers concerned an analysis of the accuracies of the positional systems of the two fundamental star catalogues, FK3 and GC, and the consistency between them. The FK3 or *Dritter Fundamentalkatalog des Berliner Astronomischen Jahrbuchs* [235, 236] had been prepared by the Astronomisches Rechen-Institut at Heidelberg and published in 1937; it contained 873 stars (a supplement had been published in 1938 with an additional 662 stars). The GC (Boss General Catalog) was the *General Catalogue of 33,342 stars for the epoch 1950* by Benjamin Boss, the most recent revision (now in 5 volumes) of the 1910 *Boss Preliminary General Catalogue* [237], prepared by his father Lewis Boss. There were considerable differences between the two catalogues in proper motions. These can be understood in terms of problems with the determination of the motion of the equinox and the precession constant. Oort's conclusions can be summarized by stating that for proper motions in right ascension the GC was to be preferred, while in declination the FK3 was a better source. Modern data obtained in the years 1989–93 with the astrometric satellite Hipparcos actually confirm Oort's preferences [238]. Proper motions are vital for a determination of the Oort constant B, as we have seen before (see Sect. 6.1). Since A can also be determined from proper motions (the amplitude of the cos $2l$ variation with Galactic longitude), Oort's analysis led to a correction to the precession constant and to new values of his constants, namely $A = 18$ km/s.kpc and $B = -8$ km/s.kpc.

9.5 Superluminal Motions in Nova Persei

We have seen that up to the late thirties, Oort's publications in astronomical journals had all concerned astrometry, fundamental systems, structure and dynamics of the Galaxy, and studies of extragalactic nebulae. In his list of published papers, the first clear break with this is the paper by Mayall and Oort (1942a): *Further data bearing on the identification of the Crab Nebula with the supernova of 1054 A.D.; II. The Astronomical Aspects*. In the *AIP-Interview* [4], David DeVorkin asked about his collaboration in this paper and its publication during the War.

> 'The discussion on the Crab Nebula must have started before the invasion of the Netherlands in 1940, and apparently I could get letters through to Mayall in 1941, before the real difficulties at the University started in 1942. My interest in expanding nebulae had arisen before the War from a study of the nebulosity around Nova Persei 1901 and the remarkable discrepancy between on the one hand the distance of about 100 pc derived by Kapteyn in 1902 from the expanding filaments observed around the nova in 1901 and 1902, on the assumption that the observed expansion was that of the flash of light emitted at the outburst, and on the other hand the distance found in 1934 when Humason had measured the expansion velocity of the gaseous shell that since 1916 had become visible. The combination of Humason's velocity with the angular expansion of this shell indicated a distance of 600 pc. The discrepancy with Kapteyn's distance showed that the signal of the flash had moved at an apparent velocity of more than five times the velocity of light. I concluded that this was due to the fact that the filaments considered by Kapteyn were not situated close the the nova as he had assumed, but at some distance in front of it where the 'reflection' of the flash would displace itself with a much higher velocity than that of light. As a consequence I obtained a picture of the three-dimensional distribution of the interstellar matter in the surroundings of the nova, which showed that this matter was concentrated in thin sheets. My study was never quite completed, principally because meanwhile a paper by A. Couders appeared in which he had elaborated very similar ideas. The essence of my own studies is given in an account written by J.J. Raimond in the 1943-volume of Hemel and Dampkring, based on a lecture I gave for the Dutch 'Astronomen Club'. There are two reasons why I mention these investigations. Firstly because they were never published, but particularly because exactly the same explanation has recently been revived for the apparently 'superluminal' velocities observed in nuclear compo-

nents of quasars. Anyhow, this explains my general interest in expanding nebulae, of which the Crab nebula was, of course, the outstanding one.'

The article by Jean Jacques Raimond involved work by Oort, so I have added that to his list of publications as Raimond and Oort (1943g). A. Couders, whom Oort referred to, is misspelled; it is André Couder (1897–1979). Oort must have known Couder well as he was a Vice-President of the IAU between 1952 and 1958, even though that is in between Oort's terms of General Secretary and President. However, the person that should have been referred to is someone else: Paul François Jean Couderc (1899–1981). The confusion probably arose from the transcription.

I provide a (loose) translation of Raimond's paper in Appendix B.4. Here I will present a short summary. For a modern introduction to Nova Persei 1901 and the matter of superluminal expansion see [239]. The early history of Nova Persei and Kapteyn's fundamental work on this can be found in my *JCKbiog* [7].

Nova Persei 1901, now known as GK Persei, appeared in February of 1901 as a star of magnitude 0 before it faded again. Soon a nebula was discovered around the nova, which was subsequently found to expand with a large apparent speed on the sky. This was a problem, as it would mean that the nova had to be very close indeed. Kapteyn had proposed that this was not a nebula expanding, but rather the progression of the light-front from the exploding nova, as it propagated through interstellar space illuminating interstellar clouds of dust that were present around the nova. The expansion was therefore at the speed of light or at least a major fraction of it, if seen in projection. It is this proposal that has resulted in Kapteyn being traditionally credited as the person having provided the fundamental idea behind superluminal expansion, as in [239]. Kapteyn derived on this basis a distance of 90 parsecs (300 lightyears) for Nova Persei.[2]

The problem arose when in 1916 a much smaller expanding nebulosity was discovered around the nova. This was not a light-front, but material actually having been expelled by the star during the outburst. In 1935 a spectrum was obtained, which showed that the expansion velocity was 1200 km/s, which indicated a distance of 2000 lightyears (613 parsec). From this it followed that the larger expanding features Kapteyn studied would not have been moving with the speed of light, but with an apparent velocity of ten times that or so! Oort had found that this could be explained when the assumption is made that the light-front actually moves on a flat semi-transparent sheet some distance (about 30 lightyears) in front of it as seen from us. If this is a uniform, thin, and flat screen perpendicular to the line of sight, the expanding, spherical light front will illuminate a circle on the screen, and if the screen is far enough away from the nova this circular image will expand at a speed exceeding that of light. Although matter cannot move faster than light—as Einstein had shown—nothing prevents an image on a screen from moving that fast. Pieter Oosterhoff, in his first-year lectures on elementary astronomy in Leiden that I attended treated this also and explained this as follows. Imagine a lighthouse and a screen at some distance. The light cone that illuminates the screen moves across this at some speed that grows when we move the screen further away from the light-

[2]The 9 pc in my *JCKbiog* [7] is a misprint and should read 90 pc.

house, and no fundamental physics prevents us from moving it out so far that the image moves faster than the speed of light.

Oort realized before he had been ready to publish these findings that Paul Couderc had independently given the same explanation and had published this in 1939 [240]. It is a rule in science that who publishes first receives all the credit and that does not change by Oort having conceived the idea independently, especially since he chose not to publish himself. Couderc is given the credit for this explanation, and rightly so.

9.6 Crab Nebula Supernova

The historical developments concerning the Crab Nebula and Oort's contributions have been described by Lodewijk Woltjer in the *LibAm80* [2]. I start with the continuation of the remarks Oort made in the *AIP-Interview* [4].

> 'I had undoubtedly discussed this with Baade during one of my visits to Pasadena, and must have been impressed by his beautiful plates of the nebula. Duyvendak, who was Professor of Chinese language and literature at Leiden, was an intimate friend of mine and was only too ready to be involved in the subject. His careful scrutiny in Chinese and Japanese sources contributed in an important way to establishing with certainty the supernova character of the object observed in 1054. The correspondence with Mayall arose from his determination of better radial velocities, and from my wish to get better data on the expansion. I interested this Chinese friend of mine in this subject. He probably looked again at it after the War after first things calmed down.'

The Crab Nebula (M1 or NGC 2952; see Fig. 9.11), situated in the constellation Taurus almost 20° north of Orion's Belt, consists of an amorphous body surrounded by wisps. It was early on found to expand, first in size on the sky and later from spectroscopic velocities. Knut Lundmark had examined Chinese chronicles and noted that there had been a 'guest star', that had appeared in 1054, not far from the Crab on the sky. John Charles Duncan (1882–1967) had measured the expansion on the sky and concluded that the age had to be of order 900 years, which made Hubble consider the identification to be secure. Early spectroscopy (first by Slipher in 1916) had shown double emission lines, which later became understood to be also due to the expansion, as particularly indicated by new spectra by Nicholas Ulrich Mayall (1906–1993) at Lick Observatory. Comparison of the expansion velocity from the spectra with the expansion on the sky led to a distance of about 1.5 kpc. The current value is still not very accurate but it may be somewhat larger (2 kpc or so). Little was known about the guest star though.

Oort had seen images of the Crab Nebula by Walter Baade, when he visited Pasadena and Mount Wilson in 1939 after the McDonald dedication. The two had started a longterm friendship and collaboration, during which Baade wrote over sixty, usually long or very long, hand-written letters to Oort, who wrote often much

Fig. 9.11 The Crab Nebula as observed with the Hubble Space Telescope [242]

shorter, typed letters to Baade. Oort happened to have a very good friend in Jan Julius Lodewijk Duyvendak, who had been appointed reader in 1919 and Professor of Chinese, Japanese and Korean in 1930 (he had obtained his Doctor's degree in 1928). Duyvendak made a thorough study of the Chinese records and provided detailed translations, which led to a much better understanding of the object. This he published in 1942 [241]. Mayall and Oort (1942a) used these translations to make an improved determination of the maximum brightness and the light curve of the star, which was established as that of a supernova with a maximum absolute magnitude of −16.5. The improvement came partly from making comparisons to light curves of actual supernovae in other galaxies. The earliest such studies of extragalactic supernovae by Baade and Zwicky had indicated that at maximum the absolute magnitudes were of order −14 to −16, which made the one that had produced the Crab Nebula the brightest one known up till then (Fig. 9.12).

Most of this must have been completed by correspondence between Oort and Mayall until it became impossible to communicate. In the AIP interview above Oort had indicated that this was still possible until 1941, after which Mayall must have arranged the publication for the two papers by Duyvendak and by himself and Oort. The last letter during the War in the Archives is from Oort to Mayall (from January 16, 1941), in which Oort mentioned that Duyvendak would write up a note on the sources and their translations and suggested that this should be published in the Proceedings of the Astronomical Society of the Pacific. He and Mayall would write a brief note to precede this. The end result was that there were two papers, the one by Duyvendak first, but indeed in the journal suggested (Fig. 9.13).

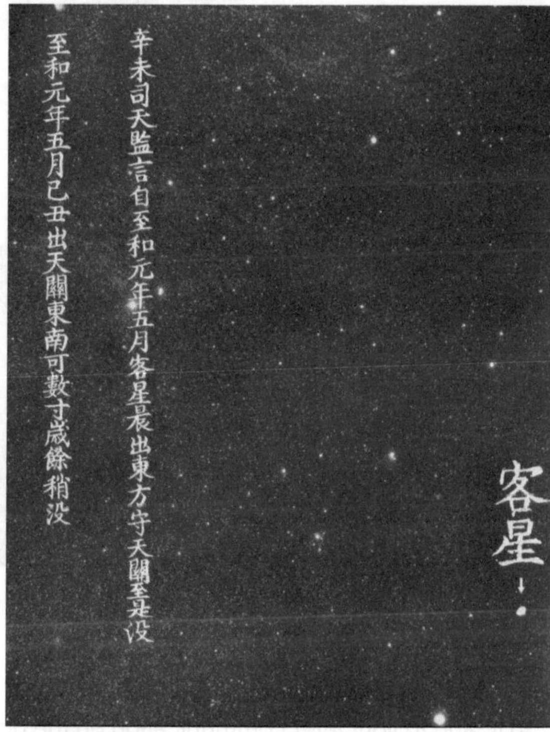

Fig. 9.12 The Crab Nebula is in the lower-right corner near the star ζ Tauri. The text at the arrow reads: 'guest-star'. The main text is (without detailed date descriptions): left column: 'On [July 4, 1054] a guest-star appeared several inches south-east of [ζ Tauri]. After more than a year it gradually became invisible.'; right column: 'On [April 17, 1056] the chief of the Astronomical Office reported that after [June 9, 1054] a guest-star had appeared in the morning in the eastern heavens, remaining in [ζ Tauri], which only now had become invisible.' Translation by Duyvendak. From *Honderd jaar Leidse Sterrewacht* [243]

The correspondence between Mayall and Oort was picked up immediately after the end of the War. On August 30, 1945, Oort wrote:

> 'It is a very good feeling to be able again to write to you. I was so sorry our correspondence was interrupted.
>
> Recently I have seen the paper by you and myself on the Crab Nebula. You have done a very good job, of which my share has been much less than might be inferred from the title! I have also shown Duyvendak his article; he says it was printed very well, there being only a few mistakes of relatively little importance.'

The letter mostly concerned an interest of Oort that he had developed in the mean time (the direction of rotation of spiral nebulae). The letter ends as follows.

> 'What have you been doing during the last years? Have you been able to do any astronomical work? I was surprised when I learned from Kuiper of the extent to which American scientists and university teachers had been drawn from their institutions. As to myself I have been able to carry on fairly actively during by far the greater part of the War. There was not much I could usefully do to help in the War.'

Mayall replied on November 3.

> 'Your welcome airmail letter of August 30 reached me in Pasadena shortly before I moved back to Mount Hamilton, and my delay in replying

Fig. 9.13 Jan Julius Lodewijk Duyvendak and Nicholas Ulrich Mayall. From Wikicommons [244] and NOAO/AURA Image Library [245]

to it, and to your later communication on the direction of rotation of M33, is due to the business of moving and trying to pick up the threads of astronomical activities suddenly dropped nearly $3\frac{1}{2}$ years ago. During all that interval, I was unable to do any astronomical work, and I had little opportunity to keep in touch with contemporary celestial progress. For one year I was in Cambridge, Mass., where I worked at the MIT Radiation Laboratory on radar navigation problems. Then I returned to California and worked for 6 months at the Mount Wilson Observatory on problems in aerial photography and fixed gunnery for fighter aircraft. Finally I joined the California Institute of Technology War project and for almost 2 years worked first on rockets and then on atomic bomb developments. Although there was a large number of astronomers at CIT [CalTech], they were all so busy on War work that they seldom had a chance to get together for astronomical discussions. In fact the degree to which American astronomers participated in the War effort is a very bright spot in an otherwise dark period. The experience was for me a very rich one, and I return to astronomy with a much better idea of science progress in this country than I could have received in any other manner.'

Mayall also wrote that he had just sent 60 reprints of the the Crab Nebula papers, to share with Duyvendak.

Before returning to Oort's astronomical work during the final years of the War and immediately after it I will first describe the situation of his almost three years of exile to a place in the middle of the Netherlands, hiding from the Germans to avoid being taken hostage.

9.7 Potbrummel, Hulshorst

When in May 1942 the Germans had arrested a large number of prominent citizens as hostages, and when subsequently more of Oort's friends were included, he obviously became worried. All of the fifteen persons that might have formed the Kleine Krans (i.e. the participants of the 1941 Dies dinner, see Sect. 9.1) had resigned from their professorships, most on or near June 1, 1942, or a few weeks later [246]. Of the ones I mentioned, Escher, Duyvendak and Kollewijn were taken hostage; Huizinga eventually was also taken hostage in August, but released for reasons of health in October, and Casimir had moved to the Philips Physics Laboratory in Eindhoven; Philips was protected because the Germans considered it as strategically important. The Royal Philips Company had started as a producer of light bulbs in 1891, but had evolved into a major international technology company dealing in lightning, consumer electronics, domestic appliances, healthcare and other areas. The Physics Laboratory, known as the Philips NatLab, was one of the largest (industrial) research establishments worldwide, concentrating much on fundamental rather than product-related research in areas ranging from electronics, physics and chemistry to more recently computer and information technology. It is less prominent now after it has been mostly disbanded in 2001. Casimir headed the Philips NatLab between 1946 and 1972.

Oort arranged through friends and acquaintances that he could use a holiday home in Hulshorst near Nunspeet in the rural part of the country (see Fig. 1.1). The house was owned by Henri Bienfait (1900–1990), a chemist that had studied and obtained his Ph.D. at the University of Amsterdam. He worked almost all his career at the Philips NatLab, eventually becoming Adjunct-Director. Bienfait was married to Bertha Elisabeth van Osselen (1902–1977), and the home was located on the property belonging to her parents, who lived in the main house. These were Jan Rudolph van Osselen (1869–1948) and Johanna Henriëtte Spakler (1872–1955). Van Osselen had served first in the Navy, but had become a technical engineer and later still Director of a mortgage bank before retirement. The street address was Potbrummel, and the home is usually referred to by that also. 'Brummel' is dialect for the Dutch 'braambes' or blackberry (cf. the English 'bramble') (Fig. 9.14).

In Oort's pocket diaries Hulshorst is first mentioned on Monday July 20, 1942, when apparently he traveled by train to Apeldoorn. There are reminders to inquire about the voltage used ('for radio, toaster and possibly vacuum cleaner') and return times the next day via Utrecht. Preparations continued, such as a visit to Rotterdam on July 31, when he walked through the 'Maastunnel', the tunnel under the 'Nieuwe Maas' that connects the northern and southern parts of Rotterdam.

Fig. 9.14 The 'Potbrummel' at Hulshorst near Nunspeet, the house where the Oorts lived through the second half of World War II, and surrounding land. Photograph provided by M.J.A. (Marc) Oort

Construction of this oldest such structure in the Netherlands had been started in 1937, continued after the War had begun and had just been opened in February 1942. The reason for the visit to Rotterdam becomes clear from an address that has been added and correspondence in the Oort Archives. The President of the supervisory Board of Directors of the 'Nationale Levensverzekeringsmaatschappij' (a large life-insurance company) had arranged, at the request of Oort's father-in-law Willem Graadt van Roggen for an introduction of Oort to the Director. The appointment was on July 31 and another meeting took place on August 20. It resulted in a letter of August 25, in which Oort was informed that this company would make use of his mathematical services in case the need would arise. After all, Oort had resigned and was in need of income to support his family. Apparently Oort had already done some work, since another letter of the same date awarded him Dfl. 40 for his services. For this Oort used an address in Oegstgeest, presumably that of his parents.

But before this they took a vacation; on July 30 Oort traveled to Wageningen where his brother John lived. He had moved to the Wageningen Agricultural University, where he had worked in the Laboratory for Plant Physiology and had obtained a Ph.D. in 1930. Since 1933 he had worked at the Laboratory for Mycology and Potato Research (he would become a professor in 1949). In 1929 he had married Titia Margaretha Martha (Pivy) van Mesdag (1908–1991). Oort and his brother were rather close, their families often spending vacations or public holidays together. The next day Oort, John, Coen and John's son Arend went on a sailing trip to the west on the Rhine, staying overnight to the south of Utrecht, where the next day they went north following the river Vecht until a village called Breukelen, somewhat to the north of Utrecht. There they were met by Pivy Oort and they stayed overnight in nearby Tienhoven on the borders of the lake area Loosdrechtse Plassen, sleeping on straw in a barn that the (John) Oorts had rented. The next day they traveled to Zeist, just to the east of Utrecht, where Mrs. Oort, Marijke and Bram were already

Fig. 9.15 A closer view of the 'Potbrummel' at Hulshorst near Nunspeet. Photograph provided by M.J.A. (Marc) Oort.

staying. There they stayed for a week or so, and moved for a few days to John and Pivy Oort's house in Tienhoven before returning to Leiden (Fig. 9.15).

The next months were full of preparations, Oort making a few more visits to the Potbrummel, until in early November when they were ready to move as a family. From Oort's pocket diaries (the children were now almost 14, 11 and 8 years of age):

> 'Tuesday November 3. Mieke and I pay many farewell visits.
>
> Wednesday November 4. Departure for Hulshorst. Mieke, Coen and Bram by taxi, Marijke and I on the tandem bike to the railway station for the 9 o'clock train. Forgot the suitcase of Marijke and Bram! Still have 8 pieces of luggage which with difficulty, and the help of two gentlemen, we get into the train at Woerden and out of it in Utrecht. Marijke and Bram stay behind in Utrecht [must be with the Graadt van Roggens] and Coen comes along with us. Much delay as a result of a derailing in Amersfoort and bombs near Stroe; we arrive at Hulshorst, 5 hours late, in the dark. Fortunately we can eat and sleep with Lies Tinbergen. Thursday morning we go in cold weather to the Potbrummel, that has been made most pleasant with flowers and fruits by Mr. and Mrs. van Osselen. We feel at home immediately. Coen and I spend quite some time transporting luggage and blankets. Miss van Hasselt introduced Mieke to the house. In the afternoon, Mieke and I go to Nunspeet with Mrs. van Loon to consult about a school for Marijke and we do some shopping.'

Stroe is somewhat off the direct route; they might have traveled through Barneveld (Kapteyn's birthplace) from where the natural bus route would lead through Stroe.

Nikolaas (Niko) Tinbergen (1907–1988), the later Nobel Prize winner in physiology or medicine, was a biologist, in 1940 appointed reader in Leiden. He had grown up in Den Haag, where his father was a HBS teacher of Dutch, but his parents rented a cottage every summer in Hulshorst for the family of five children, which included Jan Tinbergen (1903–1994), who was to study mathematics and physics and also win a Nobel Prize (in 1969 for economics in this case). Niko Tinbergen had come back to Hulshorst every summer since 1934 and organized summer camps there for his biology students. He had become fascinated by the behavior of the beewolf, an bee-killer wasp that lived in a nearby sand-drift. Tinbergen had been taken hostage to Sint-Michielsgestel on September 9, 1942. His wife Elisabeth (Lies) Rutten (1912–1990) (and their son Jaap) were living near Hulshorst, close to Vierhouten (somewhat less than 10 km away). Niko Tinbergen was released on September 3, 1944.

> 'Friday morning cycling in the surroundings with Lies Tinbergen, who introduces us to some of her acquaintances among the farmers. All day rain and mud! In the evening visit the van Osselens.
>
> Dreesje Braambos comes to help us in the morning. On Saturday the luggage arrives in a sealed wagon. Mighielsen calls us and transporter De Zwaan drives it all to the Potbrummel in two cars, where all of it is in the house by noon. Coen does not feel well and stays in bed all day. I lug crates and hammer them open. We had no less than 19 crates and suitcases +10 bags of anthracite and the piano.
>
> Tuesday November 10. Father and mother G.v.R. arrive with the children. Coen for the first time to the Lyceum, from which he returned very depressed.
>
> Wednesday November 11. At nine we bring Bram to the Christian school of Mrs. Zandbergen, where he goes to grade four. Grades 4 and 5 are taught jointly. He likes it and immediately has friends and does well with arithmetic, but still is given homework from the schoolbook to catch up. Nobody else has homework at this school, except for a child in kindergarten. Jaapje Tinbergen is in third grade and comes regularly to drink coffee. Mieke and I with Marijke visit our very nice farmer-neighbors, Hazen, van Klomp, who is feeding a baby, and Mol with a fireplace and a black kettle. No place is more beautiful and tidy as the Bonenstro's at the Groot Mulards Erf. After visiting Mrs. Kruijthof with Marijke we decide to send her to the village school as well, where she can start on Thursday in grade 7, although she is 3 arithmetic schoolbooks behind. She will have individual lessons. She also likes it the first day.
>
> The autumn mist in the forests late in the afternoons these days is beautiful. On Thursday I will start seriously with my work.'

A lyceum is a secondary school with both gymnasium and HBS streams. Jaap Tinbergen (1934–2010), son of Niko and Lies Tinbergen, eventually came to Leiden to work in astronomy. He emigrated with his parents to Oxford in 1949, obtained a B.A. in physics at Cambridge, worked on ionospheric studies at the Antarctics (where he spent an uninterrupted three summers and two winters!), and an M.Sc. in information engineering at Birmingham, before returning in 1960 to the Netherlands. At Leiden

Fig. 9.16 The Oort family in front of the Potbrummel. The picture is dated winter 1942/3. Photograph provided by M.J.A. (Marc) Oort

Observatory, he specialized in instrument building, in particular polarimetry and obtained a Ph.D. on precison spectrapolatimetry with Fjeda Walraven as supervisor in 1972 (Fig. 9.16).

Already on Tuesday November 17, Oort returned to Leiden and got back 'heavily packed' to Hulshorst on Thursday.

It took a bit longer for Oort to resign from his functions in Leiden. There is a notification in the Archives, dated November 3, 1942, that Oort had been awarded 10 months leave of absence from his duties as Adjunct-Director with his salary continuing in order to complete a textbook on the dynamics of our Galaxy. On November 10, he followed this up with a request in relation to his change of domicile to be relieved of the deduction on his salary for the occupancy of the house at the Observatory, which was granted on December 12 by the Secretary General of the Ministry of Education, Science and Culture-protection (probably a German invention). And dated January 28, 1943, there are two letters from the same authority in which honorable discharge had been granted as Extraordinary Professor of astronomy and a new appointment was made as 'beheerend'—equivalent to something like interim—Adjunct-Director of the Observatory with an annual salary of Dfl. 5040 (about €25,000 nowadays). Oort was warned though, that the character of this appointment was by definition temporary, but no date of termination was given. In the *JKM-Inventory* [1], Jet Katgert-Merkelijn wrote:

'In 1942 he resigned from the Royal Academy (to which he had been elected in 1937), from his post as a Professor Extraordinary at Leiden University, and from his adjunct-directorship of the Observatory. This last function caused the most trouble; he was summoned to an interview and told that this resignation could not be accepted as he had not been given permission to resign by his superior, E. Hertzsprung.'

This 'interim' appointment apparently solved the issue of an income. According to Leiden University [246], Oort resigned 'at his request' as Extraordinary Professor on January 28, 1945, but that is incorrect and must be 1943, and was not the end of that arrangement.

Hertzsprung remained in Leiden and formally acted as director, even after he turned seventy in 1943 and would have retired. Being Danish, he kept a low profile, avoiding to do anything that might upset the Germans or attract attention. He led a somewhat secluded life, probably concentrating on his scientific work, since there were no students and no courses to teach. Little has been recorded about his life in those years; see the few remarks by Herrmann [105] and Baneke [90].

Life at the Potbrummel was not luxurious. There was no heating, no electricity or running water. There was a stove in which wood could be burned for heating and for cooking meals. They had a stationary bicycle whose dynamo could provide some electricity for illumination at night and for running a small electric motor that would fill a reservoir with water for consumption and washing. Bram Oort has put as my disposal texts of some speeches at funeral ceremonies. I quote from the one made by Marijke Oort on November 29, 2006 on the occasion of the passing away of her brother Coen.

'[Coen] was a restless and impatient child that made a lot of noise. Our gentle father, who appreciated peace and quiet, sometimes found this difficult. Wherever Coen was, there was excitement and fun. Mother and I enjoyed this immensely. Mother told us that when she picked up Coen from school she could always locate him by looking for a group of romping and fighting boys.

My clearest memories go back to the war years, during which Coen and I were very close. We had to go from Hulshorst to Harderwijk, where our lyceum was located, on a very heavy tandem bike, which took almost an hour. Whoever had to prepare for a test paper was allowed to sit in the back, with the schoolbook attached to the clamps for hockey-sticks to perhaps pick up just a few more things. When this was my turn (we studied by speaking out loud), Coen every now and then turned around his head and shouted 'ridiculous', meaning I had to learn the most useless things. He also regularly shouted from the front: 'You're not working the pedals!'

Often, when English fighter aircraft were attacking the road, we had to hide together with the tandem in the ditch. After a year our parents kept us at home, because they felt it was too dangerous. We then sat very cozily across from each other at the table, making homework near the wood stove. In the evening we had to take turns riding the bicycle to provide light for the things the members of the family were doing. Everyone who was available to give us lessons was summoned to keep our schooling going as best they could. Coen was told to teach me French, but that did not work out of course.'

Especially the last year was very difficult, when food was scarce. In the mean time Duyvendak had also fled from Leiden and gone into hiding in nearby Ermelo (some 15 km by road) [247]. Although he had been arrested in March 1943, he was released in August. Duyvendak, Tinbergen (after his release in September 1944) and Oort saw each other regularly. To this was added, also in September 1944, Oort's brother John and his family. Their house in Wageningen had been damaged during the Battle of Arnhem in September 1944 and was uninhabitable. The younger Oort family moved into the Potbrummel and in early October Oort and his family moved to a part of the main house (Groeneveld) on the premises of the van Osselen couple.

I also reproduce some recollections written in English by the younger son Bram. I feel no need to correct the small amount of overlap (Fig. 9.17).

'My parents and their three children then moved to Hulshorst, a small village in the rural, central area of the country. They rented a summer cottage in Hulshorst, where we spent the rest of the War. Nobody in our area had electricity so we used a stationary bike to pump up water from a well and to provide light at night. My mother cooked simple meals on the wood stove; I remember that at crucial moments my older brother stopped to paddle, the milk overflowed resulting in a nauseating smell of burning milk. Several of my fathers colleagues had also moved to our area. Together they organized seminars to keep their spirits up. I remember sitting on the back of my fathers bike while cruising over uneven, sandy bike paths through the pine woods. Usually there were no German soldiers on these paths. We often returned in the dark although officially one was not allowed to go out at night. New tires were not available – some people rode on noisy solid tires. We often had to stop to repair a punctured tire; most of the tires were covered with patches. At times visitors from the western, more populated side of the country passed by, such as students and Jewish friends who were in hiding. All men above 18 years had to register for the German army.

It was easier here than in the big cities to obtain food from neighboring farmers, even during the hunger winter of 1945. I remember the constant stream of people from the big cities in the west, walking eastward along the main road with small carts to search for food. The German Army also used the same road mostly at night to avoid being attacked by English fighters. One distinct image I remember was the all day long, constant roar of English and American bombers flying at high altitude over Holland in the direction of Germany. They released little strips of aluminum to deflect the German radar – a constant stream of silver strips coming down from the sky. I loved to collect these strips. Finally in early May 1945, Canadian and American armored trucks entered Hulshorst along

Fig. 9.17 Another picture of the Potbrummel, with Mrs. Oort, Marijke and Bram on the porch. The picture is dated September 1943. Photograph provided by M.J.A. (Marc) Oort

the main road; first covered but later open. It was such a relief to see and welcome these soldiers – they shared goodies I had never tasted – like chewing gum and chocolate. A few months later in the summer of 1945, our family moved back to Leiden.

Often life was not bad at all in Hulshorst. In his diary notebooks, Oort wrote for example:

March 25, 1943. It is strange that I feel so happy at the Potbrummel in spite of all misery. It must be the harmony of the simple and peaceful existence, the absence of the confusion from relations with other people.

April 20. It is spring! Outside the singing of the birds and the buzzing of the bumble bees, a great joy it seems in animal world outside the window of Marijke's room where I sit and work.'

But in 1944 things got progressively worse, especially the availability of food in the winter of 1944/5 gave rise to widespread famine and many deaths from starvation. The western provinces ('Holland') were hit hardest. There are quite some texts in Oort's diary notebooks and pocket diaries. I have chosen a few parts of these texts to give an impresson of what it had been like in the war years in Hulshorst. Some notes in the private Archives contain the following:

'Since September '44 there are no trains. Many travel by bike, others on foot because of the bicycle claims ['vorderingen' or requisition of bicycles by the Germans]. We no longer have any contact with Holland. [...] The 'Grote Trek' [like the Great Trek of the Boers in South Africa]. [...] So many children on the road. Some die en route. Like grasshoppers. We also go around the farmers begging for food.'

I quote a few longer passages from Oort's private diary notebook.

Monday afternoon 13 November [1944]. At Groeneveld, where we have been living since 3 October, while Jo and Pivy are in the Potbrummel, and where we fortunately feel a bit at home also. [...]. We are being indulged with delicious apples and pears. Groeneveld has far too many and nothing can be transported to Holland.

26 December. Boxing Day 1944. [...] Then there are the worries about food. In the cities in Holland and Utrecht the bread ration now is 1 kg [per person per week], so that there is much hunger. We still have 1400 g and 3 kg potatoes, but the last few weeks no more butter.

In the mean time worries keep growing. What will happen to the population in Holland? There is no prospect of speedy improvement soon. Trains are not running as a result of the strike in September. There are few cars except German ones, ships cannot move in the rivers because of freezing.

It is sad to see so many people are traveling haphazardly ['op de bonnefooi'], often with children hoping to find something to eat on the way. For example, I saw a man with a girl of not even 14 stumbling behind a pushcart; they came from Meppel where they had walked to from Utrecht to find some potatoes. The girl was exhausted and had only rags on her feet. You see women with their feet wrapped in cloth and also children by themselves! One night we had two children at Groeneveld, 14 and 10 years old, that had walked for 5 days from Hilversum with a small cart and these 2 children had swapped the 2 dogs they had for 10 kg of potatoes.

Mieke and I went begging for food for one day in Oldenbroek, Marijke and I the next day in Doornspijk and Oldenbroek, but we came home with flat tires, that cannot be repaired anymore at the cycle shop because of lack of the special glue and time. [...] Another day Coen and I cycled 45 km to Genne on the river Overijselsche Vecht [beyond Zwolle] to get potatoes. On the way back we have a ruptured tire. We sleep at a farmhouse, as all 'trekkers' do if possible. Many have no sleeping place and are put up in schools. But then they do not get any food. The attitude of the farmers is generally admirable. They let the townspeople into their homes and feed them. In case they do sell they hardly ask more than the official government price for their rye and do not want to take advantage of their countrymen.

29 December '44. Yesterday I went to the road and again saw the procession of people going north. Many found a place on a truck, which they climbed onto together with their bikes. One sees automobiles full of people wrapped in blankets and children blue with cold. Not only Dutch cars, but also Germans are prepared to take people along for a while. Some have a pushcart or a baby carriage tied to a horse-drawn carriage.

28 Jan [1945]. The situation gets more and more precarious. In addition to the hunger that forces people onto the roads (as a result of the frost

there the 1 kg of potatoes per week now has also been canceled some weeks ago in most cities; even here the last 14 days no potato vouchers have arrived and no potatoes are available. The central kitchen [providing food for people that have no other means to be fed] provides only some rather thin soup, mixed with some carrot or beets). In the last month an increasingly sharper cold has arrived and in addition snow that makes travel extremely difficult. In spite of that the 'trek' has increased enormously. Two independent counts found that 15000 and 16000 people pass by us each day in both directions. The road is completely full so that you have trouble all the time overtaking others and sometimes in villages the road is completely blocked. The sight of returning people, of which half are on foot, are a dismal sight. Often they fall on the slippery roads and can only get back up with the help of others who also help to pick up the bicycle.

It is remarkable that so few die. Most look terribly pale. We were told of two cases of deaths the last few weeks, one a woman of 45 who suddenly dropped dead and a child of 14 who was on the road with his brothers of 17 and 16, who have taken his dead body back home.

Although real hunger has not yet arrived, most of the day my thoughts are occupied with food and the possibilities to collect more for the difficult months ahead. This problem is also the subject of most conversations.

10 Febr. '45. The bread ration in Holland and Utrecht is down to 500 g (at least during the frost). Now back to 800 g per person per week. The shortages are hopeless, although they are sometimes supplemented by sugar beets and also tulip bulbs that are eaten instead of potatoes. They are not too bad, remind one of chestnuts and have more nutrients than potatoes. A few days ago, when I was in Leiden for two days, I had all sorts of dishes with these, but they are rather expensive and hardly available any more.

24 Febr. '45. Made another tour north with two aims: firstly to bring 15 mud of potatoes [a 'mud' is a volume unit of 100 liters, which with potatoes amounts to some 70 kg] that Jo had ordered in Dedemesvaart for our families and that of v.d.Brink to Hattemerbroek and secondly to try myself to buy in Dedemesvaart 25 mud for the personnel of the Observatory and to have it transported there.

The transport of the potatoes [...] is much more difficult than expected. It is impossible to lug more than 5 mud up the hill to get it to the control road near the bridge, and that only with local help. Also the German soldiers relief of their conscience, helping us pushing when necessary. The older sentry on the bridge pushes every cart. Everyone helps others and that is one of the bright spots in this otherwise heavy work that directs all thoughts to a single subject, [...]'

For this expedition they had to cross the river IJssel near Zwolle. Dedemsvaart, where the patatoes came from is some 25 km north-east of Zwolle and Hattumerbroek is near Zwolle on the south-western side. Apparently they could store the potatoes in the latter place with a farmer—this would still be about 30 km from Hulshorst.

> 'This morning English or American planes came over the road and shot 3 'trekkers' dead and wounded a few others. [...] This strongly indicates a serious lack of responsibility of these, presumably young pilots.
>
> 3 March. In the last days of February I and Jan Duyvendak have been on expedition to Zwolle to bring potatoes for them and the Tinbergens across the bridge. Strange experience: the quiet calm with which Duyvendak approaches this. [...] I am nervous about the approaching closure of the IJssel and the apparent impossibility to get 20 mud across the bridge. [...] Jan D. recites poems in his mind when making long trips.'

The outcome of all of this is not mentioned.

> '5 March. In a few days the bustle that was worse than before on the roads has disappeared now that the IJssel has been closed on March 1. Only a lost cyclist or pushcart passes, where only recently there was one uninterupted procession. One of the strange things in these times is that money appears to have lost so much of its importance.'

Finally, from loose notes about the 'adventures of Leiden astronomers' in the private Archives:

> 'In the house where the assistants lived during the last 2 years [the illegal newspaper] 'Trouw' was printed. After 'Mad Tuesday' this publication was printed in ±2500 copies at the Observatory for Leiden and surroundings (once per 14 days) and distributed from there, in which Steinmetz helped. When the Allies had entered the south of the country he wanted to join them but was captured in Antwerp. Hins played an important role as reserve officer and had been hiding weapons in his own cellar.'

On Tuesday September 5, 1944 ('Dolle Dinsdag' or Mad Tuesday) rumours spread that the Allied forces were about to liberate the Netherlands and people started (prematurely, as it turned out) to celebrate in the streets. Carl Hermann Dino Steinmetz (±1920–<1999) was an astronomy student in Leiden; he left Leiden and astronomy at the end of 1947. In 1958 he informed Oort that he now was a Professor in physics in Medan (Sumatra, Indonesia) about to be tenured, but a year later that he would have to leave Indonesia, because of the Netherlands—New Guinea conflict and asked Oort to be available to write a reference letter for a job with UNESCO. There is no further correspondence.

Hertzsprung was kept completely unaware of all these activities. Although it seems inlikely that he was not aware of Oort visiting Leiden every now and then, the fact that the latter secretely taught courses in the dark room of the Photographic Telescope was kept from Hertzsprung.

9.8 Interstellar Medium

During most of his time at the Potbrummel, Oort was preoccupied with problems of the interstellar medium. According to the terms of leave of absence he was supposed to work on a textbook. He had seriously been planning to write such a book. In the Archives there is a hand-written, undated outline. Elsewhere there are also some remarks and suggestions on this in letters in 1941 by Kuiper and by van Rhijn. The hand-written outline has eight chapters. The provisional title of the book was *The Galactic System*. Of the first two chapters, 'General Introduction' and 'Data of Observation' there are detailed drafts. Further chapters on the nearest stars, Galactic clusters, stellar motions, distance criteria, interstellar material, the structure of the system and the dynamics have nothing more than titles. There were further attempts that I will return to below.

What occupied Oort in the war years were problems of interstellar material, eventually to become the subject of his George Darwin lecture (Oort, 1946e) for the Royal Astronomical Society. Oort was particularly interested in the processes that were responsible for the formation of molecules and dust particles in the interstellar medium. But first there was interstellar turbulence, on which there is correspondence at various places. The best person with whom to discuss this was his old friend Johannes Martinus (Jan) Burgers, who at that time was a Professor in Delft in the laboratory for Aero- and Hydrodynamics. This did not lead to anything concrete by way of results; with a letter of March 2, 1943, Burgers sent an unfinished manuscript that cuts off with an equation with which nothing is done. Burgers admits that indeed he had not come to any clear result. In the same letter he expressed his disagreement with Oort's withdrawal from the Koninklijke Nederlandse Akademie van Wetenschappen in protest of the German influence on it.

On the subject of the formation of molecules and 'smoke particles' in interstellar space, Oort had an excellent colleague in Hendrik Anthony (Hans) Kramers (1894–1952), Professor of physics in Leiden. Kramers worked closely with his student Dirk ter Haar (1919–2002), who was to write a thesis on the formation of the Solar System with Kramers as supervisor and defended it in Leiden in 1948. Ter Haar eventually became a fellow of Magdalen College in Oxford and a British citizen. Actually, it had started earlier with a prize competition that had been announced by the 'Rector and Senate' of Leiden University (as was customarily done every few years). In fact it comprised a set of competitions, all with a deadline of May 1, 1942. One concerned the question 'whether and to what extent during a time of $\pm 10^9$ years, during which roughly the Stellar System presumably has been in existence, solid particles could have formed in the interstellar gas clouds. Also a discussion is requested about the expected composition and the distribution of sizes of these particles.' It is likely that this text, and the idea behind it, actually came from Oort

Three essays were received and these were judged (without knowing the names of the authors) by Kramers and Oort. No prize was awarded, but two essays received an honorary mention. One of these turned out to be from ter Haar. His essay first

Fig. 9.18 Left: Hendrik Anthony (Hans) Kramers around 1928. Right: Henk van de Hulst as a young man. From Wikimedia Commons [250] and Archives Leiden Observatory

discussed an extended piece of research that he published as an astronomical paper *On the origin of smoke particles in the interstellar gas*, as Kramers had urged him (and reported to Oort in Hulshorst on a postcard in May 1943). Ter Haar's paper was published with identical text in both the Dutch journal BAN and an American one [248]. The latter had been arranged by Bart Bok. In addition Kramers and ter Haar wrote a paper on *Condensation in interstellar space* in the BAN [249]. They followed ideas by Lindblad, who had speculated that at low temperatures (when speeds are low) atoms can stick to condensation centers. Ter Haar and Kramers looked into the possibilities of the formation of molecules in space that would become condensation centers in a crystalline form. The molecules they investigated were CH and CN; carbon, hydrogen and nitrogen were known to be abundant in the Universe (Fig. 9.18).

Henk van de Hulst was the other contestant to get an honorary mention. I quote from an interview Jet Katgert-Merkelijn, the *JKHulst-interview*, had with him. She asked when he first met Oort:

'I think at the first Astronomers Conference at the Maarten Maartenshuis in '41. Minnaert had organized that and there was a good set of staff and students from first-year students on. That was fun and many gave presentations and so did Oort and I think this was the first time I met him. I had just done my Candidaats [bachelors] while on leave from military service. I had been drafted before the mobilization, just before my Candidaats and according to the rules I could ask for a leave to do the exam. So I came with a very recent Candidaats, but with a year of experience in the army, to the the observatory in Utrecht, working on line profiles in the new Solar Atlas. [...] I worked for about a year when there was this announcement of a prize competition at Leiden University. [...] Oort was in hiding [this was not the case when the prize was announced!] so this was a strange situation. That announcement was posted together with a closing date and I had seen it a couple of times in passing and then Minnaert said; 'Isn't that something for you?' And I started to look into it and – if there was a possibility to work on this at all – I had to find by myself literature in the chemistry and physics laboratories and the Observatory. [...] So, I sent in my essay and there were three contestants, none of which received a prize. I had an honorary mention I think. Kramers felt I had been too restrained in my discussion on the formation of the initial cores, but I had studied meteorological literature and that is full of horror stories about how complicated it is and how some incidental causes can lead to either the formation of cores or not at all, from which then water drops form. I was much impressed, so I had said let us just assume this happens one way or the other and Kramers disagreed. Looking back I think he did have a point. Well, and then Minneart arranged for me with support from the Kapteyn Fund that I should go for a few months to Leiden. That was a bit dangerous; you could always be arrested on the road. So in Leiden I slept in the bed that belonged to Wesselink who had gone to South Africa before the War and had not been able to return. [...]

My essay in the prize competition was only concerning the growth and composition of the particles as a result of the gradual coalescence, while ter Haar's story was in fact restricted to the formation of the cores. In that sense we were entirely complementary. A few years later, Oort and I worked some more on the growth and published two articles about it. But I never worked with ter Haar.'

In the interview van de Hulst was asked about Oort and his visits from Hulshorst to Leiden and how often he came.

'Not very often really. [...] It cannot have been more than four months or so and in that time he might have come to Leiden three to four times. Hertzsprung was kept in the dark about this, as he was a bit of a fearful person and in addition he was a Danish national and he was never sure someone at the Observatory would give away information about him. The lectures were given in the dark room under the astrograph telescope.'

Van de Hulst went back to Utrecht after a few months, but he and Oort stayed in close contact. In his contribution to the *LibAm80* [2], van de Hulst described how one day Oort visited him in Utrecht while going to Leiden. He had a flat tire.

'How he managed to have a flat tire on his bicycle at such a convenient place, halfway the 120-km route, I do not know. We sat in my father's study, discussing ways in which the interstellar particles could be kept from growing too big. [...] Oort sat there, trying out various catastrophes that might happen to a grain in this short time, and this later led to our joint paper on colliding clouds. Later we went into the garden to repair

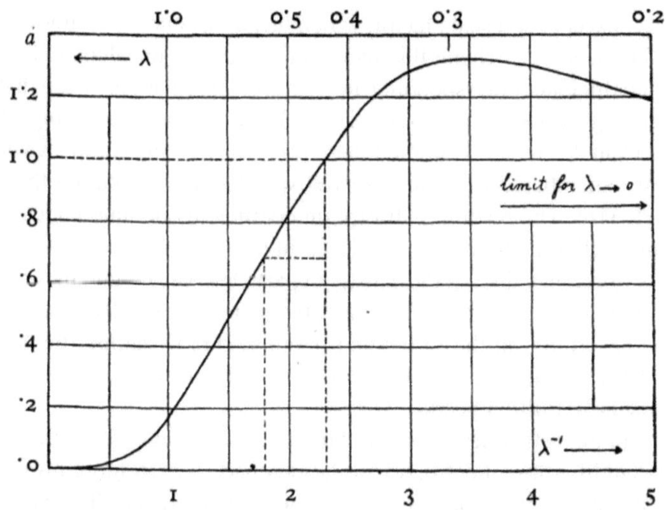

Fig. 9.19 Absorption coefficients in magnitude per kpc as a function of $1/\lambda$ (λ is wavelength in microns) from the model for interstellar smoke in Oort and van de Hulst (1946b)

> the flat tire. This, of course, was a very common job and every Dutch boy knew you must let the glue dry for ten minutes before you press the seal on the tube. Yet I had never seen anyone who actually had the patience and confidence to wait that long. Oort did, which impressed me more than his timing of interstellar collisions.'

The third contestant had been Adriaan van Woerkom, whose essay was about interstellar ionization and the charge of the solid particles. In a joint publication ter Haar, van de Hulst, Oort and van Woerkom (1943e) in the Dutch periodical of the Physics Society summed up the essays. In his paper van de Hulst makes the point that 'smoke' is the more appropriate word to be preferred over 'dust'. The real outcome of the whole exercise was the paper by Oort and van de Hulst (1946b) in the BAN. The manuscript was written before the end of the War, and according to a footnote to the title completed in February 1945. In this paper they studied the mechanisms that govern the growth and destruction of dust particles. The co-existence of gas and 'smoke' means that there must be a mechanism to evaporate gas atoms from the particles; otherwise there would be no particles or gas left. This could be by normal evaporation or through photo-dissociation when a light photon hits the particle. The first process is inefficient and the second would only work near bright, hot stars, where photons have more energy. The third possibility that they studied in this paper, is destruction or volatilization of particles in mutual collisions. The densities and velocities did not rule this out.

They next developed a model in which the size distribution of particles remained unchanged with time, the destruction being compensated exactly by the process of formation and growth of the particles. This resulted in a certain distribution of particle sizes, except for a scale factor, which meant the curve was normalized with respect to the mean size. Using this they then were able to calculate the resulting absorption coefficients corresponding to that size distribution, other than a scale factor depending on the average size. This mostly involved scattering of light by particles and van de Hulst had studied these processes. Actually, this is what van de Hulst is best remembered for. The observed wavelength distribution then fixes the mean particle size. The resulting absorption coefficients in magnitudes per kpc in the mean interstellar medium are reproduced in Fig. 9.19. It can of course be scaled vertically for different values of the density of particles. They found the mean density amounted to one particle per 7.5 million cubic meters (1.3×10^{-13} per cm^3). One can find the absorption in the visual and photographic by taking the dashed lines (visual on the left). The particles have an average radius of 0.15 micron and their mean lifetime is 50 million years.

This was an important development, but the contribution of van de Hulst to the essay prize and his subsequent period at Leiden had an even more profound effect and marks the beginning of radio astronomy in the Netherlands. The excellent book by Woodruff Turner Sullivan III (b. 1944), *Cosmic noise: A history of early radio astronomy*, presents a comprehensive discussion of the history of radio astronomy and the archive of the background research [252], including interviews, is extensive [251] (see also Sullivan's chapter in the *Legacy* [6] for early Dutch radio astronomy). The article *The beginnings of radio astronomy in the Netherlands* by Hugo van Woerden (b. 1916) and Richard Gordon Strom (b. 1944) [253] and of course the study *EarlyRadio* [9] by Astrid Elbers are excellent introductions too. I

Oort had been impressed by two publications. One was by Karl Guthe Jansky (1905–1950), who published his results, *Directional studies of atmospherics at high frequencies*, first in a journal for electronic engineers in 1932 [254], followed by papers in journals more accessible to astronomers in the next year [255] (The seminal 1932 paper is reprinted and available on the Web, see [256] for details). The other was by Grote Reber (1911–2002), published in 1940; its title was *Cosmic Static* [258, 259]. Actually already in 1941 Oort had sent the relevant journal issue to Herre Rinia (1905–1985), who worked at the Philips NatLab. Rinia later was to become one of the Adjunct-directors of NatLab. The picture of the radio telescope must have caught Oort's imagination (see Fig. 9.20). On March 12, 1941, Rinia wrote to Oort:

> 'Professor!
> Hereby I return your copy of the Astr.[ophysical] Journal with thanks. I have read the article by Reber with interest, and I have seen that there is at least one reasonable theory about the origin of this kind of radiation. It is rather nice that it would be possible in this way to 'see' a spiral nebula during the days and with clouded skies.'

I continue with the interview of Henk van de Hulst (Fig. 9.21):

> 'At some time Oort mentioned the interacademic colloquia that were always held in the mornings of the meetings of the Astronomer's Club. These had been held for a few years. And Oort had seen these papers by Reber and Jansky and said, 'we should take these'. I have always admired Oort very much, since there were all sorts of other astronomers

Fig. 9.20 The home-built antenna that Grote Reber used to detect 'cosmic static' on the grounds of his home in Wheaton, Illinois, must have captured Oort's imagination. The photograph in Reber's 1940 paper [258]—that Oort saw first—is of poorer quality; this is a different photograph that appeared in a second paper in 1944 [257]. Oort probably did not see that paper until after the War. From Wikicommons

that looked upon this as a mere curiosity that held little prospect. So it was very simple. There had to be a colloquium in half a year and Oort wanted a technical person for which he thought of Bakker from Eindhoven to speak on the techniques of receiving [radio signals] and he wanted someone from the astronomical side to discuss the origin of that radio radiation, and all I had to do in principle was discuss these 2 articles.

But then he said in passing that it would be great if one could find a [spectral] line. For me that was of course rather far from my expertise, that is to say I had done all sorts of things with spectral lines, but never at radio waves. But in that respect it was rather different, while for Oort it was very close as he after all had written a thesis on differential Galactic rotation [This is incorrect!] and he realized that these radio waves were

Fig. 9.21 Re-enactment in 1953 of the colloquium at Leiden Observatory in 1944. In the front row are Houtgast, Bakker and Oort. From the Leiden Observatory Archives

a sort of radiation that for the first time could be observed very well from the farthest parts of the Galaxy, contrary to optical emission, and he said that if one could combine this with the measurement of radial velocities you had something really great and this worked splendidly in 1951. I have been very lucky with the in some respects psychological fact that I was still a student, and that it took seven years before techniques had advanced to do the observations; you could not have arranged it better. [JKK-M: Why is that?] In some respects the fact that it took that long before it was observed, is of course a psychological boost when it is actually found. Then it is clear it was something not too obvious. And this week there is a symposium in Moscow honoring Shklovsky's eightieth birthday, where I did not go since I did not feel like traveling to Russia. He is the only other author that worked on the 21 cm line at about the same time. He had written a short article about it. I don't think it was completely independent, but it is an article in its own right. And a third man who thought of it independently, but did not know the astronomical literature, was Purcell at Harvard. I learned of this around, say, 1950–49 or 50—for the first time from Gorter, who knew Purcell and he told me that Purcell had some ideas but for Gorter this was somewhat vague. Further inquiry showed that indeed it concerned the same line. The sort of thing lying ready to be picked up; but you would actually have to pick it up.'

Fig. 9.22 Re-enactment in 1953 of the colloquium at Leiden Observatory in 1944 from another angle. From the Leiden Observatory Archives

Iosif Samuilovich Shklovsky (1916–1985) was a Russian theoretical physicist, who worked extensively on mechanisms of radio emission in celestial objects. This symposium van de Hulst referred to took place in 1996 and honored Shklovsky after his death. Edward Mills Purcell (1912–1997) was a physicist at Harvard, who as we will see, was involved in the first detection of the line emission, and Cornelis Jacobus Gorter (1907–1980) a physicist in Leiden. The other speaker at the colloquium, Cornelis Jan Bakker (1904–1960), was a physicist who worked at Philips NatLab and would play an important role in the rise of radio astronomy in the Netherlands; he was also deeply involved in the founding of the particle physics organization CERN (originally Conseil Européen pour la Recherche Nucléaire) near Geneva. He became its Director-General in 1955, but unfortunately died in a plane crash in 1960.

The rest of the story is well-known and best presented here in van de Hulst's own words in his 1998 autobiographical article [260] (Fig. 9.22).

'So I started to study spectral lines and spent several months systematically exploring all kinds of possibilities. Incidentally, I made a substitution error in estimating the broadening of the 'recombination lines' by the Stark effect, arriving at the erroneous conclusion that they would all be effaced by broadening. Later, one afternoon, my systematic exploration brought me to the hyper-fine structure line arising from a turnover of the electron spin in the field of the proton, all in the ground state of atomic hydrogen. By putting the magnetic moment of the proton – recently measured by Kopferman – into a theoretical formula derived by Fermi, I concluded that an interesting line should exist with wavelength of 21 cm. The intensity of this line seemed very uncertain. But by assuming that the magnetic dipole strength would be of the order 1 and making a somewhat liberal guess about future instrumental development, I estimated that some time in the future, this line might become observable in the Galaxy. The dramatic history is that it took another seven years before, in 1951, Ewen & Purcell first detected this line, followed rapidly by confirmations in Holland and in Australia.'

Spectral lines at radio wavelengths have to originate from transitions in atoms (or molecules), where the jump of an electron from a particular energy level to a lower level involves much less energy than for spectral lines at optical wavelengths. Since hydrogen is by far the most abundant element in the Universe, and therefore presumably also in the interstellar medium, it is natural to first look for that element. An hydrogen atom can be ionized, that is lose its electron which in interstellar space could be by absorption of a photon from hot stars or from being hit by a high-energy particle (known as cosmic rays). Hot stars are young stars of spectral type O or B and cosmic rays are thought to originate from supernova explosions. Of course an electron can also be just lifted to a higher energy level rather than being completely detached in an ionization process. When now the atom recombines with a free electron, the latter will at first settle at any energy level of the atom, the chances of which being prescribed by the energy distribution of the levels. What will happen next is that the electron will jump down to a lower level, giving rise to a spectral line in emission, the time interval and will jump down to a lower level, to which lower level again dependig on the atom's structure of energy levels.

Suppose a hydrogen atom has an electron in a high energy level. Every time it drops to a lower level a photon is emitted. If this happens at a high level to another level only a few steps lower, the energy difference is small and the photon corresponds to a long wavelength. These are the recombination lines that van de Hulst referred to. This process also occurs at lower levels, where the energy differences are larger and then the lines occur at infrared or optical wavelengths. The bright hydrogen lines in HII-regions around hot stars (see Sect. 7.2) in the optical are 'recombination lines', the Balmer-lines, where the electron drops to the second level from the levels above that. When it involves the lowest level the lines are those of the Lyman series in the ultraviolet.

Van de Hulst had incorrectly treated the Stark effect, named after Johannes Stark (1874–1957), which broadens the lines (or splits them up really) under the influence of an external electric field, and he had thereby reached an incorrect conclusion regarding the recombination lines. What became the promising transition is the one resulting from a splitting of the lowest level in the hydrogen atom due to the spins of the electron and the hydrogen nucleus (proton). These spins, which can only be either parallel or anti-parallel are associated with magnetic interactions between the two particles, so that these particles are slightly less bound when their spins are parallel. In that case the energy level is slightly higher. The difference corresponds to radio emission with in vacuum a frequency of 1420.4 MHz or a wavelength of 21.1 cm.

The presentations of Bakker and van de Hulst were published as two parts of one paper in the journal of the Dutch Physics Society [261] (for a translation with the minutes of the meeting see [262]). What had to come next was to build a radio telescope of sufficient size and a receiver with sufficient sensitivity to first detect the line and subsequently map the Milky Way. That is exactly what Oort set out to do when the War was over.

Fig. 10.0 Oort in 1946. From the Oort Archives

Chapter 10
Opening Up New Avenues

It is through science that we prove,
but through intuition that we discover.
Jules Henri Poincaré (1854–1912)

Maybe some people don't feel scared
when they think about comets and supernovas.
Maybe they think it is wonderful.
Lydia Netzer

During the first dozen years after the end of the Second World War, Oort opened up a large number of new avenues. He started radio astronomy in the Netherlands, establishing a first radio observatory at Kootwijk, co-discovering there the 21 cm line and describing the spiral structure of the Galaxy. In addition he studied the continuum radio emission from the Galaxy. Radio astronomy was a very important new tool, but it did not replace Oort's research efforts in other areas and with older techniques. He continued efforts attempting to describe the distribution of stars in external galaxies with photographic surface photometry. But he also continued to work on the formation and growth of interstellar dust particles and taking on such diverse subjects as the origin and motions of interstellar clouds and associations of young stars, the interaction of expanding shells with the interstellar medium, the origin of comets resulting in the concept of the famous Oort Cloud, and the origin of the radiation in the Crab Nebula and cosmic rays.

As far as observational techniques in the optical goes, Oort was actively involved in observations of the polarization of light in the Crab Nebula and in applying photoelectric photometry to galaxies to enlarge the potential of the photographic work. While rebuilding society on the ruins of the devastation of the War required major efforts, and funding for science was not a very high priority, Oort succeeded in securing funding for the second Kenya expedition to address the problem of absolute

French mathematician, physicist, engineer, and philosopher.
Lydia M. Netzer, née Jenkins, American author. (b. 1972) [263].

P. C. van der Kruit, *Jan Hendrik Oort*, Astrophysics and Space Science Library 459, https://doi.org/10.1007/978-3-030-17801-7_10

declinations and for starting a new branch of science, radio astronomy. He even was able to secure government funding for the construction of what for seven months would be the largest radio telescope in the world. Those twelve years up to 1957 also saw the first steps towards a proposal to build a European Southern Observatory, of which Oort was one of the founding fathers.

10.1 Director of Leiden Observatory

World War II ended for the Netherlands on May 5, 1945 with the German capitulation (except of course for the East Indies where it ended with the capitulation of Japan on August 8, 1945). Oort wrote in his diary notebooks on Friday May 4:

> 'In the evening the news comes that the Germans in the Netherlands, North-Germany and Denmark have surrendered. We first celebrate at Groeneveld with a wine toast, then at Dreesje's house, where we find Jo and Pivy and 2 Canadian soldiers, with cherries and finally at the Potbrummel with cigars and a bottle of Bourgogne'.

'Dreesje' was a lady that helped Mrs. Oort with the housekeeping, and probably also that of Oort's brother John and his wife; she lived nearby on the main road through Hulshorst. And on May 5:

> 'This morning at 8 o'clock the surrender became formal. Marijke, Erica and Bram go to Nunspeet in the afternoon for a flag procession.'

Erica is one the children of Oort's brother John.

On May 7, Oort went to Zwolle with among others Niko Tinbergen to try and obtain a pass to travel to the west of the country. On Thursday May 10 (Ascension Day), Oort, Duyvendak and Tinbergen left on a tandem and a bike to the west. At 2 in the afternoon they got to De Bilt near Utrecht, where Oort's parents-in-law lived. However, he noticed that there was no flag flying on their house. His father-in-law, Willem Graadt van Roggen had died on April 18. Oort talked with his his mother-in-law for a long time. Duyvendak and Tinbergen continued on to Leiden on the tandem, but Oort decided to return to Hulshorst to deliver the bad news. In a moving statement in his diary notebooks he wrote:

> 'When I arrived the three children were outside [...], all three adorned in orange to go to a children's party [...]. When they saw me all three ran up to me joyfully. Oh, their sad faced when I told them these was something very sad. I went upstairs to Mieke. Later we brought them also upstairs and told it to them. We sit a while on Bram's bed, a very close family. Bram cries quietly, Marijke is terribly sad, Coen tries to control his emotions, but is very pale. Marijke says she knew when I went upstairs. We read grandmother's letter and Mieke is so sweet.'

It turned out that at the funeral of Graadt van Roggen, his widow had given a very short speech and had asked that the *Wilhelmus*, the national anthem of the Netherlands, be played. This was strictly forbidden as the War still raged, but apparently done anyway. Oort went back to De Bilt next Wednesday. The weather was fine for cycling. On Thursday May 17, Oort set off again, leaving at 12:30 in the afternoon' and arriving in Oegstgeest at '±7'. This is fast, as the distance to be covered in these 4.5 h is some 65 km. He talked to various persons, in particular Oosterhoff, van Herk and others. He attended a meeting with the Faculty on May 22 and returned to De Bilt in the evening; 'beautiful evening, downwind'. After visiting Utrecht Observatory (talking with Minnaert on IAU matters and van de Hulst on his work), he went home arriving late because of flat tires. In his pocket diary he wrote on May 24: 'wedding anniversary, also for Jo and Pivy. I brought roses from the terrace at the Observatory [Leiden], that are still beautiful and smell deliciously.'

Oort's second visit to Leiden was on June 5, after having traveled by bike to De Bilt with Mrs. Oort who stayed with her mother. The Oorts returned to Hulshorst independently a few says later and then started making preparations to return permanently to Leiden as a family. There was a problem obtaining a pass for Mrs. Oort to cross the Grebbe Line (see Sect. 9.1). This defense line had been restored by the Germans and renamed Pantherstellung, but had not been cleared yet of mines. She was given a declaration of political correctness and the pass was obtained. After many farewells, Oort traveled by bike back to Leiden, leaving in the afternoon of June 18. The rest of the family apparently traveled by other transport. He visited his mother-in-law in De Bilt and van de Hulst, who was ill, at his home in Utrecht and went on 'in nice weather to Oegstgeest, where I arrived very tired'. But he added after this in his pocket diary: 'From June 18, I have been appointed 'Beheerder' [manager] of the Observatory.'

The next day, June 19, he wrote:

> 'First day as 'boss' of the Observatory, where for the moment people only work half days. In the afternoon I have a long conversation with Kuiper.'

Gerard Kuiper, who had been naturalized as an American citizen in 1937, had taken first a leave of absence during the War from Yerkes Observatory (1943–45) to join the Radio Research Laboratory at Harvard to do war-related radar research, and early in 1945 came to Europe as part of the ALSOS Mission, which investigated and collected intelligence on the German atomic, biological and chemical weapons research. He has been reported to have staged a daring rescue mission into eastern parts of Germany, during which he took famous physicist Max Planck into Allied care hours before the Soviets arrived. He traveled around a lot and undoubtedly had come to Leiden to see how the Observatory had survived the War.

On Wednesday June 20, Oort gave a speech to the full personnel. From Oort's pocket diaries:

> 'We have our '11 o'clock coffee' together. Coffee is surrogate provided by Mieke, sugar from the shipment by van Hoof [Armand van Hoof from Leuven, Belgium] and cigarettes from Kuiper! In the afternoon again a long talk with Kuiper.'

The text of the speech survived in the Oort Archives. A translation is provided in Appendix B.5. Oort immediately started to reorganize the working relations in the

Observatory, to which purpose he held talks with everyone personally and made an inventory of the wishes for support and career matters.

Ejnar Hertzsprung had been acting as director of the Observatory during the War, even though he had reached the age of retirement (70) in 1943. He left Leiden in September 1945, returning to his native Denmark. He returned for a formal farewell on June 25, 1946. Oort's farewell speech was published in 'Hemel & Dampkring' (Oort 1946d). In this he compared Hertzsprung to Tycho Brahe, also Danish and working abroad, but sharing the belief that progress requires accurate observations.

Although Leiden Observatory had come out of World War II without much loss of personnel, the Dutch astronomers in the East Indies had suffered much under the Japanese occupation. Three astronomers had died in Japanese prison camps (for obituaries see Oort 1946h). Oort's old friend Aernout de Sitter, after having obtained his Ph.D. degree with Hertzsprung, had moved as Leiden observer in 1937 to the Union Observatory at Johannesburg. In 1939 de Sitter moved on to the Bosscha Observatory at Lembang, Java, to become its Director. De Sitter was taken prisoner and died in a camp near Palembang on the island of Sumatra. Oort (1946c) described him in an obituary (written in Dutch) in 'Hemel & Dampkring' as follows:

> 'Although not particularly blessed with intellectual talents, neither possessed by the one-sided passion for research of nature that drives others to important achievements, or urged by ambition, he would, as a result of his solid professionalism that he had obtained with his unrelenting desire to serve the good cause, and in particular his clear insight of what was useful and what not, have been a Director that would have taken the Observatory to excellent performance and have brought it to great prosperity ...'.

De Sitter's wife, Henriëtte Johanna (Hette) Zoetelief Tromp (1906–19??), and their four children survived and came back to the Netherlands. The Oorts adopted them as if family as a token of the special friendship between de Sitter and Oort.

The second victim, Willem Christiaan Martin, who as already mentioned, obtained a Ph.D. under Hertzsprung in 1937, had moved to Johannesburg as the Leiden astronomer at the Southern Station in 1939. In February 1941 he had left for Lembang where he had been appointed at the Bosscha Observatory. Not long after his arrival he was taken prisoner and eventually died in June 1945 of starvation in the same camp as Aernout de Sitter. Finally, Jan Uitterdijk (1907–1945) had been a student in Leiden, but had resigned in July 1939 to become a teacher in Batavia (capital of the Dutch East Indies). He was also taken prisoner and drowned when a Japanese transport was sunk off the coast of Sumatra on September 18, 1944.

The reason for working initially only half days was that people were undernourished and in poor shape and health. Eventually things improved when more food became available. However, lecturer Jan Woltjer died in January 1946. As Dirk Brouwer wrote in an in memoriam in 'Hemel & Dampkring': 'During the 'Hongerwinter' [the famine of the 1944–45 winter] his health had been undermined so much, that he eventually died of the effects of this.' So, the loss of life due to the War was not restricted to those working in the East Indies. Housing was also

Fig. 10.1 Leiden Observatory photographed from the air around 1950. The picture is taken from the north-west. The canal is part of the original Leiden town defenses and is called 'Witte Singel'. The main telescopes are the 10 in. and 6 in. refractors on the main building, the 13 in. photographic telescope to the left (in the position before it was relocated in 1957) and the 18 in. reflecting 'Zunderman' Telescope (operated since 1947) at the right. At the left of the picture the botanical gardens of the University. From the Archives of Leiden Observatory

a problem; according to Blaauw in the *JKBlaauw-interview*, the Duyvendak family occupied for some time the upper floor of the Oort residence.

The actual move of the household of the Oorts from Hulshorst back to Leiden took place on July 10. Oort was not re-appointed until August 28; the interim arrangement ('Beheerder' of the Observatory) would have allowed the Oort family to move back into the Director's house. The formal appointment of Professor at Leiden University and Director of the Observatory was made on November 8, 1945. Also the same date, Pieter Oosterhoff was appointed lecturer and Adjunct-Director (Oosterhoff would be appointed extra-ordinary (honorary) Professor in 1948) (Fig. 10.1).

In September Oort went on a trip to England, arranged by Kuiper, who accompanied him during part of the stay. Kuiper had felt that Oort needed some better food to regain his health and strength. But there was another reason too, as we will see shortly: the International Astronomical Union. The visit was to have taken place in June, but visa and other formalities could not be completed in time. The trip started on Monday, September 23. Oort first stayed in London at the Society for Visiting Scientists, where he had to share a bedroom with three others. He slept poorly and did not feel at ease at all in London. One activity he mentioned in his pocket diaries was a meeting of the Faraday Society (for physical chemistry) with Anton Eduard van Arkel (1893–1976), Professor of chemistry in Leiden. On September 27 Oort traveled to Cambridge, where he was put up at Caius College. He dined with the fellows that lived there. He visited the Cambridge Observatory, where his host was Hermann Alexander Brück (1905–2000), a German-born astronomer who fled Nazi Germany and at that time was assistant Director of the Observatory. Oort also visited the Cavendish Laboratory, where he talked with J.A. Ratcliffe who told him about their plans for shortwave radio radiation. This is John Ashworth Ratcliffe (1902–1987), who had worked on telecommunications research during the War, together with future radio astronomers Alfred Charles Bernard Lovell (1913–2012), Martin Ryle (1918–1984), and Antony Hewish (b. 1924). Lovell was to found the Jodrell Bank Radio Observatory near Manchester, and Ryle and Hewish moved to Radcliffe's radio-physics group at Cambridge and in 1974 shared a Nobel Prize for their radio astronomical work.

On September 30, Oort returned to London, where he met Kuiper, who among others took him to the Royal Greenwich Observatory and Harold Spencer Jones, former Director of the Cape Observatory and now Director at Greenwich and Astronomer Royal. A major reason for Oort to visit Spencer Jones was the IAU. Adams at Mount Wilson, one of the Vice-Presidents of the Union, had acted as General Secretary during the War. And President Eddington had died in 1944. So measures had to be taken as soon as possible. Oort's visit to England had the urgent aim to discuss with Spencer Jones how to proceed (see next section). On Friday October 5 there was a large dinner at Oort's Leiden Observatory home for the 76th birthday of his mother with most of the family, which he then had to miss. He was to return the next day, but fog prevented the airplane from taking off from Croydon. In the end it took off on Sunday. In the afternoon he joined the family in Oegstgeest to celebrate his mother's birthday. Oort later that week wrote in his pocket diary that he still did not feel well; still the next weekend his mother came to stay with them at the Observatory.

The year 1945 ended joyfully and happy. On Sunday December 23 Oort mentioned in his pocket diary that they celebrated that on that day it was 20 years since he and Mrs. Oort had been engaged to be married. This must not have been the formal engagement with the exchange of rings; on that day they must have agreed to become husband and wife. They went walking with their children through Rhijngeest. During the fall the Oorts visited Hulshorst with their son Abraham. Oort noted in his pocket diaries that he made a short visit by himself just before Christmas to take some pictures and movies.

10.2 Recovery of the IAU

One thing that urgently needed attention from Oort when the War ended was the International Astronomical Union (see Box 3.3), of which he had been General Secretary since 1935. Briefly, the UK and France declared war on Germany following the invasion of Poland in 1939. The USSR then also invaded Poland. In the spring of

Fig. 10.2 Changing of the guards at the IAU in 1948. President Spencer Jones and General Secretary Oort (second and third from the left) were succeeded by respectively Lindblad (left) and Strömgren. From the Oort Archives

1940 followed the invasion of Denmark and Norway, the Netherlands, Belgium and France. Italy joined the war that same year, but on the German side. In the middle of 1941 the USSR followed after having been invaded by Germany and towards the end of the year the USA declared war following the Japanese attack on Pearl Harbor (Fig. 10.2).

Now the Executive Committee of the IAU found themselves divided between these groups of nations. President Eddington and Vice-President Adams were in Allied countries, Oort and Vice-President Maurice Paul Auguste Charles Fabry (1867–1945) from Paris in occupied territory, Vice-Presidents Bergstrand from Sweden and Brunner from Switzerland in neutral countries, and Giorgio Abetti (1882–1982) from Arcetri Observatory near Firenze in one of the Axis nations. In June 1940, Oort had written to as many of these as he could to arrange the transfer of the IAU secretariat to Adams at Mount Wilson, with agreement and help from Eddington. In 1940 he had sent around a full accounting of the IAU finances; bank accounts in dollars and pounds existed in addition to the one in guilders in Leiden.

When the War had ended, Oort was keen on putting the IAU onto a firm footing. What follows has been described in much more detail in Blaauw's History of the IAU, [113]. Not long after Eddington's death in 1944, Harold Spencer Jones as Astronomer Royal and former IAU Vice-President had taken the initiative to ask

Walter Adams to assume the presidency. Adams replied to Spencer Jones on March 23, 1945 as follows [264]:

'Your suggestion that I undertake the Presidency of the Union is very deeply appreciated but I have a much better one to offer. This is that you not only return to the Executive Committee but also take over the Presidency until the next meeting of the Union which is probably several years in the future. The arguments for your doing so seem to me most conclusive. In the first place you will bring to it a fund of experience in international organizations which will be invaluable in difficult times like these. Again you are strategically located with relation to nearly all the members of the Executive Committee, and will be close to Oort as Holland is liberated. This seems to me extremely important since the most difficult problems of the IAU will arise in Continental Europe and may require frequent meetings of the Executive Committee, or, in any event, of the President and the Secretary. The reorganization of the finances of the Union after the War is a case in point.'

Adams further noted that the USA wished another Englishman to fill Eddington's place. He noted that at sixty-eight and soon to be retired he hesitated to take on new responsibilities,

'especially of a character for which I am not well fitted either through ability and experience. In fact I should feel that was doing a disservice to the Union if I were to take on the Presidency at a time such as the present.'

Spencer Jones wrote back on May 16, accepting Adams' suggestion and asking him to write to the Executive Committee members to ask for their consent, which Adams did on June 30, 1945. He also wrote to Oort on that same date, notifying him as well of what was happening.

'We have thought a great deal about you during these terrible years, and it is an immense relief to us that you have been able to come through all your privations and suffering successfully. [...] Please remember me most cordially to your colleagues at the Observatory and offer them our deep sympathy for all the troubles which they have encountered. [...] You will doubtless wish to recover from your experiences for a little time to come before undertaking once more the work as General Secretary.'

The Executive Committee concurred enthusiastically with Adams to appoint Spencer Jones as President.

As we saw, Oort had visited Spencer Jones as soon as possible in September of 1945 in London. As it was not possible to hold a General Assembly on very short notice, they organized a special meeting of the Executive Committee together with senior astronomers from most countries. Unlike after World War I here was no extreme bad feeling against astronomers from the defeated countries (Abetti would be reconfirmed as Vice-President), but no astronomers from Germany or Japan were invited. Germany was not a member and Japan's membership was considered suspended. Formal membership would be possible only if these countries would regain independence and have their own governments. Adams, and some others as well, felt uneasy having friendly meetings with astronomers from these countries. The conference was held in March, 1946 in Copenhagen. Spencer Jones was reconfirmed as President and Oort as General Secretary. Adams and Abetti continued as Vice-Presidents and three new ones were appointed. The next General Assembly was planned for 1948 in Zürich, on the neutral territory of Switzerland. Individual astronomers from Germany and Japan were welcome; it was assumed that these had been scrutinized by then for possible war crimes.

This seventh General Assembly was held in Zürich in August 1948. It was attended by 279 participants, among which eight leading German astronomers, but none from Japan. At this meeting Lindblad took over as President and Bengt Strömgren as General Secretary. New was that Spencer Jones and Oort were consultants to the Executive Committee for the next period, while Oosterhoff assumed the new office of Assistant General Secretary. The final meeting gave Oort a standing ovation for his work in getting the IAU back on track (see Sadler in the *LibAm80* [2]).

Getting the IAU back in shape and organizing these Copenhagen and Zürich meetings must have taken much of Oort's energy and time. He had also worked hard to prepare the (re-)adherence of Germany and Japan, which indeed formally happened at the next General Assembly in Rome in 1952. During his tenure of the office of General Secretary he had been ably assisted by Heleen Kluyver at Leiden Observatory.

10.3 Physics of the Interstellar Medium

Before the War Oort had received some honorary appointments to learned societies (see Appendix A.5). In mid-January 1942 he had received a letter, dated December 4, 1941 that he had been awarded the Bruce Medal of the Astronomical Society of the Pacific, named after Catherine Wolfe Bruce (1816–1900), benefactor of astronomy. Oort noted on the letter, written by President Arthur Scott King (1876–1957) from Mount Wilson Observatory, that one day later than its date (Dec. 7) war broke out between the USA and Japan. The Annual Report of the secretary-treasurer of the Society for 1941 (presented January 20, 1942), recorded that Oort had been notified but no reply had been received. A year later it was reported [265]:

> 'The Catherine Wolfe Award Medal for 1942, which was awarded to Professor Jan H. Oort, of Leiden, the Netherlands, could not be delivered to him, and its casting and engraving will be held in abeyance for the time. Professor Oort has been advised of the honor bestowed upon him by the Society for his distinguished services to astronomy.'

Actually the Bruce Medal would not be awarded after 1942 until after the War had ended in 1945. On December 31, 1945, Oort was notified by the secretary that the medal was forwarded through the Dutch Consul in San Francisco to Prof. Pannekoek with the 'request to arrange for its formal presentation to you, with a fitting ceremony.'. This was done at the meeting of the Netherlands Astronomers Club on May 4, 1946 (a Saturday as customary), when Pannekoek read a statement from Arthur S. King [266].

On January 12, 1946 Oort received a telegram that he had been awarded the Gold Medal of the Royal Astronomical Society. The ceremony related to this award was on May 10, the next Friday after the ceremony for the Bruce Medal. So, Oort had to leave almost immediately for London (see Sect. 11.8 for some of Oort's personal impressions). I have referred to the address by the President of the Royal Astronomical Society on that occasion in connection with the sub-title of this book [23]. The actual presentation of the medal was preceded by the George Darwin Lecture, which Oort had given the title *Some phenomena connected with interstellar matter* (Oort 1946e). This lecture is named after George Howard Darwin (1845–1912), fifth child and second son of Charles Robert Darwin (1809–1882). George Darwin was a barrister, but later became a professor of astronomy in Cambridge. His interests focused chiefly on the interaction between the Earth and the Moon and the formation of the latter.

Oort started the lecture by pointing out that the interstellar medium is rather uneven, but that in general terms dust (sometimes called 'smoke' by Oort) and gas co-exist as evidenced by interstellar absorption lines. He emphasized a peculiar fact, namely that in the direction of the constellation Cygnus (which is near the Milky Way on the sky) there are four faint, unreddened Cepheids, discovered by Baade and observed in multiple colors by Oosterhoff, that had distances between 8 and 24 kpc. Their radial velocities confirmed these distances when compared to Galactic differential rotation. So, there is 'a stretch of at least 10,000 parsecs close to the Galactic plane [that] contains at most as much interstellar material as is normally contained in a column of 1000 parsecs.'

Oort then turned to the formation of dust particles in the interstellar medium. For this he summarized the essays in the Prize Competition and his work with Henk van de Hulst in their joint paper (1946e). The absorption must be by particles with sizes of 0.1 to 1 micron in order to explain the dependence of the absorption coefficient on wavelength. The important result, obtained with van de Hulst, was that 'the formation of smoke particles would be proportional to a high power of the gas density, of the order of ρ^8. This meant a restricted range of gas densities where absorption could be expected. When the gas density was too small the particles would grow to too large sizes, while for high densities the particles will not get a chance to grow because of frequent collisions, resulting in evaporation. The favorable range was between 2 and 50 × 10^{-24} grams per cc.

He spent quite some time on a discussion of absorption in external galaxies. First he explained the absence of conspicuous absorption in elliptical galaxies: the gas density needed to be only 1/3 or less of that near the Sun. He returned to NGC 3115 (see Fig. 8.26 and the text there), the galaxy for which he discussed the dynamics at the McDonald dedication (Oort 1940a), referring to the high matter density found there in comparison to the light. He now inferred from the absence of clear indications of dust that the interstellar medium probably would be depleted by the formation of stars. There then must be large numbers of faint dwarf stars there.

For the spiral galaxy NGC 4594 (the Sombrero Nebula, see Fig. 7.6), he presented the curve of the surface brightness along the minor axis. This indicated that the dust is concentrated in a thin ring, the central parts of the galaxy being mostly devoid of dust. This would then not be the result of a central lack of gas, but of such high gas densities that in the inner parts no dust particles could grow. Oort linked the high gas densities in the outer parts of the galaxy to the phenomenon of spiral structure, just like in the sequence of morphological types the occurrence of spiral structure correlated with the presence of a peripheral ring of dust and not with the flattening of the system. The origin (and maintenance as we will see) of spiral structure had been a major interest of Oort and here we see his early speculations. Oort mentioned that there might be other ways to explain the absence of dust in the inner parts of NGC 4595. He added:

> 'In fact, one of the motives why the study of the formation of solid particles from interstellar gas was instigated in Holland was that the easily observable smoke in extra-galactic nebulae might teach us something about the *un*observable gas in these systems. We were interested in the *gas* because it might play a dominant part in determining the field of force.'

Oort's archives on NGC 3115 contain a manuscript of a paper 'A possible explanation of the dark rings in early-type galaxies'; it is unfinished (and was never published) and undated, but from the context and the references it contains it is clear that it has been written in 1944 or 1945. He presented the arguments just reviewed and probably decided against publication when he had made the point in the Darwin Lecture.

The major subject of the lecture was the issue of collisions of gas clouds and the interaction of nova or supernova shells with the interstellar gas, which was most on Oort's mind at the time. This was rather preliminary: 'I feel considerable hesitation in discussing this subject, inasmuch as my discussion does not reach out above a reconnaissance of the field, and only a first stumbling beginning has been made with the theory of the phenomena concerned.' In this subject he collaborated not only again with Kramers but particularly with his old friend Jan Burgers. We have seen that Oort and Kramers had had some correspondence during the War on subjects like turbulence in space. One of Oort's questions was what would happen if a shell thrown out by a star moves through the gas around it. In this so-called 'snowplow' outward motion would be retarded, the temperature raised and the density in the shell increased. He presented a very interesting deduction of the spectrum of gas excited by such interactions versus that by stellar radiation, where the gas has been ionized.

In such regions one sees emission lines from hydrogen, that result when an electron recombines to form a hydrogen atom, followed by cascading down the energy levels in the atom. But there are also lines of other elements, in particular (ionized) oxygen and nitrogen. These are so-called forbidden lines: they do not occur in the laboratory, since for these lines the electrons have to stay for a relatively long time in the upper energy level and for terrestrial densities atoms/ions are usually hit by another atom/ion before the line can be emitted. Now, in the case of radiation excitation (as in a planetary nebula), cooling takes place through encounters of free electron with

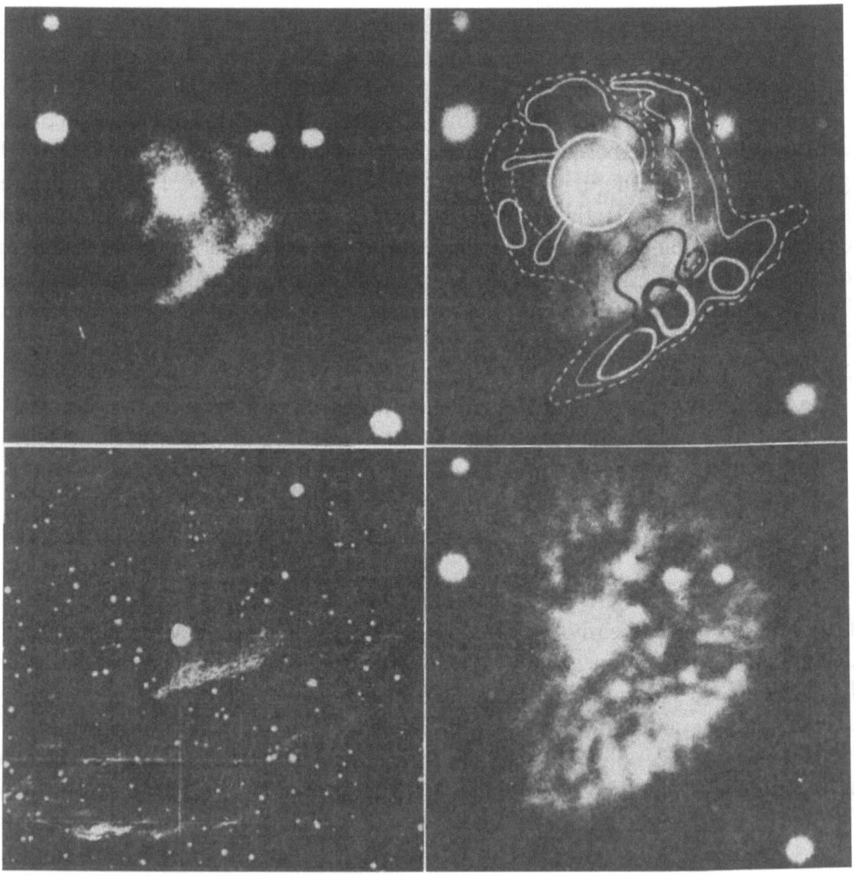

Fig. 10.3 Nova Persei 1901. Upper row: November 15, 1917 (left) and October 3, 1934 (right). Lower row: November 13, 1901 (left) and September 28, 1943 (right). On the top-right the contours represent the lower-right picture. The lower-left panel is a drawing. Oort's caption to this figure has the year of the lower-right panel wrong (1934, should be 1943). Observations by Ritchey, Hubble and Baade. Adapted from Oort (1951g)

oxygen atoms, which radiate away the energy gained from the electrons. This will keep the temperature of the electrons at a certain temperature (roughly ten thousand (10^4) K). But in the case of interstellar friction the timescale of cooling is too long and the temperature of the electrons will rise, so during collisions they can more easily excite the higher levels of the forbidden lines. This increases the strengths of the forbidden lines of ionized oxygen and nitrogen over the recombination lines of neutral hydrogen. This, Oort concluded, can help distinguish the two types of excitation, a diagnostic still used today.

Oort then went into a detailed discussion of the expansion of the shells around Nova Persei 1901 (GK Persei). The expansion could be deduced from photographs in the course of time (see Fig. 10.3). It could be used to estimate the energy released

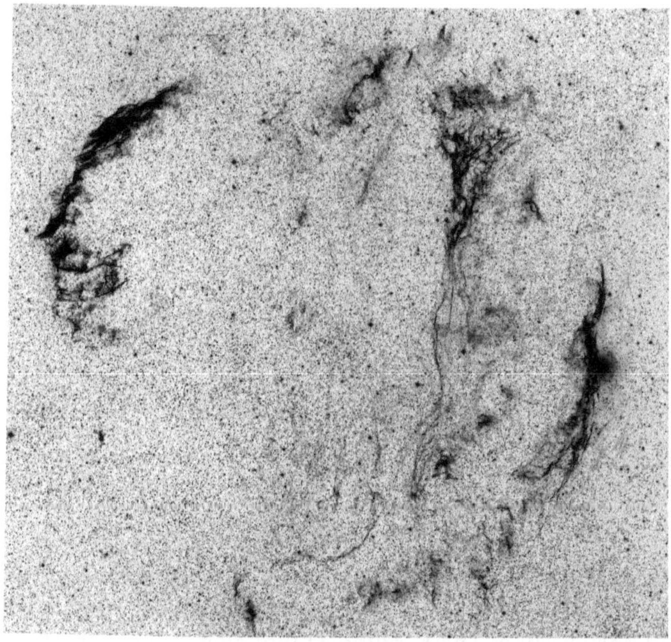

Fig. 10.4 Cygnus Loop and Veil Nebula. The arcs are known as the Veil Nebula; the circular structure that they outline is known as the Cygnus Loop, a supernova remnant. The size of the image is roughly 2.7 by 2.5 degrees on the sky. This picture has been scanned from the 103-E emulsion (red) print of the National Geographic Society–Palomar Observatory Sky Survey

per second in the deceleration and this seemed in general terms equal to that radiated at optical wavelengths. But the spectra showed enhanced forbidden lines, showing that it was impossible that the nova itself had excited the gas. In fact numerical estimates showed that the electron temperature must have risen to thirty thousand (3×10^4) K.

There were two further cases. One was the Cygnus Loop, shown in Fig. 10.4. The filaments are usually indicated as the Veil Nebula. If it resulted from a supernova, the expansion of the shell must have slowed down. One question was how the shell could remain so thin, and Oort suggested that this showed that it had to result from compression by the interstellar gas. The problem was, of course that there was no central star to excite the nebula and this all pointed to only one conclusion (already drawn by Zwicky) that it was a supernova remnant. From the observed current expansion of 0.03 arcsec per year and 75 km/s, Oort estimated a distance of 500 pc, which would mean that it had a radius of about 10 pc. Current estimates are about a factor three higher. The supernova occurred on the order of 6000–7000 years ago.

The final case considered was the Crab Nebulae (see Fig. 10.5). Oort noted:

'The mass is so large in this case that no deceleration can be expected, and indeed none has been observed. If the interstellar density were about normal the nebula might have been heated to the temper-

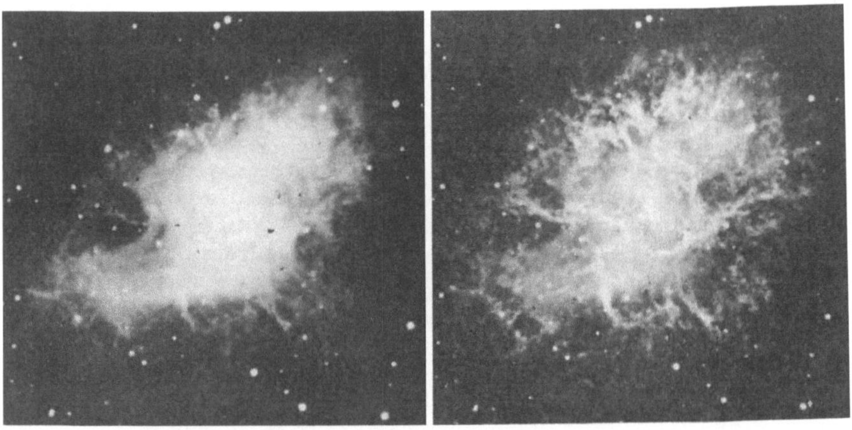

Fig. 10.5 Photographs of the Crab Nebula by Baade. On the left he blue continuum and on the right the light from the Balmer emission line Hα of hydrogen. Adapted from Oort (1951g)

> ature of the order of a few hundred thousand degrees which would be required to explain why the inner amorphous nebulosity shows no hydrogen and helium lines. The hypothesis that also in this case the density distribution in the shell and the radiation are determined by the interaction with the interstellar medium is attractive, primarily because it might explain how the exploded star can have been transformed into a shell of such surprising homogeneity as the main amorphous mass of the Crab nebula. However, the detached filamentary outer structure, whose relation to the inner mass has been so beautifully revealed by Baade's photographs in various wavelengths, seems to put serious difficulties in the way of this hypotheses.'

Oort would return to the Crab Nebula and its radiation mechanism later.

The collaboration with Jan Burgers continued for a number of years. In the end it did not result in any co-authored papers. Burgers performed a few studies of gas flows and shocks in the interstellar gas. The collaboration resulted in the organizing, together with van de Hulst, of an international symposium on *Problems of Cosmical Aerodynamics*, sponsored jointly by the International Union of Theoretical and Applied Mechanics and the International Astronomical Union, in Paris in 1949. The proceedings were published with Burgers and van de Hulst as the editors; Oort contributed an introduction (Oort 1951f) and a paper on 'the interaction of nova and supernova shells with the interstellar medium' (Oort 1951g), showing again that the 'collision of cosmical masses of gas are actually observed in our Stellar System'. The main immediate effect of Oort's interest in this area was to open up a new line of research and to involve others to bring it to fruition.

10.4 Back to Kenya

The first Kenya Expedition had been only a qualified success (see Sect. 8.3; for what follows I again refer to the article by Jet Katgert-Merkelijn on the Kenya Expeditions [140] for a more detailed description). It had been clear that the period used (the First Kenya Expedition lasted from December, 1931, to February, 1933) had been too short to obtain sufficiently many observations as was required for the objective. An extended mission would require more funding and astronomers on site for a much longer period. Whereas for the first expedition two astronomers (Hins and van Herk) had gone by themselves to Kenya (Hins was married and van Herk single), this time it was decided that the astronomers would take their wives and children along for the planned three years of the expedition. For this mission the leader would be Gijsbert van Herk, and he would bring his second wife Aartje (Attie) Konijnenbrug (1897–1975) and the two children from his first marriage (Joke and Wouter, ages 8 and 10) along. The other person was an unexpected volunteer, Willem Haske van Zadelhoff (1904–1964), Lieutenant first Class with the Royal Dutch Navy, who was seconded to Leiden Observatory by the Navy for the purpose of the mission. He had an interest in astronomy and had become aware of the upcoming Kenya Expedition through Marcel Minnaert from Utrecht. Van Zadelhoff's mother, Maria Ludovica Minnaert (1884–1966), was a first cousin of astronomer Minnaert. Van Zadelhoff was also to take his wife, Henriëtte Adrienne Naber (1906–1980), and their two children along to Kenya. Some of the persons that were involved in the second Kenya Expedition can be seen in the group photograph in Fig. 10.6.

Oort for a long time had a deep interest in astrometry, and he had taken over responsibility for the first expedition when de Sitter had died and Oort had become Adjunct-Director of the Observatory. When the results of the first expedition became evident, he had started planning for the second mission, but the second World War intervened. That was not the end of the plan, however. Actually, during the war, in 1943, the Pieter Langerhuizen Lambertuszoon Fonds, which had also supported the first expedition—funding the telescope, which was to be used again—had allocated funds to be used whenever that would become possible. At that time the island of Sumatra in the Dutch East-Indies was supposed to be the place to go for the second expedition. The funds would prove highly insufficient, so when plans became more firm after the War, Oort was able to find other sources of money, among which the IAU, four trust funds of the US National Academy of Science, various Dutch funds (among which again the Bachiene Foundation) and individuals. There is a long list in Oort's introduction to the final publication. Jan van de Bilt had also taken an important share in the planning and organization of the expedition. It must have been a major effort in terms of time to make funding applications etcetera, while administering the finances was also a complicated affair. It was done through the Sterrewacht Fund, which had been founded to administer the funds from the Rockefeller Foundation for the astrograph in South Africa. Oosterhoff managed the fund and also took control of the cash flows for the second Kenya Expedition.

Fig. 10.6 Part of a group photograph of the staff of Leiden Observatory in 1947, not long before the second Kenya Expedition. In the front Oort, Willem van Zadelhoff and Pieter Oosterhoff. Behind Oort to the right Fjeda Walraven and above him Gijsbert van Herk. In the top-left corner P. de Haan, who went along to Kenya as a carpenter for the construction of the buildings. The person in the upper-right is Adriaan Blaauw, who was in Kenya for part of the Expedition. From the Archives of Leiden Observatory

In 1947 the Dutch government, after extensive discussions starting in 1945, had founded a provisional 'Organisatie voor Zuiver-Wetenschappelijk Onderzoek' (English name: Organization for Pure Research), abbreviated ZWO, to strengthen the infrastructure for pure research in the Netherlands and to stimulate such research. In 1950 the ZWO took its definitive form and the name became Netherlands Organization for the Advancement of Pure Research; in 1988 it was transformed into a funding agency for all scientific research (the 'Pure' being dropped from its name) and it became the current NWO (Netherlands Organization for Scientific Research). For a history of ZWO, see [267, 268]. The Director of ZWO was Jan Hendrik Bannier (1909–1995), who happened to have the same first names as Oort did, but was in daily life addressed as 'Henk'. Bannier played a vital role in reviving Dutch research and was heavily involved in European science as well. He was trained as a physicist, but also had some background in astronomy; as a student he he had published a two-page article with Marcel Minnaert on lines in the solar spectrum [269]. He had worked at the Ministry of Education, Arts & Sciences when the provisional ZWO came into being and was appointed as its Director. He had not obtained a Ph.D., but was made *Doctor honoris causa* by the Catholic University of Nijmegen (now Radboud University) in 1974. A special volume appeared on the occasion of his seventieth birthday, to which Oort contributed a chapter (Oort 1979d). In 1947 there had been eighteen applications to ZWO of which twelve had been successful. Oort's request for funding of a small part of the Kenya Expedition was one of the successful ones. Including further allocations in later years, also to extend the expedition, the total contribution of ZWO to the second Kenya Expedition grew to about half of the total, so the existence of the funding agency had been a major factor in bringing it about. The cost of the expedition eventually rose to in round numbers Dfl. 115,000, which corresponds to about €0.9 million nowadays.

Eventually it was decided that the expedition would go to the same general area in Kenya that had been the scene of the first one. The actual site was chosen 3 km to the northwest of that of the previous expedition, on Timboroa Hill. This was at an altitude of 2880 m, only 36.″2 (1.1 km) north of the equator. When planning was far advanced (December, 1946, so a bit more than half a year before departure), the Observatory at Quito, Ecuador, reported much better seeing conditions than the previous time and invited the expedition to come there. This came too late to change the plans. The telescope to be used was the same horizontal one, although significant structural changes were made to it. As we have seen above (Sect. 8.3) the telescope had suffered from 'hysteresis', resulting from small movements of internal parts when turning it around, depending on the direction of turning. For this reason the option of reversal had been abandoned. The observing at different altitudes was also abandoned and the telescope was fixed at 7° above the horizon. For some impressions see Figs. 10.7 and 10.8, taken during Oort's visit to the Kenya Expedition in 1949. The two families, plus P. de Haan from the Observatory who would 'act as carpenter' in the construction of the buildings, sailed on August 30, 1947 again with the Holland-Africa Line which once more had provided financial support, and reached Mombasa on November 2.

The construction of the buildings and organization of the experiments were much more complicated than during the first visit. Roads on the site between the actual Observatory and the living quarters etc. had to be cleared by the expedition members themselves. Building materials were difficult to find so shortly after the War and the same held for qualified local workers. Carpenter P. de Haan left already in March of

Fig. 10.7 The horizontal telescope at Timboroa Hill, Kenya, its operation being demonstrated by G. van Herk during daytime. From the Oort Archives

1948 for private reasons, well before construction had been finished. It actually was not completed until October 1948, more than a year after the group had left from the Netherlands. Although the principles of the observational program were the same as before, the practical execution was very different in many respects. For example the alignment of the optics, the collimation, could not be determined now that they used a non-reversible telescope. But since observations were always subtracted from one another this was no fundamental issue. The reading of the telescope position with the micrometer was no longer done by eye, but by photographing the scales. Part of the films turned out useless because of lack of resolution.

The intention was to observe all 1193 stars between declinations +60° and −50° in the FK3, the *Dritter Fundamentalkatalog*, prepared by the Astronomisches Rechen-Institut at Heidelberg and published in 1937. The restrictions had to do with pillars, that had marks on them and were used for calibration of the azimuths; although placed at some distance they still blocked parts of the horizon. In addition another 300 stars from the GC, the *Boss General Catalogue*, also known as the *Boss Catalogue*, were observed. To some extent these were stars that were used to accurately determine the latitude of the observing station, using the zenith telescope that had also been brought along. This was accomplished reasonably completely, except that some data had to be discarded due to the unmeasurable films.

Early in 1949, Oort paid a visit to the Southern Station in South Africa and on his way back made a stop at Kenya in order to inform himself of the circumstances and progress. Figures 10.8 and 10.9 show Oort during this visit. The reason for traveling to Johannesburg was the increasing problems with light pollution from the city that threatened the Union Observatory and the work with the Rockefeller Astrograph, and therefore plans were made to relocate the Observatory away from the city of Johannesburg.

Fig. 10.8 Oort and van Herk, examining the telescope at Timboroa Hill, Kenya, during Oort's visit there in 1949. From the Oort Archives

Not long after the start of the expedition tensions between members of the team developed, which made living and working there difficult. In addition the work was hard and demanding for two observers (one of which was not a professional astronomer). To help improve the system of reading off the micrometer circles and to actually do some of the observing, Adriaan Blaauw came to Kenya between December 1949 and April 1950. Blaauw had first moved to Leiden after obtaining his Doctoral degree with a thesis under van Rhijn in Groningen in 1946. After that he had spent some time at the Yerkes and McDonald Observatories, and had been

Fig. 10.9 Oort during his visit to Kenya in February 1949. From the Oort Archives

appointed lecturer in Leiden in October 1948, at the same time that Oosterhoff had been promoted to Extraordinary Professor. Blaauw brought some new lenses for the photographic recording of the micrometer readings; this replacement improved things considerably (Fig. 10.10).

When the end of the planned three years for the expedition drew closer, it became clear that the objectives could not be met without an extension, and funds for this were requested from ZWO, and obtained. However, van Zadelhoff had by now been seconded to Leiden Observatory for four years (including the preparations in Leiden) and felt he should return to the Navy and pursue his career there. He and his family left Kenya in July of 1950. A student from Leiden, Maarten Schmidt (b. 1929), came to replace van Zadelhoff in September of 1950. Van Herk and van Zadelhoff had done almost all of the observations with the zenith telescope (with shares of respectively 45% and 55%). For the azimuth telescope van Herk had done the major share of the observing (in the end 54% of the observations), but Schmidt in spite of his relatively short stay—and being out of action for a few months, see below—still ended up having done 24% of them. Van Zadelhoff (13%) and Blaauw (9%) did the remainder.

Life was not easy at this desolate and remote place; in this small community tensions and friction between expedition members arose easily. There had been some friction between van Herk and van Zadelhoff, especially in the early stages. A Professor of physics at the University of Amsterdam, Jacob Clay (1882–1955), an expert

Fig. 10.10 Van Herk and van Zadelhoff. Both men and their wives and children arrived in Kenya in November 1947. The van Zadelhoffs left in July 1950, but the van Herks stayed until October 1951. From the Oort Archives

on cosmic rays, wanted to use the site to perform high-altitude experiments. Cosmic rays are energetic elementary particles, electrons and atomic nuclei, originating from supernovae in the Galaxy and some from the Sun (mostly protons). Most interact with particles in the Earth's atmosphere, but at high altitudes a good fraction can be observed as they were when they arrived from interstellar space. A young student, W. Derksen, first came to do the experiments, but severe tensions arose between him and the astronomers (especially van Zadelhoff). When he was replaced by Jan Strackee (1924–2009), relations improved and his presence even became helpful for the astronomical workers. One way, as Maarten Schmidt remarked to me, was that since they now were four, they could play bridge when the weather prevented observing. Strackee obtained a Ph.D. degree in 1952 from the University of Amsterdam on the basis of his work at the Timboroa site (Fig. 10.11).

There were dangers also. Van Zadelhoff reported on a bush fire on the neighboring farm that fortunately did not damage the Observatory. In the rainy season the road could become treacherous. At some stage the van Zadelhoffs set off with the truck to bring their son to the railway station in nearby Equator to travel to his school in Nairobi, but the truck slid off the road and turned over. Fortunately nobody was seriously hurt. Van Zadelhoff noted that: 'when we were lugging back up the hill, the school-bound boy did hear his train moving in the distance through

Fig. 10.11 Some of the buildings at the site of the Kenya expedition; picture probably taken by Oort in February 1949. From the Oort Archives

the mountains and rejoiced, despite the shock, in the fact that the next train would be departing only three days later'.

Schmidt had a serious accident with the car. Van Herk wrote to Oort;

'It will be about two months ago that I had a serious talk with Schmidt about his recklessness. He had almost had an accident with the sedan [...]. He promised to be more careful, but in view of his youth it is not surprising that his wildness evident in his work, speech and driving had the upper hand. On Sunday afternoon February 11 he drove the truck near Tomboroa Saw Hill on a road wide enough for a truck and without significant potholes, when he hit a tree branch the width of an arm, maybe not even as long as that, on the left-hand side. He did not see the branch and hit it at full speed with his left front-wheel. He lost control of the car, hit the shoulder and the car turned over. It happened quickly and his passenger, Strackee, could not make a better reconstruction.'

Strackee was not hurt, but Schmidt shattered his right ankle. He himself attributed the accident to shadows of trees on the road so that he did not notice the tree trunk. (see Schmidt's own account [270]). It was first thought that his right foot was pinned between the door that had opened and the car, but later it was thought that his foot then would have been cut off completely, but rather that he stuck his foot through the open window. The official doctor's report by R.W. Johnson, M.D., said (Fig. 10.12):

'Mr. M. Schmidt was admitted to hospital on the 11th of February with fractured lower right fibula and laceration of the ankle following a car accident. The skin was sloughed away at the side of the laceration thus making the fracture compound. The fracture has been twice manipulated and may require a further manipulation shortly. It will not be feasible to

Fig. 10.12 Gijsbert van Herk and his wife in Kenya; picture probably taken by Oort in February 1949. From the Oort Archives

fit a walking iron for probably four weeks and it is doubtful he will be fit for work before four months.'

Schmidt indeed was out of action for some time, being able to do deskwork after a walking iron had been fit (according to van Herk's letters in March). He recovered completely, but incidents like this put a lot of pressure on the expedition and on the progress of the observational work.

Another incident occurred when van Zadelhoff made a mistake in securing the movable hut belonging to the zenith telescope to the rails, which then was slammed into the telescope by a gust of wind. The telescope could not be repaired; the exact date is not recorded, but must have been halfway through 1950. It had been functioning for three years and the project was not seriously compromised.

In the course of September 1951 the observing was put to an end. Although funding for a further extension was applied for, ZWO did not honor this. Preparations were made for the sale of properties and transporting back instrumentation. At this stage two occasions of tension between van Herk and Oort developed. The first had to do with what was referred to as the zodiacal light experiment. Zodiacal light results from dust in the solar system, which reflects sunlight and can be seen by the naked eye under very clear circumstances as a faint band of light around the ecliptic, roughly the plane of the planets. Since the ecliptic runs through the constellations of the zodiac, it is called zodiacal light. Now, van Herk (and also van Zadelhoff) had an interest in measuring its properties. The remote, high-altitude Kenya site was ideal for this, with few sources of diffuse light from the atmosphere and human civilization. To this end a special instrument had been built at Leiden Observatory (see Fig. 10.13).

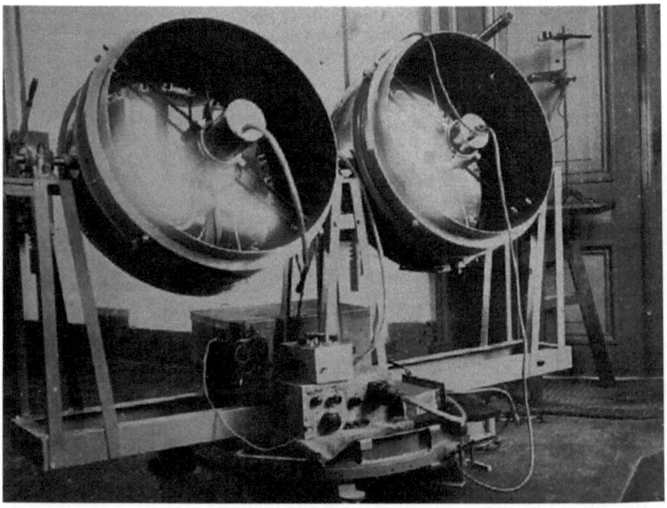

Fig. 10.13 The special instrument built at Leiden to study zodiacal light from the Kenya site. It was based on two German search lights that focused light onto the two (one red and one blue sensitive) cameras in the foci. From the Oort Archives

It was based on two German search lights to locate airplanes, that focused the light from an area of about 1° diameter on the sky onto cameras in the foci. Both observed the same area of sky, one in blue and one in red light. This program had a low priority with Oort, who wanted above all to finish the program of fundamental declinations. Van Herk managed to do some observations towards the end of the expedition, but Oort apparently had left him little time to do so. To his complaint, Oort replied:

> 'I want to say that relating to the expedition I could have done essential things better, that I felt in too much of a hurry during the preparations and to have been too much occupied with other things, both then and later. But the slow progress of the work of the expedition is not entirely my fault. It is not fair to say that the impossibility to complete the zodiacal work is to be blamed on those that had to provide for the money. ZWO will certainly not appreciate that.'

There had been some talk about transporting the instrument to Egypt to carry out the experiment. On his way back after visiting Kenya, Oort had paid a visit to the Helwan Observatory near Cairo, Egypt and again a longer visit in December 1950 and January 1951. In the end nothing came of it and the instrument was sent back to Leiden. The few data taken never led to any publication.

Then there was some friction about the return trip of the van Herks. Oort suggested they travel with the Union-Castle-Line and that Mrs. van Herk and the children would travel ahead of van Herk and Schmidt; this would be better for the children and Mrs. van Herk's health. Mrs. van Herk wrote him in a note accompanying a letter by van Herk of February 29, 1951:

'I would very much prefer to travel by Holland-Africa Line. Why so complicated going first to England and then to Holland? What a tiresome affair. Of course we want to take our dog and that then gives a lot of hassle. We do not of course expect the Observatory to pay for the dog's travel. We have lived here for almost four years in great isolation, often far from good company. I would say it is not asking too much. And I also should say that for the van Zadelhoffs taking the H.A.L. was not an issue at all. It is so much more convenient to disembark in Amsterdam.

You proposed I travel ahead of my husband with the children. But when packing and preparing for departure my help is necessary here. Concerning the children's schools, they will have lost their first year back anyway. They need to catch up a lot. When we return Wouter is 15. It will be difficult to find a suitable school for him. For my health it makes no difference that I would stay her an extra three months.

I hope you take my plea seriously concerning the Holland-Africa Line.

Are you and your wife OK? Will you give her my best wishes. What a dismal affair with Schmidt, fortunately he is a very cheerful person.'

It must have been hard on the families indeed. I was able to trace van Herk's daughter Joke, who gave me some information, but declined to provide more insight into her experience as a child in Kenya. Oort gave them permission to travel back by Holland-Africa Lijn by return mail. But it seems he had not always fully realized how difficult it all had been for the families to spend such a long time in Kenya. Oort arranged that the Observatory would try and find a suitable home for them for when they were back in Leiden, for which van Herk was grateful. In any case, the van Herk family left in October of 1951, arriving in the Netherlands on December 15, 1951, They had been away for 4 years and four months. Schmidt left somewhat earlier, traveled through Africa, ending up in Capetown, from where he sailed back via London, arriving in the Netherlands also on the same day.

For the reductions, van Herk received help from the Mathematisch Centrum in Amsterdam, an institute founded under the auspices of ZWO. The final results of the expedition were published in 1957 [271]. The results were given in a table which had the corrections necessary to the FK3 in eleven zones of declination between -50° and $+60^\circ$ with 11 correcting values ranging from -15 to $+20$ arcsec. The impact was disappointingly minimal. Probably astronomers were too conservative to accept it as superior to the traditional approaches. The discussion in Jet Katgert-Merkelijn's paper [140] actually shows that in a comparison van Herk's corrections are consistent with more recent catalogues. Van Herk's paper has zero(!) citations in the NASA Astronomy Data System ADS (see Sect. 6.1). It is true that for that period the ADS is incomplete, as references often occurred in footnotes rather than lists at the end of a paper. Two authoritative reviews of the history of astrometry, [272, 273], disappointingly do not even refer to van Herk's paper and the heroic Kenya expedition! This is really sad. I have found Oort never referring to the Kenya expeditions after their completion, while he could, as initiator of the expedition and director of the observatory where the work had been carried out, have promoted the work more at international meetings; he probably was occupied too much with radio astronomy for this. All of this is now of course superseded by astrometric satellites such as the European Space Agency ESA's Hipparcos and Gaia. Van Herk should have been given more credit and recognition than he received. He was a modest man, who was sometimes cynical about the lack of credit given to others who deserved it, while he himself never received the full appreciation due for this work. The azimuth and zenith telescopes are now, together with the Leiden meridian circle, at the National Museum Boerhaave for the history of the natural sciences.

10.5 The Oort Cloud

At the end of the War, two students in Leiden were preparing a doctoral thesis. We have already seen that Leendert Binnendijk had been working under Hertzsprung on a study on photometry of the star cluster the Pleiades. After Hertzsprung had left Leiden in 1945 for Denmark, Oort had assumed guidance of the project, and Binnendijk defended his Ph.D. thesis in 1947 with Oort as supervisor ('promotor'). The other one was Adriaan van Woerkom, who had already collaborated with Oort on the vertical force field in the Galaxy, resulting in a joint paper (Oort and van Woerkom 1941b; see Sect. 9.4). His thesis was the result of a suggestion by Hertzsprung concerning the origin of comets and the dynamics of their orbits. Since this was also the competence of Jan Woltjer, it is not inconceivable that Woltjer had been involved intimately with van Woerkom's work; he was thanked profusely in the preface to the thesis. Since Hertzsprung had left Leiden after the end of the War, and Woltjer had died as a result of the sufferings in the 'hunger winter', it was natural that Oort would act as van Woerkom's 'promotor', which resulted in a Ph.D. thesis defended in 1948 (for both theses see Appendix A.4).

Comets are small bodies in the Solar System. The best known example is Halley's Comet, which has returned to the inner regions of the Solar System every 76 years or so, most recently in 1986. Comets are made of rock and dust and frozen gases (mostly water, but also carbon oxides, ammonia and methane). When they come close to the Sun the gases evaporate and the radiation pressure of the Sun pushes them out, taking along part of the dust. Reflected sunlight then makes this visible as diffuse, long tails, pointing more or less away from the Sun and not delineating the trajectories of the comet (see the broad tail of Kohoutek's comet in Fig. 10.14). In addition charged particles (ionized by the Sun's ultraviolet light) are ejected and these interact with the solar wind, the matter, consisting also mostly of charged particles flowing out of the Sun, and this forms waves and other structures, pointing more closely away from the Sun. So-called short-period comets (of which Comets Halley and Kohoutek are examples) have elongated orbits, which according to Kepler's first law are ellipses with the Sun in one of the foci. The orbits are highly elongated and the distances of closest approach to the Sun (the perihelion) are rather small; in the case of Halley it is about 0.6 Astronomical Units (AU; the mean distance of the Earth from the Sun) and the largest distance, the aphelion, some 36 AU. Generally speaking comets can be observed when they are not more than about 2 AU from the Sun.

Fig. 10.14 Kohoutek's comet photographed by the author on January 21, 1974 with the 48 in. Schmidt Telescope at Palomar Observatory. The stars are short trails; the telescope was moved so as to follow the comet. The length of the tail is about 6° on the sky

The problem was their origin. After a limited number of passages near the Sun, the comet will have evaporated or disintegrated. There needs to be a constant supply of new comets. One of the leading theories at the time was that comets come from interstellar space and are intruders in the Solar System. This hypothesis had been put forward by Laplace in 1804. Other suggestions were a local origin, such as the idea that comets are the result of explosions on the large planets, by Italian-born, French mathematician Joseph-Louis Lagrange (Giuseppe Lodovico Lagrangia; 1736–1813). In the case of Laplace the major planets, in particular Jupiter, were supposed to force some of these incoming comets into the elliptical orbits that we observe. Evidence for this was the fact that the distribution of aphelia of comets showed a pronounced peak near the mean distance of Jupiter from the Sun and some weaker peaks for Saturn, Uranus and Neptune. Van Woerkom had studied the properties and systematics of the perturbations of Jupiter.

Oort had been fascinated by all of this and had decided to look further into it (For another excellent discussion of Oort's work on comets, see M. Schmidt's contribution to the *LibAm80* [2]). Now various workers had calculated back from the orbits of comets and of Jupiter, what the original (in-coming) orbit had been. This could not be observed since new comets on their first entry into the central parts of the Solar System do not become visible until they are within c. 2 AU from the Sun; by this time the perturbations by the major planets have already taken place.

One of the characteristics of a cometary orbit is half its major (longest) axis of the ellipse, usually denoted by a. Usually one uses not a itself, but its reciprocal $1/a$. Some orbits may have values of a that approach infinity, so then they come from outside the Solar System. When this is the case, the orbit is a parabola rather than an ellipse, and then $1/a$ equals zero. Such objects have a very small velocity with respect to the Solar System when at a large distance. But it is also possible that the velocity relative to the Sun is larger than required for a parabolic orbit, in which case it becomes a hyperbola and $1/a$ then is negative. If comets were to come from interstellar space outside the Solar System, one should expect most of them to come into the Solar System at slightly hyperbolic orbits. In terms of eccentricity e (related to the ratio of the short and long axes of an ellipse), we have for a parabola $e = 1$ and $e > 1$ for hyperbolas.

Van Woerkom had studied the effects of Jupiter on comets that fall into the Solar System on parabolic orbits. The perturbations force the comets into different orbits and what those orbits will look like depends of course on the angle of approach and the closest distance of the comet to Jupiter. From his study van Woerkom found that *on average* the effects of a passage through the inner Solar System results in a change in the reciprocal major axis $1/a$ by an amount of ± 0.0005. Thus the comet entering the Solar System on a parabola will be forced into a long ellipse, with on average $1/a = 0.0005$ or a major axis of 4000 A.U, or into a hyperbola with $1/a = -0.0005$ and then it will leave the Solar System again. If on the elliptical orbit, subsequent passages will result in further changes of the property $1/a$. From a constant supply of new comets on parabolic orbits (so from outside the Solar System) this is a process that looks much like diffusion, and it will eventually give rise to a more or less uniform distribution of $1/a$. But this is very much unlike observations, so comets could not come from outside the Solar System. Van Woerkom also concluded that the distribution could not be caused by various other mechanisms that he could think of. He already noted an absence of hyperbolic orbits among apparently 'new' comets, which could only arise when the Sun and the interstellar reservoir move at the same speed. There was no satisfactory solution to the problem if young comets came from interstellar space. Expulsions from the major planets therefore did not lead to any successful explanation of the distribution of $1/a$.

After obtaining his Ph.D., van Woerkom moved to Yale Observatory, where Dirk Brouwer was Director, and there he worked on numerical integrations of orbits in the Solar System. Eventually he left astronomy and worked on combat systems and other things related to submarine warfare, first in a commercial company and later at the Naval Underwater System Center, all in Connecticut. The problem of the origin of comets was left unsolved. Oort's intuition told him there was more to be deduced from the data.

The theory of a cloud of comets that became known as the Oort Cloud, was presented first as a scientific paper (Oort 1950b), but it was also—very fittingly—the subject of Oort's Halley Lecture, delivered on May 1, 1951 at Oxford (Oort 1951j), which—he noted—was the first Halley Lecture on comets. Oort had at his disposal reliable initial orbits of 19 comets. These showed a very strong peak in the distribution of $1/a$ for very small values; ten of them were between zero and 0.00005

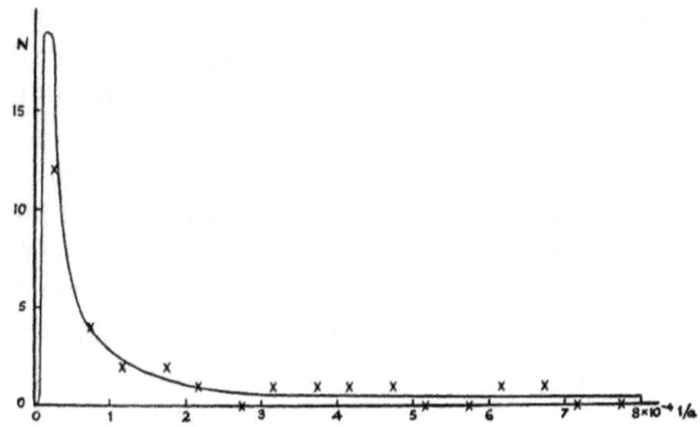

Fig. 10.15 Distribution of the reciprocal semi-major axes $1/a$ of long-period comets with major axes $2a$ in excess of 2500 AU. From Oort (1951j)

and four between that and 0.00010. None were negative, so hyperbolic orbits seemed to be absent, as van Woerkom already had concluded. The average major axis of the ten comets in the first bin was 110,000 AU (about 0.5 pc). In the BAN-article (Oort 1950b), he presented this as tabular material, but in the Halley Lecture he showed a graph (Fig. 10.15). From this he deduced (Oort 1950b)

'... that a sensible fraction of the long-period comets must have come from a region of space extending from a distance $2a = 20{,}000$ to distances of at least 150,000 AU from the Sun; that is, almost to the nearest star [which is at $\pm 270,000$ AU]. This does not mean that they are interstellar. They belong very definitely to the Solar System, because they share accurately the Sun's motion. Yet, the prevalence of these very large major axes has led several astronomers to investigate the question, whether the comets could not be of interstellar origin. It is evident that they cannot *directly* come from interstellar space, for in that case there would have to be many more outspoken hyperbolic orbits rather than nearly parabolic ones. So far no comet has been found for which the eccentricity exceeds 1 by an amount large enough to be considered as real.'

That region of space then is the Oort Cloud. But this result did not really solve the problem, because what had to be understood now was the origin and the mechanism that gave rise to the comets. The disturbances of Jupiter (and to a minor extent of Saturn) on comets that are already in the inner Solar System may result in forcing some of them on longer orbits. It follows then that in twenty million years or so all such comets will be removed from the Solar System. So the first challenge to Oort was to answer the question, what the mechanism was to get some of the comets present in the cloud to be directed rather accurately towards the Sun (so that the reservoir of the ones that can reach us would not be quickly exhausted). And, secondly, to explain how this cloud had originated in the first place.

To address the first question, Oort considered perturbations by passing stars. The effect on the velocity of a comet in the cloud depends on three parameters: the mass of the star, its velocity with respect to the Sun and the minimum distance it would have passed the comet had the latter not been deflected by the star's gravity. This was expressed in a formula that approximated the change in velocity of the comet with sufficient accuracy. This could be integrated over all possible values for the three parameters. Putting in some reasonable distributions for these then led to an estimate that the cloud will not be disrupted for distances from the Sun less than 200,000 AU, which happens to be comparable to the outer radius of the hypothesized cloud.

Oort therefore looked at the disturbances of passing stars that may 'direct' comets in the cloud to the inner Solar System, for which he takes a radius of 1.5 AU. For an assumed distribution of comet velocities at 50,000 AU (which is not very critical) he then estimated the rate at which comets undergo this process and from that he concluded that there must be a reservoir of some 2×10^{11} (200 billion) comets in the cloud. The total mass of all these objects was estimated between a tenth and a hundredth of that of the Earth.

For the origin of the cloud, Oort hypothesized that they came from the belt of asteroids (or minor planets) between Mars and Jupiter. In the early Solar System there must have been very many more asteroids— the current ones do add up to a total mass only somewhat less than one thousands of that of the Earth. If the asteroid belt is a failed planet then there must have been a large factor more of them. Oort now proposed that Jupiter (and maybe Saturn) had forced almost all of the original ones into orbits with large aphelia (maximum distance from the Sun), where disturbances by passing stars had put a fraction in less elongated orbits and that these now constitute the Oort Cloud.

The theory explained the observed random distribution of planes of comet orbits. These are not confined to the region around the orbit planes of the planets (which are the same to within a relative angle of a few degrees) while comets seem to orbit the Sun in all possible planes, with no pronounced preponderance of direct (comets going around in the same general direction as the planets) or retrograde orbits. The equilibrium distribution of cometary orbits depends on the rate at which they evaporate and disintegrate under the influence of the solar radiation and heating.

With the limited data available, Oort estimated the equilibrium distribution of the values of $1/a$'s of comets, which agreed well except for the smallest values. There were too few 'new' relative to 'old' comets. He asked Maarten Schmidt to collect information on the properties of comets, which resulted in an extensive study (Oort and Schmidt 1951c). Once comets have had one or more perihelium passages, their evaporation becomes faster and they brighten more quickly. This was also evident from the fact that new comets have more diffuse, continuous spectra (reflected sunlight) than older ones which show emission lines from excited gas. Predictions of spectacular new comets that enter the inner solar system for the first time often cause major disappointment as a result of the slow development of such new comets, as in the case of the publicity hype about comet Kohoutek in 1973— see Fig. 10.14—for which both Oort (1974e) and Schmidt correctly stated (before and afterwards) that it would not be as spectacular as predicted. By the way, comet Kohoutek was forced into a hyperbolic orbit during its passage throught the Solar System and will never return.

The reception of the model of the Oort Cloud was generally positive. One exception was Raymond Arthur Lyttleton (1911–1995). When Oort discussed the origin

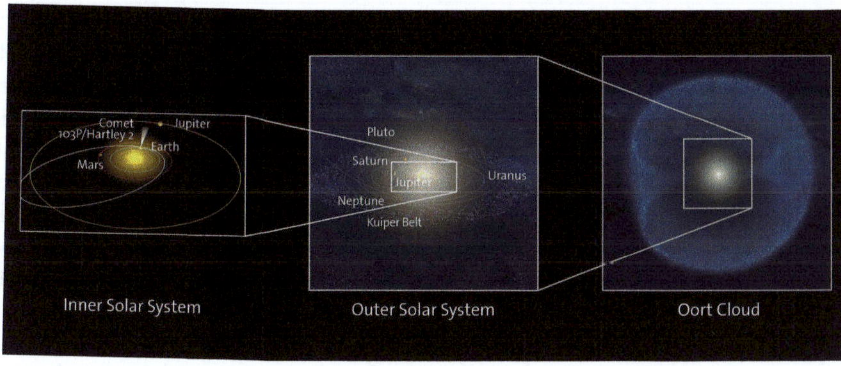

Fig. 10.16 Artist impression of the Oort Cloud. The Kuiper Belt, named after Gerard Kuiper, contains objects like dwarf planets, of which Pluto is an example. Copyright: ESA/AOES Medialab [276]

of comets during his lecture course on Stellar Dynamics in Leiden in the academic year 1967–68, a talk was published that Lyttleton had given to the Royal Astronomical Society, criticizing the Oort model (lecture summarized in the meeting's report [274]). Two of my fellow students, Rudolf Samuel le Poole (b. 1942) and Peter Katgert (b. 1944), wrote a rebuttal [275], pointing out that Lyttleton's mathematical analysis had incorrectly assumed that there was a maximum in the distribution of $1/a$'s. All Oort needed was to note was that these cluster between 1/20,000 and 1/50,000 AU^{-1} without introducing a maximum.

Oort was sometimes teased by family members with having something named after him that is quite invisible; for an impression of what the Oort Cloud looks like, we have to resort to artists impressions such as the one in Fig. 10.16.

10.6 Discovery of the 21 cm Line

The discovery of the 21 cm line of neutral hydrogen (HI) started with the first radio observations by Grote Reber, Oort's inquiry into a possible spectral line at radio wavelengths to study the motions and distribution of the gas compared to stars and how this fitted in with Baade's concept of Stellar Populations, and van de Hulst's prediction of the observability of the hyperfine transition in the neutral hydrogen atom (see Sect. 9.8). The important references on this subject are the book by Woodruff T. Sullivan, the paper by Hugo van Woerden and Richard G. Strom (referred to earlier; see [251, 253]), and *EarlyRadio* [9] by Astrid Elbers. Sullivan's chapter in the *Legacy* [6] has a detailed discussion of the issue and further interesting information, as does the introductory chapter in the *JKM-Inventory* [1] by Jet Katgert-Merkelijn.

Already on August 30, 1945, Oort wrote to Grote Reber, through Otto Struve at Yerkes, which was the nearest observatory (Reber lived in Wheaton. Illinois). He explained he planned to erect a 'large mirror and receiving set for ultra-short waves' and asked him to give 'a rough idea of the expenditure for your 31 feet mirror'. Reber wrote back on September 17, quoting an initial (1940) expenditure of $ 676.96 for the structure. Later he spent $ 500 on 'electronic apparatus', $ 553.98 for improved equipment, $ 250 on 'improved power supplies and signal generator' and $ 1000 for 'development of electronic equipment'. In addition there had been 'a very large amount of labor'. He had in the mean time spent $ 2000 for 'development work at a frequency of 480 megacycles', but had only detected the Sun and 'cosmic static from the Milky Way has so far eluded me at this higher frequency'. The value of the dollar in the early forties is equivalent to about ten dollars today. Reber ended the letter with:

> 'Your interest in this work is greatly appreciated by me because for a very long time now, it seems I have been the only enthusiast and it has been rather lonesome. Please keep me informed of your progress and if I can be of any aid please write.'

In the summer of 1945, Oort again contacted Cornelis J. Bakker (who had spoken together with van de Hulst at the 1944 meeting), at Philips about receiver issues. Bakker had an interest in ionospheric research and he had during the War published with Minnaert in Utrecht, Pannekoek in Amsterdam and Jan Veldkamp (1909–1994). Veldkamp was a geophysicist who worked at the Royal Netherlands Meteorological Institute (KNMI) in Den Bilt near Utrecht. Oort also involved his old friend Jan Burgers at Delft, whose interest in the project concerned the construction of the telescope structure. Burgers immediately used his contacts at 'Werkspoor', a machine factory and builder of railway infrastructure, bridges and similar structures, about the structural aspects of the dish. The mention of Werkspoor is interesting; Oort originally must have thought the instrument could be built by the workshop of Leiden Observatory (after all, Reber had built his telescope by himself), but quickly realized that this was in fact well beyond its capabilities. Oort also wrote Balthasar van der Pol (1889–1959), who also worked as a physicist at Philips and was an expert in the transmission of radio waves. Shortly after the War expertise on radio receivers was not very developed in the Netherlands (there were radar activities at Philips, but these were not revealed), and what there was of it was concentrated on longer wavelengths than 21 cm.

The contacts of Burgers with Werkspoor quickly resulted in a very preliminary design ('avant projet') of a radio telescope of diameter 25 m, sent to Oort on September 15, 1945. The problem of course was money. Oort decided to contact directly the then Prime Minister, which between June 1945 and July 1946 had been Willem Schermerhorn (1894–1977). Schermerhorn was a Professor of civil engineering in Delft. Oort wrote to him on November 8, 1945; he must have known him well; he was on familiar terms with him (he addressed him as 'je', rather than the formal 'u' and praised him for his courage to take on this job). This led to a visit of Oort to the Prime Minister on November 16.

In doing so, Oort bypassed in this the official channels. The formal way was would have been through the Curators of the University, but the Government Accounts Act ('Comptabiliteitswet') of 1929 was very bureaucratic. In that law the financial management of the University institutes had been centralized away from the institute Professors/Directors to the Curators, who often were lawyers and had no idea of urgency or importance of scientific initiatives and had little understanding of the daily practice of scientific research. Turning directly to the ministry was quite common practice, especially in Leiden which is located so close to Den Haag. After the War the Prime Minister had instituted a committee to advise on a new structure of higher education, resulting in a structure in which eventually (1960) the universities controlled their own income and capital [277]. But professors kept turning directly to the Ministry. Officials at the Ministry started to block this route increasingly during the late 1940s, but it never disappeared completely [278]. Turning to the Prime Minister certainly was not an act of arrogance or disdain of university administrators on Oort's behalf. It was simply efficient, common practice. Actually, Gerard Kuiper also visited Prime Minister Schermerhorn and impressed on him the need for investment in scientific research (see discussion in [279]).

This all led to the preparation of a proposal, which Oort first sent to the Royal Academy (KNAW) for endorsement and then to the Minister of Education, Arts & Sciences, Gerardus van der Leeuw (1890–1950), a Professor of theology in Groningen. The amount required was on the order of Dfl. 100,000 and such funds were not available.

In the next episode an important role was played by Herre Rinia, at the time one of the Directors of the Philips Physics Laboratory NatLab. He had made Minnaert aware of the existence of remnants of the German defense radar antennas, part of the Atlantikwall. These 'Würzburg-Riese' (Giant Würzburg) antennas had a diameter of 7.5 m and were left at many places in the Dutch dunes. Anthonet Hugo de Voogt (1892–1969) (see [280] for an article on de Voogt), head of the Radio Service of the Dutch Post, Telegraph & Telephone Company (PTT), was interested in long-distance radio transmission and the effects on it by the ionosphere and the relation to activity on the Sun. PTT had acquired a number of the antennas and de Voogt was keen to use these for his ionospheric research. Because of this he had already been in contact with Minnaert in Utrecht and his Adjunct-Director Jacob Houtgast, who were also interested in ionospheric (and solar) research. Ultimately, in 1946, the PTT through de Voogt made one of the 'Würzburgs' available for Dutch astronomy. This was a much more realistic approach than directly funding the construction of a 25 m dish, when the 21 cm line had not yet even been discovered.

This left the problem of how to fund the operation of the telescope and how to obtain a receiver. We have seen (Sect. 10.4) that in 1947 ZWO had been founded in a preliminary form to fund fundamental research, much along the lines Schermerhorn had envisaged, and that Bannier had been appointed its Director. Oort had realized, following a suggestion by Jan Burgers, that since Leiden Observatory would not be able to handle the construction of a radio telescope, a new organizational form was necessary and in 1949 this resulted in the founding of the 'Stichting Radiostraling van Zon en Melkweg' SRZM (Foundation Radio Radiation from Sun and Milky Way, which in English became know as the Netherlands Foundation for Radio Astronomy, NFRA); currently the Foundation is named ASTRON. SRZM brought together various players in the field, which proved to be vital to get radio astronomy from

Fig. 10.17 Photograph from 1951 of the Würzburg-Riese 7.5 m antenna that had been erected at Kootwijk as the first radio observatory in the Netherlands. *Courtesy* CAMRAS/ASTRON [281]

the ground, by appointing these on the Board. Oort was the chairman; astronomical members were van de Hulst (who had been appointed Lecturer in Leiden in 1948), and Minnaert and Houtgast from Utrecht. Philips Natlab was represented by Rinia and Frans Louis Henri Marie Stumpers (1911–2003), who was a radio engineer. From the PTT de Voogt was appointed and from the KNMI two geophysicists, Jan Veldkamp (1909–1994), who in addition to (paleo)magnetism had an interest in studies of the ionosphere and Felix Andries Vening Meinesz (1887–1966), at that time Director, and an expert in gravity measurements. This provided a useful base. Leiden was interested in Galactic research and Utrecht in solar and ionospheric studies. The Kapteyn Laboratory in Groningen was not represented. Under van Rhijn it had gone into a decline and there was no interest in radio work. Elbers (in *EarlyRadio* [9]) reported that in his correspondence with Oort, van Rhijn complained bitterly about his exclusion from the SRZM Board. Later this was remedied and in 1953 van Rhijn joined the Board of SRZM (Fig. 10.17).

Blaauw said a few things about this in the *JKBlaauw-interview*.

'Of course I knew van Rhijn well. After the War our contacts were few at first, because he had contracted tbc (tuberculosis) during the War and had been in a sanatorium the last years. [...] I think he could have retired but I always thought that he felt obliged to continue in order to keep his salary since he still had young children. So he was a man of very limited energy and that early phase of radio astronomy was very new. I think his mental and physical condition kept him from enthusiastically joining in, but he also has said that he found it difficult to justify to ask for so much money for an undertaking that in first instance was not so useful for the rebuilding of the nation when at the same time so much was required for just that. He was not the only one and I think Minnaert felt the same way. But at some stage van Rhijn became a member to show it was a national undertaking. But van Rhijn was not very active at all in this role.'

In 1949, the Board of SRZM applied to ZWO for Dfl. 200.000 for the construction of a 25 m radio telescope, but it was not approved. The annual budget of ZWO was about 1.25 Mfl, so it was 16% and therefore certainly a significant part of it (but expected to be spread over several years). Perhaps this refusal should not come as a surprise, since after all, the 21 cm line, which was a major pillar of the scientific case, had not even been observed. But with the acquisition of one of the Würzburg antennas the search would be possible. What was needed was a receiver and an engineer to operate it. The physicist Cornelis Jacobus Gorter (1907–1980) of Leiden University provided an adapted radar receiver that had been left as surplus American stock. And a student of electronics was found to install the receiver and search for the 21 cm line. This person, by the name of H. Hoo, turned out not to have the competence required. According to a letter of Oort to him in June 1951, he actually had an engineer's title, so he did at some stage finish his studies at the precursor of Delft Technical University. But then, in March 1950 a small fire destroyed most of the receiver before any observations could be done, which resulted in a serious delay.

Before continuing a few words need to be said about techniques of radio astronomy. The beamwidth of the telescope, the smallest structure that can be seen on the sky, is proportional to the wavelength and inversely proportional to the diameter of the dish. For a dish like the Würzburg antenna of 7.5 m, the beamwidth for observing at 21 cm wavelength is about 1.7°. This is a fundamental limitation and derives from the way the image is formed in the focal plane. The dependence holds for electromagnetic radiation at all wavelengths, so at optical wavelengths and with a reasonably sized telescope it is less than an arcsec.

One way of describing radio radiation is as an electromagnetic wave, in which a periodically changing electric and magnetic field is propagating through space. A radio telescope has a dipole in the focal plane, which basically is a rod made of some conducting material. The electric field of the wave moves the electrons in the dipole, so that radiation moving perpendicular to the dipole with a wavelength of twice the length of the dipole produces in the dipole a coherent alternating current. When the dipole is halfway cut in two and the ends connected to a receiver, a coherent alternating current in the dipole will flow through that receiver. In modern radio telescopes this is then 'pre-amplified', but this was not done in the early days (Fig. 10.18).

The signal needs to be amplified before being recorded, but this is more difficult at higher frequencies, so the signal is first converted to a lower frequency, the so-called Intermediate Frequency (IF) by 'mixing' it with a signal with a slightly different frequency. This is called a heterodyne technique, originally super(sonic) heterodyne, which was invented during World War I. Much like the 'beats' one hears when two poorly tuned musical instruments are heard together, the mixing produces a signal with a frequency equal to the difference between the two. The signal introduced for the mixing is produced by a 'local oscillator', a noise source of well-defined

Fig. 10.18 Lex Muller descending from the Kootwijk Radio Telescope. From the Archives of Leiden Observatory

frequency. This heterodyne technique was standard procedure. In the Kootwijk case the IF was around 30 MHz. The gain in the receiver is usually very unstable and needs to be monitored continuously. The solution was the 'Dicke switch' technique, developed by Robert Henry Dicke (1916–1997) during World War II, while working on microwave radar, in which the signal is compared to a stable noise source in the receiver by rapidly switching between the two signals and recording the difference. In the case of the 21 cm line detections, the signal was mixed once more with a signal from a second local oscillator to tune the frequency of the observation, before being amplified and detected. Now we return to the Kootwijk developments.

The turning point came with three new developments. First, Hoo was taken off the job (according to the letter of Oort to him of June 1951 he seemed to have gone to Indonesia on a new job) and in December of 1950 a new engineer was hired, Christiaan Alexander (Lex) Muller (1923–2004). Muller had just finished his studies at the (what we now call) Technical University Delft. Secondly, Stumpers from Philips (and a Board member of SRZM) provided equipment for the development of a new receiver. Muller, with the help of an assistant, set out to put a working receiver together. The third development was that van de Hulst early in 1951 had been appointed as visiting professor at Harvard to teach a course on radio astronomy. It is generally accepted that Bart Bok, who had realized the potential of radio astronomy

Fig. 10.19 Receiver used for the Kootwijk detection of the 21 cm line. The person sitting in the chair is Lex Muller. From the Oort Archives

for studies of the Galaxy and wanted to go into it as well, had been the person that had arranged this. At Harvard, van de Hulst found that a search for the 21 cm line was on there as well. Edward Mills Purcell (1912–1997) was a physicist interested in electrical engineering, who in 1946 had co-discovered nuclear magnetic resonance for which he would share the Nobel Prize in physics in 1952. He had learned about the 21 cm line from the publications of Shklovsky and when his student Harold Irving Ewen (1922–2015) was looking for an Ph.D. thesis subject, he suggested to search for the line (Fig. 10.19).

The story of the discovery of the line is an excellent case of cooperation rather than competition in science. The technique was to measure at the frequency of the line (1420 MHz) and then switch to a different frequency where no signal was expected, and subtract the outputs. Now Muller had realized that the receiver would be much too unstable for that, so that some kind of Dicke-type switching would be required. In the *LibAm80* [2], the Liber Amicorum for Oort's 80th birthday, he wrote:

'Our first experiments with part of the receiver had clearly shown that the stability of a simple non-switching receiver would be insufficient, and from the literature I had available, it was clear that it would be necessary to use some form of Dicke receiver in which the input was switched periodically between the antenna signal and the signal from a noise source. A synchronous detector would then measure the difference between the two signals. Then the letter arrived [from van de Hulst] with the proper solution to our problem: the noise-source signal could be replaced by an antenna signal at a different frequency. The construction of the frequency-switched receiver did not take too much time.'

So Ewen had come up with the solution to do the switching *in frequency*, comparing the signal at the frequency of the line to that of the sky at a nearby frequency where no line emission would be expected, rather than to a signal from a local noise source. However, Ewen and Purcell had no idea about astronomy and were looking for a line with a very narrow frequency width such as one would see in the laboratory. Van de Hulst on the other hand told Ewen that the reference frequency had to be much further away from the line than he had assumed, and on the other hand told Muller about the approach of switching in frequency. In an interview with W. Sullivan, Oort mentioned that the PTT arranged a long telephone conversation between him and van de Hulst at Harvard during which 'we exchanged the news about the 21 cm line' [282]. Intercontinental telephone calls at that time would have cost a fortune.

In the *JKHulst-interview* with Jet Katgert-Merkelijn, van de Hulst noted:

'The three of us were in one way or the other concerned with the 21 cm line that had not been found yet and Ewen was the man who apparently had worked a number of summers at the [Harvard cyclotron] as a technician and he knew all the modern techniques. That part he knew, but he had no idea of astronomy and neither did Purcell. They used an estimate for the width of the line that was too small, I think, by a factor of 20, since you have to know what velocities exist in the Galaxy, so I recommended him to shift his comparison frequency much further out. This way I picked up things that I forwarded to Muller. Maybe you should do things this or that way. It was a trick Ewen had thought of and it is no accident that [the line] was found about the same time from those different places.'

Ewen and Purcell detected the 21 cm line on March 25, 1951. Muller received the letter from van de Hulst early March and it took him six weeks to adapt the receiver to perform frequency switching. He found the line on May 11, 1951. As it turned out, Ewen and Purcell had detected only part of the line. Their bandwidth (the frequency range they observed) was 17 kHz, which at 1420 MHz corresponds to only 4 km/s. The switching was done by a frequency step of 75 kHz (15 km/s), while the range of velocities expected for hydrogen gas in the Galaxy is tens of km/s or maybe even more than one hundred. Ewen and Purcell also suffered from the fact that their beamwidth was larger than the width of the Milky Way. They used a horn sticking out of a window of their laboratory, which produced a beam of 12° full-width-half-maximum (FWHM) and this significantly diluted the signal. For Muller the numbers were more favorable. His beam was $2^{\circ}.8$ and the bandwidth of the line channel 25 kHz (6 km/s) and the Dicke switch offset 110 kHz (23 km/s). Still this was not really adequate to measure the full extent of HI (neutral hydrogen), which as a result of differential rotation can range over 150 km/s or even more in some lines of sight through the Galactic disk. The American receiver was more sensitive, but this was offset by its properties as summarized here.

Figure 10.20 shows early (June 4) recordings from Kootwijk. What is recorded is the difference between the 'on-line' channel and the comparison channel. Every

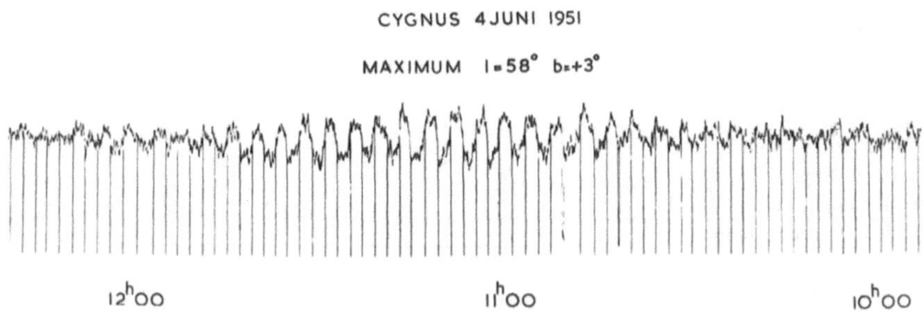

Fig. 10.20 Strip-chart recording at Kootwijk in June 1951, weeks after the first detection. During more than 2 h the Milky Way (around the constellation Cygnus) drifted through the antenna beam of the (stationary) telescope. Every 2 min (the vertical lines) the receiver switched interchanging the on-line frequency and that of the comparison channel. From the Oort Archives

2 min the two are interchanged and the difference signal changes sign. The Dicke switching itself between these two channels was done 30 times per second.

At Harvard there was also a radio engineer from Australia, Frank John Kerr (1918–2000), who had come to Harvard for a year to learn astronomy. He notified Wilbur Norman (Chris) Christiansen (1913–2007) in Sydney, who together with James V. (Jim) Hindman (1919–19??) of the Radiophysics Laboratory of the Commonwealth Scientific and Industrial Research Organisation (CSIRO) in Sydney, Australia, in an great hurry put together a receiver and were able to confirm the line towards the end of June, using a 'narrow-beam aerial', a directional antenna system developed for other purposes [283, 284].

The results were published jointly in the journal *Nature* (see [285] for reprints of the collection), first an article by Ewen and Purcell [286], followed by one by Muller and Oort (1951h), and the text of a telegram by Joseph Lade Pawsey (1908–1962) on behalf of Christiansen and Hindman. The publication date was September 1, 1951. A more detailed announcement of their observations by the latter two was published in 1952 [284].

The Ewen and Purcell paper did not attempt any astronomical analysis, but commented on excitation of the line and its opacity. Of course, Oort's chief interest had been the implications for Galactic structure. Fig. 10.21 (published in the *Nature* paper) shows results on the hydrogen distribution in scans perpendicular to the Galactic plane. This shows widths of a several degrees, and if the gas layer is thin the gas observed will have to be relatively nearby (a few to several hundred pc). In the Outer Galaxy, the layer thickens so much that it subtends quite a few degrees, but is many kpc distant. The lowest profile (at a longitude corresponding to 30° from the center) is thinner on the sky and the gas presumably more distant. Oort interpreted the velocity data to indicate that the rotation velocity of the Galaxy at about halfway from the Sun to the center was roughly 190 km/s.

Obviously, this was only a first indication and Oort would have wanted to immediately start a complete survey of the 21 cm line emission from the Milky Way as far as observable from the Netherlands. However, this had to wait; in the mean time Muller had started to overhaul and improve the receiver (Fig. 10.22).

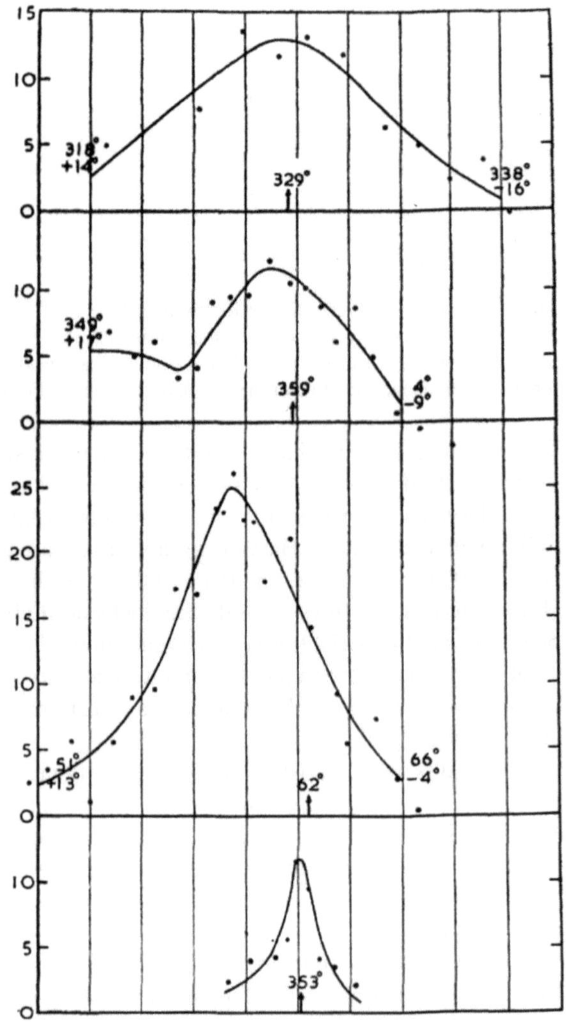

Fig. 10.21 First measurements of the HI distribution in the Galaxy from the Kootwijk Telescope. It shows the amount of neutral hydrogen perpendicular to the Galactic plane at four crossings. These are from drift measurements with a stationary telescope. The arrows show the crossings of the plane and the corresponding longitude; the end-points indicate Galactic coordinates. The vertical lines are 20 min of time apart. This figure was also included in Muller and Oort (1951h). From the Oort Archives

10.7 Radio Continuum

The 21 cm line of HI proved indeed to be an excellent tool to study the structure of the Galaxy. But there was also the more general radio emission that was seen at many frequencies and which Jansky and Reber actually had first observed. The mechanism by which this radiation was produced was not known, and Oort of course wondered if it could be used to provide more information on Galactic structure. If it were associated with or had a similar distribution as the common stars, it would be a powerful tool for tracing the distribution of stellar mass in the Milky Way. He set out to work on this with a student, Gart Westerhout (1927–2012). Westerhout had obtained his

Zaterdag 21 Juli 1951 ELSEVIERS WEEKBLAD

VERDERE TERREINWINST IN HET HEELAL

MELKWEG OP DE KORTE GOLF

Nieuw hoofdstuk in de sterrenkunde

Waterstofatomen, werkend als miniatuurzenders, vergemakkelijken onderzoek van melkwegstelsel

Fig. 10.22 National magazines and newspapers in the Netherlands reported the news of the discovery of the 21 cm line. On the left: 'Hydrogen atoms, working as miniature transmitters, facilitate research of the Galaxy'; on the right: 'Further progress in the Universe. The Milky Way at shortwave radio'. *Courtesy* Ard Hartsuijker, from the Archives of ASTRON Dwingeloo

Candidaats (Bachelor) in 1950 and was set to work on the radio continuum radiation from the Galaxy (Westerhout and Oort 1951d). Westerhout would continue to work in radio astronomy and was involved in major developments in Leiden, including work on the distribution of HI in the Outer Galaxy and a survey of continuum emission from the Milky Way which constituted his Ph.D. thesis in 1958. He moved to the USA in 1962, a move that would prove to be definite. Westerhout is one of the persons figuring in Fig. 10.23, which comes from the group photograph of the Netherlands Astronomers Conference in 1949. This picture also shows Wilhelmina (Wil) Mengerink (1922–2018) who had married van de Hulst in 1946. She had started as a student in astronomy, but switched to psychology. It also shows Hendrik Gerard (Henk) van Bueren (1925–2012), who worked on astronomical projects with Oort (such as a study of proper motions in the Hyades star cluster) before writing a Ph.D. thesis in physics in 1956. Van Bueren worked at Philips NatLab, (see Sect. 9.6), but later became Professor of astronomy in Utrecht, specializing in high-energy and laboratory astrophysics.

Radio astronomy had started at much lower frequencies than the 21 cm line. Jansky had first found his cosmic static at 20.6 MHz (14.6 m wavelength). Reber had tried much shorter wavelengths, starting at 9.1 cm or 3300 MHz, which was the highest frequency technically feasible at the time, finding nothing. Next he tried 33 cm or 910 MHz, but again was not not able to detect anything, and finally succeeded to

Fig. 10.23 Group of persons taken from the group photograph of the 1949 Netherlands Astronomers Conference. Oort is flanked by Henk van de Hulst and his wife Wilhelmina Mengerink. In the second row are from left to right Henk van Bueren, Lukas Plaut, Gart Westerhout, and Marcel Minnaert. From the Oort Archives

detect cosmic noise at 1.87 m (160 MHz). It was immediately clear that it was much too bright to be thermal black-body radiation of material at very low temperatures.

Another possibility was free-free radio emission, also referred to as thermal radio emission. Such radiation is emitted by regions of ionized gas. These contain mostly hydrogen nuclei (protons) and electrons that have been detached from the hydrogen atoms by energetic photons from hot stars. Now, if an electron interacts with a proton two things may happen. The electron can be captured again (often in a high energy level), and then cascade down to the lowest levels, giving rise to the emission lines in optical spectra. But it is also possible that the electron remains unbound; in that case there is an exchange of energy, the electron moving from one unbound to another unbound state. This gives rise to the emission of a photon, predominantly in the radio regime. This free-free transition does not occur in discrete steps as in the bound states of an electron and therefore the free-free radiation has no characteristic wavelength. But of course, it does have a specific wavelength *dependence*, corresponding to the energies the electrons have (the electron temperature).

Westerhout and Oort started out by looking into the possible contribution from this free-free radiation to the observed map of the Galaxy's radio emission. For this they used the survey by John Gatenby Bolton (1922–1993) and Kevin Charles Westfold (1921–2001) of CSIRO Radiophysics in Sydney, who mapped the southern sky at 100 MHz at a resolution of 17° [287]. The disparity between a model and the data both in distribution and intensity, led Oort and Westerhout to conclude that the contribution from the interstellar gas is negligible at this frequency. As they acknowledged, this had actually been concluded by others before them.

O- and B-stars could be eliminated on the same grounds. So Westerhout and Oort looked at objects that have the same spatial distributions as the common stars. A model for the Galaxy consisting of a few superimposed ellipsoids could provide a reasonable fit to the observations. It was based on previous work Oort had published, in particularly Oort and van Woerkom (1941b). The model fitted well at low Galactic latitudes, especially in the hemisphere around the direction of the center of the Galaxy, but in the opposite hemisphere and far from the Milky Way the radio map showed distinctly more extended emission than could be accounted for by the model. They tried to estimate what the properties of these 'radio stars' could be, making various assumptions of the distribution of luminosities. In the meantime the Crab Nebula had been found to be a discrete source of radio emission, but it was much too bright to be an example of the population required to explain the Galactic radio-continuum distribution. Finally Westerhout and Oort pointed out that in their model the Galaxy would be some seven times brighter than the Andromeda Nebula.

In 1951 Oort was invited to deliver the prestigious Henri Norris Russell Lecture of the American Astronomical Society, which he did in Cleveland, Ohio, on December 27. The title was *Problems of Galactic Structure* (Oort 1952b). In this lecture Oort first described what he called the first two great developments in investigations of the Galactic System. The first started with Herschel and had its culmination in the work of Kapteyn, who 'succeeded in determining the structure in those parts of the system that are more than about 20° from the Galactic plane, [...] and in the course of this work discovered the first dynamically important phenomena in stellar motions.' The second great development started with the work of Shapley on the system of globular clusters and culminated in Oort's own discovery of differential rotation and dynamics of the System. The third phase 'has begun by the successful reception of radiation at radio frequencies'.

Oort spent some time on showing how all determinations of the longitude of the Galactic center, from star counts, globular clusters, differential rotation, high-velocity stars and recently radio radiation maps, agreed and gave a combined value of 327° ± 1°. Next he reviewed the parameters of Galactic rotation, which gave a combined set of: distance to the center of 9.4 ± 1.2 kpc, local rotation velocity 259 ± 32 km/s and Oort constants $A = 20.6 \pm 1.4$ and $B = -6.9 \pm 0.7$ km/s.kpc. The 21 cm hydrogen line would improve that shortly and already indicated somewhat lower values for the distance and rotation velocity. Although there is no specific reference to this, the values for the Oort constants are close to those derived by Herbert Rollo Morgan (1875–1957) and Oort in a critical discussion of the constant of precession and those of Galactic rotation (Morgan and Oort 1951e). Morgan was

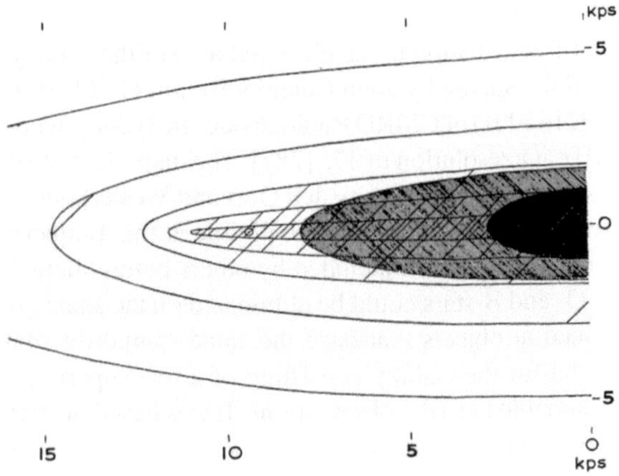

Fig. 10.24 Model of the Galaxy based on superimposed ellipsoids. The small central one is 2.8 × 1 kpc and the density 3.13 times that near the Sun. The next one is 8 × 1.6 kpc and has density 2.15. Then there are five outer ellipsoids, the innermost 11 × 0.3 kpc and density 0.446, the outer one respectively 20.3 × 4.9 and 0.003. From Oort (1952b)

an astrometrist at that time working at Yale Observatory, and this paper was the result of an international meeting in 1950 in Paris on *Constantes fondamentales de l'astronomie*, where they were asked to prepare this note.

Oort presented his best 'schematic' model of the Galactic System in Fig. 10.24. It consisted of superimposed ellipsoids that would represent the distribution of mass in the Galaxy (for details see the caption of the figure). This was essentially the model he and Westerhout had used to derive the distribution of radio radiation on the assumption that the 'radio stars' responsible for it were distributed like the common stars. This Oort presented next. But he added a figure, that displayed what we now would call the rotation curve (Fig. 10.25), which displays the rotation velocity as a function of galactocentric distance. This could of course be expected to be improved considerably when the Galaxy had been mapped in the hydrogen 21 cm line. His vision for the future was summarized as follows:

> 'The ionized regions also emit continuous radiation at radio frequencies, which should have a well-measurable intensity if observations are made near the Milky Way and with instruments of sufficient resolving power. In the surveys made so far, all with low resolving power, it is not possible to discern how much of the measured intensity is due to continuous radiation from ionized interstellar clouds and how much comes from the distant point-sources. The interstellar radiation is likely to be relatively more important at lower wave lengths.
>
> For wavelengths less than a few meters the absorption coefficient of the gas is so small that we can probably neglect the absorption, even for paths of the order of the diameter of the Galactic System. This circumstance should render it possible to get a survey of the density distribution of ionized hydrogen through the entire system as soon as sufficiently large antennas have been built, if a way can be found to separate the interstellar radiation from that of other sources.'

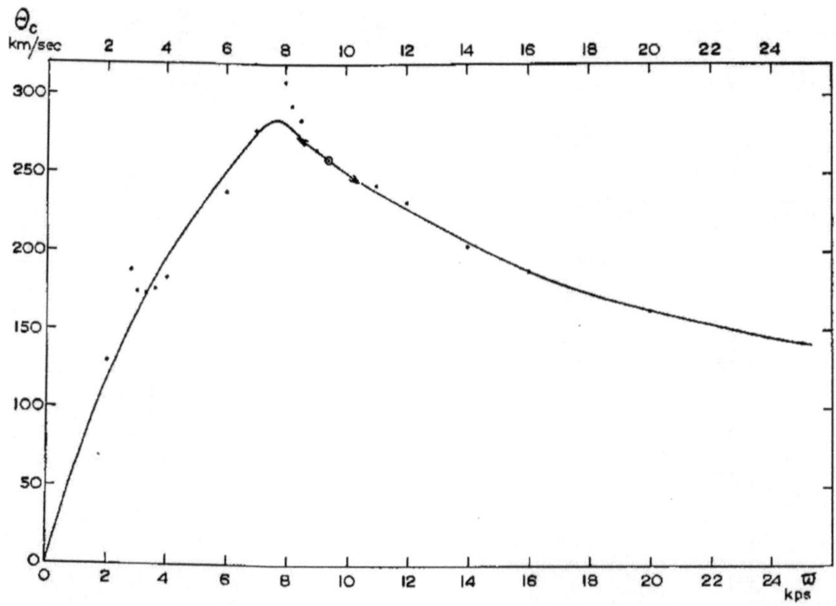

Fig. 10.25 Rotation curve of the Galaxy corresponding to the mass distribution of Fig. 10.24 and the adopted dynamical parameters. The Sun is indicated by the small circle at distance 9.2 kpc and rotation speed 269 km/s. The arrows at that position give the slope of the curve according to differential rotation and the Oort constants. The scatter in points is due to the discrete ellipsoids. From Oort (1952b)

The result Oort was looking forward to was to delineate the spiral structure. He presented some hints from other sources, namely groups of young O- and B-stars, so-called OB associations, that were being studied by Adriaan Blaauw. Proper motions had shown that these associations are expanding, which suggested ages of no more than one to a few million years. Since such entities occur frequently in late-type spiral galaxies and delineate the spiral structure, they may be used to determine the local spiral structure of the Galaxy. It remained, Oort mused,

> 'to discover the relation that should exist between the density variations and that other striking deviation from a smooth state that has long been known, namely, the difference of about 20° between the vertex of the star-streaming of the brighter stars and the longitude of the Galactic center.
>
> It appears probable, however, that, with the accumulation of more, and especially more accurate, data, the near future will bring a solution of these problems, as well as of the intriguing problem of the relation between the distribution of the common stars and the Galactic spiral arms.'

For this he first needed a more sensitive receiver for the 21 cm line at Kootwijk to survey the Milky Way and next a radio telescope with a 25 m dish to improve the resolution at 21 cm from the 2°.8 of the Würzburg dish to 0°.5.

Fig. 11.0 Oort in 1948. From the Oort Archives

Chapter 11
Spiral Structure, Interstellar Medium and the Crab

We should do astronomy because it is beautiful and because it is fun. We should do it because people want to know. We want to know our place in the Universe and how things happen.
John Norris Bahcall (1934–2005)

Young people are drawn toward a scientific career; large-scale facilities are made available that can eventually be used for basic research. Public attention is drawn toward science; inertia, if nothing else, will leave the purses open for some time.
Jan Hendrik Bannier

11.1 All in the Family

When the Oort family had returned to Leiden after the end of World War II, the children continued their secondary education. The older son Coenraad (Coen), after completing school, enrolled in Leiden University to study economics, which was a specialization within the Faculty of Law. He continued to live at the Observatory, occupying the upper part of the Director's residence. After obtaining his Leiden degree, he decided to continue his studies in Chicago, majoring in economics; he moved there in 1951. Marijke went to the Conservatory in Den Haag to study singing and piano, pursuing her musical interests. Abraham (Bram) finished secondary education in 1952 and also enrolled in Leiden University, to study physics (Fig. 11.1).

There are some documents that have been made available to me by Oort's younger son Bram, from which I will quote to give some background to Oort's private life (I changed some incorrectly indicated years). These are reminiscences of him and his sister and brother that give some impression of the family life when the Oort

In *Science* (1951; [268]), commenting on the aftermath of World War II and the science budget.

P. C. van der Kruit, *Jan Hendrik Oort*, Astrophysics and Space Science Library 459, https://doi.org/10.1007/978-3-030-17801-7_11

Fig. 11.1 The Oorts with their two of their children, Marijke and Abraham Hans (Bram) in 1956. There are two versions, slightly different (the other with rather obvious fake smiles). The other son Coen lived in the USA at the time. From the Oort Archives

children were growing up. I start with some notes that Bram Oort wrote about his older siblings.

'Let me go back to my memories of my brother, Coen. After receiving his degree in economics at Leiden University, Coen went to Chicago in 1951 to continue his education. Probably he was influenced by Milton Friedman in Chicago and must have belonged to the Chicago school – being for free trade and less government. While in Chicago, Coen met a young student, Marianne Lissy, whom he later married. Both of them came back to the Netherlands and stayed for several years in an apartment on the third floor of my parents home in the Sterrewacht (Observatory) in Leiden. Two sons, Marc and Maarten, were born there. It was a great blessing for my mother to be able to take care of the grandchildren. My brother received his PhD in economics and got a job as Professor in economics in Utrecht University where they lived until 1971. Later he became Treasurer General of the Netherlands, 1971–1977, and finally he ended up as a top banker in the ABN AMRO Bank. As a banker, he earned a good salary. For him, money was never a problem.

My sister, Marijke, admired her three-year-older brother Coen very much. They were close, and I often felt excluded as the younger brother. In high school, Marijke loved languages and became interested in singing and piano. She pursued a career in singing and piano at the Music Conservatory in Den Haag. She taught many amateur singers and piano students. Later on, to her great regret, she lost much of her hearing. Hearing problems are common in the Oort family; my Dad, Coen, Marijke, and myself have been handicapped by this. Fortunately, I have had some benefit from recent developments in hearing aids. Eventually my sister expanded her musical studies and went on to play harp.'

The next citation is part of the speech (translated from the original Dutch) that Marijke gave at Coen's funeral in 2006:

'When we returned in 1945 to our home at the Observatory we all had considerable problems adapting. Coen went to university and lived on the top floor of our house. Every thirty minutes or so he rushed down the stairs, making a lot of noise, to drink a cup of coffee or do something else. Mother and I asked ourselves all the time if he did actually work on his studies, because he turned up so often. Probably short periods of intensive study were sufficient for him.

In those days he won a prize in a competition by a broadcasting organization for an essay about Europe, and the prize was a trip for two to Rome. Until a week before we were left in the dark about who he would select to accompany him. Only at this short notice did it turn out that I was going to be the lucky one, and not some unknown, imagined girlfriend. We had a great time, attended a reception as odd ducks at the embassy and were received in audience by the Pope.

From the States, where he went to university for a year, we were told on the last page of a long letter, that he had a girlfriend that he intended to marry and would bring back to Holland with him. Contrary to all wise advice Marianne and Coen married in America, actually in the very same week when Tom and I got married: they in America, we in Leiden. They lived at first on the top floor of our house on the Observatory. Marianne enjoyed life in Leiden and we loved her immediately.

The four of us had a great time together, almost every weekend we visited each other. Marianne cooked the greatest dinners and we played all sorts of games, pretending we did that so that Marianne would learn Dutch.'

The essay contest which Coen had won was the ERP Prize; it had been organized by one of the Dutch broadcasting associations. It concerned European integration and 'international understanding'. The ERP is the European Recovery Program, popularly known as the Marshall Plan. The prize was a trip for two to London, Paris or Rome, and Coen had chosen the latter. According to the newspaper clipping from which Fig. 11.2 has been taken, he and his sister were met at Rome Airport by the local official of the ECA, the Economic Cooperation Authority, which had missions in each European capital. The winner was announced in the spring according to the newspaper, while the trip took place in December, so Coen had kept everyone in the dark about who would accompany him for more than half a year.

Fig. 11.2 Oort's elder son Coenraad (Coen) and daughter Marijke arriving in Rome in December 1950 on a KLM DC6 flight. Coen had won a trip for two in a prize essay competition and had elected to take his sister. The newspaper from which this clipping was taken from has not been recorded. From the Oort Archives

At Marijke's funeral in 2008, Bram said among other things (my translation):

'As a small girl Marijke was very good at sports; she loved water, jumped off the highest diving board, and swam as if she were an otter. Later we went sailing in our small cabin sailing boat, the 'Bounty'. I remember a sleepless night with, it seemed, continuous thunderstorms, together with Marijke on the shore of the Braassemer Lake, with buzzing mosquitoes, terrible rains and lightning. Another adventure was to the IJ in Amsterdam, where we slept in the harbor near the red light district. The next day father sailed with us to Enkhuizen and we crossed the IJsselmeer – quite a dangerous undertaking with a boat that is not seaworthy. Many a summer we went sailing, often with the Nauta family in their 'Boek-Boek'.

After we had returned to the Observatory, you went to the HBS for girls and put your talents for languages and literature to good use. Your love for music also developed there. I have fond memories hearing your singing and playing piano in the Observatory house, and of your later performances at family parties. About that time another phase started when Tom entered our family. I remember the enormous, beautiful bouquets with which Tom was courting Marijke. Your long period of engagement and your marriage in '54. Quite unexpectedly Coen just beat you by a few days in Chicago with his love, Marianne.'

And Bram about his mother:

'One aspect of my mother was her love for poetry and arts. Her father was a poet. It looks as if she developed her love for the arts by herself. Because my father worked a lot, she had much time to read and study. After WWII they took a course in arts by H. Bremmer (the painter who brought the paintings of van Gogh to the attention of Kröller-Müller; Bremmer was one of the discoverers of van Gogh's genius). Many of van Gogh's paintings can still be seen at the Kröller-Müller Museum in the Hoge Veluwe [National Park]. An anecdote: my grandmother knew the van Gogh family; she was offered some of his drawings. But she felt they were not too attractive and refused them – to the chagrin of her descendants!

Bremmer advised my parents to buy paintings and sculptures that they liked. In our house [in Vermont, USA] we have two small bronze sculptures my parents possessed; one of a deer running away – made by Raedeker. It is the first study leading to a life-size sculpture in stone that is now located in the center of the Kröller-Müller park. My parents loved it. We have also two paintings painted by R. Martinez in 1934, and a painting of Katwijk by Zoetelief Tromp. His children were very good friends of my parents; one of them, Hette, married an astronomer friend of my father, Aernout de Sitter (son of the well-known Director my father worked for). He corresponded a lot with Hette and Aernout. Aernout died working on the Burma Railroad during the Japanese occupation of Indonesia. Hette and her four children came back after the War and were 'sponsored' by my parents.

My mother was in many respects the 'soul' of the Observatory. Astronomers loved to have dinner with us or come to enjoy a cup of tea. My mother always was prepared to offer tea, biscuits, etc. Many friends loved her for her cheerful and warm character. My father learned a lot from her; without her he might have been a scientist living in an ivory tower.'

Of the persons mentioned, Vincent van Gogh needs no elucidation. The (one of the first major female) art collector Helene Emma Laura Juliane Kröller-Müller (1869–1939), who founded the Kröller-Müller Museum with her businessman husband Anthony George (Anton) Kröller (1862–1941), was advised by Hendricus Petrus (Henk) Bremmer (1871–1956). Bremmer was a painter himself but better known as an art critic, educator, collector and dealer. The sculptor is Johannes Anton (John) Rädecker (1885–1956); for some of his sculptures, among which 'Hinde' referred to, see [288]. Figure 15.6 shows the Oorts in the 1980s with the sculpture. The first painter mentioned must have been Cuban-born painter Raoul Martinez (1876–1974), who then lived in Paris, but spent his last 30 years in the Netherlands. He is also the painter of a portrait on the occasion of Oort's retirement (Fig. 14.0). The other is Johannes (Jan) Zoetelief Tromp (1872–1947). One of the things to add maybe is that

Fig. 11.3 Coen Oort on his way to defend his PhD thesis in 1958 in Leiden. In the carriage with him are his wife Marianne and his first son Marc. Photograph provided by Marc Oort

the Oorts seemed to have read works of William Shakespeare (1564–1616) aloud to each other in the evenings.

Here follows a very brief summary of the lives of the Oort children (Fig. 11.3).

In 1951, Coen Oort went to Chicago on a Fulbright Fellowship, where he obtained a Masters degree in 1954, and where—as mentioned—he met his future wife Marianne Stazie Lissy (b. 1935). They married in Chicago in 1954. After returning from Chicago, he worked on a PhD thesis in Leiden, living again at the Observatory on the top floor of the house of the Oorts. There his older two children were born, of which the first, Marc Jan Anton (b. 1958), would eventually obtain a PhD degree in astronomy from Leiden. In 1958, Coen Oort obtained a PhD from Leiden University, defending a thesis with the title *Decreasing costs as a problem of welfare economics*, after which he took a job at the Ministry of Economic Affairs in Den Haag. In 1960 he was appointed Professor of economics and statistics in the Faculty of Law at the University of Utrecht, where he stayed until 1971 when became Treasurer-General at the Ministry of Finance. This lasted until 1977 after which he held various administrative functions, particularly at a major bank in the Netherlands. His wife Marianne had studied liberal arts at the University of Chicago and in Leiden English language and literature (Candidaats/Bachelor) and Sanskrit and the history and culture of South Asia (Doctoraal/Masters in 1980). She worked as librarian in the department of the History and Culture of South Asia.

Marijke Oort married Jacobus Thomas (Tom) de Smidt (1923–2013) in 1954. Marijke as we saw, was interested in music and had studied at the Conservatory in Den Haag. Tom de Smidt had studied law and had a special interest in Dutch legal history. After obtaining a doctorate in 1954, he worked briefly for the Dutch Parliament until in 1956, when he was appointed Professor of law at the University of Amsterdam with the specialty 'introduction to jurisprudence and Dutch legal history'. From 1965 he combined that with a similar professorship on Dutch legal history at Leiden University, which he held until his retirement in 1989.

Bram Oort studied physics in Leiden. After his doctorandus (Masters) degree and military service he moved in 1961 to the Massachusetts Institute of Technology to study meteorology and oceanography until 1964, in which year he obtained a PhD at Utrecht University with a thesis *On the energy cycle in the lower stratosphere*. He married Bineke Pel (b. 1937), who had studied history. In 1966 they emigrated to the USA, where he worked in the Geophysical Fluid Dynamics Laboratory, first at Washington, D.C., but after two years at Princeton. He was visiting Professor at Princeton University, published papers on the observed energy and water cycles in the atmosphere, summarized in the monograph *Physics of Climate*, written with the late Jose P. Peixoto of the University of Lisbon. His wife taught at the Art Museum of Princeton University, studied social work at Rutgers University, and practiced Rubenfeld Synergy Therapy. After retirement in 1996 they moved to Vermont, where Bram taught Macrobiotic Shiatsu at the Kushi Institute.

11.2 Spiral Structure of the Galaxy

We have seen that after the discovery of the 21-cm line, Muller first developed a more sensitive and stable receiver. before starting a real survey of the Milky Way. In his words (from the *LibAm80* [2]):

> 'It was not until a year later, in June 1952, that at Kootwijk we started the first hydrogen-line survey with almost new equipment, helped by many students and staff members of the observatories who did the actual observations. Observing was not an easy job in the pre-automation era. During each two-hour observation of a line profile the telescope had to be moved by hand every few minutes, using two handles in a corner of the telescope cabin and reading the azimuth and declination scales through a small window in the floor and a side-wall of the cabin. In addition there were several receiver functions which had to be watched and adjusted, including a box at the back of the $7\frac{1}{2}$-m reflector. The first survey lasted one year, then the receiver was revised and extended. The second survey, which covered a strip along the Galactic plane with a width of 20° in latitude, started in November 1953 and lasted until August 1955, [...].
>
> It was not until many years later that I realized how difficult the long delay must have been for Jan Oort. I still appreciate very much the fact

that he never showed any impatience or annoyance, but just had confidence in us and gave us his full support.'

Observing at Kootwijk was indeed a tedious business. It has been described very well by one of the Leiden students involved in the execution of the survey, Gart Westerhout, in an after-dinner speech in October 2001 at a conference in Penticton [289]; it covers the first five years of Dutch radio astronomy and contains many interesting anecdotes.

The results of the first survey were published as a long (33 pages) paper in the BAN by van de Hulst, Muller and Oort (1954c), which had the title *The spiral structure of the outer part of the Galactic System derived from the hydrogen emission at 21 cm wavelength*. It started with a bit of history and acknowledgments to those that had been vital to the success of the project, and continued with a detailed description of telescope, receiver, methods of observing and data reduction. The focus contained a dipole and a reflector disk, a setup that resulted in an elliptical beam, measuring 1°.7 × 2°.8 (resp. horizontally and vertically) at half the central sensitivity (full-width-half-maximum, FWHM). It concerned mainly observations at zero Galactic latitude, although some observations were taken along a few lines perpendicular to the plane to check the symmetry.

Box 11.1 HI kinematics

Let the rotation velocity in the Galaxy at a distance R from the center be $V(R)$; near the Sun at $R = R_\circ$ this is $V_\circ$. Then along a line of sight at longitude l the radial velocity with respect to the Sun becomes

$$V_{\rm rad} = R_\circ \left[\frac{V(R)}{R} - \frac{V_\circ}{R_\circ} \right] \sin l.$$

For reasonable rotation curves the highest $V_{\rm rad}$ corresponds to the subcentral point (N in Fig. 11.5). In the case that the rotation velocity is constant with R we have

$$V_{\rm max} = 2AR_\circ(1 - \sin l) \sin l.$$

For $l \lesssim 90°$ or $\sin l \sim 1$; this can be used to put a constraint on $AR_\circ$.

Figure 11.4 shows the main outcome of the survey. This representation became know as a longitude-velocity diagram (or in short an 'l–V diagram'). The horizontal axis is radial velocity with respect to the solar neighborhood, ranging from −110 to +130 km/s. The vertical axis is Galactic longitude in old coordinates; in modern longitudes it runs from 257° at the top of the upper figure to 127° at the dividing line and −7° (or 353°) at the bottom. The Galactic anticenter is roughly halfway up the upper panel where the distribution is narrowest because most of the hydrogen moves parallel to us; the center is just above the bottom of the figure. The other parts of the

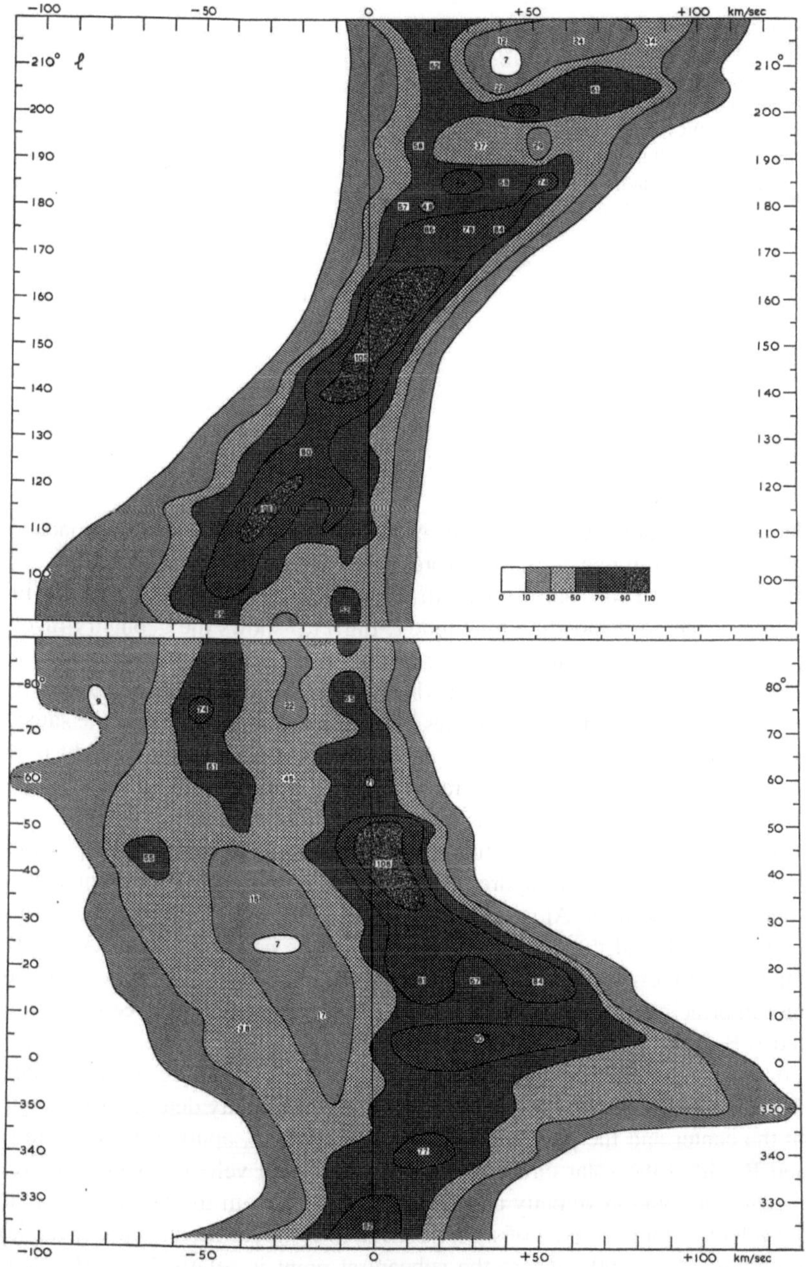

Fig. 11.4 First survey of neutral hydrogen along the plane of the Galaxy. The horizontal axis is radial velocity with respect to the Sun. The vertical axis Galactic longitude. The Galactic anti-center is roughly halfway up the top panel and the center just above the bottom of the figure. The shading corresponds to the brightness of 21-cm line emission. From the Oort Archives, published in van de Hulst, Muller, and Oort (1954c)

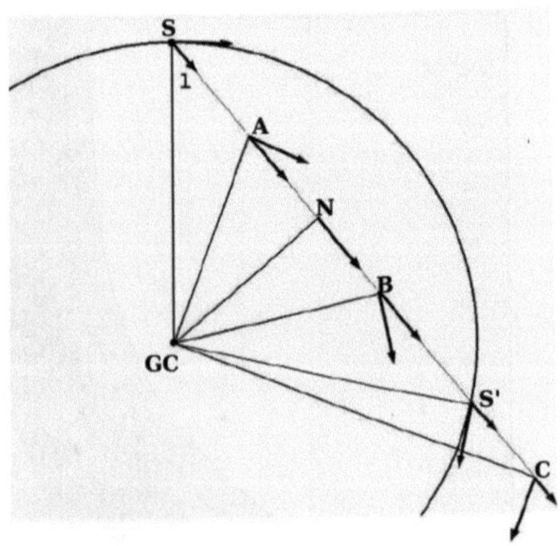

Fig. 11.5 Geometry of the HI observations. The neutral hydrogen at the various positions indicated has different velocities in the line of sight with respect to the Sun. Figure by the author

Galactic plane cannot be seen from the Netherlands. The shading corresponds to the brightness of 21-cm line emission, expressed in antenna temperature.

Each horizontal line corresponds to a line of sight through the Galaxy in the direction of the corresponding longitude *l*. Fig 11.5 shows the geometry involved. The Sun is at position S and longitude *l* is the angle with the center C. The point labeled N is closest to the center and has been called the subcentral (or tangent) point. The figure shows the rotation velocities (assumed the be more or less the same). At point S', which is at the solar circle, the projection onto the line of sight is equal to that at the Sun and the measured radial velocity is zero. At point A the projected rotation velocity is larger than for the solar position and there is a positive radial velocity left, and at point N, the full rotation velocity is in the line of sight and the radial velocity is largest. The hydrogen at point A can therefore be distinguished from the hydrogen at point N. At point C, beyond the solar circle, the projected rotation velocity is small and the radial velocity can will be negative. Points A and B are symmetric with respect to N and the observed radial velocity then is the same, so we cannot in principle distinguish solely basis of the kinematics, between hydrogen at A and at B. Box 11.1 gives the mathematics.

So, if we know the rotation velocity as a function of distance from the Galactic center (the rotation curve) we can for each observed velocity determine the distance from the center and the position along the line of sight, apart from the ambiguity A and B within the solar circle. Since at N the radial velocity is largest, we can determine the rotation velocity up to the solar radius from the observed maximum radial velocity along the lines of sight (referred to as the 'terminal velocity method'). For longitudes near 90°, where the subcentral point is relatively nearby, it turns out that the maximum observed radial velocity can be used to measure the product $AR_{\circ}$ (see Box 11.1 for details). With the Oort constant A then known from other measurements, van de Hulst et al. derived our distance from the Galactic center to be 8.2 kpc and with Oort constant B this indicated that the rotation velocity $V_{\circ}$ would be 216 km/s. There was an interval in which the IAU-recommended values were 10 kpc and 250 km/s; in 1982 these were revised to 8.5 kc and 220 km/s; however there is still considerable debate about this and the uncertainties are about 0.5 kpc and 25 km/s.

Van de Hulst, Muller and Oort then attempted to determine the distribution of hydrogen gas in the Galaxy from the data in Fig. 11.4. Because of the ambiguity inside the solar circle they restricted themselves to the region outside that radius. For the rotation curve they used the one corresponding to Oort's model for the mass distribution that we have seen before (see Sect. 10.7). The result is in Fig, 11.6. The cones towards the Galactic center and anti-center are empty; distances there cannot be determined, since the rotation velocities are nearly parallel over long lengths of the lines of sight, so there is no radial component. The authors concluded (in their Abstract):

> 'The points of maximum density are seen to be arranged in spiral arms, which seem to be only little inclined. Wherever they are inclined the direction is such that they wind up during the rotation.'

The major arm that can be seen in the figure has been referred to as the 'Perseus arm', which is now understood to be one of the two major spiral arms in the Galaxy (the counterpart, seen more clearly for the southern hemisphere, is called the 'Scutum-Centaurus arm').

It should be noted that the analysis had made a vital assumption and that is that the intensity variations in the observed profiles correspond to density variations along the line of sight, or that a peak in the profiles corresponds to the location of a spiral arm. It was found later that the velocity changes along the line of sight also have a significant (sometimes dominant) effect on the shape of the profile. This will be discussed in more detail later (see Sect. 13.4.3) (Fig. 11.7).

This discovery of spiral structure in the Galaxy was of course a monumental accomplishment. Actually the authors had been nominated for the Nobel Prize in physics. These nominations are public up to 1966 [290]. Oort, van de Hulst and Muller were nominated for the 1956 Nobel Prize by Lindblad, who was joined by Swedish solar (and stellar) physicist Karl Yngve Öhman (1903–1988) and physicist Hendrik Berend Dorgelo (1894–1961) from (now) Delft Technical University. Lindblad and Öhman repeated this in 1956, 1957, 1958 and 1961. But astronomy was not considered sufficiently close to physics. In fact, Hubble had been nominated in 1953 for the expansion of the Universe, but was not selected. Other nominations of Oort (without the other two) were unsuccessful too: three times (1958, 1959 and 1960) by Otto Struve, and in 1956 by Abraham Cornelis Sebastiaan van Heel (1899–1966), a Professor of optics in Delft, and in 1960 again by van Heel together with seven other physics professors in Delft. Even brute force would not help (Fig. 11.8).

Van de Hulst, Muller, and Oort also concluded that the 21-cm line emission was optically thin. When the density along a line of sight is high, atoms can in principle absorb emission from other, more distant atoms, shielding the distant parts from observation. Differential rotation distributes the HI-emission over a range of frequencies so that except maybe in special places, the line remains optically thin and can be used to trace the hydrogen throughout the whole Galaxy. The random motion of the gas clouds was estimated from the observed line profiles to be 8.5 km/s on

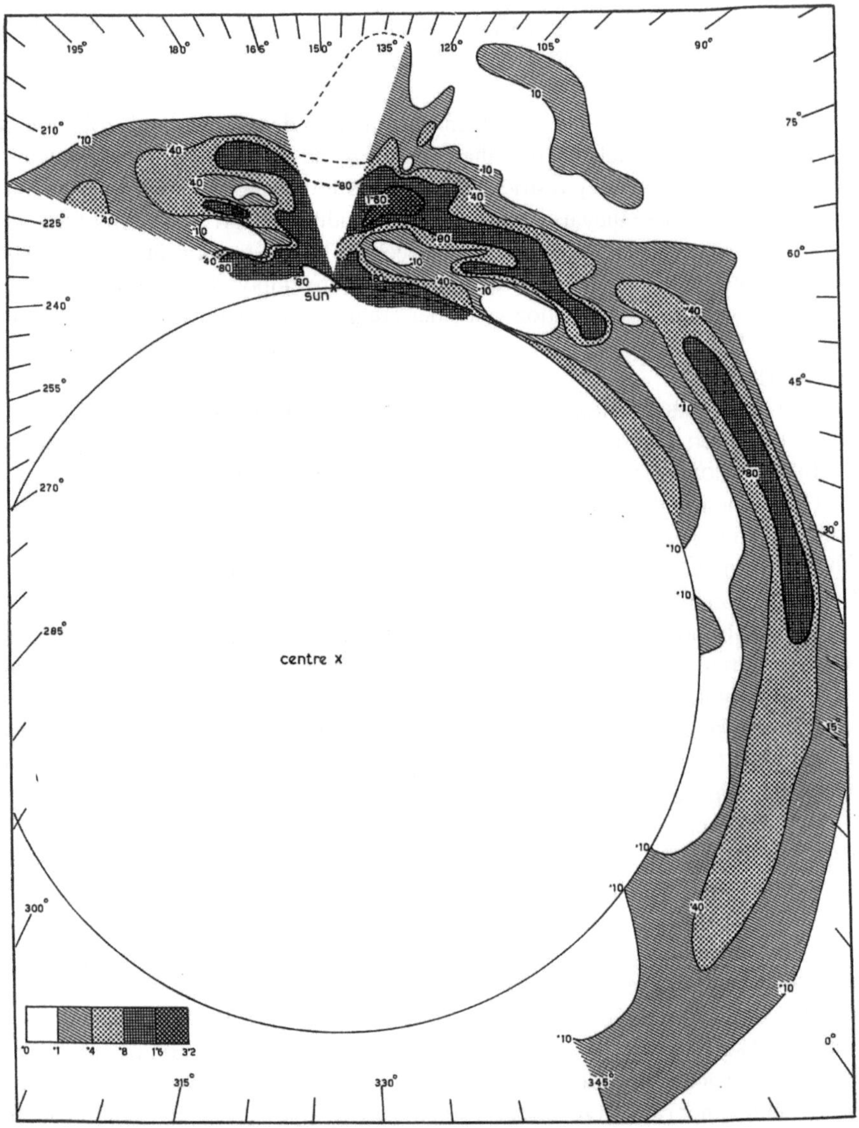

Fig. 11.6 Distribution of neutral hydrogen in the Galaxy beyond the solar circle, derived from the first Kootwijk survey (Fig. 11.4). From the Oort Archives, published in van de Hulst, Muller, and Oort (1954c)

average. In the spiral arm the density was estimated to be one to a few hydrogen atoms per cubic centimeter, and negligible in between the arms. It should be warned again that this statement is completely unwarranted in view of the incorrect assumption

Fig. 11.7 Oort and van de Hulst at the time the first observations of spiral structure in the Galaxy were found with the 21-cm line and the Kootwijk Radio Telescope. The blackboard shows a sketch to explain the method and to show the results presented in van de Hulst, Muller, and Oort (1954c). From the Archives of Leiden Observatory

that the intensity variations in the profile correspond uniquely to density variations along the line of sight (see Sect. 13.4.3).

The part of the observations referring to the Galaxy inside the solar circle was analyzed separately by Kwee, Muller, and Westerhout [291]. Kwee Kiem King (b. 1927) was another graduate student at Leiden, who worked for some time on radio observations, but eventually obtained his PhD with a thesis on variable stars supervised by Oosterhoff, and remained on the staff of Leiden Observatory working in that field. Kwee et al. used newer observations (late 1953 and early 1954) with an improved instrument. With this they determined the rotation of the inner Galaxy from the 'terminal velocity method' (see Sect. 11.2), and claimed to have found evidence for two spiral arms in the inner region of the Galaxy.

This was the beginning of the second Kootwijk survey, which lasted from November 1953 to August 1955. At the time it had been finished—as we will see below—the construction of the 25-m Dwingeloo Radio Telescope was underway and Muller and his group were relocated to Dwingeloo. Radio astronomy was then terminated at the

Fig. 11.8 Maarten Schmidt (left) and Gart Westerhout at the Kootwijk Radio Telescope. Courtesy Ard Hartsuijker, from the Archives of ASTRON Dwingeloo

Kootwijk site. The second survey was as comprehensive as the remaining time at Kootwijk would allow. It covered about the same extent in longitude as the first survey (Fig. 11.4), but in addition reached out to 10° on either side of the plane, the whole in a grid with a spacing of 2°.5. With a beam of 1°.7 × 2°.8 this means almost complete sampling. The effort was enormous: 7500 h of observing and 23,000 h of data reduction, involving staff members and students. The result was almost 800 'profiles' (the run of intensity of 21-cm emission with frequency at a particular position).

The results were published in a special issue of the BAN, No. 475, dated May 29, 1957. It contained five papers; interestingly none of them had Oort as one of the authors [292]. The first paper, by Muller and Westerhout, presented the results after describing the instrument, observations and data reduction. The second was a short paper by Ollongren and van de Hulst, discussing the way the data could be corrected for the finite resolution both on the sky and in frequency (or radial velocity of the gas). Alexander (Alex) Ollongren (b. 1928) was an astronomy student, who would later defend a thesis on numerical methods to calculate three-dimensional stellar orbits in the Galaxy, with van de Hulst as supervisor. He became a professor of computer science in Leiden and is now seen as one of the founders of that discipline in the Netherlands. The third and the fourth papers were the substance of the program. Westerhout gave a detailed analysis of the hydrogen in the outer Galaxy (beyond the solar circle), while Schmidt treated the inner Galaxy. The closing paper was by

Ernst Raimond (1932–2010), son of Jean Jacques Raimond. Both Westerhout and Raimond were graduate students. It concerned a special set of observations of two OB-associations, but these failed to reveal any neutral hydrogen.

This work had been preceded by the PhD thesis of Maarten Schmidt in March 1956. This concerned the distribution of matter in the Galaxy and was published as the paper *A model of the distribution of mass in the Galactic System* in the BAN (see Appendix A.4). In this paper Schmidt extended and improved the rough mass model that Oort presented in his Henry Norris Russell Lecture. There Oort used a few homogeneous ellipsoids, Schmidt developed the theory of the potential of inhomogeneous ellipsoids and fitted a model of the distribution of mass with such ellipsoids that used all information available on the constituents of the Galaxy. It has been in use for many years. The rotation curve following from the model was used for the analysis of the Kootwijk HI-profiles in BAN 475 to derive the distribution of neutral hydrogen in the Galaxy.

Schmidt's paper in BAN 475 attempted to use the vertical extent of HI features to break the ambiguity in the distances on either side of the subcentral point, that I discussed in Sect. 11.2. However, it is not clear from the paper how he actually did this. From the observations at or near the subcentral point he noted that the typical width perpendicular to the Galactic plane of the hydrogen gas is of the order of 220 pc for distances in excess of 3 kpc from the center. It seems that often he could not make the distinction between the two sides and simply put half the gas at the two points, which may cause the appearance of almost circular spiral arms.

Westerhout analyzed the outer parts using Schmidt's rotation curve. This, together with Schmidt's analysis of the inner Galaxy, resulted in a map of the distribution of neutral hydrogen (HI) gas in the Galaxy. The result of this enormous effort was graphically presented by two maps, both of which are reproduced here (Fig. 11.9). This is what Oort envisaged when he saw Reber's map of the radio emission that made him wonder if a spectral line would exist in the radio domain that could be used to map the Galaxy. The gray-scale presentation at the left shows the spiral structure he had been looking for. There are three major arms, from the outside inward respectively the Perseus arm, the Orion arm and the Sagittarius arm. The Sun is located just inside the Orion arm, which was in agreement with the distribution of nearby OB-associations. In the panel furthest to the right, the distribution of the HI gas near the Sun is compared with that of O-associations, as determined at Yerkes Observatory by William Wilson Morgan (1906–1994) and collaborators [293], generally accepted as the first convincing evidence for spiral structure in the solar neighborhood (see [294]). There is reasonable agreement. All three arms appear to be trailing with respect to Galactic rotation (that is in the sense of winding up). Already in a letter to Baade of July 14, 1952, Oort noted that the spiral structure is rather tightly wound (actually tighter than the OB-associations suggest), indicating a Hubble type of Sb rather than Sc (see Sect. 2.6) for the Galaxy, which means that it looks in that respect more like M31 than M101 (see Fig. 7.9).

In addition to locating the hydrogen as a function of distance from the center, Westerhout also investigated the three-dimensional distribution. He found that the gas in the outer regions systematically deviated from the plane defined by the hydrogen in

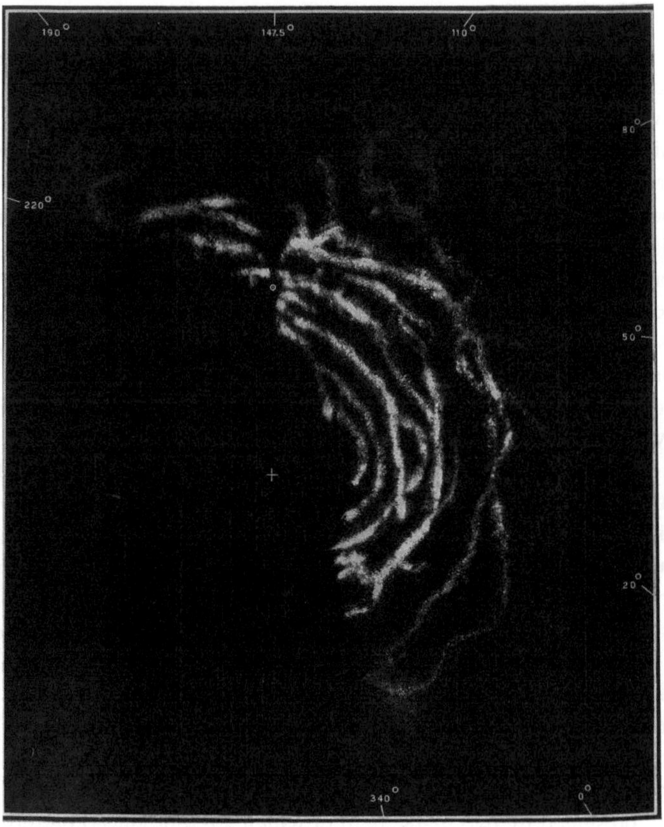

Fig. 11.9 Distribution of neutral hydrogen in the Galaxy determined from the Kootwijk survey. In the gray-scale figure on this page three major spiral arms can be distinguished. The color diagram on the next page indicates the space density; the dark areas have 1.6 atoms per cm^3 or more. In the cone around the center-anticenter line the gas has no radial velocity with respect to us and distances cannot be determined. From the Oort Archives, published in BAN 475 [292]. On the far right (next page) the area near the Sun with the open circles nearby O-associations. From the Oort Archives, published in van de Hulst, Muller, and Oort (1954c)

the inner regions. Generally speaking the gas layer on the right-hand side in Fig. 11.9 is becoming thicker and bending upward away from the plane, reaching a distance of about 300 pc at the outermost radius (which is some 14 kpc from the center). On the left-hand side (where only a small part can be mapped from the Netherlands) the bending is in the opposite direction. Later, when it became possible to observe HI in external spiral galaxies, observations of many of the ones we happen to see edge-on also show this phenomenon, which is referred to as galactic warps.

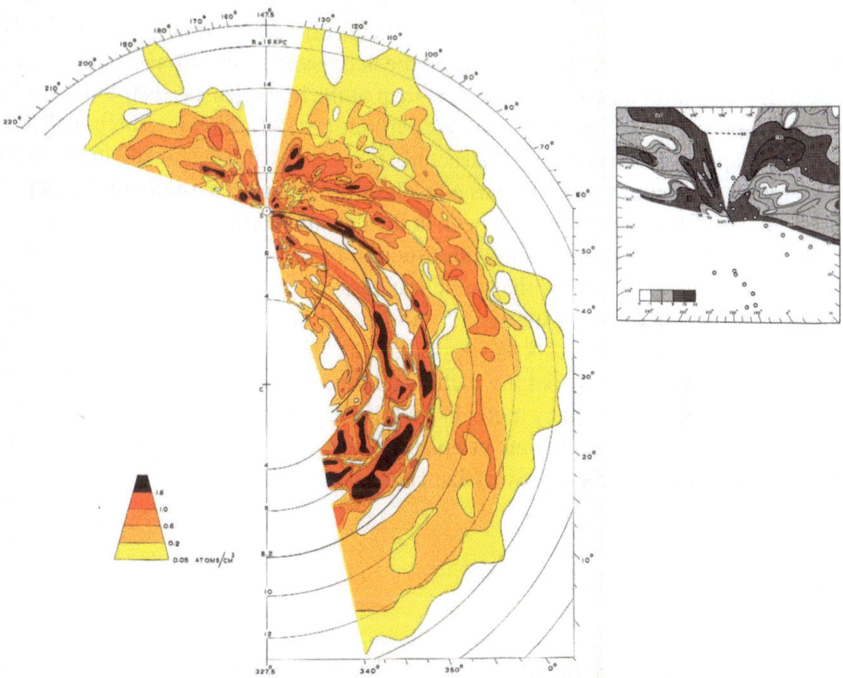

Fig. 11.10 Young Leiden astronomers, all graduate students, in 1953. From left to right: Kees van Houten, Gart Westerhout, King Kwee and Maarten Schmidt. From the Oort Archives

The main thing that remained to be done was to fill in the left-hand part of the map. For this observations in the southern hemisphere were required. One did not have to wait long for that; Frank Kerr did the necessary observations from his native Australia. The joint outcome, a complete map of the hydrogen gas distribution of the Galaxy, appeared in Oort, Kerr, and Westerhout (1958d). This is usually hailed as the first map of the Galaxy's spiral structure; however as already noted, this map was later found to be completely false as a result of the incorrect assumption that a peak in the line profile corresponds to a peak in the density distribution (see Sect. 13.4.3).

In 1945, Oort had proposed the construction of a 25-m radio telescope to detect the 21-cm line and determine the spiral structure of the Galaxy. As it turned out, all of this was accomplished with the 7.5 m German radar antenna. Yet on April 17, 1956, about a year before BAN 475 appeared, the 25-m Dwingeloo Radio Telescope was inaugurated. So the work for which it had originally been conceived had already been accomplished; yet, many more discoveries and surprises were in store now that radio astronomy had become an inseparable part of astronomy and an important, lasting contribution to the legacy of Jan Hendrik Oort (Fig. 11.10).

11.3 Travels

We have seen that only months after the liberation of the Netherlands, Oort had already traveled to England. In the following years he would travel extensively. In 1946 he went to London to receive the RAS Gold Medal by himself, where he wrote to Mrs. Oort from RAS's Burlington House on Thursday about his arrival and visit to the Royal Gallery. He reported on the proceedings the morning after it, Saturday May 11, 1946, in a letter to his wife and children, written immediately after breakfast.

> 'At four-thirty we went to the meeting hall, where some administrative matters were taken care of, and I was given the floor at five for the lecture. It lasted a bit long, as I tried to speak clearly, which slowed it down. A little over six-thirty I had finished. The chairman, Plaskett, thanked me and said among others that this had been the most original George Darwin Lecture ever! After the unavoidable praise of my work I was presented with the golden medal, and I had to give a speech to thank them.
> I felt I deserved some relaxation, so I fully enjoyed the dinner at the Royal Astr. Soc. Club that was offered afterwards. [...] Usually, guests there have to give another speech. I told the story of Erasmus's experience in 1500, when as he returned from England the customs officers in Dover took his complete fortune from him, because gold cannot leave the country. I will have to face the same problem with my gold medal.'

Desiderius Erasmus Roterodamus (1466–1536) wanted to leave England in 1500 carrying a large sum of money in gold coins that friends and supporters had donated him to make a visit to Italy possible. Customs had taken away almost all from him because of the limit on gold. His friends protested and claimed their money back, but the treasury kept the windfall.

Fig. 11.11 Oort and Oosterhoff in 1950 at Leiden Observatory, probably discussing administrative matters. From the Oort Archives

After the summer of 1947, Oort left for an extended trip to the USA. Already before the war, at the time of the McDonald dedication, Otto Struve had invited him to come to Yerkes as a visiting Professor, which Oort had declined in order to go observing at Mount Wilson. Struve renewed that invitation, but in the end it was Kuiper, who arranged the details. Struve resigned in 1947 from Yerkes and became chair of the Astronomy Department in Berkeley, while Kuiper succeeded him at Yerkes. Immediately after the War Oort had been too preoccupied with the directorate in Leiden and with the IAU to be able to leave Leiden for any extended period of time. Pieter Oosterhoff (see Fig. 11.11), Adjunct-Director since 1945 as we have seen, took an increasingly larger share of the burden of the directorate of the Observatory upon himself, making it possible for Oort to make longer visits to foreign institutes and observatories. This situation continued until Oort's retirement in 1970.

Oort left Leiden on September 13, 1947 and stayed at Yerkes during October and November. He arrived on October 2, after an inevitable visit to Jan and Jo Schilt in New York (who now lived in New Jersey) and to New Haven. In Chicago he was met by, among others, Blaauw and van de Hulst, both of whom were on leave from Leiden for longer periods, Chandrasekhar and Strömgren (Kuiper was at McDonald). Blaauw had resigned from Groningen to take up an appointment at Leiden in the fall of 1945. Oort kept a pocket diary in which he noted all social events, such as who had him for dinner or took him on a walking or swimming trip. In addition to giving lectures, he talked to many astronomers, including, in addition to the ones just mentioned, William W. Morgan, who then worked on classifications of stars and galaxies, and Georges Achille Van Biesbroeck (1880–1974), the Belgian-born expert on binary stars and Solar System objects. Oort of course visited Chicago with Kuiper more than once, and with Blaauw the astronomy department at the University of Wisconsin-Madison, which is not much further from Yerkes than Chicago (Yerkes at Williams Bay, although belonging to the University of Chicago, Illinois, is located just across the border with Wisconsin).

The diary notebook records various stories and anecdotes. Here are a few examples to illustrate the kind of things recorded. Kuiper told him about Robert Maynard Hutchins (1899–1977), who became President of the University of Chicago at a very young age (actually 30, and not 27, as Oort wrote down) and his wife novelist Maude (McVeigh) Hutchins (1899–1991):

> 'At a reception an older professor was talking to a young girl he thought was a student, saying that 27 was really much too young to become President. She: 'Do you know who I am? I am Mrs. Hutchins.' He: 'Do you know who I am?' 'No.' 'Thank God!' Exit.'

This must have been the kind of humor Oort liked. About Chandrasekhar, whom intimates called 'Chandra', Oort noted in an entry for November 26:

> 'Talked to Chandra this afternoon. He told me about a man with a broad interest who came to listen to one of his lectures. Afterwards his wife came up to the man excitedly and asked how the lecture had been. Chandra heard him say: 'Just like an American professor!
> Typical for Chandra also is his story about another lecture, where someone stood up and asked whether all of this supported the belief in God. He groped for an answer, and then the man said; 'You are quite spoiled by Western civilisation'.'

In his diary notebook Oort also noted:

> 'Tuesday evening 25 Nov.
> This afternoon I gave my last lecture. Quite a relief that I am no longer burdened by this, although I very much enjoyed preparing these seminars well; just very time-consuming. Morgan was surprisingly appreciative. 'He had never in his life experienced anything like this, never had a lecture course in which he learned so much.' I must say this was a great support.
> I did not really do Chandrasekhar much justice, I think. He had wanted to attack and disprove everything in my last lecture, Morgan told me, but was taken by surprise by the developments regarding the well-behaved vertex of distant stars.'

But Oort did not just come to lecture at Yerkes. He had corresponded with Kuiper about observing at McDonald also, again with the aim of photometry of galaxies. He was assigned time on the 82-in. Telescope in January. A younger staff member at Yerkes was William Albert Hiltner (1914–1991), who—following a suggestion by Otto Struve—had embarked on a project developing photoelectric photometry, which means measuring brightnesses directly with a photocell or equivalent detector at the telescope, avoiding all the pitfalls and complications of the photographic emulsion. Even though Hiltner was primarily interested in stellar spectroscopy, he and Oort set out to try this kind of work on extended objects, galaxies.

Oort's plans, however, were to visit Mount Wilson first in December after finishing his lecturing at Yerkes. By that time the big 200-in. Telescope at Palomar Mountain would be nearing completion. He arrived in Pasadena on December 11 and his pocket diary shows that he had dinner at the Hubbles and the Humasons and lunch with the Baades the following days. He spent many days talking to Mount Wilson astronomers and he gave some seminars.

> 'Friday evening 19 December. This afternoon I gave an introduction about problems of Galactic Structure. I had dreaded that somewhat, since they might expect a lot in view of the enormous amount of money they are supplying for my travel and subsistence and the possibly not so satisfactory colloquium presentation Wednesday morning. I did not feel in top condition after a poor night's sleep. Fortunately it went well and there were various astronomers that were rather enthusiastic about the discussion.'

The new 200-in. Telescope at Palomar Mountain was nearing completion and on December 22, Hubble and Humason drove Oort up the mountain. The situation there was the following. After the allocation of a grant from the Rockefeller Foundation, site preparation had started in 1936 and the dome and telescope structure had been completed in 1939. However, the mirror was delayed enormously, for a variety of reasons, partly World War II. The grinding took more than 11 years and the mirror had arrived at Palomar on November 19, 1947, weeks before Oort's arrival in Pasadena. When Oort came to Palomar it had just been mounted in the telescope.

> '22 December. Trip to Palomar (130 miles each way) with Hubble and Humason, where, together with Hubble, I was one of the very first to see the image of a star with the 200-inch mirror. Yesterday they did the first tests: mirror was very smooth but there was astigmatism (2 lines of foci 2 mm apart). This evening it was corrected and the mirror seemed as good as the 100-inch on many occasions! Knife-edge test with large magnification on a star image. Bowen and Anderson were there as well, and Porter. In addition Rule, the designer and builder of the mechanical part and Brown and Johanson, who had done the optical work under the guidance of Anderson (started 18 years ago with tests of quartz mirror!). Stars float in the focal plane of the mirror. Impressive trip going up with the platform and then descending into the cage, where you can sit with three persons.'

Ira Sprague Bowen (1989–1973), a nebular spectroscopist, was at that time Director of what now had become the Mount Wilson and Palomar Observatories, John August Anderson (1876–1959) had been responsible for the optical design, Bruce H. Rule (1908–1989?) had been the telescope engineer and Russell Williams Porter (1871–1949) did the design of the dome and the telescope together with Francis Pease. Porter is known for his detailed cutaway drawings [296].

Fig. 11.12 The 200-in. (5-m) Hale Telescope at Palomar Observatory. The tube hanging at the top in the telescope structure is the prime focus cage. From Wikimedia Commons [295]

The mirror was not made of quartz, as originally intended, but of pyrex, a low-thermal-expansion glass. A knife-edge test concerns the accuracy of the mirror surface. Suppose a small part is not correct and does not concentrate the light in the focus, but close to it. Point the telescope at a star; looking without an eyepiece with the eye near the focus then shows the illuminated mirror. One can hold a sharp edge (of a knife for example) right next to the focus to block the stray light and then the incorrect part of the mirror shows up as a dark area.

The 'Big Eye', as the 200-in. was fondly called, was not ready for observing yet when Oort was there, it was not even ready at the time of the formal dedication in June, 1948, when it was christened the Hale Telescope. Actually, the mirror was being perfected for about another year before it was aluminized; Hubble took the first official science exposure in January of 1949. At the time of Oort's visit the reflection of the mirror was off the glass itself. Indeed, those (like myself) who have observed from the 200-in. prime focus, can testify that the trip at night on the rising platform along the open slit of the dome really is very impressive. When it has reached the top of the slit, it allows one to step into the cage when the telescope points at the zenith (see Fig. 11.12). The cage's diameter is about 1 m, so it offers significant space for the observer (although it would be cramped with three persons and inconvenient when the telescope were pointed well away from the zenith), but it blocks only some 4%

of the 5-m diameter primary mirror. Having been one of the first to look through the 200-in. Telescope must have been one of the most imposing astronomical experiences in Oort's life.

After Christmas (at the Baades), Oort traveled to McDonald Observatory in Texas to observe at the 82-in. Telescope with William Hiltner. The idea was to investigate to what extent accurate surface photometry was possible with photoelectric techniques. The publication resulting from did not appear until 1954, the authors, van Houten, Oort, and Hiltner (1954b), apologizing for this at its end. Cornelis Johannes van Houten (1920–2002) was a student of Oort in Leiden, who eventually (1961) would write a PhD thesis on surface photometry of galaxies based on Oort's photographic plate material. The aim of the McDonald experiments was described in the van Houten et al. paper as follows:

> '.. as the difference in brightness between the nucleus and the outer parts of an extragalactic nebula is considerable, a very precise determination of the gradation curves of the photographic plates was necessary. To avoid this difficulty, it was planned to measure photoelectrically the intensity distribution in a number of extragalactic nebulae.'

The gradation curve is now called the characteristic curve (see Sect. 7.2 for a discussion of photographic surface photometry). Hiltner had published many papers on photometry, including in the ultraviolet and infrared, and on polarimetry, but never on galaxies. Oort wanted some independent check on the photographic procedures. As van Houten would describe in his thesis, calibration of the photographic material using recordings of sensitometer spots of known intensity ratios and extra-focal exposures of stars in Kapteyn Selected Areas gave inconsistent results (Fig. 11.13).

The observing was done using a photomultiplier, the output of which was amplified and recorded on a chart recorder. The light came through a diaphragm that was moved in the focal plane (the Cassegrain under the mirror; see Sect. 7.3) in a straight line by a synchronous motor. By rotating the set-up any direction on the sky could be selected. There were various diaphragms and they could move at one of three selectable speeds. In addition filters could be inserted to select wavelength ranges. A drawback was that guiding the telescope during the scan was not possible.

From the tabular data it can be deduced that he radial extent of the scans of NGC 3115 were quite similar to those presented at the McDonald dedication symposium (Oort 1940a). The authors stated that the 'general conclusion of this paper is that the attempt to measure the intensity distribution of the nebulae photoelectrically has succeeded, although for a variety of reasons the accuracy of the measures is less than that anticipated from photoelectric observations.'

After the observing session at McDonald Observatory, Oort started the return journey back home. After celebrating Jan Schilt's birthday on February 3, 1948 in New York, Oort embarked on February 7 and was back in Leiden on February 16. He had been away for four months.

In the summer of 1948 the IAU General Assembly took place in Zürich, and Mrs. Oort accompanied her husband. According to the pocket diaries, they had a great time, arriving in Zürich on August 6. A high point Oort recorded was the dinner on August 18, the day of the closing of the meeting, with Bertil Lindblad and his wife

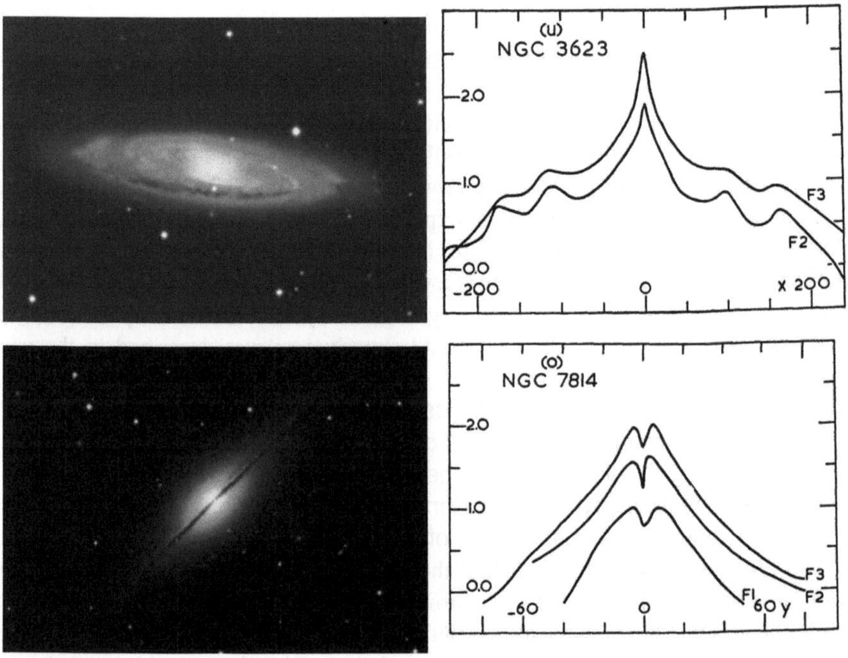

Fig. 11.13 Images and photoelectric scans of two galaxies. At the top NGC 3623, scan direction major axis. Bottom NGC 7814, minor axis. F1 and F2 indicate the use of an ultraviolet filter, F3 is in the visual. Pictures from the Digital Sky Survey, [203]. Drawings from van Houten, Oort, and Hiltner (1954b). From the Oort Archives

Dagmar, and Bengt Strömgren. Apparently their daughter Marijke (now 17 years of age) had stayed with her grandmother Graadt van Roggen, while the younger son Bram (almost 14) visited Hulshorst, camping with Jaap Tinbergen (they were born in the same year). Both children were put on the train by their grandmother in Utrecht and traveled by themselves to Switzerland to have a vacation as a family.

Already before the end of the year, on December 7, Oort started another long journey, this time primarily to South Africa to visit the Leiden Southern Station. This must have been his first intercontinental trip by air. From his diary notebook:

'On 7 Dec '48 I leave by DC6 from Schiphol for Johannesburg, seen off by Mieke, Coen and Marijke. A strange experience in that largest and fastest of airplanes to climb to 5 à 6000 meters with a cruising speed of 262 miles per hour, or 420 km/hour, length 30 meters, width 35 meters, total weight 43 tons of which 23 tons for the empty plane, to be over the Alps in as little as over an hour [in fact at that speed the roughly 650 km would have taken somewhat more than that], after twilight over the Mediterranean and after 4 hours landing in a dark Tunis. In the night we fly over the desert, where I can just see waves of sand straight below, and with stars all around us. [...] I enjoy the quiet semi-dark night in the airplane. In

> Kano, Nigeria, where we have to disembark in the middle of the night, my neighbor, geologist Mekel, leaves the flight.
> In Johannesburg, 28 hours after departure from Schiphol, I was met by Wesselink, Wim van de Bos, Jan Schilt, Erica and Wiedie. It is immediately rather hot. We drive to Joh.bg [Johannesburg] with the photometer in small crates, everything in the car of J.J. Wessels, amateur astronomer and photographer.'

Adriaan Wesselink was the Leiden representative and observer at the Southern Station, and Willem van den Bos, former Leiden observer, was now Director of the Union Observatory. The Yale Telescope facility had been joined in 1943 by the astronomy department at Columbia University, of which Jan Schilt now was Director as well as of the Yale-Columbia Telescope. In 1946 Wesselink had replaced Hendrik van Gent, who had been the Leiden observer, the final of three periods including the War. Van Gent had moved back to Leiden, but had died unexpectedly earlier in 1947 at the young age of 47 (Oort 1947c, had published an obituary in Dutch). 'Erica' was Erica Maria Zoetelief Tromp (1902–1989), wife of Jan van den Berg (1899–1982), ambassador of the Netherlands to South Africa, and Wiedie presumably a daughter. Oort knew van den Berg well (their correspondence shows that they must have been long-term friends). They were almost the same age and van de Berg had grown up in Leiden (his father was an elementary schoolmaster in nearby Hazerswoude), where he had studied Chinese language at the University. Erica was a sister of Henriëtte Zoetelief Tromp, Aernout de Sitter's widow. Before South Africa van den Berg had held various diplomatic posts at a number of places in China, before he had been appointed to Pretoria, the seat of the executive branch of the South African government. Oort stayed with the van den Bergs part of the time.

These three Dutchmen, Wesselink, van de Bos, and Schilt (see Fig. 11.14), represented the three organizations that operated telescopes in the Johannesburg area. Due to the growing city, light pollution had become a major concern and an important reason for Oort's visit was to discuss what to do about this. It was a more general problem in South Africa. Oxford's Radcliffe Observatory near Pretoria and also the Cape Observatory suffered from the same problem. During his month or so in South Africa, Oort spent much time on taking stock of the situation, talking to astronomers and officials and considering possible options for moving the station, not only around Johannesburg, but also nearer to Pretoria. He made a train trip up and down to Capetown to talk to people there also.

Oort wrote to his wife that no definite conclusions had been reached. He summed up the reasons for the trip in a letter to her as, (1) orienting himself on the situation, (2) talking to officials through van den Berg to get permissions and maybe even contributions, (3) involve local Dutchmen, (4) through contacts with Knox-Shaw and Thackeray coordinate with Radcliffe Observatory, (5) act in unison with Schilt and Yale-Columbia, (6) talk to mayors of places, maybe not too far from Pretoria, and (7) with the help of van den Berg have an interview with the secretary general of the Ministry for Education.

Although no conclusion was reached at this stage, the first steps were taken, and as will be described further on, the Leiden Southern Station was relocated to Har-

Fig. 11.14 Oort arriving in Johannesburg in December 1948. From left to right Jan Schilt, Willem van den Bos, Oort and Adriaan Wesselink. From the Oort Archives

tebeespoortdam near Pretoria with the move of the Rockefeller Astrograph and the Franklin-Adams Telescope in 1954 (and a new Leiden telescope, the 'Lightcollector'). Since I will not come back to this, I mention for completeness that in the end South Africa would concentrate *its* efforts on a new site, merging the Union and Cape facilities at the South African Astronomical Observatory at Sutherland in the Karoo and that the Yale-Columbia Telescope was moved to the Mount Stromlo Observatory near Canberra, Australia, where it was subsequently destroyed in the bush fire of January, 2003.

After this visit, Oort returned via Kenya where, as we have seen, he visited the expedition there. This trip was adventurous. First by small plane to Vereeniging, a town south of Johannesburg, where a reservoir lake is located from which larger seaplanes of the British Overseas Airways could take off. One of these seaplanes took him to the north to a place where the Victoria Falls descend into the Zambesi river, and after an overnight stay on to Kampalo on the northern side of Lake Victoria. From there after two days waiting there was a train to Timboroa near the Kenya Expedition, arriving at 4 o'clock in the morning. 'I hope I wake up in time! Van Herk writes me that he train stops only briefly.', he wrote to his wife. It must have gone well.

After this Oort traveled to Egypt, flying now with a 'land plane' from Nairobi with the same airline (BOA), and after an overnight stop at Khartoum, Sudan, on to

Fig. 11.15 Oort during his visit to the Royal Observatory at Helwan near Cairo in 1951. The other persons have not been identified, but the one on whose shoulder Oort's hand rests probably is his host Mohammed Reda Madwar. From the Oort Archives

Cairo. Here he spend a few days at the Helwan Observatory in connection with the zodiacal light experiment that had been planned for Kenya (see Sect. 10.4). He was back in Leiden on February 24, 1949 (Fig. 11.15).

It seemed from his pocket diaries that Oort had by then decided not to travel for extended periods of time. However, the visit to Helwan Observatory required a follow-up. Mohammed Reda Madwar (1893–1973) had first studied civil engineering in Edinburgh and, interrupted by a few years back in Egypt, he obtained a PhD in astronomy at Edinburgh. He became observer at Helwan Observatory, Director in 1934, and Professor of astronomy at Cairo University. After the brief visit in February 1949, Oort had written up some suggestions for work that could be done at Helwan, including monitoring of variable stars, structure of comet tails, and (unavoidably) photography of extragalactic nebulae. Madwar had visited Leiden in August, 1950 and he and Oort had discussed the possibilities for Helwan Observatory joining an expedition to Khartoum on the occasion of the solar eclipse of February 25, 1952. The expedition was organized by Bernard Ferdinand Lyot (1897–1952) of the Observatoire de Paris-Meudon. Lyot was to die from a heart attack in Cairo on his way back at the end of the expedition.

The idea was to obtain electronic photometry of the intensity and polarization of the light of the solar outer atmosphere, the corona, where it merges with the zodiacal light. This means some 20 radii out from the Sun. Polarization is the property of electromagnetic radiation, such as light and radio waves, when interpreted as a wave phenomenon, to oscillate in a particular direction. Such radiation can be understood as a periodically varying electric field and a magnetic field perpendicular to it. The plane of the electric field is that of the polarization. In principle a beam of light contains a random distribution of all polarizations, but when it has a preferred direction, the light is said to be polarized. This can for example come about when it has been scattered by dust particles that are not spherical, but elongated.

The instrument would be built at the Leiden Observatory workshop and would consist of two refracting telescopes, one directed towards the dark moon and one scanning the sky in a circle with a radius that could be set from 2 up to 20 solar radii. Walraven would assist in the design and Helwan would second a staff member, Dr. I.H. Abdel-Rahman, to Leiden for a year to complete the design and oversee the production. The costs, Dfl. 4500 for the mechanical parts and Dfl. 3600 for the electronics, would be born by Helwan Observatory. This was all done in due course, although some unavoidable bureaucratic obstacles made the timescale critical towards the end.

At the same time Oort was invited to Egypt as a visiting Professor. Oort seemed to have been reluctant to travel, but in the end agreed to come for no more than one month. He proposed to give a course of twelve lectures on Galactic structure and dynamics. Oort left after Christmas 1950 and stayed for four weeks.

The experiment was announced by Madwar in a publication of Helwan Observatory in 1952, mentioning the collaboration with Oort and Leiden Observatory. No report on the experiment or scientific publication from it could be found. There are three related publications [297–299]. Only in a general article by Madwar in 1955 was a short report on the eclipse included, mentioning the experiment, without giving any further details or any reference to Leiden or Oort. A 1958 article on the Helwan Observatory makes no mention of it; Madwar is mentioned to have been retired, Abdel-Rahman is not included in a list of staff members.

Oort did considerable traveling. In addition to what was mentioned he took trips within Europe, such as to the symposium on *Problems of Cosmical Aerodynamics* in 1949 in Paris (Oort 1951f, g) and the one on *Constantes fondamenteles de l'astronomie*, also in Paris the next year and resulting in the paper by Morgan and Oort (1951e).

11.4 Oort–Spitzer and Interstellar Motions

In the course of 1950, Oort started looking into the possibility of his next extended visit to the USA. As always he was interested to come to Mount Wilson Observatory, now that the Mount Wilson and Palomar Observatories were running the telescope facilities jointly on behalf of both the California Institute of Technology (CalTech) and the Carnegie Institution of Washington (which according to some by symmetry should be abbreviated as CarWash). Palomar had not only the 200-in. telescope, but also the wide-angle 48-in. Schmidt Telescope conducting a photographic survey of the whole northern sky, the *National Geographic Society—Palomar Observatory Sky Survey*. In a letter of May 15, 1950, to Baade, Oort mentioned a small project he wished to conduct with the 48-in., which had to do with his work on comets. He felt that there would be a good chance that Jupiter and maybe also Saturn had caught some of the incoming new comets, which would give rise to a sharp increase in the number of satellites with smaller size. Making long exposures of these planets while tracking them so that they did not move on the plate, might reveal many faint new satellites.

In a letter of October 19, 1950, Lyman Strong Spitzer Jr. (1914–1997) of Princeton University invited Oort to come for a few months as visiting Professor. They had met at the aerodynamics meeting in Paris and had started corresponding on problems concerning interstellar matter, a field in which Spitzer was a leader, and became very good friends. Spitzer had succeeded Russell as Director of Princeton University Observatory. This had on its staff Martin Schwarzschild (1912–1997), son of Karl Schwarzschild, and scientifically they ran the observatory together. Schwarzschild was Jewish and had had to flee Germany; he was an authority on stellar structure, and Princeton was therefore a very attractive place for Oort to go.

CalTech had been entrusted by the Rockefeller Foundation with the ownership of the 200-in. Hale Telescope and it was keen to set up an astronomy group and a graduate program in astronomy. In 1949 Jesse Greenstein, who had been on the staff of the Observatories since 1948, as Professor and head of the Department of Astronomy. He wrote to Oort on November 20, 1950 inviting him as visiting Professor to CalTech in combination with the Princeton visit. Greenstein and Spitzer had actually discussed the combination before the offer was made.

Oort had become interested in the question of the motions of the gas in the interstellar medium and was keen on working on that with Spitzer while visiting Princeton. Studies of interstellar absorption lines in stellar spectra as well as the 21-cm HI observations at Kootwijk indicated that gas clouds in interstellar space had random velocities of order 7–10 km/s. The question was how this could be maintained; after all, gas clouds collide and their relative kinetic energy would then be transformed into heat and radiated away. So there had to be some replenishment of kinetic energy and Oort wished to investigate the possibility that this came through heating of gas around hot O-stars.

Oort also made plans for his Pasadena period. On October 27, 1951, he wrote a long letter to the Director of the Mount Wilson and Palomar Observatories, Ira Bowen, in which he inquired into the possibility of obtaining observing time. His first priority concerned NGC 3115, for which he proposed to take a long-exposure spectrum along its major axis at the Mount Wilson 100-in., to measure the stellar velocity dispersions from the widths of the absorption lines. He mentioned that Adriaan Blaauw had tried this successfully at McDonald Observatory on M32, a small companion of the Andromeda Nebula. Two more suggestions concerned astrometry. The first was to take second epoch plates on the Mount Wilson 100-in. of fields of distant galaxy clusters to measure proper motions of foreground stars with respect to the small nebulae. The other was to take second epoch plates of RR Lyrae stars and similar objects and of planetary nebulae for which plates already were in the plate archives. These would be pilot programs to study the feasibility for a larger effort. Finally he repeated his idea of using the Palomar 48-in. Schmidt to search for faint satellites of

Jupiter and Saturn. Bowen replied, pointing out that some material already existed and 'that new equipment, which will be available at Palomar in a year or so, may make antiquated results that could be obtained this winter on Mount Wilson.' He suggested Oort should discuss things with the 'nebular group' when he was in Pasadena.

While the Oorts were making their travel plans (Mrs. Oort was to accompany her husband), Oort received the invitation to give the Henry Norris Russell Lecture, which he gave at the winter meeting of the American Astronomical Society held in Cleveland, Ohio, on December 27, 1951 (see Sect. 10.7). The ocean crossing had been rough and Oort had some difficulty preparing in time for the Russell Lecture. As usual they first visited Jan and Jo Schilt before taking the train to Chicago, where they arrived with hours of delay, but were still met at the railway station by their son Coen, who was studying there at the time. After the lecture they traveled on to Pasadena, flying to Los Angeles on January 1, 1952, delayed by 'fog, heavy traffic and headwind', so that 'we had to make an extra stop at Las Vegas to take on more fuel.' The notes in the pocket diaries are very detailed. They visited many people and at the end of January they moved to the summit of Mount Wilson, where they stayed in the Kapteyn Cottage during an observing run on the 60-in. Telescope. Oort was assisted by staff member Rudolph Minkowski (1895–1976), who was born in Germany as Rudolf Leo Bernhard Minkowski. Minkowski had worked with Baade at Hamburg, but when Hitler came to power Minkowski—who though a Christian had Jewish parentage—had to flee Germany. Baade had arranged for him to join Mount Wilson Observatory in 1935 as a research assistant. Minkowski became a well-known spectroscopist at Mount Wilson and Palomar. Oort does not specify any details of his observational program in the pocket diaries.

The Oorts had an great time in California. There was a trip to Death Valley and a further observing run at Mount Wilson with the 100-in. Telescope, which was plagued by poor weather. On March 13 Struve picked them up 'in his University car' to take them to Berkeley. 'He drives fast, often 80 miles, but the road is quiet'. They met the Shanes, who showed them San Francisco and Stanford University campus, and stayed with them at Mount Hamilton (Lick Observatory). Upon their return to Pasadena, the Oorts started 'our grand journey to the East' by car (a Ford, which they appear to have bought) on March 23. In Arizona, they picked up Coen, who had come from Chicago by train, visited the Grand Canyon into which they descended all the way to the Colorado River. They went on to Williams Bay (Yerkes Observatory), where they stayed with the Strömgrens. Eventually they arrived in Princeton on April 11, having driven 4537 miles since leaving Pasadena 'in our Ford, which has served us very well. The only problems were two flat tires' in Arizona. Oort specified in detail that he had spent 45$ for two new tires and one inner tube, 7$ to replace lights and 1$ for repair and test of a water hose to the engine. They used 308 gallons of gasoline, which corresponds to an amazingly little 14.7 miles per gallon!

The problem of random motions of the gas in interstellar space that Oort was eager to work on with Spitzer was not solved in the two months at Princeton. It resulted eventually in two papers, one by Oort himself on an *Outline of a theory on the origin and acceleration of interstellar clouds and O associations* and one with Spitzer on the

Acceleration of interstellar clouds by O-type stars, resp. Oort (1954d) and Oort and Spitzer (1955b). The latter paper has a footnote that the order of the names is chosen to be alphabetical. So, this took a long time of germination and maturation before it could be published. Oort presented the work at a conference on *Gas Dynamics of Cosmic Clouds* in Cambridge, England, in July 1953. The published contribution (Oort, 1955e) was very different—he noted himself in the text—from that presented at the meeting. This conference was a follow-up of the one in Paris in 1949 on *Problems of Cosmical Aerodynamcs* and was the second symposium sponsored by the International Astronomical Union; the proceedings were edited by van de Hulst.

The idea was the following. O-stars are hot and emit large amounts of ultraviolet radiation, which ionizes the gas around them. This had been studied in detail by Strömgren and had led to the concept of a 'Strömgren sphere'. Suppose the O-star is surrounded by a large volume of hydrogen gas. This then becomes ionized, but the free electrons every now and then encounter a proton (hydrogen nucleus) and recombine. Thus an equilibrium results in a volume around the star in which as many hydrogen atoms are ionized by the radiation field as are recombined from protons and electrons. This forms a sphere, which for the case of uniform gas density is called the Strömgren sphere. The radius depends on the gas density and the amount of ionizing radiation emitted by the star.

An example of a Strömgren sphere is the Rosetta Nebula in Fig. 11.16. It is an approximately spherical nebula bounded by the extent over which the four O-stars in the central star cluster NGC 2244 can ionize the gas. The bubble in the center is filled with extremely hot gas from the stellar winds of these same stars. Strömgren found that the gas inside his sphere (HII region) will have a temperature of about 10,000 K. This gas will then expand and push out the cold gas outside the sphere. The density in the HII region will decrease so that the radius of the corresponding Strömgren sphere will increase. But the radius of the Strömgren sphere will increase faster than

Fig. 11.16 The Rosetta Nebula and the central star cluster NGC 2244. The colors are false and correspond to emission from different atoms or ions. The light from an emission line of hydrogen is coded red, that from (doubly) ionized oxygen green and from (singly) ionized sulfur blue. From NOAO [300]

that of the expanding HII region and more ionization will take place, pushing the cold gas out even more.

The process was studied in more detail in the Oort and Spitzer paper. They considered three representative cases. The first is that of a cold hydrogen cloud near to which an O-star is formed. The radiation impinging onto the cloud will heat the part facing the star to 10,000 K or so. This upper layer will become transparent to the ultraviolet radiation so that this will penetrate deeper into the cloud. The heating of the cloud will increase the pressure and the gas will expand, but mostly in the direction of the star where the density is lower, i.e. in the direction away from the center of the cloud. This will result in a repulsive reaction on the cloud as a kind of 'rocket'-effect. The acceleration of the cloud by this Oort–Spitzer effect will increase the random motions in the interstellar medium. The other cases Oort and Spitzer considered were those of a gas cloud approaching a stationary O-star, and that in which an O-star forms within a large gas cloud. Accelerations were found up to typical values of 20 km/s.

The Oort (1954d) paper puts things in interstellar context. Oort started out by estimating the amount of energy in complexes like the Rosetta Nebula generated by O-stars and converted into kinetic energy by the Oort–Spitzer mechanism. He then estimated the energy lost in collisions between interstellar gas clouds, estimating their sizes and velocities and assuming that in a collision half the kinetic energy (since collisions are in general not head-on) is converted into heat and radiated away into space. Both energies turned out to have the same order of magnitude. He then reviewed the evidence from interstellar absorption lines that indeed expanding shells exist around associations containing O-stars and concluded that this confirmed his theory. It also could explain high-velocity O-stars. which often have velocities up to 50 or 70 km/s as forming inside an accelerated gas cloud (Fig. 11.17).

Fig. 11.17 Oort in his office in 1953. From the Oort Archives

Finally there were the expanding O-associations. Around 1950, the Armenian astronomer Victor Amazaspovich Ambartsumian (1908–1996) from the Byurakan Observatory near Yerevan in Armenia had advanced the theory that there were expanding groups of young stars. This was important as it would indicate that star formation was still going on in the Galaxy and that stars are born in groups, both of which we now know to be true. Blaauw had conclusively shown that indeed associations of O-stars were expanding with a typical velocity of some 15 km/s. Oort put forward the idea that these expanding motions would have their origin in the cold regions surrounding the ionized nebula around a newly born O-star or group of O-stars. These primary O-stars were then formed by some unknown mechanism and the stars in the association in the subsequent cool clouds around the HII region would then be moving outward from the initial O-star(s).

These papers show the typical approach Oort had to science. Everything is first treated in very general terms; the Oort (1954d) paper is full of concepts like 'rough estimate', 'tentative picture', 'incomplete theory', 'approximate estimate', etc. At the end he wrote: 'The foregoing article is evidently nothing but a first and very incomplete sketch. Although it can be, and partly has been, elaborated in various points, real progress would seem to be possible only on the basis of more detailed observations than are now available'. No need to do more theoretical work when there are no observations against which to test detailed predictions. And working out details is less interesting than developing the idea.

On July 7, the Oorts sailed from New York back home. However, Oort was soon off again, flying to South Africa, spending a week in Cape Town and discussing among other matters *Expanding groups of B-type stars* at a colloquium at Radcliffe Observatory on July 25, 1952 (Oort 1952g). Although no progress was reported, he also must have been discussing the options for the move of the Leiden Southern Station, which was not to take place until 1957.

11.5 Back to the Crab

Ever since his work on Nova Persei and the identification of the Crab Nebula supernova (see Sect. 9.6), Oort had been interested in the subject. The Crab also is a bright radio source. In the early days of radio astronomy, the brightest discrete sources were designated with their constellation. Since the Crab Nebula is situated in Taurus, the radio source that was found to be associated with it was called Taurus A. Other bright sources were Sagittarius A (the Galactic center), Virgo A (the giant galaxy M87 with a jet in the Virgo cluster), Centaurus A (a nearby colliding pair of galaxies), etc. Cygnus A was found to be a very distant galaxy, that was only identified with the help of the Palomar 200-in. It was also thought to be a colliding pair, but later studies indicated that it is a galaxy with a very active nucleus. Bright radio sources seemed associated with energetic events. The source Cassiopeia A turned out a nearby (3 kpc or so) supernova remnant that due to interstellar absorption was not recorded at the time (it should have been visible around 1670, according to the measured expansion of faint optical wisps that were found). Actually, in the late forties Oort had asked Duyvendak to examine Eastern records, but that had no success.

The Crab Nebula consists of an amorphous inner region and filamentary structures around it that are seen in certain spectral lines (see Fig. 10.5). Oort became interested in the possibility of measuring the expansion of the amorphous body of the Crab Nebula and turned to Théodore (Fjeda) Walraven to build an instrument for the purpose of obtaining such observations. Walraven had developed into an extremely innovative and capable instrument builder. Although he had worked extensively with the Rockefeller Astrograph at the Leiden Southern Station, he had also been heavily involved with a new telescope at Leiden, the Zunderman Telescope. This is a 46-cm refractor telescope, named after Hermanus Zunderman (1880–1969), instrument-maker and chief of the workshop of Leiden Observatory, where the telescope was built. It had been put into operation in 1947 with a photoelectric photometer constructed by Walraven (see Fig. 10.1 for its location at the Observatory). In view of the success of this instrument, it is no wonder Oort entrusted him with developing and building more optical instrumentation.

The instrument Walraven developed for measurements on the Crab Nebula was a photometer that could record changes in the total brightness as well as in surface brightness of the amorphous body as a result of the expansion. It was mounted in

Fig. 11.18 The photographic telescope of Leiden observatory, used by Oort and Walraven to measure polarization in the Crab Nebula. In many respects, including the 'English Mount', it resembled the telescopes used for the Carte du Ciel. These had a quite similar aperture (33 cm), but a much shorter focal length (3.4 versus 5.1 m). From the Archives of Leiden Observatory

the plate holder of the Photographic Telescope of Leiden Observatory, which has been shown in Fig. 11.18. It is a double, refracting telescope with apertures (primary lenses) of 33 and 15 cm, and a focal length of a little over 5 m. The larger aperture was used for photography, the smaller for guiding. It was acquired in 1897 as an instrument with a long focal length and consequently large 'plate scale' at the focus to facilitate measurement of proper motions. The Photographic Telescope has an unusual mount, called 'English Mount', in which the telescope is supported in a large yoke that is pointing at the celestial pole. This construction has the advantage that it makes the telescope easy to move, but has the disadvantage that the pole cannot be observed. The best-known telescope with this arrangement is the Mount Wilson 100-in. Telescope.

The instrument was an electronic photometer, the electronic detector being a photomultiplier tube, in which the light falls on a photocathode releasing electrons that then are accelerated by crossing an electric potential (90 V in this case) and fall onto another electrode releasing a number of electrons. This is then repeated a number of times (19 in this case) and at the final electrode is an anode, where the strongly ($10^6 - 10^7$ times) multiplied number of electrons results in an electric signal that can be recorded. This photomultiplier was produced by André Lallemand (1904–1978) of Paris Observatory. The set-up further included a diaphragm to select a small area of the sky and a rotating wheel with color filters that could be rotated into and out of the light path.

Before the program was well underway, Oort and Walraven heard that astronomers in the USSR had found that the light of the Crab Nebula was highly polarized. They quickly adapted the instrument, replacing some of the color filters with pieces of polaroid in three different orientations with relative angles of 60°, so that light in one of three polarizations reached the photomultiplier. With this set-up they mapped the light distribution through each of the three polaroid orientations, which then could be translated into a direction and a degree of polarization (zero if no polarization at all and 100% when completely polarized, i.e. the plane of the wave is restricted to only one orientation). For this plane the convention is to use that of the electric field, so that the (projected) direction of the magnetic field is perpendicular to that. Figure 11.19 shows the result of the observations through a half-arcmin diaphragm, as presented in Oort and Walraven (1956b). This is a remarkable observation, the more so when one realizes that it was obtained with a small telescope in the poor observing conditions pertaining to the middle of the city of Leiden.

The polarization served as an important factor towards the answer of what the physical mechanism was by which the radiation from the Crab Nebula was produced. For an explanation I have to backtrack a little to the Galactic radio continuum. Westerhout and Oort (1951d) had modeled this under the assumption that it reflected the distribution of stars. They had considered the possibility of it being free-free emission from HII-regions of ionized hydrogen (see Sect. 10.7), but had had to reject that. What they were not aware of at that time was, that in 1950 there had been a suggestion that it actually was so-called synchrotron radiation, resulting from high-energy electrons spiraling in an extended magnetic field. This hypothesis had been

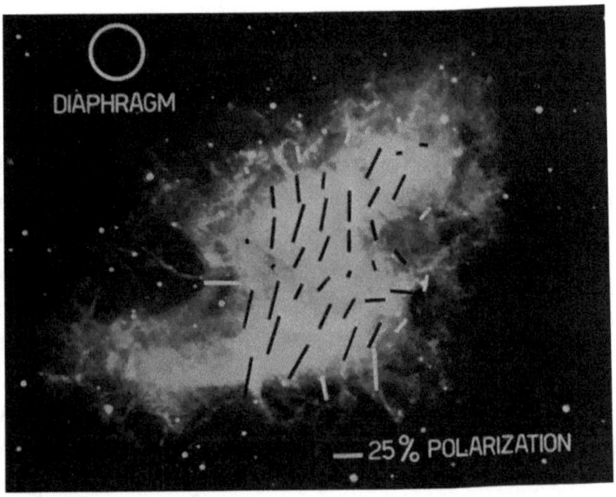

Fig. 11.19 Observed polarization of the light of the Crab Nebula through a diaphragm of 0.5 arcmin. The lengths of the lines indicate the percentage of polarization and the directions correspond to that in which the wave in the electric field has the largest amplitude. In the nebula the projection of the magnetic field would be perpendicular to this. From Oort and Walraven (1956b)

put forward by Swedish (later Nobel Prize laureate) Hannes Olof Gösta Alfvén (1908–1995) and Axel Nicolai Herlofson (1916–2004).

In early particle accelerators, cyclotrons, there is a constant magnetic field in which charged particles (e.g. electrons) are kept in circular orbits. For a given magnetic fieldstrength and particle (mass and charge), the radius of these orbits depends on the energy or speed, but the *period* is constant. In a cyclotron in some position in the orbit an electric potential is applied that accelerates the particle. This is then repeated with the electric field reversed half a period of the orbit later, so that on the opposite side a similar electric field is switched on half an orbital period later, but now with reversed field direction, so that it is again an acceleration. With increasing energy the particle then spirals outward, keeping the same period around the cyclotron. However, when the particle (say an electron) starts to acquire speeds not insignificant compared to that of light, the period of the orbit starts to depend on energy and the synchronization is lost. So physicists had to resort to accelerators that allowed for this and in which the magnetic field varies synchronously with the period to keep up with the particle in its orbit, so that it then will continue to be accelerated. Now in these synchrotrons it was noted that the electrons emitted a glow (found later in cyclotrons as well). This 'synchrotron' radiation was the result of the highly accelerated electrons being forced into a curved orbit by a magnetic field. A characteristic property of this radiation was that it was highly polarized in the plane of the curved orbit, which is perpendicular to the magnetic field.

It turns out that most continuum emission in radio astronomy is actually synchrotron radiation. In the Galaxy there is an extended magnetic field, and interstellar space is filled with high-energy electrons as a major part of cosmic rays. These cosmic rays are the high-energy particles—mostly protons and electrons—that we can detect on Earth and which in part come from the Sun and partly from Galactic and even extragalactic sources.

Now, USSR theorist Iosif Shklovsky, who we already met as independent predictor of the 21-cm line (see Sect. 9.8), had suggested that the radiation from the amorphous body of the Crab was optical synchrotron radiation. This nicely explained why that light is strongly polarized. Calculations by Oort showed that the magnetic induction can be estimated to be 10^{-3} Gauss[1] and the mean energy of the electrons that produce the synchrotron radiation would then be of order 2×10^{11} electron Volt (eV). The total energy in the electrons within the nebula thus inferred from the observed synchrotron emission is then about 3×10^{41} J. Oort compared this to the energy released if the Sun would convert all its hydrogen into helium ($\sim 10^{45}$ J) and regarded the former as not unreasonable. He also estimated that the electrons lose their energy in about 200 years, so the remains of the supernova rather than the explosion itself had to be the source of the energetic particles.

We now know that the remains of a star that has become a supernova can be a neutron star (if the mass of the original star is not too small or too large; see Sect. 2.6). Because it will contain a part of the original angular momentum of the star, it rotates very rapidly with rotation periods of milliseconds to seconds. This compact object may develop a strong magnetic field, which may in turn generate electric fields and give rise to beamed electromagnetic radiation along the magnetic axis, which often does not coincide with the stars rotation axis. The result is a rapidly rotating 'lighthouse' beam. When the Earth happens to be in the path of that beam, regular pulses may be observed. Such pulses were first seen at radio wavelengths 1967 by Susan Jocelyn Bell Burnell (b. 1943) and Antony Hewish (b. 1924) at Cambridge, UK, and the phenomenon became known as pulsar. As it happened, I had an appointment with Oort to discuss the progress of my thesis research, when he had just received the copy of *Nature*, in which the discovery of pulsars was announced. He had just convinced himself that such a rapid rotation of a neutron star appeared to be not impossible in terms of angular momentum compared to that of the Sun. When I do the comparison now, I find that a pulsar neutron star (say 2 $M_\odot$ and radius 15 km) with a rotation period of 1 second (about that of the first pulsar) has an amount of angular momentum one thousandth of that of the Sun. This must have been the quick calculation Oort had done. He was extremely excited and lectured me all about the Crab Nebula research he had done. Not much later, in 1968, a radio pulsar with a period of 33 milliseconds was detected in the Crab Nebula (as luck would have it, visible from the Earth), and in 1969 the same pulsar was also detected at optical wavelengths. It is now generally accepted that pulsars are sources of high-energy particles being fed into their immediate surroundings. Many energetic particles are of course produced during the supernova explosion itself; these have escaped long ago into the interstellar medium and are a major component of what we observe as cosmic rays. Without the new supply of energetic electrons by the pulsar the synchrotron radiation from the Crab would have faded long ago.

There were two further papers concerning the Crab Nebula published at about the same time [301, 302]. Westerhout had used the Kootwijk antenna to search for polarization at radio wavelengths, but found none, which Oort explained by the large radio

---

[1] Oort used indeed magnetic *induction*, of which Gauss is the unit, while modern astronomers loosely (but formally incorrectly) quote magnetic *fieldstrength* in Gauss. For energy Oort used ergs, which I have converted into Joules.

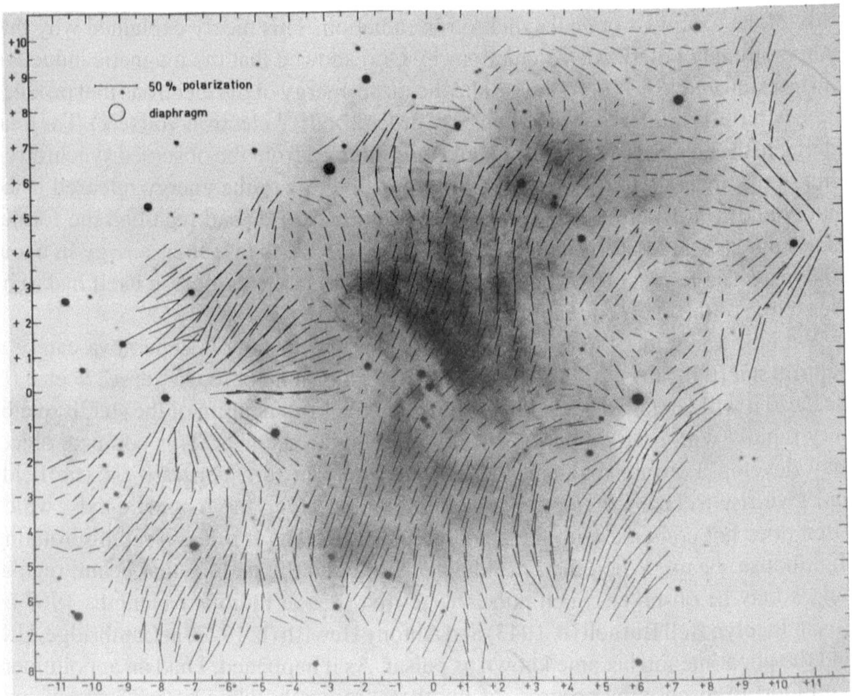

Fig. 11.20 Map of the polarization of the light of the Crab Nebula, obtained by Lodewijk Woltjer through scanning plates taken by Baade at the the Palomar 200-in. telescope through various polaroids with different orientations.. From [304]

size and the absence of polarization in the outer parts that dominated the radio emission. Baade took photographic pictures of the Crab Nebula with the Palomar 200-in. Telescope through polaroids in various orientations and found that the nebula looked very different at different polarization angles, confirming the optical polarization pattern found by Oort and Walraven.

There were two immediate follow-ups to improve the map of the polarization [303, 304]. Walraven obtained time at the 80-cm Telescope at the Observatoire de Haute-Provence in the French Alps, some 100 km to the north of Marseille; using another photometer that he had built derived a much improved polarization map of the Crab Nebula. But an even more spectacular map (see Fig. 11.20) was obtained by Lodewijk (Lo) Woltjer (b. 1930) by carefully scanning the plates Baade had taken. As Woltjer noted when he sent prints of this and related images to Baade: 'One of the most striking features is the high degree of regularity in the polarizations.'

Lodewijk Woltjer was a young student in Leiden; he was the son of Jan Woltjer, the staff member of Leiden Observatory who had died shortly after the end of the war. Like his father, Lo Woltjer had a strong theoretical inclination, but was not afraid to work on observational material.[2] He had started studying geology in 1948, but switched to astronomy in 1950 and obtained his doctorandus (Masters) degree in 1954. This polarization map actually was a prelude to his PhD thesis entitled *The Crab Nebula*, which he defended in 1957 (see Appendix A.4).

I will not discuss the thesis in detail. Woltjer used Baade's photographs and his his polarization map, and new optical spectra made available by Mayall at Lick Observatory. From these he derived (or estimated) various properties of the nebula like abundances of chemical elements, mass and temperature of the amorphous nebula and the filaments, etc. He derived motions of the filaments. Then he modeled the synchrotron origin of the emission, deriving the energy in the relativistic electrons (10^{41} J) and the magnetic fieldstrength in the center of the nebula (10^{-8} T). For comparison the magnetic field at the Earth's surface is some 5×10^{-5} T. The magnetic field resulted from turbulence in the aftermath of the supernova explosion. Woltjer also concluded that the electrons in the Galaxy that give rise to the radio continuum emission must have come from supernovae.

11.6 President of the IAU

At the 1948 General Assembly of the International Astronomical Union Oort had stepped down from the burden of General Secretary. However, together with retiring President Spencer Jones, he was retained as consulting member to the Executive Committee. The first responsible officers were from the Scandinavian countries Sweden and Denmark: Lindblad as President and Strömgren as General Secretary. The fact that they came from smaller countries (and that Sweden had been neutral in World War II) helped in resolving some of the issues that were sensitive in terms of international and global politics. As soon as 1949 Strömgren and Oort took the initiative to come to the re-adherence of Japan. This was taken up by the Executive and the matter was resolved at the next General Assembly. Although there appears to have been little actual involvement of Oort, the adherence of Germany finally took place at the same meeting.

An issue that certainly involved Oort as an adviser was that of the 1951 General Assembly, which was to take place in Leningrad. Various international developments, in particular the escalations leading to the Cold War, the political situation in the USSR, and the Korean War, made this a difficult matter. Finally, the Executive decided to postpone the meeting—and Oort must have been involved in this in his advisory role—, as 'the present international situation is likely to obstruct a representative international participation in the General Assembly and Symposia' (see Blaauw's History of the IAU [113] for this and other matters in this section). The next General Assembly was held in Rome in 1952. Oort thereafter had

[2]Some theoreticians refer to obtaining observations as 'getting your hands dirty'. Olin Eggen, an observer (see below), once reported at a meeting I attended on theoretical work he had done and referred to that as 'getting my mind dirty'.

no formal or advisory role in the IAU, but he will have been informed of developments or informally asked for advise by the President and General Secretary appointed in Rome, respectively Otto Struve and Pieter Oosterhoff.

Oort had been much involved in the matter of individual membership of the IAU. The 1931 bylaws had introduced individual membership for members of national and standing committees, although astronomers with recent PhDs were still not welcome as members. In 1948, when Oort was still General Secretary, it was decided that adhering organizations could also nominate individual members, which was triggered by unequal memberships from the various countries. After the Rome General Assembly, a small subcommittee was appointed to formulate new rules for individual membership and this was chaired by Oort. The new proposal was to appoint members (by the Executive assisted by a Nominating Committee) solely on the basis of their achievements as astronomers and not on membership of national or standing committees. Adhering organizations would propose membership for all those qualifying in this manner, which essentially meant active participation in astronomical research. This fundamentally changed the character of the Union. It was adopted by the 1955 General Assembly (in Dublin) and became effective at the 1958 one (in Moscow). In 1958, individual membership grew to over 1100.

At the 1952 General Assembly in Rome the IAU started holding special symposia during the meeting. But there were also initiatives to hold IAU-sponsored symposia in between assemblies. In 1949, the first of these took place in Paris, as we have seen already, on *Problems of Cosmical Aerodynamcs*, sponsored jointly with the International Union of Theoretical and Applied Mechanics. The first, IAU-Symposium No. 1, was held in 1953 and had as subject *Co-ordination of Galactic Research*. No. 2 has also already been mentioned; it was the one on *Gas Dynamics of Cosmic Clouds* held at Cambridge, England, in 1953. Organizing symposia rapidly grew to be a major activity of the IAU; it still continues up till today, reaching No. 349 during the General Assembly in Vienna in 2018 (Fig. 11.21).

IAU Symposium No. 1 was an initiative by Oort in his capacity as President of Commission 33 of the IAU on 'Stellar Statistics' (the name of Commission 33 changed in 1955 to 'Structure and Dynamics of the Galactic System') Its title was *Co-ordination of Galactic Research*. The aim of the meeting was to identify areas of Galactic research beyond Kapteyn's Plan of Selected Areas. Adriaan Blaauw acted as secretary. At the invitation of Pieter van Rhijn (Director of the Kapteyn Astronomical Laboratory), it was held at the Vosbergen Conference Center just to the south of Groningen, which at the time was mostly operated for use by the University of Groningen. The meeting took place from June 22 to 27, 1953. The organizing Committee, chaired by Oort, included some of the leading researchers in the field, some of which were long-time friends of Oort, such as Bertil Lindblad and Walter Baade. Actually Baade had visited Leiden for about two months preceding the conference. Oort had invited him to visit for three-and-a-half months and even obtained financial support for that from ZWO, but the visit was shortened, to a large extent due to Baade's request to American authorities for a re-entry visa having gone missing. The rationale, distributed to potential participants, most likely formulated by Oort, was the following:

Fig. 11.21 Victor Ambartsumian, Bertil Lindblad and Oort at the Vosbergen Symposium in 1953. From the Oort Archives

'During the first third of this century an important concentration of work on Galactic structure and motions has been promoted by the Plan of Selected Areas, initiated in 1906 by Kapteyn. Although the scheme outlined in 1906 has not lost its significance, it is widely felt that further research into structure and dynamics of the Galaxy should be extended beyond the original Plan. At the same time it is felt by many that some kind of co-ordination of effort remains highly desirable, [...]
[...] it is desirable to try to formulate the principal problems in which we are most interested. The emphasis will no longer be confined to general overall structure, but more on structural features displayed by special types of stars, on relations between kinematical and distributional features, on spiral structure and the phenomena connected with the evolution of stars and the Galactic System.'

The format of the conference was unusual. It consisted mostly of discussions and only a few introductory presentations summarizing recent progress in various areas. The proceedings also had an unusual format. They were written entirely by the editor, Adriaan Blaauw, and summarized the outcome of the discussions without any contributed texts from introductory reviewers or from the participants.

Blaauw noted:

'As was to be expected, the conference proved to be very useful for the formulation of the principal present problems. Concrete proposals for research with detailed specifications have been made only for a few of the many branches of Galactic research, [...]. In so far as the discussions stayed on the level of a general orientation of the main problems and the practical possibilities of attacking them, the conference served as a basis for more specific planning in the future.'

In addition a few proposals for the formation of sub-committees of the IAU were made, such as a sub-committee to Commission 33 for co-ordination of observational programs, chaired by Oort and with Blaauw as secretary. Members were among others trusted colleagues such as Baade and Lindblad, but also van Rhijn.

An important long-term program that came out of the symposium was what later became to be called the Palomar-Groningen Variable-Star Survey. The conference had noted that soon the Palomar 48-in. Telescope would finish the *National Geographic Society—Palomar Observatory Sky Survey* and the telescope would become available for large-scale surveys. During WWII, using the Mount Wilson 100-in. Telescope, Baade had found a few areas on the sky not far from the direction to the center of the Galaxy, in which the interstellar absorption was unusually small. In those areas one had as a result of the uneven distribution of dust the lucky circumstance that a long line of sight was relatively free of absorption, so one could see stars out to well beyond the central regions of the Galaxy. The largest of these is Baade's Window, only about 4–5 degrees from the direction to the center. It has a diameter of about one degree and is roughly centered on globular cluster NGC 6522. This cluster is at a distance of about 7.7 kpc from us, somewhat closer than the center itself.

The aim of the survey was to take multiple plates and search for variable stars, particularly pulsating RR Lyraes for which the distance could be estimated from the period of the variability. This could be used to study the distribution of stars in the central part of the halo of the Galaxy. The survey was undertaken by Lukas Plaut (see Sect. 9.3 and Fig. 11.22). I will come back to this survey later, because in the end Oort became involved in the analysis (Fig. 11.23).

The conference eventually had a follow-up in IAU symposium No. 7, held from 17 to 22 June, 1957, in Saltsjöbaden near Stockholm, following more or less the same format. This time the burden of writing up the proceedings was shared among six participants, Blaauw doing the coordinating and final editing. As we will see, it was overshadowed by a historic and seminal meeting on the structure of the Galaxy, that had taken place just before the Stockholm symposium, the famous Vatican Conference on *Stellar Populations* held in Vatican City May 20–28, 1957 under the auspices of the Pontifical Academy of Sciences. But this is for a later chapter (Fig. 11.24).

We saw that at the 1955 Dublin General Assembly of the IAU, a fundamental change had been adopted on the general, individual membership of the Union. During the first plenary session Oort reported in extenso on the proposal for membership and this was adopted unanimously. The change would be effective at the next General Assembly, which was held in Moscow in 1958. Oort had no further administrative duties at the Dublin meeting and could concentrate on scientific issues. Otto Struve was President of the Union, and André-Louis Danjon (1890–1967) was elected as the next president.

The report of Commission 33 on Stellar Statistics, of which Oort as its President was the author, showed very clear signs of its authorship; much attention was given to the progress in the study of Galactic structure with the 21-cm line (although

Fig. 11.22 Some participants of the Vosbergen Symposium. On the left Bertil Lindblad and Walter Baade. In front of Oort stands Lukas Plaut, with William Morgan next to him. Between the latter two we see Adriaan Blaauw. From the Oort Archives

Fig. 11.23 Oort and Harold Spencer Jones on a boat trip during the 1953 Vosbergen Symposium. From the Blaauw Archives

Fig. 11.24 Mrs. Oort with Strömgren and Lindblad during the boat trip for the participants of the Vosbergen Symposium. From the Blaauw Archives

formally nothing to do with *stellar* statistics). It emphasized the problem of spiral structure. It mentioned the work of William Morgan and collaborators, [293], of the discovery of spiral structure in the solar neighborhood using O-associations, and reported extensively on the Kootwijk results. The important problem that was and would be a major focus of attention by Oort, was the long-term stability of spiral structure. The report said:

> 'The evidence so far obtained indicates that the Galactic spiral rotates such that the convex sides of the arms precede. The observation of spiral arms in the Galactic System, where the differential rotation is known and strong, emphasizes the problem of the origin and continued existence of these arms. With an average spacing of about 2 kpc and an average derivative of angular velocity with respect to distance from the center of -9.5 km/sec.kps, the present arms would rapidly dissolve, unless they be constantly rejuvenated by a process not yet understood. Without such a process they could not be older than 300 million years.'

The matter of the direction of rotation of galaxies compared to the spiral arms, was a serious one. Lindblad had put forward theories that spiral structure resulted from dynamical instabilities and had predicted that the arms would be unwinding ('leading'), while Oort for many years had collected information that seemed to support the fact that the arms were actually winding up ('trailing'). The latter could be checked in external galaxies. This was done by selecting galaxies that are inclined highly enough to see from the projection of the dust in the disk on the central bulge, which was the near side, but still not too much inclined so that the spiral structure could be delineated. Spectroscopic measurements would then tell the direction of rotation.

New at the meeting were Symposia and Joint Discussions, the latter treating subjects across the borders of Commissions. There was one symposium before the Assembly, at Jodrell Bank Radio Observatory (near Manchester), where Oort presented his and Walraven's results on polarization of the light from the Crab Nebula (see next chapter). In Dublin there was a one-day symposium on *Comparison of the large-scale structure of the Galactic System with that of other stellar systems*, where Oort (1958c) compared the rotation curves of the Galaxy with those of the Andromeda Nebula M31 and the other small spiral in the Local Group, M33. The latter rotated much more slowly, but the Galaxy and M31 seemed comparable. In a footnote Oort referred to a very recent result with the new Dwingeloo Radio Telescope that indicated that the rotation curve in M31 'remains practically constant from 8 to 22 kpc from the centre' [353]. He noted that all systems that have spiral structure contain gas and all systems that contain gas show structure ('systematic arrangement'). In one of the three Joint Discussions, on *Fundamental Stars*, van Herk reported on the second Kenya expedition. Adriaan Blaauw became President of Commission 33, while a sub-committee on the coordination of Galactic research was established and Oort was elected chairman.

In 1958 the General Assembly was held in Moscow. There had been some difficulty accepting the Russian invitation (after all the one planned for Leningrad in 1951 had been canceled). Donald Harry Sadler (1908–1987), in the *LibAm80* [2], attributed much of the easing of tension between Russia and the IAU to the efforts of Oort and Oosterhoff, during the celebrations marking the reconstruction of the Pulkovo Observatory in 1954 (Fig. 11.25).

There is no published scientific contribution by Oort associated with this General Assembly. At the end of the meeting, Oort was elected President. His General Secretary was Sadler, who was the superintendent of His Majesty's Nautical Almanac Office at the Royal Greenwich Observatory. The two had a very fruitful collaboration (see Sadler's contribution to the *LibAm80* [2] just referred to).

There were four important issues that Oort faced during his Presidency of the Union (see also Blaauw's History of the IAU; [113]). Oort's handling of these show that he possessed admirable diplomatic skills. The first was China. The People's Republic of China, founded in 1949, considered itself to be the only China and therefore sole member of the IAU on behalf of all Chinese. The Republic of China on the island Taiwan applied for IAU membership not long before the Moscow IAU. This was a thorny issue for the IAU Executive Committee, in which two of the six Vice-Presidents were from communist countries. The two were Boris Vasilyevich Kukarkin (1909–1977) from the USSR, and Bohumil Sternberk (1897–1983) from Czechoslovakia. The others were Leopold (Leo) Goldberg (1913–1987) from the USA, Otto Heckmann from West-Germany, Robert Methven Petrie (1906–1966) from Canada and Richard Hugh Stoy (1910–1994) from South Africa. Kukarkin and Steinberk argued that there was hardly any astronomy in Taiwan to justify membership, but Oort strongly felt that since the Statutes of the IAU did not mention anything of the kind, the membership could not be denied. In the end (after much deliberation and discussions) Oort prevailed and the Executive voted along the lines mentioned, five against two, to admit Taiwan. The Peoples Republic immediately

Fig. 11.25 Oort during the celebrations of the reconstruction of the Pulkovo Observatory near St. Petersburg in 1954. From the Oort Archives

resigned from the IAU and did not re-enter the IAU until the 1989 Montreal General Assembly under the wise leadership of Adriaan Blaauw.

The second issue was triggered by a proposal from two French astronomers, Jean-Claude Pecker (b. 1923) and Evry Schatzman (1920–2010), involving a major restructuring of the commission structure of the Union. Their criticism was that the commissions became too large and that they employed no activities in between General Assemblies. They proposed a two-dimensional structure grouping also by 'methods, techniques or technical subjects'. Sadler and Oort took this very seriously and had various discussions with them (also involving Oosterhoff) and exchanges of letters and memoranda. It was a much debated issue. In the end the one-dimensional structure remained, but Commissions would have 'organizing committees' consisting of a President, Vice-President and a small group of members. These would take care of business in between meetings, prepare the report to the General Assembly, advise on IAU sponsored meetings, propose new members, etc.

The third major worry was Project Westford. This was an initiative by the US Military to launch a satellite into orbit around the Earth filled with small (1.78 cm by 25.4 μm) needles. Once they had spread around the orbit, these 480 million copper dipoles would be useful for telecommunication by the Military from almost any location (and be independent of cables on the ocean floors). The length of the needles was tuned to a frequency of 8 GHz used by a communication station located near Westford, Massachusetts (the project was coordinated from a laboratory of the

Massachusetts Institute of Technology MIT in Cambridge, Mass.). The IAU and other (radio and optical astronomical) organizations and agencies protested sharply, expecting that it would interfere severely with their astronomical work. In 1961, the launch proceeded anyway, but the needles failed to disperse. A second attempt in 1963 was 'successful', but the technology was found to be inferior to using communication satellites. The needles disintegrated soon in the Earth's atmosphere, while their presence proved unimportant in affecting astronomy. It showed the weak position of astronomy vis-a-vis military priorities during the Cold War.

But the most difficult issue must have been the fourth one, which concerned the protection of the radio bands important for astronomy. Since telecommunication was on the rise (involving more and more space probes after Sputnik had been launched in 1957), there was a problem of a conflict of interests of science and society, concerning commercial use of radio waves. As IAU General Secretary, Oort had been actively involved with ICSU, the International Council of Scientific Unions. In 1958, the ICSU had established the Committee on Space Research (COSPAR) for the 'promotion of scientific research in space on an international level, with emphasis on the free exchange of results, information, and opinions, and providing a forum, open to all scientists, for the discussion of problems that may affect space research'. Actually, Oort had asked Henk van de Hulst to attend its first meeting, who returned to Leiden having found himself elected COSPAR's first President.

In 1865, the International Telecommunication Union UTI (acronym from the French name) had been formed as the International Telegraph Union, and in 1947 had become an agency under the United Nations to coordinate the worldwide use of telecommunication. In 1959 they organized a 'treaty conference' in Geneva, where it is based, to discuss and allocate frequencies for radio and telecommunication. Oort, as President of the IAU, felt that it was imperative that radio astronomers and space scientists should attend and try to arrange that frequencies for science were given protected status. It was vital for radio astronomy, especially the part of the radio spectrum that contained the 21-cm line, that such protection be awarded, because otherwise the young science would be doomed to extinction. With the support of General Secretary Sadler, he arranged for a continuous presence of a representative of the IAU to lobby and argue the case for radio astronomy and basic science in general. This meant from mid-August until mid-December; the UTI had to consider some 6000 applications! Oort spent some time there himself as well, giving a public lecture on radio astronomy to point out its importance for an audience that was hardly aware of basic science as a user of radio frequencies. The lecture was later published in the *American Scientist* (Oort 1960g). The result of all these efforts was that radio astronomy was granted the 'undisputed use of the 1400 to 1427 Mc/s band of frequencies' (see Oort 1960g, postscript) and some other bands as well.

To safeguard the allocation and defend the interests of radio astronomy and space research, Oort convinced the IAU, URSI (the International Union for Radio Science; acronym again from the French name) and COSPAR to jointly set up and support a standing committee, the Inter-Union Committee on Allocations of Frequencies IUCAF, later renamed the Scientific Committee on Frequency Allocations for Radio Astronomy and Space Science, retaining the acronym.

Fig. 11.26 Opening ceremony of the IAU General Assembly at Berkeley in 1961. From the left Ira Bowen, Oort, Adlai Stevensen, Leo Goldberg and Donald H. McLaughlin. See text for details. From the Oort Archives

Oort presided over the next General Assembly, which was held at Berkeley in 1961. Figure 11.26 shows a picture taken during the speeches at the opening ceremony. The speaker is Leo Goldberg, chairman of the US National Committee for Astronomy. Adlai Ewing Stevenson II (1900–1965), seated to the right of Oort, was US Ambassador to the the United Nations. At the far right, Donald Hamilton McLaughlin (1891–1984), the Regent of the University of California, which hosted the meeting. Oort addressed the assembly at the opening session with a speech in English (Oort 1962c), which he subsequently summarized himself (not much shorter though) in French, honoring the fact that both languages were the official ones of the IAU. In this address he emphasized the importance of the West of the USA for the development of observational astronomy, praising of course first the Lick Observatory (see Sect. 4.7) of the University of California with its new (1959) 120-in. telescope, which in 1978 would be named the C. Donald Shane Telescope. Oort also honored George Hale for establishing the Mount Wilson and Palomar Observatories. But then he quickly turned to radio astronomy, whose telescopes 'when sufficiently large, permit observations penetrating much deeper in the Universe then is possible with optical means'. Oort referred to the fact that some of the brightest radio sources had been identified with very distant galaxies. He touched upon the need for frequency protection and on Project Westford. And than presented space research as the new area of enormous progress (Fig. 11.27).

Fig. 11.27 Oort with Donald Shane and his wife Mary Lea Heger in 1961 during the Berkeley IAU General Assembly. From the Oort Archives

The IAU was growing in size and Oort felt that was a cause for concern. He reflected: ' I have been in close contact with the Union throughout my life. I may, in fact, be the only astronomer who has been at all eleven General Assemblies, although in 1922 I attended only by accident and quite unofficially.' He foresaw that in the near future there should be separate assemblies for different subject parts of astronomy. Fortunately, that never happened. During the General Assembly in Brighton in 1970, U.K., Oort repeated this worry in the Assembly's newspaper, concluding that the solution would be more symposia and colloquia on special topics and shorter General Assemblies. Actually Oort attended IAU General Assemblies until the 17th in 1979, held in Montreal. For the 1982 Patras (Greece) meeting he had written to the General Secretary, Patrick Arthur Wayman (1927–1998) from Ireland, apologizing for not attending, which was acknowledged by his successor Richard Martin West (b. 1941), who expressed his understanding that Oort instead attended two IAU Symposia: No. 100 on *Internal kinematics and dynamics of galaxies* in Besançon, France and No. 104 on *Early evolution of the Universe and its present structure* on Crete. At the latter meeting he gave a keynote review (Oort 1983b).

The Executive's handling of the question of the membership of China and the proposal for the new Commission structure, as discussed above, were endorsed in the first business session. At the end of the Assembly the IAU elected Victor Ambartsumian from the USSR as the next President and to ensure continuing activity in the area of the restructuring of the Union Jean-Claude Pecker as Assistant General Secretary (Sadler served another term as General Secretary, after which Pecker took over this position).

Fig. 12.0 Painting of Oort by Pieter Herfst (1911–1983), likely from the 1960s. From the collection of the Academy Building, University of Leiden, see [208]

Chapter 12
Dwingeloo, ESO, the Vatican and the Cross-Antenna

You can always amend a big plan, but you can never expand on a little one. I don't believe in little plans. I believe in plans big enough to meet a situation which cannot possibly be foreseen now.
Harry S. Truman (1884–1972)

Obstacles are those frightful things you see when you take your eyes off your goal.
Henry Ford (1863–1947)

Even before the Dwingeloo Radio Telescope became operational, new plans were made. The most serious one resulted from the discussions Oort had had with Baade, when the latter visited Leiden for two months in 1953 before the Vosbergen IAU Symposium, and that concerned the establishment by the—astronomically speaking—leading European countries of a major optical observatory in the southern hemisphere, presumably South Africa. The other matter discussed was to build a successor to Dwingeloo, which in the earlier phases was envisaged as a collaboration with Belgium to build and operate a 'Benelux Cross Antenna' close to straddling the Dutch/Flemish border. Eventually these two plans resulted in the European Southern Observatory in Chile and the Westerbork Synthesis Radio Telescope not far from the Dwingeloo Observatory.

But Oort had other things on his mind as well, in addition to teaching and research and supervising students using the Dwingeloo Radio Telescope. A major share of the he time-consuming directorship, also involving the issue of the future of the Leiden Southern Station and the relation with the Bosscha Observatory in what was now Indonesia, the educational program and interface with the University, was taken up by Pieter Oosterhoff. He was Secretary General of the IAU between 1952 and 1958,

P. C. van der Kruit, *Jan Hendrik Oort*, Astrophysics and Space Science Library 459, https://doi.org/10.1007/978-3-030-17801-7_12

which must have taken some investment of time and energy as well. When Oosterhoff came to the end of his term, Oort was elected President of the IAU; Oosterhoff stayed on for one more term as adviser to the Executive and thus was able to assist Oort in IAU matters during Oort's Presidency.

12.1 The Dwingeloo Radio Telescope

In 1950 ZWO, the Netherlands Organization for Pure Research, had been formally founded. Its provisional predecessor had rejected all requests for the funding of a 25 m dish for astronomical research. By that time the annual ZWO budget had grown to one million guilders. The original plan had been for this to increase to five million by 1955, after which it was supposed to ramp down and ZWO dissolved, as it was assumed that the universities and research institutes then would have restored their research capabilities from the devastation of the War. In the end ZWO continued and even grew, but this was not foreseen at the time. When the 21 cm line had been discovered and ZWO officially founded, the Board of SRZM had submitted a new proposal. This time ZWO approved the request; the cost of the telescope had increased considerably, now approaching half a million guilders, as a result of improved designs and costing by Werkspoor, the production company.

The history of the construction of the Dwingeloo Radio Telescope has been treated in greater detail in *EarlyRadio* [9] by Astrid Elbers and in van Woerden and Strom [305]. There were some vital roles for two engineers, without whom Oort might never have obtained a telescope working as well as it did or maybe not even a working telescope at all. The first of these was Muller, who, as in Kootwijk, would be responsible for the electronics. The other one was a mechanical engineer, Bernard G. (Ben) Hooghoudt (1924–1995). He was not employed by SRZM. but rather by the Post, Telegraph, and Telephone Company PTT, but SRZM reimbursed part of his salary. Hooghoudt acted as an intermediary between Werkspoor and SRZM and this resulted among others in less dramatic increases in the budget than would have been incurred otherwise. Hooghoudt's involvement also had significant positive effects on the design of he telescope. Hooghoudt continued his career by making vital contributions to the design of radio telescopes all over the world.

There have been many delays and cost increases that must have caused Oort serious headaches. The role of Bannier, Director of ZWO, should also not be underestimated. In a festschrift for Bannier on the occasion of his 70th birthday, Oort (1979d) noted: 'The numerous meetings [of the SRZM Board] were almost without exception attended by Bannier, who was therefore in an ideal situation to explain our problems to his Board.' ZWO did approve increases in the budget, which might very well have been more difficult to obtain without Bannier's personal involvement. The personal relations between Board members and Oort (and possibly with other board-members such as Minnaert and the non-astronomers) will have helped also. After all, it was a small world. The total investment the telescope amounted to about Dfl. 0.6 million, or about €1.3 million in current purchasing power. When operations

Fig. 12.1 Aerial photograph of the Dwingeloo Radio Telescope around 1960 with on either side two Würzburg antennas mostly used for solar research. The buildings are to the top-right hidden in the woods as conservationists had demanded. Photo credit: Foto AVIODROME Lelystad

started in 1956 it was the largest radio telescope in the world, to be overtaken the next year by a significant factor by the 250-foot dish at Jodrell Bank, England; this is now the Lovell Telescope and it has a diameter of 76.2 m.

Now, was the expenditure on radio astronomy very much out of proportion? From Kersten's history of ZWO [267] it can be deduced that the total budget of ZWO between 1951 and 1960 was Dfl. 51.6 million. About half of this actually went to the Foundation for Fundamental Research on Matter (FOM)—'a cuckoo's egg in the ZWO nest' according to Kersten. During the 1950s FOM operated two research institutes, one for nuclear physics and one for mass spectroscopy (in fact uranium enrichment). Although branching out into other fields of physics during the 1960s, FOM was at first nuclear physics related, which was socially very relevant because then there were great expectations for nuclear energy. Since SRZM was allocated an operations budget when the Dwingeloo dish was working (of no less than Dfl. 365,000 in 1959), the total share SRZM received from ZWO's total budget over this period of ten years might have been some 5% and may not be that outrageous. It is high for the field like astronomy, but note that this is 5% of only the expenditure on *pure* research. It was the result of a considered and deliberate choice of the ZWO Board; and still nuclear physics received about 10 times as much (Fig. 12.1).

The Dwingeloo Radio Telescope has an alt-az mounting (see also Sect. 5.1), which means that in order to point the telescope at a particular position on the sky one has to convert the right ascension and declination into azimuth and altitude using the sidereal time at that moment (see Sect. 2.1). Nowadays this is done digitally in a computer; in the Dwingeloo telescope it was done with an analogue computer, a clever construction of moving yokes and axes and quartz clocks that did the conversion.

This 'pilot' was used until 1993. The radio emission was received at the focus of the dish by the 'feed', which contained two perpendicular dipoles (to be sensitive to all polarizations) and was mounted on a pole which had its base at the center of the dish. The beam was circular with a diameter (full-width-half-maximum, FWHM) of 34 arcmin at 21 cm. The surface of the dish consisted of conductive wire and its mesh had to be significantly smaller than the wavelength of the radiation to be observed in order to assure complete or almost complete reflection; in Dwingeloo it was 15 × 15 mm.

The site selection was another thorny issue, described in detail in Elbers' studies *EarlyRadio* [9]. Initially, for a short period there was a possibility of Belgian participation and a site near the Netherlands-Belgian border directly to the south of Eindhoven was considered. When this fell through, the north of the country, the province of Drenthe south of Groningen, came forward as an ideal location, also because a car-free area around the Observatory was required to minimize interference from automobile and other engines. There was a protracted period of opposition from city councils, apprehensive about negative effects of the car free zone on tourism, and from farmers and conservationists. Again Bannier played an important role in this. In the end a site near the village of Dwingeloo and the hamlet Lhee

Fig. 12.2 Queen Juliana of the Netherlands and Oort after the opening of the Dwingeloo Radio Telescope on April 17, 1956. From the Oort Archives

Fig. 12.3 The Dwingeloo 25 m not long after construction. The long pole to the focus where the feed is located, has been replaced after a few years by a tripod, later by a quadrupod. From the Archives of Leiden Observatory

was selected and approved by Staatsbosbeheer, the Dutch government organization for the management of forestry and nature reserves, and the responsible Ministry of Agriculture and Fisheries (see Fig. 1.1). It was located at the northern edge of a large nature reserve area, the Kraloër and Dwingeloosche Heide, now the largest part of the National Park Dwingelderveld. From the chosen site there is a clear view across heathland towards the southern horizon. The latter was important since at Dwingeloo the Galactic center, an important subject of study for the radio telescope, does not rise more than some 8 degrees above the horizon in the south.

The official inauguration took place on Tuesday April 17, 1956, by Queen Juliana of the Netherlands (see Fig. 12.2). The moving of the telescope could not yet be done automatically, so the button she pressed at Oort's invitation only lit a bulb in the control room and the movement of the telescope was started manually. At first the telescope started moving in the direction opposite to what had been planned, but the day was saved by quick intervention of Gart Westerhout (see [289]). The Leiden newspaper 'Leids Dagblad' ran an almost full page coverage (although on page 7) in its April 17 edition with heading; 'On the quiet heathland of Drenthe – The Netherlands pricks up its ears to hear the whispering of the Universe' [306]. At a gathering in the village of Dwingeloo before the inauguration, in the presence of the Queen, the Minister of Education, Arts, and Sciences,[1] Jozef Maria Laurens Theo Cals (1914–1971), conferred upon Oort a Royal Distinction, viz. Knight in the Order of the Netherlands Lion. This order is the highest civilian one in the Netherlands. Oort is reported to have stated in his speech, that 'The sky is opening up to us in all directions. Because in addition to radio telescopes we have V2s, aerobees and other rockets that climb beyond the atmosphere. Soon these will be succeeded by artificial satellites. From those satellites we will for the first time observe far-ultraviolet and X-ray radiation from celestial bodies.'

Among the people I consulted there is a clear consensus that Oort's most important contribution to astronomy has been the realization of the potential of radio waves for the study of the Galaxy and his subsequent efforts leading to the work at Kootwijk. The Dwingeloo dish was designed from the start entirely for astronomical application. With its completion astronomers in the Netherlands owned a world class observing facility and would no longer have to rely for a large extent on foreign telescopes. This was a turning point in Dutch astronomy and a crown on Oort's work. The challenge was to use it to produce first class science and to make the next step to a bigger facility to remain at the forefront.

On behalf of the SRZM, Herman Kleibrink, the then photographer at Leiden Observatory, produced a movie of about half an hour on the building of the Dwingeloo Radio Telescope (Fig. 12.3), which is available on the Web [308].

12.2 An Observatory in the South

The observing conditions due to light pollution at the Leiden Southern Station were deteriorating (see Sect. 11.3). During Oort's visit to Johannesburg in 1948, he had discussed this with Adriaan Wesselink, the Leiden observer at the time, and with Willem van den Bos, the Director of the Union Observatory. At the time of that visit the Yale facility was still in South Africa; it had been joined in 1943 by Columbia University (so that it now was the Yale-Columbia Telescope). In 1956 the telescope was moved to Mount Stromlo Observatory in Canberra, Australia. Wesselink had moved to Radcliffe Observatory in 1950 and had been replaced by Fjeda Walraven; the latter was married to Johanna Helena (Jo) Verlinden (1921–1989), who participated in his astronomical work throughout her life. The Walravens moved back to Leiden in 1955 (where the polarization work on the Crab nebula was performed) and Andreas Bernardus (André) Muller (1918–2006) came to Johannesburg. With Oosterhoff as supervisor Muller had written a thesis on a study of the variable star XZ Cygni, an RR Lyrae type variable star, and had obtained his doctorate in 1953.

Another player in the field was the Boyden Station of Harvard College Observatory, which in 1929 had been moved from Arequipa in Peru to Mazelspoort, 20 km from Bloemfontein, Orange Free State, South-Africa. An important development at

[1]The name is now 'Education, Culture, and Science', ignoring the fact that the sciences—like the arts—are part of culture.

Boyden had been the construction of the Armagh-Dunsink-Harvard Baker-Schmidt Telescope, brought about to a large extent by Eric Lindsay (1907–1974) of Armagh Observatory in Northern Ireland with the help of German-born Hermann Alexander Brück (1905–2000), Director of Dunsink Observatory at Dublin (in 1957 Brück would move to Edinburgh and become Astronomer Royal for Scotland). This was a 36 inch telescope with a large field of view, based on the principle of the Schmidt telescope (see Sect. 3.3), but modified by James Gilbert Baker (1914–2005) with a secondary mirror, so that the image was not formed in the prime focus but at the Cassegrain focus at the back of the primary mirror. It had been installed in 1950. The ADH Telescope never was a success due to alignment problems and it was scrapped in the mid-1970s. When in 1953 Harvard threatened to withdraw from Boyden, Lindsay and Brück found additional support from Armagh and Dunsink, but also from observatories in Belgium, Germany and Sweden, to continue operations.

There had been the possibility that Leiden would share in the Mazelspoort facility near Bloemfontein, but in the end the Leiden Southern Station was moved to Hartebeespoortdam near Pretoria. Pretoria is some 50 km north-east of Johannesburg, Bloemfontein somewhat more than 300 km to the south-west. The decision to go to Hartebeesport Dam was taken in 1953. On November 11, 1953, Oort wrote to the South African Minister of Education, Arts and Science:

> 'When the plans to move our instruments to Hartebeespoortdam were first discussed, the expectation was, as was that of the Director of the Union Observatory, that the Franklin-Adams Telescope would be moved there in 1949 and that this would enable us to obtain complete certainty about the question if the situation at Hartebeespoortdam would meet all astronomical requirements and that therefore it would be justified to also move the larger instruments there. The urgency to take a decision on our move led us to have our Johannesburg observer perform special observations of a preliminary nature in the second half of 1952 with instruments erected near Hartebeespoortdam, especially in view of a comparison with conditions elsewhere in South Africa and particularly Mazelspoort near Bloemfontein. These observations gave very satisfactory results for the conditions in Hartebeespoortdam.'

The main cause for the letter came in the next paragraph, when Oort informed the Minister that the decision to move to Hartebeespoortdam had been taken and thanked him that he 'particularly through the very generous allocation of funds for the accommodation of our station in Hartebeespoortdam had encouraged the continuation of our collaboration'.

Oort explained that the decision had been held up by the uncertainty about the resolve of Harvard University to close the Boyden Station and the possibility for Leiden to take a share in it.

> 'As a result of this, Leiden Observatory has taken an initiative for a much more extensive plan that aims at a large cooperative observatory in the southern hemisphere (preferably South Africa). This Observatory will aim at extending those observations that are essential for our understanding of the Universe, and that at present can only be performed in

the northern hemisphere, to the southern parts of the sky. This observatory would be financed jointly by about six European countries. Although it appears from the first, unofficial discussions that there is much enthusiasm for this plan, it cannot be sure yet that it will actually be executed.'

How had this come about?

Baade's visit in 1953 had been a very productive one. Originally Oort had intended it to be for three and a half months, but it had to be curtailed to two months (see Sect. 11.6). Oort had urged Baade to come 'incognito', that is without letting others know, because they might invite him to come to their institute as well and take away from the Leiden visit. We have seen that Oort and Baade prepared for the Vosbergen Symposium during that period. But another matter came up in discussions between them that would have profound consequences, the founding of the European Southern Observatory ESO. Adriaan Blaauw has written an excellent book on ESO's early history, to which I refer for details [309].

Baade flew from the USA on May 2, 1953. A year later, in a letter of April 17, 1954, Baade recalled:

'I shall never forget the beautiful last stretch of the trip from Glasgow to Amsterdam, mostly across the North Sea until we hit the Dutch coast and banked down for the landing, everything in the most beautiful sunshine. But you and Mieke topped it off when you picked me up at the airport for a drive through the flowering bulb fields [the Keukenhof]. The walk through the beautiful park where we had a cup of 11.am. coffee was a real welcome treat to me and I still see before my eyes the magnificent beeches in their first green garlands. It was a wonderful time I had in Holland.'

Baade and Oort must have discussed various matters of common interest, but a southern observatory was an important part of the discussions. Now, Oort had a meeting in Leiden about the Boyden Station on May 12. Oosterhoff wrote minutes of the meeting, but I have not been able to locate these. However, at a further meeting on June 12 (see below) this gathering has been summarized as follows:

'... held on May 12th at the instigation of Prof. Brück and Dr. Lindsay, in which the possibility was discussed of a co-operative management of the Boyden Station at Mazelspoort near Bloemfontein. This problem had become very urgent since the Harvard Corporation announced that for financial reasons it would liquidate its obligations for maintaining the southern station in July 1954. According to the scheme discussed a number of independent stations could be set up at Mazelspoort, each station having the full responsibility for its own instruments and staff. [...] Useful and practical as such a project may be, it never would lead to the erection of a really large telescope.'

This was part of the exercise by Lindsay and Brück that in the end successfully prevented closure of the Boyden Station, although without Dutch participation and for a limited period. Baade might have attended the meeting. In any case, Oort will have discussed it with him—and it seems that among themselves both had been discussing the prospects of building a major telescope in the south and saw the Mazelspoort problem as an opportunity to expand further on such a plan.

The outcome of the Oort-Baade brainstorms was to try and whip up support for the idea among European colleagues that were attending IAU Symposium No. 1 on the *Co-ordination of Galactic Research* at Vosbergen (see Sect. 11.6). Oort invited a number of participants from European countries in Leiden to meet on the subject the day before the Vosbergen Symposium would start. This meeting on Sunday June 21, 1953 formed the informal start of the European Southern Observatory ESO. The minutes of this meeting, written by Oosterhoff, record that Belgium was represented by Paul Eugène-Edouard Bourgeois (1898–1974), France by André-Louis Danjon (1890–1967), Germany by Heckmann, and Sweden by Lindblad. Spencer Jones from the UK could not make it, but attended the Vosbergen Symposium, where he was brought up to date. From the Netherlands Adriaan Blaauw was present; he and the Kapteyn Astronomical Laboratory, of which he would become Director a few years later, would play an important role in the preparations, construction and early operation of ESO. Most importantly, Bannier from ZWO took part; he was president of CERN Council (in the notes designated as the European Council for Nuclear Research) on which eventually ESO would be modeled. Bannier had a decisive role to play.

The meeting started with the reporting of Oort cited above on the discussions with Brück and Lindsay on the future of the Boyden Station, which was presented as a promising starting point for the necessary infrastructure, at a time when several European countries were considering putting small or moderate telescopes in the south. It was noted that all reflectors with apertures in excess of 80 inches were in the north, while some of the most interesting regions of the sky are in the south (the central parts of the Galaxy, the Magellanic Clouds, etc.). The most important part of the proceedings was taken up by Baade's detailed ideas on what telescopes it would be reasonable to aim for. A copy of Palomar's 200 inch telescope was beyond the technical and financial possibilities, but a 120 inch reflector, like the one at Lick Observatory, was a realistic option. Baade felt that with such a telescope, together with an 48 inch Schmidt similar to the one at Palomar, 'all the important problems could be tackled. The support of the mirror would be easy and no difficult engineering problems would be expected'. The total investment was estimated at \$ 2.5 million and the running budget would ramp up to a level of \$ 100.000 per annum (respectively €20 million and 0.8 million today). These would be shared among the participating nations in a ratio corresponding to the respective numbers of IAU members. A general organizational structure was suggested with a Director, but otherwise a minimum of scientific and technical staff. There would be an executive council with one vote per country.

In his book on ESO's early history Adriaan Blaauw has included some photographs taken during a boat trip on the IJsselmeer for the participants of the Vosbergen Symposium. On this trip the possibilities of a southern observatory must have been one of the important topics of conversation (see Fig. 11.24, but also Fig. 11.23).

In the beginning the work proceeded quickly. Another meeting of astronomers from Belgium, France, Germany, Great Britain, the Netherlands and Sweden was

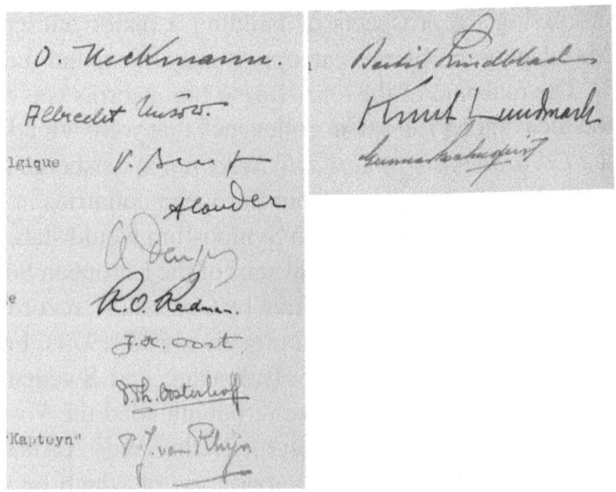

Fig. 12.4 Signatures under the ESO Declaration, drawn up in the Senate Room of Leiden University on January 26, 1954. The signatures are usually reproduced as a single column; however he ones on the left are at the bottom of the second page, the ones on the right at the top of the third. From the Oort Archives (at ESO)

held in Leiden on January 25–27, 1954, where the plans were worked out in some more detail under the chairmanship of Lindblad. Support from the astronomical communities in the countries had been easily obtained. A special committee was formed, with an office in Leiden, to oversee a seeing (quality of the images) expedition. Walraven was assigned the tasks of looking into the matter of the seeing in South Africa and working out how to go about choosing a site (Mazelspoort was not the only place considered), and more thought went into defining a budget, including estimates for the investments. The improved budget that was adopted, showed a total investment of £ 1.2 million and operations eventually leveling at £ 160.000 (nowadays respectively € 27 and 1.0 million). Operations started at £ 10.000, viz. the cost of the seeing expedition.

This meeting, which was held at the Senate Room of Leiden University, is considered the first of the ESO Committee, which would be a precursor to a formal Council. The occasion was marked by the signing of a declaration by the participants (see Fig. 12.4); this is now displayed at a plaque outside the doors of the Senate Room. The original text of the declaration is in French [310]. The signatories were Heckmann (Hamburg), Unsöld (Kiel), Bourgeois (Brussels), Couder (Paris), Danjon (Paris), Roderick Oliver Redman (1905–1975) (Cambridge), Oort (Leiden), Oosterhoff (Leiden), van Rhijn (Groningen), Lindblad (Stockholm), Lundmark (Lund) and Karl Gunnar Malmquist (1893–1982) (Uppsala). Two persons present at the meeting did not sign and these were Bannier and Funke. Gösta Werner Funke (1906–1991) was the equivalent in Sweden of what Bannier was in the Netherlands (Fig. 12.5). Both were member of the Council of CERN, knew each other and the relevant diplomatic world very well. Both actually have been presidents of both the Council of CERN and that of ESO (Funke the second after Oort, Bannier the third) after it was established, and played a very important role in bringing ESO about (see Fig. 12.6). Like Bannier, Funke was a physicist, but Funke held a doctor's degree (as we have seen, Bannier was promoted to doctor *honoris causa* after his retirement).

Fig. 12.5 Oort and Otto Heckmann in 1953. From the Oort Archives

Fig. 12.6 Two important diplomatic players in the establishment of ESO were Bannier (left) and Funke, Directors of the research councils of the Netherlands and of Sweden. This picture was taken at CERN in 1967 [312]

The next meeting was held in Paris in November 1954, where Bannier and Funke presented a draft version of the convention and much discussion took place on the site-selection activities. It quickly became obvious that they had to go beyond the existing sites and special expeditions needed to be organized. For the next years until the signing and ratification of the convention, preparations for these formal steps to establish ESO and extensive site-testing were the main activities of the ESO Committee (see also Oort, 1956f). The road to the signing of the convention was a tortuous one, and Oort took full leadership in this. It must have taken a large fraction of his time; the number of letters in the Oort Archives on ESO in those years is enormous. Where the fourth, fifth and sixth meetings had been chaired by the hosts (respectively Heckmann, Lindblad and Bourgeois), the meeting of the ESO Committee in October 1957 at Leiden chose Oort as the permanent chairman. He remained in that position of leadership until 1965; by that time the convention had been ratified and the provisional Committee turned into a Council.

There were many problems, as described in much more detail by Adriaan Blaauw [309]. First the United Kingdom decided to leave the project in order to build a southern hemisphere telescope in the framework of the British Commonwealth. This became eventually the 3.9 m Anglo-Australian Telescope at Siding Spring in Australia. At the same site, the UK built the UK-Schmidt Telescope, which is a copy of the 48 inch Schmidt Telescope at Palomar Observatory. Eventually, ESO with its 1 m Schmidt Telescope and the Anglo-Australian Observatory shared the work on the southern survey. After 1957 the UK did not attend the ESO Committee meetings, and it formally notified withdrawal from the project in 1960.

There were also problems due to the unstable political situation in France, with rapid succession of governments. A related factor, which I appreciate from my own experience as later President of Council, is the fact that, contrary to other countries, in France Council members need a mandate from three ministries (Science, Finance and Foreign Affairs), which makes life complicated (see [311] for my memoirs as President of ESO Council).

Oort approached the Ford Foundation in 1957. This foundation had been established in 1936 by Edsel Bryant Ford (1893–1943) and Henry Ford (1863–1947) and had as one of its aims promoting international collaboration. Initially Oort had not met with a positive attitude towards supporting ESO. However, a new attempt in 1959 was successful. The Foundation would allocate $ 1 million, as soon as four of the five prospective members (Belgium, France, Germany, the Netherlands and Sweden) had ratified a convention. This amount was one fifth of the total sum of $ 5 million required according to budget estimates at the time and it indeed proved to be the push needed; on October 5, 1962 the five countries in the end signed the convention. The Netherlands, Sweden and Germany ratified in 1963, and France followed suit on January 17, 1964 so that the Ford Foundation's conditions were fulfilled. Belgium did not ratify until 1967, a few months after Denmark had joined.

In addition to Oort, the driving forces behind ESO had been Lindblad, Danjon and Heckmann. The latter became the first Director General, and he provided the daily leadership. However, before all this happened much activity had been going on in the area of site selection, in which Oort had been heavily involved. Other persons that played an important role were Charles Fehrenbach (1914–2008) of the Observatoire de Haute-Provence in southern France, and Adriaan Blaauw.

Blaauw had left Leiden in October 1953 to take up a position at Yerkes Observatory, but was lured back to the Netherlands in 1957 by Oort and the prospects of radio astronomy. In that year van Rhijn had retired at the age of 65 and Blaauw was appointed as his successor at Groningen. At Yerkes, Blaauw had by this time been appointed as associate Director (Bengt Strömgren was Director), and it had generally been expected he would remain in the USA.

In the *JK Blaauw-interview*, Blaauw said about this:

'Oort was the boss, and Oosterhoff controlled things in the other building [the stellar photometry group] that we called Astro and one had the feeling one could easily stay here for another 20 or 25 years, but the world is so large and there are so many interesting things to see, so I resigned from my position as Reader and accepted a temporary appointment at the University of Chicago and the Yerkes Observatory; in hindsight a strange gamble and I think few people would do that these days, give up a permanent job for one in America that I think paid less than Leiden, but I must have felt that either I stayed – and I would get holed up in one place, although with very nice company –, or look for adventure. That's the way I am. But others have done the same thing. Gart Westerhout had the same feeling and there have been more.'

Before accepting the Groningen Directorship, Blaauw had made some important conditions. One of these was a staff increase. This was first of all to get involved in radio astronomy, now that Dwingeloo was operational. But there was also another development relevant to ESO. Again from the *JKBlaauw-interview*.

'Then there are persons like André [Muller], who were not treated very nicely by Jan. André had been in South Africa with his family. It must have been around '57, when André came back with his wife and he said to me that he had been told to look for a job elsewhere. That was very painful and had to do with Oort feeling that Walraven was the better man to be there. André and his wife Louise had a large family. They had 6 children, although I don't know they had already 6 then. [...] In any case André had been in Africa for quite some time and looked after things very well, but he was simply dismissed and I felt that was very unjust.

Well, André then was given a job in Groningen. I had convinced people to accept him, since the situation in Leiden was very poor. And then the developments in the framework of ESO demanded that someone would supervise things that developed and André with his experience in South Africa was the right person. And slowly he shifted from a job in Groningen to one at ESO. It all worked out well for him, but he and his family sacrificed a lot when he worked in Chile. That has been very difficult.'

Muller and Walraven did not go along well, so that one of them had to go back from South-Africa to the Netherlands. Walraven preferred to stay, but there was no suitable work for Muller in Leiden. Oort felt Walraven was the most useful person on the Observatory staff, which has been confirmed by Walraven's further work. Yet, Muller deserved better treatment after his return from South-Africa than Oort gave him. A well-written article was published by Richard West at the time of Muller's death, which describes his work and personality very well [313]. André had been vital to ESO's success.

Fig. 12.7 Oort on horseback in Chile in 1963. Together with a small ESO group he visited the site Cerro Morado, which was a possible choice for ESO. Is is located near Cerro Tololo, where the Americans were planning a southern observatory. This picture is cut from one that shows the full party. Credit: ESO/F.K. Edmondson [315]

In the years up to 1963 site testing continued in South Africa, as described in much detail in Blaauw's book on the history of ESO [309], generally moving south away from large cities into the Great Karoo. Many astronomers participated, in addition to Muller, Michael Grewing (b.1940) from Germany and Fehrenbach from France should be mentioned, nor should major contributions from a number of young astronomers from the various ESO countries be ignored.

From 1962 onward, while the site testing was still proceeding in South Africa, ESO's attention was turning to Chile. In the USA, AURA, the Association of Universities for Research in Astronomy, founded by seven universities in 1957 to coordinate their efforts in optical astronomy, had been looking for sites in South America, particularly Chile. In 1962 the Board of AURA chose a site in the general vicinity of the city of La Serena, some 500 km north of Santiago, for its southern observatory. Jürgen Stock (1923–2004), a student of Otto Heckmann then working for the University of Chicago, had been the leader of extensive site selection activities in Chile and AURA based its decision on his work. Stock was to become the founding Director of the Cerro Tololo Inter-American Observatory (CTIO). News had come to Europe of Stock's results and in 1960 Oort contacted Donald Shane at Lick and asked for information. Shane was keen to provide this and suggested that much profit could be gained from a choice to build the American and European observatories not too far apart (the one planned by the Carnegie Institution would end up at Las Campanas, a site not too far from Cerro Tololo). As a result, European astronomers, among them André Muller joined Stock's team in 1962. In 1963, the ESO Committee meeting was held in Chile and on that occasion a group visited a prospective site at Cerro Morado near Cerro Tololo (see Fig. 12.7).

ESO decided for Chile instead of South Africa in November 1963 because of the superior climate for astronomical observations. The convention had been ratified and Heckmann, as Director General, led the negotiations with Chile to establish an observatory. The selection of La Silla followed in May of 1964. The final result was that there were three new observatories lying roughly on a north-south line, CTIO at Cerro Tololo about 80 km mostly east and somewhat south of La Serena, ESO's La Silla Observatory 160 km to the north of Cerro Tololo and the Carnegie Institution's Las Campanas Observatory a further 30 km north.

From then on Oort's active involvement slowly lessened and leadership was provided by Heckmann as Director General and Adriaan Blaauw as his successor between 1970 and 1975. The dedication of ESO took place on March 25, 1969; various smaller telescopes (some owned by partner nations but operated for astronomers from other member states as well) were operational. The 1-m ESO Schmidt Telescope became operational in 1971, although it took a number of years (and much time of André Muller) to carry it to perfection. In 1976 the 3.6 m Telescope became operational, fulfilling the foresight and dreams of Oort and Baade in 1953 and crowning the dedication and perseverance of Oort and many others in the project.

12.3 Hartebeespoortdam

Well before the move of the Leiden Southern Station to Hartebeespoortdam was decided and accomplished, plans had been made to extend it with a new reflector for photoelectric photometry of stars. Oort had tried out photoelectric photometry of galaxies during his visit to Yerkes and McDonald in 1947, and now it had also become clear that photographic stellar photometry, as had been done for many years with the Rockefeller Astrograph at the Leiden Southern Station, was no longer state of the art. In 1947, the 46 cm Zunderman Telescope had become operational in Leiden at the Observatory with a special photometer developed by Walraven, and there was a strong desire in Leiden, certainly on the part of Walraven, to do photoelectric stellar photometry from the south.

In a letter to Adriaan Wesselink at the Union Observatory on January 5, 1948, Oort wrote:

> 'We are making plans (but for the moment no more than that) to build a 90 cm mirror [reflector] for Johannesburg especially with a view towards photo-electric observations. Greenstein and Henyey have made an interesting design for a 100-inch for Yerkes, of which we could build a smaller version. It would consist of a spherical 90-cm mirror and a special Cassegrain. The length of the tube then would be no more than 3 meters, while the effective focal length would be about 12 meters. This combined with a multiplier and amplifier should get us down to 15th magnitude. The optics could be produced for us by Yerkes. But even if we could find the money soon, it still would take some four years before such an instrument could be operational'.

The Rockefeller Telescope was formally owned by the Leiden Observatory Fund, so the relocation of the Southern Station and any new telescope would have to be

Fig. 12.8 The 90-cm Lightcollector of Leiden Observatory's Southern Station at Hartebeespoortdam (it was later relocated to ESO at La Silla). The photometer is situated below and to the side of the telescope. It was quite unique in that it had very fast and accurate pointing with preset coordinates. Photograph supplied by Jan Willem Pel

done under the auspices of that Fund. In the minutes of the Board meeting of May 12, 1948 it has been recorded:

> 'Prof. Oort also advocated the purchase of a 90 cm reflector for the Southern Station. The optics for this could be produced at the Yerkes Observatory for the cost price of about 4000 $. The instrument itself could then be produced in the workshop of the Observatory. It was decided to postpone this plan.'

Oort must have arranged this option while he was visiting Yerkes in 1947. In the meeting of September 17, 1948, the secretary reported that the price Yerkes had determined amounted to $ 4500 and that a subsidy for an equivalent had been requested from ZWO. It resulted in an item of Dfl. 12,000 on the Fund's budget of the next year, which in the meeting of March 31, 1949 was explained as necessary for the purchase of 'two mirrors of a special type' for the new instrument for the Southern Station. Yerkes would not make a profit on this. The ZWO request was pending.

At the March 9, 1950 meeting Oort reported that the contract still needed approval from the University of Chicago and ZWO (so the request must have been granted), and by February 22, 1952 that the contract had been signed and an advance of $ 4500 had been transferred, being three quarters of the agreed price. In the meeting of December 4, 1951 (Fig. 12.8):

> 'Prof. Oort reports on the matter of the construction of the 90-cm 'light collector'. The optics, which will be produced at the optical workshop of

Yerkes Observatory will probably arrive soon. In the meantime an order has been placed with the Company Rademakers in Rotterdam for those parts of the telescope that cannot be produced in our own workshop. The costs will be around Dfl. 34,000. From extra budget made available by the University a part amounting to Dfl. 10.000 was received in 1950, in 1951 this was Dfl. 4,000 and in 1952 Dfl. 10.000. The remaining Dfl. 10,000 had also been requested for 1952, and if that does not prove possible for 1953.
The aim is to have the telescope finished some time during 1953. It was suggested that we should consider to have the optics transported directly from the USA to South-Africa.'

For comparison, an amount of Dfl. 10,000 in 1952 corresponds to a purchasing power at present-dat prices of an amount of about € 36,000. The willingness of Leiden University to support this project may have profited substantially from the fact that the President-Curator of the University and the chairman of the Leiden Observatory Fund were the same person, Jacob Evert (Jaap) Baron de Vos van Steenwijk. As mentioned above, he held a Ph.D. in astronomy which he had obtained in 1918 under the supervision of E.F. van de Sande Bakhuyzen. De Vos van Steenwijk was President-Curator from 1945 to 1962 and had been chairman of the Leiden Observatory Fund since 1950 (until 1964).

The actual year the telescope, that had been dubbed the Lightcollector, came into operation turned out to be 1958. Fjeda Walraven had built an ingenious instrument with five simultaneous photo-electric channels for a five-band photometric system. This was used in the first publication by the Walravens to study O- and B-stars [314]. The Station at Hartebeespoortdam was officially opened on September 9, 1957 in the presence of de Vos van Steenwijk and Oort. The Rockefeller Astrograph and the Franklin-Adams Telescope had been moved there also.

The special optics Oort referred to in his letter to Wesselink is a form of what is called the Dall-Kirkham type of telescope, after British telescope builder Edward Stafford Dall (1901–1986) and American amateur astronomer Alan Robert Kirkham (1909–1968). This modification of the classical Cassegrain mirror optics employs a concave ellipsoidal primary mirror and a convex spheroidal secondary. For many years the Lightcollector was used at Hartebeespoortdam for photometry in the 'Walraven System' of photometric bands, but due to deteriorating conditions it was relocated to the ESO site at La Silla in Chile in 1978 [316].

12.4 Stellar Populations at the Vatican

In 1957 a seminal conference took place at the Vatican, the *Semaine d'Étude* on *Stellar Populations* of the Pontifical Academy of Science and the Vatican Observatory. This rather small conference in terms of number of participants was so important because it brought together leading experts in most fields of astronomy, in all of which significant progress had been made since World War II, and the meeting synthesized all information into a consistent picture of the formation, structure and evolution of the Galaxy and through that of galaxies in general. It was based on Baade's concept of Stellar Population that he had introduced in 1944. I will summarize Baade's work here as a background to a discussion of that meeting, but refer for a more detailed treatment to the excellent biography of Baade that has been written by Donald Edward Osterbrock (1924–2007) [317].

During the War, Baade—being a German citizen—was restricted to the Pasadena area, but had special permission to go to Mount Wilson, where, as many astronomers were recruited to other places and duties because of the war effort, he had much telescope time for himself. He had figured that he should have been able to resolve the brightest stars in the Andromeda Nebula and its small elliptical companions. Baade had noted that on ordinary (blue-sensitive) photographic emulsions he was unable to resolve the amorphous central bulge of M31 into its brightest stars. This may sound contradictory, since after all Hubble had found Cepheids in the Andromeda Nebula to derive its distance. But this was in the sparse, outer regions and Baade's effort concerned resolving the dense inner parts into individual stars. Baade had studied some dwarf galaxies (called Sculptor and Fornax after their constellations in the directions of which they are located) and had found that the brightest stars had the colors of K-giants. Such small systems are spread throughout the so-called 'Local Group' of galaxies, of which M31 and the Milky Way are the main components.

Not only did Baade have much observing time on the Mount Wilson 100 inch Telescope, but due to the War the city lights of Los Angeles were subject to partial blackouts and the sky at Mount Wilson was unusually dark. With yellow-sensitive emulsions and his great care and observing skills, he was in 1942 able to see that he was on the verge of resolving the brightest M31 stars, so he went to the best red-sensitive emulsions that the Eastman Kodak Company, which had long been responsible for the most advanced photographic emulsions used in astronomy, had available. This proved successful during the fall of 1943. Baade concluded that there were two populations: Population I, in which the brightest stars were blue as in the solar neighborhood, and Population II, in which these were red such as in globular clusters. This distinction was reminiscent of that among stars in the Galaxy, as Oort had pointed out long before. Among the Galactic high-velocity stars there were no O- and B-stars in contrast to the lower-velocity stars. However, the full meaning of the existence of distinct populations (as introduced in general terms on Sect. 2.6) remained to be elucidated. Baade published the results in 1944 [318].

One result that followed from this discovery is Baade's 'doubling of the size of the Universe'. The distance scale among galaxies was set by the distance of the Andromeda Nebula and Hubble's expansion law. One critical element is the rate of expansion, expressed as the Hubble constant. We have seen (Sect. 7.2) this was estimated early on to be of the order of 400–500 km/s Mpc. If the expansion is linear with time, this would mean that the expansion had started some 2 billion years ago. By 1950, the adopted value for the Hubble constant, and the estimated age of the Universe, was still of that order. However, there was evidence that the age of the Earth—and therefore of the Sun—was of the order of 3, maybe 3.5 billion years, which was a glaring inconsistency. The distance to the Andromeda Nebula had been determined from pulsating stars; Cepheids, whose period-luminosity relation had been calibrated with local examples of such stars. It was known that there were two types of Cepheids, which could be distinguished by their light curves, the classical ones of which δ Cephei was the prototype and another type with prototype W Virginis. The slopes of the period-luminosity relations might be different. The zero point in use at the time was tied to the absolute magnitudes of a third type of pulsating stars, the RR Lyrae stars of the globular clusters. As it turned out, W Virginis and RR Lyrae stars were part of Population II and the classical Cepheids, used for the calibration of the distance of M31—and hence of the extragalactic distance scale—belonged to Population I. From a comparison of the brightest stars in Galactic globular clusters and those in M31s Population II stars Baade concluded that the zero-point of the period-luminosity relation for classical Cepheids had to be 1.5 magnitudes (a factor 4) brighter. This doubled the distance to the Andromeda nebula and therefore the scale of the Universe and solved the age discrepancy between the Solar system and the Universe.

Baade had also devised a method to estimate the distance of a pulsating star from the variations of its color and radial velocity (of the surface, showing the pulsation). If we know the color we can deduce the temperature and therefore the amount of radiation per unit area of the stars surface. Doing this at two times in the pulsation cycle and calculating the difference in radius between the two instants (in km) from the radial velocities would give the radius in km and therefore the surface area and the absolute magnitude. This method, later improved by Adriaan Wesselink, has become known as the Baade-Wesselink method. The results strongly supported the difference in zero point between the two types of Cepheids. The Populations concept thus indicated fundamental differences among stars.

At the time of Baade's discovery of Stellar Populations, very little was known about stellar evolution, differences in chemical composition and the connection to kinematics, all of which proved vital to understand the true significance of the concept of Stellar Populations. Some relevant reviews have been presented in a half-day history section at the beginning of IAU Symposium **164** on *Stellar Populations*, that I organized with Gerard Francis (Gerry) Gilmore (b. 1951) at the 1994 Den Haag General Assembly, celebrating the 50th anniversary of Baade's seminal paper [319]. The opening paper by Owen Jay Gingerich (b. 1930) described the progress of the understanding of stellar evolution up to 1950 (Fig. 12.9).

It took a long time to realize that the Sun was almost entirely made up of hydrogen and some helium. The problem of the energy production was solved only shortly before WWII by Hans Albrecht Bethe, (1906–2005) and Carl Friedrich von Weizsäcker (1912–2007), who showed that this happened by fusion reactions of hydrogen into helium. As a result, a slowly growing core of helium forms at the stars center. This central helium core should be able to support the outer layers of the star and Màrio Schönberg (1914–1990), a Brazilian physicist, and Chandrasekhar found that this central core then could contain no more than about 10% of the star's mass, named after them the Schönberg–Chandrasekhar limit.

But what happened after that only became clear in 1952, when Allan Rex Sandage (1926–2010) and Martin Schwarzschild were able to show that as it reaches this

Fig. 12.9 Walter Baade (right), his wife and Martin Schwarzschild in 1956 at an unknown location. From the Oort Archives

limit the core contracts, accompanied by an expansion of the envelope, which in the Hertzsprung-Russell diagram (see Sect. 2.6) corresponds to a move from the Main Sequence to a position on the giant branch. Subsequently, Fred Hoyle (1915–2001) and Schwarzschild were able to calculate the evolution all the way up to the tip of the giant branch, where helium burning sets in. This explained the Hertzsprung-Russell diagram of globular clusters and the evolution of stars away from the Main Sequence when the core had exhausted its hydrogen. The latter meant that the age of a cluster could be obtained from the position where the Main Sequence ended, the more massive and brighter stars having evolved away. For more background see the papers in the history session at IAU Symposium 164 [319].

In the mean time much work had been done on the formation of the chemical elements (nucleosynthesis). George Antonovitsj Gamow (1904–1968) advanced the theory that it had all happened in the very early Universe, which was hot, by repeated capture of neutrons, building up the whole range of elements in the periodic system from hydrogen onward. Eventually this became the '$\alpha\beta\gamma$'-theory, referring to a paper by Ralph Asher Alpher (1921–2007), Bethe and Gamow. This theory failed for elements beyond helium because of the absence in nature of stable atoms with mass numbers 5 and 8. Synthesis of the heavier elements inside stars became a viable alternative, when it was realized that a semi-stable state of beryllium-8, formed from two colliding helium nuclei, could exist long enough in the core of helium-burning giants to be hit by another helium nucleus before decaying. This provided a means of synthesizing carbon and from that heavier elements up to iron in late stages of stellar evolution and the synthesis of even heavier elements in supernova explosions. The work of experimental nuclear physicist William Alfred (Willie) Fowler (1911–1995) provided the necessary input on nuclear reaction rates. The full theory of nucleosynthesis was reviewed in a famous, often-cited paper (referred to as B^2FH) by Eleanor Margaret Burbidge (b. Peachey; 1919), Geoffrey Ronald Burbidge (1925–2010), Fowler and Hoyle in 1957, *Synthesis of the Elements in Stars* [320].

Hoyle was interested in stellar nucleosynthesis because of his belief in a Steady-state Universe, in which the dilution by the expansion is compensated by creation of matter. If this newly created matter would be in the form of hydrogen there had to be a way to transform part of it into heavier elements and where else could that happen than in stars. Fowler eventually (1983) won the Nobel Prize for this work, shared with Chandrasekhar; Hoyle never got a share in the Prize and his sympathy for the Steady-state Theory and other unconventional ideas might have played a role in that.

Another very important discovery that would throw light on the significance of Stellar Populations was the relation between abundance of heavy elements and kinematics. Astronomers often refer to the abundance in elements heavier than helium as the '*metal abundance*', since many of the more readily identifiable chemical elements in stellar spectra are metals. I will adopt this terminology for what follows. Much of the relevant work was done by Nancy Roman, who has summarized all this in a presentation to the history session of IAU Symposium 164 [319]. In the early 1950s she was involved in the classification of stellar spectra and noted that these could be significantly different in terms of the strengths of lines of heavy elements with respect to those of hydrogen and that all 'weak-line' stars happened to have higher space velocities with respect to the Sun. Roman also found that high-velocity stars were deficient in heavy elements. Similar spectroscopic work supporting this point of view was done by William Morgan and collaborators at Yerkes (the place where Roman also worked) and by Russian-born Alexander Nikolayevich Vyssotsky at Leander McCormick Observatory of the University of Virginia in Charlottesville. A correlation between metal abundance and kinematics appeared, Population II stars being 'metal-poor' as well as old and having different velocity distributions than the stars in the disk.

After the War, the Director of the Vatican Observatory, Daniel Joseph Kelly O'Connell (1896–1982), was making plans to organize a *Semaine d'Étude*. This had actually been proposed by former Director Johannes Stein, but in this case the War also intervened and the plans had not been realized when Stein died in 1951 (three such study weeks in other areas had been held by then). Jason John Nassau (1893–1965) suggested the subject of Stellar Populations to O'Connell. Nassau, born in Turkey to Greek parents, had studied in the USA and Scotland and now worked at the Case Institute of Technology and was Director of the Warner and Swasey Observatory of Case Western Reserve University in Cleveland, Ohio. He had met Baade at the McDonald dedication and they had become very good friends. Baade had advised him to put the funds he had available into a Schmidt telescope to survey the Milky Way for red stars; this became the 24 inch Burrell Schmidt Telescope (Burrell was an engineer at the Warner and Swasey Company that also built the McDonald 84-inch). The telescope was outfitted with an objective prism, so that he could effectively and efficiently search for M-stars and similar objects. Nassau had become an expert in Galactic structure using late-type stars.

His suggestion resulted in an Organizing Committee, which in addition to O'Connell and Nassau consisted of Baade, Oort and Schwarzschild. It met during the 1955 Dublin IAU General Assembly. The Study Week would be held May 20–28, 1957 at the Vatican under the auspices of the Pontifical Academy of Science and the Vatican Observatory. The correspondence in the Oort Archives on the planning and organizing of the meeting is not very extensive. I visited the Huntington Library in San Marino, California, where the Baade papers are kept and found that whereas Baade was very sloppy in archiving letters (there are only two from Oort, and for one of these there is only the second page, while they exchanged well over one hundred letters that are in the Oort Archives) there is a large collection of letters with O'Connell, who together with Baade appears to have done most of the organizing.

The attendance was limited, but would include most of the important players in the field. The (well-known) official group photograph has not been reproduced in full, but I have enlarged two parts of it. The top panel in Fig. 12.10 shows Jan Oort and Walter Baade in the foreground, with Lindblad to the left of Oort. The person at the left edge is Nassau. He and Baade somewhat look alike; Osterbrock, in his Baade biography [317], wrote:

> 'The two men had the same general build, height, hair style and color, prominent hawk nose, and liked to dress alike at scientific meetings which they both attended. Frequently in group pictures they stood at opposite ends of the front row, often striking very similar poses, which they must have rehearsed or at least discussed in advance.'

The person next to Nassau is Daniel Chalonge (1895–1977) of the Observatoire de Paris, whose specialty was spectral classifications. Next we have Martin Schwarzschild, Adriaan Blaauw and at the right Otto Heckmann. In the back from left to right we see Edwin Ernest Salpeter (1924–2008) from Cornell University, who worked on energy production in stars and their formation and evolution, Bengt Strömgren and Lyman Spitzer.

The bottom panel shows in the front from left to right Father O'Connell, Giuseppe Armellini (1887–1958) from Rome, and famous cosmologist Georges Henri Joseph Édouard Lemaître. Arminelli and Hermann Brück (not in the figure) were astronomer members of the Pontifical Academy. In the second row are Fred Hoyle, William Morgan, George Howard Herbig (1920–2013) from Lick Observatory, expert on young stars, and Allan Sandage. In the back Willie Fowler and Andrew David Thackeray (1910–1978) from Radcliffe Observatory, who was to report on stellar populations in the Magellanic Clouds.

I have gone into all this detail in order to stress the fundamental breakthroughs in stellar structure and evolution and in galaxy structure and dynamics, that were achieved in the first decade after the end of the War. Essentially all of the players attended the Vatican Conference! An obvious omission is Chandrasekhar. He might have declined to attend, but this is not clear from the documentation. When I was preparing for IAU Symposium 164, I asked Allan Sandage to be a member of the Scientific Organizing Committee and to be one of the invited speakers. He wrote back (July 21, 1993), that he would not be able to come to Den Haag for the IAU meetings. He noted:

Fig. 12.10 Participants to the *Semaine d'Étude* in 1957 at the Vatican on Stellar Populations. The original group photograph has for clarity been cut into two parts (and some Vatican members at the edges deleted). Individuals have been identified in the text. From the Oort Archives

'But neither was stellar evolution, nor the age dating problem, nor chemical composition differences understood in 1940s when Baade resolved M31, etc. So neither age differences, nor chemical composition, nor disk–halo connections were part of Baade's initial concept. All of these essential elements of the modern view came gradually, all later than 1944 when he defined his concept in ApJ 100. These later, now cen-

> tral concepts, developed in the decade of the 1950's when the method of age dating and chemical differentiation came to the front. Of course, kinematics was a strong part of his globular cluster – open cluster separation, based on the known RR Lyrae high velocities and the Oort (1926) high velocity field star results in Groningen Pub. #40. Stellar evolution became central in this history only after 1950.
>
> [...] The age dating breakthrough came only in 1952, together with Schwarzschild's explanation of the step beyond the Schönberg–Chandrasekhar limit as evolution off the Main Sequence. The role of chemical composition then began with Nancy Roman's discovery of the UV excess connection with kinematics [...] of individual high velocity stars.'

In that fairly long letter Sandage suggested that Chandrasekhar, although he was not at the Vatican meeting, should be asked to review the development of stellar evolution before 1950. When I invited him, he wrote back on November 19, 1993:

> 'I am sorry that I will not be able to participate in the IAU symposium 164. Besides, I have no interest in recalling incidents of 50 or more years ago. I am indeed surprised that anyone should have thought otherwise!'

Sandage also suggested Martin Schwarzschild should review stellar evolution after 1950; however, I had already in 1992 invited him, as a potential participant, to support my proposal to the IAU to hold Symposium 164, which he did but then already mentioned he would not participate. He wrote to me (April 20, 1992):

> 'It strikes me as historically most timely and as an active modern topic – as defined by your list of subtopics. So I happily applaud your plan. '[...] I realize that there are few of us left of the 'Vatican' generation. But one should be more active and inventive than I am now, to properly represent that sparkling conference.'

Adriaan Blaauw presented a highly recommendable review of the Vatican Conference at IAU164. Nancy Roman reviewed her work. With Osterbrock's review of Baade and the discovery of populations, the history session of IAU Symposium 164 presents a good introduction to appreciate the importance of the Vatican Conference.

It is interesting to note that year before the journal 'Scientific American' had invited a number of leading astronomers to review in a special issue (of September 1956) recent developments in the study of the Universe and there is significant overlap with those attending the Vatican Conference. Although there were articles of prospective Vatican-goers on other issues than addressed at the Vatican (on the Steady-state Universe by Fred Hoyle and on the redshift by Allan Sandage), there were also issues closely connected with that meeting, by Fowler on the origin of the elements, Baade on the content of galaxies and Oort on the evolution of galaxies. The latter paper (Oort, 1956c) ascribed the different types of galaxies to a relative importance of Baade's two Populations, an idea that was to a significant extent, but certainly not entirely, attributable to Oort, and presented a growing consensus among astronomers. Oort described the formation of galaxies as a process in which the older population II formed early on in a more or less spherical volume, while the remaining gas dissipated its random motions in collisions and contracted subsequently into a flat, rotating disk where star formation proceeds up to today. The various Hubble types are then a sequence from pure Population II in elliptical galaxies to a progressively more dominating Population I in spirals from types Sa to Sc.

The synthesis arrived at during the Vatican Conference resulted in the definition of five Populations, presented in slightly different forms in tables in the papers by Schwarzschild and by Oort. These populations formed a sequence in flattening, age and chemical composition. It envisaged the Galaxy being made up of a more or less spherical Halo Population II, which contained old stars with low metal abundances and little rotation, down to Extreme Population I, which was very flat and rapidly rotating, contained gas and very young stars and was relatively metal-rich. It showed the evolution of the Galaxy in time, becoming more flattened and metal-rich in the course of time.

Oort, as one of the leading figures of the meeting, was assigned to summarize the meeting from the observational point of view. The theory-related summary was covered by Fred Hoyle, who nicely summed up all the accomplishments and successes in the theory of stellar structure and evolution. Oort definitely played a very important role in the Vatican Conference and this reconfirms my choice of the subtitle of this book, Master of the Galactic System.

Some years after the Vatican Conference, in 1962, a seminal paper appeared by Olin Jeuck Eggen (1919–1998), Donald Lynden-Bell (1935–2018) and Sandage that had the title *Evidence from the motions of old stars that the Galaxy collapsed* [321]. This famous paper, often referred to as ELS, showed that there was a very clear relation among over 200 dwarf stars in the solar neighborhood between their so-called ultraviolet excess and motions in the Galaxy. Ultraviolet excess is the difference in color index between an ultraviolet and a blue wavelength band as observed and as expected from the spectral type. It is a measure of metal abundance or chemical composition. Stars with fewer heavy elements, or low metal abundance, are systematically brighter in the ultraviolet compared to the blue for two reasons. Metal-poor stars have a somewhat different structure, since the internal energy transport depends on the atomic nuclei present, and this results in a higher surface temperature, so a bluer color. Secondly, the stellar spectra show absorption lines from elements in the star's atmosphere, which take away light from the star, and when there are fewer metals, there are fewer lines. Now the relevant spectral lines occur in larger numbers at ultraviolet than at blue wavelengths, so that metal-poor stars are brighter in the ultraviolet (this effect is called 'line-blanketing'). Now it turned out, as ELS concluded, that the stars with larger ultraviolet excess had larger velocities perpendicular to the plane of the Galaxy and in the plane had orbits that deviated more from circular. The collapse of the Galaxy was preserved in the motions of these stars. ELS estimated that the collapse happened fast (much less than a billion years) and that the collapse was a factor 10 in the radial and 25 in the vertical direction. This nicely confirmed the picture developed at the Vatican, in which Oort played such a central role (Fig. 12.11).

I quote one final paragraph from Sandage's letter of July 21, 1993 to me:

'One final point on the history is, I believe, important. The Vatican Conference was not only important because of its later influence; it was also highly important re scientific venture. This, even as late as 1957, was the

Fig. 12.11 The Oorts and Baade after the Vatican Conference relaxing for a weekend at the Italian coast at Amalfi. From the Blaauw Archives

first major conference after the War which was truly international. Look at the list of participants and the countries they came from. I remember, for example, Heckmann so very well as we became good friends during the meeting. The most astonishing thing about that meeting that is still so strong in memory, is that the central idea of stellar evolution, of age differences among open clusters and the moving off the MS instead of up it (Gamow's incorrect dogma of 1938) seemed so new to Heckmann and to Oort, although it was 5 years old by then in the West. Some flavor of the newness of the idea can still be gained by reading the Vatican Conference report, especially in Oort's various questions and especially his criticisms of age differences because of the absence of enough 'old' ('if that was, in fact what they were') open clusters. So, the Vatican Conference was the healing conference after the War, along with whatever else it accomplished.'

The detailed Vatican picture has broken down somewhat, mostly as a result of the realization that the correlation between age and metallicity is much more complicated than simply one-to-one. Nowadays the halo is thought to have formed out of smaller units in which stars formed early on and went through a protracted merging process. But that is beyond the scope of this book.

12.5 BCAP: The Benelux Cross Antenna Project

Oort did not sit still after the Dwingeloo Telescope was inaugurated. The need for higher resolution was obvious, especially when one would want to study other galaxies. There was also a new field that had captured Oort's interests and that was cosmology and the counts of radio sources.

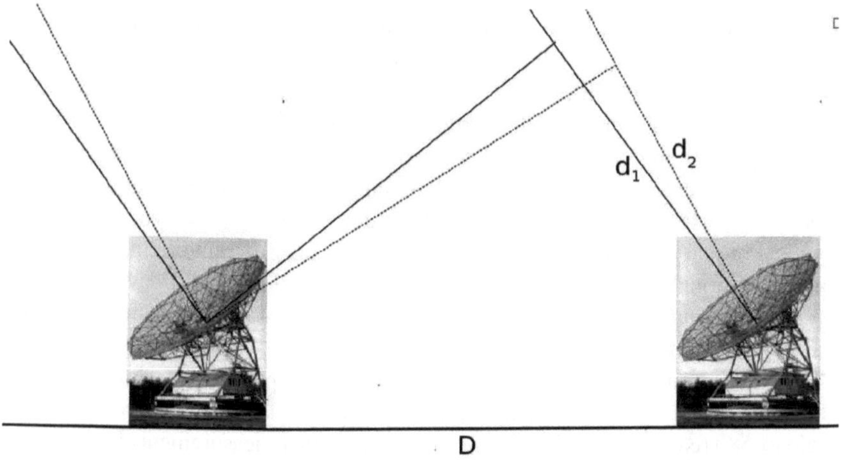

Fig. 12.12 Signals in two telescopes at a distance D correspond to different pathlengths. For two directions on the sky these are d_1 and d_2. Figure by the author

As mentioned before, the resolution of a telescope (optical, radio or any other wavelength) is the smallest structure that can be seen on the sky. In radio astronomy this is called the beamwidth, which is proportional to the wavelength and inversely proportional to the diameter of the dish. For a dish like the Dwingeloo antenna of 25 meters, the beamwidth for observing at 21 cm wavelength is about half a degree. When observing a point-source in the sky, this is the diameter of the circular area at which the response of the telescope is half of that at the center. Therefore it is also referred to as the full width at half maximum (FWHM) beam. This beam shape results from the interaction of all waves that reach the focus after reflection from the dish. The beam center is the point where all reflected waves arrive in phase. The wave amplitudes add up here, resulting in maximum intensity. The edge of the beam corresponds to those points on the sky for which all reflected wave amplitudes in the focus cancel. This is the case for points at an angular distance from the bean center equal to (in radians) the distance between the telescopes expressed in wavelengths. The beamwidth, and therefore the resolution, is thus set by the ratio between wavelength and dish size.

Oort was thinking of a resolution of order one arcmin, which at a wavelength of 75 cm (chosen as a possible reference wavelength for the new telescope) would mean a dish of some two-and-a-half km diameter. This is obviously not realistic, but such a large telescope dish can be broken down into smaller pieces. Take two radio telescopes at a distance D (see Fig. 12.12) and combine ('correlate') the signals they observe at the same time from the same part of the sky. For the center of the beam (full-drawn paths) these arrive at the two telescopes with a pathlength difference d_1, which results in a time difference that can be allowed for by delaying the signal from the telescope on the left. This delay compensation can be achieved by sending the signal through a longer cable; nowadays it is done electronically. Then transmit the signals from the telescope to a correlator through cables of equal lengths, so that there is no time difference left. The two waves will then add up. For a different sky position the pathlength difference is d_2. If this differs from d_1 by half a wavelength the two signals when entering the correlator will be 180° out of phase and therefore cancel. A point source can then be located on the sky (in the direction parallel to D) with an accuracy of the corresponding angle between the two paths. For such a 'single interferometer' it is not possible to determine the structure of a complex source. For that different distances D are required (and different orientations for the two-dimensional structure to be recovered). If D is 2.5 km and the wavelength 75 cm then the smallest dimension measurable is 1 arcmin and radio sources smaller than this will be seen as discrete, unresolved sources.

The interferometer technique was pursued at various places. At Cambridge (UK), Martin Ryle and Antony Hewish in 1955 used four aerials at the corners of a rectangle of 580 (east-west) and 52 m (north-south) so that measurements in these both directions could be made. The wavelength was 3.7 m, and the beamwidth was 0°.38 × 4°.3 (east-west and north-south respectively). With this 'Cambridge Telescope' they detected 1936 radio 'stars' in the available sky. This was analyzed by Ryle and his student Peter August Georg Scheuer (1930–2001), a German by birth, and they found an increase in the number of sources as they went fainter that was steeper than expected from a uniform distribution in space [322, 323]. The distribution on the sky was isotropic so that the population was most likely outside the Galaxy.

According to them this had profound consequences. Hoyle's Steady-state Universe, in which the intrinsic luminosity distribution has to be the same at all distances, required things not to change with time (and place) and therefore 'attempts to explain the observations according to Steady-state Theories offer little hope of success', thus escalating a controversy between Hoyle and Ryle that became increasingly ugly and public. According to Gart Westerhout [324], Oort had written on top of the Ryle & Scheuer preprint: 'This is the most fascinating paper I have ever seen'. Oort, who had kept up a regular correspondence with Ryle, was very much impressed by this result and wrote him on June 1, 1955:

> 'I have just read your and Scheuer's article on the radio stars, and do not want to wait another hour to congratulate you with this splendid discovery you have made. To me it is the most fascinating discovery I have ever seen.
>
> Your explanation seems plausible, and forms the first indication [...] that we are really looking back into the young stage of the universe.'
>
> If I were you I would at once proceed to build an instrument of at least twice the size of the one you have used for these observations. A more worth-while scientific investment of a relatively small amount of money can hardly be imagined.'

So, at the start Oort was not necessarily thinking of an instrument for Galactic studies using the 21 1960 cm line for his next-generation radio telescope, but had source counts (as this became known) at the forefront of his mind.

The history of the Benelux Cross Antenna Project (BCAP) leading up to the Westerbork Synthesis Radio Telescope (WSRT), has been described in much more detail than required here. In particular I refer to the book and thesis *EarlyRadio* [9]. And a chapter on *Four decades of Dutch radio astronomy: twenty-five years Westerbork Telescope* by Ernst Raimond in a volume that appeared on the occasion of the WSRT's silver anniversary in 1996 [325].

In 1951, a young, American radio engineer, Charles Louis Seeger (1912–2002), who was spending some time in Sweden to help develop radio astronomy there, visited Leiden where, according to an interview with Woody Sullivan [326],

> 'I gave a seminar at Leiden and Oort said: 'How about coming down here?' And so, I packed up in October 1951 and went down there for a year. And one year stretched into another and another and in the end I was there ten years. It was like being in the center of the universe in radio astronomy.'

In 1958 Oort asked Seeger to make a concept design for a telescope with an angular resolution of one arcminute. Seeger followed the idea developed in Australia of a cross, such as the Mill's cross built by Bernard Yarnton Mills (1920–2011) at Fleurs near Sydney, which had two arms, east-west and north-south, that together produced a 'fan beam' (elongated in the north-south direction due to projection of the corresponding arm as seen from the sky); see [327] for more on Bernie Mills and this telescope. Seeger's conceptual design, also presented in 1958, had two arms of 3 km length, which when used at 75 cm wavelength would give the required resolution of about 1 arcmin. The arms consisted of a reflecting screen (made of wires) with dipoles hanging above it to record the radiation.

The cross design was supported by the developments in Australia. The Mill's Cross had also been used to do source counts (catalogues of radio sources) from Australia, which turned out to be irreconcilable with the Cambridge results. This was due to 'confusion': the fact that in the Cambridge data the resolution was too poor to resolve nearby sources from each other, which were then counted as one. Ryle and his co-workers quickly changed the frequency of the receivers on their telescope from the 81.5 MHz (3.6 m) to 159 MHz (1.9 m) to improve the resolution. The resulting 3C (third Cambridge) catalogue has been a fundamental catalogue ever since.

The Cross-Antenna project was too expensive for the Netherlands to pursue alone, so Oort asked Belgian colleagues to join. The name 'Benelux' refers to a customs agreement signed in 1944 in London between the governments of Belgium, the Netherlands and Luxembourg. This name was adopted for the Benelux Cross Antenna, even though Luxembourg was not part of it.

For the project quite a few more foreigners spent an extensive period in the Netherlands and were of great importance. For example, there was Chris Christiansen, who had built his own 'Chris Cross' in Australia. He was seconded during 1960–1961 to Leiden by his boss, Joseph Lade (Joe) Pawsey (1908–1962) of the Radiophysics Division of the CSIRO in Sydney. The second, Jan Arvid Högbom (b. 1929) was Swedish, but had obtained a Ph.D. in 1959 in Cambridge with Ryle. And the third was the American William Clarence (Bill) Erickson (1930–2015), who was hired between 1962 and 1963. Major contributions, of course, came from the Netherlands,

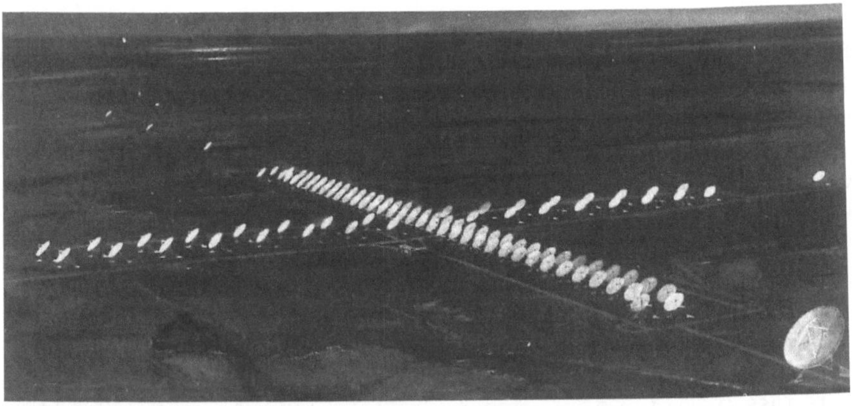

Fig. 12.13 The second design for the Benelux Cross Antenna of 1961. The arms were about 1.5 km long; the main cross consisted of some 100 dishes with a diameter of 30 meters. From the Archives of Leiden Observatory

in particular Lex Muller, Ernst Raimond, Ben Hooghoudt, Wim Brouw have been invaluable.

The conceptual design of Seeger was developed into a full design by Christiansen and Högbom and presented to the Board of SRZM early in 1961 The two arms were 5 km long and the operating wavelength would be 75 cm (frequency around 408 MHz). The costs came out much too high. A second design was developed in 1961, now with observations at 21 cm in mind. It was presented by Högbom and Erickson in the fall of 1962. It consisted of over 100 separate 30 m dishes along two arms (see Fig. 12.13) and a single 70 m dish for total brightness measurements. For a wavelength of 21 cm the arms could be shorter than in the first design and were envisaged to be some 1.5 km long. A possible site near the Belgian-Dutch border had been identified. This design was much too ambitious and would have been very expensive to build; to operate it would require complicated electronics and extensive reduction techniques that were not then available.

In the mean time Ryle had adopted a new technique for his next telescope at Cambridge (the One-Mile Telescope). Remember that in a single dish the image in the focus is formed by the total of all interferometric combinations from two elements of the dish, not only covering all lengths up to the diameter, but also the full 360° of orientations. Now with an east-west set of telescopes one can observed a given part of the sky for 12 h, while the rotation of the Earth carries this baseline, as seen from the source, in all directions (because of symmetry only 180° needs to be observed). If one then also makes the component telescopes movable so that their relative distances can be changed, the full required range of distances can be covered. Actually, Högbom had considered the idea already while a Ph.D. student of Ryle. Ryle set out to test the concept [328] using an existing telescope, operating at 178 MHz, that had antenna sections along an east-west line of about 1 km length. Individual elements could be used to make an interferometer. However, making use of the Earth's rotation required tracking the source for 12 h, which it could not do. So he observed the north celestial pole, which is stationary in the sky. This resulted in a map with a radius of 4° around the pole, showing many point sources (in fact 87 with detection limits varying with polar distances). This map captured Oort's imagination. Peter Scheuer reported in 1984 [329] (see also Elbers in *EarlyRadio* [9]) how Oort looked for a long time at the map displayed during a conference at the Royal Greenwich Observatory at Herstmonceux Castle in January of 1963. The paper [328] had been published in 1962 and Oort would not have missed this publication and must have seen the map before that.

Ryle received half the 1974 Nobel Prize for this concept of Earth-rotation aperture synthesis (Hewish the other half for discovering pulsars, in which Jocelyn Bell did not share). It is not clear how the concept was adopted in the Netherlands for the BCAP design (see the discussion in *EarlyRadio* [9]), but the deep impression that Ryle's map made on Oort must have played an important role. And so must have been Högbom's presence in Leiden. The third design for BCAP, produced in 1963 mainly by Högbom and Hooghoudt, reduced the cross to a single east-west line. To avoid political difficulties it was presented as a reconfiguration rather than a change of concept. It should be said that Högbom probably never received the credit he deserved for this.

The Belgian-Dutch collaboration in BCAP was not without difficulty (see *EarlyRadio* [9]). Oort tried to find funding from the Organisation for European Economic Co-operation (OEEC), the precursor of the Organisation for Economic Co-operation and Development (OECD), through its Governing Committee for Scientific and Technical Personnel. Initial contacts were made in 1958 for an amount of Dfl. 100,000 (some € 300,000 today), but that took time for it to be supported by both governments (the Netherlands and Belgium had to match each award by contributing a quarter of the amount, eventually adding up to contributing a half of the total). Bannier was very important in this effort. In 1960 a Belgian-Dutch Council for the Benelux Cross Antenna Project was formed, chaired by Oort. The cost of the design and preparation of BCAP had gone up in the mean time from the estimated Dfl. 100,000 to something more like Dfl. 255,000. The agreement with the OEEC was finally signed in June 1960, contributing somewhat less than Dfl. 150,000. However, Belgium still had not officially confirmed it would contribute to the Cross Antenna.

The OEEC actually became a divisive issue, as documented in *EarlyRadio* [9], when the organization wanted to further internationalize the project by including more partners, which the Belgians liked but which the Netherlands and Oort strongly opposed so as not to have to share the observing time with too many others. Part of the political game was a conference that the (in the mean time) OECD had organized in December, 1961 in Paris on the issue of the construction of large radio telescopes [330]. Oort made four contributions (1962j-m). The first was an introduction, in which he gave the credit for first suggesting the Cross Antenna to Charles Seeger 'long ago' (Seeger had been in Leiden since 1951). Next a long article on ways of studying the Universe, which concerned extragalactic radio sources including source counts, which required resolutions of order one arcmin. Oort discussed the 23 brightest radio sources in the full sky, of which at that time 7 definitely belonged to the Galaxy, 8 were extragalactic and 7 unidentified. These extragalactic ones were distant and luminous. As an example Oort chose Cygnus A, one of the brightest in the sky, which had been identified with what then was thought to be a collision between galaxies (nowadays we believe it is a single galaxy with an active nucleus) by Walter Baade and Rudolph Minkowski at an enormous distance then taken to be 220 Mpc [331]. Studying these was interesting in itself, counting them as a function of brightness might give clues on the large-scale structure of the Universe.

It was followed by a short note on 21 cm line studies of nearby galaxies—Oort mentioned M81 (see Fig. 2.24) and M51 (see Fig. 8.12), which required resolutions of order one arcmin. The Andromeda Nebula M31 (see Fig 7.9) is highly inclined, so that a similar resolution is required along the minor axis; however in this system the spiral arms are more difficult to trace because of the inclination. Oort's final contribution concerned possible programs for a large radio telescope. where he again mentioned faint extragalactic radio sources first, while nearby galaxies (also in the continuum) and our Galaxy came in second and third place.

The meeting was also attended by government representatives. For the Netherlands this was Johan Henri Eliza Ferrier (1910–2010), a Adviser to the Minister of Education, Arts and Sciences (and former Prime Minister of Suriname, at that time an autonomous country within the Kingdom of the Netherlands), with Oort as adviser to him. This group produced a few recommendations, concluding that there was no competing proposal to the Benelux Cross, that smaller instruments could also do useful work and that an even larger project on an international scale should be considered. They were very clear in recommending frequency protection and urged that action be taken on the recommendations of IUCAF. A special expert group should be set up, but no real action on this appeared to have been taken.

In the end the Belgian National Council advised its government that for participation in the cross antenna definite conditions would have to be honored concerning internationalization and contracts for Belgian industry. There is no need to document all details here of the strenuous relationship that ensued, since it is well covered in *EarlyRadio* [9]. The collaboration between Belgium and the Netherlands seemed doomed, since the differences between the two were fundamental. The Netherlands placed a much stronger emphasis on fundamental—pure—research, while Belgium stressed much more the social role of the scientific endeavor and applications. Astronomy was also much less prominent in science policy in Belgium than it was in the Netherlands; in 1960 the Belgian government decided the focus should be on molecular biology, space research and human genetics (*EarlyRadio*, [9] page 147). As it turned out, another collaboration between the two countries, the Antarctic expeditions, provided a final push towards an end to BCAP. During the International Geophysical Year (1957–58) Belgium had been involved in establishing scientific research stations in Antarctica and in the wake of this it started organizing further expeditions. The Belgian government had hoped to organize more Dutch-Belgian expeditions. Although this had no great priority in the Netherlands, a joint expedition had been undertaken, with financial support from ZWO, starting in December 1963. Subsequently other Dutch-Belgian crews moved to the Roi Baudouin base in Antarctica, until ZWO decided to withdraw from the project, which forced the

Belgians to close the base in early 1967. This, rather than a choice for Belgian astronomers between ESO and BCAP, which is sometimes suggested, forced the end of the Cross Antenna Project; Belgium formally withdrew in 1967.

The third design of the BCAP had changed already from a cross antenna to a single east-west array of telescopes employing Earth-rotation synthesis, as we have seen, and this eventually would lead to the Westerbork Synthesis Radio Telescope. But I will come back to that later. When examining the Oort Archives on these developments, one is struck by the enormous amount of work and time, he must have invested in this BCAP episode. In addition, at least in the beginning, the developments concerning ESO had put a heavy demand on his time too, while between 1958 and 1961 he also had been President of the IAU. It is a miracle he kept up scientific research, but he did and I will return to that presently.

12.6 More Dynamics

Although radio astronomy and other matters had captured Oort's imagination, stellar dynamics had not disappeared from his mind. Fundamental astrometry remained an important interest as well, so that when 'The Second astrometric conference' was organized in Cincinnati in 1959 to address 'problems of improving astrometric research in the southern hemisphere', Oort attended. His contribution was modest; it was aimed at a recommendation to use 'fundamental systems to improve the positions of some stars of particular interest' (Oort 1960b).

A fundamental study he undertook concerned the dynamics of globular clusters, specifically of Messier 3. It was undertaken in collaboration with Gijsbert van Herk (Oort and van Herk 1959c). In 1958 the latter had moved to the United States Naval Observatory (USNO), since meridian-circle work had become impossible from Leiden [332]. He briefly returned to Leiden towards the end of 1960, but moved on to Copenhagen Observatory after a few months, from which he returned to Leiden permanently in 1962.

Globular clusters are very regular, smooth, spherical collections of stars. M3, the subject of the Oort and van Herk paper, is at a little more than 10 kpc distance from us and contains on the order of half a million stars. The half-light radius (the radius of the circle that encloses half the total brightness) is about 11 pc. It is a relatively large globular cluster. The globular cluster M53, illustrated in Fig. 3.5, is somewhat smaller (and further away), but otherwise very similar. The investigation started when Oort visited the Mount Wilson and Palomar Observatories in 1952. Allan Sandage had improved and extended counts of stars as a function of radius by using plates with the 200 inch Telescope at Palomar; these had never been fully published, but Sandage put them at Oort's disposal.

The smoothness of globular clusters had usually been attributed to two-body encounters. In the same way as molecules in a gas become distributed evenly by frequent collisions, the stars in a star cluster also become smoothly distributed by encounters—they do not collide, but gravitationally deflect each other in their orbits during close passages. In a gas the velocities of the molecules become distributed in an equilibrium distribution, the 'Maxwell-Boltzmann distribution' named after James Clerk Maxwell (1831–1879) and Ludwig Boltzmann. The same process brings about 'equipartition of energy', which means that on average kinetic energy is distributed equally over the stars, and therefore on average the massive ones have lower velocities. It was expected that the stars in a globular cluster would end up with the same distribution as in a gas. The question is then, how long it takes for such an equilibrium to be reached. The time this takes is the 'relaxation time', which is defined as the time in which the cumulative exchange of kinetic energy of an average star is equal to the average kinetic energy of the stars.

Oort proceeded to make an improved estimate of relaxation times and found that roughly speaking the central 8 pc were 'relaxed'. They presented a model for the cluster, in which the stars in the outer parts are on progressively more elongated orbits. The smooth transition from the relaxed inner regions to the outer parts suggested an exchange of stars between these regions. The equipartition resulted in massive stars concentrating at smaller radii and Oort and van Herk estimated that when these lost mass by stellar evolution and had become white dwarfs, the cluster had lost 40% of its mass and expanded by a factor of about two.

This was not the final answer to the question of the structure of globular clusters, but paved the way for further studies. The now canonical models were developed by Ivan Robert King (b. 1927) in the 1960s [333] and these are roughly spherical collections of stars that have truncated isothermal velocity distributions; isothermal meaning that the average velocity is the same everywhere and truncated because stars of sufficiently high velocities have escaped from the cluster as a result of the tidal effects of the gravitational field of the Milky Way.

Oort also returned to the vertical force in the Galaxy and the local density in the Galactic disk. The work was done by a student from Australia, Eric Richard Hill (1927–2016). Joe Pawsey of the Radiophysics Laboratory of CSIRO in Sydney, whom we met as one of the confirmers of the detection of the 21 cm line, intended to have him join his radio astronomy group at CSIRO after studying with Oort in Leiden. Pawsey felt that the group not only needed radio engineers, but in addition someone with experience in astronomy. In his letter of September 11, 1952, Pawsey reported to Oort that Hill had left Australia:

> 'I do not have definite ideas as to what he should do in Leiden other than that he should attempt to acquire a general background in astronomy and learn something of the spirit of the approach to research which you and your people demonstrate so effectively. I prefer to leave you to decide on details in the light of these very general ideas.'

Pawsey thanked Oort and his wife for arranging living quarters for Hill and his mother, who accompanied him to Leiden. It turned out that Hill did a detailed study of a new determination of the vertical force field near the Sun, trying to improve on what Oort(1932a) and Oort and van Woerkom (1941b) had done. Hill used new data on K-giants at high Galactic latitudes and the work was intended to become a Ph.D. thesis to be awarded to him before his return to Australia.

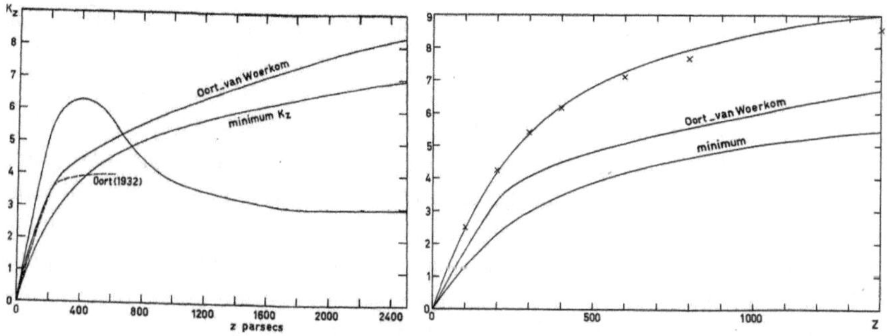

Fig. 12.14 The vertical force K_z in the Galaxy near the Sun, according to Hill (unlabeled curve at left) and to Oort (top curve and points on the right). The labeled curves are from Oort and van Woerkom (1941b) and a minimum estimate from all known constituents of the disk. The left panel also displays the curve from Oort (1932a). The vertical scales are the same; the extents in z (horizontal axis) are very different: 2.5 kpc (left) and 1.4 kpc (right). From [334] and Oort (1960c); the Archives of Leiden Observatory

Remember that the technique was to use the vertical space distribution of some collection of stars together with their velocity dispersions to deduce what the attractive force had to be for these stars to remain in equilibrium with that force. This is an application of the continuity equation in stellar dynamics (see Sect. 7.1). Hill carefully derived the best distributions of K-giants near the Sun and their velocity dispersions. For this he used Nancy Roman's data on three groups of stars (metal-strong, metal-weak and high-velocity) and their space velocities, plus radial velocities of high-latitude K-giants, to derive the run of vertical velocity dispersion with height above the plane. He then used photographic star counts and corrected these to K-stars and used the *Bergedorfer Spektral-Durchmusterung* (see Sect. 8.4) to estimate what fraction of the star counts were K-stars. Proper motion distributions and known distributions of absolute magnitudes among K-stars (from van Rhijn, among others) were then used to estimate the fraction of K-stars that were giants at various levels of apparent magnitude.

From these data Hill derived the vertical force K_z from the continuity equation and found the curve in Fig. 12.14 (left-hand panel). It rises faster than what was found by Oort and van Woerkom (1941b), but dips down under this curve at higher z. Of course the force will at first increase, but will indeed eventually have to decrease when one moves away from the disk. However, the turnover would be expected to occur at higher z; in the Schmidt mass-model of the Galaxy it occurs at 3 or 4 kpc, much higher than Hill found. So, Hill's curve seemed to be incorrect and Hill's analysis of faint K-giants was probably unsuitable for a useful determination of K_z above about 500 pc. This analysis makes no use of Poisson's equation that relates the gravitational forces to the space distribution of gravitating matter. Hill therefore calculated the 'minimum' force to be expected from all known constituents that could contribute to the vertical force using Poisson's equation and that curve is of course lower than Oort & van Woerkom. Above 800 pc Hill's curve dips down even further and therefore must be wrong. Hill realized this and attributed it to uncertainties in the faint star counts. His end result was that he only had a K_z that was reliable only below 500 pc or so; this was still an important result since the slope near the plane could be used for a determination of the local mass density. This came out as 0.13 $M_\odot$ pc^{-3}, which is about 1.5 times higher than the value Oort had found in his 1932a paper. Hill published his results with considerable delay in 1960 in the BAN [334].

Hill's BAN paper is followed by Oort (1960e), in which Oort amended Hill's analysis. Oort started out by stressing the uncertainties in the fraction of K-giants in general star counts at apparent magnitudes fainter than about 13 (photographic). In fact he supposed that this drops quickly, which gives much lower counts at faint levels than Hill had found (and used), based on the assumption that the fraction would not decrease that fast. Oort's main point, however, was that one could use Poisson's equation with the assumption that the total mass would have the same vertical distribution as the K-giants. This gave the shape of the K_z curve at distances larger than 500 pc and the brighter K-giants could then be used to derive the multiplicative factor. This resulted in the right-hand panel of Fig 12.14, which Oort believed would be a reliable determination of the force up to larger distances. From the slope near the origin, Oort derived a mass density near the Sun of 0.15 $M_\odot$ pc^{-3}.

Now we have to go back to December 9, 1955, when Oort wrote to Pawsey that Hill had decided to return to Australia without having 'concluded the work and received the degree which he deserves'.

> 'The principal cause of this was apparently that he has apparently become perturbed by the head office of the CSIRO in Melbourne and has become worried that he might lose his job if he did not return quickly. Other reasons were that he has stranded on some difficulties in the final discussion of his results and that these have unduly discouraged him.'

The thesis was practically written. Oort promised to sacrifice his Christmas holidays (he had been planning to work on the Crab Nebula) and work with him on 'the best way of dealing with the final results and about the best form in which these should be stated.' Pawsey quickly sent a telegram that Hill's immediate return was not required at all, so Hill remained in Leiden.

On March 29, 1956, however, Oort wrote to Pawsey:

> 'I am distressed to have to inform you that Hill has decided to return to Australia without having got his degree. [...] you must feel that we have let you down, in particular after the liberal way in which you have several times complied with requests to let Hill stay. [...] We arranged for a date for his promotion on April 18; we gave him all necessary help for preparing the stencils, so that the thesis could actually be ready in time. Apparently, however, the strain of having to finish the last things in a hurry have been too much for him, and he told me a few days ago that he wished to cancel the arrangements, because he did not wish to get his degree on a thesis with which he was not satisfied himself. Van de Hulst, Oosterhoff and I, as well as Schmidt, who besides myself has been in closest contact with his work, are in complete agreement that, objectively, there is no justification for this unfavorable judgment on his own work'.

Oort did worry about how the useful work might be published. However, as we saw, with considerable delay it eventually appeared in the BAN in 1960, albeit after many letters between Oort and Pawsey (and apparently unanswered ones from Oort to Hill himself).

In the *JKBlaauw-interview*, Adriaan Blaauw said about Eric Hill:

> 'I know about him. We lived at the Observatory and Eric visited us regularly. [...] Eric Hill lived for a long time together with his mother in Leiden. Well, Eric told me sometimes about his work [...] and I do know that when this seemed to come to some kind of conclusion and that Jan told him that it was not sufficient for a thesis, which was of course very disappointing and that he then had gone back to Australia.'

Blaauw was rather negative about the way Oort had conducted himself here. One should, however, remember that at this time (late 1955, early 1956) Blaauw had left Leiden and worked at Yerkes Observatory, where he had moved to in 1953 (indeed after Hill's arrival in Leiden). Blaauw's account is at best second hand, and he could indeed easily have been ill informed. The way Oort wrote to Pawsey about all this is much more concerned with Hill than Blaauw's story would imply. And if indeed Oort had refused to approve the thesis, Hill would have told Pawsey upon his return and there would have been more discussion between Pawsey and Oort. To the contrary, the letters indicate that both Oort and Pawsey urged Hill to complete the work as a paper. The continuing delay by Hill to complete the paper in spite of insistence by Pawsey and Oort and his failure to answer letters to Oort give more the impression of a person with some mental problems. Jet Katgert told me that computer C.J. Kooreman has mentioned to her that his major problems in Leiden were caused by an all-pervading homesickness. In any case, Blaauw's account seems to be based on incorrect hearsay.

The third development in stellar dynamics that Oort had a strong interest in, is that of the problem of the third integral (see Sect. 6.2). I will start by giving a brief discussion what the problem is (see also [164]). Stars in their orbits have conserved quantities. The most obvious one is the total energy, the sum of the potential and kinetic energy. The orbits derive from three differential equations of the second order (accelerations in three coordinates), and integrating these would result in six constants. Now most of these are only implicit in the solution and cannot be interpreted as a physical property with useful meaning. The total energy also is a so-called *isolating* integral. It is most useful to consider this in the context of phase space, which is a six-dimensional space in which the coordinates are three spatial directions (usually designated R, z and θ by Oort) and the velocity components (Π, Θ, and Z) in these three directions. Any point in phase space corresponds then to a position in space in the Galaxy and a particular velocity vector. The distribution function then is the number of stars with those six coordinates. Now the total energy is the sum of the potential energy (which follows directly from the position) and the kinetic energy (which follows directly from the velocities). For any point in phase space there can therefore be only one total energy. Conversely, a value for the total energy corresponds then to the collection of all points where that total energy applies and a star with that total energy can only be at those points. So the total energy isolates part of phase space. It does not mean of course that such a star will be able to visit all points of this restricted space in its orbit; that depends also on other integrals of motion. For every potential the total energy is an isolating integral.

In the case of spherical symmetry the three components of the angular momentum are all isolating integrals. In the case of axisymmetry only the angular momentum about the symmetry axis is an isolating integral. A consequence of that is that at each point in the Galaxy, the velocity dispersion in the radial and in the vertical direction should be equal. This is so, because these two velocities appear in precisely the same manner in the set of two integrals (the tangential velocity appears in the same way in the total energy, but in addition in the angular momentum integral). We have met this requirement before. In the case of the dynamics of NGC 3115, Oort (1940a) actually made used of it. In the case of K_z (Oort, 1932a) this contradicted observations and Oort assumed that in the case of the very flat Galaxy it would not apply.

Box 12.1 Integrals of motion

The **total energy** is always an integral of motion and is always isolating, since for any value it restricts the star to a (six-dimensional) volume in phase space. In the axisymmetric case:

$$I_1 = E = \Phi(R, z) + \tfrac{1}{2}(V_R^2 + V_\theta^2 + V_z^2).$$

The second isolating integral in axisymmetric potentials is the **angular momentum**:

$$I_2 = J = RV_\theta.$$

Since V_R and V_z appear in the same way in I_1 and I_2, the dispersions in those two directions will have to be the same. For V_θ the dispersion will be different. When the potential can be separated in R and z

$$\Phi(R, z) = \Phi_1(R) + \Phi_2(z),$$

the **decoupled** z**-energy** then is the Oort-Lindblad third integral

$$I_3 = \Phi_2(z) + \tfrac{1}{2}V_z^2.$$

Since V_R and V_z now appear differently in all of the three integrals, the dispersions in those two directions can be different. This integral is only isolating when the frequencies of oscillations in R and z are in a ratio that is a rational number.

In the Galaxy for small z the radial force parallel to the plane hardly changes with z so that in excellent approximation the vertical motions only depend on K_z and are essentially decoupled from the ones parallel to the plane. In that case there is an approximate third integral and that is the total energy in the z-direction. Mathematically it means that for small z the potential can be written as the sum of two functions, one with R as the only variable and one with z as only variable. This total energy in the z-direction is sometimes called the 'Oort-Lindblad third integral' which they discussed in 1932 and 1933. But it is an approximate integral and no longer applies when we go up to higher z. So, the question Oort (and others) had was whether or not there actually was a third integral for somewhat larger motions.

This approximate third integral can be isolating but does not have to be so. When we look in the (R, z)-plane the star osculates in the R-direction (in its epicycle) and in the vertical direction. This results in an orbit that resembles the well-known Lissajous-figures, named after Jules Antoine Lissajous (1822–1880), of combined harmonic oscillators. It is well-known that when the two frequencies are related to each other as a rational number (a quotient of two integers) the orbits are closed and the object cannot reach every part of the rectangle enclosed by the two oscillations. If this happens in the Oort-Lindblad third integral, the integral is isolating; otherwise it is not. For some mathematical background see Box 12.1.

Alexander Ollongren was a student at the time and Oort suggested that as a Ph.D. thesis project he should try and numerically calculate orbits of stars in the Galaxy. This was a new approach that had only been attempted for stars with low velocities near the Sun by George Contopoulos (b. 1928) of the University of Thessaloniki, later University of Athens. Contopoulos did this work partly in Sweden in collaboration with Per Olof Lindblad, the son of Bertil Lindblad, before returning to his native Greece. They found evidence that almost all orbits had a third integral. Oort suggested that Ollongren calculate orbits of higher velocity to see if these had a third integral as well. For this he first used the ARMAC computer of the Mathematical Center MC in Amsterdam (ARMAC stands for Automatische Rekenmachine MAthematisch Centrum or Automatic Calculating Machine of the AMC), with financial support from ZWO. This was the fastest computer the Netherlands had; it had been become operational in 1956. Important initial calculations were done, but ARMAC was not good enough for the real purpose. So Ollongren turned to the Swedish BESK (Binär Elektronisk SekvensKalkylator or Binary Electronic Sequence Calculator), which had been completed in 1953 and for a short time had been the fastest computer in the world. Ollongren was of partly Swedish descent; with the help of Oort, again involving a grant from ZWO, he arranged for an extended stay at Lund Observatory in Sweden in 1959, where he had access to BESK (located in Stockholm). With the help of Ingrid Torgård (1918–2001), an astronomer on the Lund staff, he programmed BESK to calculate stellar orbits in the Galaxy. Eventually Ollongren obtained his Ph.D. with a thesis on this work, with Henk van de Hulst as the supervisor, in 1962 [335] (see Appendix A.4); it was published both in the BAN and the Lund Observatory Annals [335].

Ollongren used the gravitational field corresponding to the mass model of the Galaxy from the thesis of Maarten Schmidt. The calculations were really state-of-the-art, but time-consuming, so Ollongren had to restrict himself to a carefully chosen set of initial parameters at the start of the calculation of the orbits. An important class

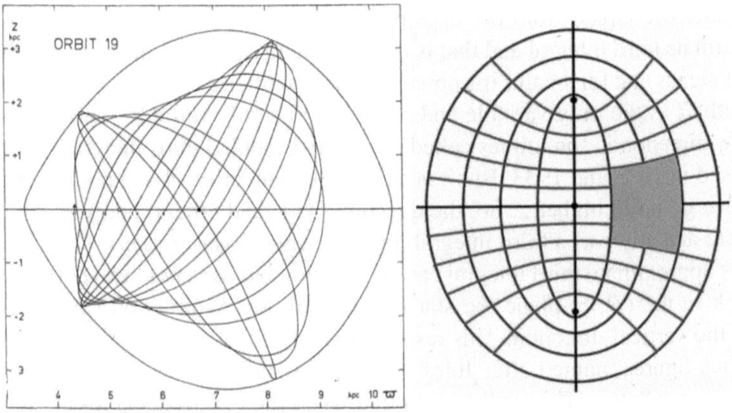

Fig. 12.15 Ollongren's Box orbit 19 in the *R*-*z* plane (left). On the right confocal ellipsoidal coordinates. From the Archives of Leiden Observatory and adapted from Wikimedia Commons [336]

of orbits, called box orbits, were topologically distorted Lissajous figures in the *R*-*z* meridional plane. One example has been shown in Fig 12.15. The envelope in the form of a thin oval (in the real meaning of egg-shaped) curve represents the 'zero-velocity curve'. On this the total energy is entirely in the form of potential energy and therefore the kinetic energy and velocity are zero. No star then can cross this line. But the calculations show that the actual orbit does not completely fill this area and that is evidence some third integral must apply.

Contopoulos had found both analytically and by numerical integration, that low-energy orbits had trapezium-shaped boundaries [337]. Ollongren studied various types of coordinate systems that could be used to describe this behavior for his orbits with higher energy and the most successful case was that of confocal ellipsoidal systems. In this the two-dimensional case is described by sets of ellipses and hyperbolas that share their foci (see right-hand panel in Fig 12.15). It would mean that in these coordinates the potential is separable; that is, can be written as a sum of two functions, each dependent upon only one of these coordinates. This was worked out by Henk van de Hulst in a local approximation, in which the positions of the foci of the coordinate system depend on where the fit is made [338]. To this end he applied Stäckel potentials, named after Paul Gustav Samuel Stäckel (1862–1919), which are described by separable functions in confocal ellipsoidal coordinate systems.

The fit of the potential field to separable functions is only approximate and breaks down for orbits with even higher velocities. Eventually there isn't even a third integral, when the force field of the disk has become insignificant at larger distances from it. Eventually using the concept of confocal ellipsoidal coordinate systems and studying third integrals was taken to three dimensions and applied to (tri-axial) elliptical galaxies in the Leiden Ph.D. thesis of Pieter Timotheus (Tim) de Zeeuw in 1984. The way the coordinate system is generalized to three dimensions by (a) rotating the coordinate system on the right of Fig. 12.15 along the vertical axis through the foci

Fig. 12.16 Oort and some persons who played an important role in his life, cut from a group photograph of staff and visitors at Yerkes Observatory in 1947. From the left Adriaan Blaauw, Gerard Kuiper, Oort and Otto Struve. From the Oort Archives

to provide ellipsoids for surfaces of constant first coordinate and hyperbolas of two sheets for the second one, and (b) rotating along the horizontal axis to give rise to hyperbolas with one sheet for the third coordinate. It is of interest to quote from the notes that Oort made for himself when preparing questions to be posed during de Zeeuw's thesis defense (Fig. 12.16).

'I have been almost since my birth as an astronomer been interested in the problem of the 'third integral' that has been solved so elegantly in this thesis. There was a general theorem by Jeans that the velocity distribution in the Π- and Z-directions had to be equal. But all determinations of the velocity distribution of stars in our neighborhood showed clearly that the dispersion in Z was much smaller than in Π. The large dispersion in Π led to Kapteyn's discovery of the so-called Star Streams. The disagreement between theory and observation was not very disturbing then, because the z-motions in such a thin system had to be almost independent for those in the other two directions. [...]

I am deeply impressed by the elegant manner in which you have been able to bring a problem that has confronted astronomy ever since the advent of stellar dynamics to the complete solution that is now lying in front of us. I cannot discuss with you the manner in which this was accomplished. But I would like to ask two questions that occurred to me in the context of your thesis.

[The first concerned the application of this work to determine the Galactic potential, for example when better data than we have today on the velocity distribution of RR Lyrae variables become available].

2. In your introduction you suggest that tri-axial density distributions are 'natural'. To be able to pronounce judgment on that one has to ponder the question how galaxies have been formed and evolved in their early stages. Concerning the last point, Lynden-Bell has some time ago presented an interesting idea that he called violent relaxation. Do we have

to conclude from the existence of tri-axial galaxies that violent relaxation did not play an important role?'

We have seen from the discussion of the dynamics of globular clusters that these become 'relaxed' by two-body encounters between the individual stars. The timescale for this in the case of elliptical galaxies is much longer than the age of the Universe, so that cannot work there. Lynden-Bell's violent relaxation poses that an elliptical galaxy is collapsing quickly with a timescale much shorter than the typical timescale of stellar orbits in the system. The relatively rapidly changing gravitational field strongly affects the stellar orbits. Since there are no two-body gravitational encounters there is no transfer of energy or angular momentum, so that in violent relaxation the integrals of a stellar orbit are conserved. So it will not wipe out tri-axiallity or for that matter force it upon the system.

12.7 Stars and Stellar Systems V

During the War, while hiding in Hulshorst, Oort had worked on a textbook on the Galaxy. Actually, his salary from Leiden University had been continued with this in mind. We have seen though that nothing more that a list of chapter titles and drafts of two chapters came out of it. Oort tried again in 1956, when he gave a series of 'Waynflete Lectures' at Magdalen College of Oxford University. William Waynflete (c.f. 1398–1486), born William Patten, had been not only Provost of Eton, Bishop of Winchester and Lord Chancellor of England, but also founder of Magdalen College. Oort's course consisted of six Wednesday lectures, three in February and March, and three more in April and May of 1956. The announcement said: 'The Lectures will be open to members of the University in academical dress and, if space permits, to the general public.' Oort had agreed with Clarendon Press in Oxford to publish the lectures and this must have been a continuation of the original attempt during the War to write a textbook on Galactic structure and dynamics. Although Oort reported on September 20, 1956 to the President of Magdalen College that he was making progress (and had written up two lectures), and although there are typed-up texts with annotations by others of two chapters, and some handwritten notes on a third, the project was never completed. An important factor in the delays might have been that Oort all the time he encountered things that he felt first needed to be sorted out better, before he could continue.

Blaauw was rather negative about this in the *JKBlaauw-interview*:

'Well, yes, there are these aspects of his personality. What I never understood was the ease with which he did not keep his promise to write a book on the Galaxy, which as far as I know he had promised Clarendon. He had given a series of lectures and I am almost sure he told me that the idea was to summarize these in a book, and that he had indeed started to write that. But why did he not complete this; after all he had promised Clarendon! It was an obligation, you see, and I can only conclude he was a bit of a spoiled little boy then. He could have thought: 'Jan, you should

Fig. 12.17 Oort with Bart Bok in 1963. The location could be the 64 m Parkes Radio Telescope in Australia. Bok would then be pointing at the 'Master Equatorial', which controls the pointing of the telescope. From the Oort Archives

force yourself to do that, it should be finished because you promised it.' But no, then his desire to do the things he enjoys, comets or whatever, take precedence. It really is a pity, because there is all this stuff that with only a little bit of additional effort could have been turned into something very valuable to all of us.'

There must have been a further attempt in 1957; in August of that year the Noord-Hollandsche Uitgevers Maatschappij (later part of Elsevier publishers) sent a contract for the publication of a book. This also came to nought. Eventually, the effort paid off as a chapter in Volume V of *Stars and Stellar Systems* (Oort 1965a).

This compendium was organized by Gerard Kuiper; actually it had been preceded by one on *The Solar System*. Kuiper had organized that series in 1950 and it consisted of volumes with chapters contributed by experts in particular areas. The first volume on *the Sun* had appeared in 1953 and the second on *The Earth as a planet* in 1954. Originally it was planned to proceed quickly with the volume IV appearing in 1957, but serious delays developed. Volume III on *Planets and Satellites* was not published until 1961 and then with the help of an associate editor, Barbara Mary Middlehurst (1915–1995). Middlehurst had moved from her native Britain in 1959 to Yerkes Observatory and later to Steward Observatory in Tucson, Arizona. Kuiper left Yerkes and the directorship of McDonald Observatory in 1960 and founded the Lunar and Planetary Laboratory also in Tucson. The fourth volume of the Solar System compendium, also edited by Kuiper and Middlehurst, appeared in 1963. Its subject was *The Moon, meteorites and comets* and Oort contributed a short (9 page) chapter on comets (Oort 1963c) (Fig. 12.17).

Kuiper had from the beginning (1950) of the project intended to also edit a compendium on stellar astronomy and wrote to Oort about that in November of 1954 (when the Solar System project seemed nearing completion) for further advice. At that time it was envisaged to be a six-volume set of books. He would have separate editors for each volume to distribute the work load. Kuiper worried about the overlap with a *Handbuch der Physik* that publisher Springer in Germany was contemplating. Oort supported Kuiper's idea as being much more comprehensive than the German one, but warned he was much too optimistic. And indeed, the set was extended later to nine volumes and the ninth volume appeared only in 1975 (volume IV never appeared). The series was a success though, particularly Volume V on *Galactic Structure*, which appeared in 1965 and was edited by Adriaan Blaauw and Maarten Schmidt. Oort wrote a chapter on *Stellar dynamics* (Oort 1965a). It turned out an excellent chapter; when I attended Oort's course on that subject in 1967, he distributed with obvious pride a reprint of the chapter among his students (he had ordered no less than 200 of them). There were corrections that were indeed applied in further printings, but he had these added by hand in the reprint he handed out. And he dictated new values in one of the tables for us to add ourselves by hand. I have consulted this reprint regularly during my research and teaching career. Oort's notes for that course ('Jan-May 1967, started after Christmas because of a stay at Columbia in the fall of 1966') list 13 students attending for the first time and two for the second time. We crowded on the first rows since he talked softly, often turned towards the backboard, while his chalk clicked on the board when 'improving' his drawings of circles and other sketches.

Adriaan Blaauw replied as follows in the *JKBlaauw-interview* to the question whether it was difficult to get this chapter out of his hands:

> 'No, but of course in this case I sort of mentored him in the following sense. We knew each other, I knew he had done a lot of work on a possible book, so I could say, isn't this an excellent occasion to write down what you wanted to say in the book, or at least a part of it. And now is the moment to bring it to some kind of fruition. It was not difficult, I actually believe he enjoyed it a lot, to do this, since it was a chapter in a book, where it was preceded by chapters, where many subjects already were treated by others. So he could concentrate nicely on what he liked most writing about. And he is maybe not at his very best here, but certainly he had really warmed up to it. You can feel he wrote it with pleasure; he really was in his element.
>
> And there is a very nice review of that chapter by Bart Bok. Bart Bok wrote about the book in Sky & Telescope. One of the remarkable things is that Bart is not a contributor to the volume. He had a reputation of being an expert on Galactic research, so you would think he would have been an author too. Mart and I could not think of a suitable subject, since you would not ask him to write on spiral structure, since that could of course be done by others, like Pawsey and Kerr and so. Analysis of star counts was something for Mart and me, so we really had no good subject for Bart Bok. Bart never blamed us for it, but said later: Since I am not an author I can write a glowing review of that book, so I will take it along on a boat trip, from Australia to Europe or so I think, and then he says: I am

Fig. 12.18 Geoffrey and Margaret Burbidge (middle-left), and Oort (top-right) in the audience at the 1958 Moscow IAU General Assembly. The person in front of Oort is Bart Bok. From the Oort Archives

on such a long boat trip and then I will carefully read that book and write a review about it. And he indeed does write a review and then he notes that this is a chapter where you see Jan at his best. That is still the case. But in this way it was not difficult at all.'

One other activity to Oort's credit, related to educating astronomers in Galactic astronomy, was the organization of a top-level Summer School in 1960 at the Nyenrode Castle at Breukelen near Utrecht. The venue is now the seat of the Nyenrode Business University. This three week course on *Present problems concerning the structure and evolution of the Galactic System* was financed by NUFFIC, the Netherlands Universities Foundation for International Cooperation. It involved lectures by top scientists, although Oort had hoped to enlist a number that in the end did not come to the Netherlands. He had tried in vain to persuade Lyman Spitzer, Martin Schwarzschild, Fred Hoyle, and Bengt Strömgren. Walter Baade was to be another star at the school; he had enthusiastically indicated he would come, but Baade suddenly died on June 25, 1960, a little more than a month before the summer school was to take place (Fig. 12.18). Allan Sandage at first accepted, writing to Oort on August 10, 1959, that he would bring his wife, whom he had just married. She was 'an astronomer (née Mary Connelley of Harvard)', he wrote, 'and is interested in Galactic structure problems.' But on February 9, 1960: 'My wife and I are to be parents in April [...] I, of course could come alone. My wife has urged me to do so. However, after thinking a great deal about this, I am convinced this would be most unfair. At a time when Mary will be busy with a new and unknown problem, I feel I must stay here.'

This was of course a setback. Lecturers from outside the Netherlands included Margaret and Geoffrey Burbidge, and Per Olof Lindblad. Oort, Henk van de Hulst and Adriaan Blaauw took a large share of the course. For some yorung people that later became well-known astronomers it was a very special experience, for example Vera Florence Rubin (née Cooper) (1928–2016) and Donald Lee Morton (1934–2014) [339, 340]. Rubin's paper was distributed widely, to a large extent by Oort himself. The Nyenrode Castle meeting was an important success!

Fig. 13.0 Oort during the ceremonies where he received honorary doctorates at Leuven in 1955 (left) and Cambridge in 1960 (right). From the Oort Archives

Chapter 13
Galactic Structure and the Dwingeloo Telescope

... an autocratic gentleman of about seventy,
with a balding skull and a sharply etched face.
[...] *That was the director, he said* [...],
– who had not only shown that the Milky Way system was
rotating, but also that it had spiral structure.
Harry Mulisch (1927–2010)

With their radio telescopes they can capture
wisps of radiation so preposterously faint that the
total amount of energy collected from outside
the solar system from all of them together
since collecting began (in 1951) is less than the
energy of a single snowflake striking the ground.
Carl Edward Sagan (1934–1996)

13.1 Private Affairs

When the younger son, Abraham (Bram) Oort, and his wife emigrated to the United States in 1966, Bram's parents frequently visited them. Bram worked at the Geophysical Fluid Dynamics Laboratory, first located in Washington, DC, and later on

Harry Kurt Victor Mulisch was a prominent Dutch writer. The quote (my translation) is from his novel *The Discovery of Heaven* [341]. One of the main characters, Max Delius, is—just as I was—a Ph.D. student in Leiden around 1970, studying the Galactic center in Dwingeloo and preparing to use the Westerbork Telescope.

The numerical justification is questionable, see [342]. The comparison is sometimes incorrectly attributed to William McGuire (Bill) Bryson (b. 1951), who actually does quote it with proper reference in *A short history of nearly everything* [343].

P. C. van der Kruit, *Jan Hendrik Oort*, Astrophysics and Space Science Library 459, https://doi.org/10.1007/978-3-030-17801-7_13

Fig. 13.1 Oort delivering his congratulatory speech at the wedding of his younger son Abraham and Bineke Pel in 1961. From the collection of Abraham H. Oort

at the Forestall campus of Princeton University. The first two years Bram and his family lived in Alexandria, VA, and then moved on in Rocky Hill, New Jersey, a mile north of Princeton, NJ. Bram had married Bineke Pel in May, 1961. On that occasion Oort had given his congratulatory speech dressed up as a magician/astrologer (see Fig. 13.1). He had collected character traits from their astrological zodiacal signs and the positions of the planets, concluding that all indications promised an ideal marriage. In particular he concluded that their love for good food and drink promised a surprisingly good fit. Bram Oort wrote:

> I now realize better what we and the children have missed in the fifty years since our immigration in 1966. Of course, we first of all missed our Dutch family. Fortunately, my parents were able to visit us frequently since my father was invited to work at the Institute for Advanced Studies in Princeton, and was offered an apartment at the Institute. Almost every year they tried to visit us until they became too old to travel. Their last visit was in 1985. They loved to watch our children grow up and invited them often to stay in their apartment at the Institute. It was clear to all of us how much my parents loved and cared for their children and grandchildren both in the Netherlands and the US.

Figure 13.2 shows the Oort generations in the early 1960s, including Mrs. Oort's mother. About the latter, Bram Oort mentioned:

> My maternal grandmother, Heleen Graadt van Roggen, was a strong and inventive woman. She had a small car in the garage in De Bilt during World War II. She could keep the car because of the ill health of my grandfather. After the death of her husband and the end of the War, she was allowed to keep her car to drive Prof. F.A. Vening Meinesz, for several

Fig. 13.2 Four generations of the Oort family in the 1965 or 1966. In the back in the middle Oort's mother-in-law Heleen Graadt van Roggen-van de Water. At the top-right daughter Marijke and son Coen, and in the front-left the other son Bram. in the back at the left their spouses Bineke Pel, Marianne Lissy and Tom de Smidt. Three more grandchildren (two girls and one boy) would be added later to the six boys and one girl in the picture, one each by the three Oort children. The boy in the front is Marc Oort, who later earned a Ph.D. in astronomy in Leiden. From the collection of Abraham H. Oort

years from his home to the KNMI when he was the Chief Director. After his retirement, she drove Prof. Wouter Bleeker (1904–1967), Director of Research, to the KNMI. Much later in 1964 Prof Bleeker became my PhD thesis adviser at the University of Utrecht.

Felix Andries Vening Meinesz was a geophysicist, known for his method to precisely measure Earths gravity, which was used to study gravity anomalies above the ocean. He was a Professor at Utrecht and later at Delft, and served as Director of the Royal Netherlands Meteorological Institute (KNMI) between 1945 and 1951. Meteorologist Wouter Bleeker had worked at the KNMI since 1928, in various leadership functions, becoming Director in 1965.

My parents liked flowers; they usually had a bouquet of flowers on the coffee table and on my fathers desk. In the summer, we had flowers from our garden in the Sterrewacht, and in other seasons my parents took their bikes to buy flowers on the Saturday market in Leiden. My mother did most of the work of weeding and planting. I remember the abundance of red and pink roses climbing high up along the trellises in front of my fathers official study next to our dining room. My father had the habit of cleaning up dead or almost dead flowers in bouquets, not just in our own house but also in friends' homes – a source of enjoyment for all of us. Later on after their retirement and the move to their first own home on the President Kennedy street in Oegstgeest, they maintained a beautiful garden. They both enjoyed so much working in the garden.

And a bit more on family life from Bram Oort;

Coen and I had little in common and became more distant after our emigration; we were different in character and style of living. He was extravert, noisy and talkative; while I was more introvert, quiet and silent. He liked to live in a complex, luxurious way while I liked to live simply. When I was very young, I remember how he could tease me when I was absorbed in my pretend games, and how he would ridicule me when I was playing teacher or minister with my puppets. On the other hand, I admired how well he could play the cello – I loved to listen to him. I also enjoyed his help in translating Latin texts from Caesars De Bello Gallico. He was a good teacher.

I enjoyed Marijke's singing and playing piano in the Sterrewacht while I was in high school, and later when we visited Holland. She performed often during family celebrations. She and a close friend of Bineke sang 'Sound the Trumpet' by Purcell during our wedding in The Hague in 1961. Marijke resented that we emigrated to the US; she had always wanted to have a sister like Bineke. The last time we saw Marijke in optimal form was during the celebration of her 75th birthday in 2006. Two years later she passed away unexpectedly due to a fast-growing abdominal cancer. I was present during her last days – she recognized me but could not speak anymore. It was a sad moment when the last of my siblings passed away.

Oort's mother, Ruth Hannah Faber died in November 1957. There are no documents related to this in the Oort Archives. Like for her husband, there had been an announcement in the 'Leids Dagblad', the local newspaper. Although her husband Abraham Oort had been buried in Oegstgeest (in 1941 during the War), Ruth Faber was cremated at the Westerveld cemetery and crematorium in Driehuis near Velsen (see Fig. 1.1). This was the first crematorium in the Netherlands, operated since 1914, although cremation was illegal (but tolerated) and only legalized in the Netherlands in 1955. Westerveld, founded in 1888, was the first crematorium open to all religions and prominent Dutchmen have been buried (or cremated) there, among whom Jacobus C. Kapteyn.

The Oorts for many years organized parties at the Observatory for all staff, hoping to improve the relations and contacts between the academic staff and other personnel (particularly the computers and the instrument builders in the workshop). Mrs. Oort usually took the initiative and Oort with some help selected a theme. One was held in the winter and one during the summer months. The floor in the hall of their house was used for a 'human chess game', where individuals acted as the chess pieces. For this purpose tiles had to be covered with pieces of black cardboard. The tradition ceased in the 1960s (Fig. 13.3).

Fig. 13.3 The Oorts, probably in 1956. They liked reading aloud to each other in the evenings, and this maybe the setting of such an occasion. From the Oort Archives

13.2 Honors and the Vetlesen Prize

Oort has received many honors. A full list of these can be found in the *JKM-Inventory* [1] by Jet Katgert-Merkelijn and in Appendix A.5.

There are 22 (honorary) memberships of academies and learned societies, among which the KNAW (Royal Academy) (1937), but also some prestigious foreign ones such as the American Academy of Arts and Sciences (1946), the (USA) National Academy of Science (1953), and the (UK) Royal Society (1959).

Oort received 10 honorary doctorates, some from the most prestigious universities, such as Oxford, UK (1951), Harvard , USA (1957) and Cambridge, UK (1960).

Of the prizes the most prestigious ones probably are the Bruce Medal of the Astronomical Society of the Pacific (1942), the Gold Medal of the Royal Astronomical Society (1946), the Vetlesen Prize of Columbia University (1966), and the Balzan (1984) and the Kyoto Prizes for Astronomy (1987) (see Sect. 15.4).

Two Royal distinctions were bestowed upon Oort by the Queen (Juliana in Oort's case) of the Netherlands. The first is that of Knight in the Order of the Netherlands Lion in 1956 and the second Commander in the Order of Orange-Nassau in 1970. The first was on the occasion of the inauguration of the Dwingeloo Radio Telescope (see Sect. 12.1) and the second of his retirement. The Order of the Netherlands Lion was founded by King William I (1772–1843) of the Netherlands in 1815, immediately upon the proclamation of the Kingdom after Napoleon's defeat. It is the highest civilian order in the Netherlands and awarded only in highly exceptional cases (significantly less than 1% of all Royal awards). There are three grades, of which Knight is the lowest. Oort's second Royal distinction is Commander in the Order of Orange-Nassau. This order was founded in 1892 by Queen-regent Emma (1858–1934) on behalf of her under-aged daughter Wilhelmina (1890–1962). It is a somewhat lower order; it is meant for 'everyone with special merits and services for society', while for the Netherlands Lion 'of a special nature' is added. Oort's rank of Commander, though, is the third highest of six and is considered one step higher than

Knight in the Order of the Netherlands Lion, according to the 'wearing order', which determines the order in which different distinctions by a person are worn on official occasions.

In 1955, Donald Robinson, 'former chief historian for the U.S. Military Government, aided by over 100 governmental agencies, foreign missions and private organizations' compiled a list for Look Magazine of the 100 most important people alive. In the category 'Science' there were eleven names, among which Oort. Others were atomic physicist Edward Teller (1980–2003), computer scientist John von Neumann (1903–1957), and Nobel Prize winners Linus Carl Pauling (1901–1994), Niels Henrik David Bohr (1885–1962), and Subrahmanyan Chandrasekhar. The Dutch newspaper 'De Telegraaf' followed suit and proclaimed Oort 'Man of the Year' 1955 in the Netherlands (see Fig. 13.4).

The Vetlesen Prize of Columbia University is awarded 'for scientific achievement resulting in a clearer understanding of the Earth, its history, or its relations to the Universe' and designed to rank as honorable as the Nobel Prize. It has been awarded roughly every two years since 1960, when it was instituted by the G. Unger Vetlesen Foundation. This foundation that was founded by Norwegian-born American shipbuilder Georg Unger Vetlesen (1889–1955). Except for Oort and in 1973 Willie Fowler it has been awarded exclusively to geophysicists and geologists (KNMI Director Felix Vening Meinesz shared the prize in 1962) and in these fields it indeed has a status similar to the Nobel Prize. The Lamont–Doherty Earth Observatory at Columbia administers the prize. Lodewijk Woltjer, who worked at the time at Columbia University, must have taken the initiative to award the prize to Oort, in a (not really successful) attempt to broaden its field to include astronomy. The prize

Fig. 13.4 On the left, part of an article in the Dutch newspaper 'De Telegraaf' of December 27, 1955, that proclaimed him 'Man of the Year'. On the right, Oort in 1962, wearing his ribbon bar of Knight in the Order of the Netherlands Lion. From the Oort Archives

carries a cash award of $ 250,000 (and a gold medal). The cash amount is equivalent to about € 1.3 million in today's buying power.

The award ceremony took place on October 28, 1966 and was followed by a one-day symposium on *Galaxies and the Universe*. At the ceremony Bengt Strömgren presented an 'Appreciation of Jan Hendrik Oort'. He noted

> 'If today we have a clearer understanding of the history of the Earth and its relation to the Universe than we had three or four decades ago we owe this partly to astronomers who through their research work provided some foundations for the investigations in question. Detailed knowledge of the structure, kinematics and dynamics of our Galaxy pertaining to stellar as well as interstellar components has been obtained. Insight has also been given into the nature of the forces of evolution at work in every part of the Galaxy. Among the astronomers who have contributed to this knowledge, Professor Oort in the opinion of the astronomical community ranks highest.' [...] Professor Oort's outstanding contributions made it clear to those who made the choice for the Vetlesen Prize that he should be the first astronomer to receive the Prize. On this occasion I would like to mention a few other features of the whole picture that we are so happy about this evening. Professor Oort not only contributed through his own research, but has been outstanding as a leader in the field. His contributions to the international organization, the International Astronomical Union, are well known. I think I do not exaggerate when I say that Professor Oort, as General Secretary and as President of the Union, has contributed more than anyone else to transforming it into a Union in which research and collaboration in research come first. But his influence in contemporary astronomy goes much further than that. Whenever there is an important astronomical gathering, Professor Oort's advice is likely to be sought; and I think that through his influence, directly on his students and collaborators, and indirectly on a large number of astronomers, he has made a contribution that is also outstanding.'

The Symposium involved four invited speakers; the participants that formed the audience have not been recorded. Oort started out by reviewing *Radio astronomical studies of the Galactic System* (Oort, 1968e). This review included the results from the Kootwijk and Dwingeloo Radio Telescopes that interested him most: spiral structure, the Galactic center and high-velocity clouds. These are the subject of three subsections of the next section. The three other speakers were C.C. Lin on theories of spiral structure, Bruno Benedetto Rossi (1905–1993), who at the Massachusetts Institute of Technology pioneered X-ray astronomy, and Allan Sandage. Rossi mentioned the detection of the first discrete sources of X-ray emission, which would turn out to have to do with compact stellar objects in close binary systems. Sandage's presentation on *The time scale of Creation* concerned age determinations of the oldest stars, of the chemical elements and of the Universe from its expansion. The ages of oldest stars could be determined utilizing progress in the understanding of stellar evolution and Hertzsprung–Russell diagrams of globular clusters and old stellar populations in the Galactic disk; ages of the chemical elements followed from

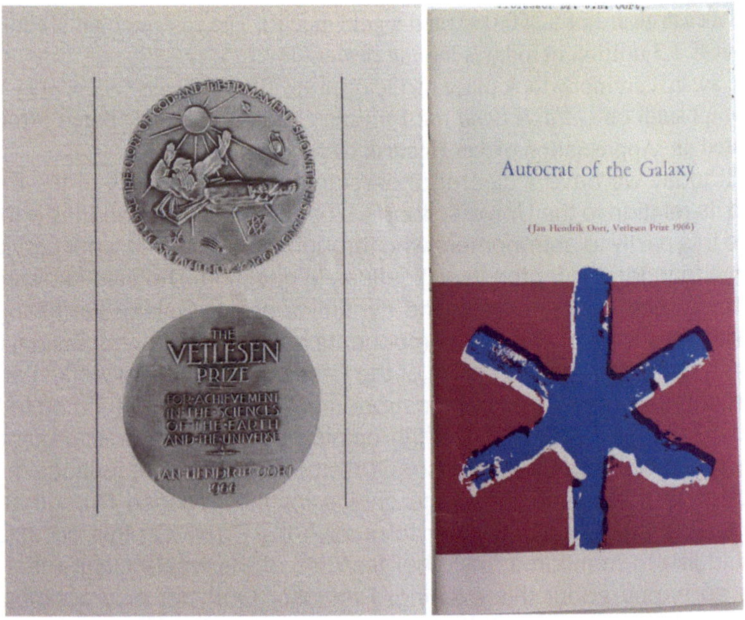

Fig. 13.5 On the left the Vetlesen medal as it was displayed on the menu for the dinner after the award ceremony. On the right the cover of a small booklet published by Radio Nederland Wereldomroep on the occasion of a radio program celebrating the Prize. From the Oort Archives

relating the abundance of particular isotopes of elements in meteorites to radioactive decay. These subjects must have interested Oort enormously.

Figure 13.5 shows the gold medal that was associated with the Prize and the cover of a small booklet published by Radio Nederland Wereldomroep (Radio Netherlands Worldwide) which produced radio programs for a worldwide audience of listeners, accompanying a program dedicated to Oort's Vetlesen Prize. Until 2012 RNW transmitted programs in the Dutch, English and Indonesian languages.

Oort never boasted about his many prizes. Yet, he would have been proud to receive such honors. In the case of the Vetlesen Prize he organized a major party at his home for all the Observatory staff to celebrate what he also must have regarded as a joint recognition.

13.3 Galaxies: Photometry, Dynamics and Supernova Rates

We have seen that Oort had been interested for a long time in the dynamics of external galaxies, for which measuring the distribution of stars was a prerequisite. We saw that he unsuccessfully spent a few months at the Perkins Observatory in 1932, and then had some plates available from Mount Wilson, obtained by Pieter Oosterhoff in 1936 on three galaxies, and that he himself collected more plates in 1939 at the Mount Wilson 60 and 100 in. telescopes. At the McDonald dedication in 1939 he had presented surface brightness measurements from the 1936-

plates and a preliminary dynamical analysis, which involved first deprojecting to the 3-dimensional distribution and subsequently using rotation velocities measured by Milton Humason at Mount Wilson (see Sect. 8.6).

In the 1950s the number of galaxies with reliable, 2-dimensional surface photometry maps was very limited (see Sect. 7.2), so Oort must have felt it was important to pursue this and further reduce the plate material he had available. As described already (see Sect. 11.3), in 1948 Oort had pioneered with William Hiltner the approach of photoelectric photometry at the McDonald 82 in. telescope in order to improve the calibration of the photographic material. His student Kees van Houten first reduced the photoelectric data (van Houten, Oort, and Hiltner, 1954b). In the end van Houten derived two-dimensional surface brightness maps of 20 galaxies, based for three on the material obtained by Oosterhoff and the rest on the plates Oort had taken. These were published as part of his Ph.D. thesis in 1961 (see Appendix A.4).

The calibration of the plates had been attempted in two ways, by a set of dark spots on the plate exposed with known intensity ratios and by extra-focal exposures of stars in Kapteyn Selected Areas (of many of these the apparent magnitudes were known). The problem remained that these two methods did not give similar results, so van Houten spent some time at Yerkes in 1955 for a second set of calibration measurements at McDonald. For seven of the galaxies he obtained photoelectric calibration, which showed that the method of extra-focal exposure worked best. Van Houten used a self-recording microphotometer, making one-dimensional tracings across the photographic plate and recording the measurements on a strip chart. For more or less round nebulae the tracings were through the nucleus in various position

Fig. 13.6 The Professors of Leiden Observatory in 1961: Oort, van de Hulst and Oosterhoff. The telescope stems from the eighteenth century, see [344]. From Wikimedia [345]

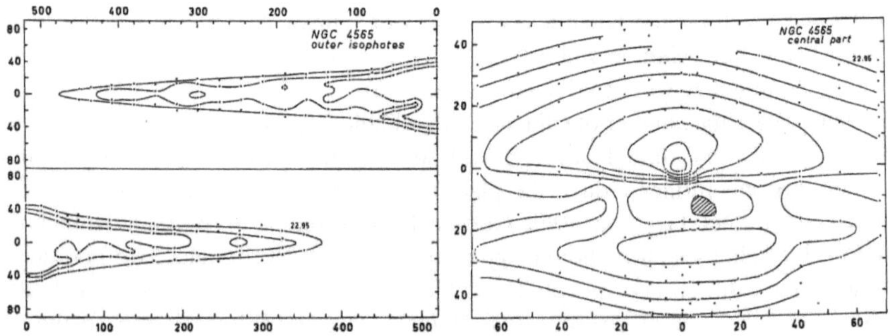

Fig. 13.7 Isophote map of NGC 4565 (see Fig. 8.10); in two halves on the left, and the central parts blown up on the right. The plate was broken during transport, but that cannot be seen in the end result. The isophotes are half a magnitude apart; the faintest is at about 23 magnitudes per square arcsec in the photographic band. From van Houten's thesis (see Appendix A.4)

angles, highly inclined systems were scanned parallel to the major and minor axes (Fig. 13.6).

The twenty galaxies in the sample contained eleven disk galaxies seen (almost) edge-on, like NGC 4565 (Fig. 8.10)) and the Sombrero Nebula NGC 4595 (see Fig. 7.6), but also NGC 3115 (Fig. 8.11). The analysis concerned mostly the edge-on disk systems. Figures 13.7 and 13.8 show isophote maps for two of these. The remaining ones were elliptical galaxies and some spirals (not edge-on).

The analysis chiefly concerned the dust lanes and effects of extinction by scattering of light out of the line of sight, but also by (forward) scattering into it, e.g. of the light of the bulge of the Sombrero Nebula as it grazes the dust layer. The light distributions in the edge-on systems were understood as consisting of two components, a bulge and a disk system. This has been confirmed in numerous later studies, the bulge often being understood as the central part of Population II (although van Houten did not mention this). The disks have sometimes inner edges, which has not been corroborated by later studies. Color measurements were also reported, but these were at the limit of observational possibilities.

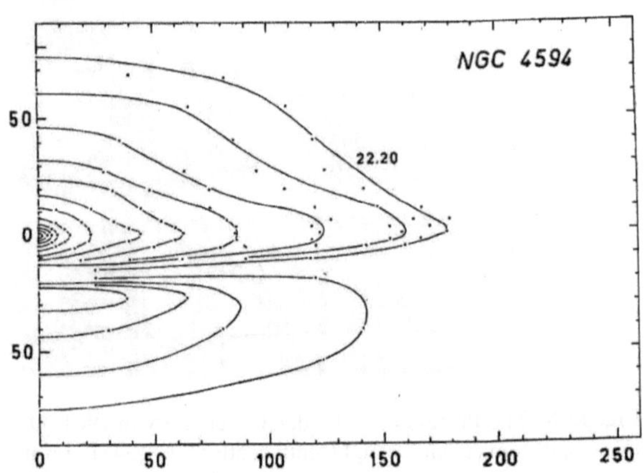

Fig. 13.8 Isophote map of the Sombrero Nebula M104 or NGC 4594 (see Fig. 7.6). The two halves of the galaxy (left-right in this orientation) have been averaged to improve the accuracy. The isophotes are half a magnitude apart. The faintest is at 22.2 magnitudes per square arcsec, which corresponded to about 30% of that of the night sky itself during the observations. From van Houten's thesis

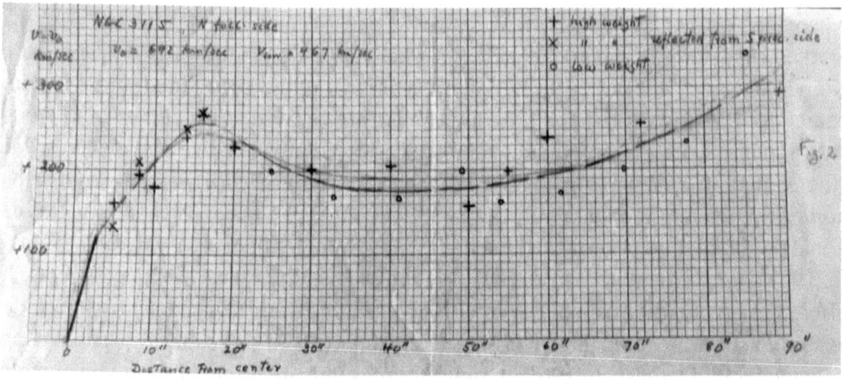

Fig. 13.9 Plot of the rotation curve Minkowski obtained for NGC 3115 around 1960. The pluses are high-weight points and circles low weight on the north-following (higher right ascension) side, crosses mirrored from the south-preceding side. The systemic velocity is 692 km/s with respect to the Sun, 467 km/s after correction for the Galactic rotation. From the Oort Archives

In the meantime Oort had not given up on dynamics of external galaxies. There has also been correspondence in 1942 and 1945 between Oort and Lindblad on the direction of rotation of spiral galaxies. They used the dust lane to judge which side is the near side, the minor axis being unaffected by dust on the far side. No definite conclusion was reached, but we now know that spiral arms always appear to by winding up rather than unwinding as Lindblad's ideas required.

Oort was keen to improve the work on NGC 3115 that he had reported at McDonald (Oort, 1940a). The corresponding file in his archives is extensive. Correspondence on NGC 3115 with Minkowski started in 1958. Apparently Oort had called upon him to try and determine a better rotation curve than the few points he had available from Humason that he had used in his 1940 paper. Minkowski used the 200 in. telescope, but even with this instrument it turned out to be a difficult task. In the end he provided a rotation curve and an estimate of the central velocity dispersion. He presented these as preliminary results in two reviews at meetings, in which he discussed the problems of nebular spectroscopy [346, 347]. The rotation curve was rather different from what Oort had available in 1939, where it rose to 316 km/s at 32″.2 and 447 km/s at 45″.5. Instead (see Fig. 13.9) it now rose to 250 km/s at 20″, then slowly turned down to about 190 km/s at 50″, and then rose again to 300 km/s at 80″. The central velocity dispersion was some 200 km/s. Oort insisted on trying to improve the outer points and push the observations to even larger distances. At about the same time, Maarten Schmidt, who now was a Professor at CalTech and a staff member of the Mount Wilson and Palomar Observatories, had tried this, undoubtedly at Oort's request, also using the 200 in., but the spectrum (according to Minkowski) was rather poor. Dark skies are essential for the faint outer parts, but observing conditions at Palomar were severely affected by the eruption of Mount Agung on Bali in February 1963, from which dust released into the atmosphere dispersed all around the Earth, making the background night sky unusually bright.

Kees van Houten suggested that NGC 3115's light distribution was really the superposition of a spheroidal component and a disk and Oort had accepted that suggestion. Van Houten had moved to the Landessternwarte Hêidelberg-Königstuhl, where he and his German-born wife Ingrid van Houten-Groeneveld (1921–2015) worked on asteroids. They had met at Yerkes while they both were working under Kuiper and were married in 1955 before Kees returned to Leiden and Ingrid accompanied him. After van Houten's thesis defense they moved to Heidelberg, but would return again to Leiden to pursue studies of asteroids for the rest of their lives. Van Houten had promised to do the calculations for the deprojection of the light distribution of the two components in NGC 3115, and after that the corresponding gravitational potential, but progress had been very slow. When Michael Moore (Mike) Davis (b. 1938), who attended Oort's course on stellar dynamics, heard of the problem, he offered to do at least some of the calculations and he did. Davis had come to Leiden to study astronomy, and after working with van de Hulst on scattering problems, he defended a Ph.D. thesis in 1967 on radio sources that are bright at short wavelengths with van de Hulst as supervisor. Davis returned to the USA and worked there in radio astronomy for the rest of his career, most recently at Arecibo Observatory at Puerto Rico, administered by Cornell University, and the SETI Institute (Search for Extraterrestrial Intelligence) at Mountain View, California.

The Oort Archives contain a number of drafts of a publication entitled *Rotation and dynamics of NGC 3115*, authored by Minkowski, Oort, van Houten and Davis, but none has ever been submitted for publication. Satisfactory solutions were never found, suffering always from high values of the mass-to-light ratio in the outer parts. This is inherent in the rising nature of the rotation curve near the end of the measurements. Oort was able to reproduce Minkowski's rotation curve only by invoking a large velocity dispersion (200 km/s or so) in the region where it dips to somewhat lower values, but that was not a physically acceptable solution (see Oort's remark in the discussion after Minkowski's 1960 paper in [346]). Oort never felt sufficiently confident of the result to publish the paper (for more discussion see also my remarks in [348]).

Instead he tried to convince others to measure the rotation curve and velocity dispersion in NGC 3115. The archives contain letters to among others Margaret Burbidge and Sandra (Sandy) Faber (b. 1944), both of whom worked on spectroscopy and kinematical and dynamical studies. In the end, Maarten Schmidt did obtain good quality data and had his student Theodore Burton Williams (b. 1952?) reduce and publish the data [222]. It showed a quick, initial rise to 260 km/s at 20 arcsec, like Minkowski found, but then stayed flat at that value out to 120 arcsec from the center. This would have changed the conclusions for the dynamical structure of this galaxy considerably. But by then it was 1975 and Oort never came back to NGC 3115, his interests having turned elsewhere. In fact many others have conducted studies of the dynamics of NGC 3115 (e.g. [349]; these authors cite Oort's (1965a) review of stellar dynamics, but not his seminal 1940 paper).

In the 1960s much work was done on rotation curves in spiral galaxies using emission lines from HII regions. This was mainly done by Margaret and Geoffrey Burbidge, working at the University of California at San Diego (La Jolla), using the 82 in. Telescope at McDonald Observatory. In this they cooperated extensively with dynamicist Kevin H. Prendergast (1929–2004) of Columbia University. After some time Vera Rubin also joined them. During this period in particular the Burbidges spent part of their summers in Leiden working on their observations and discussing matters with Oort. Many students of my generation will remember them walking on the Observatory lane with their daughter Sarah (she chose not to follow her parents into astronomy and became an attorney) to the nearby shops. More detail on this episode can be found in the chapter the Burbidges wrote in the *LibAm80* [2]. Vera Rubin also became an occasional visitor to Leiden, as she recounts in her 2011 autobiographical article [350], and she and her husband became very good friends of the Oorts, who stayed with them a few times in Washington D.C.

The Crab Nebula and a few other bright radio sources were examples of remnants of recent (a few hundred years), nearby supernovae in our Galaxy. The question arose how often a supernova goes off in a galaxy. Fritz Zwicky had conducted surveys for supernovae since the 1930s, mostly using an 18 in. Schmidt Telescope erected for that purpose at Palomar Mountain in 1936, well before the 200 in. Telescope had been completed. Over the years he discovered over one hundred supernovae by comparing photographic exposures of galaxies at different times. From this he estimated that per galaxy the supernova rate was one in a few centuries. It was student Peter Katgert, who wondered how, if that were true, it could be that after 65 years of surveying there were a few galaxies that had shown multiple supernovae; in fact there were even five that had three occurrences, which would be 'utterly improbable' if the frequency was as low as Zwicky had suggested. He and Oort set out on a statistical study (Katgert and Oort, 1967a), of the available data and concluded that the rate was more like one every fifty years or so (currently it is believed to be more like thirty). There were two very distinct types of supernovae according to their light curves; type I was exclusively seen in spiral galaxies, while the type II occurred in equal numbers in spirals and ellipticals (and S0 systems). So there seemed to be a relation to Stellar Populations. We now know that type II corresponds to massive stars having come at the end of their evolution (and thus are Population I) and type II supernovae the other ones triggered by mass exchange in lower mass binary stars, so they occur in older stars, Among these are (sub-)Type Ia supernovae that more recently have been used as a very accurate standard candle leading to the discovery of the acceleration of the expansion of the Universe.

The result was corroborated by those radio sources that were local supernova remnants. In addition to Taurus A (the Crab Nebula; 1054, ~2 kpc) and Cassiopeia A (~1670, ~3.5 kpc), two more were known within a few kpc; one corresponding to a very bright supernova recorded in 1006 (SN1006, ~2 kpc), and one discovered in 1572 by Tycho Brahe (~3 kpc). The one discovered by Johannes Kepler in 1604 is 7 kpc or more from us and probably part of the central bulge of the Galaxy. Four supernovae within 4 kpc (not counting Kepler) in the last thousand years implies a rate for the whole Galaxy of one per 25 years.

13.4 Science with the Dwingeloo Radio Telescope

The late 1950s and the 1960s were the golden years of the Dwingeloo Radio Telescope. Most Galactic research was done in Leiden by members of the staff or particularly young students. In *Dwingeloo—the golden radio telescope* [305], van Woerden and Strom presented an excellent overview of the work with the Dwingeloo dish. I will here discuss some of the studies concerned, but refer to their paper for the complete journal references. I will present first a short overview and then concentrate on the issues that Oort was most involved with.

The first months of the operation of the telescopes were devoted to a continuum survey (at a wavelength close to that of neutral hydrogen) of the region around the Galactic plane as far as observable from the Netherlands. Eventually this became the Ph.D. thesis of Gart Westerhout, which he defended in 1958 with Henk van de Hulst as supervisor (see Appendix A.4). This study located about 80 discrete sources (now identified with a W-prefix), almost all of which had radio spectra (brightness—or in technical terms flux density—as a function of wavelength) indicative of free-free emission of 'thermal' electrons in regions of ionized hydrogen (see Sect. 10.7). The survey also found an extended background from the Galaxy that was found to contain components of both thermal free-free emission and non-thermal synchrotron radiation from high-energy, relativistic electrons in a magnetic field (see Sect. 11.5).

Dwingeloo has been involved extensively in mapping the full sky at various radio wavelengths. The first paper on this was authored by Charles Seeger, Gart Westerhout and Henk van de Hulst, and published in 1956. These authors had mapped the full sky from Dwingeloo at a frequency of 400 MHz, very different from the hydrogen line at 1420 MHz. It turned out that there was radio radiation coming from directions well away from the Milky Way in structures that were described as loops and spurs. One of these structures was a major loop on the sky with a radius of order 60°, the North Polar Spur. Other observers at different telescopes were finding the same results. Further continuum studies were undertaken, which involved students, such as Louise Marie Jeanne Sophie Volders (1935–2017), later Gelato-Volders, Mike Davis, and Jet Merkelijn, later Katgert-Merkelijn). Louise Gelato-Volders later obtained a Ph.D. in mathematics from the University of Bologna, where she worked from 1961 to 1965 in radio astronomy, and after a short period in Leiden at the Computer Center she moved to Geneva (Fig. 13.11).

Fig. 13.10 Staff of Leiden Observatory in 1960. From the Oort Archives

Fig. 13.11 Some persons that play a role in this narrative, taken from the 1960 group photograph of Fig. 13.10. From left in the front: Dini Ondei, King Kwee, Oort, Pieter Oosterhoff, Gail (Bieger-)Smith, and Gart Westerhout, between Kwee and Oort Louise (Gelato-)Volders, and between Oosterhoff and Smith, Ben Hooghoudt. From the Oort Archives

Fig. 13.12 Top: A few more persons that play a role in the story of early radio astronomy in the Netherlands, again from Fig. 13.10: Jan Högbom, Whitney Shane, Jaap Tinbergen, Alex Ollongren, Ernst Raimond, and Wim Brouw. Bottom: Some more persons that played a role in Dutch radio astronomy in the 1960s. From left to right: Jet Merkelijn, Luc Braes, Hugo van Woerden, Elly Berkhuijsen, Butler Burton (behind him Kees van Houten), and Wim Rougoor. From group photograph of IAU Symposium No. 31 at Noordwijk in 1966. From the Oort Archives

Fig. 13.13 Some attendants at the 1963 IAU-URSI Symposium in Canberra, Australia, after arrival. In the left panel in front Bertil Lindblad and Frank Kerr with Richard Stoy in between them and at the right Maarten Schmidt. In the middle Mrs. Oort and Oort. Behind Mrs. Oort to the left Gérard de Vaucouleurs and behind Oort, Andrew Thackeray and Adriaan Blaauw. At the right Bart Bok and Fjeda Walraven. From a group photograph in the Oort Archives

And then there were surveys for not only the distribution of radio brightness, but in particular its polarization at various frequencies (400, 610, 820 and 1415 MHz). This would be expected for synchrotron emission and could be used to map the distribution of magnetic fields in the Galaxy. This was undertaken by a series of astronomers in addition to Seeger and Westerhout (both of whom left Leiden around 1961-2), namely Elisabeth Mabel (Elly) Berkhuijsen (b. 1937), Willem Nicolaas Brouw (b. 1940), Jaap Tinbergen and Titus Adrianus Thomas Spoelstra (1945–2010). Berkhuijsen and Spoelstra were to write Ph.D. theses under Oort (see Appendix A.4); Berkhuijsen in 1971, in which she identified the loops and spurs as remnants of ancient, nearby supernovae, and Spoelstra in 1972 with Harm (Harry) van der Laan (b. 1936) as additional supervisor. Spoelstra modeled the spurs as old supernova remnants; the radio structure of supernova remnants had been the subject of van der Laan's Ph.D. thesis research in Cambridge under Martin Ryle. Brouw went on to write a thesis on the software for the Westerbork Synthesis Radio Telescope under Alex Muller and Tinbergen went into optical polarimetry and obtained his Ph.D. in 1972 with Fjeda Walraven as supervisor (Fig. 13.12).

However, it was not Oort, but Henk van de Hulst, who reviewed the polarization data and the implications for the properties of the magnetic field of the Galaxy, in an extensive article in 1967 in the 'Annual Review of Astronomy and Astrophysics', [351]. The continuum work was definitely very interesting, but there are three subjects that we find repeatedly being reviewed by Oort at international conferences and in review papers by Oort and those involved the neutral hydrogen 21 cm line. These concerned the Galactic center, Galactic spiral structure and high-velocity clouds. Before turning to these I will first say a few words on external galaxies (Fig. 13.13).

13.4.1 External Galaxies

The beam of half a degree for the Dwingeloo Radio Telescope was of course too coarse to study external galaxies in any detail. The Andromeda Nebula M31 has a diameter in the optical of about 3 degrees, so only a crude study is possible. The 21 cm line emission is weak, so it could only be detected by observing a field away from the galaxy as a reference signal to subtract from the data. The first attempt to observe M31 was made not long after the Dwingeloo Telescope came into operation, and it proved successful. The study was published by van de Hulst, Raimond and van Woerden in 1957 [352], with a follow-up by Schmidt [353]. Raimond had been involved in Kootwijk and the BCAP studies, as we have seen. Hugo van Woerden (b. 1926) had started as a volunteer at Leiden Observatory in the summer of 1943; he and his father had visited Oort in Hulshorst in the summer of that year, van Woerden senior being anxious to find out his son's prospects after studying astronomy. Hugo had returned home (in Arnhem) in August 1944, because his father felt that was safer, and returned to Leiden as a student in 1945.

The Dwingeloo study [353] of the Andromeda galaxy produced a rotation curve of the galaxy, rising to 278 km/s at $0^\circ\!.5$, falling of only slowly to 221 km/s at the edge of the hydrogen distribution at $2^\circ\!.5$. This is more extended than the optical image. M31 is larger and more massive than our Galaxy. Schmidt's article concerned a model for the distribution of mass, which turned out rather similar to that of the light.

A few more systems were observed, such as the other Local Group spiral M33 by Ernst Raimond and Louise Volders. The resolution of the Dwingeloo dish was clearly insufficient. It only produced a radial velocity of the system and a total mass in neutral hydrogen from the integrated (over velocity) signal. Louise Volders also observed the large, nearby spiral M101 with the same result. The early work showed the promise for mapping HI in external galaxies, but that would have to await instruments with much better angular resolution. It must have strengthened Oort's resolve to push for a large radio telescope.

Throughout the 1960s there were three persistent research topics concerning our own Galaxy that kept Oort's attention. Illustrative is Oort's opening review at a well-attended symposium in March of 1963 in Canberra, Australia, on the subject *The Galaxy and the Magellanic Clouds*. It was IAU symposium no. 20, held under the auspices of both the IAU and URSI. It high-lighted radio astronomical research and the relation to optical work, especially on the Galaxy. The scene was set by Oort by identifying the current problems both as far as radio observations as well as optical data were concerned (Oort, 1964b). He spent most time on introducing the state of affairs in the areas of the gas layer and the spiral structure, the central regions and gas in the halo. Two further contributions (Oort, 1964c, 1964d) discussed the last two subjects.

13.4.2 Galactic Center

Already in the observations at Kootwijk a large range of velocities was found to exist in the HI in directions near that of the center of the Galaxy, while from Galactic rotation little is expected in the line of sight. The discovery of these expanding structures was first reported on the basis of Dwingeloo observations by van Woerden, Rougoor, and Oort (1957c). There was an indication of an arm inside the solar circle, that was seen in absorption against the radio source in the center of the Galaxy (Sagittarius A) so that it had to be on this side of the central parts, expanding away from the center at 53 km/s. It could be traced over an extent in longitude, that would indicate a distance of 3 kpc from the Galactic center. It was dubbed the '3-kpc arm' (see Fig. 13.14).

Gerrit Willem (Wim) Rougoor (1931–1967) was to pursue the issue and would write a Ph.D. thesis about it, which he defended in 1964. The structure of the central parts of the Galaxy was described by him to consist of the following components. Within about 750 pc from the center there was a rapidly rotating nuclear disk, reaching rotation velocities at its rather sharp outer edge of 250 km/s. In his contribution to the Vetlesen Symposium, Oort (1968e) made the point that 'we see the hydrogen rotating at quite high rotation velocities; at 100 pc from the center the time of revolution is about 3 million years, that is, almost one hundred times faster than in our vicinity.' Then there were expanding arms, on our side of the center the 3-

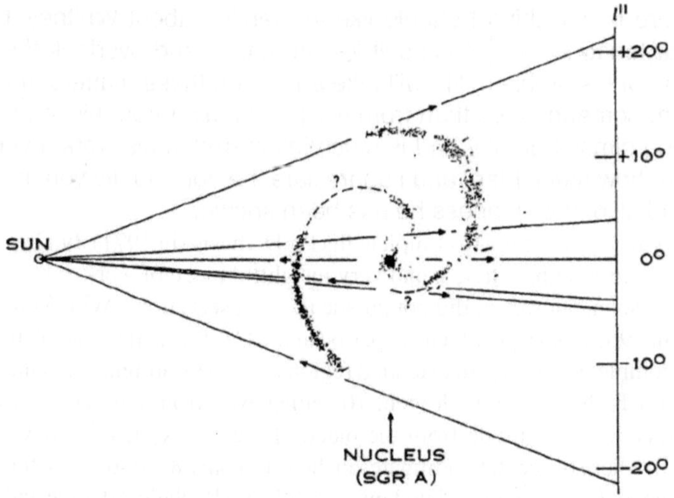

Fig. 13.14 Oort's schematic drawing of the neutral hydrogen in the central part of the Galaxy that he used in various publications and presentations. The 3-kpc arm on the left is seen in absorption and moves towards the Sun at 53 km/s in front of the central radio source Sagittarius A (Sgr A), while on the opposite side there is no absorption from the arm moving at 135 km/s. From the Oort Archives

kpc arm, moving at 53 km/s from the center, and on the opposite side an arm moving outward at 135 km/s; this arm was not seen in absorption against Sagittarius A.

I quote further from Oort's Vetlesen lecture:

'The total mass involved in this rapid expansion is of the order of $4 \times 10^7 M_\odot$. Unless there is some way to replenishment, the whole central region up to 2 or 3 kpc would be emptied by this expansion in 20 to 50 million years. [...] It is difficult to judge whether the phenomena might be caused by an asymmetry in the Galactic gravitational field or by a large-scale pulsating motion of the gas in the Galaxy. It seems doubtful that the very large outward motions close to the center could be maintained in such a way. Similarly, estimates by Woltjer and others have indicated that it is unlikely that magnetic fields could be of sufficient strength to cause the observed motions. At the moment the least unlikely explanation would seem to be an explosive activity in the nucleus.'

Oort further made mention of gas with high velocities just outside the plane of the Galaxy on either side of the nucleus moving with velocities of over 100 km/s in directions opposite to what the rotation of the Galaxy would prescribe. These observations were ascribed to another student, William Whitney Shane (b. 1928), the son of Donald Shane, Director of Lick Observatory and longtime friend and colleague of Oort, and his wife, Mary Lea Heger (1897–1983), also educated as an astronomer. Whitney Shane had met Oort when the latter visited Lick Observatory in 1958, while Shane was thinking about a thesis project involving comparison of Kootwijk data with optical observations. Oort had then invited him to come to Leiden to work on Dwingeloo data. In reply to some questions I asked him, Whitney noted that he had enormous help to settle in Leiden from King Kwee to the extent of even how to use his coal stove. Whitney's father wrote to Oort in a hand-written letter of October 15, 1958:

'There is one thing I should like to mention about Whitney. He has a great tendency to get himself lost in details and overlook the larger aspects of his problem. He will take advice on these matters much better from someone else than from me [...] Judging from his letters he is enjoying himself greatly and is much interested in his work. I want you to know how much Mary and I appreciate the opportunity you are giving him and all of the kindness he has been shown.'

Whitney took some time to complete his Ph.D. thesis (in 1971), but it was indeed an extremely thorough, detailed, and very insightful piece of work.

Whitney Shane, however, did not pursue these observations. Wim Rougoor, after returning in 1966 from a postdoctoral period at the California Institute of Technology set out to completely survey an extended region on the sky around the Galactic center to search for high-velocity hydrogen, 10° either way from the center in longitude and 5° either way in latitude from the plane of the Galaxy (as far as visible from Dwingeloo, where the center culminates in the south only 8° or so above the horizon). As it happened, I, had just completed my Candidates (Bachelor) degree and as part of my Doctoraal (Master) curriculum would assist in this research. However, Rougoor had been diagnosed with leukemia, and in spite of undergoing various experimental treatments he died in 1967 at the age of only 36. I was to continue this work under the supervision of Oort and to some extent Whitney Shane's. In the end the Dwingeloo survey would evolve into my Ph.D. research, when early in 1968 Oort urged me to obtain my Doctoraal as quickly as possible and formally start the Ph.D. research before the end of the year. Eventually I defended my thesis in 1971 with Oort as thesis supervisor. I was in fact Oort's last student to start a thesis project (see Appendix A.4).

The way Oort supervised me was by approaching me informally when in the morning all staff and students gathered in the former meridian room for morning coffee. Some played billiards (usually in vain trying to beat Oosterhoff) or table tennis. He used to have a minuscule scrap of paper with the names of the persons he wished to talk to. Others, who worked on subjects that held his interest less directly than the Galactic center, have told me that they saw even less of him. However, in the later stages he made very extensive notes and comments on drafts of chapters and papers of his students and discussed these at length.

Originally in Dwingeloo the data were recorded at the telescope on strip chart recorders and reduced manually. By the time my survey started this had been replaced by recording on punched paper tape, which was transported to Leiden where it was fed into the computer of Leiden University, an Electrologica X1, the first digital solid-state (transistorized) computer designed and produced in the Netherlands. It was programmed in ALGOL (Algorithmic Language). Producing the paper tape was not without problems, because the punching machines at the rate they were used in Dwingeloo needed regular repair and attention. There were three of them and they were made to be in operating order on Friday afternoon, which often, but not always, was just sufficient to make it until the next Monday morning. Around 1968, the Leiden computer was replaced by an IBM 360/50, for which the programming was done in PL/1 (Programming Language 1). Oort did not involve himself personally with any of that. The computer transition involved a period of months when no

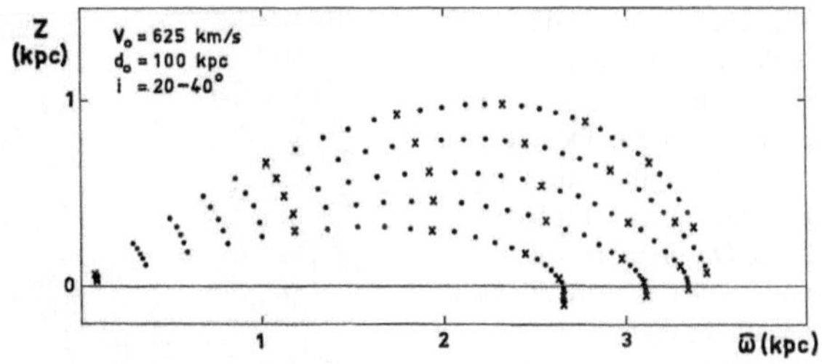

Fig. 13.15 Orbits of gas clouds, expelled from the Galatic center, falling back into the plane and being braked by gas already present. For expanation see text. From the author's thesis (see Appendix A.4)

computer reduction was possible. Oort was keen to report results of my project and 'asked' me to perform preliminary reductions by hand using printed date directly from the paper tapes and a noisy electric table calculator. I spent weeks doing that.

Some of Shane's observations were published by Belgian astronomer Pierre Cugnon (1940–2004), who had, while a student in Liège, worked at Leiden for some time with Whitney Shane. Cugnon worked as a solar physicist for the rest of his career. When in Leiden he found a cloud of neutral hydrogen outside the plane at a distance of 11°, moving at 50 km/s away from us opposite to rotation in the inner Galaxy. If near the center of the Galaxy it would have a mass in hydrogen of a few million solar masses. My survey showed a corresponding feature, which I designated 'feature XII', situated opposite the center from Cugnon's cloud and moving towards us also at about 50 km/s. I suggested that these would be clouds in some way expelled from the nucleus of the Galaxy in two opposite directions.

One reason why this seemed a viable explanation was that there was a class of spiral galaxies, named after Carl Keenan Seyfert (1911–1960) that had bright, point-like nuclei, in which optical spectroscopy had shown evidence for clouds, probably of similar mass, moving at high velocities (500 km/s or more). Our Galaxy could be a later stage of that same phenomenon. In addition bright radio sources in galaxies and quasars often showed a radio structure that suggested expulsion in two opposite directions. Oort was very happy with this result. Not long after I had found this in 1969, he had Donald Edward Osterbrock (1924–2007) as a guest staying with him and 'asked' me to come in on Saturday morning (since the early 1960s, the Netherlands had gone to a 40 h workweek and Saturday was not a workingday anymore) to show the results to Osterbrock, who had worked as a spectroscopist among other subjects on Seyfert nuclei. Oort described my work enthusiastically and I remember Osterbrock was much impressed and encouraged me to publish the work soon.

The question arose of whether or not the expanding arms *in* the plane of the central region could be the result of an earlier, more energetic phase of such expulsions. Without being more specific, Oort suggested that I should 'do some calculations', so I used the new IBM 360/50 computer of Leiden University to numerically integrate orbits of gas clouds expelled at an angle from the nucleus, which would then fall back into the disk, subsequently being braked by the existing gas (see Fig. 13.15). For this I first constructed a model for the mass distribution along the lines of Maarten Schmidt's method. The orbits in the figure are for an expulsion velocity of 625 km/s at 0.1 kpc and angles with the plane of 20°–40°. The points are about half a million years apart in the orbit. The parameter $d_{\circ}$ describes the effect of braking by the existing gas (for aficionado's: it is the ratio of the initial gas surface density in the expelled

Fig. 13.16 Oort and myself at the reception at Leiden Observatory after my Ph.D. thesis defense in October 1971. When the astronomy department was still located at the Observatory site, the young doctors were allowed a walk through the Hortus Botanicus (established in 1590, the oldest botanical garden in the Netherlands) from the Academy Building to the Observatory. Candidates defend their theses in dress suit with tailcoats (ladies usually long dresses). Professors wear their academic gowns, but Oort had taken his off before joining the party in the walk. The tube at the lower right contains the formal diploma. From the author's collection

clouds to the space density of gas in the plane). The simulations indicated that the expulsion required to explain the expanding arms involved 5–10 million solar masses of gas, expelled about 12 million years ago, at an angle of 25°–30° with the plane. The asymmetry between the near and far side from the center then would result from a small asymmetry in the angles involved on the two sides of the expulsions. This work was published as two papers; it constituted my Ph.D. thesis (see Appendix A.4), which I defended in 1971 with Oort as 'promotor' (see Fig. 13.16). In a review at IAU Symposium 38 at Basel in 1969 on 'The spiral structure of our Galaxy', Oort (1970b) summarized my work (somewhat prematurely, since the calculations were at that time unfinished) as follows:

> '..., we might imagine that strong nuclear activity is a thing that comes in bursts, that we are now in a relatively quiet period, and that the motion of the 3-kpc arm is due to much stronger activity at an epoch a few times 10^7 years ago. The circumstance that the streams appear to be ejected at a considerable angle with the Galactic plane makes it understandable that the small rotating nuclear disk found by Rougoor and Oort might have survived the period of violent activity.'

Other explanations, involving dynamical effects, for the 3-kpc arm and its counterpart at the opposite side have been put forward and are now more likely possibilities, but the gas clouds outside the plane of the Galaxy do not fit these propositions in a straightforward manner. I will return to this later (see Sect. 14.5).

Box 13.1 Dynamical relations

Jeans instability. In a homogeneous medium with density ρ and velocity dispersion $\langle V^2\rangle^{1/2}$, the maximum radius for which a sphere is stable against gravitational collapse, is (G is the gravitational constant)

$$R_{\text{Jeans}} = \left(\frac{3\pi}{32}\frac{\langle V^2\rangle}{G\rho}\right)^{1/2}.$$

A similar relation exists for a flat disk.

Toomre's criterion for local stability. This depends on the surface density of the infinitesimally thin disk σ, the velocity dispersion $\langle V_{\text{R}}^2\rangle^{1/2}$ and the epicyclic frequency κ. The criterion then says that Q should exceed unity for stability, where

$$Q = \frac{\langle V_{\text{R}}^2\rangle^{1/2}\kappa}{3.36G\sigma}.$$

Inner Lindblad resonance. The rotation velocity in a disk V_{rot} translates into an angular velocity $\Omega = V_{\text{rot}}R$. In a frame that rotates with an angular velocity Ω_{ILR} stars describe two epicycles in the time of one revolution around the center, where

$$\Omega_{\text{ILR}} = \Omega - \frac{\kappa}{2}.$$

Lindblad was the first to note that in the Galaxy over a large range of galactocentric distances this angular velocity was almost constant.

13.4.3 Spiral Structure

Spiral structure has been a longterm puzzle for Oort. In the 1930s the suggestion that some dynamical instability gave rise to it, made by Lindblad, was behind Oort's attempts to understand the dynamics of NGC 3115. Lindblad's theory predicted that the spiral arms should be leading, i.e. that the rotation of the galaxy would be in the direction of unwinding the spiral structure. Oort had corresponded with American astronomers like Edwin Hubble, Nick Mayall and others on the issue. The trick is to take galaxies that are close enough to edge-on so that the spiral structure can still be delineated, while the projection of the dustlane onto the central bulge can be used to tell which is the galaxy's near side. Spectroscopy then is needed to find which half of the galaxy is approaching and which receding. From this, a picture had been emerging that the spiral structure is not leading, as Lindblad required, but trailing.

Dynamical instability as a subject has a long history. Already there has been a discussion about this in an exchange of letters between Isaac Newton and theologian, critic and general scholar Richard Bentley (1662–1742). How can the Universe be stable when every body in it attracts every other body? Why doesn't it collapse? Newton's answer was that matter was distributed evenly throughout space so that all attractions on a body in it will compensate one another. For flat galaxies this problem was addressed by Alar Toomre (b.1937) in a very mathematical paper in 1964. For small scales he resorted to the velocity dispersions (in both stars and gas), in which the kinetic energy in the random motions compensates the gravitational attraction. However, as James Jeans had shown long ago, for a certain density of matter and velocity dispersion (or temperature) there is, according to the virial theorem, a *maximum* radius for stability against gravitational collapse (the 'Jeans radius' or correspondingly the 'Jeans mass'; see Box 3.1). In a galaxy disk at larger distances it is the shear from the differential rotation that counteracts gravity, and here there is a *minimum* radius beyond which it will work. Toomre defined a criterion in which the galaxy disk is just stable on all scales when this minimum radius equals the Jeans radius. What was known of our Galaxy near the Sun indicated that it was just stable in this sense, making gravitational instability unlikely as the cause of spiral structure. In more recent times, measurements, particularly by myself and some of my students, have shown that disks in galaxies are also marginally stable in Toomre's sense.

Already at a meeting held at the Institute of Advanced Studies in Princeton in 1961, Oort (1962i) had considered persistence of spiral structure, which might arise if the gas in a galaxy does not move in circular orbits, but is streaming along the arms. The expansions in the central region then add gas to the arms in their inner edges, which is again lost at the outer ones. In a suitably chosen rotating system the arms would have a fixed shape. In this connection there was a curious property of the Galaxy, Oort must have felt intuitively to have some fundamental significance.

We have seen that the orbit of a star that has a velocity not too different from circular rotation can be described by an elliptical epicycle superimposed onto the circular rotation. It is always possible to define a rotating frame in which the rotation around the center is equal to twice the period in the epicycle. In this frame the stellar orbit is a closed one of a quasi-elliptical form. This is called the inner Lindblad resonance. Now Bertil and his son Per Olof Lindblad had noted that in the Galaxy the angular velocity of this frame was almost the same over a wide range of galactocentric distances, at least between 3 and 10 kpc but probably also beyond. Rotation curves of external galaxies M31 (the Andromeda nebula) (see Fig 7.9) and M81 (see Fig. 2.24) showed a similar behavior. It was this property that Oort suspected to provide a clue to the problem of spiral structure, since it suggested that there was a rotating frame in which any pattern in the stellar distribution over most of the disk would be maintained.

Bertil Lindblad introduced the concept 'dispersion orbit'; if such a closed orbit is populated with stars that have small departures from the required velocities, they will in the course of time disperse along the orbit in a band with finite thickness. Per Olof Lindblad carried out numerical experiments, using the same computer that Ollongren used for his stellar orbits, see Sect. 12.6, to calculate the time development of a few such rings (see his paper in the conference proceedings preceding Oort, 1962i) and found some spiral structure developing.

It is possible to think of an arrangement in which the dispersion orbits have a rrelative orientation that produces a spiral pattern. Agris Janis Kalnajs (b. 1937) produced such an picture in 1973 [354] and I used it to model the spiral galaxy NGC 3310 [355] (Fig. 13.17). The orbits come close together and in this way produce density enhancements in the form of spiral structure, in a pattern that remains the same in the corresponding rotating frame.

The problem remained, though, of how to *maintain* such a structure. This was solved in the 'density-wave theory' proposed by Chia-Chiao Lin (1916–2013) and Frank Hsia-San Shu (b. 1943) in 1964. In broad terms their idea was the following. A wave in the form of a spiral pattern of density enhancements exists in the stellar distribution, which propagates in such a way that it conserves its pattern in some rotating frame (slower than galactic rotation itself). This means that stars in this rotating frame approach the spiral arms from the inside. The enhancement in front of them increases their random motion, but once the stars have moved throught the arms they will be decelerated by the enhancement now behind them. Stars will then 'pile up' in the arms and in this way the wave will maintain itself. Their collaborator Chi Yuan (1937–2008) applied the theory to the Galaxy, with some very promising results.

The idea of density waves must have been very convincing to Oort. Of course the increased density of stars in the arms would be very difficult to observe, since what we see as the outlines of the spiral arms are the young, luminous stars and the HII regions of ionized hydrogen around them. And the 'streaming motions' due to the density wave in stars would also be undetectable. But such motions in the gas, particularly in the neutral hydrogen at 21 cm, could be studied very well. For a

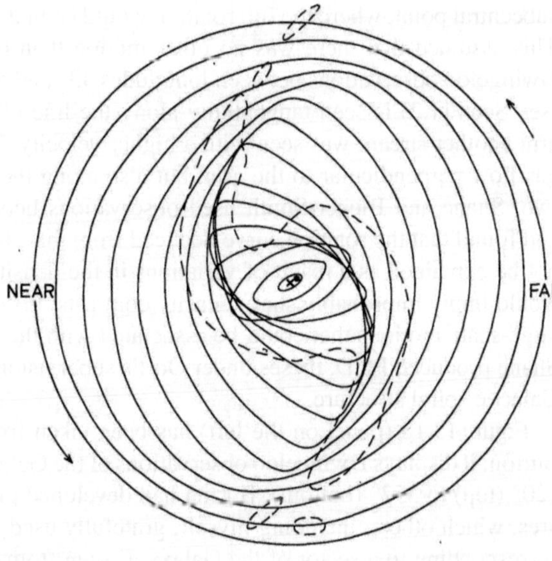

Fig. 13.17 Arrangement of dispersion or Inner Lindblad resonance orbits in the corresponding rotating frame to model the spiral structure and motions in NGC 3310. The dashed lines indicate the spiral structure in an optical photograph. From van der Kruit (1976), following Kalnajs (1973) [355]

galaxy like our own the magnitude of these streaming motions was expected to be 5–10 km/s.

A first test of the presence of streaming motions concerned the indication that the Sun, together with stars in our neighborhood, was moving outward with respect to the center of the Galaxy at a speed of 7 km/s. This had been found by Frank Kerr from Australia. In Leiden Lucien Lucas Eduard Braes (b. 1936), a student from Belgium, used the Dwingeloo dish to check this, but found that there was no conclusive evidence for it. Around Christmas 1966 Oort and Lin had an exchange of letters about this issue. Systematic motions along spiral arms became an important ingredient of the Dwingeloo studies of the Galaxy in the 21 cm line. Most of the work in the 1960s was done by three students from the US: Whitney Shane, Gail Smith and Butler Burton. Whitney Shane had come to Leiden in 1958, as we have seen. Gail Patricia Smith (b. 1939) had come to the Netherlands also in 1958 and had obtained a candidates (Bachelor) degree in 1960. Butler Burton (b. 1940) had studied with Peter van de Kamp, coming to the Netherlands in 1962, where he obtained a doctoraal degree in 1965. After a few years, Gail Smith had married and changed her name into Bieger-Smith. In a telephone interview in September 2017, she mentioned that originally Oort had given his lectures in English (according to her among others Whitney Shane and Mike Davis had followed these as well), but one day announced that he would switch to Dutch the following week. Maybe he felt they should learn the language of the country they worked in. This meant they had to learn Dutch quickly. So the American students started speaking Dutch among themselves to practise (see also Sect. 13.4.4).

In 1966, there were two important papers, one by Butler Burton and another by Whitney Shane and Gail Bieger-Smith on the question of streaming along spiral arms [356, 357]. The Galactic rotation curve inside the solar circle could be determined from selecting the highest radial velocity, which would correspond to the subcentral point, where the full rotation would be in the line of sight (see Sect. 11.2). This assumed that there was no other motion than circular rotation. Burton used Dwingeloo observations between longitudes 43° and 56°, where the Sagittarius arm (see Sect. 11.2) is seen tangentially along the line of sight. Apart from this spiral arm another stream was seen with a higher velocity. This could be interpreted as a gas-flow perpendicular to the arm, but also along the arm or a combination of the two. Shane and Bieger-Smith used observations between longitudes 22° and 70° and found that the rotation curve deduced from that showed irregularities that could not be explained as a result of variations in the density of hydrogen gas, since that would imply improbably sharp density contrasts. Instead they had to be ascribed to large-scale motions that could be associated with the spiral arms. Both Burton and Shane produced Ph.D. theses under Oort's supervision, addressing the nature of the Galactic spiral structure.

Figure 13.18 (panel on the left) has been taken from the Ph.D. thesis of Butler Burton; it displays Dwingeloo observations of the Galactic plane between longitudes 120° (top) to 352° (bottom). Burton had developed plotting programs for such figures, which others, including myself, gratefully used. The figure is an l-V diagram corresponding to a sector of the Galaxy as seen from the Sun from just outside the solar circle to just beyond the center. The horizontal axis is the observed radial velocity V (from −160 to +180 km/s) and the contours show intensity of 21 cm emission. When we look at longitudes less than 90°, positive velocities correspond to material along the line of sight inside the solar circle, negative velocities outside that.

We have seen that the rotation curve of the Galaxy within the solar circle could be determined by selecting the largest radial velocity at each longitude, which would correspond to the subcentral point. The rotation curve of the Galaxy thus corresponds to the measurements along the high-velocity envelope of the l-V diagram. The dots in the right-hand panel of Fig. 13.18 show these terminal velocities as a function of longitude, als determined by Shane and Bieger-Smith. The full-drawn line shows the same, but now determined from a theoretical l-V diagram that corresponds to a density wave model of the Galaxy. The wiggles can all be explained by this theory.

From the Kootwijk days on the simple interpretation had been that the intensity in the profiles simply corresponded to the density of hydrogen gas. However, it turned out not to be that simple. If along the line of sight the variation in radial velocity is small, signals add up and would give an increased 21 cm line intensity. Furthermore, when there are streaming motions and/or a velocity dispersion in addition to the circular rotation these would also affect the intensities. Figure 13.19 shows model calculations first published by Butler Burton in a review article by him and Whitney Shane at an IAU Symposium in 1970 [359], when both were in the later phases of

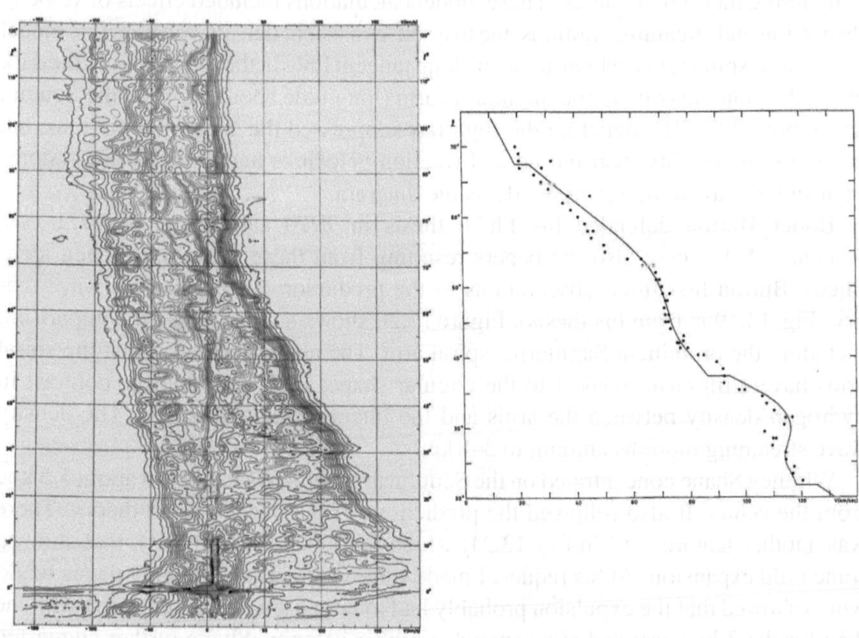

Fig. 13.18 Longitude-velocity diagrams and terminal velocities of a part of the Galaxy. See text for explanation. From the Ph.D. thesis of W. B. Burton and from Burton and Shane (1970) [358, 359]

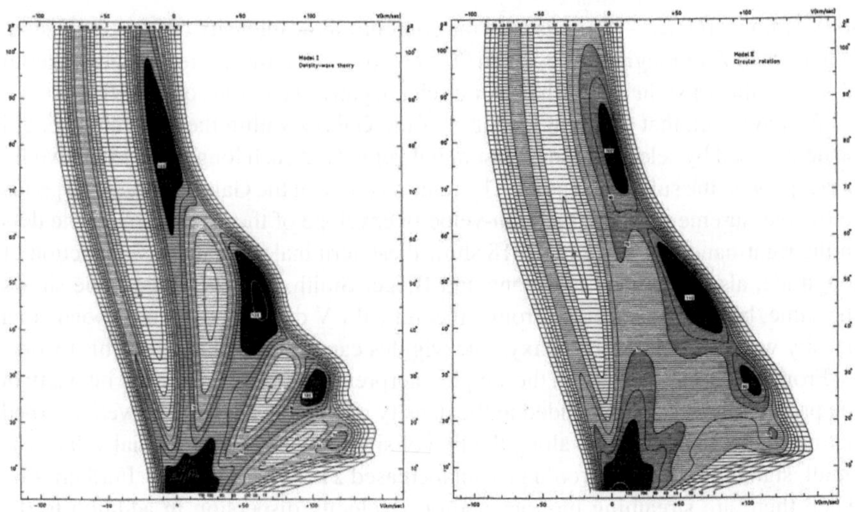

Fig. 13.19 Theoretical l-V diagrams for the case of a density wave (left; Model I) and for the same density distribution but without the streaming motions (right; Model II). From the Ph.D. thesis of Butler Burton (see Appendix A.4)

completing their Ph.D. theses. These model calculations included effects of velocity dispersion and streaming motions due to a self-consistent density wave. In this model we see four spiral arms, which are seen along tangent lines at the longitudes of the dark areas. The middle two are the Sagittarius arm (longitude about 50°) and the Scutum arm (about 30°). The panel on the right has suppressed the streaming motions, but has the same density distribution and rotation velocities and velocity dispersions. Inclusion of streaming severely affects the diagram.

Butler Burton defended his Ph.D. thesis in 1970 and Shane in 1971 (see Appendix A.4, where also the papers resulting from these theses have been identified). Burton first fitted observations to the predictions of the density wave theory (Fig. 13.19 is from his thesis). Figure 13.20 shows a more detailed comparison, including the prominent Sagittarius spiral arm. The analysis showed that the spiral arms have a tilt (with respect to the circular shape) of 7° and that the contrast in hydrogen density between the arms and the interarm regions is 3–1. The density wave streaming motions amount to 3–8 km/s.

Whitney Shane concentrated on the Scutum arm, which is located at about 5.5 kpc from the center. It also followed the predictions of the density wave theory. There was another feature ('X' in Fig. 13.21, which is from Shane's thesis), that showed some mild expansion. At his request I made some calculations as in my thesis work, which showed that the expulsion probably had to be at a smaller angle with the plane than for the 3-kpc arm and at a somewhat higher velocity. Shane further suggested that as an alternative to the nuclear expulsion hypothesis, the 3-kpc arm could be a dispersion orbit, in which gas is distributed along an orbit that was an inner Lindblad resonance (see Sect. 13.4.3). The orientation of feature 'Y' then had to be such as in Fig. 13.21, so that the parts where the gas is moving outward from the center

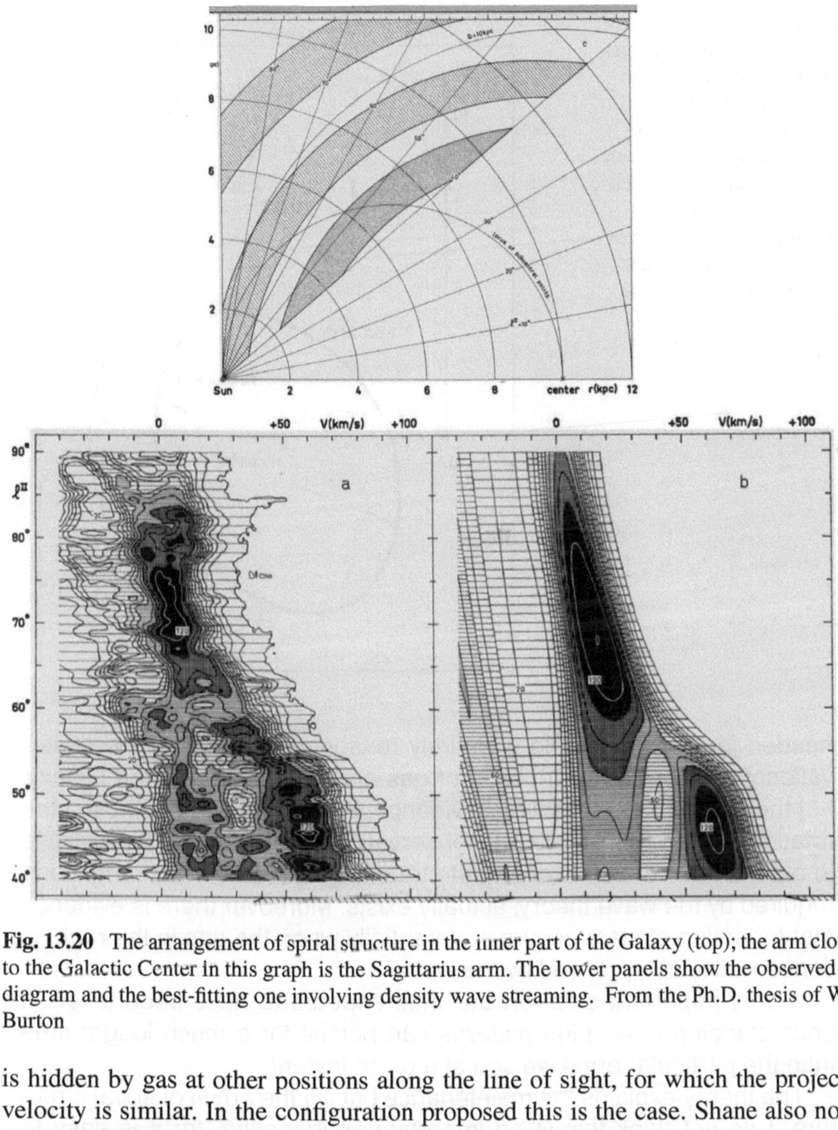

Fig. 13.20 The arrangement of spiral structure in the inner part of the Galaxy (top); the arm closest to the Galactic Center in this graph is the Sagittarius arm. The lower panels show the observed l-V diagram and the best-fitting one involving density wave streaming. From the Ph.D. thesis of W.B. Burton

is hidden by gas at other positions along the line of sight, for which the projected velocity is similar. In the configuration proposed this is the case. Shane also noted that this could explain some aspects of the observations, but not all.

In the review at the Basel meeting referred to before, Oort (1970b) summarized the situation regarding spiral structure as follows:

> 'It was suggested many years ago by Bertil Lindblad that spiral patterns could be maintained for a long time by a sectorial wave system. This idea has in recent years been further worked out by Lin, Shu and Yuan, who showed among other things how such a density wave could be sustained by its own spiral gravitational field superposed on the general axisymmetric field of the galaxy. The density excess in the arms

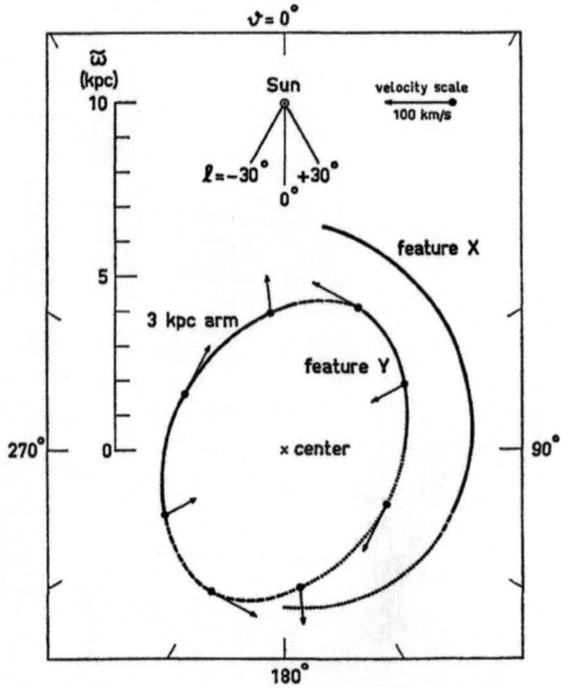

Fig. 13.21 The arrangement of structure in the inner parts of the Galaxy, according to Whitney Shane. The elliptical structure (feature 'Y') is a dispersion ring with streaming motions along it as indicated by the arrows. It constitutes an explanation of the 3-kpc arm alternative to it resulting from an nuclear expulsion. From the Ph.D. thesis of W.W. Shane

needed for this spiral field is entirely reasonable in view of the observational data on the distribution of gas and young stars in the Galaxy. [...] there is considerable direct evidence, especially in the humps of the rotation curve found from 21-cm observations in our Galaxy, but likewise in other 21-cm observations, that streaming of the kind and amplitude required by the wave theory, actually exists. Moreover there is evidence that formation of stars occurs preferentially when the gas in the regions concerned is compressed by the wave.

A very important step forward thus appears to have been taken in understanding how spiral patterns can persist for a much longer time than the particular arms we see at a given instant.

The theory explains the maintenance but *not* the *origin* of spiral structure. I do not think this is an important shortcoming, for it is easy to conceive of processes which would *start* a spiral structure. Indeed, any mechanism for formation of a rotating galaxy is likely to produce a system of large scale initial asymmetry in the plane of rotation, which will almost automatically develop into a two-armed spiral structure.'

It should be stressed that much more research has been performed at Dwingeloo concerning the hydrogen in the Galaxy, some of which resulted in Ph.D. theses. Without going into detail, I mention work by Hugo van Woerden, which resulted in a Ph.D. thesis on the gas in the region of the Orion association under Adriaan Blaaaw in Groningen in 1962, and by Ernst Raimond on the gas around the Monoceros association in a Ph.D. thesis under Oort, defended in 1966 (for both see Appendix A.4). Per Olof Lindblad paid an extended visit to Leiden on Oort's invitation and found that the Outer Arm, the most distant arm that could be recognized in the 21 cm maps, moved inward at some 30 km/s, while it had been found by Westerhout from the Kootwijk survey that this arm and the outer disk of the Galaxy were warped (curved away from the inner flat plane). Harm Jan Habing (b. 1937) found that the warp was extended even more. All these studies must have interested Oort very much as well, even though he was not directly involved.

13.4.4 High-Velocity Clouds

The riddles of the expansions near the Galactic center and the nature of spiral structure were the two subjects that figured from the beginning of radio astronomy as major interests for Oort, but a third one was added when the high-velocity clouds were discovered. This started with a prediction by Spitzer, published in 1956, but undoubtedly discussed extensively with Oort when working on the Oort and Spitzer (1955b) paper, that the Galaxy should be surrounded by a hot gaseous corona. Spitzer based this on arguments of pressure equilibrium both between the gas in the spiral arms and in between—which would require a high-temperature gas in the interarm regions—but also between that in the disk and in the volume around it. Spitzer quoted interstellar absorption lines in high-latitude O and B stars, which would be more than one kpc from the Galactic plane, as some evidence for this corona, which could contain $10^8 M_\odot$ or so. Oort suspected that if this were indeed the case, there would also be neutral condensations that could be observable in Dwingeloo. Searches for hydrogen 21 cm line emission were conducted as early as 1959, but were only successful after a number of years when more sensitive receivers became available. Most of the early work on these so-called high-velocity clouds or HVCs was done by Ernst Raimond and another student, Adrianus Nicolaas Maria (Aad) Hulsbosch (b. 1941). The most striking finding was that the gas everywhere had high *negative* (infalling) radial velocities and was concentrated in a relatively small part of the sky, viz. an area of 90° or so in Galactic longitude and 30° in latitude, more or less centered on what became known as HVC Complex A at about Galactic longitude 155°, latitude 40°, velocity −175 km/s. Complex A had a diameter of a few degrees and a mass of 2500 solar masses assuming a distance of 1 kpc (but of course the distance was completely unknown and the mass scales with its square). Some other material was found as well at lower latitudes and 'intermediate' velocities, in particular in work at Groningen by Adriaan Blaauw, Hugo van Woerden, American postdoctoral fellow Charles R. Tolbert (b. 1937) and Harm Habing, at that time a student in Groningen. More modern work has revealed many HVCs, widely spread over the sky, especially those in the category of compact HVCs, with substantial positive velocities.

Oort (1966b) wrote an extensive article in the BAN exploring the possible origins of this high-velocity material. He first discarded the possibility that they were parts of nearby supernova shells, since at least some of them would be so nearby that

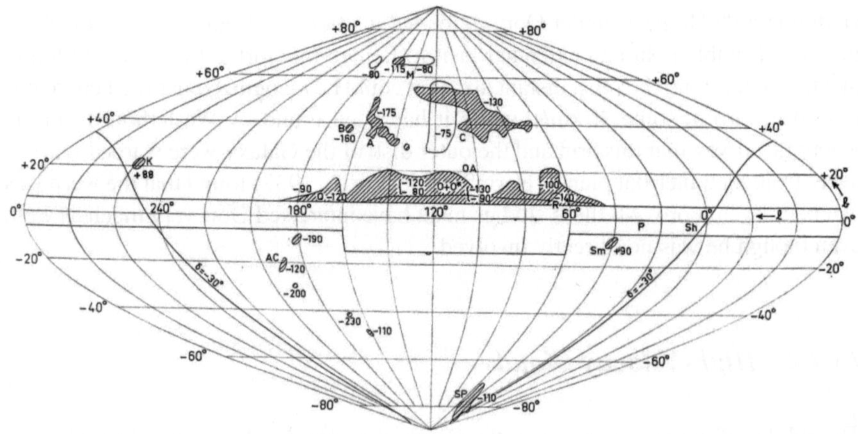

Fig. 13.22 Plot of the distribution of high-velocity clouds over the sky. The horizontal central line is the Milky Way plane, with longitude 120° in the middle. Velocities with respect to the Sun are indicated. The material near the plane in the middle is intermediate velocity material, probably associated with the outer parts of the Galaxy's disk. The complex in the middle above the plane is Complex A. The Smith Cloud is halfway to the right just below the plane, labeled 'Sm'. Figure from Oort's Nuffield Lecture in July 1969 (Oort, 1969b). From the Oort Archives

absorption lines would be visible in some suitably situated stars. Condensations in the Galactic corona was an option, but Oort argued that these would not be able to acquire the observed high velocities and that the masses would be too large in view of pressure equilibrium and evaporation time scales. If the corona were a transient phenomenon related to expulsion of gas from the Galactic nucleus, the velocities would probably not be that large at the apparent distances from the center, and should not show a preference to have negative velocities either. The same held for satellite systems while these would not have coherent structure over such a large area of sky either. Another possibility considered was to postulate superexplosions in the disk, in which large masses of gas were ejected. Oort did not rule this out, but he prefered the possibility that remained, namely that it was infalling gas into the Galaxy from intergalactic space. This resulted in a paper (Oort, 1970f) on the formation of galaxies, in which the HVCs would show the late stages (Fig. 13.22).

In later years the explanation involving a 'Galactic fountain', worked out in detail by American astronomer Joel Norman Bregman (b. 1951), gained recognition as a viable alternative to Oort's infall model. The fountain would result from supernova explosions in the disk, in which hot gas, invisible at 21 cm, is ejected and subsequently condenses into cool neutral hydrogen clouds that fall back into the disk. This explains the predominance of negative radial velocities. More details on the history of HVC research and current interpretations can be found in the proceedings of a conference in Groningen on the subject in 2004, edited by Hugo van Woerden and collaborators [360]. There now seem to be three types of HVCs, associated with either a Galactic fountain, infall of intergalactic gas, or resulting from interaction between the Galaxy and its companions, the Magellanic Clouds.

An interesting exception to the predominance of negative velocities is a cloud, first described by Gail Smith [361], and dubbed the Smith Cloud. She worked with Whitney Shane on the Galactic rotation curve (see Sect. 13.4.3). This cloud is at Galactic longitude 40°, latitude −15° and radial velocity +90 km/s. It moves faster than the maximum radial velocity due to Galactic rotation at its longitude, and does not seems to be associated with the disk. Oort (1964d) presented it at the 1963 IAU-URSI Symposium in Canberra, speculating that it was the remnant of an old, expanding (but very massive) supernova shell. In 2008, Felix James (Jay) Lockman (b. 1947) and co-workers showed that it is a cloud of gas falling in from intergalactic space. It contains more than a million solar masses of gas and is 12 kpc away from us and 3 kpc out of the Galactic plane, moving at 300 km/s with respect to the Galaxy (corrected for our rotation velocity); it is expected to hit the disk in some 27 million years [362].

This gave Gail Bieger-Smith, who had left astronomy, some publicity in the media at the time (2008). Some reporters (e.g. Govert Schilling (b. 1956) [363] in a newspaper article claimed that she had left astronomy, because Oort had felt that there was no future for women with children in astronomy. Actually this is incorrect as she confirmed in a telephone conversation with me on September 13, 2017. After she got married and had obtained her Doctoraal (Masters) degree, she went to the USA for a year to show her Dutch husband where she came from. Before she returned she wrote Oort, informing him of her upcoming return. On March 31, 1964 Oort wrote back, asking if she was interested to come and work at the Observatory again. She replied on April 15 that she was indeed interested and Oort offered her a job (letter not in the Archives). Gail Bieger wrote back in an undated letter that she was very happy with the offer and the salary, but had discovered that she was 'unexpectedly pregnant' and would stop working in January. On August 23, Oort congratulated her and told her he was sorry that for this reason she 'was not able to stay very long with us', but said he would arrange for a 'short-time appointment' as assistant. Gail Bieger-Smith had *herself* decided not to pursue a career in astronomy (and became a scientific translator); as she told me, she very much liked studying astronomy but less so the performance scientific research. She described Oort as a very friendly man, who cared a lot about his students, helped her to find a house to live, etc., and certainly never discouraged her as a mother to pursue astronomy. Later she and her husband became good friends with Coen and Marianne Oort and they came to parties at the Sterrewacht.

Gail Bieger-Smith noted that there was a feeling among students at the time that Oort felt that it would be difficult for mothers to combine an astronomy career with the care of children. Many persons I talked to (in particular Hugo van Woerden, with whom I had an extensive interview in April 2018) confirm Oort's opinion that a career in astronomy required full dedication. But there is no evidence that he had any problems with female astronomers with or without children. After all, Oort had very good relations with Margaret Burbidge; there were a few summers in which the Burbidges and their daughter spent a few weeks visiting Oort at the Observatory. And Oort had many very good contacts with Vera Rubin, whom he actually encouraged (as she once told me) to pursue her career in astronomy at the Nyenrode Summer

School in 1960 (Vera was a mother of four; her youngest child was actually born in May 1960; one of her children was to become an astronomer also).

After publication of the 1966b paper on the origin of the high-velocity clouds, Oort pursued the matter on various fronts. His first concern was the question of the distances of the clouds. One possible way to get a handle on that was observing distant stars in the same direction and see if these had interstellar absorption lines at the same velocities. Material at low latitudes and intermediate velocity could be shown from existing publications to be part of the Galaxy, but no such material was available for the high-velocity clouds. For this he corresponded with many spectroscopists he knew, such as George Wallerstein (b.1930), then at the University of California at San Diego, Margaret Burbidge, George Howard Herbig (1920–2013) at the University of California at Berkeley (all of these had access to the facilities of Lick Observatory), Allan Sandage and John Beverley (Bev) Oke (1928–2004) from CalTech, who observed at Palomar Observatory, and a number of others. None of these was able to produce a definite result, although Wallerstein did confirm that intermediate velocity neutral hydrogen was at least one kpc away. Only in more recent years distances have been established with this method by Hugo van Woerden and collaborators (e.g. HVC Complex A is 4.5–7 kpc from us).

Oort instigated various other studies of HVCs related to what he considered the most likely origin, that of an inflow from intergalactic space. The first problem was that it was not obvious that the gas would remain neutral. When intergalactic gas hits the Galactic corona it will be at velocities of order 500 km/s and strong shocks will develop that will heat and ionize the gas. At Oort's suggestion a study was conducted by Malcolm Paul Savedoff (b.1928), who visited Leiden at the time as a postdoctoral fellow, Joachim Willem (Joop) Hovenier (b. 1936) and Bram van Leer (b. 1942), graduate students who both would obtain a Ph.D. with Henk van de Hulst as supervisor (Hovenier on scattering of light on particles and van Leer on methods for numerical schemes of shocks). They concluded that indeed strong shocks should develop and estimated cooling and recombination times for the hydrogen after the shock to become neutral again. These depended strongly on the actual gas density, so the results were not very conclusive, but the timescales could be as short as a few million years, so that neutral hydrogen would be observable.

Shklovsky had suggested that the 21 cm emission might have been amplified by a maser affect, indicating more neutral hydrogen than actually present. Maser stands for Microwave Amplification by Stimulated Emission of Radiation (in optical light this would be a laser). This works as follows. Stimulated emission occurs when a photon corresponding to a particular transition hits an atom that has its electron in the upper level of that same transition. Then this gives rise to the emission of another photon in addition the incident one, so amplifying the radiation. Normally this is countered by absorption of the same transition by the more numerous atoms in the lower level. Now, the hydrogen atom has a ground state (split in two for the 21 cm transition) and higher levels. Take the first level above the ground state (also split). The transition from any sub-level of this first one to the groundlevel corresponds to the so-called Lyman-α line with an ultraviolet wavelength of 1216 Å; it occurs in HII regions when hydrogen recombines (like the 'Balmer' lines in the optical, which are to the first

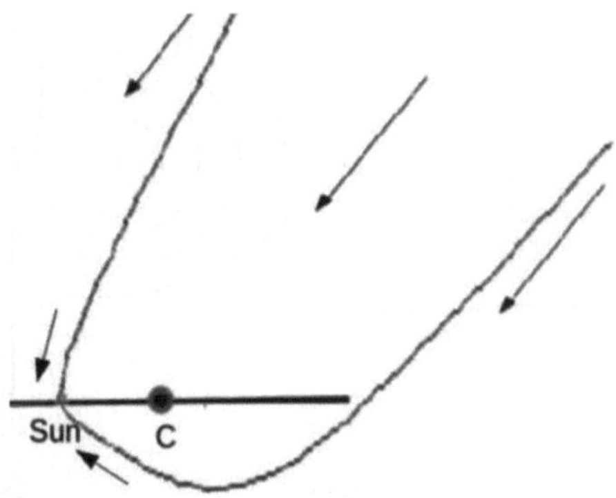

Fig. 13.23 Schematic sketch of the orbits of infalling gas into the Galaxy at the position of the Sun. There is a possibility for a direct and an indirect orbit, although the latter can be blocked by the gas in the plane of the Galaxy (thick line). After a figure from an unpublished paper by Oort and the author

level; see Sect. 9.8). This Ly-α emission is heavily absorbed in the disk of the Galaxy, but can escape in other directions and reach high-velocity clouds. So hydrogen atoms will be excited by Lyman-α photons from the ground state into the first level and these will then fall back to the ground state (emitting a Lyman-α photon). Some electrons will end up in the upper level of the 21 cm line. Now the decay from the first level to the ground state (Lyman-α) occurs fast, the average lifetime in the upper level is a few nanoseconds, while for the 21 cm transition this is of order ten million years (it is a so-called 'forbidden' transition). A process of continuous absorption of Lyman-α photons results thus in an accumulation of atoms in the upper level of the 21 cm line, and eventually there will be more there than in the lower level. This is referred to as population inversion by Lyman-α pumping. Then there is more stimulated emission than absorption and amplification of 21 cm radiation occurs. However, Henk van Bueren and Oort (1968c) considered this mechanism in much detail and concluded that the effect was unimportant.

If the Galaxy moved through an intergalactic medium, what could be said about the direction of this motion and about the internal motions in the stream? The concentration at HVC Complex A could indicate the main direction of the relative velocity of Galaxy with respect to the medium, but what more could be said about it? In 1966 I had obtained my Candidates degree (Bachelors) and for the Doctoraal (Masters) two research projects were required. The major one would be my Dwingeloo work on the Galactic center (see Sect. 13.4.2) and as the smaller project Oort suggested I would do some calculations concerning this issue. I first noted that there would be two types of trajectories (see Fig. 13.23). In addition to the direct one, there would also be an indirect one, which went through the plane before reaching the solar neighborhood. If this crossing occurred within the gaseous disk of the Galaxy, the infalling gas would be blocked by the disk gas, but farther out the HVC gas could actually reach the Solar neigbourhood.

Assuming for simplicity that all mass of the Galaxy was concentrated in the center, orbits could be calculated that would reach the Sun, and assuming a Gaussian distribution of random velocities in the medium (or clouds) I could calculate the resulting distribution on the sky. I assumed that the Galaxy moved at 50 km/s with respect to the medium (gas would be accelerated and move much faster when approaching

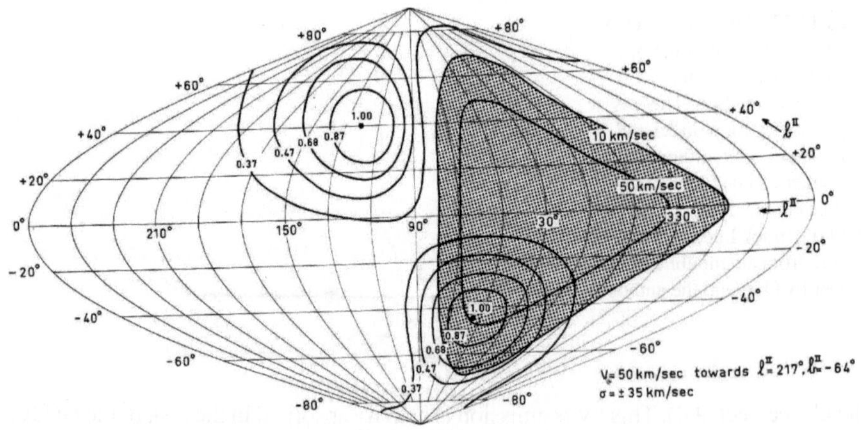

Fig. 13.24 The predicted distribution of High-Velocity Clouds over the sky in the case of infall from intergalactic space due to the Galaxy moving with respect to that medium. The shaded area corresponds to indirect orbits that cross the Galactic plane within 15 kpc. From an unpublished paper by Oort and the author

the Galactic disk), and that the position of Complex A corresponded to the main infall direction now observed, which gave the direction of the Galaxy's motion in space. I then assumed that the medium had an internal velocity dispersion of 35 km/s and the Galaxic disk a radius of 15 kpc for blocking indirect orbits. This resulted in Fig. 13.24, which showed where we would observed high-velocity clouds. The shaded area indicates where indirect orbits would have crossed the Galaxy and no clouds should be observed there (the two outlines correspond to gas with different internal motions in the medium, actually their velocities at infinity). The contours on the top-left indeed included most HVCs observed, none were seen in the shaded area (Smith's Cloud is an exception, but that is at positive radial velocity and unrelated to this general stream), and indeed some high-velocity material was observed near the south Galactic pole (the South-Polar Complex). It gave a qualitative explanation of the observed sky distribution. I did these calculations on a slide-rule while observing at Dwingeloo and when Oort was on an extended stay in the USA staying with his son in Princeton and was also visiting the Mount Wilson and Palomar Observatories in Pasadena. I had sent him a progress report towards the end of December 1967 there, but only presented him the full outcome after his return early in 1968. He was extremely pleased with it (and my performance I suppose) and a few weeks later he invited me to his office and urged me to finish my Masters before the end of the year and offered me a position to continue my Galactic center work for a Ph.D. thesis. I did my Masters in November of 1968. I wrote up this work as a joint paper, but although Oort had it typed up and made annotations and corrections, this was never completed and ended up as an unpublished manuscript in the Oort Archives. Oort referred to it in talks and some publications (such as Oort, 1970f). It gave support to the idea of an inflow of intergalactic gas.

The main work on the observations of HVCs was done by Aad Hulsbosch, who after a long delay submitted his Ph.D. thesis in 1973. He had been involved in Dwingeloo observations of HVCs since about 1964 and collected an extensive data set, contributing a major part of what was known about them at that time.

13.4.5 *Galaxy Formation*

Although Oort usually qualified any statement that the inflow of intergalactic gas was the most probable origin for the high-velocity clouds, he regarded it as strong enough evidence for the fact that such gas indeed was flowing into the Galaxy that a fundamental rethinking of the problem of the formation of galaxies was required. This resulted in two fundamental papers, Oort (1970f) on galaxy formation and Oort (1970g) on the density in the Universe. Most of the material was presented as his Nuffield Lecture on July 7, 1969 (Oort, 1969b), held at the Institute of Theoretical Astronomy of Cambridge University. This institute had been created by Fred Hoyle in 1967. The Nuffield Foundation is a charitable trust established in 1943 by William Richard Morris, 1st Viscount Nuffield (1877–1963) based on his fortune from his automobile industry. It had committed itself for some years to supplying 2/5 of the running costs of Hoyle's Institute. The chairman of the Foundation, Alexander Robertus Todd (1907–1997, biochemist at Cambridge and Nobel Prize laureate, was a good friend of Hoyle's. As it turned out, Oort's first Nuffield Lecture was the first and last of an intended series.

The HVCs indicated, according to Oort, an inflow rate of the order of 3×10^{17} H-atoms per cm^2 per million years, which corresponds to a few thousandths of a solar mass per square parsec in the disk per million years. In a billion (10^9) years, Oort estimated, the mass of the Galaxy would increase over its lifetime at this rate by 0.9 and 10% or so. This is a significant increase.

A crucial part of any theory of galaxy formation is understanding how they acquired their angular momentum. A popular idea was that galaxies resulted from density fluctuations in an early stage of the Universe when gravity became more important than the energy in the radiation from the hot Big Bang. However, Oort argued, that at the time the density was too high and the proto-galaxies too small to contain enough angular momentum to explain the current rotation. One option was that the angular momentum had been acquired later from external forces, 'tidal torques' on irregular protogalaxies from neighboring mass concentrations (attracting one side more than the opposite one). Oort estimated that this effect could not produce the angular momentum required. He therefore assumed that galaxies would detach themselves much later when large-scale turbulent cells would contain sufficient angular momentum. Some turbulent cells would have sufficient inward motion to counterbalance the universal expansion and such cells detached themself from their surroundings. Regions that do not have the required amount of angular momentum and radial motions would not collapse into a galaxy. The matter in the favorable cells will at first expand with the Universe, but slow down, collapse, and form a rotating system. The outer layers will collapse more slowly and produce an inflow, albeit at a reduced rate, up to the present.

Oort estimated that the protogalaxies collapsed relatively late (from 1/10 to 1/4 or so of the present age of the Universe), but would keep acquiring new gas as evidenced by the high-velocity clouds. Since there had to be cells that would not collapse into galaxies, there had to be much intergalactic material and in the second paper he estimated that galaxies will form only in a small fraction of space. He argued as follows: Start with a cell that could become a galaxy, assume it has some angular momentum and look at the corresponding plane of rotation. There are other streaming motions as well in this plane and along any of four perpendicular directions from the center these can be inflow or outflow. There is no a priori reason for any of these to be preferred. Along these four directions there are sixteen permutations and combinations possible, while collapse will only occur in one case, namely when all four streamings are inward. In the third dimension there is no rotation and no centrifugal force and contraction will eventually occur in almost all cases. Oort's formulation is a bit confusing because he distinguished the two opposite directions of rotation, which, however, is not essential for his argument. So no more than 1/16 of the cells will result in a contraction to form a galaxy and 15/16 of space would be filled by an intergalactic gas. This is not observed so it has to be hot and observable only in soft X-rays. The factor 16 would bring the total density in the Universe close to the critical one required to just stop the expansion in an infinite time (the Einstein-de Sitter Universe).

Our understanding of galaxy formation has developed rapidly by newer insights and much progress has been made using numerical simulations. Furthermore, the discovery in the 1970s of 'dark haloes'around galaxies consisting of mysterious 'dark matter' has led to the conclusion that the masses of galaxies are larger by an order of magnitude, and nowadays the origin of angular momentum in galaxies by the tidal torques from neighboring mass concentrations in the early Universe is generally accepted, where Oort concluded on the basis of then prevailing estimates of the masses of galaxies it would be insufficient. Oort's deduction that there is an inflow of gas from intergalactic space still stands, at least as an explanation for part of the HVC phenomenon, but his work on galaxy formation has been superseded by later developments. His prediction that the mass we see in the form of stars and gas in galaxies is only a small fraction of the total mass has largely been confirmed; however it is in the form of 'dark matter' and not in failed galaxies.

13.5 Advisory and Administrative Functions

The professors in the University were supposed to take turns in administrative functions relating to the running of the affairs (see for example [364] for a history of Leiden University). The top of the organization was the Board of Curators. All professors together formed the Senate, where important decisions on budgets and appointments were made subject to the Curators. It was chaired by the Rector Magnificus and he was assisted by a secretary (who would be the next Rector) and a Presidium Senatus consisting of representatives, usually Deans from the Faculties. The term of office for Rector Magniticus was one year and the most senior professor (as a primus inter pares) was expected take his turn as Rector. Faculties had Deans that had terms of two years and a Board of Professors, half of which was replaced every two year.

Oort served as Dean of the Faculty of Mathematics and Natural Sciences from 1955 to 1957 and was a member of the Praesidium Senatus for the same period. An important development during Oort's tenure of the Deanship was the establishment of an ad hoc national committee on funding of natural sciences, eventually known by the name of its chairman as the 'Casimir Committee' (see [365])

The economy of the Netherlands (as in the western world) prospered in 1950s and the need for young professionals with a background in natural science or engineering grew. The number of schoolchildren had grown as a result of the baby boom just after the War and this would soon result in a need to accommodate many more students at the universities. The Minister of Education, Arts, and Sciences, Jozef Cals was much aware of this and appointed a commission to investigate and advise him on what needed to be done so that universities would be able to cope with this and whether there was a need to found more universities and institutions for higher education. However, this was 1956 and the growth of the economy showed signs of slowing down and significant economies by the government were in order, including an end to the annual increase of the funding of and of new positions at universities for 1957. In particular there would be an immediate stop on hiring personnel.

Of course, the Curators and other authorities at the universities were approaching members of Parliament to avert this. However, Oort felt that this was too urgent to wait for developments, especially for the areas of natural sciences. He proposed to his fellow Deans at the other state-funded universities (Groningen and Utrecht) to send a letter directly to the Prime Minister, Willem Drees (1886–1988). Drees was a social democrat, who first served as Minister of Social Affairs (1945–1948) and then as Prime Minister from 1948 to 1958. As minister Drees introduced a general state pension and he is seen as the father of the economic recovery of the Netherlands from the devastation of the Second World War, and of the introduction of the 'welfare state'.

Oort sent a draft of the letter to the Deans at Groningen and Utrecht (the latter happened to be astronomer Marcel Minnaert) and with their approval wrote to other Deans on December 3, 1956. He stressed to them the need for breaking the ban on new appointments and offered the opinion that this could be achieved only by having the matter discussed at the Cabinet level. Since it was expected that the Cabinet would complete its proposal for next year's budget before Christmas, time was short and the letter would have to be from the Deans rather than from the Faculties. The reason for passing by the Senates and Curators of the universities was not only this pressure of time, but also that the need was most extreme in the case of Mathematics and Natural Sciences.

The letter was sent to the Minister on December 7, 1956, with the support from the Deans of the University of Amsterdam and the Free University in Amsterdam. Oort sent the same date a copy of the letter to the Minister, but in the cover letter did not try to justify why he was approaching the Prime Minister directly. Passing by the Minister was a bold step. Astrid Elbers in *EarlyRadio* [9] and David Baneke in [365] mention Oort's approach of the Prime Minister in 1945 to propose the construction of a radio telescope as a comparable (and for Oort typical) action. Why this comparison is relevant is not clear; after all, the concern about the future of the natural sciences by a group of Deans in these areas after all is very different from obtaining funds for a specific scientific instrument in a single discipline. And twice is hardly a habit. It does show Oort's genuine concern that the Netherlands not become second rate (or worse) in science in general which was a real threat to the future prosperity of the country.

The letter ran for 7 pages, including two appendices, and must have been carefully and painstakingly composed by Oort. Minnaert had suggested some numerical background to support the case. The letter stressed that the 'bottleneck' for the development of natural science and engineering in the USA was manpower, and stated that also in the Netherlands such a shortage was the most important obstacle for progress. In addition the program of building new laboratories had to be accelerated significantly. Much had been done by the government, also through ZWO, but it would be disastrous to undo this now. A budget increase, even of 6% annually, was absolutely insufficient.

On April 6, 1957 the Deans and Secretaries of the Faculties of the three state universities met with Minister Cals and his highest civil servant in the area concerned, Arie Johannes Piekaar (1910–1990). The latter had studied indology (history, culture and literature of the Dutch East Indies) and had been a civil servant there since 1934, including a period in Japanese internment during the War and a spell after the Indonesian independence. Since 1953 he had worked at the Ministry of Education, Arts and Sciences, eventually as Director-General for Science Policy. He would be very important in the realization of the growth of academic education until his retirement in 1975. In that conversation, of which Oort sent a summary to both the Minister and his fellow Deans, the Deans had presented a strong plea for the establishment of a high level advisory commission. The Minister wrote to Oort cum suis on July 1, 1957 that he had much appreciated the letter to the Prime Minister and the discussion and was prepared to do his utmost to end the hiring stop and provide funds for new buildings, but wanted a further proposal from them concerning the advisory committee they had proposed.

Oort sent such a proposal, including a list of names in October 1957 (the Oort Archives contain a draft of the letter dated July 15), and the Commission was formally established on December 3, 1957. It was designated as the Commission for the Development of Research in the Natural Sciences, but became commonly known as the Casimir Commission, named after its chairman Henk Casimir, physicist and Director of Philips NatLab (see Sect. 9.6). Members were Oort, plant physiologist Willem Hendrik Arisz (1888–1975) from Groningen, physicist Cornelis Gorter from Leiden, physical chemist Jan Theodoor Gerard Overbeek (1911–2007) of Utrecht, and Hendrik Willem Slotboom (1904–1996), chemical engineer and coordinator research at Royal Dutch Shell. They did not waste time to start their work and after a number of iterations filed their report in October of 1958.

The recommendations of the Casimir Commission were never implemented as a package, but most of them were in the end realized. The number of students increased even more sharply than the 'Centraal Bureau voor de Statistiek' (CBS; Statistics Netherlands [366]) had predicted (in the natural sciences by a factor 3 between 1957 and 1970). Budgets for academic education increased by a factor 6 or so [365] over this period and new campuses with larger laboratories were built starting in the 1960s on the outskirts of university cities. The Commission had proposed that promising students (who were offered a position at the universities) would be exempted from military service; this was never realized in a general measure across the board, but rules for deferral were made more flexible; 6 or 7 years for the phase to the doctorandus degree were not exceptional, as was a further extension by 4 years to prepare a Ph.D. thesis.

Although it is not realistic to attribute all of what had been accomplished to the Casimir Commission, the initiative of Oort to write the Deans' letter to the Prime Minister left a vital mark on the conducting of (natural) science in the Netherlands. Approaching the Prime Minister directly was an exceptional procedure, which Oort must have considered carefully and which in his judgment would not (unlike probably in most cases) be counterproductive. His judgment again proved right.

In the same period Oort's turn came to be Rector Magnificus of Leiden University for one academic year. This system had been in use for a long time, although there certainly have been professors that took various terms as Rector. But Oort objected strongly. On October 16, 1957, he wrote to the incumbent rector, physician and pharmacologist Samuel Elzevier de Jongh (1898–1976):

> 'My competence is in the first place research and I have the feeling this is a period in which I could do quite some useful research. In practice this is only possible for a very small fraction of my time. I am involved in a continuous battle to keep some time open for my own research. At this time there are so many tremendously interesting things to find out about the Universe that I would have to decide to turn away from, but would not be able to justify.
>
> This opinion is reinforced by the following: I believe that the one-year rectorate is an institution that is not profitable for the University. It is no longer appropriate for the extensive tasks of the board of a Senate. [...] For this development a Board of the Senate needs a permanent Rector, freed for a large part of his other professorial duties. [...]
>
> It has been remarked that it is in the interest of the University to have a Rector with some international renown. I have never subscribed to that opinion, since I believe I could enhance the standing of the University more by an extra year in research than a year as Rector. [...]
>
> I am convinced that a year as Rector would, next to the worries, be a year of rich experiences, which by themselves I find attractive. I also see the value of the tradition of a one-year rectorate and the regular succession based on seniority. Had I had the time for it without harming other interests I would gladly have accepted.'

The University leadership, maybe in part in response tot Oort's arguments, decided to abolish the old system and Oort would never assume the office of Rector Magnificus.

The Casimir Commission had an ad hoc nature and Oort favored its continuation on a structural basis. This did not happen, but it continued in another form. In a new law on higher education in 1960, following up on a suggestions that had been made since 1945, a body was instituted, the Academic Council, to advise the Minister on all matters concerning the universities and academic affairs, and for the universities for mutual consultations of the universities, and coordination of their efforts. It was organized into sections and Oort was asked for the section on exact sciences. He declined, but briefly was interim chairman until a chairperson could be found. A main issue in the first years was to justify how many staff were needed in each discipline and university in view of what the number of students would be, etc. After four years Oort decided it took too much of his time. He resigned by letter of July 6, 1964 to the chairman of the Council, quoting the pressure on his time by the Directorship of the Observatory and international duties, which left him too little time for research.

> 'This is of course an inefficient arrangement of my life, since I am in the first place a researcher and it is in research that I am best.
>
> I feel very responsible for the general development of scientific research in the Netherlands. That is why I accepted the membership of the Commission Exact Science which is important in my opinion. In practice I have not been able to devote sufficient time to it to make my membership worthwhile.'

Oort had tried during the 1960s to free more time for research by resigning from international bodies. He presided over ESO Council until the end of 1965. However, he remained as chair of the Board of the Foundation SRZM, undoubtedly because he wanted to see the Westerbork Telescope construction to a good end. And as Director of Leiden Observatory he took an active part in the founding of the European journal 'Astronomy & Astrophysics' (A&A), which was a merging of European national astronomical journals, some major and some with more modest influence, including the 'Bulletin of the Astronomical Institutes of the Netherlands' (BAN). The UK 'Monthly Notices of the Royal Astronomical Society' did not join this effort. The journal's first issue was published in 1969. Oort remained a member of the Board of Directors until 1972.

After Oort's formal retirement the universities were restructured in response to the student protests of 1968 demanding more democracy in the decision process. The Directorship of Leiden Observatory was replaced by a council with representation from all workers, which was chaired by Henk van de Hulst. At the same time, Oort resigned as chair of the SRZM now that the WSRT had come into operation, and again the position went to van de Hulst. However, Oort stayed on as an honorary member of the Program Committee (PC), selecting observing programs from proposals for the WSRT.

Oort had been involved in many advisory commissions outside astronomy or the university. For example he spent some effort on the National Museum Boerhaave for the history of the natural sciences in Leiden, for which he was a member of the Board of Advisers between 1966 and 1969. He seemed to have greatly enjoyed his chairmanship of the Board of the Leiden Academic Arts Center (LAK) between 1946 and 1951, which organized concerts and exhibitions. Oort and his wife were much involved in the visual arts section. Their contacts with Hendricus Bremmer would have stimulated this, while it has probably helped organize an exhibition *From Fantin Latour to Picasso* in 1950, where many paintings from the collection of the Kröller-Müller Museum were displayed in the Museum de Lakenhal in Leiden. In the framework of the LAK, Oort put much effort in organizing an exhibition reviewing the works of prominent Leiden painter Jan Josephszoon van Goyen (1596–1656), but to his great disappointment this failed due to lack of cooperation from the museum.

A less successful period was Oort's membership of the Council of Curators of the Royal Netherlands Meteorological Institute (KNMI). The file in the Oort Archives contains quite a number of letters excusing him from attending meetings, starting in January 1959. The formal appointment in the Council by Royal Decree of Queen Juliana is dated December 1961. Oort resigned from the Council again by Royal decree in August of 1968, proposing Henk van de Hulst as his successor.

13.6 Learned Societies and the Gouden Ganzenveer

Appendix A.5 lists Oort's honors, in his own handwriting. A complete list has been published earlier in the *JKM-Inventory* [1] by Jet Katgert-Merkelijn (item 333 starting on page 149). He was elected to 22 academies or learned societies from countries all over the world. The two academies in which Oort was most active, are the Dutch Royal Academy (Koninklijke Nederlandse Academie van Wetenschappen, KNAW) and the Pontifical Academy of Sciences (Pontifica Academia Scientarium).

At the KNAW, of which he became a member in 1937, he very regularly reported on the recent astronomical developments he was involved in, actually giving presentations more often than the rotation roster for members scheduled him. From his full list of publications (again in the *JKM-Inventory* [1], appendix, starting on page 161) the ones that were published by the KNAW have been listed. No publication by Oort (nor its English translation) has been included in the KNAW Digital Library. Oort's full publication list reveals that in most years after the War there was one article of a few pages, often reporting on recent developments at Kootwijk, Dwingeloo or Westerbork. After 1972 this became less, his last such contribution dating from 1982. Oort also contributed a few obituaries of astronomers in addition to the ones in Dutch he provided for the amateur periodicals 'Hemel and Dampkring' and 'Zenit'.

Oort's association with the Pontifical Academy stemmed from the organization of the *Semaine d'Etude* on Stellar Populations in 1957. He was elected as a member in 1961. From the correspondence in the Oort Archives it is clear that he very regularly corresponded with the Academy advising on matters such as the election of new members. He often visited the annual meetings, according to the correspondence usually at the expense of the Academy and accompanied by Mrs. Oort. His opinion was highly valued. Oort was very much involved in the organization of two *Semaines d'Edude*, one on *Nuclei of Galaxies* in 1970, and one on *Astrophysical Cosmology* in 1981. Oort attended both of these and gave presentations.

I have already alluded to some of Oort's honors and prizes, in particular the Bruce Medal of the Astronomical Society of the Pacific, the Gold Medal of the Royal Astronomical Society and the Vetlesen Prize of Columbia University. An

Fig. 13.25 Oort receives the 'Gouden Ganzeveer' (Golden Quill), from the Minister of Education, Arts, and Sciences, Jozef Cals. From the Oort Archives

other interesting one he received was the 'Gouden Ganzenveer' (Golden Quill) of the Royal Dutch Association of Publishers, which he was awarded in 1960. This prize is awarded at intervals of 1–5 years, to a person or institution for their exceptional significance for the written and printed word in the Dutch language. The selection is done by a jury consisting of members from the areas of culture, politics, science and industry. I quote from the laudation at the award ceremony, which was held on October 13, 1960. It starts by pointing out the prominent place astronomy has enjoyed internationally which is out of proportion to the size of the country, and in spite of the atmospheric conditions. This is exemplified by men like Kapteyn, Julius, de Sitter and Pannekoek (Fig. 13.25).

> 'Among the living Dutch astronomers [...] Oort undoubtedly ranks first. [...] Much of his early research was conducted related to the distribution of stars in our Galaxy and their velocities, through which the rotation became evident. Later he turned to other spiral galaxies. At an early date a general understanding of the dynamics of galaxies resulted from his studies.
>
> After World War II the development of radio astronomy provided a new tool, that made it possible to explore the whole Galaxy, which due to absorbing matter was not open to optical methods.
>
> By promoting radio astronomical research in the Netherlands through the construction of the Dwingeloo Radio Telescope, Oort, in collaboration with van de Hulst, has revealed and investigated the spiral structure in our own Galaxy. His work has received high international acclaim. [There then is a partial summing up of memberships in learned societies, honorary doc-

Fig. 13.26 Oort and Mrs. Oort at the Boekenbal talking to Queen Juliana. In the background Harry Mulisch, who later wrote *The Discovery of Heaven* (see quote at opening of chapter 13). From the Oort Archives

torates, roles in the IAU, ESO and the Benelux Cross.] Not only through his own publications, but also through the researches of his associates, Oort has contributed to the fame of Dutch astronomy. Many great observatories abroad are being directed by Dutchmen, and most of these have been influenced by Oort's views during their studies.

In addition to his many duties, Oort always paid much attention to making the newest results accessible to a wider audience. Examples of his efforts are his more popular presentations in the form of books and review articles in periodicals such as 'Hemel en Dampkring' and the 'Nederlands Tijdschrift voor Natuurkunde'.

All these facts make Prof. Oort the person, who according to the jury is the most qualified to be awarded the Gouden Ganzenveer.'

Actually, in addition to regularly writing popular articles Oort gave many popular talks and lectures. In the Archives notes or texts for some eighty of these between the mid-1920s and mid-19890s are present and there probably have been many more. Nowadays the Gouden Ganzenveer would never have been awarded for popularization of science.

Since 1932 there has been annually a week in which books and literature in general are promoted in the Netherlands. During this so-called boekenweek anyone who spends more than a given amount on the purchase of Dutch-language books is given a free copy of a novella that has been specially written for the occasion. Since 1947, it has been preceded by a gala party, the 'Boekenbal', held in the Concertgebouw, the famous concert hall in Amsterdam. Oort and Mrs. Oort were invited to attend; Fig. 13.26 shows them talking to Queen Juliana. The picture in the Oort Archives gives the year as 1960; however, the Boekenweek (and the Boekenbal) is always held in March and if the invitation of Oort would be related to him receiving the Gouden Ganzenveer, this is most likely 1961.

Fig. 14.0 Painting of Oort presented to him on the occasion of this retirement in 1970 by colleagues, friends, and students. It was painted by Raoul Martinez (1876–1974), a painter of French-Cuban descent, who lived for a long time in the Netherlands. From the collection of the Academy Building, University of Leiden, see [208]

Chapter 14
Westerbork, Retirement and Beyond

Near Westerbork a gigantic Synthesis Radio Telescope was being built, consisting of twelve dishes, [...] It was going to be the biggest instrument in the world and he expected a lot of it.

Harry Mulisch

I will never be an old man. To me, old age is always 15 years older than I am.

Francis Bacon (1561–1626)

In the after-dinner speech at the IAU Symposium 74 on *Radio astronomy and cosmology* held in Cambridge, UK, in August 1976, Oort told the following story (from my personal recollection): A traveler came through a village, where he noted an old man sitting in front of his house. 'Have you lived here all your life', the traveler asked him. The old man replied: 'Not yet.' Oort remained active in research after his retirement on September 1, 1970 (the start of the academic year succeeding his seventieth birthday) and his attendance of this symposium showed his continued interest in the large-scale structure of the Universe.

The year 1970, and especially the month of April, was a very busy time. On April 13–18 a *Semaine d'Étude* was held at the Vatican on *Nuclei of Galaxies*. And on April 20–22 a symposium was held on *Radio sources and the Universe* at the Kapteyn Observatory in Roden near Groningen, to celebrate his 70th birthday and the completion of the WSRT. During this symposium Oort gave a special address in the main auditorium of the University of Groningen (Oort 1970h), the place where in 1926 he had defended his PhD thesis. And then his actual birthday, April 28, was celebrated with a big party at the Leiden Observatory. In June the actual inauguration of the new radio telescope at Westerbork followed. Oort's valedictory lecture took

In: *The discovery of Heaven.*

P. C. van der Kruit, *Jan Hendrik Oort*, Astrophysics and Space Science Library 459, https://doi.org/10.1007/978-3-030-17801-7_14

place in Leiden on September 18. And during this time the Oorts had to leave their house at the Observatory and move to Oegstgeest. But I need to turn first to the construction of the Westerbork Synthesis Radio Telescope.

14.1 The Westerbork Synthesis Radio Telescope

Again I refer for more details on the subject of this section to *EarlyRadio* [9] by Astrid Elbers and Ernst Raimond's book on 25 years of WSRT [325].[1] I pick up the story with the third design of the Benelux Cross Antenna Project (Sect. 12.5) from which Belgium formally withdrew in 1967. This BCAP third design had changed from a cross antenna to a single east-west array of telescopes employing the Earth-rotation synthesis technique. This third design envisaged the construction of 28 steerable telescopes arranged east-west over a length of 1.5 km with an additional set of 6 dishes to the east (or the west) that would give a total 'baseline' of 3 km. This would result in a resolution of order 10 arcsec at 21 cm. Initially a separate 65 m dish was included but this gradually disappeared from the discussions. The cost was projected at Dfl. 30 million. After Belgium resigned from the BCAP project, the design was made more modest and affordable and presented by Muller in 1964. It had 9 (a little later 10) dishes of 25 m diameter on an east-west baseline that was to be 1200 m long with an additional dish on a railtrack of 300 m length. The costs would be about half of what had been projected for the third design and since the Netherlands had committed to half the budget, the project could go ahead and it went to tender in 1965. Later that year it was decided to add a 12th movable dish on the same railtrack, since the contract and budget allowed this addition (in 1975 two more dishes were added on another railtrack 1.5 km to the east, so that it became a 3 km array).

The mid-1960s were marked by a short period of poor relations between Bannier and Oort. It took place around New Year 1966. The construction of the telescope had finally begun, but Oort was worrying about possible significant delays (which actually occurred). He was concerned that other countries were moving ahead faster and was in a hurry. In her *EarlyRadio* [9], Astrid Elbers drew some far-reaching conclusions about Oort's character. The letters in the following paragraphs were examined by Elbers in the archives in NWO (successor of ZWO) offices in Den Haag. Unfortunately NWO has since destroyed its older records, so this policy does not allow re-examination. These letters are also absent from the Oort Archives and that of SRZM in Dwingeloo. Oort possibly felt embarrassed by them and destroyed them (Fig. 14.1).

On December 8, 1965 Bannier had written to Oort complaining about his lack of attention to formal deadlines for budget requests. Oort, being under much pressure, had become a bit lax with this, but always got away with it. Bannier threatened to no longer consider requests that arrived after the deadline. Oort wrote back a week later (December 15) explaining the reason for the delay, but also he noted that he did

[1] While the present book was nearing completion a commemorative volume was published celebrating the approaching 50th anniversary [367].

Fig. 14.1 The Westerbork Synthesis Radio Telescope, photographed from the west, not long after its inauguration in 1970. The building on the left is the construction hall for the antenna's. The operations building is located halfway along the array, hidden in the forest to the right. The telescope consisted of ten dishes of 25 m diameter and two movable ones on a railtrack at the top, for a total length of 1.5 km. Later two more dishes along the same line beyond the top of this picture to make it a 3 km array. At the top left the remains can be seen of the Westerbork concentration camp, at that that time still occupied by Moluccans. From the Archives of Leiden Observatory

not like the tone of the letter and was not prepared to be treated 'like a school-boy'. Bannier also was irritated about the way Oort ran the SRZM Board (letter of Bannier to Oort of January 4, 1966), inviting an engineer to the meeting or without prior notice putting a case before the board of paying an astronomer's trip to the USA for an observing session associated with a Dwingeloo project. Both of these seem perfectly normal to me, and I might have acted similarly when I was chair of the same Board. But Elbers concluded that Oort was authoritarian, too much used to get things his way, and only acting along lines of his own interests. To strengthen her point, Elbers also took the two cases, where Oort directly approached the Prime Minister, ignoring the official channels. The one case was in 1945, before ZWO existed, when Oort approached Schermerhorn about a radio telescope (see Sect. 10.6) and the second had to do with the Casimir Committee (see Sect. 13.5). Neither of these involved ZWO (and were completely acceptable in my opinion), although Elbers claimed he made it 'a habit' of 'not going through ZWO'. Oort might have felt that dealing with the Minister directly would speed things up; however Bannier correctly stated in another letter to Oort[2] that ZWO's involvement had in fact expedited the progress.

Oort did indeed determine the scientific directions almost by himself, but at that time there really wasn't anybody else to provide this leadership. I have heard no complaints from astronomers that worked with him at the time that 'Oort's idiosyncratic way of arranging things started to annoy his colleagues more and more' (quote from Elbers, who doesn't remember whose remarks she based this on). Hugo van Woerden told me in an interview, that Oort made him (as a student) responsible for the scheduling of the 21 cm observations in Dwingeloo and later readily agreed to his proposal to have a program committee of peers to select the priorities among proposals for WSRT observations. Colleagues I talked to who knew Oort well, invariably describe him as absolutely not authoritarian. Harry van der Laan, who joined the Leiden Faculty in 1966, told me he was 'definitely not authoritarian, but he was tenacious'. Hugo van Woerden characterized Oort as 'charming and friendly. Humble is a word that definitely applies to him'. Oort has been described to me as soft-spoken, unassuming, helpful, caring, understanding and open to opinions and ideas of others, but certainly not stubborn, arrogant, dictatorial, dominant or idiosyncratic. It is true he demanded dedication and could be impatient. My sources agree with me that the conflict with Bannier must have been an isolated incident, the latter as a bureaucrat feeling he had to insist on at least some strictness. Van Woerden also mentioned that as secretary of the SRZM Board between 1965 and 1970 (and at Board meetings earlier he attended as program leader of the Dwingeloo 21 cm observations) he observed Oort's and Bannier's relationship to be 'always warm and amicable' (see also Richard Strom's review of *EarlyRadio* [9] on this issue [368]). I agree with these descriptions of Oort's character from my own experience. It must be said however that although Oort probably was not an authoritarian man, he maybe not always appreciated how much authority his opinion carried. And there have also been cases, where he could have supported staff working for him more than he did, such as in the cases of Gijsbert van Herk and André Muller (see Sects. 10.4 and 12.2).

Humbleness should not be confused with modesty. Bannier wrote in the *LibAm80* [2]: 'A friendly word here and a small prod there, together with a superb feeling for finding the right people, able, willing and persevering enough to do the spade work, have enabled him not only to reach these goals but also to make lasting friendships in the process.' And: 'I simply do not hold with modesty. If a man [...] knows that he has a sound idea [...] he should say so openly and without restraint. If Jan Oort had been modest, [...], i.e. if he would not really have known what he wanted, then he would have kept his mouth shut. And astronomy and the Netherlands would have been the poorer for that.'

As in the case of the Dwingeloo radio telescope, there was the problem of selecting the location. It was not easy to find a piece of uncultivated land, 1.5 km long and with the possibility to establish a car-free zone of a few km around this to avoid

[2]Elbers attributed this to a letter of Oort to Bannier of December 15, 1965; her footnote 139 (171 in the PhD thesis version). Elbers agrees it actually was from Bannier to Oort on December 8.

Fig. 14.2 Element dishes of the Westerbork Synthesis Radio Telescope. This photograph is taken halfway along the array looking east. Courtesy of ASTRON, Dwingeloo

electrical interference from ignition spark plugs. Next to the Dwingeloo dish was impossible in view of the conservation of the Kraloër and Dwingeloosche Heide nature reserve. Most other parts of the country were too densily populated. The site that finally appeared was in the province of Drenthe some 20 km to the north of Dwingeloo. It was a forestry (the Boswachterij Hooghalen), in which during World War II a concentration camp had been located where Dutch Jews and some others were interned before being transported to the concentration camps of Auschwitz and others. The camp was named after the nearby (10 km) village of Westerbork rather than Hooghalen. After the War it had been used as a place for internment of collaborators with the German occupation, but that had ended after a few years. The forest also offered the possibility to extend the array to the east by another 1.5 km, which indeed happened a few years after the telescope became operational (Fig. 14.2).

At that time the camp was put to use as a temporary housing facility for a large group of 'Ambonese', who had to flee Indonesia (having been sympathetic to the Dutch colonialists) after its independence in 1949. In 1950 these inhabitants of the South Moluccas, a group of islands in the east of Indonesia (between Sulawesi and New Guinea), had declared an independent 'Republic of South Molukka' (RMS from Republik Maluku Selatan). This happened on one of the islands, Ambon, so that name became associated with it as well. It was quickly occupied by force by the Indonesian government. Many men had served in the Royal Dutch Indonesian Army (acronym KNIL from the Dutch words) and were transported to the Netherlands with their families (a total of ca. 12,000 persons) temporarily—until they could return to

Fig. 14.3 Some of the element dishes of the Westerbork Synthesis Radio Telescope seen from the eastern side with the first two movable dishes on a railtrack. Courtesy by ASTRON, Dwingeloo

the RMS, which of course never happened. Now the Dutch government had put them up in the remaining buildings of the Camp Westerbork (and other former camps such as in Vught in the south of the country). Some 1200 still lived there when the camp area was designated as the site of the new radio telescope. Oort had to use all his government and other bureaucratic contacts and all his diplomatic skills to have the authorities put pressure on them to move. Relocating them fitted the Dutch policy of discouraging hope to return to a RMS in the South Moluccas and trying to integrate the Moluccans with Dutch society. In the end the project was not affected.

The new telescope became known as the Westerbork Radio Synthesis Telescope, acronym WSRT. This was an issue since the name Westerbork was associated with the concentration camp, but common sense prevailed over sentiment. A large monument and memorial center has been constructed at the site of the camp (Fig. 14.3).

The construction of the telescope was not without problems and delays, and it must have put a heavy burden on Oort in his final years before retirement in 1970. The 25 m dishes had to be put exactly on an east-west line and at exactly the same relative distance from each other (144 m). Gijsbert van Herk was charged to perform observations of the stars to determine this direction. This turned out to be less trivial then expected due to atmospheric conditions; I remember van Herk mentioning at the time that he was never happy with the accuracy. Then there was the construction of the dishes, each of which had to be identical. For this part of the project concerning the mechanical and electrical aspects, Ben Hooghoudt was appointed project manager. For the design and testing of the dishes (like the Dwingeloo radio telescope built by Werkspoor), Hooghoudt and engineer Jacob Wilhelm Martin (Jaap) Baars (b. 1937) oversaw the process and tested the results. The construction phase is very well described in a paper by these two [369]. The project manager for the electronics was Lex Muller; the electronics were partly developed and built in Dwingeloo, where the laboratory and workshop were expanded, and partly at Leiden Observatory. The latter group worked under the leadership of Jean Luc Casse (b. 1934), an engineer from Belgium who would stay with the radio observatory in his further career and would be instrumental in developing new electronics and receivers for the WSRT. This Leiden group moved to Dwingeloo to be integrated with the rest of the observatory staff during the 1970s, on the instigation of Harry van der Laan, who first became member of the SRZM Board and then 'delegated board member' (exercising in fact the role of chairman), after Oort had retired and Henk van de Hulst had nominally taken up that position. A difficult decision Oort still had to take is to relieve Lex Muller from his position with the Foundation; he, as Harry van der Laan has documented to me, and his junior staff had developed poor relations and Harry, at that time a new member of the Board of SRZM, had advised Oort to let him go. Oort agreed, in spite of a very good friendship between him and Muller, arranging an appointment as professor for him at the Technical University Twente at Enschede.

And, not trivial at all, was the fact that there had to be ways of using the combined interferometric recordings to produce maps of intensity on the sky. This was a major effort in computer programming and development of reduction techniques and routines, performed in Leiden almost single-handedly by Wim Brouw, still an astronomy graduate student at the time. And throughout the process Ernst Raimond, who had been associated with BCAP almost from the beginning, played a vital role in coordinating, providing the online computer control of the telescope and receivers, and contributing to the computer reduction software. Raimond became the first Director of the WSRT. Of course very many more people contributed, but those just mentioned were probably the most crucial ones.

The construction of the mechanical structure was completed in late 1968. Test observations started in 1969 and lasted into 1970. The total expenditure on the telescope depends a bit on what is counted as investment, but was somewhat over Dfl. twenty million (€ 25 million in current purchasing power). At the time the design was made the budget was half of that of the BCAP third design, so half of Dfl. 30 million. The support of ZWO and Bannier had been vital. The inauguration took place on June 24, 1970, as it happened only months before Oort's formal retirement.

14.2 Retirement

As mentioned above, the first part of 1970 was an extremely busy time for Oort. On his birthday (April 28) a big party had been given at Leiden Observatory. There was classical music played by members of the Observatory, followed by drinks, dancing and conversation, and later in the evening an entertaining performance by Jaap Tinbergen.

Oort was a member of the Scientific Organizing Committee of the 1970 *Semaine d'Etude* on *Nuclei of Galaxies*. The plans for this had been made during the General Assembly of the IAU in Prague in 1967 by Daniel O'Connell of the Vatican Observatory and Martin Ryle, Allan Sandage and William Morgan. It was held April 13 to

Fig. 14.4 Maarten Schmidt, Lo Woltjer, and Oort on their way to an audience with the Pope during the *Semaine d'Étude* in 1970. In the background Martin Francis McCarthy S.J. (1923–2010) of the Vatican Observatory. From the Oort Archives

18, 1970 at the Vatican. Oort was to review the centers and nuclei of both our Galaxy and the Andromeda Nebula (Oort 1971b). In this he discussed the similarity in the distribution of near-infrared radiation from the central parts of both galaxies, which was probably a tracer of stars and from which a dense central core was evident. He also reviewed the data on the expanding motions in the central region, presenting my as yet still unpublished calculations of gas clouds expelled by the nucleus giving rise to the expanding arms. The published proceedings are available on ADS, [370] (Fig. 14.4).

Margaret Burbidge reviewed observations and her first four conclusions were:

'1. It has become increasingly clear that a large fraction of galactic nuclei have various degrees and forms of activity.
2. Matter is ejected from active nuclei and QSO's.
3. There is strong evidence that massive clouds of gas, relativistic particles and perhaps coherent objects are ejected.
4. Very high star densities exist in the nucleus of M31 and probably in the nucleus of our Galaxy.'

QSO's, quasi-stellar objects, now called quasars, are stellar objects at enormous distances related to very energetic activity in nuclei of galaxies, probably driven by black holes. They were discovered in 1963 by Maarten Schmidt. The first one discovered has a redshift of 0.158, meaning that spectral lines have a wavelength of 1.158 times those in the laboratory. At the time of writing quasars are known to redshifts up to 6 or so. Burbidge's conclusions certainly reflected the thinking of the time and also Oort's opinions. The evidence for ejection of matter from nuclei of galaxies like our own is no longer considered so strong and in this respect Oort does not seem to have had the intuition he has displayed in other matters. The meeting did not have the enormous impact the *Semaine d'Étude* on Stellar Populations had (Fig. 14.5).

Fig. 14.5 Oort and Bengt Strömgren meet Pope Paul VI (Giovanni Battista Enrico Antonio Maria Montini, 1897–1978) during the 1970 *Semaine d'Étude*. From the Oort Archives

The symposium *Radio sources and the Universe* on the occasion of both Oort's seventieth birthday and the inauguration of the WSRT, was held at the Kapteyn Observatory of Groningen University in Roden from April 20–22, 1970, immediately following the Rome Study Week. The topics were very similar to what had been discussed at the Vatican and in the end no formal proceedings were published. Part of the program was a formal address by Oort in the main auditorium of the Academy Building of Groningen University, the actual place where he had defended his PhD thesis in 1926 (Oort 1970h). He reviewed his ideas of the formation of galaxies and the origin of their rotation, the explosive events in nuclei and the formation of large scale structure in the Universe. This address was more speculative than usual. He apologized for this at the beginning, recalling first that he was standing very near Kapteyn's Astronomical Laboratory, which brought back many memories. The laboratory indeed stood very near the Academy Building, but was demolished in 1988 after a fire (see Sect. 2.2). His notes used for his introduction are in his Archives; in these he reminded the audience that Kapteyn

> 'always impressed above all on his students, the need to 'always look through things', as he expressed it. His pupils and his pupils' pupils have all to a greater or lesser extent been indoctrinated with this attitude. While thinking about this I began to wonder what Kapteyn would have said if he could have listened to the speculative aspects of the lecture I have prepared for today. But he was an extremely benevolent man, and he might have forgiven his pupil for relaxing the critical attitude towards speculation on which he always put so much emphasis.'

The symposium was attended by many of Oort's colleagues, collaborators and students (see Fig. 14.6). Persons are identified in the caption by the numbers. For further persons in the photograph see [112] (Fig. 14.7).

1 2 3 4 5 6 7 8 9 10 11 12 13 14 15 16 17 18 19 20 21 22 23 24 25 26 27 28 29 30 31 32 33

34 35 36 37 38 39 40 41 42 43 44 45 46 47 48 49 50 51 52 53 54 55 56 57 58

◄**Fig. 14.6** Group photograph of the attendees of a symposium at the Kapteyn Observatory near Groningen, on the occasion of Oort's 70th birthday and the inauguration of the Westerbork Synthesis Radio Telescope. From the Oort Archives. In the first place there were persons that had been very much involved (as astronomers or engineers) in realizing the Westerbork Telescope, both then and earlier on (with the numbers in the figure): Chris Christiansen (1), Ernst Raimond (4), Wim Brouw (6), Ben Hooghoudt (10), Jean Casse (16), Johannes Petrus Hamaker (b. 1937) (26), Bill Erickson (36), Kelvin J. Wellington (40), Lex Muller (41), Jaap Baars (42), Charles Seeger (45), and Jan Högbom (50). Old time (and more recent) collaborators and colleagues included Wallace Leslie William Sargent (1935–2012) (2), Lukas Plaut (3), Peter Scheuer (11), famous cosmologists John Archibald Wheeler (1911–2008) (13) and Martin John Rees (b. 1942) (18), Willie Fowler (22), Henk van Bueren (28), Otto Heckmann (29), Margaret and Geoffrey Burbidge (37, 49), Gart Westerhout (38), Harry van der Laan (51), Fred Hoyle (53), Lo Woltjer (55), Maarten Schmidt (56), Adriaan Blaauw (57), and Hugo van Woerden (58). Finally, students and young staff at the time, that would soon become early Westerbork users or related researchers, included Vincent Icke (8), Paul Ronald Wesselius (b. 1942) (14), Renzo Sancisi (b. 1940) (15), Ronald John (Ron) Allen (b. 1940) (27), Jet (Katgert-)Merkelijn (31), Peter Katgert (32), Rudolf le Poole (43), Arnoldus Hendrikus (Arnold) Rots (b. 1946) (47), myself (Piet van der Kruit), and Walter Joseph Jaffe (b. 1946) (52)

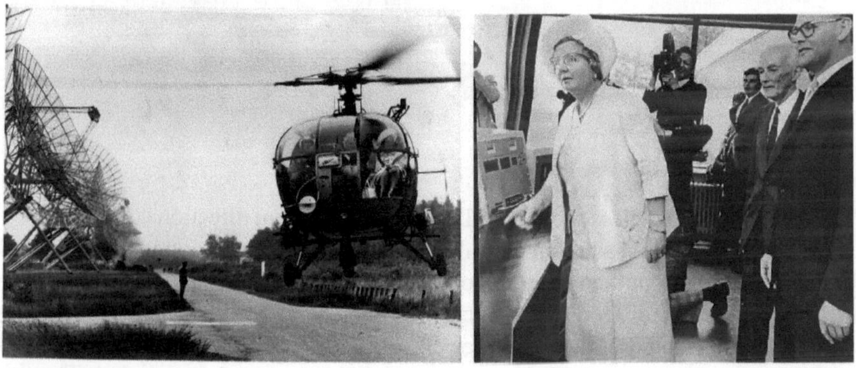

Fig. 14.7 Left: Queen Juliana arriving at the Westerbork Synthesis Radio Telescope for the inauguration on June 24, 1970. Right: The Queen in the control room. The person on the right is Ernst Raimond. From the ASTRON and Oort Archives [372]

The inauguration of the Westerbork Telescope was—just like the one at Dwingeloo—performed by Queen Juliana. The date was June 24, 1970. This was only some two months after the 25th anniversary of the liberation of the Netherlands from the Nazi German occupation and the Queen had at that date dedicated a monument in the former concentration camp not far from the radio telescope, commemorating the people that were transported to their death from here by the initial part of the rail track and 10,200 small stones, one for each of them. The Queen arrived by helicopter (see Fig. 14.7). Most guests were collected in the large construction hall (see Fig. 14.1), where the dishes were assembled. As Oort grandson Marc Oort recalls:

> 'The helicopter the Queen arrived in was so noisy that the speeches that were being made at that time were barely audible. This to the great consternation of the organizers; strangely enough this is what I—as then 12-year-old—remembered most.'

Fig. 14.8 Oort and Mrs. Oort at the ceremony of his valedictory lecture on September 8, 1970. From the Oort Archives

The elements of the telescope were all pointing at different directions on the sky, but after the Queen pushed a button they all moved to the same position in the sky. None of the persons present then that I talked to remembered what that position was.

The formal date of Oort's retirement was September 1, 1970. It was marked by a traditional 'afscheidscollege', a farewell or valedictory lecture, which took place on September 18. Figure 14.8 shows Oort and Mrs. Oort at that occasion. The title was *To the horizon of the Universe*, and it was a tour along the highlights of his career. It was published in Dutch in the magazine 'Zenit', the successor of 'Hemel and Dampkring', Oort (1970j). Although an English transalation has been published, this is not easily accessible to the English-speaking world and I have provided an independent translation in Appendix B.6. As Oort has done on many occasions he started out by recalling his student years, giving much credit to Kapteyn, whom he described as 'not only an astronomer of genius, but also an exceptional teacher'. In his *AnnRev81* [3], the autobiographical essay he wrote in 1981, Oort wrote that if he were to write a book on his life, he would call it *Looking Ahead in Wonder*. The valedictory lecture actually did look back in much detail, especially his account of the work on the Crab Nebula (to the extent that he admits he had considered to dedicate the full lecture to this object); at the end he looked forward towards solving the problems of the structure of the Universe and expressed the hope that the dreams of the time of his farewell would not turn out to be nightmares for his successors.

The lecture was followed by a surprise visit by the Minister of Education and Science, Gerard Veringa (1924–1999), who bestowed the high Royal Distinction of Commander in the Order of Orange-Nassau upon Oort.

Oort's retirement also meant he and his wife had to move out of the Director's residence at the Observatory, where they had lived since 1935 (excepting the almost three years during the war in the Potbrummel in Hulshorst). They had seen this coming, of course, and had bought a comfortable, newly-built house in Oegstgeest, the town very near Leiden where Oort had grown up (see Fig. 14.9). The house is located on a canal that connects the lake district north of Leiden ('de Kaag') to Katwijk. From numerous financial contributions from friends, colleagues, family, students, in fact astronomers worldwide, the Oorts had been given three significant presents on the occasion of his retirement. One of these was the painting that is reproduced in Fig. 14.0. The other was a sundial in the form of one of the elements of the Westerbork telescope and has been erected in the garden of the new Oort residence. It can be seen in both panels of Fig. 14.9. The third present was a small sailing boat. The Oorts had organized a housewarming party, which was held in their garden (the weather actually cooperated) and the boat arrived during this cocktail party. Oort spent a significant part of the party trying out the sailing boat in the canal. Geoffrey and Margaret Burbidge had organized the effort of collecting money for a present from the attendants of the Roden symposium and the former wrote on May 12, 1970 that 'Margaret and I certainly hope we shall have an opportunity of visiting you and sailing with you in the not too distant future.' Whether they did sail (the boat was not that large and Geoffrey not too light) is not recorded. Oort wrote a letter on July 9, intended for all contributors at the Roden symposium, thanking everyone and noting:

> 'It has taken some time of looking at different types of boats before we could decide which we liked best; we have now found the ideal: a beautiful four-meter long wooden 'zeilschouwjte'.
>
> [...] I have always been particularly attracted by ships because of the adventure of which they hold a promise, or perhaps still more the illusion that they could sail to unattainable horizons. In recent years we have done little sailing; it is a wonderful idea that now we shall soon be regularly on the lakes again.'

Oort and his wife now also had more time to spend in their second home. They had had this built in 1960 in Haamstede on the island of Schouwen-Duiveland in the province of Zeeland in the southwest of the country to replace the one they had owned at Katwijk. They also named this new home Sandy Hook.

It must have been a relief when all celebrations and changes of environment and responsibilities were over. The first results the new radio telescope at Westerbork had begun to provide captured Oort's interest and attention.

14.3 Westerbork and Nearby Galaxies

When the WSRT was being constructed and Oort's retirement approaching, the question arose who would succeed him and who would take charge of the scientific exploitation of the instrument. Formally it was not up to Oort to search for his suc-

Fig. 14.9 The house in Oegstgeest the Oorts moved to from the Director's residence at the Observatory following Oort's retirement. Bottom picture: Oort in the garden of the house with the sundial he had been given as a present for his 70th birthday. From the collection of Abraham H. Oort

cessor, but Henk van de Hulst was the natural person to lead that effort. Oort would have liked to have one of his most successful students to assume leadership of the Observatory, particular Maarten Schmidt and Lo Woltjer. and these were approached some time in 1969, most likely by van de Hulst. Adriaan Blaauw was another option. Woltjer decided he wanted to remain in New York at Columbia University, and Blaauw was expected to be appointed Director General of ESO and would at least temporarily give up his directorship of the Kapteyn Laboratory in Groningen. Van de Hulst was too busy with Space Research, for which he had started a group in the physics department (the 'Cosmic Ray Working Group'), and would not be able to concentrate on radio astronomy. Oort's preferred option was Schmidt, to whom he wrote extensively in a long letter of June 26, 1969, following up a first approach by van de Hulst. Oort mentioned Woltjer's position and that Blaauw's ESO appointment was still being negotiated, but expected to be announced soon. Work on radio sources was a primary objective for the new telescope at Westerbork, and Schmidt as discoverer of the quasars would be the natural choice to lead this. Schmidt had access to the 200 in. telescope at Palomar Observatory, but—Oort argued—Leiden would soon have a major optical telescope at ESO.

In the end Schmidt also declined; and he wrote to Oort to explain his position in a letter of August 7, 1969. The load of the Directorship and the prospect of having to give up his work at Palomar, which he described as making a real sacrifice, were the primary reasons, although house prices and the general circumstances of life in the Netherlands compared to California played a secondary, but significant, role. The actual salary and the Dutch weather hardly mattered.

Faced with this the Observatory had to look for a new possibility. Fortunately some years earlier Oort had hired a new staff member who could take over leadership of the radio astronomy program. This was Harry van der Laan, who at age 16 had emigrated with his parents to Canada ('I have three sons, so I need three farms and that is not possible in the Netherlands', his father had said), and had come to Cambridge, UK, on a Commonwealth fellowship to work on a PhD thesis in mathematics or physics. After talking to various professors, he had been attracted most to Martin Ryle and switched to astronomy, producing a thesis on the physics of radio emission from supernova remnants, and returned to Canada as the conditions of his fellowship required. But he had not remained unnoticed in the Netherlands and his background was ideal in the context of the new Westerbork facility. He actually got unsolicited offers from Amsterdam, Groningen and from Oort in Leiden. He had met Oort during the Herstmonceux conference described above in the context of Scheuer's remark on Oort's fascination of the Cambridge map of the north polar area (see Sect. 12.5). Even though the offer from Leiden concerned the most junior position of the three, he accepted that one and came to Leiden in 1967. So, in 1970 van de Hulst took over the leadership of the Observatory (although not as director due to democratic reforms), and van der Laan was expected to lead the scientific use of the WSRT.

Harry van der Laan prepared the junior staff, including PhD students, for Westerbork by organizing literature discussions on various aspects of radio astronomy. Many of us, graduate students, actually worked at Westerbork testing the telescope when it was in the final stages of construction. Oort wrote a widely-read Internal Technical Note (of the Foundation SRZM) concerning programs to be carried out at Westerbork (Oort 1968d). The first four programs all concerned radio sources in the Universe; counts of faint sources, structure and polarization in brighter sources and detailed mapping of some of the very brightest ones. The fifth category was radio emission from nearby, large spiral galaxies.

It turned out that nearby galaxies would provide the first spectacular results. One reason was that source counts take much preparation and careful analysis. Another was that during the 1960s it had become clear that the counts were saying more about the cosmological evolution of radio sources than about the geometrical structure of the Universe, the reason why Oort had become interested in these counts. In the mean time quasars (see Sect. 14.2) had been discovered and the study of violent activity in galaxies itself had become a major source of interest.

One of the earliest programs at Westerbork, in which Oort took a major interest, was carried out by myself and concerned observations of the brightest Seyfert galaxies and for comparison some bright normal spiral galaxies. Seyfert galaxies (see Sect. 13.4.3) had bright optical nuclei and large motions in them and would indicate activity as was also deduced to have happened in the nucleus of our own Galaxy. The results showed radio emission for the nuclei of the Seyfert galaxies, while the normal galaxies in general showed no central radio source or only a faint one. The publication of the paper describing these results [371], actually was held up, so that the first scientific paper from the WSRT could be one from Oort. This was a quick observation of the nearby galaxy Maffei I (Oort 1971e), an elliptical galaxy discovered in the infrared by Paolo Maffei (1926–2009) very close to the Galactic plane and obscured by interstellar dust. No emission from Maffei I was detected. My paper appeared a few months later.

An impressive result, often seen as the first major Westerbork breakthrough, was the observation of the Whirlpool galaxy M51 by Donald Seaforth Mathewson (b. 1929), myself and Wim Brouw. Mathewson spent a sabbatical year at Leiden from Mount Stromlo Observatory in Australia; he was an expert in studies of the Galactic magnetic field using polarization measurement in both the optical and the radio. The usual procedure was that around coffee time the computer plots. etc. from overnight computer runs for the Westerbork reductions were brought into the Observatory and everybody, while gathering for morning coffee, would have a look. The first map of M51 was brought in late in January 1971. It was in the form of the left hand panel of Fig. 14.10 and left a very strong impression. The radio map showed spiral structure.

The radio emission coincided with thin, prominent dust lanes at the inner edges of the spiral arms and provided a beautiful confirmation of the density wave theory. When the gas approaches the arms from the inside (see Sect. 13.4.3), the velocities are supersonic (the speed of sound in the interstellar medium is a few km/s) and strong shock compression in the gas and dust occurs. This supposedly triggers star formation, which after moving into the arms on a timescale of ten million years or so, gives rise to the prominent HII regions that delineate these arms. Moreover, the magnetic field (and the relativistic particles) are compressed along with the interstellar gas, giving rise to enhanced synchrotron emission. Actually, hardly any polarization was detected.

There were other noteworthy results too that fascinated Oort. One was the detection of X-ray sources, which were one type of objects Oort had proposed would gain insight from Westerbork observations. These sources result from mass exchange in binary systems, in which one component is a white dwarf, neutron star or black hole. Observations by Luc Braes and by the Irish postdoc (and later staff member in

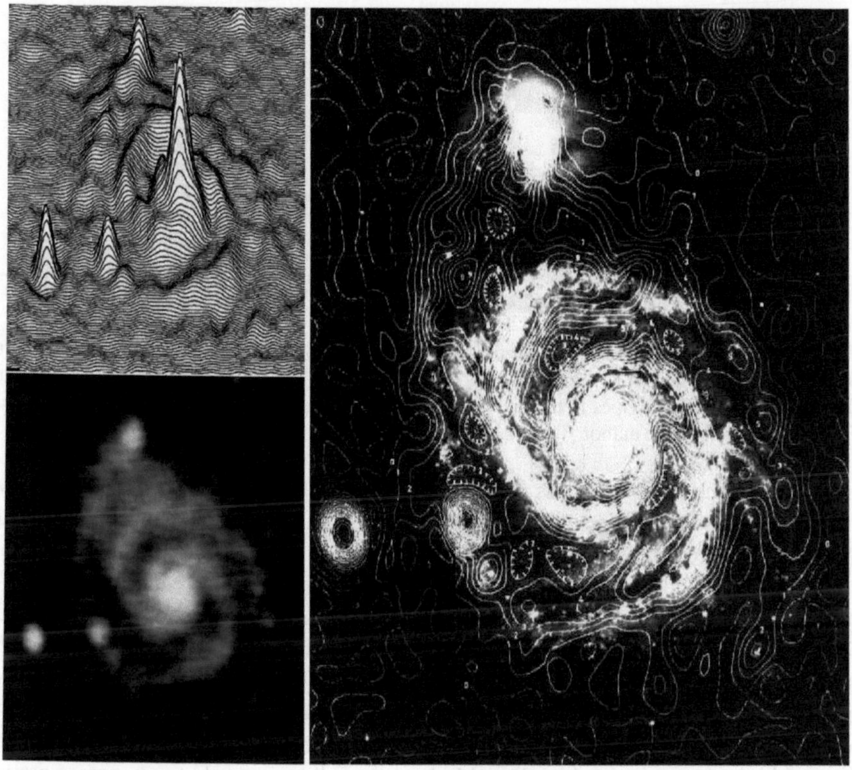

Fig. 14.10 Westerbork observations of the continuum radiation of the Whirlpool galaxy M51. On the top-left a so-called ruled surface plot and below that an intensity representation produced by Walter Jaffe on a cathode-ray tube. On the right brightness contours superposed onto an optical image [373]. From the archives of the author

Leiden) George Kildare Miley (b. 1942) showed strong radio emission in addition to the X-rays. Another noteworthy result that excited Oort, was the study of radio sources in the Perseus Cluster by George Miley, postdoc Giuseppe Cesare Perola (b. 1943), myself and Harry van der Laan. The Perseus Cluster is a rich cluster of galaxies at a distance of about 75 Mpc. Galaxies that are strong radio sources leave behind trails of radio emission as they move through the intra-cluster gas, giving the appearance of 'head-tail' sources. Another strong source (3C 129) in another cluster even showed a 400 kpc long trail (see Fig. 14.11).

After his retirement Oort had to give up his impressive office at the Observatory and was moved to a smaller office (a corner room on the first floor of his former home) comparable to other staff members. Don Mathewson was given the office and he and I spent many hours there working on the M51 observations and further galaxies we observed. Oort kept his collection of bound journal volumes there and often (almost daily in my recollection) knocked on the door, coming to check some article in a journal. This probably was often a pretense to have a look at our results. He

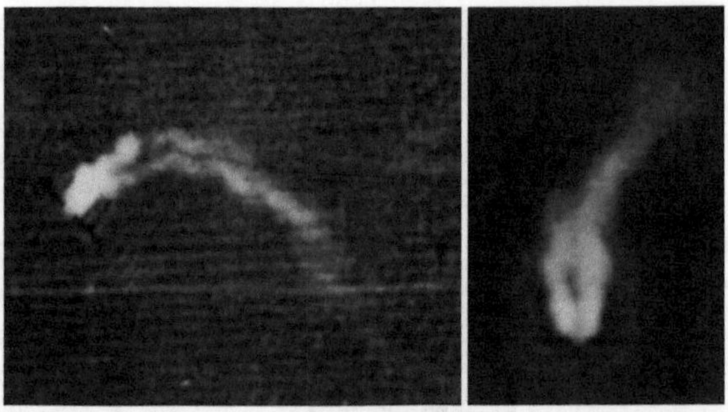

Fig. 14.11 Westerbork observations of the continuum radiation of 'head-tail' sources. On the left 3C 129 and on the right NGC 1265 in the Perseus Cluster [374]. The optical galaxies are located at the bright 'heads'. The stripes in the picture on the left are imperfections in the cathode-ray tube. From the archives of the author

then spent some time to discuss matters with us, which of course we welcomed very much. The spiral structure of M51 impressed him, but he was even more fascinated when we received the observation of NGC 4258 (see Fig. 14.12).

This galaxy had 'anomalous' arms that curved in the same direction as the 'normal' spiral arms, but seemed a completely independent phenomenon. I soon found from a literature search that the arms had also been seen in the light of ionized hydrogen (Hα). When he first saw these data Oort collected from his room all journal volumes that contained articles on this galaxy before he went off for the weekend to his vacation house in Haamstede. After returning he asked with some hesitation if he could join the project, which Don Mathewson and I were only too happy to agree to. In the end it resulted in the paper by van der Kruit, Oort and Mathewson (1972b). We came to the conclusion that the anomalous arms probably were the result of an enormous explosion in the nucleus, now not at an angle with the plane but *in* it, so that these expelled gas clouds plowed their way through the disk. The clouds followed curved trajectories as they swept up material already present, thus gaining angular momentum. The very sharp outer edges of the anomalous arms were the shockfronts while this still was happening. Following Oort's remark 'any mechanism for formation of a rotating galaxy is likely to produce a system of large scale initial asymmetry in the plane of rotation, which will almost automatically develop into a two-armed spiral structure' at the 1970 Basel meeting (Oort 1970b; see Sect. 13.4.3), we speculated that this was a way to form spiral structure.

Spectra had taken a few years later with the 200 in. telescope while I was a Carnegie Fellow, together with some spectra Maarten Schmidt had put at my disposal, I confirmed that the anomalous arms moved more slowly than the disk itself and that the emission lines indicated ionization in shocks. In 1978 Geert Dick van Albada (b. 1961) produced a thesis based on 21 cm line observations at Westerbork of NGC 4258 (see Appendix A.4). Dick is a son of Gale Bruno van Albada (1912–1972) and Elsa van Dien (1914–2007), both astronomers that had worked for many years at the Bosscha Observatory at Lembang and later at the University of Amsterdam. Dick van Albada found evidence for an effect of a bar in the system. He concluded that in general his data strengthened the ejection hypothesis. Somewhat disappointed, Oort

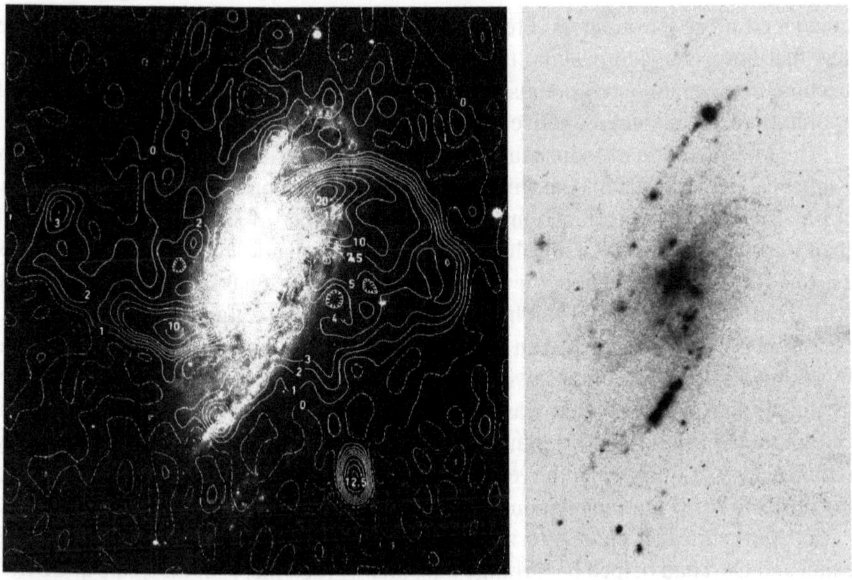

Fig. 14.12 NGC4258. On the left an optical image with the radio continuum contours superimposed and on the right a photograph taken in the light of ionized hydrogen (Hα). From the collection of the author

(1978j) remarked at a meeting on quasars and active nuclei in Copenhagen: 'Observations of neutral hydrogen have given interesting new information, but have not yet led to a satisfactory model.'

More recent models link the nuclear activity to jets of relativistic material that emerge from the nucleus, which in turn may drive gas motions. The anomalous arms are also seen in X-rays (see [375] for more discussion, also on alternative models).

14.4 Other Scientific Research

Oort kept coming to his office at the Observatory quite regularly, biking from Oegstgeest to Leiden. He still wrote many letters and his longtime secretary, Dini J. Ondei-Beneker (~1940–2008) (see Fig. 13.11) assisted him for many years with support from the Observatory Fund. She deserves enormous credit for preserving the Oort Archives.

The determination of counts of faint radio sources had been the most prominent argument to build the WSRT; in the years before it became operational, Oort invariably listed this as the first item. Yet, as we have seen, his involvement in the early Westerbork years concerned radio emission from nearby, normal galaxies. The field of cosmology and radio sources had changed rapidly. There was no widespread support for the Steady-State Theory of the Universe (see Sect. 12.4); the argument of Ryle and others based on radio source counts, the discovery of the Cosmological Microwave Background and the Big-Bang Theory of the origin of the Universe had convinced most astronomers. Fred Hoyle and Geoffrey Burbidge were among the few that never stopped pushing the steady state. As mentioned already, it had also become clear that source counts were a powerful tool to study the cosmological evolution of radio sources, active galactic nuclei and quasars.

The interpretation of radio source counts bears some resemblance to the methods Kapteyn and Oort used to determine the spatial distribution of stars from star counts. It involves a luminosity function, which is the number of sources as a function of their intrinsic luminosity or absolute magnitude. In Galactic studies this function was assumed constant throughout the system, in cosmology it has to be assumed to vary with distance, or redshift. The PhD thesis Jet Merkelijn (Katgert) produced in 1970 (see Appendix A.4) concerned an early determination of that function. This was done by taking a sample of sources over a large part of the sky (the Universe was always assumed isotropic, so that it was the same in all directions) complete down to a certain flux density (radio brightness) limit. These then had to be identified optically with a galaxy or quasar and the redshift had to be measured, so that the distance and the intrinsic radio luminosity could be derived.

Jet Merkelijn had chosen to spend some time in Australia after her Doctoraal exam and Oort had helped her arrange this. She worked there for 2.5 years at CSIRO in Sydney with John Bolton. The work involved source counts obtained with the 64 m Parkes Radio Telescope in Australia and optical identifications, which allowed measurements of the redshift and distance. In cases where no redshift measurement is available one can estimate this from the apparent magnitude of the galaxy assuming these to be in a small range in absolute magnitude (unidentified would mean very large distance). Merkelijn compared the luminosity function at two redshifts, 0.10 and 0.17 (one tenth and one sixth of the age of the Universe back in time). She determined the luminosity function using two radio frequencies, relatively low at 400 MHz and relatively high at 2.7 GHz. The result was that at least the more powerful source population has to evolve significantly with cosmic epoch.

Producing source counts at Westerbork was a difficult observation, requiring much care in calibration and date reduction. Quite some undertaking at Westerbork was dedicated to this, culminating in a few PhD theses, of which the first one was in 1977 by Peter Katgert and eventually in 1987 by Oort's grandson Marc Oort (for details see Appendix A.4). Oort must have been following these developments with much interest; he discussed them sometimes in presentations before the Royal Academy, and in other popular reviews and contributions. At the same time quite some work with the WSRT was undertaken in the area of the detailed structure of radio sources and quasars. This also captured Oort's interest, for example he co-authored an article on giant radio galaxies in 'Scientific American', Strom, Miley, and Oort (1975k), but did not appear as co-author on any refereed papers on these subjects.

Early in the 1970s Oort returned to some old work that had remained unfinished and unpublished. Gerrit Pels had been working on the Hyades star cluster. This cluster covers an area of sky centered on the constellation Taurus. The brightest star, (α Tauri), Aldebaran, the red eye of the bull, does not belong to the Hyades, but most of the other brighter stars in the Taurus constellation do. Because of its proximity the Hyades are not easily recognized as a star cluster simply by looking at sky images.

Fig. 14.13 Oort during a television recording in Groningen at the time of the celebration of the centenary of Kapteyn's professorship at Groningen (1978). The painting is the well-known one by Jan Veth, which resides in the Kapteyn Room in the Kapteyn Astronomical Institute where the photograph was taken (see Fig. 3.7). The table contains a box with plates for the Cape Photographic Durchmusterung and the plate holder used by Kapteyn to measure star positions, which reside in the Groningen University Museum. From the Oort Archives

However, this has not prevented astronomers starting from the ancient Greeks to list it as a cluster (along with the Pleiades). In Greek mythology, the Hyades were the daughters of Atlas and sisters to the Pleiades. Its proximity (about 48 parsecs, roughly 150 light years) makes it a stepping stone for studies of stellar distances and beyond. Henk van Bueren had studied the cluster before; actually it had been the subject of extensive studies by Kapteyn and his student Herman Albertus Weersma (1877–1961), as described in detail in my *JCKbiog* [7]. Pels had completed a proper motion survey in 1962 and extended the known membership, but then had fallen ill and died in 1966. Together with Pels' widow, Heleen Kluyver, Oort set out to finish this work, which resulted in a detailed paper (Pels, Oort, and Pels-Kluyver, 1975h). The structure of the cluster (by them assumed to be at 40 pc) showed clear 'mass segregation', the brighter, more massive stars being strikingly more concentrated within 3 pc, but outside 4 pc the distributions for different masses were similar. Stars beyond 10 pc or so are too loosely bound to the cluster to avoid being detached from the cluster by the Galaxy's gravitational field (Fig. 14.13).

Another extensive program that Oort involved himself in was the completion of the Palomar-Groningen Variable-Star Survey (see Sect. 11.6), that had been initiated in discussions with Walter Baade at the 1953 Vosbergen Symposium and had been carried out by Lukas Plaut. It involved mostly RR Lyrae stars, for which the distance can be obtained from the period of the light variation. Plaut had published the results in a series of six papers between 1966 and 1973. Oort and Plaut (1975g) used these data for a discussion of the distance to the Galactic center and the Galactic constants. They found a sharp peak in the density of RR Lyrae stars (260 per kpc^3), corresponding to a distance of the center of 8.7 ± 0.6 kpc, and a value for the Oort constant A of 16.9 km/s kpc. Constant B remained uncertain due to the deviation of the vertex, but the rotation of the Galaxy near the Sun was determined to be 226 ± 2 km/s. Up to 2000 this paper was cited regularly, but then the much more accurate results of the Hipparcos satellite came available. Oort and Plaut's work gave rise to a study of the distribution of planetary nebulae and of the gravitational field near the Galactic center (Oort 1976d).

An object that intrigued Oort very much was NGC 1275. This is a giant galaxy in the center of the Perseus Cluster. It is at a distance of around 75 Mpc. This cluster showed the first examples of head-tail radio sources (see Sect. 14.3). NGC 1275 is a strong radio source (Perseus A) and has two sets of optical filaments, one at more or less the velocity of the elliptical galaxy (and was assumed to be gas expelled by its nucleus) and one that moves 3000 km/s faster. The latter had been interpreted also as expulsion at high velocity, but also as a late-type spiral colliding with the elliptical galaxy. Oort had been intrigued by this object and he corresponded extensively on it with many astronomers, among whom Allan Sandage and in particular with Vera Rubin (16 letters between them). She agreed to take some more spectra of the object and in the end this resulted in the paper by Rubin, Ford, Peterson, and Oort (1977a). They partly mapped the velocity field of the high-velocity system and concluded that it is a spiral galaxy seen accidentally superimposed onto the elliptical galaxy, but not as part of a collision.

The file in the Archives also contains an exchange of letters with Sidney van den Bergh (b. 1929), who questioned the superposition hypothesis. Van den Bergh was born in the Netherlands as a grandson of one of the Unilever company co-founders, but after a brief period at Leiden University he left for the USA. He subsequently obtained a PhD in 1956 from the University of Göttingen, Germany, on polarization in the Pleiades area. He worked most of his career in Canada. Van den Bergh had proposed that the two-galaxy hypothesis was false but Oort strongly disagreed, which he expressed in a letter of April 1, 1977. They addressed each other formally. Oort ended his letter by saying that the one thing he *did* like about Sydney's paper was the Winnie-the-Pooh quote at the start (*'The feeling on me grows and grows / that hardly anybody knows / if those are these or these are those.'*).

14.5 Galactic Center Revisited

After his retirement, Oort's publication output became prolific; he had 82 publications (papers, reviews, popular articles, obituaries, historical, etc.) between 1971 and 1985, after which the frequency decreased significantly. Of these 82 publications, 23

Fig. 14.14 Oort in 1980 in his office at the Observatory after it had moved to the Huygens Laboratory on the northern side of Leiden. From the Oort Archives

were in refereed journals. Up to the end of 2018 these have been cited about 1500 times. The refereed papers concern subjects such as comets, star clusters, globular clusters, high-velocity clouds, active galactic nuclei, quasars, large-scale structure and cosmology. The most impressive and probably most labor intensive ones were two large review papers in the 'Annual Review of Astronomy and Astrophysics' in 1977 and 1983. Over the last ten years (three-an-a-half to four decades after their appearance) these still are cited twice per year each. The first one is the subject of this section (Fig. 14.14).

The editor at that time of the Annual Review, Geoffrey Burbidge, had invited Oort in a letter of May 21, 1974 to write a review article of 20 pages on the Galactic center. Oort answered on June 10 that he was hesitant to accept because he had been 'collecting too many obligations' and experience in the last few years had taught him that there were 'always several that I cannot fulfill anywhere near the time promised'. Burbidge did not take no for an answer, since he felt that it would be 'exceedingly valuable' (letter of August 6). Only on January 2, 1975 Oort wrote in a postscript to a letter on a different subject:

> 'I have not yet answered your letter of August 6. This is one of the many things I feel ashamed of.
> I had two long articles very much on my mind, and just did not want to

> undertake a new obligation (which would turn out to be rather heavy) before I had finished some articles. One, on the RR Lyrae variables, is now at last ready for the printer. The other, on the Hyades, is in the process of being revised, and will presumably be finishod this month. Then there is an Encyclopedi [(del Novecento, see below)] article on Interstellar Matter which has been due for more than a year, and which I must do next. And there is still the paper on NGC 3115 which has been lying around in an almost finished form for years, because of worry about Minkowsky's observations; but it must now be prepared for the printers.
>
> However, I am deeply interested in the problems around the nucleus of our Galaxy, and would find it an interesting challenge to try to put all the data together. That is the reason why I did not send you a negative answer, as I perhaps should have done.'

We have seen that the paper on NGC 3115 never was submitted, but it is interesting to see that Oort at least still thought about that in 1975. Burbidge took this for an acceptance and Oort promised to free some time over the summer for the exercise. But he did not finish the manuscript in time, of which he informed Burbidge on September 30, 1975. The latter was sorry of course but offered to reserve space in the next volume, if Oort agreed to an absolute deadline of October 15, 1976. However, on November 3, 1976, Oort sent a manuscript for 'the first part of the article', which would run already about 25 pages. Burbidge hoped that he would receive the rest by December 10. Oort sent another ('rather long') section on December 16, but still the paper was incomplete. It took until February before the full manuscript was finished, but it would run 57 pages (Oort's own estimate) and had about 120 reference; the invitation had been for 20 pages.

Oort accepted too many obligations; but he also underestimated the time it would take him. A major factor in that must have been that when concentrated on a subject he encountered many aspects that he felt he had to look into. New papers and preprints appeared all the time and needed to be included. This meant adding significantly to the work he had anticipated. Indeed the Galactic center review treated all possible aspects of the subject, including recent new approaches at different wavelengths or angular scales on the sky. The paper did get published in the 1977 Review (volume 15), actually running to 68 pages (Oort 1977c). For Burbidge, having been extremely considerate, this must have been gratifying. Actually, this was not the worst case. Oort's contribution on *Interstellar matter and radiation* to the 'Enciclopedia del Novecento' was solicited (and accepted by Oort) in 1970 and did not appear until 1979, after many reminders that the manuscript was overdue. In the end it ran for 14 pages and was completed only after Groningen astronomer Stuart Robert Pottasch (1932–2018) contributed about half the text (Oort and Pottasch 1979f).

The Galactic center review started with the statement that it was partly a review and partly a research paper. As a prelude it noted the position on the sky of the center of the Galaxy and its distance from us. Although the RR Lyrae stars indicated a smaller distance (8.7 kpc) Oort used 10 kpc for consistency with the standard value most astronomers used. Then he turned to the mass distribution, as could be determined from near-infrared mapping—this gave the distribution of old stars, much less affected by absorption than in the optical—and the rotating hydrogen of the nuclear disk. For the mass distribution Oort used similar maps of the center of the Andromeda Nebula as a guide. A major fraction of the article concerned the expanding motions. The features observed at 21 cm were at that time thought to be the result of an expulsion of gas from the nucleus, but the 3 kpc arm and the corresponding 135 km/s arm could also be part of resonance orbits. There were more expanding features observed at other wavelengths closer in to the nucleus: molecular clouds from CO-lines at radio waves within 300 pc, ionized gas from thermal radio emission (see Sect. 10.7) and recombination lines (see Sect. 9.8) within 50 pc, and high-velocity streams (from an infrared line of ionized neon) within 1 pc. In the latter region he described a 'remarkable concentration of discrete sources' heated by old stars and young O- and B-stars. It all came down to an 'extremely compact radio-emitting body' containing 'the engine providing the energy and mass required for the expulsion phenonema'. He noted that the central mass of 4×10^6 (4 million) solar masses could be concentrated into a black hole, which then would have a diameter of only 0.1 Astronomical Unit. Since then quite convincing proof for the existence of such a black hole has been accumulating.

At the present time (October 2018) the ADS (see Appendix C.1) shows the article has been cited 347 times and has accumulated 822 reads. Remarkably, even in the last few years it still has enjoyed 25 to 30 reads per year!

14.6 Literature

The NASA Astronomy Data System ADS (see Appendix C.1) at the Harvard-Smithsonian Center for Astrophysics (CfA) currently is the most important tool astronomers use to access the literature. It is open to everybody from anywhere, lists essentially all publications in astronomy and astrophysics and provides electronic scans of most of these back to the nineteenth century, or sometimes earlier still. For more recent publications it provides links to WWW-sites from where these can be downloaded (often free of charge in the framework of free access). Together with modern preprint servers (astro-ph of arXiv.org at Cornell University [376]), it has made the availability of a library with printed material unnecessary for keeping up with journal articles, observatory publications and conference proceedings. ADS further lists all publications referred to in each article with its ADS designation, making bibliometric investigations possible. The completeness of ADS is excellent, at least for recent years (e.g. [377]). However, in the first part of the twentieth century it was customary to put references in footnotes in the texts rather than collecting these at the end of an article. ADS has only listed cited articles when there are such lists at the end and therefore is very incomplete for those years. This is relevant for Oort's publications since the change is in the middle of the 1950s; for example the decision to list references at the end of papers—for clarity, especially for papers that were cited more than once—was made by the 'The Astrophysical Journal' with volume 120 halfway through 1954, and by the 'Bulletin of the Astronomical Institutes of the Netherlands' in 1957 in the middle of volume 13.

Table 14.1 The top dozen Oort papers in terms of the total number of citations and those between 1-1-2013 and 1-1-2019. According to the ADS

	Citations total		Citations 2013–2018	
1	Oort (1950b) Origin of comets, Oort Cloud	521	Oort (1932a) Vertical force K_z	105
2	Oort (1932a) Galactic center review	352	Oort (1950b) Origin of comets, Oort Cloud	74
3	Oort (1977c) Vertical force K_z	347	Oort and Spitzer (1955b) HI spiral structure in the Galaxy	34
4	Oort (1970f) Galaxy formation	307	Oort, Kerr and Westerhout (1958d) Acceleration interstellar clouds	32
5	Oort (1960c) K_z revisited	254	Oort (1960c) K_z revisited	21
6	Oort (1983a) Superclusters review	245	Oort (1927c) Galactic differential rotation	20
7	Oort and Spitzer (1955b) Acceleration interstellar clouds	217	Oort (1970f) Crab Nebula polarization	17
8	Oort and Walraven (1956b) Crab Nebula polarization	188	Oort and Walraven (1956b) Properties of interstellar dust	13
9	Oort and Plaut (1975g) RR Lyrae stars in Galactic bulge	184	Oort (1940a) Dynamics, NGC 3115	13
10	van der Kruit, Oort, and Mathewson (1972b) NGC 4258 Westerbork	179	Oort (1966b)	12
11	Oort (1966b) Origin high-velocity clouds	169	Oort and van de Hulst (1946b) Galaxy formation	11
12	Oort (1954d) Motions interstellar medium	163	Oort (1983a) Superclusters review	11
	Total	3126	Total	363
	All 116 refereed	5309	All 116 refereed	445

These days bibliometric measures play a major role in employment, science funding and policy. The parameters used are usually an 'impact factor', which is the total number of citations received by a set of papers divided by those of the median papers in the same journals in the same years. It is easier to find one's *h-index*, defined by Jorge Eduardo Hirsch (b. 1953); it has a value *n* if one has *n* papers that have been cited *n* times or more. To correct for bias by publications as part of large consortia and to honor first authorship, the best usage would be to supplement the *h-index* by those for normalized citations (divided by the number of authors) and for first-author publications only [378]. That the these indices depend on discipline or branch of science is usually ignored. By definition the impact factor should be unity for the average astronomer, and for the three *h-indices* these would be the medians in their distributions, which are respectively 17, 9 and 8, according to [378, 379].

From ADS (in particular the ADS-Bumblebee version) one can easily derive publication and citation related properties. Oort has 116 papers in refereed journals. These have been cited 5309 times up to the end of 2018. In agreement with what has been described above, he has only 58 citations up to the end of 1954, or only a few per year. This is due to the incompleteness just described. But then it quickly increases and it has peaked at 130 or so per year in the 1980s, the time of his last refereed

Table 14.2 The top dozen Oort papers in terms of the number of reads between January 1, 2013 and December 31, 2018. The number of citations over the same period has been listed for comparison. According to the ADS

	References	Reads	Cits	Subject
1	Oort (1932a)	2453	105	Vertical force K_z
2	Oort, Kerr, and Westerhout (1958d)	1397	32	HI spiral structure in the Galaxy
3	Oort (1950b)	1329	74	Origin of comets, Oort Cloud
4	Mayall and Oort (1942a)	547	8	Crab supernova identification
5	Oort (1927c)	521	20	Galactic differential rotation
6	van de Hulst, Muller, and Oort (1954c)	328	10	Kootwijk 21-cm survey
7	Oort (1940a)	370	10	Dynamics, NGC 3115
8	Muller and Oort (1951h)	332	23	Detection 21-cm line
9	Oort and Spitzer (1955b)	293	31	Acceleration interstellar clouds
10	Oort and van de Hulst (1946b)	249	14	Properties of interstellar dust
11	Oort (1928b)	230	4	Local Galactic dynamics
12	Oort (1977c)	209	10	Galactic center review

publications, but over the last years (2013 to 2018) Oort still has 70 or so citations per year (plus 20 or so for unrefereed publications). For completeness, Oort's *h-indices* are 33, 29 and 29, but in the light of what has been said above these are meaningless.

Table 14.1 lists Oort's dozen most cited papers over all times and those cited most often over the years 2013–2018. The paper on the Oort Cloud gets cited very often when the origin of comets is addressed, while the paper in which the Oort constants have been introduced is seldom referred to when these constants are used. There are also few citations to the work on high-velocity stars or the identification of the Crab supernova. The K_z analysis of 1932 is frequently cited. Oort always stressed that this built on Kapteyn's 1922 paper [70], where he introduced the method. For a long time this Kapteyn paper was cited once or twice per year at most, while in recent years there has been a definite increase: it was cited 41 times in the period 2016–18.

Citations are not very useful for seeing what Oort at the present time is mainly remembered for. More relevant in this context might be the frequency with which his (refereed) papers are being consulted. ADS also keeps track of so-called 'reads', which is the number of times either the full bibliometric information, including the abstract if available, or the electronic versions of the full texts of papers have been accessed through an ADS search page (the latter condition in order to exclude access by 'robots', etc.). The results for the period between January 1, 2013 and December 31, 2018 (6.0 years) for the top dozen Oort publications are collected in Table 14.2. The citations over this same period are given also and this shows that some classical papers are read often but cited rarely. Most reads concern Galactic structure, dynamics, and the interstellar medium, including even Oort's papers between 1927 and 1932 on this subject, supporting the choice of the undertitle of this book. But of course Oort's important papers on comets and the Crab supernova remnant are also being examined still. The fundamental chapter on *Stellar Dynamics*, Oort (1965a) is missing; it scores 87 reads, but many ADS users may not be aware of the fact that the full text can be downloaded directly.

The revival of Kapteyn's paper of 1922, in which he studied the dynamics of the Sidereal System and laid the foundations for Oort's 1932-work, is also evident in the reads. It has 1168 reads in 2013–18 (a record 360 in 2016). This is also reflected in Kapteyn's total citations (32 in 2016, while the average in the 1980s and 1990s is 5 per year). Kapteyn has scored 283 reads per year in the three-year period 2016–1018, up from 103 in the previous ten-year interval. Oort would certainly have been gratified by this recognition of 'my inspiring preceptor'.

The growing volume of scientific publications is reflected in the increasing difficulty of keeping up with the literature. This raises the question of how Oort coped with this. Oort's Archives contain seven hardcover, large-size (varying, but of order 20 by 35 cm) notebooks, of which two sample pages are shown in Fig. 14.15. The pages have been cut on the right-hand side to provide tabs for browsing. Each tab corresponds to a general area ranging from solar system subjects to cosmology. On the pages are notes for each article read in Oort's neat but tiny handwriting, usually restricted to up to four or five lines per paper. Although he read articles in English, the notes are in Dutch.

Oort was evidently very conscientious in keeping up with the literature. The earliest entries are from 1945, the latest from 1983, after which he must have given up following the astronomical literature systematically. It should also be mentioned that for a long time he had personal subscriptions to major journals, which were bound per volume in hardcovers. These occupied many meters of shelf space in his office at Leiden Observatory.

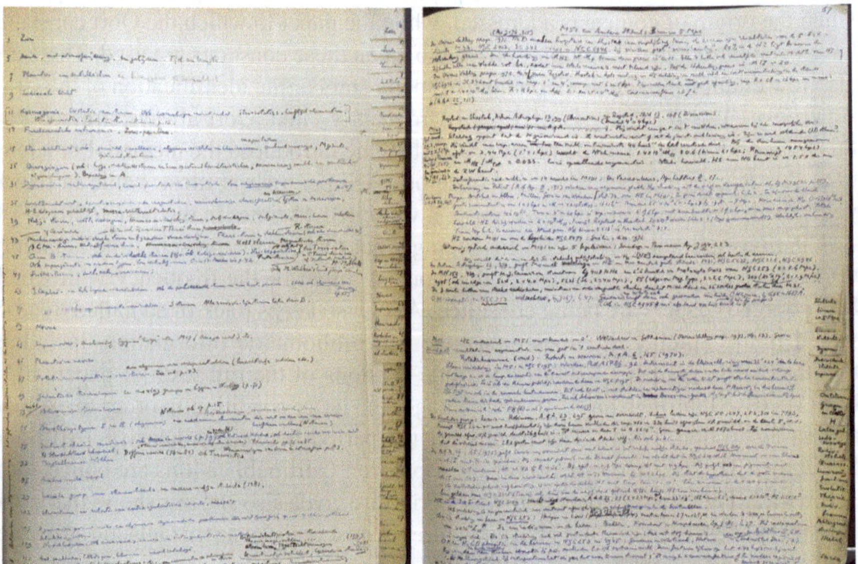

Fig. 14.15 Two pages from one of Oort's literature notebooks. On the left a table of contents; note the cut-out tabs on the right. On the right a sample page (not intended to be readable). From the Oort Archives

14.7 Pocket Diaries and End-of-Year Highlights

Starting with 1964 Oort had the habit to add to his about to go out-of-date pocket diary a list of noteworthy things that had happened during the almost completed year. A few years earlier he had often included a list of the donations and the costs he had incurred of entertaining guests and visitors, possible to be used when filing his income tax forms. From 1964 onward he also listed the books he had read.

To give an impression of the kind of books he read: In 1967 (a busy year so time for reading must have been limited) he read (the order of his list) *The Prize* by Irving Wallace (1916–1990), *Sparkenbroke* by Charles Langbridge Morgan (1894–1958), *War and Peace* by Lev Nikolajevitsj Tolstoj (1828–1910), *Thousand Cranes* by Yasunari Kawabata (1899–1972), *Ballad of the Sad Café* by Carson McCullers (1917–1967), again Tolstoj (*The Cossacks*), and *The Valley of the Dolls* by Jacqueline Susann (1918–1974). In later years there are also lists of books that he and his wife had read aloud to each other. For example in 1969 these include *Ulysses* by James Augustine Aloysius Joyce (1882–1941), *The Hunchback of Notre-Dame* by Victor Marie Hugo (1802–1885), *Northanger Abbey* by Jane Austen (1775–1817) and *Books of Small Souls* by Dutch author Louis Marie-Anne Couperus (1863–1923). After retirement they more often read books to each other, the mix of authors remaining similar. For example in 1977 the authors of both books read together and by himself are William Shakespeare (1564–1616), William Makepeace Thackeray (1811–1863), George Eliot (Mary Anne Evans; 1819–1880), and Henry James (1843–1916), but also Dutch author Jacob van Lennep (1802–1868), and the more modern authors John Ernst Steinbeck (1902–1968) and Chaim Potok (1929–2002), Heinrich Böll (twice) (1917–1985), George Papashvily (1898–1978) and Helen Waite Papashvily (1906–1996), James Albert Michener (1907–1997), but also Agatha Mary Clarissa Christie (née Miller; 1890–1976) and Nicholas Meyer (b. 1945) (a book entitled *The 7% Solution* about Sherlock Holmes).

As for important events and happenings for the year 1964 he mentioned that Dini Ondei had been working for him since February 10; as already mentioned she would work as his secretary for many years after Oort's retirement. Mrs. Oort in February hurt her ankle so badly that she had to stay down for a long time and still had not fully recovered by the end of the year. His mother-in-law Hélène Graadt van Roggen-van de Water, suffered a serious car accident on his birthday (April 28), and Mrs. Oort had to stay with her mother for a few weeks with her hurt ankle (Mrs. Graadt van Roggen recovered remarkably well). On September 28 his 40-years association with Leiden Observatory had been celebrated and he was presented with a wind vane and prism binoculars. Oort and his wife spent quite some time in their second home Sandy-Hook in Haamstede (during the next year, 1965, they lived there for 65 days and Mrs. Oort even 8 days more than that). For 1964 he ended by noting: 'At the end of the year I feel tired and getting old'. The lists also mentions the research he had done, his meetings and travels, notes on his students and various matters concerning the children and grandchildren (for the latter births, exams, schools, etc.) (Fig. 14.16).

Occasions when Oort acted as formal supervisor ('promotor') at PhD award ceremonies were recorded also, although originally only as a one-sentence note in his

Fig. 14.16 The Oorts with Oort's brother John and sister-in-law Pivy. This photograph dates from 1984. From the collection of Abraham H. Oort

pocket diary. Appendix A.4 lists his students. It is remarkable that Oort had so few of them. The first three he actually inherited from others (van Herk from de Sitter, Binnendijk from Hertzsprung, van Woerkom from Woltjer). Of his 15 students, seven defended their theses after his formal retirement. Sometimes the notes in the pocket diaries on the actual dates are rather extensive. For example in the case of Whitney Shane (June 16, 1971) he recorded the other professors and guests present, who were the paranymphs (ceremonial assistants to the candidate) and the party in the evening. The paranymphs were often fellow students, as in Whitney's case Wim Brouw and Renzo Sancisi. There was an excellent dinner in a restaurant near Noordwijk. In my case (October 6, 1971) Oort noted only one paranymph (Peter Katgert), the other was unknown to Oort; he was a friend of mine, Herman Frederik Johan (Manfred) Horstmanhoff, who majored in classical languages. Mrs. Oort had returned from the USA in the morning and only joined at the reception. In the evening 'we go back to the Observatory, where in our [sic!] hall there are tables lighted by candles and a meal of cheese fondue. It is extraordinarily enjoyable. [...] I am given a beautiful vane for my schouw and Mieke a few bottles of sherry'. Peter Katgert was my office mate, as was Vincent Icke who had prepared the meal and arranged the vane. In the case of Titus Spoelstra (really a student of van der Laan) Oort was very understanding of the fact that Spoelstra could not afford a party as he needed all his finances for a study he had started in Kampen at the Theological University. The last time Oort acted as promotor was for the defense of Aad Hulsbosch (January 10, 1973); he only noted that there was a dinner in a nearby restaurant. Actually Hulsbosch had started on his thesis research almost ten years earlier. After myself in December 1968 Oort had not taken on any new students (except agreeing to jointly supervise Spoelstra with van der Laan); according to university rules he could only act as promotor up to five years after his retirement.

In addition to astronomy there is mention of family matters, music, books, art, theater. And of course an increasing number of deaths of family, friends and colleagues. Also important national or international events are mentioned in the annual summaries, such as the (Dutch) Royal wedding of 1966, student revolts of 1968, devel-

Fig. 14.17 Plaque Mrs. Mieke Oort gave her husband Jan Oort on the occasion of their fortieth wedding anniversary in 1967. See text for explanation. From the collection of Abraham H. Oort

opments in the Middle East, the end of the Cold War, etc. Although Oort recorded important international events, he did not comment on them. He and his wife were not active in politics or members of a political party. It is not known how they voted, but grandson Marc Oort suspects they voted for the liberal party (VVD, Party for Freedom and Democracy), which stems from the Thorbecke liberal movement of the nineteenth century and is considered somewhat conservative to the right of the political spectrum.

In Haamstede there was for years no television and telephone to keep life quiet. Until Mrs. Oort won a television set at a lottery and there was no other (better) destination for it. The disappeared within a few years in a burglary. But by then the tone was set and there a new one was bought. But in 1969, when the landing on the Moon of Apollo XII took place, there still was no television; so Oort went out and rented a television set to witness the first steps by a human on the Moon.

During the periods the Oorts spent in Haamstede, they often were joined by Oort's brother John and sister-in-law Pivy. This intensified in 1968 when these bought a house in Haamstede. John Oort turned 65 that year and by the end of the year he retired from Wageningen Agricultural University.

From the long list of noteworthy happenings I select the following.

In 1967 the Oorts had celebrated their 40th wedding anniversary. On that occasion Mrs. Oort had presented her husband with a plaque, which is shown in Fig. 14.17. It is a quotation from Dutch poet Pieter Corneliszoon Hooft (1581–1647): 'I dedicate myself to your service, my love, my light, my life.'

In January 1974, Oort's mother-in-law Hélène Graadt van Roggen (fondly called Oma Lène in the family) died at the age of 93. She had lost her only son and her husband many years before, but was a strong woman that set an example to all in the family. In his speech (written together with his wife), Oort said:

> 'A new emptiness engulfed her when father died towards the end of the war, and with him many relationships and friends. Somewhat later however, in Utrecht and De Bilt a new family center developed. Coen first accepted a job at the Jaarbeurs [... and later he] became a professor in Utrecht and Coen, Marianna, Marc and Maarten came to live near her and in the end Bram and Bineke settled in De Bilt.
>
> How nice would it be if grandma would meet her husband and son again so that they could talk and catch up. It was not grandma's nature to allow such speculations. The only thing she ever said like it, was at

the funeral of grandpa on April 21, 1945, which none of us could attend because of the war. She said among others: 'Dying is not like a cup falling and breaking into fragments and dying is also not like a string breaking and all the beads being lost, but dying is: the small brook ending up flowing into the infinite waters; the unifying of the part with the whole'.'

In 1974 the staff of Leiden Observatory moved to the new complex of laboratories and institutes that Leiden University was building on a campus at the northern edge of town. The astronomy department was located in a tall building, named the Huygens Laboratory, and Oort of course moved there too. Obviously there was absolutely no space for him to take along his extensive library in the office he had occupied for so long at the 'Sterrewacht'. He moved part of it to his private home in Oegstgeest and some to his new office, but the size of the collection prompted him to decide to donate most of his library to the European Southern Observatory. The first part was offered not long after the move; Director General Lo Woltjer acknowledged the offer in January 1976. Oort had stated that at that time he would donate only a part of it, the second part being reserved for when he would stop doing active research. Woltjer was very pleased and promised he would make it a separate part of the new library after the move of ESO. At that time ESO was using space at the CERN premises in Geneva, while the office of the Director General was located in Hamburg; a new building to bring all parts together was being constructed at Garching near München. The second installment was acknowledged by Woltjer in June 1982 from Garching. This in spite of the fact that Oort had not yet stopped doing active research. The Leiden astronomy department moved in 1998 to an adjacent building that was named after Oort. The Oort Building is somewhat futuristic in that the facade reclines ten degrees from the vertical.

The annual lists of important happenings end with the pocket diary of 1979. For many years the pocket diaries had a week on two opposite pages with for each weekday eight or so lines to write on. Many of the early ones are from 'de Jaarbeurs', the organization his father-in-law had been Secretary General of and which organized annual fairs. Later they were of different make, in which each month fills two opposite pages and in which every day has been allocated a single line. They have the logo of the KLM, the Royal Dutch Airlines, of which Oort had for many years been a regular customer (Fig. 14.18).

The 1979 pocket diary in the old style records (Wednesday January 3):

'On the canal people skate in spite of the snow. I take out my skates to check out the ice. It is difficult because of the snow, but I can skate under the bridge and under the railway bridge to the Leede, but on the Leede skating is almost impossible. On the Dwarssloot [also named the 'Groote Sloot'] the snow has been blown off the ice and skating is nice, although every 15 meters you have to step over a ridge of snow. On the Zijl it is crowded with skaters and the ice is reasonable. Just as I was about to return I fell on the back of my head and only woke up when I was being helped to the bank of the river supported by a female student, Sytske Bant. Standing on the bank I had a strange 'dream' experience,

Fig. 14.18 Oort in front of the Huygens Laboratory in 1984, where Leiden Observatory had been based since 1974. The building has actually eleven floors, and the astronomy department was mostly located on the fifth and sixth (European notation). From the collection of Abraham H. Oort

in which I first thought it was unreal that there was ice and that I had put on my skates at home. After 10 to 15 minutes the girl's boyfriend, Mr. Groothuis, came to take me home in his car, where I arrived at 4:30 pm. An enormous mass of blood under my ice cap and with primitive bandaging I go to bed. Henk Bannier is coming to dinner. At 7:00 pm physician Held arrives and puts a large plaster bandage on my head. He orders me to keep to my bed for 4 or 5 days and have few visitors. I do talk a bit to Henk anyhow, but retire with [a fever of] 38.5.'

Skating to the small river 'Zijl' is about 5 km, but Oort did not record how far he skated on it, but in total he probably had skated no more than 10 km or so before deciding to return. The next day Oort remained quiet, but still read a few articles and wrote two letters to the authors. On Saturday (January 6) physician Held came to check up and Harry van der Laan to visit. Harry came on skates (he lived some 15 km skating away) and told about a car accident of his son Vincent and the funeral of his father in Canada. There is lots of skating although the thaw had started. Sunday is Mrs. Oort's birthday and Coen and Marijke with their families visit and they sit near the fire place, except for Coen who leaves after lunch as he has to travel to Mexico (by Concorde, the supersonic airliner). Oort noted that at the end of the day he was rather tired. On Monday he still feels a bit light in the head, but seems OK again on Tuesday.

Oort never skated by himself again; Mrs. Oort had ruled that they would only skate together, which they did occasionally. However, he kept up rowing until he was about ninety, but stopped using his sailing boat at about eighty because it involved so much effort to prepare.

Fig. 15.0 Painting of Oort that is displayed in the Senate Room in the Academy Building of Leiden University. It stems from around 1980 and has been painted by Gerard de Wit (1931–2010). From the collection of the Academy Building, University of Leiden, see [208]

Chapter 15
The Last Horizon

It is terrible to become eighty,
but not to become eighty is even more terrible.
Harry Mulisch

At the end, the horizon is reached;
it turns out to be incomprehensible.
Jan Hendrik Oort

15.1 Eighty Years

When in 1980 Oort reached the age of eighty a celebration was organized in Museum de Lakenhal in Leiden, with which the Oorts had dealings before in the framework of the Leiden Academic Arts Center, LAK (see Sect. 13.5). But the most tangible was the *Liber Amicorum* that was presented to him. This volume, to which I have referred in this biography as *LibAm80* [2], had been edited by longtime collaborators Hugo van Woerden, Wim Brouw, and Henk van de Hulst. Adriaan Blaauw, who wrote the opening chapter reviewing Oort's work, noted (Fig. 15.1):

> 'Among those whose signature is engraved on the astronomy of this century, Jan Oort's is one of the prominent ones. That familiar signature, just four characters preceded by two initials, simple and usually without any underlining, is always clearly written, be it on a postcard with greetings from a conference, under a letter about his ideas or criticisms, or under a weighty document. It symbolizes the clarity of his insight and the style of his presentation with which he has let us share in the fruits of his many talents.'

From the cover letters Oort sent to friends presenting them a reprint of the Zenit issue with the Dutch translation of his Kyoto Prize Lecture.

P. C. van der Kruit, *Jan Hendrik Oort*, Astrophysics and Space Science Library 459, https://doi.org/10.1007/978-3-030-17801-7_15

Fig. 15.1 Oort in 1980, probably at the celebration of his eightieth birthday. From the Oort Archives

Harry van der Laan reviewed Oort's work after the age of seventy in an article entitled *Meritus emeritus*, stressing his prolific output during that decade. An interesting chapter was written by Henk van de Hulst on *Style of research*:

'... the best antidote to perfectionism: to remain open to and get inspired by an even more exciting problem. Oort manages to do so quite well and this point alone, a very genuine interest in the marvelous Universe, may very well be the key to his style of research.'

A few sentences from Oort's letter of thanks to the contributors:

'The book reflects everywhere this same spirit and warmth that has meant so much to me, and that began with two astronomers who died long ago, but whom I want to recall at this moment: J.C. Kapteyn and W. de Sitter. Both very remarkable men in their work and scientific attitude, but no less in their personality. [...]

Adriaan Blaauw, my nearly life-long friend who shared with me the same interests and the same manner of research; nobody could have better reviewed my work. [...]

Bart Bok and Per Olof Lindblad, whose articles call up scenes of the early years in Leiden, with Bok, Kuiper and Oosterhoff, and the strong background of Ejnar Hertzsprung, and outside Leiden the beginnings of a very precious tie, the friendship with Bertil Lindblad and his family. [...]

Fig. 15.2 Oort at the scientific colloquium on the occasion of the inauguration of the new ESO headquarters in Garching-bei-München, Germany in May 1981. From the European Southern Observatory

> Henk van de Hulst, as a so-highly esteemed and appreciated companion at the Leiden Observatory, for his imprint on the book's character and his interesting and well-chosen article on Styles of Research.'

There are two further characterizations van de Hulst has given in his article. One is his attitude that 'carefull assessment of observational data is the root of all astronomy', and 'basically he remains an ardent observer'. Indeed, Oort was not a theoretician. The foundations of Galactic dynamics were laid by Jeans and Eddington, not by Oort; he had the insight to apply these to observational data to describe Galactic rotation and derive local properties of the Galaxy. Or as he said himself about his great example Kapteyn (quoted from Oort, 1981d): 'Two things were always prominent: first the direct and continuous relation to observations, and secondly to always aspire to, as he said, look through things and not be distracted from this clear starting point by vague considerations.'

The other remark van de Hulst made is (Fig. 15.2):

> 'Oort once said:'you can do only one thing well at a time'. I think he follows this maxim himself. But he has managed to do many things well at different times. In this respect his style of work must be classed as somewhat nomadic, although he returns very frequently to his preferred pasture, the structure of our Galaxy and it's mysterious center.'

15.2 Superclusters

In 1964 the then General Editor of the Annual Review, Leo Goldberg corresponded with Oort on a chapter on *Interpretation of counts of radio sources*, a subject that was very much on his mind. The efforts required for the BCAP were obviously in the way, and on October 1965 Oort wrote that he was not able to produce the manuscript: 'My wife is ever again reproaching me for my time-optimism. She is certainly right; I wonder if I should try to change this.' Oort suggested that Goldberg might ask someone else. But he actually asked Oort not to give up on it. However, in 1967 Oort was informed that Martin Ryle was preparing an article on the counts of radio sources, which indeed appeared in 1968 [380]. But in 1968 Goldberg asked Oort again to write on counts of radio sources for the 1970 volume. Oort thought about this, but was hesitant to commit himself and suggested that they ask Harry van der Laan. Later he changed his mind and suggested he write the article together with van der Laan. This was accepted; however in December 1969 Oort wrote that the Westerbork Telescope would take longer to become operational, but since source count observations would have the highest priority, results were expected very soon. He asked for an extension by a year. But that deadline again passed without a manuscript. It was suggested that Peter Katgert, who would be using the WSRT to prepare a thesis on the subject, would be a third author. In October 1971 Oort reported that the three authors felt that observations that would be obtained at Westerbork over the next few years would make a review now obsolete in one and a half to two years, and he withdrew from the assignment.

Oort kept in contact with the new chief editor Geoffrey Burbidge, suggesting authors and subjects for new chapters. And Oort wrote the review on the Galactic center (Oort, 1977c), followed by his autobiographical contribution a few years later (Oort, 1981d). That invitation had come in May 1979. Oort accepted 'with hesitation', noting that 'I am not good in history, have no vivid memory of the past, and am much more interested in the future. I wish the Editorial Board had thought of a, to me, more intriguing subject (such as, for instance, the prospect of measuring the curvature of the Universe)'. Burbidge wrote in reply that after the prefatory chapter he could 'then go on and write on the topic that you're are most interested in. I shall be delighted to publish whatever you send us.' Oort took it upon himself to write the prefatory chapter and then one on superclusters.

The autobiographical article was written in 1980 and appeared in 1981. He wrote to Burbidge (December 1980) that while writing on the subject of superclusters, he had become more and more interested in it, 'but time is closing in'. At about the Brück that the Pontifical Academy had decided to organize another *Semaine d'Etude*, this time on Cosmology and Fundamental Physics. As a member of the Academy Oort could easily arrange to be invited and he proposed that he would make a presentation on superclusters ('if indeed it was felt that this was appropriate'). He also wrote a number of letters with the aim that Russians would also be invited, such as Yakov Borisovich Zeldovich (1914–1987). After having been involved with the Russian atom bomb program and elementary particle physics, Zeldovich had made important contributions to cosmology and Oort had been in contact with him regularly. The Study Week was held from September 28 to October 2, 1981, but no Russians attended. Oort indeed gave a presentation on *The largest structures in the Universe* (Oort, 1982b). But progress with the Annual Review paper was slow,

Fig. 15.3 Oort meeting Pope John Paul II, Karol Józef Wojtyla (1920–2005), during the *Semaine d'Etude* on *Astrophysical Cosmology* at the Vatican in 1981. From the Oort Archives

possibly because of the meeting at the Vatican. There are again letters with apologies for not making the deadline (September, 1981) and in December he submitted a first part and the rest in January 1982. However, Burbidge reported to Oort that the article was very long, while he had more manuscripts than he could accommodate. The paper was deferred. During 1982 Oort rewrote it, extensively discussing many new publications and submitted it in December 1982, well beyond the October 15 deadline. When published (Oort, 1983a) it ran 56 pages (Fig. 15.3).

Superclusters are the largest known structures in the Universe, up to ten times larger than the usual large clusters. They are irregular, contain several clusters and in their longer dimensions have crossing times (the time needed for a galaxy with an average speed to cross it from one side to the other) longer than the age of the Universe. A typical size is 350 Mpc and a typical mass is $10^{15} - 10^{16} M_{\odot}$ (of order ten thousand times our Galaxy). In a way Oort returned to his fascination of the distribution of galaxies that he had already shown in his inaugural lecture of 1936 (see Fig. 8.9); indeed Fig. 1 of the review is the same one. Oort concluded that these would have formed as flat structures, 'pancakes', from collapse in the gaseous stage of the Universe prior to the formation of galaxies.

After 1980 Oort published (or co-authored) only four more research papers in refereed journals. One of these (Schwarz and Oort, 1981c) concerned Westerbork observations of parts of the most prominent high-velocity cloud Complex A. With its high resolution the WSRT showed much small-scale structure and turbulence. There is no discussion of the implications for the formation of high-velocity clouds.

In the same year two more papers were published; these were on superclusters. The first (Oort, 1981a) concerned gas in superclusters. Quasars showed in their spectra absorption lines at redshifts different from that of the object itself. In particular the fundamental hydrogen line Lyman-α (see Sect. 9.8) is prominent in this respect. This line has a wavelength (in the laboratory) of 1216 Å, but in a quasar it will be shifted into the optical domain by the expansion of the Universe. Also material between us and the quasar will show up in absorption through this line. Now Oort, who in the paper called the discovery of these lines (by Wal Sargent and collaborators) 'one of the most interesting studies of the last quarter century', showed that the characteristics and distribution of the lines indicate that it is 'pancake gas' that has remained uncondensed, and that the lines do not originate in galaxies.

The other study (Oort, Arp and de Ruiter, 1981b) concerned the distribution of quasars. The second author, Halton Christian (Chip) Arp (1927–2013), had spent decades trying in vain to convince the astronomical community that redshifts in quasars are not related to the expansion of the Universe, but result from some unknown physical property. This paper is one of the very few Arp co-authored in which it had been assumed throughout that redshifts are 'cosmological'. Arp worked for many years at the Mount Wilson and Palomar Observatories, but from the late 1970s onward lived in München. Hans Rudolf de Ruiter (b. 1946) is a graduate from Leiden, working at that time at ESO. Arp had been looking for associations of quasars with brighter, more nearby galaxies (from which he speculated they had been expelled) and in that process had found pairs that were relatively close in the sky and had very similar redshifts. The twelve pairs Arp had found would have physical separations of 5 to 30 Mpc (assuming that their redshifts *are* cosmological), the scale of superclusters. The analysis showed that in all likelihood all quasars lie in and are associated with superclusters, indicating that superclusters formed before the epoch of the formation of quasars (and since quasars presumably are located in galactic nuclei, also that of galaxy formation).

In his last refereed paper (Oort, 1984a) he again reviewed the association of quasars and superclusters. The Lyman-α lines indicate that the gas in superclusters must be highly ionized and Oort assumed that the strong ultraviolet radiation from the quasars is responsible for this. Quasars have a very pronounced maximum in density at a redshift of 3.5 (corresponding to an epoch of 15% of the present age of the Universe); if they are related to the formation of pancakes and superclusters, as Oort surmised, the gas at that time would have to be neutral and in principle observable in the 21 cm line. Such observations are actually possible nowadays.

15.3 The Return of Halley

Oort's paper on the vast swarm of comets that had become known as the Oort Cloud received much attention in 1986, when Halley's comet made another pass through the inner Solar System. The media attention was enormous and the Volkssterrewacht 'Public Observatory' Simon Stevin near Breda had organized seven flights with a charter airline company, in which 240 persons could view Halley from an Airbus-310 aircraft at cruising altitude. Oort was invited to come along on one of these on January 9, 1986. This was before the comet actually made its closest pass of the Sun. He described it as an enormous experience. In the article from which Fig. 15.4 has been taken (in the newspaper 'De Telegraaf') Oort told the reporter:

Fig. 15.4 Destination Halley's Comet. Oort being accompanied by stewardesses of the charter company Martinair to board a flight to view Halley's comet from an altitude of 10 km. The date was January 9, 1986. From the Oort Archives [381]

> 'At the time of the last passage of comet Halley in 1910 I did see a comet. My father, a physician at Oegstgeest, took me outside one evening and there was THE comet. I remember a beautiful tail. Later I came to realize that my father really did not know anything about astronomy. On the evening in question it was dark early and cold. Halley's comet was not to appear in full splendor until May and could be seen and admired in the sky around midnight. Something was wrong and in the course of time I became convinced that in 1910 I did not see Halley's comet but the 'Johannesburg Comet' that crossed the sky in the first months of the year.'

This was the 'Great January Comet' of 1910, also referred to as the 'Daylight Comet', designated C/1910 A1. As the name reveals it was so bright that it could be seen before the Sun had set.

The European Space Agency ESA had launched a spacecraft to make a close passage of the nucleus of the Halley's comet. The probe was named Giotto after the medieval painter Giotto di Bondone (1270–1337), who had depicted the Star of Bethlehem as a comet on one of his famous frescoes. Mauritania had issued a series of postage stamps to commemorate the return of the comet and the Giotto fly-by, one of which bore an image of Oort (see Fig. 15.5).

The space craft Giotto actually was part of a much larger, worldwide effort of five space probes (collectively referred to as the 'Halley Armada'). Giotto—the first to encounter Halley—provided a very good idea of the position of the cometary nucleus for the other probes. The closest approach of Giotto to the comet occurred on 14 March 1986 just after midnight in Europe, and Dutch television had an extensive covering of that event starting late in the evening of March 13. Oort had been invited to the studio to comment and explain his 'Oort Cloud' theory before the program switched to ESA Control Center in Darmstadt. Oort drove the presenter of the first part of the show, Petrus Lambertus Lucas (Piet) Smolders (b. 1940), to despair by giving only superfluous answers to his questions on comets and turning the subject every time quickly to cosmology and the large-scale structure of the Universe.

The organizers of the annual Oxford Halley lecture decided in 1986 that the return of Halley's comet and the encounter with Giotto warranted a second Halley Lecture

Fig. 15.5 Commemorative stamp issued by Mauritania on the occasion of the return of Halley's comet to the inner Solar System and the fly-by of ESA's Giotto spacecraft. From the Oort Archives

by Oort, which he delivered on May 6 (Oort, 1986a). He discussed observations of Halley's comet and reviewed his theory with the reservoir of comets in a distant cloud, but the lecture did not present fundamental new insight into the origin of comets. It was a fitting tribute to Oort's insight and particular his remarkable intuition to realize that there was something fundamental to learn from the distribution of semi-major axes of comets.

Oort actually did some more research on comets. In 1990 he published an update on the statistics of orbital parameters among comets, which he had started working on in preparation for the Halley Lecture. In 1991 he submitted a very short note (the text is one A4-page) to 'Astronomy and Astrophysics' (*Evidence that the long-period comets are covered by a layer of extremely volatile material*), but the referees felt that the ideas were not really different from what he had been published earlier (Oort and Schmidt, 1951c) and Oort withdrew the manuscript.

15.4 Balzan and Kyoto Prizes, the Jan Hendrik Oort Fund

In 1984 Oort was awarded the Balzan Prize. The International Balzan Prize Foundation annually awards prizes for outstanding achievements in four fields, among which natural sciences. The Prize is named after Eugenio Balzan (1874–1953), an Italian newspaper owner. His daughter Angela Lina Balzan (1892–1956) inherited his fortune and before her death she left instructions to use it to establish a Foundation. This Balzan Foundation has two seats: the administration is located in Milan, while its Headquarters are in Zürich. The Prize was first awarded to the Nobel Foundation in 1961 and there were a few individual recipients in 1962. However, a serious disagreement developed within the Board over the appointment of new members on the Board by its chairman, Father Enrico Zucca, a Franciscan friar. The Swiss and Italian governments disagreed with his choice and the enormous endowment was blocked so that the Prize could not be awarded. However, the Board announced the 1964 Peace Prize anyway, which would be awarded to the United Nations. Apparently this was done by Father Zucca and a few of his supporters without notifying the full Board. The Secretary General of the UN, U Thant (1909–1974), actually was handed a cheque that was not validated and which he returned. The Board was dismissed and replaced by two special commissioners, on behalf of Italy and Switzerland, to which Father Zucca of course vehemently objected, taking the view that this was nationalizing a private institution against the will of the benefactor. The resulting stalemate lead to a legal case that went on for years.

In spite of this, the dismissed Board announced in February 1965 that Prizes would be awarded that year in five areas; among the winners were Oort and economist Jan Tinbergen, the brother of Oort's close friend Nico Tinbergen. This of course was blocked by the governments. In March 1965, Oort wrote to Tinbergen, congratulating him, but expressing the view that they should not expect too much of all this. Indeed, the 1965 Prize was not awarded in the end; the only correspondence of this is in Oort's Archives is a letter from Father Zucca informing Oort of the necessity of legal procedures. In interviews at the time, Oort attributed the disagreements in the Board to the fact that under Father Zucca the prize money that had been awarded was more than the interest on the endowment.

The case took a long time to be resolved. In the end the governments prevailed and in 1978 a single award (to Mother Teresa of Calcutta) was made and more followed. Oort received the Balzan Prize in 1984, but Jan Tinbergen never did, although he lived to 1994. Tinbergen shared the Nobel Prize in 1969.

The Prize was 250,000 Swiss Franks, which is in current purchasing power about €215,000. Oort accepted the Prize in Rome on November 15, 1984. In his acceptance speech he noted:

> 'The pleasure of receiving this is particularly enhanced by the fact that it comes from Italy, the country where Galileo made his great contribution to our insight in the vastness of the world in which we live, [...]
>
> On the occasion of a high award one's thoughts naturally turn to the sources of one's scientific interest. The main inspiration in my life has undoubtedly come from Professor Kapteyn in Groningen, who was one of the earliest explorers of the Milky Way System. [...]
>
> In recent years my work has carried me to the largest existing structures and their evolution in the course of time. I wonder sometimes whether this work would have the approval of my former professor, strongly interwoven as it is with its speculative aspects. But on this point I derive comfort from a verse by Browning [Robert Browning (1812–1889)]: 'But a man's reach should exceed his grasp, or what is heaven for?'.'

In 1987 Oort received the Kyoto Prize. I quote from an interview I had with Harry van der Laan (my translation).

> 'So they were in Haamstede and I received a phone call from the Japanese Embassy [...] and they told me Jan had won the Kyoto Prize, I think half a million Guilders or so. I knew they would return in two days, so I thought 'It is no use bothering them now, they will be packing for the return trip anyway'. So I called them in Oegstgeest [...] and Mieke answered the phone. I told her I had something to tell them but preferred not to do that over the phone. [...] When I arrived they had tea prepared, and for me it was only a quarter of an hour by bike. 'Well Harry', Mieke said, 'what is it you have to tell'. I said: 'I had a call from the Japanese Embassy'. She said: 'Jan, what do you have to do with Japan?' 'Well', Jan said : 'I have a few friends there, but nothing special.' And I said: 'According to the ambassador, or his spokesman, you have been awarded a major prize. It is called the Kyoto Prize and it means a large sum of money, about half a million guilders.' Jan just looked, and Mieke said: 'Oh, then we can finally have this leaking roof repaired.' [...] I said: 'There is a complication in the conditions and that is that you have to pick up the prize personally.' Now Mieke sat in a wheelchair, because of problems with her knees. 'Well', Jan said, 'That should be possible. [and to her] And you have to come along.' Coen [their older son] worked for the ABN Bank and also was chairman of the Supervisory Board of KLM. So he arranged that KLM would sponsor this. So, Coen, his wife, and Jan and Mieke, went to Tokyo first class.'

The Prize was accompanied by a workshop, for which Oort prepared a review on *Origin and structure in the Universe* (Oort, 1988a). The Kyoto Lecture, *Horizons*, has been published in Dutch (Oort, 1989b). In Appendix B.7, I reproduce the English text he probably read out in Kyoto.

I again quote from my interview with Harry van der Laan:

> 'One morning Oort shuffled into my office, closed the door, which he never did, and said a little nervously: 'Harry, can you spare a little bit of time for me to talk to you?' [...] We sat down and he started immediately: 'What should I do with all that money?'

Harry suggested that he put it in a fund that would finance a special visiting professorship, which Oort did after talking it over with his children. It resulted in the Jan Hendrik Oort Fund, which established an Oort Professorship, which includes a workshop for staff and students and a public evening lecture, the Oort Lecture (following the example of the Jansky Lecture at the National Radio Astronomy Observatory NRAO,). The first Oort professors were Lo Woltjer in 1990 and Maarten Schmidt in 1991. Leiden Observatory posted photographs of all Oort Professors.

15.5 Personal Matters

The Oorts very much enjoyed their family life. One annual festivity in the Netherlands is 'Sinterklaas', celebrating the anniversary of a legendary figure based on Saint Nicholas of Myra (probably 270–343), patron saint of children. People give each other presents, often elaborately packaged in a creative way, referred to as a 'surprise' (from the French). These are accompanied by humorous poems from Sinterklaas, often bearing personal messages, usually teasing the recipient with poor habits or other character traits. Marc Oort recalls:

> 'Sinterklaas indeed was always a highlight. Lots of poems and surprises (no draws, just everyone made something for everyone else) and it regularly required working until after midnight. And no one was spared; the lesser qualities came on the table without restraint. Grandpa and Grandma felt it was a great party. They had two sessions; first with us and usually the day before or after with aunt Marijke. Any excuse to get the whole family together was allowed. They wanted to have everyone around them as much as possible.'

We have seen that the Oorts shared a love of arts. From a note by Bram Oort:

> 'Together my parents attended a volleyball group, a Shakespeare discussion group, and art courses. Following the advice of their teacher they bought some contemporary paintings and original sculptures from well-known Dutch artists. My wife, Bineke, and I have inherited a few of these, enjoying them now in our home in Hartland [Vermont]. I must have gotten my love for art from my parents, especially from my mom.'

In Fig. 15.6 we see the Oorts with the sculpture 'Hinde' by John Raedecker located in the center of the Kröller-Müller park in the Hoge Veluwe (see also Sect. 11.1). The picture was taken by Bram Oort when

> 'my parents and I had a pick-nick there. My parents previously had bought a first study of the deer in bronze. It used to be on the coffee table at my parents' home but is now on our coffee table in Hartland.'
>
> 'During one of their visits to Princeton in the 1980s, they patiently posed for a clay sculpture I made of them. These portraits are now exhibited at the Observatory in Leiden in a room dedicated to my father.'

Figure 15.7 shows these clay sculptures together with pictures of the Oorts at that time. In October 1982, Mrs. Oort posed for a sculpture session in the apartment the Oorts occupied at the time on Einstein Drive in Princeton, where they were staying during one of their visits. The sculpture Bram Oort made of her head, was fired in 1983. Bram had a cast made of the side view and offered it in January 2000 to the Oort Room at Leiden Observatory (in the Oort Building at the Leiden Campus to the north of the city). In December 1984, Bram worked on a bust of his father. This took a much longer period to complete with several sessions. These took place again in the apartment of the Oorts during their visits to Princeton. Much later, when Henk van de Hulst visited Bram Oort in Rocky Hill, New Jersey, where they lived at the

Fig. 15.6 The Oorts posing with the sculpture 'Hinde' by John Raedecker located in the center of the Kröller-Müller park. The Oorts owned a first study for this sculpture in bronze. From the collection of Abraham H. Oort

time, and saw the bust, he suggested that Bram should donate it to the Oort Room, which he did. It still resides there.

In 1977 the Oorts celebrated their fiftieth wedding anniversary. I quote from the speech Oort gave during the family celebrations, in which he reflected on their life together and the role Mrs. Oort (to whom Oort referred as 'mother') played in this (Bram Oort's and my translation):

> 'Probably my life has been too focused on my profession with always too many daily obligations that never were ready in time so that I could not consciously enjoy the family. But while I was immersed in my work there was always the background of mother, the joy of a real home that is the basis of our life. [...] That home was mother's creation: without much fuss, without ever boasting about it (to the contrary; she herself always felt that her presence did not matter and sometimes was unhappy about that). But it was she who, almost unknowingly, shaped the family, formed our circle of friends at the Observatory and the university, welcomed our guests and was the source of the fascination in our lives. Intuitively she felt what is beautiful and good, always thinking of others rather than herself. [...] In the beginning, but later also, we talked about our ideals, what one wants to accomplish in life. For mother this was simple: the wish to be pure. For me it was more complicated and unattainable. But

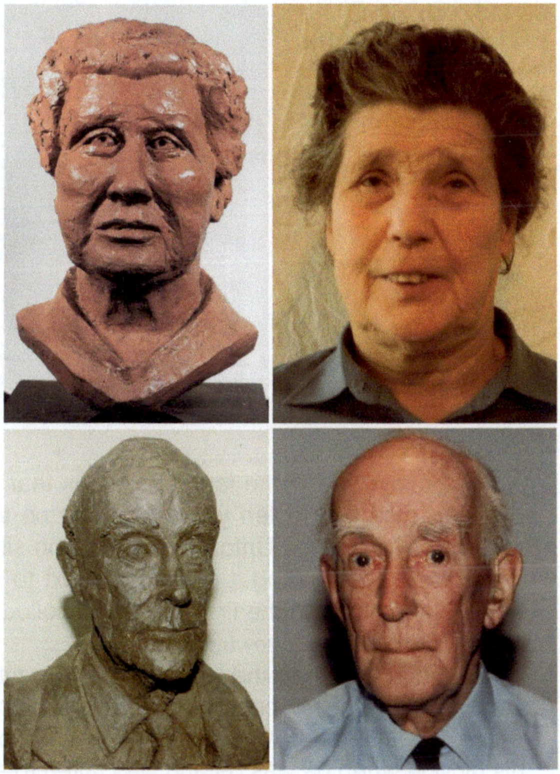

Fig. 15.7 Clay sculptures Bram Oort made of his parents and photographs while made posing for this. Top: Mrs. Oort; the clay model was produced in 1982 and it was cast in 1983. A cast of the side view resides in the Oort Room in Leiden. Bottom: Oort; the production took several sessions starting in 1984. It resides now in the Oort Room. From the collection of Abraham H. Oort

> in reality we enjoyed the same things and generally loved the same people, especially our children.'

In 1987 Coen Oort said at their sixtieth wedding anniversary (supplied by Bram Oort) (Fig. 15.8):

> 'In the foreground but also in the background of my memories, the most vivid is the warmth you both gave and the examples you set (although I often protested) and the togetherness of the family. What I especially remember is that I never felt uneasy; there was always support, self-evident and therefore not oppressive. You were 'modern' parents, very tolerant and permissive, which we realized and understood only later, when we raised children ourselves.'

It must have been a major milestone and also a source of pride when in 1987, Oort's grandson Marc obtained a Ph.D. in astronomy. During the festivities around this Oort has made a speech and I quote from notes he left.

> 'It was a great time, all those years together at the Observatory. Years in which we worked on cosmological subjects and were interested in the same things, even though our actual daily work was different. [...] I treasured the thought that you would continue the work and that you

Fig. 15.8 Oort with his grandson Marc in 1984. The latter was to obtain a Ph.D. in astronomy in 1987. From the collection of Abraham H. Oort

would be involved in the fantastic things that will happen in astronomy over the next five to ten years and that so to speak all of this would stay in the family. [...] Unfortunately the job situation in astronomy is so unfavorable that finding a position is next to impossible, so we will no longer sit together in the first row at the colloquia.'

Marc Oort provided the following reminiscences:

'Grandpa cycled to the Huygens lab [the Observatory offices] until the very last (one of his complaints about the Johannahuis [in Wassenaar, where they moved in 1991] was that it was no longer possible). He fell off his bicycle when he was 89 and was taken to a first aid post. The doctor asked (to see if he had a concussion) 'what is 7×7' and immediately got the right answer. The next question was 'and 7×17', upon which he also got an immediate answer. After long thought and calculation the doctor came to his surprise to the conclusion that answer was also correct. No concussion (but several abrasions). Just like with skating he (almost) stopped cycling.

When I came to the Observatory after my Candidaats [Masters] it became a tradition that every week after the Thursday afternoon colloquium I cycled with him to Oegstgeest to have dinner there. The rule (Grandmother's) was that astronomy should not be discussed. He found this very difficult and it rarely worked. We kept that up for at least three years (then I got a girlfriend and changed my priorities).

Because our family is quite informal with each other (always 'jij' [the informal version for 'you'] to Grandpa and Grandma and say things straight as they are), I have always benefited much from our time together at the Observatory. He always told me when I was wrong or when I was being sloppy (at my thesis defense I got a question that came down to the fact that he found a certain conclusion absolute nonsense) and vice versa. I always joked with him. He was very sorry that the postdocs and Ph.D. students always treated him with such respect. He would have preferred to sit at the lunch table with them and talk on an equal footing. But it rarely worked and he missed that. Especially after his hearing deteriorated. We shared an office for two months (after my Ph.D. until I got a job at Fokker Space). His was the tidiest desk in the universe, mine the most chaotic.

My thesis defense was indeed a unique experience. My uncle Tom (who as a professor of law had also talked himself in the committee) recalled that there were 5 generations of Oorts present. From Henricus Oort (in a painting on the wall) to our (then unborn) eldest son. Afterwards I heard that he had been thinking for days about how he would address me. Eventually it became the neutral 'mister candidate'.'

Oort was also concerned with humanitarian issues. In the 1980s he signed petitions demanding the release of political prisoners in countries with dictatorial regimes such as Chile, Turkey and the USSR. He wrote personal letters to high Chilean officials including president Pinochet (1987) and the prime minister and president of Turkey (1988). In particular he was concerned with the treatment of Russian dissidents, including physicists Andrei Dmitrievich Sakharov (1921–1989) and Yuri Fyodorovich Orlov (b. 1924). Similarly as others he did not give up his membership of the Russian Academy, feeling that as a member he might carry more weight than as a private citizen. Both physicists were allowed to leave the country.

From the correspondence Oort had during the last years of his life, I select one final issue. Oort had agreed on the phone to participate in a series of interviews by Bernardus Hendrikus Rootmensen (1939–2013). Rootmensen was student-pastor or minister for the students at the University of Amsterdam. The interview was to take place much later. It was followed up by a letter (January 6, 1988), in which Rootmensen explained a little more about the way he was going to conduct this. The aim was 'to bring young people into contact with the life experiences of the elderly'. The series was called 'Old enough and wise enough'. The description did not mention religion as a subject per se. One of the questions was 'If you put your life on a sieve and you shake it, what do you feel remains as essential, as precious?'. In July Oort was reminded of the event, but then he started to have seconds thoughts. He wrote back on July 28, saying that he had thought it concerned

'a kind of informal conversation in a small group and not something coming close to a lecture. I am not really good at conveying my life experience in an adequate manner to young people. My thinking has been dominated by astronomical problems and only very little on important human questions as you mention so that I cannot have an interesting conversation about them. [...] When I showed your letter to my wife her reaction also was that the subject of your interviews, interesting as they may be, are too far removed from my thinking for me to provide a useful contribution.'

Oort seemed reluctant to enter into matters of religion or philosophy of life. He also claimed that his hearing problems were in the way of a useful conversation. Rootmensen wrote back explaining in more detail what was expected of him, saying he would not ask about his private life, but would pose questions like why astronomy

Fig. 15.9 The Oorts in Franeker in 1990 on the occasion of the revealing of the plaque on the house where Oort was born (see Fig. 1.3). From the Oort Archives

is interesting, what developments he was involved in, what his current scientific interests were. So, Oort agreed in the end to come and according to a letter by Rootmensen after the interview it was a success.

Oort increasingly more often complained in letters and conversations about his hearing problems. Before the meeting of the Program Committee, which ranked observing proposals for the WSRT, in October 1986, he came to me (I was chairman of the PC at the time) saying that he wished to resign. He was not able to follow conversations and was not in a position to usefully contribute. We agreed that he would keep refereeing proposals and send in reports and, as we had no chance to prepare for this, in the near future come to a meeting once more so that we could formally mark his departure. Later he indicated that he would not want to come to any further meetings and did not feel there was a need for a formal goodbye. He had been a member of the PC from the beginning, seldom missing a meeting (Fig. 15.9).

15.6 The End

Bram Oort put at my disposal a few letters Oort and Mrs. Oort had in the last years of their lives written to him and his wife Bineke, for which I am very grateful. From a letter dated February 1, 1986 (Oort started it, Bram Oort's and my translation):

> 'Mother has a pain in one of her legs and has difficulty walking. We hope it will disappear as quickly as it came up. Her physiotherapist has been on a winter vacation in Austria but she will come here tomorrow.

[...] Your visit for mother's 80th birthday made her very happy. She has the fondest memories of that day; your presence was the high point. We saw on television the explosion of the Challenger and were very much shocked but were impressed by the speech President Reagan delivered after the accident.'

Mrs. Oort continued thanking her daughter-in-law Bineke Oort-Pel, who had made an audio recording for Mrs. Oort's 80th birthday that Bram had delivered.

'I always experienced your going to America as something very painful, although I do agree with it now. It would never have worked for Bram in De Bilt. In the US it has been possible for Bram to flourish and he is now an authority in his field!! In the beginning of our marriage we went to the USA for some time, especially to find out if we could settle in the USA. Father had more and more offers for a position in the USA, and he wanted to find out if I would like it there. My conviction grew that I would be very happy to live in Holland but probably would not adapt well in America. It might be a drama as happened in the emigration of our friends, the Strömgrens. I would also find raising children very difficult there. You did that very well! Hats off! I have been lucky that father's position improved (and his recognition) just in time so that we decided definitely to stay in Holland. The financial side was never very important for us. Although the first Sterrewacht years in that large house were difficult and the salaries of a professor were lagging behind. But we also were lucky; we did not have to spend a cent on extra lessons for the children, except for music lessons. All of you went through school without any problems; for years we never had worries about any of you three!

It is a pity I cannot reread the sweet things Bineke said on the tape she sent me for my 80th birthday; it was much too much. Yes, I think I was flexible, I believe, but it all was worth the trouble to fight for (I know this sounds melodramatic) against the background of the great love for each other which dominated everything. In the beginning I had, like you, problems with the seriousness and great appetite for work in the Oort family and only now father and I enjoy each other more than in all preceding years. It is such a deeply rooted thing in the Oort family that I had to find a modus vivendi to cope with it, but I have always got what I needed! I have never been deprived of anything, because father is in the end a very sweet human being—always seeing the best in others, always optimistic; very special qualities that not always come to the foreground because of the seriousness from the outside—difficulty expressing his feelings—his lack of time to concentrate on other things than his work.'

Bram Oort said in an email:

'As far as emotions are concerned, my mother was more'in touch' with her emotions. In case of tragic events my father often seemed not to know what the right emotions were—he often looked at my mother to see what was going on. Maybe this also had to do with his hearing problems. He was interested in everything that was new to him, he was

Fig. 15.10 Oort at his ninetieth birthday reception with three Dutch Directors General of ESO. From the left: Lo Woltjer (DG 1975–1987), Adriaan Blaauw (1970–1974) and Harry van der Laan (1988–1992). From the European Southern Observatory

naturally curious! I remember an occasion when my parents were visiting us in Rocky Hill, NJ; they had an apartment at the Institute for Advanced Study in Princeton, NJ. Bineke and I and our children were involved in the EST Training [Erhard Seminars Training]. Our oldest son, Pieter Jan, moderated a group about 'communications' and invited the four of us to join. My father sat in the front row (my mother decided not to attend). I don't know how much he understood. My fathers main reaction was his surprise that many participants so easily expressed their (deep) feelings in public. Later, at our home he listened intensely to tapes by Werner Erhard.'

Later during the 1980s Oort came less often to the Sterrewacht (as to his delight the astronomy department was still officially called). But he kept up with literature and kept coming to colloquia. I quote from a newspaper cutting of July 1987 [382]:

'It is not easy to go through all the astronomical journals every month. But apparently at a very advanced age one can still remain curious, have the urge to find out many things. I still concern myself with the largest structures in the Universe. It is amusing to conclude that in the 1920s my borders were determined in space, by those of the Milky Way Galaxy that were at 20.000 lightyears. Now, my horizon is determined in time, 15 billion years, the age of the Universe. It is a pity that before too long I will have to die and then can no longer see the new developments.'

In 1990, when Oort turned ninety, a large reception was held at which guests came from all over the world. It was held at the Kasteel Duivenvoorde, in the village of Voorschoten, some ten km to the southwest of the center of Leiden. Duivenvoorde Castle dates back to 1226, making it one of the oldest castles in Holland (the provinces). During this reception the photograph in Fig. 15.10 was taken in which Oort posed with three Directors General of the European Southern Observatory ESO, two former (Adriaan Blaauw and Lo Woltjer) and the then current one (Harry van der Laan).

In the summer of 1991 the Oorts were no longer able to live in their own home and moved to the Johannahuis, Schouwweg 72 in Wassenaar. The town of Wassenaar is between Leiden and Den Haag and is known for its luxurious housing due to the fact that many well-to-do people live there. The Johannahuis is an assisted-living residence for senior persons, so it included daily care. The move was the result of a number of incidents that showed that living on their own was no longer feasible. As their grandson Marc Oort told me, at one stage one had a fall in their living room, and the other in trying to help had also fallen; neither was able to get up. They did have alarm buttons to wear around their necks, but headstrong as they could be they did not wear these and had been lying there for hours. Coen Oort more or less forced the move upon them because he felt they found it too difficult to make that decision themselves.

In the Johannahuis they occupied a spacious apartment with a large living room, a bedroom, bathroom and kitchen, so it was very comfortable. They could not take all their possessions, but Oort took most of his books (also astronomical ones) along. Marc Oort took the sundial in the form of a radio telescope that is seen in Fig. 14.9 and it is still in his garden. They did not have much contact with the other tenants, and were not very keen on joining the social program. They had good relations with their neighbors, though. From a letter of the Oorts to Bram and Bineke in the USA, early 1992:

> 'Father half the time in bed, half the time out of it. New Years Eve quietly together, went to bed at 11 pm with a plate with all sorts of goodies brought by the Snieders. Very nice neighbors! [...] The home is always very pleasant and we are taken care of very well! What else would one want?'

Their children Coen and Marijke visited weekly, the grandchildren (with in the meantime great-grandchildren) came every now and then. They did not leave their home very often. One last occasion when Oort attended a formal academic ceremony was the thesis defense of Marinus Adrianus Maria (Rien) van de Weygaert (b. 1963) in September 1991. This (see Fig. 15.11) was very likely the last time ever he wore a formal academic gown.

The end came late in 1992. Oort took a bad fall in his apartment and broke his hip. Two days later, on November 5, 1992, Jan Hendrik Oort died from complications.

Bram Oort:

> 'My parents had been married for 65 years when my father died in 1992. My mother, my wife, Bineke, and myself were at his bedside when he passed along peacefully. For me, it was the first time to be fully present with a dying person. Although it was sad, it was also a special gift to have been there and to notice the transition between life and death. It was clear that something—his soul/spirit?—had left; what remained of my father's body was just a body without life. My brother and sister were not aware that the time had come and had left.

Fig. 15.11 Oort's probably last public appearance in academic gown. The date is September 25, 1991 and the occasion the Ph.D. thesis defense of M.A.M. (Rien) van de Weygaert. The person behind Oort is Vincent Icke, who completed a thesis in Leiden in 1971 (see Appendix A.4). Collection Rien van de Weygaert

The funeral ceremony was held on November 10 at the crematorium Westerveld near Velsen (see Sect. 13.1), where also Oort's mother had been cremated. In addition to the family, many astronomers and friends attended.

Not much more than seven months later, Mrs. Oort also had a bad fall and she died of the complications on June 7, 1993. She was also cremated at Westerveld.

On the funeral card, the children quoted a poem by Frederik Willem van Eeden (1860–1932) [383]:

I love the immaculate waterlily white,
so pure, so silent
as she unfolds her crown,
in summer's light.

Rising out of the lightless chilly pond,
she's found the summer light
and has gladly unlocked her golden heart.

Now she lies pensive on the water's plane
and wishes no more...

– –

Many persons have expressed their feelings about Oort, as a family member, teacher, scientist, administrator, etc. I have selected four excerpts to include here. But I first present some remarks of my own.

I have experienced Oort as an inspiring professor, not because of superb teaching or exciting conversations as a thesis supervisor. He instilled a feeling of fascination for astronomy in his students and a deep marvel at the beauty of the Universe and love for science. His courses on the Solar System and the Galaxy left me with a great admiration for the works of Kapteyn and Kepler and their approach to scientific inquiry, and a special affection to the field of stellar dynamics. Oort's actions as director or administrator were always driven by what was good for astronomy and personal matters of his subordinates were secondary. Some persons suffered from this (e.g. Gijsbert van Herk and André Muller). Personal sympathies sometimes played a role (e.g. his efforts to find a job for A. de Sitter), but could be put aside when required (e.g. Alex Muller). He pursued his own interests and priorities with vigor, but left less room for those of others, which led to the departure from Leiden of some astronomers (e.g. Westerhout and Blaauw). His intuition of what were promising new lines of research was remarkable; examples are his early realization of the potential of radio astronomy and his work on the Crab Nebula. But it also sometimes failed, as in the case of the expulsion of matter from galactic nuclei. He understood the need for international collaboration, as his efforts for the IAU and the establishment of ESO show, but where possible chose to minimize partnership, which can be seen in the radio facilities at Dwingeloo and Westerbork and the termination of the Benelux Cross Project. He was a brilliant scientist, basically an observational, interpretive astronomer, not a theoretician, which derives to a large extent from Kapteyns influence and his insistence to concentrate on observations. Oort was soft-spoken, helpful, kind and interested in others, but also persistent and sometimes impatient. I was lucky to have him as professor, mentor and supervisor.

I quote Gart Westerhout from a letter to Oort on August 9, 1977, not long after he had been appointed scientific director of the US Naval Observatories.

> 'In my years as a student in Leiden, especially when I was a doctoraal [Masters] student I always enjoyed the Oort lectures. In those years, when my astronomical knowledge and background was formed (my formative years) I was always impressed by the systematic manner in which you approached a problem, and emphasized how the many aspects of astronomy are interrelated. In the lectures on dynamics, galaxies, statistical astronomy, etc. you always stressed how fundamental measurements in fact influenced the whole building of astronomy, and even how modern astronomy and astrophysics would be impossible without continued attention to fundamental astronomy. [...] You stressed the importance of proper motion studies for Galactic astronomy and this impressed me greatly. As a result of that I always followed the developments in the area of astronomy that we call 'fundamental' and tried—actually without much success—to pass this on to my students.
>
> I wanted to write you this to thank you for the enormous influence you had on my career and my love for our profession that you taught me.'

Oort's granddaughter Yvette de Smidt said at the funeral of Oort (supplied by Bram Oort, translation by me and checked by Bram):

'We all have been in the dome at the Observatory when we were children and experienced the excitement of the starry sky opening up just for you alone. Later we understood—one maybe more than the other—his passion for the Universe. A year ago he decided to convene his grandchildren so that he could tell of the latest developments. And as everyone knows, when grandpa made a plan it had to be executed. So this spring we experienced his last lecture as a very special happening.

The Haamstede saying: 'O future generations, see this, and remember to stay united', is a wish that has been fulfilled thanks to the strong bond between grandpa and grandma.

Beloved grandpa, thank you for everything.'

I quote Whitney Shane from his written remarks to my interview questions:

'No one could claim that Oort was a good classroom lecturer. With a certain poetic license, I could describe an Oort lecture as follows: Oort would enter the lecture hall quietly and smile meekly at the eagerly assembled multitude. He would clear his throat, open his box of Wybertjes [a type of small cough drops in the form of a diamond] and slip a couple in his mouth, and begin his lecture. He would turn to face the blackboard and begin to whisper. In his right hand he would hold a piece of chalk with which he would write something in his own illegible hand, and in his left hand he would hold an eraser with which he would erase what he had just written. When the lecture was finished he would turn around and smile once again at the sleepy multitude and leave the lecture hall. You may well ask how anyone could put such a lecturer near the top of the list of his best teachers, as I do. I do not ask how skilled he was as speaker, nor how well I thought I understood the material at the time. Instead I wait 40 years (or more) and ask how his teaching has influenced me. I remember a high school English teacher with whom I did not get along well and from whom I thought I did not learn very much. Now when I write I find myself writing as he had told me I should, and when I read I find myself reading the books that he recommended. So is it with Oort. When, many years later, I had to give lectures on roughly the material that Oort had covered, I found myself modeling my course (although not, I hope, my presentation) on that of Oort. I did not think of Trumpler, whose course on approximately the same subject I had followed some years earlier and who was widely recognized as an excellent lecturer. So despite his obvious shortcomings, I have to praise Oort as a good teacher, although how he did it (as is so often the case with Oort) remains a mystery. Somehow he was able to convey something of his appreciation and understanding of the subject while leaving the details up to the initiative of the student. Perhaps this is the way it should be done.'

Fig. 15.12 Oort on the roof of Leiden Observatory in 1991. From the Oort Archives

Finally, in an oral interview Harry van der Laan said (Fig. 15.12):

"Characteristics of Oorts uniqueness', was one section of my talk at the Centenary Symposium at Leiden in the year 2000. And I say there Jan Oort was strong yet modest to the point of shyness. I never saw him push anyone to the limit except himself. He was persistent, he was tenacious. So that's illustrated by when he needed funds for the Benelux Cross he again and again tried to get an appointment with the minister. And finally the secretary of the minister called to ask Oort: How much do you really need? And Oort said: 'Well, 1.5 million Guilders for a start.' So he did not have to go to the minister. The minister knew apparently enough about him that Oort was very persistent and very tenacious. He was also very persuasive, but he was never coercive. In the program-committee of Westerbork which was the committee in which he served to his late eighties, he believed in the things he believed in, he pleaded for it, but he never threw his weight around. He was a leader, not a boss, so in that sense he was not at all authoritarian. And he certainly was not Machiavellian playing off one person against another. Divide and rule was total alien to Jan Oort ... totally alien. I also think he was very courageous. He never chose the path of least resistance, he was always looking for quality and aborting the well-advanced Benelux Cross project shows how courageous he was after he had read the Ryle-Neville North Polar Survey paper in Monthly Notices. He supported young staff and I am, of course, a prime example of a young staff member, whom he supported to the hilt, because he believed in my vision which he shared.

I would say: 'I shared his vision.' And if you look at Adriaan Blaauw's histories of both ESO and the IAU, two books by Adriaan, you can see what he accomplished by force of ... yeah, of arguments. So ... and also when Oort accepted his professorship he spoke about problems, difficulties, labor that demand all of our resources That's in his inaugural lecture. And if you look at Oort family's statements after he had died it also shows you what drove him. It was a curiosity-driven life in research.'

I close this book with the text Bram Oort used in his contribution to the *LibAm80* [2] and Mrs. Oort and the family quoted in part at the funeral announcement card of Jan Hendrik Oort. It is from American anthropologist and science writer Loren Eiseley (1907–1977), from his book *The Immense Journey*:

'Down how many roads among the stars must man propel himself in search of the final secret? The journey is difficult, immense, at times impossible, yet that will not deter some of us from attempting it. We cannot know all that has happened in the past, or the reasons for all of these events, any more than we can with surety discern what lies ahead. We have joined the caravan, you might say, at a certain point; we will travel as far as we can, but we cannot in one lifetime see all that we would like to see or learn all that we hunger to know.'

Appendix A
Curriculum Vitae, Publications, Honors, Students and Academic Genealogy

Do I ever get tired of writing?
No. I can do it sitting down, can't I?
Simon Vestdijk (1898–1971)

A.1 Curriculum Vitae

Figures A.1 and A.2 show Oort's curriculum vitae as he has written it himself. From the years quoted, it can be concluded that he compiled this after he had turned seventy.

A.2 Publications by J.H. Oort

The bibliography of Oort below is taken from the *JKM-Inventory* [1] by J.K. Katgert-Merkelijn, which contains a complete list of all Oort's publications. *The list below is not complete but contains only the papers referred to in this book.* The original list in the *JKM-Inventory* [1] (appendix, starting on p. 161) covers the period of 1922–1992 and was compiled first by Dini Ondei-Beneker, Oort's longtime secretary, and by Jay Ekers, Willem N. Brouw and Hugo van Woerden, for the *Liber Amicorum*, *LibAm80* [2], of 1980 and has been updated and revised by Dini Ondei and Jet Katgert-Merkelijn. The original designations as in the *JKM-Inventory* [1] have been kept, so e.g. Oort (1981d) is for his autobiographical essay in the *Annual Review*, *AnnRev81*

Dutch writer. No Dutch or Flemish author ever won the Nobel Prize for Literature. Vestdijk probably came closest; between 1950 and 1966, for which the Nobel archives are public, he has been nominated fifteen times.

P. C. van der Kruit, *Jan Hendrik Oort*, Astrophysics and Space Science Library 459, https://doi.org/10.1007/978-3-030-17801-7

JAN HENDRIK OORT

Curriculum Vitae

Born: Franeker, the Netherlands, on April 28, 1900.
Married: to Johanna Maria Graadt van Roggen, May 24, 1927.
Three children.

Secondary school Leyden (certificate 1917).
Student at the Groningen University from 1917 to 1921 (Professors of astronomy: J.C. Kapteyn and P.J. van Rhijn).
Doctor of astronomy at Groningen 1926.

1921-'22	Research assistant at the Laboratory of Astronomy at Groningen.
1922-'24	" " " " Yale Observatory
Since 1924	Astronomer at the Leyden Observatory
1926-'30	Lecturer at the Leyden University.
1930-'35	Assistant-professor at the Leyden University.
1935-'45	Associate professor at the Leyden University and Associate director of the Observatory.
1945-'70	Professor and Director of the Leyden Observatory.
1949-'70	President of the Netherlands Foundation for Radio Astronomy.
1952	"Visiting professor" at the California Institute of Technology at Pasadena and at the University of Princeton.
1956	Waynflete Reader, Magdalen College, Oxford.
1958	Special Fellow Carnegie Institute and Alexander Morrison Research Associate, University of California.
1935-'48	General Secretary of the International Astronomical Union.
1958-'61	President of the same Union.

Honorary degrees: Copenhagen, Glasgow, Oxford, Louvain, Harvard, Brussels, Cambridge, Bordeaux, Canberra, Torún.
Gold Medal of the Royal Astronomical Society, London.
Bruce Gold Medal of the Astronomical Society of the Pacific, San Francisco.
Janssen Medal of the Société Astronomique de France, Paris.
Janssen Prize of the Société Astronomique de France, Paris.
Vetlesen Prize of Columbia University of New York.
Prize of the International Balzan Foundation, Milan.

Fig. A.1 Oort's own text for his c.v. It must have been produced after he turned seventy

[3] There are a few additions to this list. The first one—Oort (1921)—is a Dutch article about Kapteyn that he wrote as a student in 1921 and that has been published in the yearbook (studentenalmanak) of the student association *Vindicat atque Polit*

- 2 -

Jan Hendrik Oort, curriculum vitae.

Correspondent of the Bureau des Longitudes, Paris.
Member of the Royal Academy of Sciences at Amsterdam.
Foreign Associate of the Academy of Sciences at Paris.
Honorary member of the Royal Institution, England.
Foreign member of the Royal Society, England.
Foreign Associate of the National Academy of Sciences, Washington.
Foreign member of the Academy of Sciences, Moscow.
Member of the Pontifical Academy of Sciences.
Ridder in de Orde van de Nederlandse Leeuw.
Commandeur in de Orde van Oranje Nassau.

Fig. A.2 Oort's own text for his c.v.–continued

in Groningen. Other additions are Oort (1925g), Oort (1943g) and Oort (1951l). The contributions of Oort to the OECD Symposium on Large Radio Telescopes in Paris in 1961, have been quoted incompletely in the *JKM-Inventory* [1]. Only 1962j was identified there as the conference contribution, 1962k was identified as a BCAP Memo, but not as part of this conference, while 1962l and 1962m are new additions here.

Many, but far from all, of these publications have been listed by the NASA Astronomy Data System ADS (see Appendix C.1) and usually scanned copies are available there. I have indicated the ADS code at the end of the full reference between square brackets. Another source, which does require access arrangements, is the non-profit organization JSTOR at jstor.org. The publications by the Royal Netherlands Academy of Arts and Sciences (KNAW) are made available on the *Digital Library—Dutch History of science web centre* [384].

A complete list of Oort's publications with links to scans at ADS or other '>places can be found on the accompanying Website www.astro.rug.nl/JHOort.

Oort. J. H. 1921. *Iets over het werk van Prof. J.C. Kapteyn*, Groningensche Studentenalmanak voor het jaar 1922, vier-en-negentigste jaargang, 202–215.

Oort, J. H. 1922. *Some peculiarities in the motion of stars of high velocity*. Bulletin of the Astronomical Institutes of the Netherlands, 1, 133–137. [1922BAN.....1..133O]

Schilt, J. & Oort, J. H. 1923. *The frequency of a component of the linear velocity for stars brighter than* $3^{\mathrm{m}}.8$ *of spectral types F, G, K and M, derived from the proper motions of Boss' Catalogue*. Hoitsema brothers, Groningen.

Oort, J. H. & H.M. Marsh, H. M. 1924a. *On the proper motions of stars of the thirteenth magnitude*. Popular Astronomy, 32, 559–561. [1924PA.....32..559O]

Oort, J. H. 1924b. *A comparison of the average velocity of binaries with that of single stars.* Astronomical Journal, 35, 141–144. [1924AJ.....35..141O]

Oort, J. H. 1924c. *Note on the difference in velocity between absolutely bright and faint stars*. Proceedings of the National Academy of Sciences (U.S.A.), 10, 253–256. [1924PNAS...10..253O]

Oort, J. H. 1924d. *On a possible relation between globular clusters and stars of high velocity*. Proceedings of the National Academy of Sciences (U.S.A.), 10, 256–260. [1924PNAS...10..256O]

Oort, J. H. 1924e. *A comparison of the average velocity of binaries with that of single stars*. Popular Astronomy, 32, 226–227. [1924PA.....32..226O]

Sanders, C. & Oort, J. H. 1925b. *A discussion of the determination of declinations from azimuth measures made near the equator*. Bulletin of the Astronomical Institutes of the Netherlands, 2, 201–208. [1925BAN.....2..201S] Errata: BAN 2, V.

de Sitter, W. & Oort, J. H. 1925d. *Provisional scheme for the determination of fundamental declinations from azimuth observations*. Bulletin of the Astronomical Institutes of the Netherlands, 3, 1–6. [1925BAN.....3....1D]

Oort, J. H. & Doorn, N. W. 1925f. *The solar motion from radial velocities of the absolutely brightest stars of spectra F, G, K and M*. Bulletin of the Astronomical Institutes of the Netherlands, 3, 71. [1925BAN.....3...71O]

Oort, J. H. 1925g. *Bepaling van declinatie's onafhankelijk van vertikale refractie* (abstract). Hemel & Dampking, 23, 79–80.

Oort, J. H. 1926a. *Asymmetry in the distribution of stellar velocities*, The Observatory, 49, 302–304 [1926Obs....49..302O]

Oort, J. H. 1926b. *The stars of high velocity*. PhD thesis, University of Groningen, also: Publications of the Kapteyn Astronomical Laboratory Groningen, 40, 1–75. [1926PhDT.........1O]

Oort, J. H. 1927a. *Niet-lichtgevende materie in het sterrenstelsel*, Inaugural Lecture, reprinted in Hemel and Dampkring 25, 13–21 and 60–70. *Non-Light emitting matter in the Stellar System*, translated by P. C. van der Kruit in the *The Legacy of J.C. Kapteyn*, Appendix A.

Oort, J. H. 1927c. *Observational evidence confirming Lindblad's hypothesis of a rotation of the Galactic System*. Bulletin of the Astronomical Institutes of the Netherlands, 3, 275–282. [1927BAN.....3..275O]

Oort, J. H. 1927d. *Investigations concerning the rotational motion of the Galactic System together with new determinations of secular parallaxes, precession and motion of the equinox*. Bulletin of the Astronomical Institutes of the Netherlands, 4, 79–89 (Errata: 4, 94). [1927BAN.....4...79O]

Oort, J. H. 1927e. *Additional notes concerning the rotation of the Galactic System*. Bulletin of the Astronomical Institutes of the Netherlands, 4, 91–92. [1927BAN.....4...91O]

Oort, J, H. 1927f. *Summary of the principal radial velocity data used for the results of Bull. Astr. Inst. Netherlands 3 275 and 4 79*. Bulletin of the Astronomical Institutes of the Netherlands, 4, 93–94. [1927BAN.....4...93O]

Oort, J. H. 1928a. *An investigation of the constant of refraction from observations at Leiden and at the Cape*. Bulletin of the Astronomical Institutes of the Netherlands, 4, 137–142. [1928BAN.....4..137O]

Oort, J. H. 1928b. *Dynamics of the Galactic System in the vicinity of the Sun*. Bulletin of the Astronomical Institutes of the Netherlands, 4, 269–284. [1928BAN.....4..269O]

Oort, J. H. 1928c. *Catalogue of 460 stars for the epoch and equinox 1885.0, from meridian observations made at Leiden in the years 1880 to 1897*. Annalen van de Sterrewacht te Leiden, 13, D1–D47. [1928AnLei..13D...1O]

Oort, J. H. 1929a. *Note on O. Struve's intensity estimates of the Calcium K line in early-type stars*. Bulletin of the Astronomical Institutes of the Netherlands, 5, 105–107. [1929BAN.....5..105O

Oort, J. H. 1930a. *The estimated number of high velocities among stars selected according to proper motion*. Bulletin of the Astronomical Institutes of the Netherlands, 5, 189–192. [1930BAN.....5..189O]

Oort, J. H. 1930b. *The motion of the Sun with respect to the interstellar gas*. Bulletin of the Astronomical Institutes of the Netherlands, 5, 192–194. [1930BAN.....5..192O]

Oort. J. H. 1930c. *Note on the velocities of extragalactic nebulae*. Bulletin of the Astronomical Institutes of the Netherlands, 5, 239–241. [1930BAN.....5..239O]

Oort, J. H. 1931a. *Some problems concerning the distribution of luminosities and peculiar velocities of extragalactic nebulae*. Bulletin of the Astronomical Institutes of the Netherlands, 6, 155–160. [1931BAN.....6..155O]

Oort, J. H. 1931b. *Licht-absorptie in het Melkwegstelsel*. Hemel & Dampkring, 29, 41–49.

Oort. J. H. 1932a. *The force exerted by the stellar system in the direction perpendicular to the Galactic plane and some related problems*. Bulletin of the Astronomical Institutes of the Netherlands, 6, 249–287. [1932BAN.....6..249O]

Oort, J. H. 1932b. *Note on the distribution of luminosities of K and M giants*. Bulletin of the Astronomical Institutes of the Netherlands, 6, 289–294. [1932BAN.....6..289O]

Oort, J. H. 1935a. *Obituary: Willem de Sitter*, The Observatory, 58, 22–27. [1935Obs....58...22O]

Oort, J. H. 1936b. *De bouw der sterrenstelsels*, Inaugural lecture, reprinted in Hemel & Dampkring, 34, 1–16.

Oort, J. H. 1936c. *Mean parallaxes of faint stars derived from the Radcliffe Catalogue of proper motions*. Bulletin of the Astronomical Institutes of the Netherlands, 8, 75–104. [1936BAN.....8...75O]

Oort, J. H. 1937a. *A redetermination of the constant of precession, the motion of the equinox and the rotation of the Galaxy from faint stars observed at the McCormick Observatory*. Bulletin of the Astronomical Institutes of the Netherlands, 8, 149–155. [1937BAN.....8..149O]

Oort, J. H. 1938c. *Absorption and density distribution in the Galactic System*. Bulletin of the Astronomical Institutes of the Netherlands, 8, 233–264 [1938BAN.....8..233O]

Oort, J. H. 1939a. *Motions of RR Lyrae variables.*. Bulletin of the Astronomical Institutes of the Netherlands, 8, 337–338. [1939BAN.....8..337B]

Oort, J. H. 1939b. *De nieuwe dubbele astrograaf van de afdeeling der Leidsche Sterrewacht te Johannesburg*, Hemel & Dampkring, 37, 129–133.

Oort, J. H. 1940a. *Some problems concerning the structure and dynamics of the Galactic System and the elliptical nebulae NGC3115 and 4494.* Astrophysical Journal, 91, 273–306. [1940ApJ....91..273O], erratum [1940ApJ....92..440]

Oort, J. H. & van Woerkom, A. J. J. 1941b. *The attractive force of the Galactic System as determined from the distribution of RR Lyrae variables.* Bulletin of the Astronomical Institutes of the Netherlands, 9, 185–188 (Errata: 11, 270). [1941BAN.....9..185O]

Oort, J. H. 1941c. *Note on the structure of the inner parts of the Galactic System.* Bulletin of the Astronomical Institutes of the Netherlands, 9, 193–196. [1941BAN.....9..193O]

Mayall, N.U. & Oort, J. H. 1942a. *Further data bearing on the identification of the Crab Nebula with the supernova of 1054 A.D.; II. The Astronomical Aspects.* Publications of the Astronomical Society of the Pacific, 54, 95–104. [1942PASP...54...95M]

Oort, J. H. 1942b. *A determination of the Galactic pole from stars at large distances from the Galactic plane.* Bulletin of the Astronomical Institutes of the Netherlands, 9, 324–325. [1942BAN.....9..324O]

Oort, J. H. & Oosterhoff, P.T. 1942c. *Note on the distances and motions of some extremely remote Cepheids in Cygnus.* Bulletin of the Astronomical Institutes of the Netherlands, Vol. 9, p.325–327. [1942BAN.....9..325O]

Oort, J. H. & van Tulder, J. J. M. 1942d. *On the relation between velocity-and density-distribution of long-period variables.* Bulletin of the Astronomical Institutes of the Netherlands, 9, 327–331. [1942BAN.....9..327O]

Oort, J. H. & van Tulder, J. J. M. 1942e. *Remark on the distances of long-period variables.* Bulletin of the Astronomical Institutes of the Netherlands, 9, 332–334. [1942BAN.....9..332O]

Oort, J. H. 1942f. *The constants of differential rotation and the ratio of the two Galactic axes of the velocity ellipsoid in the case when peculiar motions are not negligible.* Bulletin of the Astronomical Institutes of the Netherlands, Vol. 9, p.334–334. [1942BAN.....9..334O]

Oort, J. H. 1943a. *De twee sterstroomen: Het verschijnsel verklaard uit de rotatie van ons Melkwegstelsel*, Hemel & Dampkring, 41, 40–46.

Oort, J. H. 1943b. *Some remarks on the fundamental systems of the General Catalogue and the Dritter Fundamentalkatalog.* Bulletin of the Astronomical Institutes of the Netherlands, 9, 417–422. [1943BAN.....9..417O]

Oort, J. H. 1943c. *Tentative corrections to the FK3 and GC systems of declinations.* Bulletin of the Astronomical Institutes of the Netherlands, 9, 423–424. [1943BAN.....9..423O]

Oort, J. H. 1943d. *The constants of precession and of Galactic rotation.* Bulletin of the Astronomical Institutes of the Netherlands, 9, 424–427. [1943BAN.....9..424O] ter Haar, D. van de Hulst, H. C. Oort, J. H. & van Woerkom, A. J. J. 1943e. *De vorming van vaste deeltjes in het interstellaire gas*. Nederlands Tijdschrift voor Natuurkunde, 10, 238–257 (1943).

Raimond, J. J. & Oort, J. H. 1943g. *Nevels rondom Nova Persei: Een boeiende geschiedenis*. Hemel & Dampkring, 41, 145–163 and 200–206.

Oort, J. H. 1945a. *In memeriam Professor Frank Schlesinger*, Hemel & Dampkring, 43, 27–30.

Oort, J. H. & van de Hulst, H. C. 1946b. *Gas and smoke in interstellar space*. Bulletin of the Astronomical Institutes of the Netherlands, 10, 187–204. [1946BAN.... 10..187O]

Oort, J. H. 1946c. *In memoriam Dr. A. de Sitter*, Hemel & Dampkring, 44, 33.

Oort, J. H. 1946d. *Bij het afscheid van Professor Ejnar Hertzsprung*, Hemel & Dampkring, 44, 166–168.

Oort, J. H. 1946e. *Some phenomena connected with interstellar matter* (George Darwin Lecture). Monthly Notices of the Royal Astronomical Society, 106, 159–179. [1946MNRAS.106..159O]

Oort, J. H. 1946h. *Obituaries: W. C. Martin, A. de Sitter, J. Uitterdijk*. The Observatory, 66, 265–266. [1946Obs....66..265O]

Oort, J. H. 1947c. *In memoriam Dr. H. van Gent*, Hemel & Dampkring, 45, 159–160.

Oort, J. H. 1950b. *The structure of the cloud of comets surrounding the Solar System and a hypothesis concerning its origin*. Bulletin of the Astronomical Institutes of the Netherlands, 11, 91–110 (errata p.VIII). [1950BAN....11...91O]

Oort, J. H. & Schmidt, M. 1951c. *Differences between new and old comets*. Bulletin of the Astronomical Institutes of the Netherlands, 11, 259–269. [1951BAN.... 11..259O]

Westerhout, G. & Oort, J. H. 1951d. *A comparison of the intensity distribution of radio-frequency radiation with a model of the Galactic System*. Bulletin of the Astronomical Institutes of the Netherlands, 11, 323–333. [1951BAN....11..323W]

Morgan, H. R. & Oort, J. H. 1951e. *A new determination of the precession and the constants of Galactic rotation*. Bulletin of the Astronomical Institutes of the Netherlands, 11, 379–384. [1951BAN....11..379M]

Oort, J. H. 1951f. *Introduction*. In: Problems of Cosmical Aerodynamics. Ed. J.M. Burgers & H.C. van de Hulst. Central Air Documents Office, Dayton, Ohio, 1–6. [1951pca..conf....1O]

Oort, J. H. 1951g. *Interaction of nova and supernova shells with the interstellar medium*. In: Problems of Cosmical Aerodynamics. Ed. J.M. Burgers & H.C. van de Hulst. Central Air Documents Office, Dayton, Ohio, 118–124. [1951pca..conf..118O]

Muller, C. A. & Oort, J. H. 1951h. *Observation of a line in the Galactic radio spectrum: The interstellar hydrogen line at 1,420 Mc./sec. and an estimate of Galactic rotation*. Nature, 168, 357–358. [1951Natur.168..357M]

Oort, J. H. 1951j. *Origin and development of comets* (Halley Lecture). The Observatory, 71, 129–144. [1951Obs....71..129O]

Oort, J. H. 1951l. *Obituary: Dr. C.H. Hins*. The Observatory, 71, 243–244. [1951Obs....71..243.]

Oort, J. H. 1952b. *Problems of Galactic Structure* (Henry Norris Russell Lecture). Astrophysical Journal, 116, 233–250. [1952ApJ...116..233O]

Oort, J. H. 1952g. *Expanding groups of B-type stars*. Monthly Notes of the Astronomical Society of South Africa, 11, 91–92. [1952MNSSA..11...91O]

van Houten, C.J. Oort, J. H. & Hiltner, W.A. 1954b. *Photoelectric measurements of extragalactic nebulae*. Astrophysical Journal, 120, 439–453. [1954ApJ...120..439V]

van de Hulst, H. C. Muller, C. A. & Oort, J. H. 1954c. *The spiral structure of the outer part of the Galactic System derived from the hydrogen emission at 21 cm wavelength*. Bulletin of the Astronomical Institutes of the Netherlands, 12, 117–149. [1954BAN....12..117V]

Oort, J. H. 1954d. *Outline of a theory on the origin and acceleration of interstellar clouds and O associations*. Bulletin of the Astronomical Institutes of the Netherlands, 12, 177–186. [1954BAN....12..177O]

Oort, J. H. & Spitzer, L. 1955b. *Acceleration of interstellar clouds by O-Type stars*. Astrophysical Journal, 121, 6–23. [1955ApJ...121....6O]

Oort, J. H. 1955e. *Outline of a theory on the origin and acceleration of interstellar clouds and O-associations*. In: Gas dynamics of cosmic clouds, IAU Symposium 2, Ed. H.C. van de Hulst, North Holland Pub. Co. Amsterdam, 147–158. [1955IAUS....2..147O]

Oort, J. H. & Walraven, Th. 1956b. *Polarization and composition of the Crab nebula*. Bulletin of the Astronomical Institutes of the Netherlands, 12, 285–308. [1956BAN....12..285O]

Oort, J. H. 1956e. *The evolution of galaxies*. Scientific American, 195(3), 100–108. [1956SciAm.195c.100O]

Oort, J. H. 1956f. *A New Southern Hemisphere Observatory*. Sky & Telescope, 15, 163. [1956S&T....15..163O]

van Woerden, H. Rougoor, G. W. & Oort, J. H. 1957c. *Expansion d'une structure spirale dans le noyau du Systéme Galactique, et position de la radiosource Sagittarius A*. Comptes Rendus de l'Academie des Sciences, 244, 1691–1695. [1957CRAS..244.1691V]

Oort, J. H. & Walraven, Th. 1957d. *Polarization and the radiating mechanism of the Crab Nebula* In: Radio astronomy, IAU Symposium 4, Ed.H.C. van de Hulst, Cambridge Univ. Press, 197–200. [1957IAUS....4..197O]

Oort, J. H. 1958c. *Comparison of the Galactic System with other stellar systems*. In: Comparison of the large-scale structure of the Galactic System with that of other stellar systems, IAU Symposium 5, Ed. N.G. Roman, Cambridge Univ. Press, 69–72. [1958IAUS....5...69O]

Oort, J. H. & van Herk, G. 1959c. *Structure and dynamics of Messier 3*. Bulletin of the Astronomical Institutes of the Netherlands, 14, 299–321. [1959BAN....14..299O]

Oort, J. H. 1960b. *Very accurate positions of selected stars*. Astronomical Journal, 65, 229–231. [1960AJ.....65..229O]

Oort, J. H. 1960c. *Note on the determination of K_z and on the mass density near the Sun*. Bulletin of the Astronomical Institutes of the Netherlands, 15, 45–53. [1960BAN....15...45O]

Oort, J. H. & Rougoor, G. W. 1960f. *The position of the Galactic Centre*. Monthly Notices of the Royal Astronomical Society, 121, 171–173. [1960MNRAS.121..171O]

Oort, J. H. 1960g. *Radio astronomy—A window on the Universe*. American Scientist 48, 160–178. JSTOR: www.jstor.org/stable/pdf/27827535.pdf

Rougoor, G. W. & Oort, J. H. 1960h. *Distribution and motion of interstellar hydrogen in the Galactic System with particular reference to the region within 3 kiloparsecs of the Center*. Proceedings of the National Academy of Sciences of the United States of America, 46, 1–13. [1960PNAS...46....1R]

Oort. J. H. 1962i. *Spiral Structure*, In: The distribution and motion of interstellar matter in galaxies. Proceedings of a conference held April 1961 at the Institute for Advanced Study, Princeton, Ed. L. Woltjer. New York, Benjamin, Inc. 234–242. [1962dmim.conf..234O]

Oort, J. H. 1962j. *Some considerations concerning the study of the Universe by means of large radiotelescopes*, In: Large Radio-Telescopes, Proceedings of the OECD Symposium on Large Antennae for Radioastronomy in Paris, 35–51. [1961lrt..conf...35O]

Oort, J. H. 1961k. *Considerations concerning the minimum resolving power required for 21-cm line observations with a very large antenna. Ibid.*, 53–54. [1961lrt..conf...53O]

Oort, J. H. 1961l. *Introduction. Ibid.*, 33–34. [1961lrt..conf...33O]

Oort, J. H. 1961m. *Some suggested programmes. Ibid.*. 123–129. [1961lrt..conf.. 123O]

Oort, J. H. 1962c. *Address by the President of the Union*, Transactions of the International Astronomical Union, 11B, 7–13. [1962IAUTB..11....7O]

Oort, J. H. 1963c. *Empirical data on the origin of comets*, In: The Solar System IV: The Moon, meteorites and comets, Ed. G.P. Kuiper & B.M. Middlehurst. University of Chicago Press, Chicago, 665–673. [1963mmc..book..665O]

Oort, J. H. 1964b. *Structure of the Galaxy*, In: The Galaxy and the Magellanic Clouds, IAU-URSI Symposium 20, Ed. F.J. Kerr & A.W. Rodgers, 1–9. [1964IAUS...20....1O]

Oort, J. H. 1964c. *A large high velocity cloud at* $l^{II} = 41°$, $b^{II} = -15°$, In: The Galaxy and the Magellanic Clouds, IAU-URSI Symposium 20, Ed. F.J. Kerr & A.W. Rodgers, 130. [1964IAUS...20..130O]

Oort, J. H. 1964d. *Recent observations at Dwingeloo of the central region of the Galactic System*, In: The Galaxy and the Magellanic Clouds, IAU-URSI Symposium 20, Ed. F.J. Kerr & A.W. Rodgers, 179–183. [1964IAUS...20..183O]

Oort, J. H. 1965a. *Stellar dynamics*. In: Stars & Stellar Systems V: Galactic structure. Ed. A. Blaauw & M. Schmidt. University of Chicago Press, Chicago, 455–511. [1965gast.conf..455O; pdf download option might give wrong pages]

Oort, J. H. 1966b. *Possible interpretations of the high-velocity clouds*. Bulletin of the Astronomical Institutes of the Netherlands, 18, 421–438. [1966BAN....18..421O]

Katgert, P. & Oort, J. H. 1967a. *On the frequency of supernova outbursts in galaxies*. Bulletin of the Astronomical Institutes of the Netherlands, 19, 239–245. [1967BAN....19..239K]

van Bueren, H. G. & Oort, J, H. 1968c. *On the possibility of amplification of 21-cm radio emission in high-velocity clouds*. Bulletin of the Astronomical Institutes of the Netherlands, 19, 414–416. [1968BAN....19..414V] Errata: BAN 20, 224.

Oort, J. H. 1968d. *Survey of possible programmes with the SRT at 1415 MHz*. Internal Technical Report of the Synthesis Radio Telescope Project, No. 74.

Oort, J. H. 1968e. *Radio astronomical studies of the Galactic System*, In: Galaxies and the Universe, Proceedings of the Vetlesen Symposium, Ed. L. Woltjer. New York: Columbia University Press, 1–32. [1968gaun.book....1O]

Oort, J. H. 1969b. *Infall of gas from intergalactic space*. Nature, 224, 1158–1163. [1969Natur.224.1158O]

Oort, J. H. 1970f. *The formation of galaxies and the origin of the high-velocity hydrogen*. Astronomy and Astrophysics, 7, 381–404. [1970A&A.....7..381O]

Oort, J. H. 1970g. *The density of the Universe*. Astronomy and Astrophysics, 7, 405–407. [1970A&A.....7..405O]

Oort, J. H. 1970h. *Galaxies and the Universe*. Science, 170, 1363–1370. [1970Sci ...170.1363O]

Oort, J. H./ 1970j. *Afscheidscollege*, Nederlands Tijdschrift voor Natuurkunde 36, 321–325. Also Hemel & Dampkring, 68, 257–161.

Oort, J. H. 1971b. *Composition and activity of the nucleus of our Galaxy, and comparison with M 31*, In: Nuclei of Galaxies, Pontificiae Academiae Scientiarum Scripta Varia, 321–349. [1971swng.conf..321O]

Oort, J. H./ 1971d. *To the horizon of the Universe*, translation of 1970i, Delta 14, No. II, 33–45.

Oort, J. H. 1971e. *Absence of Radio Emission from Maffei I*, Nature, 230, 103–105. [1971Natur.230..103O]

van der Kruit, P. C. Oort, J. H. & Mathewson, D. S. 1972b. *The radio emission of NGC 4258 and the possible origin of spiral structure*, Astronomy and Astrophysics, 21, 169–184. [1972A&A....21..169V] Erratum: [1973A&A....22..479V]

Hulsbosch, A. N. M. & Oort, J. H. 1973d. *Note on Verschuur's article on high-velocity clouds and 'normal' Galactic structure*, Astronomy and Astrophysics, 22, 153–154. [1973A&A....22..153H]

Oort, J. H. 1974e. *Zwakte van Kohoutek was te verwachten*, Zenit, 1, No.2, 13–14.

Oort, J. H. 1975b. *Phenomenology of spiral galaxies*, In: Structure and Evolution of Galaxies. Ed. G. Setti. Reidel, Dordrecht, ISBN 90-277-0325-6[1975 ASIC...21...85O]

Oort, J. H. & Plaut, L. 1975g. *The distance to the Galactic centre derived from RR Lyrae variables, the distribution of these variables in the Galaxy's inner region and halo, and a rediscussion of the Galactic rotation constants*, Astronomy and Astrophysics, 41, 71–86. [1975A&A....41...71O]

Pels, G. Oort, J. H. & Pels-Kluyver, H. A. 1975h. *New members of the Hyades cluster and a discussion of its structure*, Astronomy and Astrophysics, 43, 423–441. [1975A&A:...43..423P]

Strom, R. G. Miley, G. K. & Oort, J. H. 1975k. *Giant radio galaxies*, Scientific American, 233, No. 2, 26–35. [1975SciAm.233b..26S]

Oort, J. H. 1976c. *Interpretations of radio and optical observations of NGC 1275*, Publications of the Astronomical Society of the Pacific, 88, 591–593. [1976PASP...88..591O]

Oort, J. H. 1976d. *A survey of planetary nebulae near the Galactic center*, Publications of the Astronomical Society of the Pacific, 88, 596–597. [1976PASP...88..596O]

Rubin, V. C. Ford, W. K. Peterson, C.J. & Oort, J. H 1977a. *New observations of the NGC 1275 phenomenon*, Astrophysical Journal, 211, 693–696. [1977ApJ...211..693R]

Oort, J,. H. 1977b. *The expected number density of globular clusters near the Galactic center*, Astrophysical Journal, 218, L97–L101. [1977ApJ...218L..97O]

Oort, J.H. 1977c. *The Galactic center*, Annual review of astronomy and astrophysics, 15, 295–362. [1977ARA&A..15..295O]

Oort, J. H. 1978f. *Eruptive phenomena near the galactic centre*, Physica Scripta, 17, 175–184.[1978PhyS...17..175O]

Oort, J. H. 1978g. *Remark on activity in nuclei of spiral galaxies different from that in Seyfert galaxies*, Physica Scripta, 17, 369. [1978PhyS...17..369O]

Oort, J. H. 1979c. *Great expectations*, The Messenger (ESO), 16, 17–19. [1979Msngr..16...17O]

Oort, J. H. 1979d. *Bannier's betrokkenheid bij dertig jaar sterrenkunde*, Onder de ZWO-Bannier, ZWO, 85–101.

Oort, J. H. 1979e. *Luminosity distribution and shape of the Hyades cluster*, Astronomy and Astrophysics, 78, 312–317. [1979A&A....78..312O]

Oort, J. H. & Pottasch, S. R. 1979f. *Materia e radiazione interstellare*, Enciclopedia del Novecente, ed. G. Bedeschi, Vol. IV, 19–32.

Oort, J. H. 1981a. *Superclusters and Lyman α absorption lines in quasars*, Astronomy and Astrophysics, 94, 359–364. [1981A&A....94..359O]

Oort, J. H. Arp, H. & de Ruiter, H. 1981b. *Evidence for the location of quasars in superclusters*, Astronomy and Astrophysics, 95, 7–13. [1981A&A....95....7O]

Schwarz, U.J. & Oort, J. H. 1981c. *HI fine structure in a high velocity Cloud (HVC AI)*, Astronomy and Astrophysics, 101, 305–314. [1981A&A...101..305S]

Oort, J. H. 1981d. *Some notes on my life as an astronomer*. Annual Review Astronomy and Astrophysics, 19, 1–5. [1981ARA&A..19....1O]

Oort, J. H. 1982b. *The nature of the largest structures in the universe*, In: Astrophysical cosmology, Specula Vaticana, Pontificia Academia Scientiarum, 145–163. [1982ac...proc..145O]

Oort, J. H. 1982e. *Reminiscences of Astronomy in the Twentieth Century*. Memorie della Societa Astronomica Italiana, 53, 795–801. [1982MmSAI..53..795O]

Oort, J. H. 1983a. *Superclusters*, Annual review of astronomy and astrophysics, 21 373–428. [1983ARA&A..21..373O]

Oort, J. H. 1983b. *Structure of the universe*, Keynote review in: Early evolution of the Universe and its present structure, IAU Symposium 104, ed. G.O. Abell & G. Chincarini, Dordrecht, Reidel, 1–6; Discussion, p. 6. [1983IAUS..104....1O]

Oort, J. H. 1984a. *Superclusters at large redshifts—Can protosuperclusters and the birth of 'pancakes' be observed?*, Astronomy and Astrophysics, 139, 211–214. [1984A&A...139..211O]

Oort, J. H. 1986a. *The origin and dissolution of comets*, The Observatory, 106, 186–193. [1986Obs...106..186O]

Oort, J. H. 1988a. *Origin of structure in the universe*, Publications of the Astronomical Society of Japan, 40, 1–14. [1988PASJ...40....1O] Erratum: [1988PASJ ...40..637O].

Oort, J. H. 1989b. *Horizonnen*, Zenit, 16, 124–132. [1989Zenit..16..124O]

Oort, J. H. 1990a. *Orbital distribution of comets*. In: Physics and chemistry of comets, ed. W.F. Huebner, Springer, 235–244. [1990pcc..conf..235O]

Oort, J. H. 1992a. *Exploring the nuclei of galaxies*, Mercury, 21, 57–61. [1992 Mercu..21b..57O]

A.3 Publications About J.H. Oort

A complete list of these publications with links to scans at ADS or other places can be found on the accompanying Website www.astro.rug.nl/JHOort.

Articles (in chronological order).

Plaskett, H. H. 1946. *(Presidential Address) on presenting the Gold Medal to Professor J. H. Oort*. Monthly Notices of the Royal Astronomical Society, 106, 242–243. [1946MNRAS.106..242P]

King, A. S. 1946. *The presentation of the Bruce Gold Medal for the Year 1942 to Dr. Jan H. Oort*, Publications of the Astronomical Society of the Pacific, 58, 229–232.

van der Laan, H. 1992. *Jan Hendrik Oort (1900–1992)—Looking ahead in wonder*. The Messenger, 70, 1–2. [1992Msngr..70....1V]

Blaauw, A. 1993. *Jan Hendrik Oort, 28 april 1900—5 november 1992*. Zenit, 20, 196–210. [1993Zenit..20..196B]

van de Hulst, H. C. 1993. *Levensbericht J.H. Oort*, Levensberichten en herdenkingen, 1993, Koninklijke Academie van Wetenschappen, 67–73. www.dwc.knaw.nl/DL/levensberichten/PE00002152.pdf

Woltjer, L. 1993. *Jan Hendrik Oort 1900–1992*. Journal of Astrophysics and Astronomy, 14, 3–5. [1993JApA...14....3W]

Blaauw, A. & Schmidt, M. 1993. *Jan Hendrik Oort (1900–1992)*. Publications of the Astronomical Society of the Pacific, 105, 681–685. [1993PASP..105..681B]

Woltjer, L. 1993. *Jan Hendrik Oort died 5 November 1992*. Physics Today, 46, 104–105. [1993PhT....46k.104W]

van den Bergh, S. 1993. *An Astronomical Life—Oort, J.H.*. Journal of the Royal Astronomical Society of Canada, 87, 73–75. [1993JRASC..87...73V]

Pecker, J.-C. 1993. *Jan Hendrik Oort, 28 avril 1900—05 novembre 1992*. Astronomie, 107, 60–64. [1993LAstr.107...60P]

Pecker, J.-C./ 1993. *La vie et l'œuvre de Jan Hendrik Oort (1900–1992)*. Comptes Rendus de l'Academie des Sciences, Sér. Gén. Vie Sci. 10, 535–540. [1993CRASG..10..535P]

Schilling, G. 1993. *Jan Oort Remembered*, Sky & Telescope, 85, 44. [1993S&T.... 85...44S]

Herrmann, D. B. 1993. *Jan Hendrik Oort. Zum ersten Todestag des großen Niederländischen Astronomen.* Astronomie und Raumfahrt, 30, Heft 18, 6–7. [1993A&R....30....6H]

van de Hulst, H. C. 1994. *Jan Hendrik Oort (1900–1992).* Quarterly Journal of the Royal Astronomical Society, 35, 237–242. [1994QJRAS..35..237V]

Books

L. Woltjer. 1968. *Galaxies and the universe*. Proceedings of the Vetlesen Symposium, held at Columbia University, October 19, 1966, Columbia University Press, ISBN-10: 0231031106.

van Woerden, H. Brouw, W. N. & van de Hulst, H. C. 1980. *Oort and the Universe. A sketch of Oort's research and person*, Liber Amicorum presented to Jan Hendrik Oort on the occasion of his 80th birthday, Dordrecht: D. Reidel Publishing Company, ISBN-10: 9027711801.

Katgert-Merkelijn, J. K. 1997. *The letters and papers of Jan Hendrik Oort as archived in the University Library*, Dordrecht, Kluwer, ISBN: 978079234542.

Katgert-Merkelijn, J .K. & Damen, J. 2000. *Jan Oort, Astronomer*, Catalogue of an exhibition in Leiden University Library, April 20—May 27, 2000, Leiden Univ. Lib. ISSN 0921-9293, vol. 35.

Elbers, A. 2015. *Early Dutch radio astronomy (1940–1970): The people and the politics*, PhD thesis, Leiden University. Published as *The rise of radio astronomy in the Netherlands: The people and the politics*, Springer, ISBN: 978-3-319-49078-6 (2017).

A.4 Oort and His Students: a List of Ph.D. Theses

The following are Ph.D. theses supervised by Oort.

GIJSBERT VAN HERK (1 May, 1936): *Eenige uitkomsten van de waarnemingen in de jaren 1931–1933 te Equator gedaan voor de verbetering van het declinatiesysteem van Boss* (Some results of observations made in Equator from 1931 to 1933 to improve Boss' system of declinations), eventually part of [1938AnLei..18A...1H]

LEENDERT BINNENDIJK (8 January, 1947): *A study of stars in the Pleiades region, based on photographic magnitudes, colour-aequivalents, spectral types and proper motions*, published partly also as [1947BAN....10..259B]

ADRIANUS JAN JASPER VAN WOERKOM (29 September, 1948) *On the origin of comets*, published also as [1948BAN....10..445V]

MAARTEN SCHMIDT (21 March, 1956): *A model of the distribution of mass in the Galactic System*, published also as [1956BAN....13...15S]

LODEWIJK WOLTJER (6 November, 1958): *The Crab Nebula*, published also as [1958BAN....14...39W]

CORNELIS JOHANNES VAN HOUTEN (14 June, 1961): *Surface photometry of extragalactic nebulae*, published also as [1961BAN....16....1V]

GERRIT WILLEM ROUGOOR (30 September, 1964): *The neutral hydrogen in the central region of the Galactic System*, published also as [1964BAN....17..381R]

ERNST RAIMOND (16 December 1964): *Hydrogen connected with stellar associations in Monoceros*, published also as [1966BAN....18..191R]
JEANETTE KATE MERKELIJN (18 November, 1970): *A determination of the luminosity function of radiogalaxies at 400 and 2700 MHz*, published also as [1971A&A.... 15...11M]
WILLIAM BUTLER BURTON (25 November, 1970): *The Sagittarius spiral arm and the distribution of hydrogen between 0° and 90° longitude*, published also as [1971A&A....10...76B]
WILLIAM WHITNEY SHANE (16 June, 1971): *Observations of neutral hydrogen in an interior region of the Galaxy and the structure and kinematics of the Scutum spiral arm*, published also as [1971A&AS....4....1S], [1971A&AS....4..315S] and [1972A&A.....16..118S]
ELISABETH MABEL BERKHUIJSEN (7 June, 1971): *A Study of the Galactic radiation and the degree of polarization at 820 MHz with special reference to the loops and spurs*, published also as [1971A&A....14..359B] and [1972A&AS....5..263B]
PIETER CORIJNUS VAN DER KRUIT (6 October 1971): *Evidence for past activity in the Galactic nucleus*, published also as [1970A&A.....4..462V] and [1971A&A....13 ..405V]
TITUS ADRIANUS THOMAS SPOELSTRA (20 December, 1972): *Linear polarization in the Galactic spurs at 1415 MHz* (supervisors J.H. Oort & H. van der Laan), published also as [1971A&A....13..237S], [1972A&AS....5..205S], [1972A&AS....7..169S] and [1972A&A....21...61S]
ADRIANUS NICOLAAS MARIA HULSBOSCH (10 January, 1973): *Studies on high-velocity clouds*, published also as [1975A&A....40....1H]

After 1970, when Oort had become emeritus, he had to stop taking on new Ph.D. students, but he had some that were still working on their thesis research. Of these I was the last one Oort hired as a Ph.D. student in December 1968. Hulsbosch's thesis research had started earlier and took almost ten years to complete. Spoelstra's ZWO-funded project was intended for newly appointed lector ('reader') van der Laan, but the latter wished to make it a joint project with Oort. The supervision during the writing up of the thesis was mostly by him.

After 1975, five years after retirement, Oort was, no longer allowed to formally act as supervisor/promotor according to the protocols; he kept reading theses and taking active part in the opposition during the thesis defense ceremonies.

There are some other Leiden theses that were prepared during Oort's days in which he was not one of the formal Ph.D. supervisor/promotor, but in which he must have had a strong interest and at least some involvement.
GART WESTERHOUT (8 July, 1958): *A survey of the continuous radiation from the Galactic system at a frequency of 1390 Mc/s* (supervisor H.C. van de Hulst), published also as [1958BAN....14..215W]
ALEXANDER OLLONGREN (5 July, 1962): *Three-dimensional stellar orbits* (supervisor H.C. van de Hulst), published also as [1962BAN....16..241O]
HUGO VAN WOERDEN (6 July, 1962): *De neutrale waterstof in Orion* (The neutral hydrogen in Orion) (Supervisor A. Blaauw, degree awarded in Groningen), abstract published as [1963AJ.....68S.296V]

WILLEM NICOLAAS BROUW (15 September, 1971): *Data processing for the Westerbork synthesis telescope* (supervisor A.C. Muller), published in different form as [1974A&A....33..289H]

VINCENT ICKE (13 June, 1972): *Formation of galaxies inside clusters* (supervisor H.C. van de Hulst), published also as [1973A&A....27....1I]

ELLY DEKKER (16 April, 1975): *Spiral structure and the dynamics of flat stellar systems* (supervisor H.C. van de Hulst), published in a different form as [1976PhR....24..315D]

PETER KATGERT (25 May, 1977): *Populations of weak radio sources* (supervisor H. van der Laan), published as [1978A&AS...31..409K] and [1979A&A....73..107K]

GEERT DICK VAN ALBADA (4 October, 1978): *The peculiar spiral galaxy NGC 4258* (supervisors H. van der Laan & W.W. Shane), published as [1980A&AS...39..283V] and [1980A&A....90..123V]

PIETER TIMOTHEUS DE ZEEUW (19 June, 1984): *Dynamics of tri-axial stellar systems* (supervisor H.C. van de Hulst) [1984PhDT........87D]

MARC JAN ANTON OORT (19 September, 1987): *Radio galaxies at very low flux levels* (supervisors H. van der Laan & P. Katgert), published as [1987A&AS...71...25O], [1988A&A...192...42O] and [1988A&A...193....5O]

A.5 Oort's Honors

Figures A.3 and A.4 constitute a list of Oort's honors as he listed them himself. He stops with the Balzan Prize for Astrophysics in 1984. There are two more to add to Oort's own listing, namely the Kyoto Prize for Astronomy in 1987 and the emeritus membership of the Academia Europaea in 1989.

A complete list of 'Memberships of academies and learned societies, honors, award and prizes (1935–1989)' has been printed in Jet Katgert-Merkelijn's *JKM-inventory* [1] (item 333, starting on p. 149). That list does not mention the prize-winning publication with Jan Schilt for the essay competition of the Bachiene Foundation (Schilt and Oort, 1923), described in Sect. 3.2 of the main text.

There are 22 memberships of academies and learned societies from the Netherlands, France, UK, USA, Belgium, Denmark, Sweden, Portugal, South Africa, Canada, Germany, Italy, Vatican, and the USSR. Oort received 10 honorary doctorates. These were from universities at Copenhagen, Denmark (1946), Glasgow, UK (1950), Oxford, UK (1951), Louvain, Belgium (1955), Harvard, USA (1957), Bruxelles, Belgium (1959), Cambridge, UK (1960), Bordeaux, France (1961), Canberra, Australia (1963) and Torún, Poland (1973).

The prizes, medals and lectureships are the Bruce Medal of the Astronomical Society of the Pacific (1942), the Gold Medal of the Royal Astronomical Society (1946), the Medaille Jules-César Janssen of the Société Astronomique de France (1946), the Medal of the Université de Liège (1953), Waynflete Lecturer at Magdalen College, Oxford (1956), the 'Gouden Ganzeveer' (Golden Quill) of the Royal Dutch Association of Publishers (1960), the Vetlesen Prize of Columbia University (1966),

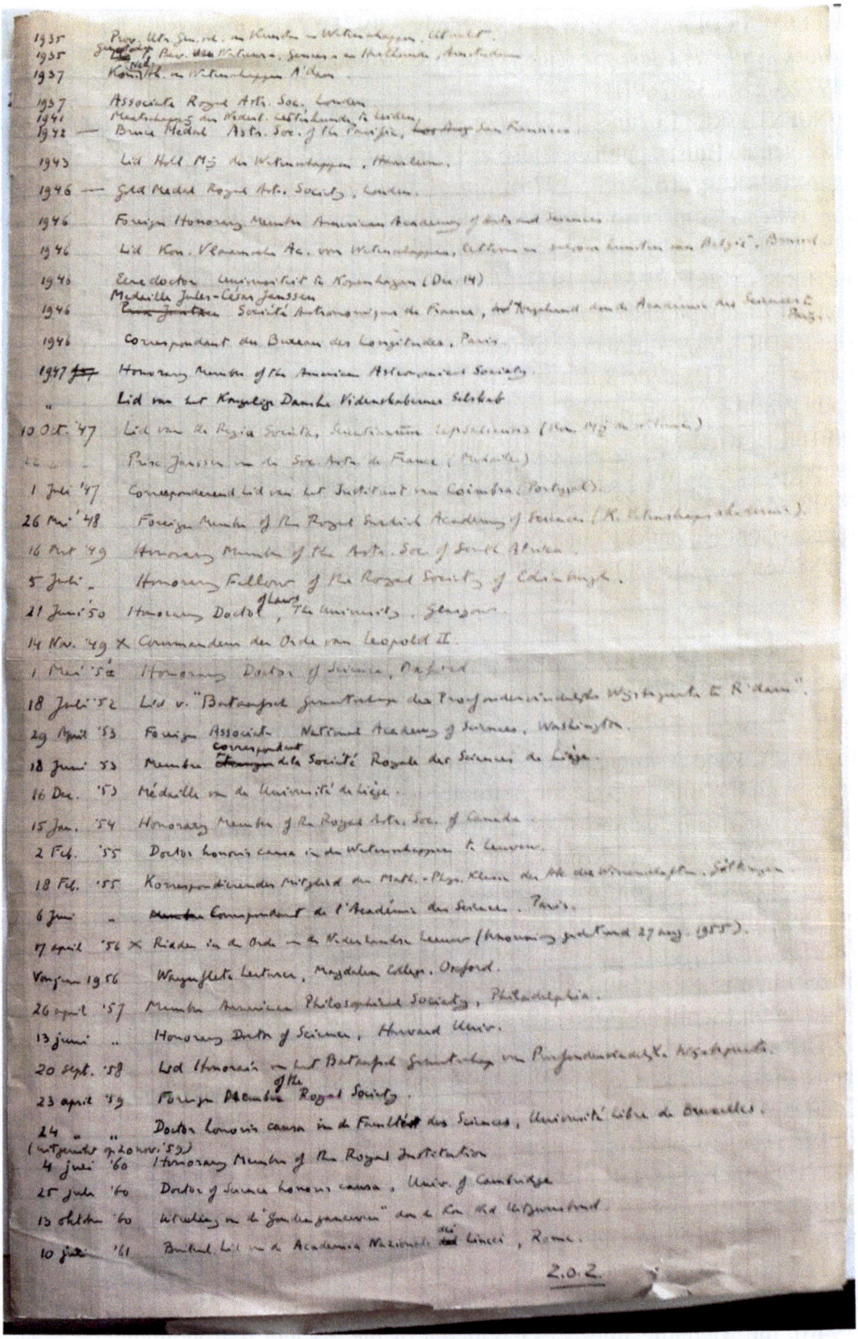

1935 Prof. [illegible], Utrecht.
1935 [illegible] Prof. [illegible], Amsterdam.
1937 Kon. Ak. v. Wetenschappen A'dam.
1937 Associate Royal Astr. Soc. London
1941 Maatschappij der Nederl. Letterkunde te Leiden,
1942 — Bruce Medal Astr. Soc. of the Pacific, ~~[illegible]~~ San Francisco.
1943 Lid Holl. Mij der Wetenschappen, Haarlem.
1946 — Gold Medal Royal Astr. Society, London.
1946 Foreign Honorary Member American Academy of Arts and Sciences
1946 Lid Kon. Vlaamsche Ac. voor Wetenschappen, Letteren en Schoone Kunsten van België", Brussel
1946 Eeredoctor Universiteit te Kopenhagen (Dec. 14)
1946 Medaille Jules-César Janssen ~~[illegible]~~ Société Astronomique de France, [illegible] Académie des Sciences te Parijs.
1946 Correspondant du Bureau des Longitudes, Paris.
1947 ~~[illegible]~~ Honorary Member of the American Astronomical Society.
" Lid van het Koninklijke Danske Videnskabernes Selskab
10 Oct. '47 Lid van de Regia Societas Scientiarum Upsaliensis ([illegible])
[illegible] Prix Janssen van de Soc. Astr. de France (Medaille)
1 Juli '47 Correspondeerend Lid van het Instituut van Coimbra (Portugal).
26 Mei '48 Foreign Member of the Royal Swedish Academy of Sciences (K. [illegible]).
16 [illegible] '49 Honorary Member of the Astr. Soc. of South Africa.
5 Juli " Honorary Fellow of the Royal Society of Edinburgh.
21 Juni '50 Honorary Doctor of Laws, The University, Glasgow.
14 Nov. '49 × Commandeur der Orde van Leopold II.
1 Mei '52 Honorary Doctor of Science, Oxford.
18 Juli '52 Lid v. "Bataafsch Genootschap der Proefondervindelijke Wijsbegeerte te R'dam".
29 April '53 Foreign Associate National Academy of Sciences, Washington.
18 Juni '53 Membre ~~[illegible]~~ correspondant de la Société Royale des Sciences de Liège.
16 Dec. '53 Médaille van de Université de Liège.
15 Jan. '54 Honorary Member of the Royal Astr. Soc. of Canada.
2 Feb. '55 Doctor honoris causa in de Wetenschappen te Leuven.
18 Feb. '55 Korrespondierendes Mitglied der Math.-Phys. Klasse der Ak. der Wissenschaften, Göttingen
6 Juni " ~~[illegible]~~ Correspondant de l'Académie des Sciences. Paris.
17 april '56 × Ridder in de Orde van de Nederlandse Leeuw ([illegible] 27 aug. 1955).
Voorjaar 1956 Waynflete Lecturer, Magdalen College. Oxford.
26 april '57 Member American Philosophical Society, Philadelphia.
13 juni " Honorary Doctor of Science, Harvard Univ.
20 Sept. '58 Lid Honorair van het Bataafsch Genootschap van Proefondervindelijke Wijsbegeerte.
23 april '59 Foreign Member of the Royal Society.
24 " " Doctor honoris causa in de Faculteit des Sciences, Université Libre de Bruxelles.
(uitgereikt op 20 nov. '59)
4 juni '60 Honorary Member of the Royal Institution
25 juli '60 Doctor of Science honoris causa, Univ. of Cambridge
13 oktober '60 Uitreiking van de "Gouden ganzeveer" door de Kon. Ned. Uitgeversbond.
10 juni '61 Buitenl. Lid van de Accademia Nazionale dei Lincei, Rome.
Z.O.Z.

Fig. A.3 Oort's own listing of his honors (continued in Fig. A.4). From the Oort Archives

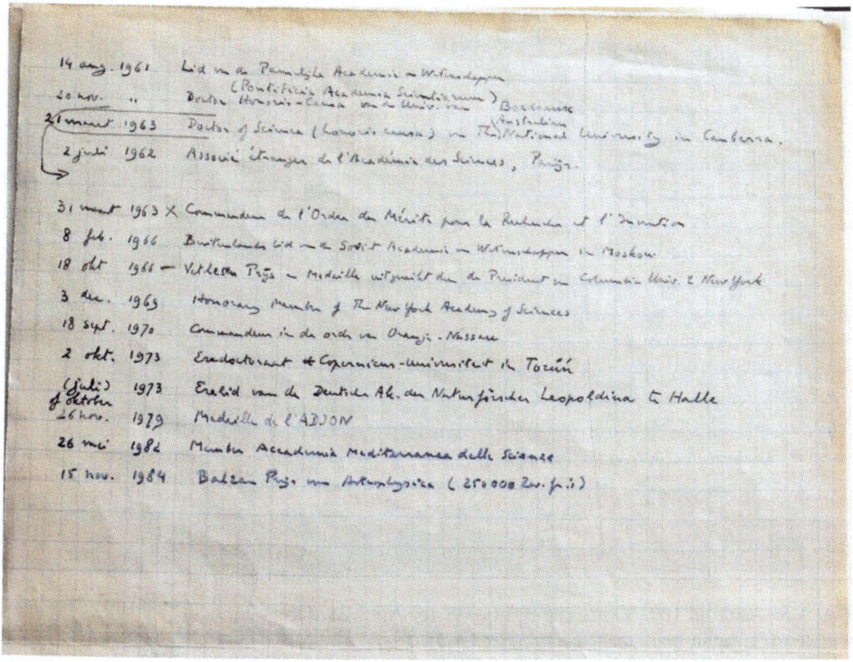

14 aug. 1961 Lid van de Pauselijke Academie van Wetenschappen (Pontificia Academia Scientiarum)
20 nov. " Doctor Honoris-Causa van de Univ. van Bordeaux
31 maart 1963 Doctor of Science (honoris causa) van The Australian National University in Canberra.
2 juli 1962 Associé étranger de l'Académie des Sciences, Parijs.

31 maart 1963 X Commandeur de l'Ordre du Mérite pour la Recherche et l'Invention
8 feb. 1966 Buitenlands lid van de Sovjet Academie van Wetenschappen in Moskou
18 okt 1966 – Vetlesen Prijs – Medaille uitgereikt door de President van Columbia Univ. te New York
3 dec. 1969 Honorary Member of The New York Academy of Sciences
18 sept. 1970 Commandeur in de orde van Oranje-Nassau
2 okt. 1973 Eredoctoraat of Copernicus-universiteit te Toruń
(juli) of oktober 1973 Erelid van de Deutsche Ak. der Naturforscher Leopoldina te Halle
26 nov. 1979 Medaille de l'ADION
26 mei 1982 Member Accademia Mediterranea delle Scienze
15 nov. 1984 Balzan Prijs voor Astrophysica (250000 Zw. fr.)

Fig. A.4 Oort's own listing of his honors (continued from Fig. A.3). It stops with the Balzan Prize (1984); the Kyoto Prize for astronomy (1987) and emeritus membership of the Academia Europaea (1989) should be added. From the Oort Archives

the Medaille de l'ADION, Observatoire de la Côte d'Azur (1979), the Balzan Prize for Astrophysics (1984) and the Kyoto Prize for Astronomy (1987).

High distinctions are Commander in the Order of Leopold II (1949), Knight in the Order of the Netherlands Lion (1955), Commander de l'Ordre de Mèrite pour la Recherche et l'Invention (1963) and Commander in the Order of Orange-Nassau (1970). In Sect. 13.1 of the main text some more details can be found.

Asteroid *1691 Oort* had been discovered in 1956 in Heidelberg by Ingrid Groeneveld (who had married Kees van Houten the year before) and a local astronomer (K. Reinmuth). After the orbital elements had been determined with sufficient accuracy they had proposed to the IAU, as is their privilege, to name the object after Oort. It has a somewhat irregular shape with diameters of 30–35 km and rotates about its axis in 10.3 h. It moves around the Sun in 5.62 years in a modestly elliptical orbit with a semi-major axis of 3.2 Astronomical Units. Its plane is inclined by some 4° to that of the Earth. The orbit with respect to those of the planets up to Jupiter and the position of the asteroid on April 28 (Oort's birthday) in 2019 has been displayed in Fig. A.5.

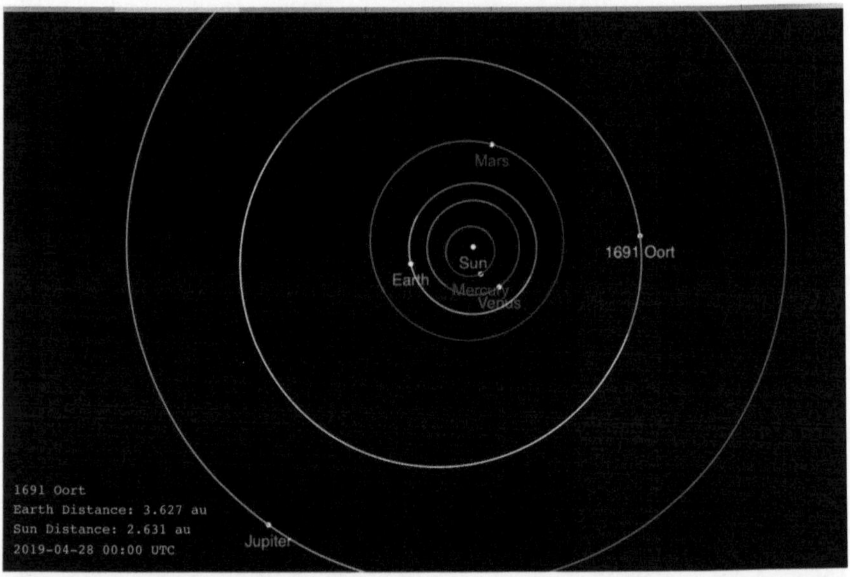

Fig. A.5 Asteroid 1691 Oort. The position is for April 28 (Oort's birthday) in 2019. Produced with the JPL small-body database browser [385]

A.6 Academic Genealogy

This following is an abbreviated version of the genealogy of Jacobus Cornelius Kapteyn in my *JCKbiog* [7].

An academic genealogy should ideally follow the Ph.D. theses and advisers. This is not always possible. In the text here a broad interpretation has been adopted. So I have accepted Johann Samuel König as a student of the Bernoullis and von Wolff, even though he did not formally submit a Ph.D. thesis; I also ignored the fact that Gerrit Moll received his Ph.D. degree *honoris causa*. Likewise I have ignored any inconsistencies between the listings in the Mathematics Genealogy Project [386] and other biographies. I start with Oort's supervisor van Rhijn and go back in time.

Pieter Johannes Van Rhijn (1886–1960)—Student of J.C. Kapteyn; Ph.D. thesis: *Derivation of the change of colour with distance and apparent magnitude together with a new determination of the mean parallaxes of the stars with given magnitude and proper motion*, University of Groningen, 1915.

Jacobus Cornelius Kapteyn (1851–1922)—Student of C.H.C. Grinwis; Ph.D. thesis: *Onderzoek der trillende platte vliezen* (Study of vibrating flat membranes), University of Utrecht, 1875.

Cornelis Hubertus Carolus Grinwis (1831–1899)—Student of R. Van Rees; Ph.D. thesis: *De distributione fluidi electrici in superficie conductoris* (On the distribution of electricity over the surface of a conductor), University of Utrecht, 1858.

Richard van Rees (1797–1875)—Student of G. Moll; Ph.D. thesis: *De celeritate soni per fluida elastica propagati* (On the speed of sound in an elastic fluid), University of Utrecht, 1819.
Gerrit (Gerard) Moll (1785–1838)—Student of J.T. Rossijn; Ph.D. thesis: the degree was awarded *honoris causa*, University of Utrecht, 1815.
Johannes Theodorus Rossijn (1744–1817)—Student of A. Brugmans; Ph.D. thesis: *De tonitru et fulmine ex nova electricitatis theoria deducendis* (On thunder and lightning according to the new theory of electricity), University of Franeker, 1762.
Antonius Brugmans (1732–1789)—Student of J.S. König; Ph.D. thesis: *Dissertatio philosophica inauguralis de phaenomeno* (Philosophical dissertation on the phenomena), University of Franeker, 1749.
Johann Samuel König (1712–1757)—Student of Johann and Daniel Bernoulli and Christian Wolff. König studied in Basel under Johann from 1730 and under Daniel Bernoulli from 1733. He did not formally receive a Ph.D. degree. In 1735 he went to study under Christian von Wolff in Marburg.
Daniel Bernoulli (1700–1782)—Student of Johann Bernoulli(?); Ph.D. thesis: *Dissertatio physico-medica de respiratione* (Dissertation on the medical physics of respiration), University of Basel, 1721.
Johann Bernoulli (1667–1748)—Student of Jacob Bernoulli; Ph.D. thesis: *Dissertatio physico-anatomica de motu musculorum* (Dissertation on the physics and anatomy of muscular motion), University of Basel, 1694.
Jacob Bernoulli (1654–1705)—Student of G.W. von Leibniz; Ph.D. thesis: *Solutionem tergemini problematis arithmetici, geometrici et astronomici* (Solutions to a triple problem in arithmetics, mathematics and astronomy), University of Basel, 1684.
Christian von Wolff (1679–1754)—Student of E.W. von Tschirnhaus and G.W. von Leibniz; Ph.D. thesis: *Dissertatio Algebraica de Algorithmo Infinitesimali Differentiali* (Dissertation on the algebra of solving differential equations using infinitesimals), University of Leipzig, 1704.
Ehrenfriend Walter von Tschirnhaus (1651–1708)—Although he did study in Leiden around 1669 to 1674, there is no evidence that he actually obtained a Ph.D.
Gottfried Wilhelm von Leibniz (1646–1716)—Student of E. Weigel; Ph.D. thesis: *De casibus perplexis in jure* (On perplexing cases in law), University of Althof, 1666.
Erhard Weigel (1625–1699)—Student of P. Müller; Ph.D. thesis: *De ascensionibus et descensionibus astronomicis dissertatio* (Astronomical dissertation on risings and settings), University of Leipzig, 1650.
Philipp Müller(1585–1659)—He was the thesis adviser of Erhard Weigel and a professor of mathematics at the University of Leipzig from 1616 onward. Müller had a keen interest in astronomy and was one of the first to accept Kepler's laws of planetary motion. He corresponded extensively with Johannes Kepler (1571–1630). Jacob Bartsch (ca 1600–1633), a pupil of Philip Müller, is known to have made a star chart based on Müller's data and later became Kepler's assistant and son-in-law.

Oort probably was not aware of this. Had he been, he would have been gratified (as I am, since it my genealogy too) that it goes back to persons so close to Kepler (Figs. A.6, A.7, B.1).

Fig. A.6 In Leiden a number of *Wall formulas* have been painted on houses. They all have some connection to science at Leiden University [387]. The one with Oort's constants is located to the west of the Observatory just across the canal (Witte Singel). The bottom part of the figure shows the formulae. The text says: '*We move at a large speed around the center of the Milky Way*'. See also [388]. Photograph by the author

Fig. A.7 The 'Oort Cloud', part of the 11Fountains project in which a fountain has been erected in each of the eleven Frisian cities. This one, in Franeker, the city where Oort was born, has been designed by French artist Jean Michel Othoniel (b. 1964) [389]. The golden, open grid sheds a mist of fine droplets, depicting the Oort cloud. From 11Fountains, with permission

Appendix B
Texts of Selected Documents

It is better to have read a great work
of another culture in translation
than never to have read it at all.
Henry Gratton Doyle (1889–1964)

Some important documents or texts cited in the main part of this book are in Dutch. Fore some I here present English translations. Note that I have already provided an English translation of the lecture delivered at the occasion of Oort's appointment of 'privaat docent' in Leiden, *Non-light-emitting matter in the Stellar System* (Oort, 1927a) as Appendix A. in the *Legacy* [6].[1]

B.1 J.C. Kapteyn to J.H. Oort

See Figs. 2.19 and 2.20 for a reproduction of the only letter of Kapteyn to Oort. Here is the full text in English translation.

Amsterdam, 10 Emmaplein, 29 Dec. 1921

Amice,

A few days ago I received a copy of the Groningsche Studentenalmanak and in that I found to my surprise your article, 'Something on the works of J.C. Kapteyn'. The sympathetic manner in which the article has been written gives me much joy, but it has given me special pleasure that it was You who wrote it.

I could not help at my departure from Groningen to feel great satisfaction. The many tributes and especially the powerful cordiality made me happy.—But, believe me in spite of that feeling of satisfaction I still have felt that much has been lacking during the completion of my duties. Not one of the lesser of these shortcomings that I have felt is that in the final

[1] Professor of Roman languages, translator, editor.

P. C. van der Kruit, *Jan Hendrik Oort*, Astrophysics and Space Science Library 459, https://doi.org/10.1007/978-3-030-17801-7

year I have not cared sufficiently for my students. I left that for the major part to van Rhijn.—The reason, you will undoubtedly know, resulted from my irresistible wish to bring my life's work to some sort of completion. I have taliter qualiter [more or less] reached that completion. You will see in what manner in two articles for the Mount Wilson Contributions that are in the hands of the printers.—I had to sacrifice much too much for that. –

It feel relieved to have been able to say this here.

Maybe I can at a later stage make up for that.—We have now bought a house in Hilversum. You could look us up there and I hope that I have then the opportunity to discuss and review all sort of things concerning our beloved science.

——

I have heard to my delight that you have started with the measuring and reduction of the Greenwich parallaxes. I hope you will make careful notes on the nature of the images; it may for example be necessary later to discard the too clearly deformed images.—The reduction will be performed such I think that also the PM [proper motions] can be determined. Or have other plates for that been requested from Greenwich? Entirely usable results cannot be expected other than on the basis of PM.—

With a warm handshake,

J.C. Kapteyn

B.2 Propositions Belonging to J.H. Oort's Ph.D. Thesis

See Sect. 5.2 for more background to the thesis and propositions.

I. There is no approximate equilibrium of energy in the Stellar System. It is therefore unlikely that stars can have existed as such for more than 10^{13} years. Likewise it is unlikely that, as Smart supposes (*Monthly Notices R.A.S.* **85**, p. 426), that the faintest dwarf stars have been formed from giants by emission of matter.

II. Pannekoek did not prove that the difference between spectroscopic and true absolute magnitude is related to the mass of a star (*Bull. Astr. Inst. Netherlands* No. 19, p. 116).

III. It should be recommended to determine the reduction curves for spectroscopic absolute magnitudes in such a way that, when ordered by spectroscopic absolute magnitude, the average values for the true absolute magnitudes become equal to the spectroscopic ones.

IV. It is important to investigate whether there is a gap between binary or multiple stars and open star clusters.

V. Aitken did not prove that binary stars are relatively more frequent per unit volume in the plane of the Milky Way than far outside of it (*The Binary Stars*, p. 260).

VI. The random errors of the proper motions of planetary nebulae measured by van Maanen and Mrs Marsh (see *Mt Wilson Contr.* No. 290) are significantly larger than they have indicated, so that it becomes doubtful whether any real meaning can be attached to the motions they found.
VII. Öpik's conclusions in *Tartu Publ.* **25**, Nos. 5 and 6 (Statistical studies of double stars) are not fully justified.
VIII. Shapley's opinion that it should be very well possible to use apparent diameters of open clusters to deduce a system of relative distances (*Proc. Nat. Ac.*, **5**, p. 345) can be contested.
IX. Kapteyn and van Rhijn incorrectly pose that the existence of a concentration of stars with large proper motion at negative Galactic latitudes is a priori improbable.
X. The 'curious' result of probability calculations, that Weber and Bauschinger find on p. 369 of the *Encykl. der Elem.-Mathematik* 2. Auflage), is incorrect.
XI. Eddington's claim that equipartition of energy is a third order effect compared to the general exchange of speed between two encountering stars, is incorrect (*Stellar Movements*, p. 250).
XII. The explanation attempted by Sommerfeld and Unsold of the intensity anomaly in the spectrum of hydrogen by meta-stability is wrong (*Z. S. für Physik*, **36**, p. 266).
XIII. The small value found for the relativity doublet in hydrogen is not in violation of the exclusion rules or the correspondence principle (see e.g. Janicki and Lau, *Z. S. für Physik*, **35**, p. 1).
XIV. The criticism Mrs. Ehrenfest-Afanassjewa[2] brings to bear on the purely empirical conception of our Euclidean imagination, is not well founded (*Physica*, 5th vol. p. 445).

B.3 Inaugural Lecture: The Structure of Galaxies

This is an English translation of Oorts inaugural lecture as professor of astronomy in Leiden on November 15, 1935. It has appeared in Dutch as Oort (1936b). The footnotes are Oort's. The figures are in the main text of the book.

The distribution of matter in the Universe is highly irregular. As far as it is present in the form of gas, we encounter it in extended, thin clouds with such a small density, that reproducing it in the laboratory is well beyond current capabilities; in addition we do encounter it with high density in enormous entities, that we call stars. Except for the gaseous form of matter, it also exists in solid form; not only as planets and satellites, but mainly as extended,

[2]Tatyana Alexeyevna Afanasejeva (or Afanassjewa) (1876–1964) was the wife of Leiden physics professor Paul Ehrenfest.

often irregular clouds of dust, scattered between the stars, with a total mass probably not much less than that of the stars.

If we wish to study more closely the form and nature of the Universe, we must restrict ourselves to the stars. The interesting questions arising from the rest of matter have only been answered satisfactorily in a fragmentary manner. One may say that the important role that the part of matter that is in the form of dust is playing, is such as to make it as difficult as possible to determine for researchers of the Sidereal System.

Looking at the distribution of stars in space we are immediately struck by the extraordinarily irregular distribution. Just like matter on smaller scales is concentrated into stars, these stars in turn have been clustered in discrete swarms, that we call galaxies and that may be properly considered to be the fundamental building blocks of the Universe. The largest of these swarms contain a few billion stars, while the smallest that we know still consist of a few tens of millions.[3]

At the time when observational data on these was almost entirely lacking, speculative thinkers have asked the question, whether these galaxies in turn can be clustered in systems of a higher order, and maybe these again in systems of still higher order, etcetera.

Even the most cursory view of the distribution of galaxies indeed shows an extremely irregular distribution in space.[4] Next to each other one sees areas where the galaxies are close together and areas where they are far apart. The boundaries of these different areas are generally very irregular and often difficult to define. In the more populous regions again sometimes clear, sometimes vague, local concentrations occur, sometimes entailing some thousand galaxies, sometimes consisting of only two or three lying close to each other. The structure of the Universe, as outlined by this distribution is difficult to describe in a few sentences; it is an unimaginable chaos, but a chaos that is completely different from what one would see had the galaxies been distributed in space by pure *chance*. Surely, there also is no question that they are in general organized in somewhat clearly demarcated systems of a higher order, that could be compared to systems formed out of stars; if one finds here or there a system that is only somewhat regular in its form, even then it still

[3] I leave the so-called star clusters out of consideration, even though, at least as far as the globular clusters are concerned, we are not sure whether these swarms of no more than a few tens of thousands of stars should be regarded as independent miniature galaxies or as satellites, and therefore parts of larger systems; the ones we know probably belong to the latter category.

[4] A good impression of this irregularity is shown in the figures published by Shapley and Adelaide Ames, which show the distribution of the thousand brightest galaxies over the sky (Annals of the Harvard Observatory, Volume 88, part 2 (1934); see Fig. 1 [Fig. 8.9]. An important part of the irregularity in these figures results from absorbing matter in our own Galaxy; this has been taken into account in the statement above on the general irregularity, as can be appreciated by looking only at those parts of the figures more than 40° away from the circle of the Milky Way. In Fig. 8.9 these are the areas inside the thick-drawn circles. That the irregularities in the distribution of galaxies in *these* areas are real and not much affected by the absorption of light, has been shown convincingly by studies of Shapley and Hubble.

seems to gradually merge into its surroundings; only in very few exceptions of very concentrated clusterings of galaxies one gets the impression of more or less independent entities, in which the members are held together by their mutual attraction. The study of the internal motions of these special conglomerations of galaxies promises the become very interesting in the near future, because they would make possible the determination of the average density of matter in space, a property that plays an important role in the theory of the Universe.

It is a remarkable feature that the irregularities in the distribution of galaxies are in general relatively small. They are seen most clearly when one compares volumes whose sizes are only a few times the mean separation of the galaxies and which do not contain more than a few hundred of them. The larger the volumes considered, the smaller, percentage-wise, the differences in the total numbers of galaxies in those volumes; when one considers volumes of space with sizes of fifty or a hundred times the average separation between two galaxies, then the distribution of galaxies over such volumes is remarkably regular. There is no question of any inhomogeneity on large scales in the beautiful counts carried out by Hubble of the Mount Wilson Observatory, stretching over areas of at least one hundred million galaxies.

Nothing in these observation should keep us from speculating on the possibility that galaxies could be clustered in hypersystems of an even larger order of magnitude than the volume considered by Hubble. In the last few years conjectures have been put forward that if one could speak of such a hypersystem, there would be only one, which, one may estimate, entails ten to a hundred thousand more galaxies then have been observed at present. About this hypersystem, the *Universe*, whose discovery and study is so closely related to the work of de Sitter, I have no time to speak further. I will, because of what follows, only mention the curious observational fact that the Universe is rapidly expanding, from which it follows that galaxies recently (*recently* to be understood on an astronomical or geological timescale) must have been much closer together than nowadays.

In this hour I wish to beg your attention for a somewhat less comprehensive topic, namely the separate structural phenomena that can be found in the Universe. I already briefly discussed the remarkable irregularities in the distribution of galaxies; the chaotic character and the forms that are difficult to define in this structure that in general has made it unsuitable for deeper study.[5] It is completely different in the case of galaxies themselves, where regular features provide sufficient cause for further consideration of their internal structure and motions.

[5] An exception here are the more compact groups, the so-called 'clusters of galaxies', that I mentioned earlier, and that are in various respects important for the study of galaxies. The most beautiful example of this is the cluster discovered by Wolf in Heidelberg in the constellation 'Coma Berenices' (see e.g. Publ. des Astrophysikalischen Observatoriums Königstuhl Heidelberg, Bd. 1, p. 173).

The star closest to our heart, the Sun, is part of a Sidereal System that in reality is not different from the other galaxies; we may hope that what we may learn from the Galaxy around us concerning its constitution, may give us also some insight into the structure of galaxies in general. This Stellar System that surrounds us on all sides and usually is designated as the Galaxy, contains all visible stars, even all those observable with the largest telescopes, and undoubtedly a much larger number of fainter stars, that are outside the reach of our optical instruments. (The stars in other galaxies in generally are not observable separately; of the millions of members of these distant swarms of stars we observe only the collective light as a nebulous spot, that is usually designated with the, in fact hardly appropriate, name 'nebula').

The number of stars that are in principle observable in our Galaxy is over ten billion. It is of course not possible to observe such a large number of objects separately and what one had to do is to confine oneself to various samples to delineate the global contours of our swarm of stars. It has been one of the greatest merits of my inspiring preceptor Kapteyn to clearly see, when only few dared to consider such a grand undertaking to determine the boundaries of the Galaxy, the importance of an early organized collaboration. With his ebullient enthusiasm he succeeded in enlisting the joint forces of a number of observatories to concentrate their work on faint and therefore distant stars in these sample fields, of which he defined some two hundred, regularly distributed across the sky.

It is obvious that one should start to make homogeneous star counts in the various areas to stars as faint as feasible. In this manner, Sir William Herschel in the eighteenth century had already tried to determine the form of the Sidereal System. But if one wants to study the variation in the 'population density' and the *dimensions* of the system better, this is not enough and one should, next to the *numbers* of stars, know the *distances* as well. One should do measurements that make it possible to determine distances of stars. The only measurement that one could perform up till then for the faintest stars was their brightness; that property has indeed been determined with the greatest accuracy at the Mount Wilson and Harvard Observatories in the so-called 'Selected Areas' defined by Kapteyn. A comparison of this brightness with relatively nearby stars makes it possible to estimate how much more distant these faint stars are. Since it has been possible to measure the distances of nearby stars using other methods, this then gives us the distance of the faintest stars. The calculation presents quite a number of difficulties that I don't want to address now; difficulties on the one side arising from the attenuation of the starlight on the path between the star and us as a result of absorption and scattering; and on the other hand from the fact, that the real brightnesses of stars vary so greatly. For both these reasons it is of the utmost importance—which nowadays seems to enter the realm of possibilities—to measure in addition to the brightnesses also the colors of these faint stars, since stars of the same intrinsic color differ much less among them in true brightness than all stars mixed together.

Distances of stars can be determined with more certainty and accuracy from the apparent motions on the sky than with measurements of brightness. Measuring these motions requires great precision, and patience to wait the many years that stars need to move so far that this displacement can be seen from here; and that is the case in particular for more distant stars. While a few years ago we had only sporadic determinations of motions of faint stars that can be used for this, in the last few years many valuable data have been collected, where special mention should be made of the motions of very faint stars in Kapteyn's Selected Areas published by the Radcliffe Observatory at Oxford, that make it possible for us to calculate distances that in some directions do almost reach the outermost stars in our swarm. The difficulty with those calculations is, that for this we need to know the distributions of the velocities of these stars expressed in linear measure, e.g. kilometers per second. This has been determined well from spectroscopic measurements for the bright stars, but for the faint stars, that are of primary interest in this case, this is based on extrapolation.

In Kapteyn's Plan the necessary restrictions of the observations are realized by concentrating on a few small areas. The same result could be obtained by directing the observations at rare sorts of stars that are especially suited for this. Such stars should fulfill the following conditions: they have to be distinguishable from ordinary stars—for example like variability of their brightness or a specific spectrum—; they will have to have a large real luminosity, so that they can be distinguished up to very large distances and finally we would need to know their real luminosity to estimate their distances, There actually are objects in our Stellar System that fulfill these requirements reasonably well. I am thinking e.g. of stars with extremely high surface temperatures, and of the variable stars of the so-called δ Cephei or RR Lyrae type for which we can determine so nicely to which class of luminosity they belong.

The characteristics mentioned are also conformed to by a different kind of object: star clusters, and in particular so–called globular clusters, that as a result of this have played an important role in the exploration of the Sidereal System.

If one would know all objects of such a sort that occur in the star-swarm (maybe a few thousand in the case of the variable stars, not mote than a few hundred in the case of globular clusters), then one could define probably a kind of skeleton for the Galaxy. In addition to the problem of the absorption of light, to which I will return below, there is the serious difficulty to select such objects to a certain level of completeness and derive their properties well. Work in this area has been started under the leadership of Hertzsprung in South Africa, and at present is being continued in Leiden using plates taken by van Gent in Johannesburg. At various other observatories such work is also being pursued. It has already provided important results on the distribution of particular kinds of stars.

The efforts to discover the boundaries and the structure of the Galaxy using the various routes sketched, have resulted in a number of important insights,—they have not led, however, to an even somewhat satisfactory solution to the central problem. They have shown that the Stellar System is much smaller in one direction than in those perpendicular to it, that is has, in other words, the form of a flat disk or a very thin automobile wheel. We see in the ring of light in the Milky Way the gleam of the distant clouds of stars in its plane. It has also been possible to conclude that the Sun is located roughly in the middle of the upper and lower boundaries of this disk—the thickness of the disk can de determined well, but it was actually not possible to derive anything from direct observations about its form and size. The reasons for this restriction are the absorbing veils spread out over the space in between the stars in this disk, which restrict—like the smoke over a large city—our view towards the distant areas. These veils probably are for the largest part made up of small dust particles with diameters of the same order as the wavelength of the light. The existence of such dust clouds, visible as dark areas that contrast with the background of the star-studded Milky Way background, and that with their irregular distribution cause the irregular impression made by photographs of some parts of the Milky Way, has been known for some tens of years. Only during recent years astronomers have realized the fatal extent of the difficulties that blocked the progress in research in the structure of our System. Not only do we have to give up hope that we will ever be able to see the central parts of the wheel-shaped Galaxy, but also for the parts that we can see our ignorance of the extent to which the light of the stars has been attenuated, makes our conclusions about the general structure extremely doubtful.

That in spite of this we do know where the center of the Galaxy is located, is a result of the existence of objects, like the globular clusters just mentioned and among others also the RR Lyrae variables, that unlike most other objects in the System, move well outside the dust-ridden, disk-like space, but are still as a result of the attraction of the distant central parts of the Galaxy grouped symmetrically around it. Just as we can locate from the sea an island full of birds, that itself has been hidden by the fog in the lower levels of the atmosphere, by the swarms of seagulls hovering over it, this distant cloud of globular clusters above (and below) our star-disk shows us the position where this central part of the Galaxy must be located.[6] That the invisible center is indeed at that location has been confirmed by phenomena related to the rotation of the Galaxy, that I will return to below.

How important the knowledge may be,that we can deduce from the facts just mentioned, it remains in fact no more than global information. We will not be able to make progress as long as the absorption remains an unknown in studies of the distributions of stars. In this area it seems that absorption is the most important thing at which to direct an attack. I will try to sketch a few avenues along which this attack could be directed.

[6] See Fig. 2 [Fig. 3.4, lower panel].

Already in 1908, Kapteyn tried to determine whether the colors of distant stars was different from those nearby. His reasoning was that, in the case of thinly spread absorbing matter in the Universe, the stars lying behind these would appear redder than normal, much like the setting Sun by absorption in our atmosphere. Kapteyn, nor other astronomers that looked into this phenomenon, found anything definite of the kind, until a few years ago, when the American astronomers Trümpler, Stebbins and Vyssotsky independently and using different methods showed convincingly the existence of such reddening. One of the reasons why it took so long to discover this is the fact that the reddening is so small in comparison with the absorption itself, in a relative sense much less than in the case of the setting Sun. The ratio of these two must be determined empirically in the Universe and that is a rather difficult problem, especially as one does not know if the ratio is the same everywhere. Once this ratio is known, an accurate determination of the colors of star-types for which the intrinsic colors do not differ much among themselves (like stars of very high temperature) will suffice to show the absorption that the light from such a star is liable to. One factor that plays a major role in these researches is a criterion that is related to the rotation of the Galaxy, which makes it now possible for us to better estimate distances of more distant stars than was possible before from the apparent motion.

There is a completely different way to approach the problems of the absorption. When I discussed the distribution of the galaxies in space I pointed out that although these systems are distributed unevenly when one considers small volumes, that irregularity more or less disappears and is replaced by a great regularity when comparing large areas, like the ones covered by the Mount Wilson counts. These counts *apparently* show a strong irregularity over large areas of the sky that in fact is the result of absorbing clouds in our System through which we look to the outside. When in a certain direction there is much absorbing matter, the light of the galaxies lying in that direction will be attenuated before it reaches us, which will mean in practice that we see fewer of them. In this way counts of distant galaxies gave us a tool to determine the total amount of absorption that light rays reaching us from different directions have undergone in the Galaxy; and also to select areas in which the absorption is small and that can be used effectively to soundings in our Galaxy.

The extent to which absorption has influenced our thinking is shown in a remarkable way in the differences in opinion that exists about the so-called 'local system'. The very common opinion had taken root that the Sun is located in the middle of a large local clustering of stars, that is in fact a local concentration in the large swarm. Preliminary studies along ways sketched above now give reason to believe to the contrary that the part of the Galaxy in which the Sun resides is significantly rarer than the surrounding areas.

It looks like that with the data that we have collected now about the absorption we can find out significantly more about the structure of the Galaxy. At the present time these studies are still for such a large part in the melting pot, that I will restrict myself to giving no more than an only somewhat clearer outline of the smoothed model, that I already sketched along the way. We know now that the center of our disk-like system is removed very far from us and that the Sun is located rather close to the outer boundary of this dislike structure. How thin the system is may be illustrated by the statement that in the neighborhood of the Sun, the thickness of the layer in which half the stars reside amount to only about a twenty-fifth of the distance of the Sun to the center.

So it is very flat, the Sidereal System resembling the Planetary System, in which the orbits of the large planets are roughly in the same plane. In another respect the two systems resemble each other as well; not only do the motions of the stars appear generally to be controlled by the attraction of the large, central mass, but the different stars—much like the planets—also move in the same direction around the center. This is often expressed by saying that the Galaxy rotates. The large attracting mass that in the Galaxy assumes the position that the Sun holds in the Planetary System, is exactly the invisible central part of the Stellar System itself. The total mass of this is about seventy billion times the mass of the Sun; the period of the rotation is some two hundred million years.

The assumption that the majority of stars move around a distant center with a common speed that is significantly larger than the relative velocities that we see in our observations, has been put forward clearly for the first time and the important consequences clearly outlined by the Swedish astronomer Lindblad. The correctness of this proposition can be shown in a rather direct manner in the systematic differences between the motions of distant stars, the so-called rotational effects, that in addition to providing support for the theory of rotation give a good method to determine the distances to these distant objects. Except in the effects of rotation, the action of the force originating from the invisible center shows itself in two different properties that the Galaxy has, that have appeared very mysterious to us when we did not know what the effects of absorption were. I mean the phenomenon of the star streams and the curious asymmetry in the motions of the stars of high velocity. The phenomenon of star streams was discovered early this century by Kapteyn. This was the first time that something special was found in the motions of the stars, something that disagreed with the picture of a perfectly random distribution and this finding has in no small way contributed to the awakening of the interest in the study of the Sidereal System. Kapteyn's interpretation of his discovery was that in our neighborhood there are two different, but mixed stellar systems that moved systematically with respect to each other. Schwarzschild showed that the observed phenomena can be interpreted as well in a picture in which the stars belong to one single system, in which the motions in that system for unknown reasons have a preference to occur in two opposite directions. Data that have been collected since then, which also show that the phenomenon occurs up to very large distances and everywhere in the same manner, have made it likely that Schwarzschild's opinion is closest to the truth. Lindblad finally succeeded in finding the most important reason for this mysterious preference in the stellar motions, when he showed that they are a direct consequence of the rapid rotation of the Galaxy.

In the case of the stars of high velocity it also was difficult to understand this phenomenon until a natural explanation was found in the new conception of the Galaxy, outlined here.

One might think that the discovery of the distant center and of the rotation of the Stellar System were the most important problems to be solved concerning the construction of the Galaxy; that what remained to be done would predominantly be improving and refining the already established conception.

I cannot ask more of your already very much taxed attention to have to listen to a recital of the different problems that are still completely unsolved, such as the remarkable consistency of the properties of the stars and their motions and distribution in the large stellar swarm; I allow myself to remark on only one more set of phenomena.

In the southern part of the Milky Way, in the constellations of Centaurus and Scorpius, there is a group of so-called helium stars, with high surface temperatures, that seem to have some connection to our subject. This has been deduced from the fact that they form a clear clustering in this area and that their motions are similar. Now the curious thing is that the different members of this group are not *precisely* the same, or at least cannot remain exactly the same in the long run due to the rotation of the Galaxy. This follows from the fact that the Galaxy does not rotate like a solid wheel in which the relative positions of the various parts remain the same, but that—like in the Solar System—the parts that are closer to the center rotate faster than those that are further out. Because of this, groups like that will be torn apart, unless the internal forces keep them together, which is very unlikely. The fact that these and other similar local clusterings exist, tells us that they can have formed only rather recently through an unknown cause, such that the rotation of the Galaxy did not have the time to tear them apart.

The very inhomogeneous distribution of the more or less rare helium stars over our System is remarkable. Maybe the enormous stellar conglomerations that we see in the Milky Way also derive their appearance of more or less loose clouds of stars not only from the contrast with the surrounding absorbing veils, but are real concentrations in the large star-swarm. We know very little about the Milky Way clouds; there is no question of any regularity in their relative appearance.

As I mentioned earlier the reasons for our incomplete understanding are both the absorbing clouds that are located in a relatively thin layer in the plane of the Milky Way, as well as the fact that the Sun is located within this layer, where the view is so restricted. Would we have been at a more favorable position, at some distance from the plane of the Galaxy, we would have very little difficulty having a good overview of the structure of our Stellar System. Not only would our view be blocked very little by the dust clouds, the structures that lie in the Galactic disk that are now from our current position seen projected on top one another in such a confusing manner, would be seen separately so that we could get a clearer overview.

Such a vantage point as for this enviable observer is taken by us in the case of the many other stellar systems, be it from a rather large distance. The architecture that is shown to us this way is surprising and interesting.

The stellar systems appear for the most part to have a regular structure: a bright center whose light falls off gradually outwards, from which at two opposite points mighty streams of stars emerge, that move out in flat spiraling lanes from the central parts.[7] This strange spiral structure, from which the commonly used name spiral nebula originates, is displayed by about three quarters of the bright nebulae. There is however a large range of appearance among the various individuals.[8] These spiral arms almost always have a rather irregular, chunky appearance, sometimes they are clearly delineated, sometimes only vaguely visible. There are systems that appear to have nothing else but widely spread out spiral arms and in which the nucleus has shrunk to only a very small part of the total image. On the other side there are systems in which the nucleus dominates and the spiral arms are insignificant. The latter systems form a gradual change towards a somewhat smaller category of galaxies that consist *only* of a very regular central part, without spiral arms.[9]

Among the systems in this latter group one finds different degrees of flattening, that probably is related to different rotation velocities of these systems. Very likely there are systems that are almost spherical, while at the same time we see flattenings that correspond to a ratio of the longest to the shortest axis of three to one. Larger flattenings do not occur among these so-called elliptical systems, but the spiral nebulae probably all have a substantially flatter form.[10] It looks as if galaxies with a faster rotation than corresponds to an axis ratio of one to three could not exist in the smooth form of ellipsoids, but were forced to develop spiral arms, that is, it seems as if stellar systems become unstable at this axis ratio.

In addition to the ellipsoidal and spiral systems there is a third kind of galaxy that has neither spiral arms nor a regular, towards the center concentrated form like the others, but are rather irregular concentrations of masses, without a center signified by a clustering of stars. To this rather rare category belong

[7] See Fig. 5 [Fig. 8.12].

[8] Among the spiral nebulae a separate group can be distinguished, in which the spiral windings do not directly emerge from the central parts, but from the ends of a more or less rod-shaped extension. In English this class of objects is often designated as 'barred spirals'.

[9] See Fig. 4 [Fig. 8.11].

[10] See Fig. 3 [Fig. 8.10].

the Magellanic Clouds, the only stellar systems that will be known from direct observation by at least some of you.[11]

I have sketched in rough terms the various kinds of galaxies that exist in the Universe. To which of these groups does our Galaxy belong? We have already seen that it is highly flattened, we also know that this extremely flat part stretches out to large distances in all directions from the Sun, while the rotational properties show that there is a strong concentration of matter in the central regions. These facts seem to exclude the possibility that the Galaxy would belong to the group of elliptical or irregular systems. It may be regarded as very probable that the Galaxy belongs to the group of systems with spiral structure and that we are surrounded by the windings of spiral arms[12] rather than by a smooth stellar system. It is a matter for the future to detect the way these spiral arms have been laid out in our System.[13]

Does the Galaxy also compare well to the spiral nebulae in terms of size, mass and composition? As far as we can deduce from the little information available there is no difference in terms of composition. Also in terms of mass and size the Galaxy is similar to other systems, although it seems to be comparable to the largest of these; but the latter conclusion is no reason to regard it as being of a different kind.

We have succeeded in explaining in general terms the observed distribution of the stars and their motions in the Galaxy. It is a completely different matter regarding the spiral structure, that we can not study because of our unfavorable position in our own System, but that appears to be one of the most general and characteristic natural phenomena in the Universe. On the one hand this is the result of the incompleteness of available measurements, that are restricted to the light distribution in some elliptical galaxies and to spectroscopic determinations of the rotation in the centers of a few spiral galaxies, but it is equally well due to the theoretical complexity of the problem.

Various attempts have been undertaken for a partial explanation. Jeans found from calculations that a rotating volume of gas that as a result of gradual contraction rotates faster and faster, will become unstable at a certain flattening, corresponding to the largest flattening observed among ellipsoidal galaxies, and he argued that under particular conditions this instability could give rise to the formation of spiral arms, without being able to say anything

[11] It is unclear whether the Magellanic Clouds may be distinguished as separate galaxies or should be seen as a kind of satellites to the Galaxy. As far as size and structure are concerned, they are not inferior to the other irregular galaxies that we see in the Universe.

[12] This does not have to be in contradiction to the observed kinematical phenomena or with the explanation that has been given by the theory of rotation of the Galaxy.

[13] A courageous attempt to do this has been undertaken by our fellow Dutchman Easton, who interpreted the structure in the Milky Way as the projection of a spiral-like, wound-up system. The better insight that has been gained for the absorbing matter in the Galaxy takes away much from this work. Furthermore, we now know that in the case that our System indeed does have spiral structure, it must be completely different from that in Easton's interesting attempt.

quantitatively. It is true that spiral nebulae are now known to be certainty not entirely gaseous, but to consist of mostly stars. Later, Lindblad showed that a similar instability can occur in a stellar system. Vogt takes the general expanding force in the Universe as an explanation of the spiral structure, which however, except in the outermost parts is not in agreement with the quantitative data.

It is difficult to pose that these attempts, interesting as they may be, have provided already a satisfactory solution to the essence of the questions.

Maybe the most remarkable characteristic of the phenomenon of the spiral nebulae is the limited lifetime of this structure, and it shares this property with the phenomenon of loose star clusters, like in Centaurus and Scorpius, that we find in our Galaxy. We have to do almost certainly with a process that runs only in one direction.

When we consider these apparently temporal forms we are led to suspect that these must have had an astronomically speaking recent origin, a more or less contemporaneous beginning and that a disturbance from outside the systems must have played an important role in this. It is natural to think of the important disturbing forces that two systems exert on each other when having a close encounter. At present these mutual distances are so large that at first sight it seems rather improbable that the majority of the systems have had an encounter with another system.

Prof. de Sitter, who unfortunately had no more opportunity than making a few exploratory calculations concerning this problem, was the first to point at a possible connection between of the formation of spiral structure and another process that develops in one direction with time, namely that of the expansion of the Universe. It shows after all that *in fact all* systems have to have been closer together in the past.

The spirals and the Galaxy in both the regularity and irregularity of their structure bear the stamp of the past. They are like stones in which fragments of inscriptions have been etched, which we suspect relate if not to their origin then at least to a very special epoch in the history of the Universe. But their language cannot be read by us yet.

B.4 Nova Persei and Superluminal Expansion

The paper *Nevels rondom Nova Persei* (Nebulosities around Nova Persei) described unpublished work Oort had done on the observed motions that appeared to indicate an expansion velocity in excess of the speed of light. Oort never published this, but he had given a presentation of his work at the meeting of the Nederlandse Astronomen Club, the professional society of astronomers, probably the one of Saturday January 10, 1942. In 1943, Jean Jacques Raimond (these names are French and pronounced as such) published an account of this in the amateur periodical 'Hemel and Dampkring'. Because of Oort's involvement I have added it to his list of publications as Raimond

Fig. B.1 Title figure of the article in 'Hemel and Dampkring' on Nova Persei (Raimond and Oort, 1943g)

and Oort (1943g). Raimond, who had obtained a Ph.D. in astronomy in 1933 on properties of Galactic absorption in Groningen with van Rhijn as supervisor, was for many years director of the Zeiss Planetarium in Den Haag and was used to speaking for a lay audience and children. He also had been a very active member of the editorial board of Hemel and Dampkring over a long period. All of this is evident in the rather pedantic style of the article. Below I therefore provide not a precise translation. The footnotes are mine.

Nova Persei appeared on February 21, 1901. At a position on the sky where before no star brighter than magnitude 12 had been seen, there suddenly was a star as bright as Vega.[14] Before 1901 there had been an irregularly variable star of magnitude 13 or 14 at its position. In about 24 h the brightness increased to magnitude 0; the nova radiated as much light as ca. 100,000 times the Sun. According to Gamov,[15] the energy production in the star had reached a critical stage where some kind of saturation occurred and the star had to open all valves to prevent worse things from happening. However we are not concerned with the star or the cause of the outburst, but with the surroundings of the star.

Figure B.2 shows the light-curve. The extreme change in brightness had been accompanied by abrupt changes in the spectrum. About a month and a half after the outburst the spectrum showed lines that we see exclusively in emission nebulae; they were called lines of 'nebulium', decades later explained as lines from (sometimes multiply) ionized atoms of oxygen and nitrogen. This showed that nebulosities had to be present.

---

[14] Vega or α Lyrae is of magnitude 0 and the brightest stars in the northern hemisphere.

[15] George Antonovitsj Gamow (1904–1968) was a Russian-born (Ukraine), American nuclear physicist and cosmologist, who addressed the question of the production of the chemical elements.

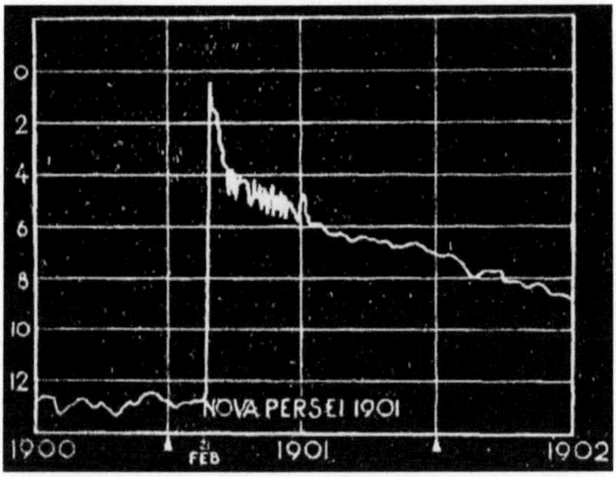

Fig. B.2 Light curve of Nova Persei 1901. From Raimond and Oort (1943g)

The problems started in August of 1901. Camille Flammarion sent a telegram to Max Wolf at Heidelberg to inform him that plates he had taken on February 22, 1901, showed a round nebula around the nova. Wolf, however, realized this was a 'pseudo-nebula'; it was an artifact that had been caused by the objective lens used for the exposure. The nova had emitted large amounts of ultraviolet radiation and the objective had not been corrected for that. The nebula was the out-of-focus image of the violet light of the nova.

The pseudo-nebula was really the cause of the discovery of the real nebula. Max Wolf found on some excellent photographs—that had been exposed for 4 h—a number of curved nebular veils. At Wolf's request, Charles Dillon Perrine (1867–1951) of Lick Observatory used the Crossley Telescope to observe the nebula. And Ritchey at Mount Wilson used the 60-in. telescope for the same purpose. On these pictures the pseudo-nebula did not appear, because these were reflectors which do not suffer from color effects. But these photographs did confirm without question the existence of the nebular veils that Wolf had indisputably confirmed. These nebulosities are shown in Fig. B.3; from September 20, 1901 these had been photographed regularly.

Perrine made the important discovery that the nebula expanded. He compared his images of November 7 and 8 to those obtained by Ritchey on September 20. It appeared that in less than two months the nebula had expanded in all directions by 1.′5. In February 1902, about a year after the outburst, the nebulosities covered an area of sky not much less than that of the full moon.

Perrine's discovery was not only important, it was also remarkable. A displacement of 1.′5 in two months is very large. It corresponds to a proper motion of 9 arcmin per year. The largest proper motion observed was 10 arc*sec* per annum. This is the proper motion of Barnard's star, which is only at only 6 lightyears from us and moves with a velocity of 90 km/s.

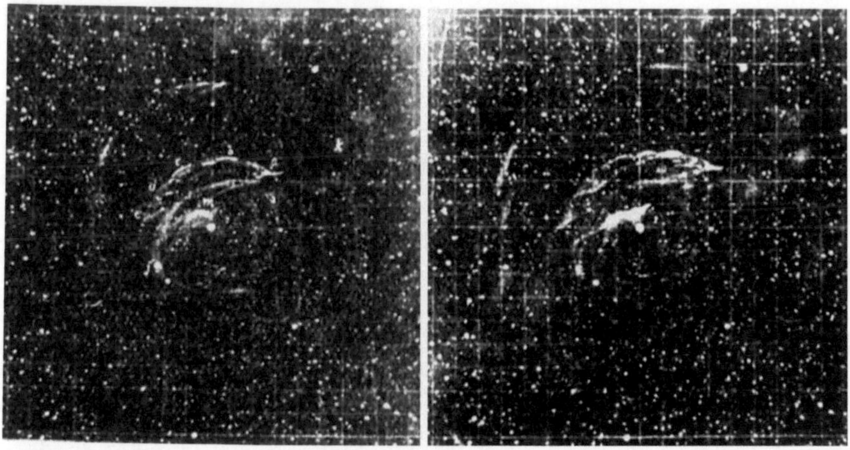

Fig. B.3 Drawings from Ritchey's photographs. On the left: September 20, 1901. On the right: November 13, 1901. From Raimond and Oort (1943g)

It was very improbable that the nova should be among our nearest neighbors. The linear velocity would have to be fantastically high. There were two options; either the nova was extremely close, or the speeds in the nebula were indeed extraordinarily high.

Our great compatriot Kapteyn gave a very satisfactory explanation. While others were thinking of expanding matter that had been expelled by the nova during the outburst, Kapteyn supposed that there were dark and therefore invisible clouds around the nova, that might move but for which the distance from us was so large that these motions were not perceptible in a few months. However, Kapteyn did not think of motions of clouds, but of the enormous amount of light that the nova had emitted during the short-lived eruption. On February 21, 1910, a flash of light left the nova and moved in all directions with the speed of light and illuminated parts of the clouds at larger and larger distances from the nova. The idea was immediately accepted by others, among them von Seeliger. Kapteyn calculated that the nova had to be 300 lightyears from us,[16] since a velocity of 300,000 km/s would give rise to a displacement of about 10 arcmin per year. So, it was not matter that moved but a light-front that illuminated the invisible clouds around Nova Persei—see Fig. B.4.

After a year the light-front had already moved to 17 arcmin from the nova and the phenomenon became increasingly fainter. Perrine was only able to see a very faint blob of light on November 1902, even though he used exposure times of 7–11 h. But Perrine provided a very neat confirmation of Kapteyn's theory. At the expense of an exposure time of 35 h, divided of course over several nights, he obtained a spectrum of the nebulosities. This turned out to be an exact copy of the spectrum of the nova during the eruption.

[16]Or 90 pc; the 9 pc quoted in my *JCKbiog* [7] on p.314 is a misprint.

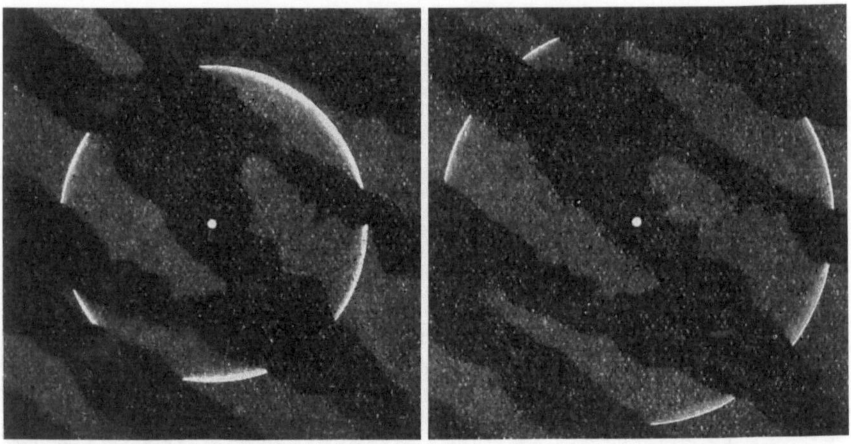

Fig. B.4 Kapteyn's explanation. The figure is schematic. On the left the light-front on October 1, on the right December 1. From Raimond and Oort (1943g)

The light from the nebulosities had a clear signature, namely that of the date February 1901; it had been emitted on that date and had in the intervening time reached the clouds and was now observed months after the outburst.

There was *one* fact that did not draw the attention it should have had in 1901 and 1902. Perrine had also seen a nebulosity on an exposure of March 29, 1901. Even though only five weeks had passed, the light-front would already have been 5 arcmin from the nova. That could have been used to prove that the nebulosities did *not* expand with a constant speed. Von Seeliger, however, seems to have noted this without attaching much attention to it.

Everything was quiet about Nova Persei until 1916. In that year, Barnard observed a very small nebula around the weak star that had remained of Nova Persei. The radius of the nebulosity was only 8 arcsec. This finally constituted the remains of the material the nova had expelled and that had now grown to observable size. During the outburst the nova expelled its atmosphere with great force and surrounded itself with nebulosities that had been seen in the 'nebulium' lines that appeared in the spectrum of the nova in April 1901. Later photographs confirmed this. The nebula—a distinctly different phenomenon from the light-front that had been observed in 1901 and 1902—expanded in all directions.. The proper motion was 0$''$.4 per annum, a motion that in itself was not remarkable. If Kapteyn's estimate of the distance were correct, it would correspond to a velocity of about 400 km/s. This is a high speed, but not really larger than would be expected from a violent eruption like the explosion of a nova.

But that was not the end. The various pictures that had been taken between 1916 and 1935 showed the expansion of the nebula, but no observer had succeeded in observing the nebula spectroscopically. This was done by

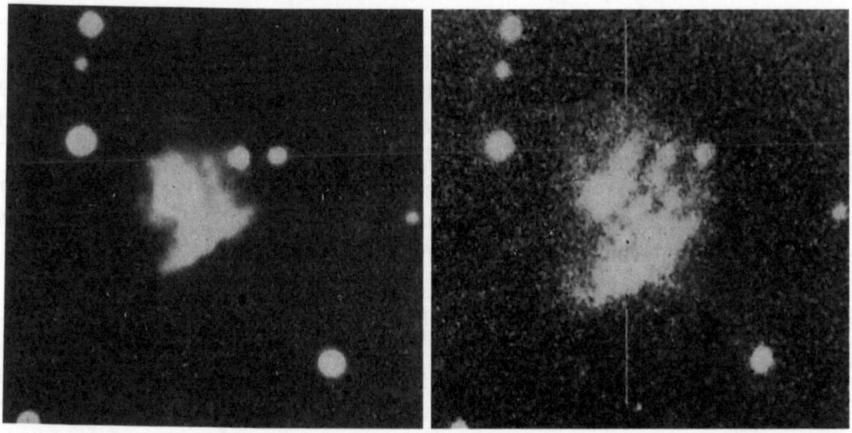

Fig. B.5 On the left a photograph of the nebula, discovered in 1916 by Barnard, obtained by Ritchey on November 15, 1917. Exposure 5.5 h. On the right; Image by Humason in 1934. The thin line gives the position of the slit of the spectrograph used for the spectrum in Fig. B.6. From Raimond and Oort (1943g)

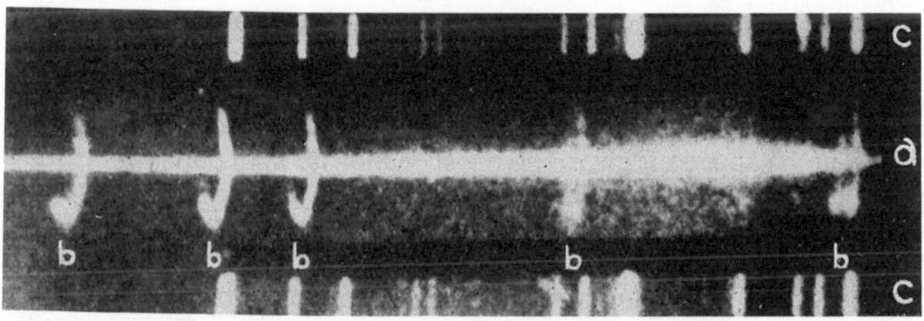

Fig. B.6 Spectrum of the nebula in 1934. From Raimond and Oort (1943g)

Humason, who had performed more difficult observations in his program to measure radial velocities of extragalactic nebulae. In 1935 he succeeded in obtaining a spectrum of the nebula around Nova Persei.

The spectrum is strange. Figure B.5 shows a direct image of the nebula and Fig. B.6 a reproduction of the spectrum. Humason used the 100-in. telescope for this. The nebula is not symmetric; part of it is bright and other parts are faint.

[Raimond then went into a long exposition on how the spectrograph worked and how the hook-shaped lines have originated. I omit this detail and also the figures that go with it. In Fig B.6 the horizontal direction corresponds to wavelength and the vertical displacement is along the slit, the nova being the long horizontal feature *a* in the center (and *c* the comparison spectrum on either side for calibration of the wavelength scale). The nebular lines are elliptical. If the nebula would be an expanding spherical shell the lines would be double at the center (from the front coming towards us and the backside away from us) and single at the edges. The nebula is not uniform and therefore the ellipses are not filled and look like hooks.]

Humason found that the front of the shell is coming towards us with a velocity of -1400 km/s and the back away from us at $+1050$ km/s. This indicated that the whole is moving towards us at 175 km/s. It is very large compared to radial velocities of other stars, but the spectrum was very fussy and the velocities rather uncertain. [Actually, the radial velocity is not 175, but $+28$ km/s]. The nebula expands in all directions at a velocity of 1225 km/s, or in round numbers around 1200 km/s. [This is still a rather good value.]

From direct photographs between 1916 and 1935 the radius of the nebula on the sky was found to grow by $0''.4$ per year. From this one can easily calculate the distance. A velocity of 1200 km/s then would mean a distance of 2000 lightyears and not the 300 lightyears that Kapteyn deduced. But from this it follows that *in 1901 and 1902 the light-front moved away from the nova at a speed that is about ten times as large as the speed of light.*

[At this point the article is broken in two parts. Raimond ends the first part by comparing it to a detective story, challenging the reader to find the solution that Oort had found, before the next installment a month later. He does acknowledge at the start of the second installment that, independently from Oort, Paul Couderc had also found this solution.]

Assume a star that is uniformly surrounded by non-light-emitting clouds. Assume that at some time $t_\circ$ the star emits a burst of light, a light-front, that moves at the speed of light c. The question then is: what parts we do see at some later time? Referring to Fig. B.7, assume that at some time t the point K_1 of the nebula is seen by us illuminated by the nova. There are other parts of the nebula we also see as long as for these the total path traveled by the light is equally long. Thus, the points K_2, K_3, etc. share the property that the sum of the total path from the nova to us is the same as for K_1. The geometrical figure connecting all these points is an ellipse with the nova and the Earth in its foci. The lightrays considered all have left the nova at the same time and have reached the Earth also at the same time. This is for the flat picture in the figure; in reality we have three dimensions and the figure becomes a revolution ellipsoid.

After a few days the luminosity of the nova diminished; let us for simplicity assume that the nova emitted light only between the times $t_\circ$ and $t_\circ + \Delta t$; Δt then is a short time measured in days. The light that left the nova at time $t_\circ + \Delta t$ must also lie on an ellipse—in the three-dimensional case on an ellipsoid. All parts of the cloud illuminated by the outburst then lie in between these two ellipsoids.

If the space around the nova were completely filled with dark nebulosity we would see around the nova a disk of light—an aureole—. In 1901 there were only light streaks so that we have to conclude that the nebula was made up of flat layers. For what follows we will assume for simplicity that there is only one layer. We will also assume that the sight lines from the illuminated parts of the cloud towards the Earth are parallel. This is justified because the

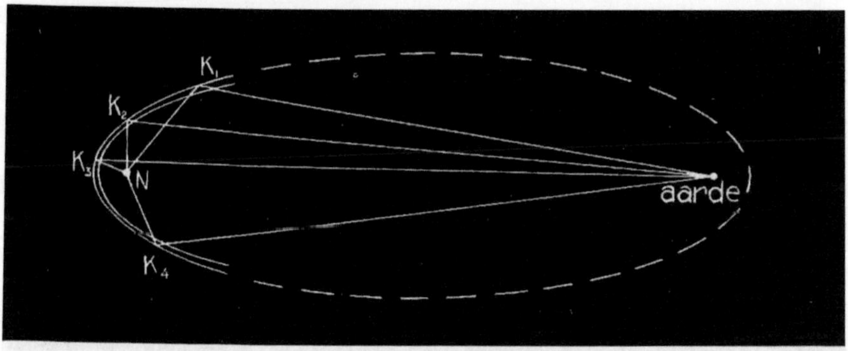

Fig. B.7 The lightrays that reached the Earth at the same time have been reflected by various parts of the nebula that are all lying on a revolution ellipsoid with the nova and the Earth in the foci. From Raimond and Oort (1943g)

dimensions of the clouds are at most some lightyears, while the distance to the Earth is 1600 lightyears. The question then is what do we observe at the Earth. We will have to distinguish three cases. The first case is that the nova is *inside* the layer; this is the case that Kapteyn had tacitly assumed. For this see the top panel of Fig. B.8.

The light that left the nova at time $t_\circ$, reached the Earth at the time $t_\circ + r$ if r is the distance of the nova to the Earth expressed in lightyears. The light that reached the point K_1 after 0.1 years, would arrive at the Earth at $t_\circ + r + 0.1$. The path it had followed via NK_1 was a tenth of a lightyear longer than the direct path. Similarly the light that reached K_2 after 0.2 years reached us at time $t_\circ + r + 0.2$. The light-front traveled along the layer with a speed c and that was reflected in the apparent expansion on the sky.

But now look at the case that the nova is behind the layer (see the middle panel of Fig. B.8). For the sake of the argument let us assume that the layer is 30 lightyears from the nova. This is the distance a in the figure. The direct signal, now through point K_1, reached us at time $t_\circ + r$. A tenth of a year later the signal via NK_1 showed us that the nova was fading. At the same time a signal reached us via NK_2, which is a path 0.1 lightyears longer than the direct one. Indeed on March 9, 1901 a picture was taken at Lick Observatory of a circular nebula, but that escaped observers at the time. The distance that the light-front had traveled through the layer follows from Pythagoras' theorem as $\sqrt{(a+0.1)^2 - a^2} = \sqrt{0.2a + 0.01}$. Similarly for other points (see the figure). When $a = 30$ the distances become $K_1K_2 = 2.4$, $K_1K_3 = 3.5$, $K_1K_4 = 4.5$ lightyears. The distance K_2K_3 is 1.1 lightyears and seen from the Earth the front had moved that distance in 0.1 years; so the average velocity was 11 times the speed of light.

On September 20—about 0.7 years after the outburst—a nebular veil was observed. If our assumptions are correct, the then observed parts of the layer had to be at a distance $\sqrt{(a+0.7)^2 - a^2}$ from K_1. The distance K_1K_7 then is

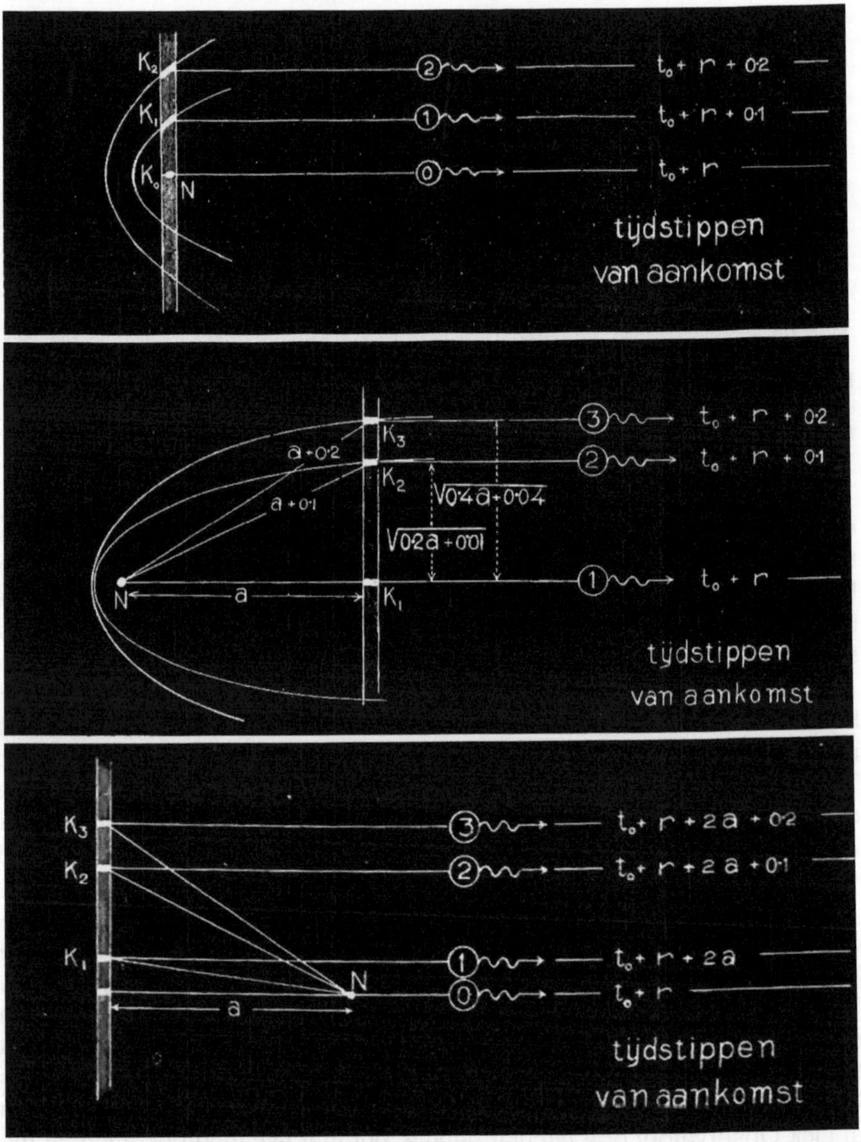

Fig. B.8 Three cases of the path of the light from the nova to the observer. At the top the nova is situated *inside* the reflecting layer, in the middle *behind* it and at the bottom *in front of* this layer. 'Tijdstippen van aankomst' means 'times of arrival'. For further explanation see the text. From Raimond and Oort (1943g)

$\sqrt{1.4a + 0.49} = \sqrt{42.49} = 6.5$ lightyears. That means a mean speed of 6.5

lightyears in 0.7 years, or 9 lightyears per year. *The light has traveled through the layer at a speed almost ten times that of light*. Of course this holds for a single layer; on September 20 there actually were other cloud veils as well.

We still have to look at the third option, that the layer is behind the nova. This is illustrated in the bottom panel of Fig. B.8. In Fig. B.3 we see a fan-shaped blob—see the letter *m* in the left-hand panel—; it was very bright and observable for a full year. The corresponding cloud must have been situated just behind the nova; it was illuminated by almost unattenuated light from the nova.

It is not likely that clouds very far behind the nova will be observable. Figure B.8 shows that a cloud that is 10 lightyears behind the nova, would only have appeared to us in 1921. However, the image would be very faint; the particles in the cloud scatter only a small faction of the light back towards the nova and therefore to the Earth. Most of the light will be scattered in directions that are not very different from those away from the nova.

Following Professor Oort we can ask the question how the fine structure in the aureole can be explained. According to the drawings of Ritchey—see Fig. B.3—the aureole consisted of a few narrow arches of light; some of these have structures that can be identified and followed in time. Their motions have been measured and checked against the theory presented. The conclusion was: there exists an irregular cloud between the Earth and the nova that consists of a number of patches. These patches are layers with a thickness of about 1.5 lightyears. Some of these are in or near a plane that has an angle of about 45° with the line Earth-nova. Two of these patches have given rise to the appearance in 1901 and 1902 of the outermost arches. The cloud layer that corresponds to the arch *abc* in Fig. B.3 (left panel) is about 10 lightyears from the nova.

[Raimond ends the article with a long discussion of factors that influence the brightness of the aureole, namely the density of scattering particles in the cloud, the characteristics of the scattering process and the fact that the ellipsoids become more and more separated as the distance to the nova increases.]

B.5 Speech at Re-opening Sterrewacht on June 20, 1945

Sterrewachters,
Now that the War is over here in the West, the professors that left Leiden have returned, and there is no longer any reason to have Prof. Hertzsprung after his formal term—which ended already September last year—carry the burden of the directorate, the management has been referred to me. On this occasion I feel the need to say a few words to you.

Since the German army invaded our country much sorrow and misery has come our way. You also have not been spared. After the for those concerned much too long time of mobilization, that some of us had to endure, after the great worries of the few days of fighting, came the deep humiliation of the quick defeat and surrender. The military occupation by the enemy, however hard, would have been possible to accept. But it soon turned out that we had to do with something more serious than a military occupation, namely that the 'Third Reich' wanted to make us support their Nazism. What this meant for the country, I will not recite. It is too fresh in our memories: hostage camps, concentration camps, executions and murders, persecution of Jews, and that at such a manner unworthy of man that it has deeply hurt all of us, mostly inactive spectators, and finally the starvation of half the population on a scale that resistance was almost broken. For Leiden University that led to closure, and it had its effect on all of us as we are here today. Kooreman [C.J. Kooreman worked at the Observatory for most of his career, starting in 1927 as computer and ending in 1967 as technical/astronomical officer] was prisoner of War, others constantly felt threatened by O.T. [the Organisation Todt that used and employed forced labor] and Gestapo [Geheime Staatspolizei]. Much of the normal work was made impossible, and life especially during the last year was made difficult by worries for food and fuel.

But in spite of all misery also something much better grew. Slowly, gradually and quietly a stubborn spirit of resistance against the moral and material suppression appeared. The desperation that we felt after the surrender was replaced by another battle. A battle in which many took part, everyone according to own character and capability, and from which in spite of heavy loss no defeat, no surrender resulted. We all know what tasks Dr. Hins as former reserve-officer took upon himself, with what idealism Steinmetz put his life on the line, and many others have actually helped to ease the life of Jews and subducers.

Many Dutchmen originally were critical of the resistance movement. They only noticed the casualties and loss of many valuable things. They lost sight of the fact that not fighting back would result in even graver moral loss.

Those of you that participated, fully or part of the time, must have felt something new came to life inside them. They might have felt the joy of the bond between comrades fighting together and of working jointly for an ideal that transcended daily life. Of course only a small fraction of Dutchmen had the opportunity to be active in the resistance movement. But there were also others that were able to break from their daily drudgery in which many lifes seemed lost.

And now we return to 'ordinary' life. It is no surprise that some of us will feel some aversion. We have been away from it, have seen something wider and do not want to return to the narrow everyday business. Some may want to accept the solution to volunteer for more fighting for freedom in the Indies, but that road is not open to all of us.

But those that stay behind do not have to return to the same life! We will bring something new into our lives and work. Something in which we recognize our ideals. The war of resistance is really not the only road to happiness, which we may experience as comrades building something together.

Let us realize and keep in mind that whoever does his work at the Observatory with real dedication, helps build something valuable for the country. As Salazar [António de Oliveira Salazar (1889–1970), Prime Minister of Portugal] said: 'Work, all work, is equally noble and worthy, when seen as a contribution according to capability of everyone to the community they belong to.'

We should realize that we all work together for the discovery of a new, grand world around us, to which we all form links in the chain of the history of human civilization.

Without being arrogant we should realize that we work at an observatory that occupies one of the most important places among European observatories. We cannot exaggerate the importance of our own institute. We can all be proud to be associated with an institution where so many important contributions to science were produced under the leadership of men that—like Prof. de Sitter and Prof. Hertzsprung—will for ever be honored in the history of astronomy as major figures. Not every staff member can be a great astronomer himself, but he has been called to do as much as possible the work for which he is best fit by talent and education. The work often is very unassuming, for many it threatens to become dull and routine because it involves repeating similar calculations. Although I will try to bring you as much variation as possible, it is unavoidable that much remains monotonous. Even the most simple work brings responsibility. All work, from complicated to simple, demands the same complete dedication. When you do your computational work with the same dedication and care with which a farmer clears the weeds from his land, you will find some of the same satisfaction when seeing the finished work as the farmer who looks his crop. And try to remember that your contribution is part of a much grander design for mankind to fulfill its calling to disclose nature. And: give it all your energy. Because then you will acquire happiness.

I very much like you to go back to work with joy. I will try to promote a closer contact between the scientific staff and the observers, instrument-makers and computers. For that purpose there will be some changes in the assignment of rooms and everyone will have some special affiliation with a staff member.

You should not expect that you or I will be able to remove in a single instant the old grass of slowness that has grown over life. We have to fight it continuously. We also will have to experiment to optimize the team work. In hindsight it might have been better if some things had been done differently, and we will therefore, especially in the beginning, have to make changes. I will always be open to your wishes and to accepting your advice in case you want changes for good reasons.

A final practical remark.—for a short period we will work half days. I expect you all to be actually present at the Observatory between nine until half past twelve. If you have a special reason not to be present you will need to discuss that with me the day before. In case of illness I ask you to send a note immediately with the character of your ailment.

With respect to holidays you will keep for this year the rights for a complete vacation, unless officials decide differently. For this I would like to set the same rules that were held under Prof. de Sitter, which means at least one long vacation.

It is my intention to make it possible that we can have a joint coffee break together either in the Lecture Room or outside when the weather permits, at 11 o'clock, for which I make available a quarter of an hour.

I end with the hope I can count in all respects on your full cooperation and dedication for the beautiful job entrusted to us, for the prosperity and blessing of our Leiden Observatory.

B.6 Valedictory Lecture: To the Horizon of the Universe

This lecture was delivered on Friday September 18, 1970, and published in Dutch as Oort (1970j) without a title. An independent, but rather inaccessible translation has appeared as Oort (1971d). There the title was given as: *To the horizon of the Universe*. Below a new, independent translation.

A valedictory lecture is a natural occasion to look back over one's working life to the dreams for the future one had at the beginning, and the extent to which these have come true. What strikes one foremost in this is how inadequate the early expectations and fantasies turned out to be. Nature always seems to have in store completely new, úndreamed-of things that are more important than the ones that were pursued.

Of what are the most intriguing subjects today, no astronomer had any idea then. Not even science-fiction writers. Really new territory is only discovered by chance, while one is searching for something else. In this lecture I want to illustrate this with a few examples from my own experience.

I have been fortunate to have known a very eminent scientist, early on in my studies in Groningen. Prof. J.C. Kapteyn was not only a genius astronomer, but also an exceptional teacher, who was able to instill in his students a spirit of research that many have been nourished all their lifes. He lectured only about classical subjects, such as the motions of the planets, and the tides of the oceans on Earth, things that interested me enormously, among other things since an early attempt at research while still a highschool student, when I tried to find a relation between the crudely measured times of high tide at Katwijk and the positions of the Moon and the Sun. But the most important thing was the *manner* in which Kapteyn approached his topics. He taught us how astronomers in the course of time came to their insights, how discovery really worked. Two things were always prominent: first that there should be a direct and continuous relation to observations, and secondly that one shoudl always aspire to, as he said, 'look through things' and not be distracted from this clear starting point by vague considerations. His lectures only concerned the planetary system. But his Laboratory Kapteyn, in the meantime, was busy with something else. He was investigating the structure of the Galaxy. That was pioneering work.

At the end of the 18th century William Herschel had investigated the shape of our Sidereal System, but the observations were too uncertain to draw more than a few crude and rather vague conclusions. Kapteyn and his collaborators were the first to seriously attempt to put their research on a solid, quantitative basis. As always in case of pioneering work, it was hard and time consuming. His co-workers not only did the brick-laying, they also had to collect the bricks. For many years Kapteyn collected them with his own hands. Because of this there was a deep humility and modesty at the Groningen Laboratory, which left a lasting impression on all who grew up there. Kapteyn, however, never was fixated by his work only. He retained his vision of the broader picture, his cheerfulness and his inspiration.

In that respect Groningen was different from the foreign observatory where I worked for two years after finishing my studies. There the careful masoning was so much more important than what had to come next that over the entrance there was an engraving: 'You who enter here must give up all hope for discovery'. But one received a thorough grounding in the practical aspects of the craft.

But let us return to the exploration of the Galaxy. Important progress had been made on the one hand by the thorough long-term work at the Groningen Laboratory, on the other by the more adventurous work of Harlow Shapley at Mt Wilson. However, the results of these two approaches were difficult to reconcile. The 'Kapteyn Universe' was almost ten times smaller than the swarm of globular clusters on which Shapley had concentrated his research. Furthermore, they had a completely different shape. His [Shapley's] was almost spherical, while the most pronounced characteristic of system investigated by Kapteyn was its disklike form, a structure that we can see as the faint glow of the Milky Way. Notwithstanding these radical differences there could not be any doubt that the globular clusters belonged to the Galaxy. The most convincing piece of evidence for this was that the center of the swarm of globular clusters was situated right in the Milky Way and moreover in one of the most prominent parts of it.

These were the first, hesitant steps towards a new astronomy in which a young astronomer was growing up. For three centuries astronomical research had been directed at explanations of the motions in the Planetary System. The discovery of the Milky Way Galaxy was a jump further into space by a factor one hundred million.

Another jump had been made at the same time. While investigating the motions of stars, with the goal of using these to derive their distances, Kapteyn had found something unexplored, namely that stars did not, as was always assumed, moved randomly in all directions in space like molecules in a gas, but had a preference for a particular direction. This discovery of the so-called Star Streams marked the beginning of Galactic dynamics. It was an area that I found very attractive, and by chance I became involved in it during my studies. A somewhat older fellow-student had mentioned the strange behavior of stars that had high velocities. I do not remember what brought me to look further into this, but through these stars of high velocity, that would later become the subject of my Ph.D. thesis under van Rhijn, I entered a field of research that was almost completely unexplored and not understood at all. I received the doctors degree, but in hindsight I did not really deserve it, because the *explanation* of the phenomena addressed in the dissertation, had not been found while in reality it was up for grabs. The explanation was proposed a few years later by a young Swedish astronomer, Bertil Lindblad. Even then I almost missed the solution because of the rather inaccessible way in which Lindblad wrote his papers. When I did understand these, I was immediately convinced of the correctness of his idea. This led to the discovery of so-called differential rotation of the Galaxy and at the same time to the start of a lifelong friendship with this Swedish astronomer.

The discovery of the rotation of the Galaxy confirmed that Shapley's swarm of globular clusters indicated the correct size of the Galaxy. It also made clear that Kapteyn and van Rhijn had only mapped a small part of the system. The population density of the stars *outside* the plane of the disk was well determined, but *in* that plane the dimensions were affected by absorption of the light of distant stars, of which they had underestimated the importance. The many billions of suns that make up the Galaxy, are predominantly concentrated into a thin disklike swarm, of which the thickness is only a few hundredths of the diameter. In the disk there is a thick fog. The fog layer is thin, so that we can look up and down almost unaffected, and in that way we can explore the world *outside* the disk almost unhindered by it. But *in* the disk a thick mist surrounds us like an impenetrable wall. Unlike to mist on Earth, the Galaxy-fog never lifts. In the thirties it looked as if we simply had to live with that, and that by far the largest portion of the Galaxy would for ever remain hidden from us.

Although the malady could not be remedied, it still seemed important to try and find out what caused it. How could small particles of ice that presumably made up the interstellar fog, have been formed and how were they destroyed? Those questions led in 1941 to the announcement of a prize competition by Leiden University. This competition became the starting point of a long series of studies, in which especially van de Hulst would take an important and lasting part. It also led to stimulating collaborations with Kramers in Leiden and Jan Burgers in Delft, of which I have the most pleasant recollections. Through these problems a wide area of research was opened concerning rarefied clouds of gas that exist between the stars, and also groups of young stars and the supernovae that cause the motions of these clouds.

In this way the *Crab Nebula* entered my life. At the position where we now see the nebula, a star exploded 900 years ago and flared up as a so-called supernova. We still see the parts of the star that was torn apart expanding at about the speed that they attained in the explosion. This Crab Nebula became the key for the solution of some of the most fundamental problems in astronomy. Of the adventurous travels I made as an astronomer, the one to the Crab Nebula has been the most fascinating. The study of this one object has affected so many fields of research that at the Massachusetts Institute of Technology last year a full extra-curricular course has been given for students both in science and in humanities. I have briefly been tempted to build this complete lecture around it, but in the end I decided not to since the subject is a little outside the work that has occupied me for the larger part of my life.

My association with this nebula started a few years after the War in a conversation with my friend Duyvendak that led him to examine Chinese chronicles of the period around 1054, where he located very interesting information on the flaring up of the supernova on July 4, 1054, and on further properties of that curious star.[17] It is remarkable that there is no record in Western countries of this striking phenomenon. A few years later, Walraven and I decided to undertake a careful investigation of the brightness of the Crab Nebula to find out how it would decrease as a result of the expansion. It was during the very cold winter of 1954/55 that I was involved in this as a sort of 'night assistant' during the observations. By complete coincidence, Walraven had also mounted a polarimeter on the telescope. And this caused us to find something completely different, infinitely more important than what we were after, namely the irrefutable evidence that a part of the nebula emits light of a completely different nature than the light with which we are acquainted in everyday life as well as in astronomy. The discovery of this 'synchrotron radiation' in the Crab Nebula provided a basis for many new contacts and friendships, especially in the Sovjet Union, where, as became evident in hindsight, the first research on this kind of radiation had been performed; outside Russia it had remained almost unknown. It also renewed contacts with the enthusiastic astronomer Baade, who subsequently found the most beautiful details about the synchrotron radiation from the Crab Nebula. Baade was a man who always had a treasure of unpublished material at his disposal, which was also the case the Crab Nebula, in which he had observed over the course of time other, equally mysterious things, that now came in the the open. All these phenomena in the Crab Nebula were very puzzling. Just as later with the quasars, nobody understood *where* the continuing activity from this star which had exploded almost a thousand years ago, came from; there was also no indication what had become of the star. Only two years ago have the remains been discovered. It turned out to have become a very faint star that revolves about its axis at a fantastic speed; a complete star going around in one thirtieth of a second. It is an example of so-called pulsars, objects that have been discovered lately by radio astronomers; these are collapsed stars with a density some trillion times larger than that of the matter we find on Earth

[17]Actually this happened just *before* the War; Duyvendak published these results in 1942, together with the Mayall and Oort (1942a) paper (see Chap. 9).

or in normal stars like our Sun.[18] This pulsar turned out to be the engine that keeps these strange phenomena in the Crab Nebula going. But that is not all. The pulsar in the Crab Nebula might be a signpost to the solution to the still more mysterious—and *enormously* more powerful—problem of what happens in the *nuclei of galaxies*.

So we are back at the Galaxy and the mist that we left a few moments ago. The mist cannot clear, but help came from an unexpected direction, namely from *radio waves*. None among the astronomers had sufficient insight into the techniques of radio reception to come up with the idea that it could be used to explore the Galaxy. Even when in 1931 an engineer of the Bell Telephone Company, Jansky, investigating the cause of interference in radio reception at short waves, discovered that these were related to the Milky Way, it still did not dawn on the astronomers what a powerful tool they had in their hands. The realization only came ten years later when a second American engineer made new observations with a radio telescope that he had built single-handedly in his backyard. These eventually, although not without difficulty, ended up in an astronomical journal.

Then it became clear that this would open tremendous possibilities to penetrate the fog in the Galaxy. The radiation at radio wavelengths travels unhindered through the thickest mist, even through a snowstorm.

In Kapteyn's footsteps we were involved in Galactic research more than anywhere else. No wonder it is that where the first plans were formulated to profit from these new possibilities. The possibilities were enhanced considerably when van de Hulst had calculated that specific radiation at a wavelength of 21 cm would be emitted by hydrogen atoms, and that this radiation would probably be emitted by the interstellar gas with a measurable intensity. All of this took place during the War.

It was a fortunate circumstance to find immediately after the War in prime minister Schermerhorn a person who was greatly interested in our plans to build a cheap large radio telescope, the 'chicken wire' telescope as he called it. In the end it was not cheap, but ZWO and its director Bannier was so interested in the project that they did not blame us for the initial low estimates of the costs! In this way radio astronomical research started in the Netherlands. Actually it started with the radar antennas left by the Germans in the dunes, the so-called Würzburg Riesen that on the initiative of Ir. de Voogt were transported to Radio Kootwijk by the PTT, one of which the newly founded 'Stichting Radiostraling van Zon en Melkweg' was given the use of.

With the extensive observations of the 21-cm line a few enthusiastic young men, among which Westerhout and the later discoverer of the quasars

[18]Oort wrote in Dutch 'triljoen'. In the Dutch language the 'long system' is commonly used, where an increase of a factor one thousand gives (starting at a million) 'miljoen, miljard, biljoen, biljard, triljoen, etc.' Then triljoen is 10^{18}. In English the 'short system' is usually used: 'million, billion, trillion, etc.' The density in a neutron star (which is what a pulsar is) is some 5×10^{17} kg per m^3, and on Earth the average density is about 5×10^3 kg per m^3. The factor thus is 10^{14}, 'honderd biljoen' in Dutch or 'one hundred trillion' in English.

Maarten Schmidt, produced the first map of the spiral structure of the Galaxy, while at the same time the rotation and the mass distribution were determined. Things of which *before* the War I would never have dreamed that they could ever be observed.

In itself this was quite important. But it turned out as we expected: The majority of galaxies show spiral structure, which is why they are called 'spiral galaxies', and we had not expected anything else than that our Galaxy also would have spiral arms. We would almost have felt the same as Eddington, when after a long and difficult series of observations he had found that the deflection of light by the gravity of the Sun, predicted by Einstein, indeed existed. He is reported to have uttered the famous words: 'Foiled again: we have learned nothing new about Nature.'

The greatest advances come about when predictions are *not* confirmed. Or from the unexpected things that you find by chance while you are looking for something else. Things you never dreamed of that they would exist, but that you run into at the corner of the street. Like when in 1960, when we were searching for a gaseous atmosphere, or halo, of the Galaxy. Indeed, after much effort—since they were very thin—we found gas clouds outside the plane of the Galaxy, that also moved with the expected velocities. However, they turned out not to move in random directions but seemed to be falling towards the plane of the disk. This probably is gas that streams in from the surrounding space so that we have to conclude that the Galaxy is still growing.

And just as unexpected, it was discovered, during the investigation into the spiral structure, that in the central parts of the Galaxy enormous quantities of interstellar gas were systematically moving outward. It was immediately clear that there was something very strange behind this. Later studies, mostly of *nuclei* of *external* galaxies, have shown that in these nuclei very strange things may happen; that—probably intermittently—enormous explosive activity may take place, activity that results in the so-called radio galaxies. In its extreme form we see this activity as the so-called *quasars*.

No one would have expected that a normal, seemingly quiet galaxy like our own would also take part in that. But it seems more and more likely to me that the enormous expanding gas masses that we have found in the central part of our star swarm have originated in the small nucleus at the Galactic center.

But how? And what does this mysterious nucleus contain?

There are only vague speculations: Fairytale-like speculations about 'black holes', holes in space, through which matter is being sucked up to become forever invisible. Rapidly rotating black holes with their strong magnetic fields could be scaled-up versions of the pulsar in the Crab Nebula. The large quantities of energy that might be liberated could be the cause of the eruptive phenomena in galactic nuclei.

There are even more eccentric speculations in which it is supposed that in the nuclei of galaxies pieces remain of the primeval fireball out of which the Universe originated ten billion years ago. In reality the study of the nuclei of galaxies has in recent years become one of the most fascinating areas of research. The study of the central parts of our Galaxy in Dwingeloo has provided an important contribution to that.

The instability, or explosiveness, of their nuclei appears to have become one of the most fundamental properties of galaxies. During these violent events it seems that invariably high-energy particles are being produced that radiate at radio wavelengths. This is why radio astronomy has shown us in particular the *unstable* aspects of the world, while classic optical observations by and large have shown us the *quiet*, stable aspect of the galaxies.

For astronomers the existence of *in*stability in the Universe is a bonus of the highest order.

Thanks to these instabilities we can now hope to determine the size of the Universe with radio telescopes. Only these eruptions can be powerful enough to be observable out to what one might call the horizon of the Universe, from which the ordinary radiation of the quiet systems is no longer observable.

One of the questions concerning the Universe is whether it is finite or infinite. Many people think it is finite. In that case *space* has to be curved like the *surface* of a sphere is curved. And like one can determine the curvature of a spherical surface—and therefore also the radius of the sphere—from the measurement of angles, we may now expect that by measurement of angles at which we see the remotest radio sources we can observe, we may determine the curvature of space and through this the radius as well as the total volume of the Universe. This is theoretically possible. Whether it will be feasible in practice will become clear in the near future. Finding an answer to these questions would be of the utmost importance for our understanding of the physical world in which we live and its history. But even if it would turn out impossible to answer these fundamental questions now, it is to be expected that the voyage to the horizon of the Universe will turn out a fascinating one during which stranger and more unimaginable things will be encountered than all the wonders that the explorers of the Earth found on their journeys in the 16th and 17th century.

When I started out as an astronomer it was the *Galaxy* that lured as an distant goal and for which explorations were prepared to the borders of the system.

Now, at my farewell, again everyone is called upon for the preparation of great expeditions. One of these expeditions is ready to go underway at Westerbork. This time we will not go to the outskirts of the Galaxy, but to those of the *Universe*.

Current expeditions are not only reaching out in *space*. This time we also will travel back in *time*, to the period during which the galaxies were born, and partly also to the early phases of the Universe.

Such an undertaking may seem like defying the gods. But it is probably not more foolhardy then the voyage of Columbus in the old days, and maybe even more promising than the jump from the Planetary System to the Galaxy in the days of Kapteyn.

For the moment these are only dreams. I hope they will not turn out nightmares for my successors.

B.7 Kyoto Prize Lecture: Horizons

Oort held the lecture on the occasion of the award of the Kyoto Prize on November 11, 1987. It was published in Dutch in the journal 'Zenit' (Oort, 1989b). The following is the English text Oort used when delivering the lecture; it is part of the Oort archives. At the request of the organizers it was to a large extent autobiographical. Text underlined in pencil in Oort's copy have been italicized and reference to figures adapted.

When we were young we may have watched how ships gradually disappeared when they were farther and farther away from us. We saw first the hulls disappear, while the rest remained visible. Finally, as they sailed away, even the tops of the masts dipped under the 'horizon'. As children we were intrigued by this horizon. What was behind it? Grown-ups have a similar curiosity. They sailed to the horizon, and found that there were new horizons behind, and that this was repeated almost eternally. Until at last they found they had circumnavigated the whole Earth, and had discovered the whole world.

But there remained the space above us: the realm of the Sun, Moon and stars. From the earliest times from which writings have come to us it has been thought that the heavenly bodies formed the outer limit of the world. As there were no ships, they were unreachable. Nobody knew their distances.

However, Greek thinkers conceived models of the world which incorporated the heavens. Among many other things they understood that the eclipses of the Moon were due to the Moon's coming into the shadow of the Earth. The observation that this shadow was round taught them that the Earth must have a spherical shape. They also used observations of this shadow to obtain a measurement of the distance to the Moon: because the Moon moved so fast, going around the sky in a month's time, while the Sun took a year to do this, it was generally, and correctly, thought that the distance of the Sun must be considerably larger than that of the Moon. The shadow of the Earth over the range concerned must then be almost a cylinder, with a radius equal to that of the Earth. By measuring the angular radius of the shadow of [on] the Moon one can therefore find the distance to the Moon, once the Earth's radius is known. The latter had been determined by Eratosthenes from a combination of the linear distance between Syene in Egypt and Alexandria with the angular distance between stars going through the zenith at the two cities. The resulting distance to the Moon was 60 times the radius of the Earth, or 400,000 km: the first measurement of a distance in space! It was a tremendous step.

Almost two thousand years had to pass before the next jump, to the distances of the planets and the Sun could be made.

The Greek astronomers had conceived a model of the world in which all heavenly bodies were attached to rotating spheres. The outermost sphere, the 'primum mobile', contained the fixed stars. Inside this was a sphere containing the planet Saturn; it rotated around its own axis which was attached to the outer sphere. Then came the Jupiter sphere whose axis was attached to that of Saturn and again had its own rotation, and so on for Mars, [the Sun], Venus, Mercury and the Moon. At the center was the immovable Earth. The spheres were crystalline and transparent, so that the inner ones did not obstruct the view to the outer planets. The model was fully described and transferred to later generations by Ptolemy in his famous book the 'Almagest, which for many centuries remained the standard textbook on astronomy. It was only in the 16th century that Ptolemy's system was seriously challenged by Copernicus. He [Nicolaus Copernicus (1473–1543)] pointed out that it was more plausible that the daily rotation of the stars was only an apparent motion; that it was the Earth which was rotating, and that the stars were at rest; that, furthermore, the Sun did not describe an orbit around us, but that we described an orbit around the Sun. An entirely new model of the planetary system was constructed by Copernicus and his followers, Tycho Brahe, Kepler and Galileo. It was in this era that by technical refinement of instruments and passionate devotion of scientists like Tycho and Kepler a second tremendous leap in measuring distances in the Universe was made. It was accomplished by measuring the direction towards a planet from two different positions as illustrated in the picture. Due to the Earth's rotation the observatory swings around, and will alternatively see the planet in slightly different directions. Knowing the radius of the Earth the observer also knows the distance between A and B: precise measurement of the angle between directions AP and BP then enables us to construct the triangle APB and thereby derive the distances AP and BP. It was a marvelous accomplishment; the angle between AP and BP, which was the essence of the measurement, being less than a minute of arc. In this way the scale of the planetary system became known and the foundations were laid for the magnificent theory of gravitation by which Newton finally crowned the exploration of the Solar System (Figs. B.9).

But what about the fixed stars? Just because of their fixedness they should be much further away than the planets. But how much further? Some original scientists had already speculated that they might be things similar in

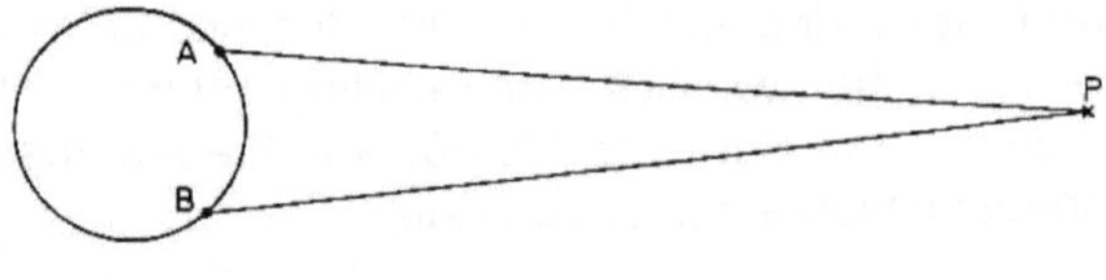

Fig. B.9 Figure from Oort's Kyoto lecture. From the Oort Archives

nature to the Sun. If that would be so their distances must be a million times larger than that of the Sun, far beyond any distance that could be measured in Kepler's time. Once more, as in Tycho and Kepler's time three centuries earlier, it was an enormous advance in precision brought about by technical developments, and in particular the construction of large telescopes, the 'space ships' of those days, which enabled man to measure the distances to the fixed stars. Just as in the case of the planets the measurements were based on trigonometry. But this time the basis of the triangle was not the diameter of the Earth, but the 25,000 times larger diameter of the Earth's *orbit*. Nevertheless the measurements were at the extreme verge of what could be accomplished. The angle between AP and BP was less than a second of arc which is the angle under which you would see a small coin at a distance of 2 km.

The observations conformed that the stars were indeed of a brilliance comparable to that of the Sun.

The measurements had also shown that the stars were not fixed but had considerable motions. A young Dutch astronomer, J.C. Kapteyn of Groningen, had even discovered a systematic trend in these motions which became known as *'star streams'*. They were the first discovery of *stellar dynamics*.

A gigantic new world had become accessible to exploration. The research was started vigorously in a few centers. One of them was the University of Groningen.

It had long been recognized that the world of stars did not extend indefinitely; but it was unknown how far it extended and what shape it had. Kapteyn had set it as his task to find out. He also wanted to investigate the motions of the stars and the forces that held them together.

It was in this early stage of reconnaissance of the Milky Way System, or Galaxy, as the star swarm was called, that I began my university studies. Drawn to Groningen by Kapteyn's fame I was soon fascinated by the inspiration which radiated from his lectures. So much so that in my first year I tried to make my fellow students in law and medicine share this inspiration. And what luck it was for a student to grow up in an environment where through hard work and enthusiasm the first traces of a new world were being revealed!

This world contained so large a number of members (in fact, some hundred thousand million) that it was impossible to study them all. For this reason Kapteyn had proposed to concentrate observations in some 200 small fields distributed all over the sky, in which observations would be made down to the faintest observable stars. He succeeded to instill this enthusiasm in his colleagues, and persuaded a number of observatories spread all over the world to join in his *Plan of Selected Areas*. The Groningen department itself had no telescopes. It made its contribution by measuring plates obtained at other observatories. But its staff was too small to cope adequately with the enormous number of measurements required. But Kapteyn was undaunted, and found a successful solution by asking and obtaining permission to get prisoners to assist in the work. [This story is sometimes told in connection with another project, the production of the *Cape Photographic Durchmusterung*, but probably is not the truth (see my *JCKbiog* [7], Sect. 6.10).]

During my student years the great project had already led to a provisional result. In this bold first model, the so-called 'Kapteyn Universe' the Sun was assumed not far from the center of the spheroidal swarm which had a diameter of some fifteen thousand lightyears, and a thickness of one fifth of this. The equatorial plane coincided with the plane of the Milky Way, which gave it its name 'Galaxy'. Of course there were no abrupt boundaries, the star density falling off gradually towards the outside. At the distances mentioned the density was about one tenth of that near the center. [Here Oort referred to the picture in the bottom panel of Fig. 3.2.]

My own work, during the years in Groningen was directed to the *motions* of stars. A fellow student had drawn my attention to an article describing a peculiar property of the motions of stars which had high velocities relative the the mean of the stars measured. The phenomenon proved to be interesting. The investigation led to my first article in an astronomical journal, and later became the basis for my doctor's thesis.

I was of course also involved in thinking about the Kapteyn System and its place in the *Universe*. In the 'school' of Kapteyn the big swarm of stars surrounding us *was* the Universe. It ended at the outer edges of the swarm.

In this same period the Universe was also being studied at the Mount Wilson Observatory in California by a younger astronomer, Harlow Shapley, in a quite different manner. Shapley concentrated entirely on the so-called globular clusters. These are concentrations of stars, with some hundred thousand members each. By studying a special set of variable stars in these clusters, and by various other means, Shapley succeeded in finding their distances. These came out to be very large; several ten thousand lightyears. If correct, the globular clusters would therefore lie well beyond the frontiers of the Kapteyn Universe. The roughly hundred clusters formed a swarm, much like Kapteyn's swarm, but of some five times larger dimension. It was concentrated towards a center some 20,000 lightyears away, far outside the Kapteyn system. What was remarkable was that this distant center lay precisely in the plane of Kapteyn's disk, i.e. the plane of the Milky Way!

Shapley imagined that the actual Galactic System was outlined by his globular clusters, and that the Kapteyn 'Universe' was one of a large group of 'islands' spread around through the much larger swarm of globular clusters. The model is illustrated in the illustration [Fig. 3.4, top panel]. At the time this picture seemed far from satisfactory; why did we not see the other islands, and why did the center of the large system coincide exactly with the plane of our local island?

The final solution came (somewhat later) through the realization that the real world was entirely different from what either the Kapteyn group or Shapley had imagined. It *did* stretch out over the entire diameter of the swarm of clusters but it did not *resemble* it. While the globular clusters formed a nearly spherical system, the *real* star swarm was a thin disk, whose thickness was nor more than a hundredth of its diameter (cf. the schematic representation in the illustration [Fig. 3.4, bottom panel]). At first sight, this model would appear to contradict the Groningen investigation. Actually, it does *not*. The Groningen astronomers had principally investigated their 206 areas distributed evenly over the sky, but in concentrating on these regions, had practically overlooked the thin band of the disk system. An important other factor was that the major part of the disk was hidden by absorption through dark interstellar clouds, whose overruling importance was not realized at the time, due to the fact that they were entirely confined to the disk. Actually, the whole of the Galactic disk outside a small circle around the Sun is invisible.

The true, complex structure of the galaxy became known only gradually, partly due to the circumstance that not all of its populations lay in the disk. It contained a mixture of various populations with varying degrees of concentration towards the disk, the extremes being the globular clusters which showed hardly any affinity to the disk.

The differences between the populations were strongly reflected in their motions. It is here that the peculiarities of the high-velocity stars which I had investigated in the years at Groningen began to fit. But I was still too much indoctrinated by the Kapteyn system to take the step to a model in which all stars around us would actually be part of a much larger continuous disk system. The step was taken in 1925 by the Swedish astronomer Bertil Lindblad by suggesting that all slow-moving stars in our surroundings would share a fast rotation around the center of the system of globular clusters. Lindblad thought of a solid rotation. I realized that because the mass of the Galactic System was concentrated towards its center, the angular velocity of rotation should increase towards he center, just like the motions of the planets in the Solar System. I had found earlier that distant stars in the Milky Way had unexplained systematic motions, and it now dawned upon me that these motions were just what one would expect in a rotating system of stars whose inner regions rotated faster that the outer parts. This led in 1927 to the discovery of the differential galactic rotation. The motions confirmed also that the point around which the System rotated lay in Sagittarius, in an exceptionally bright region of the Milky Way, coinciding precisely with the center of the system of globular clusters. It was a wonderful discovery; it showed that the same law of gravitation which had been so successful in explaining the motions of the planets in the Solar System was applicable also in the millions of times larger Galaxy, and that there was a strong analogy between the two systems.

In the following couple of years most of the characteristics of the Galaxy's structure and internal motions became understood, including Kapteyn's Star Streams.

However, one serious incompleteness remained: Due to the obscuration in the disk the major part (viz. that outside a small circle around the Sun) remained hidden behind the absorbing clouds. We could not even observe the center, which later proved to be the seat of extremely interesting phenomena.

Nor could we see whether the disk had any large-scale structure like the spiral nebulae. It was only with the advent of *radio astronomy*, almost twenty years later, that the enigmas of these regions were revealed.

But long before that time, during the first explorations of the Galaxy, a problem of much wider scope began to present itself; viz. that of the nature of the spiral nebulae. While making the famous survey of the sky with their large telescopes, William and John Herschel had found that beside stars and clusters there existed also a large number of nebulous-looking objects. Because many had a striking spiral structure they were commonly referred to as *Spiral Nebulae*. There was considerable doubt about their nature. Some astronomers, including Shapley, thought they were objects inside our Galaxy. Others, who finally proved to be correct, put them at much larger distances and pictured them as individual star systems, 'island universes', much like our own Galaxy. They would be so far away that not even with the largest telescopes individual stars could be discerned, which explained their nebulous appearance.

Already in my student days I was convinced that this was the reality and that the world of spiral nebulae opened our eyes to a *new Universe*. The second half of my lecture will be devoted to this astounding world.

But before entering upon this I must conclude the story of our own Galaxy. Quite unexpected new ways for its exploration were opened by the advent of radio astronomy. An engineer of the Bell Telephone Company, Jansky, who was searching for the cause of a disturbing noise in radio receivers, discovered that the noise was related to the direction of the Sun. It also depended on the position of the Milky Way, and specifically of the *center* of the Galaxy.

Due to a surprising lack of interest by the optical astronomers it took nearly 20 years before the significance of Jansky's discovery for the exploration of the Universe was realized by astronomers. Exploratory work was started by another engineer of the Bell Company. Around 1945 Grote Reber built in his own yard the first radio telescope for exploring the Galaxy.

The advantages of using radio waves for investigating the structure of the Milky Way System were obvious; for radiation at meter- and decimeter wavelengths penetrates unhindered through the dust clouds that obscured our view of the Galactic disk. But because of the long wavelengths, telescopes of large aperture were needed to obtain the necessary resolving power. Grote Reber's results were so promising that the Dutch astronomers who had been so deeply involved in the first exploration of the Galaxy made plans for building a radio telescope that would be sufficiently large for an adequate study of the structure of the Galaxy. We therefore made a proposal to build a radio telescope of 25 m aperture which would enable us to resolve the structure of the most distant regions of our Galaxy at a wavelength of 21 cm.

There was a a special reason why this wavelength was chosen: The Dutch astronomer van de Hulst had shown that atomic hydrogen, which, besides helium, is the principal element populating the space between the stars, will emit radiation at this wavelength. Accurate observations of the interstellar hydrogen clouds at this wavelength would enable us to determine not only their density but also their velocity. The radio telescope could penetrate to the hidden regions close to the center and measure the rotation velocity in these regions.

The large telescope was only completed in 1957, but ten years earlier we had the luck that the Postal Service put at our disposal one of the 7.5 m radar telescopes salvaged from the dunes where the retreating German armies had left them behind. Observations were started in 1952, and by 1954 a complete map of hydrogen density and velocity over the entire Galaxy as far as it was visible in the Netherlands was ready. It was *the first map of the Galactic disk*, and was quite an achievement. The work was in large part done by students who worked almost day and night, with mutually inspiring enthusiasm. It was a good time at the Leiden Observatory!

The map showed a striking *spiral-like structure*, confirming again the resemblance between our Galaxy and the spiral nebulae.

However interesting the 21 cm hydrogen radiation was, the radio emission of our Galaxy had other, totally *new*, things in store for us: A large fraction of the radiation does not come from tiny vibrations in atoms, like the light with which we are familiar, but comes instead from very-high-speed electrons describing orbits of many lightyears radius in the magnetic field of the Galaxy. This sort of radiation had been known from observations in the large accelerators used to investigate the structure of atomic nuclei. It was therefore called *'synchrotron radiation'*. One of its main characteristics is that it is polarized: the light vibrations are confined to one direction, contrasting to common light, where they show a random distribution. Synchrotron radiation had never before been observed in nature.

By a curious train of events I became involved in the early history of finding this radiation. It came about through observations of the Crab nebula, which is probably the most remarkable object in the sky. The Crab nebula was born on the 4th of July 1054 by the explosion of a faint, unknown star in the constellation of Taurus. In the first year after its explosion it became so bright that it could be seen in full daylight. The star was almost entirely disrupted. Its fragments were expelled at high velocity, and after several centuries had grown to such a size that they could be observed as a nebulous object. In 1954 I asked Dr. Walraven in Leiden to measure the rate at which its brightness decreased by the expansion. Dr. Walraven, who was a genius in refined observations, did more: he measured not only the brightness, but also the polarization. We had heard that observers in the Soviet Union had found that the light of the Crab nebula was polarized, and wanted to see if this was correct. It turned out that it was not only correct, but that the *degree* of polarization was so high that it could not be ascribed to known mechanisms of producing polarization through diffraction of interstellar dust particles. It indicated convincingly that it must be intrinsic in the light emitted by the Crab nebula, and that this light must therefore be of the synchrotron sort. This was an exciting discovery! The cold February nights I spent with Dr. Walraven at the telescope, watching the construction of the first synchrotron map ever made were probably the most wonderful times in our lives.

It was Walraven's talent which made it possible to make these revolutionary observations with a small telescope under the most unfavorable conditions in the midst of a fully illuminated city.

We now return to the exploration of the world.

We have explored the limits of the huge star-swarm in which we live, and have found that neither the Kapteyn 'Universe' nor Shapley's extended globular cluster system can be the complete Universe. They are no more than an island in an ocean that extends far beyond; an ocean that contains many other islands. The spiral nebulae are such islands. They are numerous. Thousands of nebulae had already been cataloged by William Herschel and his son John in their large survey of the two previous centuries. The nebulae, or galaxies as we shall call them from now on, are at least as numerous as the stars in our own Galaxy. On long-exposure plates, such as those taken with the 5 m Hale telescope on Mount Palomar in Southern California, we see two sorts of images: those with sharp boundaries and those with fuzzy boundaries. The first are images of *stars*, like our Sun, but at distances of several thousand lightyears; the nebulous ones are *galaxies*, swarms of at least a hundred billion suns each, lying roughly a million times further away than the sharp-image objects.

The farther we look the more galaxies we see. Is there an end? Do the galaxies form a swarm, like the stars in our own Galaxy, but one of higher order? No evidence has been found for this. There appears to be no *end* to the world of galaxies.

True, their distribution is far from random: they have a strong tendency to cluster together, in groups of all kinds of sizes, ranging from doubles, triples etc. to groups and clusters containing thousands of galaxies. [Here Oort showed Fig. 8.9], which shows the distribution of the 1300 brightest galaxies and gives an impression of the complicated structure. The picture contains one large cluster, the Virgo cluster but it is too large to show adequately.

In general the rich galaxy clusters have roundish shapes, with diameters of the order of ten billion lightyears. There exist still vaster structures, with hundred times larger diameters. They are called *'superclusters'*. Their shapes are far from round. Sometimes they are string-like. Usually they are very irregular. The asymmetrical structure of the superclusters suggests that there has been no important mixing since their birth. Apparently we see them in the stage of their formation. This is exciting: the study of superclusters may then teach us something about the manner in which large structure has originated. During the last five years I have become deeply involved in their investigation.

Meanwhile our knowledge of the Universe had undergone a radical change. The technical development in building telescopes and spectrographs had, early in this century, enabled astronomers to obtain *spectra* of galaxies of sufficient quality to measure Doppler shifts of their spectral lines and to derive their velocities. Their radial velocities revealed something very remarkable: the galaxies all moved away at high velocities, higher than any velocities observed within our Galaxy. The velocities increased the further away the galaxies were. Hubble succeeded in showing that the motions were proportional to the distances. This indicated that they were not only moving away from *us* but also from each other: *The Universe appeared to be expanding*. This remarkable phenomenon was amply confirmed by subsequent observations. It is called *Hubble expansion*; the rate of expansion is called the *Hubble constant*.

At earlier epochs the galaxies must thus have been closer together, and there must have been a time that they all started at one point. If the rate of expansion had been constant, this time (the so-called Hubble time) would have been about twenty billion years ago. Actually, the expansion velocity cannot have been constant. It must have been decelerated by the mutual attraction of the galaxies, The expansion must therefore have been faster in the past, and the time elapsed since its start must have been shorter. Present estimates are about 13 billion years. This is *the age of the Universe*.

The more distant a galaxy is, the fainter it will appear. In principle we can determine the distance from its apparent magnitude.

In observing objects that are far away in *space* we also look far back in *time*, But there is a limit beyond which we cannot look. This limit is the age of the Universe, which at present corresponds to 13 billion lightyears. It is a *new horizon*, of another kind than the one behind which ships disappear when they sail away from the coast. This horizon can only be surpassed by *patience*. A billion years from now we can see the light emitted by galaxies a billion lightyears further away than the most distant ones we can now observe. The number of observable galaxies is thus continually increasing.

Let us now look the other way, and ask how the Universe was in the past. In order to penetrate as far as possible into the past we should evidently choose the most luminous objects. Such objects can often be recognized by their exceptionally strong radio emission. Particularly powerful objects for studying evolution in the Universe are the so-called *quasars*. These galaxies are characterized by having exceptionally bright nuclei. They are so luminous, and often so distant, that the light waves we now receive from them were emitted at times when the Universe was between five and ten times younger, and between five and ten times smaller, than at present.

[The following paragraph is crossed out, but included in the Dutch translation (Oort, 1989b)] From the counts of quasars at these epochs we see that a very strong evolution has occurred in their number density per volume, if we take a volume expanding with the Universe. The density at early epochs proved to be several hundred times higher than the present density, probably implying an equal increase in their birth rate. Similarly large increases have been found for the most powerful radio galaxies. No plausible explanation for this birth rate explosion has been found. Nor is there any sign of newly forming galaxies at these early times. The subject is eagerly being followed up.

The Universe cannot always have consisted of stars and galaxies. At earlier epochs it must have been a more or less continuous medium of radiation and particles. It must have had a very high density and such a high temperature that no condensation into stars could have taken place. It gradually cooled proportionally with the expansion. When it had cooled to a few thousand degrees, formation of galaxies and stars became possible.

We cannot directly observe the earlier high-temperature stage of the Universe, but there is one valuable piece of information on the properties of the early Universe, viz. the abundance of helium relative to hydrogen.

Particle physicists assert that helium nuclei can be formed only under extreme conditions of pressure and temperature, conditions that do not generally exist in the present Universe, not even in the center of stars. The only place where temperature and pressure have been of the right order to make the 25% helium that exists today, was in the Universe shortly after its creation in the 'Big Bang'. According to the standard theory the temperature of the medium three minutes after the Big Bang would have been a thousand million degrees, and both density and temperature would have been just right for the formation of helium nuclei. Due to the rapid expansion the density and temperature fell steeply, so the interval of time when conditions were suitable was short. The resulting number of helium nuclei was determined by the length of this interval, and the exact temperature and density.

Thanks to a most remarkable discovery all three data can be computed from measurements in the Universe today. The helium abundance thus gives us a glimpse of the Universe when it was 3 min old.

The temperature, or the radiation density, continued to drop in proportion with the expansion. At the present time, roughly 13 billion years after the Big Bang, it has been reduced to three degrees Kelvin.

It is one of the big triumphs of science that astronomers have been able to measure this temperature. In 1964 Penzias and Wilson of the Bell Telephone Company [Arno Allan Penzias (b. 1933) and Robert Woodrow Wilson (b. 1936), winners of the 1978 physics Nobel Prize.] succeeded in showing that the Universe is indeed filled with radiation of 3K, and that therefore at the age of three minutes the temperature must have been 1000 million K, just what was needed for the formation of helium.

The observation of the 3 °K radiation was one of the greatest discoveries of cosmology. In combination with the helium abundance it gave us a glimpse of the Universe as it was 3 min old. But it should not close our eyes for the big enigmas that remain unresolved. For instance, where did the very large structures observed in the present Universe come from? How can the three-degree background radiation coming from opposite directions be so nearly equal if, due to the expansion, the two regions can never have been in contact? And, finally, what caused the Big Bang?

It is interesting to contemplate how cosmology has become interwoven with particle physics. Physicists have taught us to understand the evolution of the Universe in the first fractions of seconds of its existence. In exchange, the expanding Universe, in diving into energy regimes far beyond those attainable in the largest accelerators, might ultimately contribute to a better understanding of some of the deepest problems of physics.

In the very beginning the whole observable Universe was contained in a tiny space, no more than a dew drop, but perhaps comprising the solution to all enigmas, like the dew drop in Issa's haiku

A world of short-lived dew,
And in that dew-drop,
What violent quarrels!

The actual Universe would have occupied only a minute fraction of the 'dew drop', and this would have contained the seed for all the wonderful phenomena in the immense Universe which were to grow out of it.

Fig. C.0 Oort photographed in his office at the Observatory, Huygens Building, in 1987. From the Oort archives

Appendix C
Notes

Science has authority, not because of white coats, or titles,
but because of precision and transparency:
you explain your theory, set out your evidence, and
reference the studies that support your case.
Ben Goldacre (1974–present)

C.1 Conventions

Papers by Oort and collaborators are in the text referred to by authors and years of publication. These are listed in Oort's publications in Appendix A.1.

NASA Astronomy Data System ADS. Many astronomical journal papers that I refer to are available from the NASA Astronomy Data System ADS. The homepage of ADS is at ads.harvard.edu/, One can search the literature either using the classical site at adsabs.harvard.edu/abstract_service.html, or the new 'Bumblebee' version at ui.adsabs.edu. There are also mirror sites (see adsabs.harvard.edu/mirrors.html); the European mirror site is located at ESO and can be found at esoads.eso.org/abstract_service.html. Whenever available, references to entries in ADS are included in these notes (and for papers in Appendix A) *by their ADS designation.* So if the note refers to the paper *Some notes on my life as an astronomer*, published by Oort in 1981 in the Annual Review of Astronomy and Astrophysics, 19, 1–5, the annotation [1981ARA&A..19....1O] has been added. This means that this paper is listed in ADS and the full URL to access this paper is adsabs.harvard.edu/abs/1981ARA&A..19....1O in the USA node, esoads.eso.org/abs/1981ARA&A..19....1O in Europe, etc. If you prefer the new 'ADS Bumblebee' version of ADS use ui.adsabs.harvard.edu#abs/1981ARA&A..19....1O. For many, espe-

Benjamin Michael (Ben) Goldacre is a British physician and science writer.

P. C. van der Kruit, *Jan Hendrik Oort*, Astrophysics and Space Science Library 459, https://doi.org/10.1007/978-3-030-17801-7

Fig. C.1 Covers of three publications about Jan Hendrik Oort. On the left Jet Katgert-Merkelijn's inventory of the Oort Archives (*JKM-Inventory* [1]), in the middle the Liber Amicorum on the occasion of his eightieth birthday (*LibAm80* [2]), and on the right the catalogue of the 2000 centenary exhibition (*Cat2000* [5])

cially older publications, scanned versions of the papers are provided by ADS in .pdf or .gif formats, but in other (mostly recent) cases electronic subscriptions may be required to download the full text from the journal or publisher's site, depending on their free-access policy.

International Astronomical Union. The transactions of the IAU have been published online by Cambridge University Press as part of 'CambridgeCore'. For access of this a valid subscription is required. The URL is www.cambridge.org/core/journals/transactions-of-the-international-astronomical-union/all-issues. Some volumes are still missing at the time of writing.

URLs have been cut-off at the end of the lines without a hyphenation dash. The symbol ∪ in a URL indicates a blank space.

Frequently referred books and other publications have been given a special designation. These are listed as references [1–9]. The covers of three examples have been illustrated in Fig. C.1.

Interviews. Some time in the 1990s Jet Katgert-Merkelijn interviewed both Adriaan Blaauw and Henk van de Hulst and recordings of these have been made available to me by her on audiotape. I have had these transformed into electronic mp3-files at the University of Groningen and arranged for transcriptions at the Kapteyn Astronomical Institute. These interviews have been conducted in Dutch. I refer to these as the *JKBlaauw-interview* and the *JKHulst-interview*. These transcripts are not publicly available; inquiries should be directed to Jet Katgert. Likewise, transcripts of my interviews with Maarten Schmidt, Hugo van Woerden, Butler Burton, Harry van der Laan, Jet Katgert and (in writing) Whitney Shane are not publicly available without consent by these individuals.

References

1. J.K. Katgert-Merkelijn, *The letters and papers of Jan Hendrik Oort*, Astrophysics and Space Science Library no. 213, ISBN: 0-79234-542-8, Kluwer, (1997). (*JKM-Inventory*).
2. H. van Woerden, W.N. Brouw, & H.C. van de Hulst, *Oort and the Universe. A sketch of Oort's research and person*, Reidel, ISBN-10: 9-027-71180-1, (1980). (*LibAm80)*.
3. J.H. Oort, *Some notes on my life as an astronomer*. Ann. Rev. Astron. Astroph. 19, 1–5, (1981), [1981ARA&A..19....1O]. (*AnnRev81*).
4. David DeVorkin, Interview of J.H. Oort on November 10, 1977, American Institute of Physics www.aip.org/history-programs/niels-bohr-library/oral-histories/4806. (*AIP-Interview*).
5. J.K. Katgert-Merkelijn & J. Damen, *Jan Oort, Astronomer*, catalogue of the centenary exhibition in 2000 at Leiden University, Leiden Univ. Lib., ISSN 0921-9293, vol. 35 (2000). lampje.leidenuniv.nl/digitale-tentoonstellingen-voor-2007/Jan_Oort/object1.htm. ([*Cat2000*).
6. P.C. van der Kruit & K. van Berkel, *The legacy of J.C. Kapteyn: Studies on Kapteyn and the development of modern astronomy*, Springer, ISBN 0-7923-6393-0 (2000). (*Legacy*).
7. P.C. van der Kruit, *Jacobus Cornelius Kapteyn: Born investigator of the Heavens*, Springer, ISBN 978-3-319-10875-9 (2015). (*JCKbiog*).
8. A. Elbers, *The rise of radio astronomy in the Netherlands: The people and the politics*, Springer, ISBN 978-3-319-49079-3 (2017).
9. A. Elbers, *Early Dutch radio astronomy (1940-1970): The people and the politics*, PhD thesis, University of Leiden (2015), openaccess.leidenuniv.nl/handle/1887/36547. (*EarlyRadio*).
10. Ancestry, www.ancestry.com/genealogy.
11. Pondes, www.pondes.nl, Geni: www.geni.com.
12. Online Familieberichten, www.online-familieberichten.nl.
13. Stamboomzoeker, www.stamboomzoeker.nl.
14. GenVer, www.genver.nl.
15. Institute of Social History, www.iisg.nl/hpw/calculate.php.
16. Dutch Ancestry Coach, www.dutchancestrycoach.com/content/how-much-did-you-say-converting-dutch-historic-currencies.
17. Currencyconverter, www.historicalstatistics.org/Currencyconverter.html.
18. Pictoright, www.pictoright.nl
19. Catawiki, www.catawiki.nl/help.
20. Schilderijen-site, www.schilderijen-site.nl.
21. OudLeiden, www.oudleiden.nl, galerielaimbock.com.

Chapter 1. Growing up in Oegstgeest

22. *Levensbericht J.H. Oort*, KNAW, Levensberichten en herdenkingen, Amsterdam, 67–73 (1993). www.dwc.knaw.nl/DL/levensberichten/PE00002152.pdf.
23. H.H. Plaskett, *(Presidential Address) on presenting the Gold Medal to Professor J.H. Oort*, Monthly Notices of the Royal Astronomical Society, 106, 242 (1946), [1946MNRAS. 106..242P].
24. https://en.wikipedia.org/wiki/Elfstedentocht.
25. https://en.wikipedia.org/wiki/City_rights_in_the_Low_Countries.
26. *Levensbericht van Henricus Oort, 27 December 1836 – 13 December 1927*, Jaarboek van de Maatschappij der Nederlandse Letterkunde, 77–120 (1929) www.dbnl.org/tekst/_jaa003192901_01/_jaa003192901_02_0021.php.
27. *Veertigjarig Artsjubileum Dr. A.H. Oort*, Nederlands Tijdschrift voor de Geneeskunde, 80, 3207–3208 (1936). www.ntvg.nl/system/files/publications/1936132070002a.pdf.

28. www.familyberry.com/?q=87A077CFB3B70E5726C535F305B0BC2C. I have excluded names with an obvious non-Dutch background.
29. allefriezen.nl/zoeken/persons?f="search_s_register_gemeente":"v":"Franeker"&ss="q": "Oort"&page=3.
30. J. van Gijn & J.P. Gijselhart, *Jelgersma en zijn kliniek*, Nederlands Tijdschrift voor de Geneeskunde, 155:A3219 (2011).
31. H. Oosterhuis & M. Gijswijt-Hofstra, *Verward van geest en ander ongerief: Psychiatrie en geestelijke gezondheidszorg in Nederland (1870–2005)*, Bohn Stafleu van Loghum (2008), ISBN 978-9-031-35238-8, p.172.
32. *Jaarboekje 1904 - Historische vereniging Oud Leiden*, www.oudleiden.nl/pdf/1904.pdf.
33. upload.wikimedia.org/wikipedia/commons/thumb/c/c3/Exterieur_overzicht_villa,_behorend_bij_Jelgersmakliniek_-_Oegstgeest_-_20310532_-_RCE.jpg.
34. Bastiaan Willink, *De tweede Gouden Eeuw: Nederland en de Nobelprijzen voor natuurwetenschappen 1870–1940*, ISBN10 9-035-11942-8, Bakker (1998). www.dbnl.org/tekst/tekst/will078twee01_01/.
35. Bastiaan Willink, *Origins of the Second Golden Age of Dutch Science after 1860.* sss.sagepub.com/content/21/3/503.abstract.
36. Dork van Delft, *Heike Kamerlingh Onnes: de man van het absolute nulpunt*, ISBN10 9-035-12917-2, ISBN13 9-789-035129-177, Bakker (2005). www.dbnl.org/tekst/delf006heik01_01/.
37. www.erfgoedleiden.nl/ae887bbc-26bc-11e3-bc3a-3cd92befe4f8.
38. *Levensbericht van Arend Joan Petrus Oort (1903–1987)*, Koninklijke Nederlandse Academie van Wetenschappen, www.dwc.knaw.nl/DL/levensberichten/PE00002151.pdf.
39. www.erfgoedleiden.nl/ca8b3656-26bc-11e3-8138-3cd92befe4f8.
40. Amersfoorts Dagblad/de Eemlander, 30 juni 1932. archiefeemland.courant/issue/ADDE/1932-06-30/edition/0/page/6.
41. A.H. Oort (Oort's son), private communication.

Chapter 2. Kapteyn and Groningen

42. P.C. van der Kruit, *Inleiding Sterrenkunde*, www.astro.rug.nl/~vdkruit/Inleiding.html
43. P.C. van der Kruit, *Surface photometry of edge-on spiral galaxies. V. The distribution of luminosity in the disk of the Galaxy derived from the Pioneer 10 background experiment*, Astronomy & Astrophysics 157, 230–244. (1986). [1986A&A...157..230V].
44. J.C. Kapteyn, *Über eine Methode die Polhöhe möglichst frei von Systematischen Fehlern zu bestimmen*, Copernicus 3, 147–182 (1884).
45. H.M.A.A. Hertzsprung-Kapteyn, *J.C. Kapteyn: Zijn leven en werken*, Noordhoff, Groningen (1928).
46. bookhistory.harvard.edu/takenote/sites/default/files/attachments/Edward Pickering.jpg.
47. [1918AnHar.101.....1P], [1923AnHar.102.....1P], [1923AnHar.103.....1P], [1952QB819.R46......] and [1930QB6.M78........].
48. Wikimedia Commons, https://commons.wikimedia.org/wiki/File:Mutua_Fides_ca_1890.png.
49. memim.com/grosses-schauspielhaus.html or www.today-in-history.de/index.php?what=thmanu&manu_id=1659&tag=28&monat=11&year=2010&dayisset=1&lang=en.
50. *De theorie van eb en vloed.* Lecture delivered on February 27, 1909, Jaarboekje 1909–1910 van de Mijnbouwkundige Vereeniging te Delft. 5 pages + figures.
51. Arthus Koestler, *The Sleepwalkers*, Hutchinson (UK) and MacMillan (USA), (1959).
52. https://en.wikipedia.org/wiki/File:Obafgkm_noao_big.jpg. 'This file is in the public domain because it was solely created by NASA'.
53. Annals of Harvard College Observatory, 27, 1–388 (1890), [1890AnHar..27....1P].

54. www.atlasoftheuniverse.com/hr.html and https://en.wikipedia.org/wiki/File:HRDiagram.png; licensed under the Creative Commons Attribution-Share Alike 2.5 Generic license.
55. www.noao.edu/image_gallery/html/im0493.html and www.noao.edu/outreach/aop/observers/n891.html (N.A. Sharp and Dale Cupp/Flynn Haase/NOAO/AURA/NSF).
56. M.J. Way & D. Hunter, (eds.), *Origins of the expanding Universe: 1912–1932*, Astronomical Society of the Pacific Conference Proceedings, 471, (2013), [2013ASPC..471.....W].

Chapter 3. Stars of High Velocity

57. R.W. Smith, *Beyond the big Galaxy: The structure of the Stellar System 1900–1952*, Journal for the History of Astronomy, 37, 307–342 (2006), [2006JHA....37..307S].
58. J.C. Kapteyn, *Remarks on the determination of the number and mean parallax of stars of different magnitude and the absorption of light in space*, Astronomical Journal 24, 115–122 (1904), [1904AJ.....24..115K]
59. J.C. Kapteyn, *On the absorption of light in space*, Astrophysical Journal 29, 46–54 (1909), [1909ApJ....29...46K].
60. J.C. Kapteyn, *On the absorption of light in space*, Astrophysical Journal 29, 30, 284–317 (1909), [1909ApJ....30..284K].
61. disc.leidenuniv.nl/view/action/singleViewer.do?dvs=1483352434763710&locale=en_US&VIEWER_URL=/view/action/singleViewer.do?&RY_RULE_ID=10&frameId=1&usePid1=DELIVEtrue&usePid2=true or www.strw.leidenuniv.nl/album1908/index.html.
62. https://commons.wikimedia.org/wiki/File:PJvanRhijn.jpg.
63. H. Shapley, *Studies of magnitudes in star clusters. I. On the absorption of light in space*, Proceedings of the National Academy of Sciences of the United States of America, 2, 12–15 (1916), [1916PNAS....2...12S].
64. H. Shapley, *Studies of magnitudes in star clusters. I. On the absorption of light in space*, Proceedings of the National Academy of Sciences of the United States of America, 2, 12–15 (1916), [1916PNAS....2...12S].
65. P.J. van Rhijn, *On the number of stars of each photographic magnitude in different Galactic latitudes* Publications of the Astronomical Laboratory at Groningen 27, 1–63 (1916), [1916PGro...27....1V].
66. J.C. Kapteyn, P.J. van Rhijn, & H.A. Weersma, *The secular parallax of the stars of different magnitude, Galactic latitude and spectrum* Publications of the Astronomical Laboratory at Groningen 29, 1–63 (1918), [1918PGro...29Q...1.] plus [1918PGro...29...45.]
67. J.C. Kapteyn & P.J. van Rhijn, Publications of the Astronomical Laboratory at Groningen 30, 1–110 (1920), [1920PGro...30....1.].
68. J.C. Kapteyn & P.J. van Rhijn, Astrophysical Journal 52, 23–38 (1920), [1920ApJ....52...23K].
69. W.J.A. Schouten, *On the determination of the principal laws of statistical astronomy*, PhD thesis, University of Groningen (1918).
70. J.C. Kapteyn, Astrophysical Journal 55, 302–328 (1922), [1922ApJ....55..302K].
71. P.C. van der Kruit & L. Searle, *Surface photometry of edge-on spiral galaxies. I. A model for the three-dimensional distribution of light in galactic disks*, Astronomy & Astrophysics, 95, 105–115 (1981), [1981A&A....95..105V].
72. Macmillan (1914). Re-issued Cornell University (2009) ISBN-10: 1-112-5608-8.
73. W. de Sitter, *Kosmos*, Harvard University Press (1932); Van Stockum, Den Haag (1934), Amsterdam University Press (https://www.aup.nl/en/book/9789089641540/kosmos), 'Digitale Bibliotheek voor de Nederlandse letteren' (www.dbnl.org/tekst/sitt003kosm01_01/).
74. H. Shapley, *Studies based on the colors and magnitudes in stellar clusters. XII. Remarks on the arrangement of the sidereal Universe*, Astrophysical Journal, 49, 311–336 (1919), [1919ApJ....49..311S].
75. https://www.spacetelescope.org/images/potw1140a/.

76. M.A. Hoskin, *The 'Great Debate': What Really Happened*, Journal for the History of Astronomy, 7, 169–182 (1976), [1976JHA.....7..169H].
77. V. Trimble, *The 1920 Shapley-Curtis Discussion: Background, Issues, and Aftermath*, Publications of the Astronomical Society of the Pacific, 107, 1133–1144 (1995), [1995PASP..107.1133T].
78. E. Hertzsprung, *Über die räumliche Verteilung der Veränderlichen vom δ Cephei-Typus*, Astronomische Nachrichten, 196, 201–209 (1913), [1913AN....196..201H].
79. H. Shapley, *Through rugged ways to the stars: The reminiscences of an astronomer*, Charles Scriber's Sons (1969).
80. www.aip.org/history-programs/niels-bohr-library/oral-histories/4518-1.
81. *The Bachiene Institution prize essay*, The Observatory, 43, 157–158 (1920), [1920Obs....43..157O].
82. Willem J. Luyten, 1940 East River Road, Minneapolis, MN 55414, USA. 20+203 pp. (1987), [1987fsya.book.....L].
83. L. Boss, Carnegie Institution of Washington, archive.org/details/preliminarygener00carnrich, [1910pgcs.book.....B].
84. W.S. Adams & A.H. Joy, *The motions in space of some stars of high radial velocity*, Astrophysical Journal, 49, 179–185 (1919), [1919ApJ....49..179A].
85. A.S. Eddington, *Stellar movements and the structure of the Universe*, McMillan & Co (1914): https://archive.org/details/stellarmovements00eddirich/.
86. J.H. Jeans*Problems of Cosmogony and Stellar Dynamics*, Cambridge University Press (1919): https://archive.org/details/problemsofcosmog00jeanrich.
87. https://commons.wikimedia.org/wiki/File:Museum_Boerhaave_-_Snellius_quadrant.jpg.
88. G. van Herk, H. Kleibrink, & W. Bijleveld, *De Leidse Sterrewacht: Vier eeuwen wacht bij dag en bij nacht*, Leiden Observatory (1965).
89. F.P. Israel, *De Leidse Sterrewacht: Glorieus als vanouds*, Waanders/De Kler, Zwolle (1983) and Leiden University (2011).
90. David Baneke: *De ontdekkers van de hemel*, Prometheus, ISBN10 9-035-13688-8 (2015).
91. Hans Hooijmaijers, *Een passie voor precisie: Frederik Kaiser en het instrumentarium van de Leidse Sterrewacht*, Studium, 4, 195–126 (2011) www.gewina-studium.nl/index.php/studium/article/view/1545/7241.
92. F. Kaiser, Astronomische Nachrichten 60, 273–286 (1863), [1863AN.....60..273K].
93. F. Kaiser, *Verslag van de Staat der Sterrenwacht te Leiden en de aldaar volbrachte werkzaamheden, 1875–1876* [1876VeLdn..12....1V].
94. John Lankford, *American astronomy: Community, careers and power, 1859–1940*, University of Chicago Press; 1st edition (1997), ISBN-13: 978-0-226-46886-0.
95. A. Heck (ed.), *Information handling in astronomy – Historical vistas*, Astrophysics and Space Science Library, 285, 267–273, Kluwer, ISBN: 1-4020-1178-4. (2003), [2003ASSL..285..267J].
96. W. de Sitter, *Determination of the mass of Jupiter and elements of the orbits of its satellites by W. de Sitter, D. Gill & W.H. Finlay*, Annals of the Cape Observatory, 12, 1.1-1.173 (1915), [1915AnCap..12....1D].
97. J.M. Pasachoff, *Simon Marius's Mundus Iovialis: 400th anniversary in Galileo's shadow*, Journal for the History of Astronomy, 46, 218–234 (2015), [2015JHA....46..218P].
98. W. de Sitter, Monthly Notices of the Royal Astronomical Society, 91, 706–738 (1931), [1931MNRAS..91..706D]
99. Jan Guichelaar, *De Sitter en Einstein: 'Het lijkt mij dat Einstein hier een vergissing begaan heeft.'*, Studium, 9, 77–95 (2016).
100. D.M. Baneke, *Teach and travel: Leiden Observatory and the renaissance of Dutch astronomy in the interwar years*, Journal for the History of Astronomy, 41, 167–198 (2010), [2010JHA....41..167B].
101. D.M. Baneke, *Als bij toverslag, De reorganisatie van de Leidse Sterrewacht, 1918–1924*, BMGN/Low Countries Historical Review 120, 207–225 (2005).

102. D.M. Baneke, *Hij kan toch moeilijk de sterren in de war schoppen. De afwijzing van Pannekoek als adjunct-directeur van de Leidse Sterrewacht*, Gewina 27, 1–13 (2004).
103. Anton Pannekoek, B.A. Sijes, E.P.J. van den Heuvel, J.M. Welcker & J.R. van der Leeuw, *Herinneringen: Herinneringen uit de arbeidersbeweging: Sterrenkundige herinneringen*, van Gennep, ISBN10: 9-060-12350-6 (1982).
104. WS. de Sitter, *Short history of the Observatory of the University at Leiden 1633–1933, published at the occasion of the celebration of the 300th anniversary of the foundation of the Observatory*, Enschedé en Zonen, Haarlem (1933).
105. D.B. Herrmann, *Ejnar Hertzsprung, Pionier der Sternforschung*, Springer (1994) ISBN 3-540-56788-6.
106. A.R. Sandage, *Centennial History of the Carnegie Institution of Washington: Vol. 1*, Cambridge University Press (2004), ISBN-13: 978-05-218-3078-2
107. Susanne Elisabeth Nørskov, AU Library, Fysik & Steno, Institut for Fysik og Astronomi, Aarhus Universitet, Denmark.
108. rasathane.en.ankara.edu.tr/.
109. C. Güner Omay, *Prof. Dr. Egbert Adriaan Kreiken*, Ankara, ISBN 978-605-87419-0-4; www.bitav.org.tr/TR/Genel/dg.ashx?DIL=1&BELGEANAH=176&DOSYAISIM=EAK_Life_Story_Book_CGO3.pdf
110. Juus Kreiken, *Egbert Adriaan Kreiken: pionier in de Turkse sterrenkunde*, Zenit, januari 2013, 14–18.
111. De Sitter Archives, Leiden Observatory (WdS 30 – correspondence with Kapteyn).
112. www.strw.leidenuniv.nl/oortfotos/.
113. A. Blaauw, *History of the IAU, Birth and first half-century of the International Astronomical Union*, Kluwer (1994), ISBN 0-7923-2979-1.
114. https://commons.wikimedia.org/wiki/File:PSM_V84_D203_Frank_Schlesinger.jpg.
115. J. Hamaker-Nauta, niece of Oort, private communication.

Chapter 4. Via Yale to Leiden

116. Abhishek Ratna, *No parking. no halt. success non stop!*. Supernova Publishers (2015), ISBN 8-189-93098-2.
117. *Yale University buildings and grounds photographs, 1716-2004 (inclusive)*. Manuscripts & Archives, Yale University. images.library.yale.edu/madid/.
118. Dorrit Hofleit, *Yale University Observatory at the turn of the century*, Irish Astronomical Journal, 12, 104–108 (1975), [1975IrAJ...12..104H].
119. D. Gill & W.L. Elkin, Memoirs of the Royal Astronomical Society, 48, 1–198 (1885), [1885MmRAS..48....1G].
120. E.D, Hoffleit (ed.), *Astronomy at Yale, 1701–1968*, Connecticut Academy of Arts and Sciences, ISBN 9-781-878508-027 (1992).
121. https://commons.wikimedia.org/wiki/Category:Rotterdam_(ship,_1908)?uselang=nl#/media/File:Rotterdam-Ship.jpg.
122. https://commons.wikimedia.org/wiki/File:YaleOldCampus2009.jpg.
123. https://home.earthlink.net/~anne_peter/family/anne_paternal/vanleeuwen/vanleeuwen.html.
124. www.aaa.si.edu/collections/konrad-and-florence-ballin-cramer-papers-9413/.
125. F. Schlesinger, *Two historical notes: on the variation of latitude, and on Ptolemy's catalogue*, The Observatory, 59, 191–196 (1936), [1936Obs....59..191S].
126. University of Chicago Photographic Archive, apf6-04494, Special Collections Research Center, University of Chicago Library.
127. F. Schlesinger, Astronomical Journal, 34, 42–44 (1922), [1922AJ.....34...42S].

128. F.R. Helmert, *Starke Änderung der geographischen Breite in der zweiten Hälfte des Jahres 1889 zu Berlin, Potsdam, Prag und Strassburg*, Astronomische Nachrichten, 124, 177 (1890), [1890AN....124..177H].
129. J.C. Kapteyn, *Über eine photographische Methode der Breitenbestimmung aus Zenithsternen*, Astronomische Nachrichten 125, 81–86 (1890), [1890AN....125...81K].
130. F. Schlesinger, Popular Astronomy, e.g. for 1923 it is P.A. 23, 162–164, contained in [1924PA.....32..139].
131. D. Brouwer, National Academy of Sciences, Biographical Memoirs, Vol. XXIV – third memoir (1945).
132. www.truthnet.org/islam/Jewish/.
133. A.R. Sandage, R.L. Beaton, & S.R. Majewski, *Comparison of Hipparcos trigonometric and Mount Wilson spectroscopic parallaxes for 90 subgiants that defined the class in 1935*, Publications of the Astronomical Society of Pacific, 128, 064202 (2016), [2016PASP..128f4202S].
134. https://had.aas.org/resources/aashistory/early-meetings/1922-1927#31.
135. https://had.aas.org/resources/aashistory/early-meetings/1922-1927#32.
136. Centraal Bureau voor de Statistiek, *Tachtig jaren statistiek in tijdreeksen*, Den Haag, Staatsuitgeverij (1979).

Chapter 5. Galactic Rotation; Marriage

137. www.universetoday.com/30710/galaxy-rotation/.
138. NIMH, Netherlands Institute of Military History, 2017. https://commons.wikimedia.org/wiki/File:NIMH_-_2011_-_0300_-_Aerial_photograph_of_Leiden,_The_Netherlands_-_1920_-_1940.jpg.
139. J. Schilt, *Description of the new thermopile-microphotometer of the Leiden Observatory*, Bulletin of the Astronomical Institutes of the Netherlands, 2, 135–139 (1924), [1924BAN.....2..135S].
140. J.K. Katgert-Merkelijn, Journal for the History of Astronomy 22, 267–296 (1991), [1991JHA....22..267K].
141. C.H.L. Sanders, The Observatory 40, 271–273 (1917) and 42, 90–91 (1919), [1917Obs....4040..271S, 1919Obs....42...90S].
142. H. Spencer Jones, Monthly Notices of the Royal Astronomical Society, 80, 455–463, [1920MNRAS..80..455S].
143. L. Courvoisier, Inaugural-Dissertation Kaiser-Wilhelms-Universität Strassburg (1901).
144. G.B. Strömberg, *The distribution of the velocities of stars of spectral types F to M*, Astrophysical Journal, 56, 265–294 (1922), [1924ApJ....59..228S].
145. G.B. Strömberg, *The asymmetry in stellar motions and the existence of a velocity-restriction in space*, Astrophysical Journal, 59, 228–251 (1924), [1922ApJ.....56..265S].
146. J.H. Jeans, *The motions of stars in a Kapteyn Universe*, Monthly Notices of the Royal Astronomical Society, 82, 122–132 (1922), [1922MNRAS..82..122J].
147. www.oughtred.org/archive/bulletinsummer08/people.htm.
148. https://hetutrechtsarchief.nl/collectie/archiefbank/indexen/personen/,File:NL-UtHUA_A152357_000301.jpg.
149. B. Lindblad, *Star-streaming and the structure of the stellar system*, Arkiv för Matematik, Astronomi och fysik, 19a, 1–8 (1925). But see also *On the cause of star-streaming* Astrophysical Journal, 62, 191–197 (1925), [1925ApJ....62..191L].
150. www.aip.org/history-programs/niels-bohr-library/oral-histories/4518-1.

Chapter 6. Galactic Structure and Dynamics

151. C.V.L. Charlier, *Stars, motions and distances, An investigation on the motion of the stars*, Monthly Notices of the Royal Astronomical Society, 73, 486–491 (1913), [1913MNRAS..73..486C].
152. J.K. Fotheringham, *Precession, galactic rotation, and equinox correction, ibid.* 86, 414–426 (1926), [1926MNRAS..86..414F].
153. J. Schilt, *The effect of a rotation of the Galaxy on proper motions in right ascension and declination*, Proceedings of the National Academy of Sciences of the United States of America, 13, 642–646 (1927), [1927PNAS...13..642S].
154. J. Schilt, *Galactic rotation derived from the proper motions of Boss's Preliminary General Catalogue*, Astronomical Journal, 39, 17–20 (1928), [1928AJ.....39...17S].
155. J.S. Plaskett, *The rotation of the Galaxy*, Monthly Notices of the Royal Astronomical Society, 88, 395–403 (1928), [1928MNRAS..88..395P].
156. A.S. Eddington, *The rotation of the Galaxy*, Monthly Notices of the Royal Astronomical Society, 88, 329–331 (1928), [1928MNRAS..88..329]
157. A.S. Eddington, *The rotation of the galaxy, being the Halley lecture delivered on 30 May 1930*, Oxford, at the Clarendon Press (1930).
158. G.B. Strömberg, *Asymmetry in the distribution of stellar velocities*, The Observatory, 49, 279–280 (1926), [1926Obs....49..279S].
159. I. King, G. Gilmore, & P.C. van der Kruit, *The Milky Way as s Galaxy*, University Science Books, ISBN: 0-935702-62-8, p.334, [1990mwg..conf.....K.] (see also www.astro.rug.nl/~vdkruit/Books.html).
160. E.P. Hubble, *Angular rotations of spiral nebulae*, Astrophysical Journal, 81, 334–335 (1935), [1935ApJ....81..334H].
161. A. van Maanen, *Internal motions in spiral nebulae*, Astrophysical Journal, 81, 336–337 (1935), [1935ApJ....81..336V].
162. The SIMBAD Astronomical Database at simbad.u-strasbg.fr/simbad/sim-id?Ident =HD+134439 (or ...40);
163. M. Ishigaki, M. Chiba & W. Aoki, *Chemical abundances of outer halo stars in the Milky Way*, Publications of the Astronomical Society of Japan, 62, 143 (2010), [2010PASJ...62..143I].
164. P.C. van der Kruit, *Structure and Dynamics of Galaxies*, www.astro.rug.nl/~vdkruit/Beijing.html.
165. https://commons.wikimedia.org/wiki/File:Karl_schwarzschild.jpg and .../File:James_Hopwood_Jeans.jpg. Copyright expired.
166. A.S. Eddington, *The dynamics of a stellar system. Third paper: oblate and other distributions*, Monthly Notices of the Royal Astronomical Society, 76, 37–60 (1915), [1915MNRAS..76...37E].
167. J.H. Jeans, *On the theory of star-streaming and the structure of the Universe*, Monthly Notices of the Royal Astronomical Society, 76, 70–84 (1915), [1915MNRAS..76...70J].
168. B. Lindblad, *On the state of motion in the Galactic System*, Monthly Notices of the Royal Astronomical Society, 87, 553–564 (1927), [1927MNRAS..87..553L].
169. J.S. Plaskett & J.A. Pearce, *The motions and distribution of inter-stellar matter*, Monthly Notices of the Royal Astronomical Society, 90, 243–268 (1930), [1930MNRAS..90..243P].
170. *International Astronomical Union, Cambridge, 1925 June 14–22*, The Observatory, 48, 241–261 (1925), [1925Obs....48..241].
171. C.H.L. Sanders & J.J. Raimond, *On a new method for the determination of absolute declinations*, Bulletin of the Astronomical Institutes of the Netherlands, 5, 9–16 (1928), [1928BAN.....5....9S].
172. H.T. Stetson, *The Students' Astronomical Laboratory of Harvard University, 1903–1928*, Popular Astronomy, 36, 589–596 (1928), [1928PA.....36..589S].

Chapter 7. The Oort Limit and Galaxy Photometry

173. Jodi Lynn Picoult, *My sister's keeper*, Atria, ISBN: 0-7434-5452-9 (2004).
174. I. Asimov, *The intelligent man's guide to the sciences, Vol. I the physical sciences*, Basic Books (1960).
175. W.J. Luyten, *On the relation between parallax, proper motion, and apparent magnitude*, Lick Observatory Bulletin, 10, 135–140 (1922), [1922LicOB..10..135L].
176. W.J. Luyten, *Parallaxes of stars determined from spectral class and proper motion* Lick Observatory Bulletin, 11, 33–40 (1923), [1923LicOB..11...33L].
177. www.spacetelescope.org/images/opo0328a/.
178. W. de Sitter, Monthly Notices of the Royal Astronomical Society, 76, 699–728 (1916), [1916MNRAS..76..699D].
179. W. de Sitter, Monthly Notices of the Royal Astronomical Society, 77, 155–184 (1916), [1916MNRASRAS..77..155D]
180. W. de Sitter, Monthly Notices of the Royal Astronomical Society, 78, 3–28 (1917), [1917MNRAS..78....3D].
181. W. de Sitter, Bulletin of the Astronomical Institutes of the Netherlands, 5, 157–171 (1930), [1930BAN.....5..157D].
182. E.P. Hubble, *A relation between distance and radial velocity among extra-galactic nebulae*, Proceedings of the National Academy of Sciences of the United States of America, 15, 68–173 (1929), [1929PNAS...15..168H].
183. C. O'Raifeartaigh, *The contribution of V.M. Slipher to the discovery of the expanding Universe*, discussion, in: Origins of the expanding Universe: 1912–1932. ASP Conference Proceedings, 471, 49–61 (2013), [2013ASPC..471...49O].
184. A. Einstein & W. de Sitter, *On the relation between the expansion and the mean density of the Universe*, Proceedings of the National Academy of Sciences of the United States of America, 18, 213–214 (1932), [1932PNAS...18..213E].
185. K. Lundmark, *Determination of the apices and the mean parallax of the spirals*, The Observatory, 47, 279–281 (1924), [1924Obs....47..279L].
186. G. Strömberg, *Analysis of radial velocities of globular clusters and non-galactic nebulae*, Astrophysical Journal, 61, 353–362 (1925), [1925ApJ....61..353S].
187. R.L. Voller, *The muleskinner and the stars: The life and times of Milton La Salle Humason, Astronomer*, Springer, ISBN 978-1-4939-2879-8 (2015).
188. Sloan Digital Sky Survey Skyserver. cas.sdss.org/dr6/en/credits/ and www.sdss.org/.
189. C. Easton, *A photographic chart of the Milky way and the spiral theory of the Galactic System*, Astrophysical Journal, 37, 105–118 (1913), [1913ApJ....37..105E].
190. Davide De Martin (ESA/Hubble). spacetelescope.org/images/heic1112f/l (M31), www.spacetelescope.org/news/heic0602/ (M101).
191. J. Reynolds, *The light curve of the Andromeda nebula (NGC 224)*, Monthly Notices of the Royal Astronomical Society, 74, 132–136 (1913), [1913MNRAS..74..132R].
192. G.H. de Vaucouleurs, *The first half century of galaxy photometry: Methods and results*, in: Photometry, Kinematics and Dynamics of Galaxies, Department of Astronomy, University of Texas at Austin, 1–12 (1979), [1979pkdg.conf....1D].
193. B. Lindblad, *On the distribution of light-intensity and colour in the spiral nebula NGC7331*, Stockholms Observatoriums Annaler, 13, 8.1-8.32 (1941), [1941StoAn..13....8L].
194. E. Holmberg, *A photographic photometry of extragalactic nebulae*, Meddelanden fran Lunds Astronomiska Observatorium, Ser. II, Vol. 136, 1–103 (1958), [1958MeLuS.136....1H].
195. P.C. van der Kruit, *Truncations of stellar disks and warps of HI-layers in edge-on spiral galaxies*, Astronomy & Astrophysics, 466, 883–893 (2007), Appendix, [2007A&A...466..883V].
196. B.M.H.R. Wevers, P.C. van der Kruit, and R.J. Allen, *The Palomar-Westerbork survey of northern spiral galaxies*, Astronomy & Astrophysics Supplement Series, 66, 505–662 (1986), [1986A&AS...66..505W].
197. V.M. Slipher, *The detection of nebular rotation*, Lowell Observatory Bulletin, 2, 66 (1914), [1914LowOB...2...66S].

198. M. Wolf: Vierteljahrsschrift der Astronomischen Gesellschaft, 49, 169 (1914) (a single paragraph, in: Jahresbericht für 1913 of the Königstuhl Sternwarte.
199. P.C. van der Kruit & K.C. Freeman, *Stellar kinematics and the stability of disks in spiral galaxies*, Astrophysical Journal, 303, 556–572 (1986), [1986ApJ...303..556V].
200. https://observatoriesofohio.org/perkins-observatory/.
201. M. Minnaert & L.S. Ornstein, *The Heliophysical Institute of the Physical Laboratory at Utrecht: Report for the year 1932*, Bulletin of the Astronomical Institutes of the Netherlands, 7. 81–82 (1933), [1933BAN.....7...81M].
202. www.prewarbuick.com/cars/83/1925-Buick-Sedan.
203. *Digitized Sky Surveys, based on photographic data obtained using the 48-inch Schmidt Telescope on Palomar Mountain*, Space Telescope Science Institute. [stdatu.stsci.edu/cgi-bin/dss_form].
204. https://en.wikipedia.org/wiki/Gezelligheid.
205. M. Minnaert, J. Houtgast, & G.F.W. Mulders, *Photometric atlas of the solar spectrum from λ 3612 to λ 8771 with an appendix from λ 3332 to λ 3637*, Sterrewacht Sonnenborgh (1940).
206. M. Minnaert, & J. Houtgast, *Direct registration of intensities in spectrophotometric work*, Zeitschrift für Astrophysik, 15, 354–360 (1938), [1938ZA.....15..354M].
207. Klaas van Berkel, *Growing astronomers for export: Dutch astronomers in the United States before World War II*, in the *Legacy*, p.151.
208. https://hoogleraren.leidenuniv.nl/id/1855.

Chapter 8. Professor and Adjunct-Director

209. J.K. Katgert-Merkelijn, *De opvolging van W. de Sitter*, Jaarboekje voor geschiedenis en oudheidkunde van Leiden en omstreken 1997 / Vereniging 'Oud-Leiden', 128–143 (1997). www.oudleiden.nl/pdf1/1997_12.pdf.
210. Leiden Observatory Archives, directorate E. Hertzsprung. These can be accessed online through https://catalogue.leidenuniv.nl/primo_library/libweb/action/search.do.
211. H. Shapley & A. Ames, Annals of Harvard College Observatory 88, 41 76 (1932), [1932AnHar..88...41S].
212. H.C. Hins & G. van Herk, *Catalogue of fundamental declination derived from azimuth observations at a station on the Earth's equator*, Annalen van de Sterrewacht te Leiden 18, 1–79 (1938), [1938AnLei..18...68H] (see adsbit.harvard.edu/cgi-bin/iarticle_query?journal=AnLei&volume=0018&type=SCREEN_THMB).
213. H. Knox-Shaw & H.G Scott Barrett, Humphrey Milford (1934), [1934rcpm.book.....K].
214. A.D. Thackeray, *The Radcliffe Observatory, 1772–1972*, Radcliffe Trust (1972), [1972raob.book.....T].
215. E.P. Hubble, Astrophysical Journal 79, 8–76 (1934), [1934ApJ....79....8H].
216. J.L. Greenstein, *A determination of selective absorption based on the spectrophotometry of reddened B stars*, Astrophysical Journal, 87, 151–175 (1938), [1938ApJ....87..151G].
217. A.I. Eremeeva, *Political repression and personality: The history of political repression against Soviet astronomers*, Journal for the History of Astronomy, 26, 297–324 (1995), [1995JHA....26..297E].
218. *Obituary: Richard Prager*, The Observatory, 66, 186–187 (1945), [1945Obs....66..186.].
219. Otto Struve, *Dedication of the McDonald Observatory*, The Observatory, 62, 187–191 (1939), [1939Obs....62..187], and *Astronomical Symposium on Galactic and extragalactic structure*, Science, 90, 85–88, [1939Sci....90...85S].
220. S. Chandrasekhar, *The dynamics of stellar systems. I-VIII*, Astrophysical Journal, 90, 1–151. (1939), [1939ApJ....90....1C].
221. S. Alexander, *On the origin of the forms and the present condition of some of the clusters of stars, and several of the nebulae*, Astronomical Journal, 2, 95–160, as eight installments (1852), [1852AJ.....2...95A and further].

222. Williams, *The rotation curve of NGC 3115*, Astrophysical Journal, 199, 586–590 (1975), [1975ApJ...199..586W].
223. E.E. Barnard, E.B. Frost & M.R. Calvert, *A photographic atlas of selected regions of the Milky way*, Carnegie institution of Washington (1927).

Chapter 9. World War II: The Watershed

224. P.J. Idenburg, *De Leidse Universiteit 1928–1946: Vernieuwing en verzet*, Universitaire Pers Leiden, ISBN: 9-060-21425-0 (1978).
225. P.J. Idenburg, *De Leidse Universiteit tegen nationaal-socialisme en bezetting*, Brill, Leiden, ISBN: 9-004-06764-7 (1982).
226. www.news.leiden.edu/news-2013/leiden-classics-cleveringas-protest-against-the-anti-jewish-measures.
227. www.dbnl.org/tekst/clev00326no01_01/clev00326no01_01_0002.php.
228. B.J. Bok & V. Kourganoff, V. *The Committee for the Distribution of Astronomical Literature and the Astronomical News Letters*, Vistas in Astronomy, 1, 22-25 (1955), [1955VA......1...22B].
229. https://leiden.courant.nu/issue/LD/1941-05-12/edition/0/page/4 (Oort's second initial is incorrect).
230. www.astronomenclub.nl/.
231. A. Blaauw, *My cruise through the world of astronomy*, Annual Review of Astronomy & Astrophysics, 42, 1–37 (2004), [2004ARA&A..42....1B].
232. P.J. van Rhijn, *Astronomy in the Netherlands during the war*, The Observatory, 66, 258–261 (1946), [1946Obs....66..258V]
233. J.J.M.A. de Kort, *Some remarks on the system of galactic globular clusters*, Bulletin of the Astronomical Institutes of the Netherlands, 9, 189–193 (1941), [1941BAN.....9..189D].
234. J.J.M. van Tulder, de Kort, *A new determination of the Galactic pole and the distance of the Sun from the Galactic plane*, Bulletin of the Astronomical Institutes of the Netherlands, 9, 315–323 (1942), [1942BAN.....9..315V].
235. A. Kopff, *Dritter Fundamentalkatalog des Berliner Astronomischen Jahrbuchs. 1. Die Auwers-Sterne für die Epochen 1925 und 1950*, [1937VeABD..54.....1K]
236. A. Kopff, *Ibid. – II. Die Zusatzsterne für die Epoche 1950*, [1938AbhP....3.....K].
237. B. Boss, Carnegie Institution of Washington. ntrs.nasa.gov/search.jsp?R=19840015352, [1937gcse.bookQ....B].
238. A. Blaauw, *FK3 or GC – a difficult choice half a century ago* Acta Historica Astronomiae, 3, 114–122 (1998), [1998AcHA....3..114B].
239. M.F. Bode, T.J O'Brien & M. Simpson, *Echoes of an explosive past: Solving the mystery of the first superluminal source*, The Astrophysical Journal, 600, L63–L66 (2004), [2004ApJ...600L..63B].
240. P. Couderc, *Les auréoles lumineuses des novae*, Annales d'Astrophysique, 2, 271–302 (1939), [1939AnAp....2..271C].
241. J.J.L. Duyvendak, *Further data bearing on the identification of the Crab Nebula with the supernova of 1054 A.D.: Part I. The ancient Oriental chronicles*, Publications of the Astronomical Society of the Pacific, 54, 91–94 (1942), [1942PASP...54...]
242. www.nasa.gov/multimedia/imagegallery/image_feature_1604.html. NASA, ESA, J. Hester, A. Loll (ASU).
243. *Honderd jaar Leidse Sterrewacht*, Sterrewacht Leiden (1965).
244. https://commons.wikimedia.org/wiki/File:JJL_Duyvendak.jpg.
245. NOAO/AURA/NSF. www.noao.edu/image_gallery/html/im0490.html.
246. *Leidse hoogleraren vanaf 1575*, hoogleraren.leidenuniv.nl/search?keyword=Oort;docsPerPage=1;startDoc=4.

247. P. Monderaa, *Jan Julius Lodewijk Duyvendak*, Jaarboek van de Maatschappij der Nederlandse Letterkunde, 1956, 68–81. www.dbnl.org/tekst/_jaa003195601_01/_jaa003195601_01_0009.php.
248. D. ter Haar, *On the origin of smoke particles in the interstellar gas*, Bulletin of the Astronomical Institutes of the Netherlands, 10, 1–8 (1943) & Astrophysical Journal, 100, 288–299 (1944), [1943BAN....10....1T0] and [1944ApJ...100..288T]
249. H.A. Kramers & D. ter Haar, Bulletin of the Astronomical Institutes of the Netherlands, 10, 137–146 (1946), [1946BAN....10..137K].
250. https://commons.wikimedia.org/wiki/File:Kramers_1928.jpg.
251. W.T. Sullivan, *Cosmic Noise*, Cambridge University Press, ISBN 978-0-521-76524-4 (2009).
252. W.T. Sullivan, online archive, www.nrao.edu/archives/Sullivan/sullivan.shtml.
253. H. van Woerden & R.G. Strom, Journal of Astronomical History and Heritage, 9, 3–20 (2006), [2006JAHH....9....3V]
254. C.G. Jansky, Proceedings of the Institute for Radio Engineers, 20, 1920–1932 (1931), [2005ASPC..345....3J].
255. C.G. Jansky, *Electrical phenomena that apparently are of interstellar origin*, Popular Astronomy, 41, 548–555 (1933), [1933PA.....41..548J].
256. C.G. Jansky, *Radio waves from outside the Solar System*, Nature, 132, 66 (1933), [1933Natur.132...66J].
257. https://commons.wikimedia.org/wiki/File:Grote_Antenna_Wheaton.gif. It appeared in [259].
258. G. Reber, *Cosmic Static*, Astrophysical Journal, 91, 621–624 (1940), [1940ApJ....91..621R].
259. G. Reber, *Cosmic Static*, Astrophysical Journal, 100, 279–287 (1944), [1944ApJ...100..279R].
260. H.C. van de Hulst, *Roaming through astrophysics*, Annual Review of Astronomy & Astrophysics, 36, 1–16. (1998), [1998ARA&A..36....1V].
261. C.J. Bakker & H.C. van de Hulst, *Radiogolven uit het wereldruim*, Nederlands Tijdschrift voor Natuurkunde, 11, 201–221 (1944).
262. D. Hartmann & W.B. Burton, *Atlas of Galactic Neutral Hydrogen*, Cambridge University Press, ISBN-13: 978-0-521-47111-4 (1997), paperback 978-0-521-28312-0 (2012).

Chapter 10. Opening up New Avenues

263. L.M. Netzer, *How to tell Toledo from the night sky*, St. Martin's Press (2014).
264. The Adams papers, Huntington Library, San Marino, California.
265. *Annual reports*, Publications of the Astronomical Society of the Pacific, 54, 59–63 (1942) & 55, 56–61 (1943), [1942PASP...54...59.] and [1943PASP...55...56].
266. *The presentation of the Bruce Gold Medal for the year 1942 to Dr. Jan H. Oort*, Publications of the Astronomical Society of the Pacific, 58, 229–232 (1946), [1946PASP...58..229.].
267. A.E. Kersten, *Een organisatie van en voor onderzoekers: De Nederlandse Organisatie voor Zuiver-Wetenschappelijk Onderzoek (Z.W.O.), 1947–1988*, Van Gorcum (1996), ISBN-13: 978-9-023-23051-9.
268. J.H. Bannier, *Z.W.O., The Netherlands Science Foundation*, Science 113, 197–201 (1951).
269. M.C. Minneart & J.H. Bannier, *Equivalent breadth of some Fraunhofer lines in the region between 10030 and 10540 Å*, Zeitschrift für Astrophysik, 11, 392–393 (1936), [1936ZA.....11..392M].
270. M. Schmidt, *Exploring the Universe*, Annual Review of Astronomy & Astrophysics, 53, 1–14 (2015), [2015ARA&A..53....1S].
271. G. van Herk, *Absolute declinations of 1493 stars from azimuth observations*, Annalen van de Sterrewacht te Leiden, 18, E1–E58 (1957), [1957AnLei..18E...1V].Herk
272. M. Perryman, *The history of astrometry*, The European Physical Journal H, 37, 745–792 (2012), [2012EPJH...37..745P].

273. Erik Høg, *Astrometric accuracy during the past 2000 years*, Astro-ph, 28+8 pages (2017), [2017arXiv170701020H].
274. R.A. Lyttleton, The Observatory, 88, 9 (1968), [1968Obs....88....6.].
275. R.S. le Poole & P. Katgert, *The major-axis distribution of long-period comets*, The Observatory, 88, 164–166 (1968), [1968Obs....88..164L].
276. sci.esa.int/science-e-media/img/ee/SolarSystem_3frames_annotated_1280.jpg.
277. G. Vernmeer, *De macht van de tikmachine. Over bureaucratisering van de universiteit*, in: *Een universiteit in de twintigste eeuw*, red. K. van Berkel & F.R.H. Smit, Rijksuniversiteit Groningen (1999).
278. Klaas van Berkel, private communication.
279. F. Hoeneveld, *Een vinger in de Amerikaanse pap: fundamenteel fysisch en militair onderzoek in Nederland tijdens de vroege Koude Oorlog*, Ph.D. Thesis, University of Utrecht (2018).
280. R.G. Strom, *Ir A.H. de Voogt: life and career of a radio pioneer*, Astronomische Nachrichten, 328, 443–446 (2007), [2007AN....328..443S].
281. CAMRAS, the A.C. Muller Radio Astronomy Station, operates the Dwingeloo Telescope for amateur use. Courtesy Ard Hartsuijker.
282. www.nrao.edu/archives/Sullivan/sullivan_transcript_oort_1971.shtml.
283. W.N. Christiansen & J.V. Hindman, *21 cm. line radiation from Galactic hydrogen*, The Observatory, 72, 149–151 (1952), [1952Obs....72..149C].
284. W.N. Christiansen & J.V. Hindman, *A preliminary survey of 1420 Mc/s. Line emission from Galactic hydrogen*, Australian Journal of Scientific Research A, 5, 437–455 (1952), [1952AuSRA...5..437C].
285. W.T.. Sullivan, *Classics in radio astronomy*, Reidel, ISBN-13: 978-9-400-97754-9 (1982), ia600205.us.archive.archive.org/12/10.1007-978-94-009-7752-5/10.1007-978-94-items/springer_009-7752-5.pdf.
286. H.I. Ewen & E.M. Purcell, *Observation of a line in the Galactic radio spectrum: Radiation from Galactic hydrogen at 1,420 Mc./sec.*, Nature, 168, 356–358 (1951), [1951Natur.168..356E].
287. J.G. Bolton & K.C, Westfold, *Galactic radiation at radio frequencies. I. 100 Mc/s. Survey*, Australian Journal of Scientific Research A, 3, 19–33 (1950), [1950AuSRA...3...19B].

Chapter 11. Spiral Structure, Interstellar Medium and the Crab

288. krollermuller.nl/nl/page/15?q=keywords:John+Rädecker#filters.
289. G. Westerhout, *The start of 21-cm line research: The early Dutch years*, In: Seeing through the dust: The detection of HI and the exploration of the ISM in Galaxies, ASP Conference Proceedings, 276. 2002., 27–33 (2002), [2002ASPC..276...27W].
290. Nobelprize.org. [www.nobelprize.org/nomination/archive/search_people.php].
291. K.K. Kwee, C.A. Muller & G. Westerhout, *The rotation of the inner parts of the Galactic System*, Bulletin of the Astronomical Institutes of the Netherlands, 12, 211–222 (1954), [1954BAN....12..211K] .
292. *Galactic structure determined from 21-cm observations*, Bulletin of the Astronomical Institutes of the Netherlands, 13, 151–273 (1957), [1957BAN....13..151M], [..196O], [..201W], [..247S] and [..269R].
293. W.W. Morgan, A.E. Whitford, & A.D. Code, *Studies in Galactic structure. I. A preliminary determination of the space distribution of the blue giants*, Astrophysical Journal, 118, 318–322 (1953), [1953ApJ...118..318M].
294. O.J. Gingerich, *The discovery of the spiral arms of the Milky Way*, in: IAU Symposium 106, H. van Woerden, R.J. Allen, & W.B. Burton (eds.), Reidel, 59–70 (1985), [1985IAUS..106...59G].

295. https://commons.wikimedia.org/wiki/File:HaleTelescope-MountPalomar.jpg.
296. R.W. Porter, *Photographic giants of Palomar*, Westernlore Press, ASIN: B003127S18 (1952). For a description of Palomar's history, see e.g. www.astro.caltech.edu/palomar/about/history.html.
297. M.R. Madwar, *Preliminary notice on the total solar eclipse of 25th February 1952*, Helwan Institute of Astronomy and Geophysics Bulletins, 42, 3–13 (1952), [1952HelOB..42....3M].
298. M.R. Madwar, *Solar work and plans in Egypt*, Vistas in Astronomy, 1, 732–737 (1955), [1955VA......1..732M].
299. M.R. Madwar, *The scientific work of the Helwan Observatory*, Soviet Astronomy, 2, 256–259 (1958), [1958SvA.....2..256F].
300. www.noao.edu/staff/rector/digital/rosette/index.html. Credit line: National Optical Astronomy Observatory/Association of Universities for Research in Astronomy/National Science Foundation.
301. G. Westerhout, *Search for polarization of the Crab nebula and Cassiopeia A at 22 cm wavelength*, Bulletin of the Astronomical Institutes of the Netherlands, 12, 309–311 (1956), [1956BAN....12..309W].
302. W. Baade, *The polarization of the Crab nebula on plates taken with the 200-inch telescope*, Bulletin of the Astronomical Institutes of the Netherlands, 12, 312 (1956), [1956BAN....12..312B].
303. Th. Walraven, *Photo-electric observations of the polarization and surface brightness of the Crab nebula made at the Observations de Haute Provence*, Bulletin of the Astronomical Institutes of the Netherlands, 13, 293–300 (1957), [1957BAN....13..293W].
304. L. Woltjer, *The polarization and intensity distribution in the Crab nebula derived from plates taken with the 200-inch telescope by Dr. W. Baade*, Bulletin of the Astronomical Institutes of the Netherlands, 13, 301–311 (1957), [1957BAN....13..301W].

Chapter 12. Dwingeloo, ESO, the Vatican and the Cross-Antenna

305. H. van Woerden & R.G. Strom, *Dwingeloo – the golden radio telescope*, Astronomische Nachrichten, 328, 376–387 (2007), [2007AN....328..376V].
306. leiden.courant.nu/issue/LD/1956-04-17/edition/0/page/7.
307. lintjes.nl/onderscheidingen/de-orde-van-de-nederlandse-leeuw, or the Wikipedia site en.wikipedia.org/wiki/Order_of_the_Netherlands_Lion
308. H. Kleibrink, *De Bouw van een Radiotelescoop* (1956). www.astron.nl/about-astron/history/footage/historic-footage or www.youtube.com/watch?v=3SvkWdm-KOw.
309. A. Blaauw, *ESO's early history: The European Southern Observatory from concept to reality*, ESO (1991), ISBN 3-9-235-2440-4.
310. www.eso.org/public/images/eso_statement_26-01-54/ and www.eso.org/public/announcements/ann15013/.
311. P.C. van der Kruit, *Five-and-a-half years in ESO Council*, https://www.astro.rug.nl/~vdkruit/President.html (2006).
312. CERN Document Server: cds.cern.ch/record/41046.
313. R.M. West, *André B. Muller (25.9.1918-1.4.2006)*, The Messenger, 124, 37 (2006), [2006Msngr.124...37W].
314. Th. & J.H. Walraven, *A new photo-electric method of classification of luminosity and spectral type for O-and B-type stars*, Bulletin of the Astronomical Institutes of the Netherlands, 15, 67–83 (1960). [1960BAN....15...67W].
315. www.eso.org/public/images/eso0912d/.
316. J. Lub, *The Dutch telescope on La Silla*, The Messenger, 19, 1–4 (1979), [1979Msngr..19....1L].

317. D.E. Osterbrock, *Walter Baade – A Life in Astrophysics*, Princeton University Press, ISBN: 9-780-69104-9366 (2001).
318. W. Baade, *The resolution of Messier 32, NGC 205, and the central region of the Andromeda Nebula*, Astrophysical Journal, 100, 137–146 (1944), [1944ApJ...100..137B].
319. P.C. van der Kruit & G. Gilmore, *Stellar populations*, IAU Symposium 164, the Hague, the Netherlands, August 15–19, 1994, Kluwer (1995), [1995IAUS..164.....V].
320. E.M. Burbidge, G.R. Burbidge, W.A. Fowler & F. Hoyle, B^2FH, Reviews of Modern Physics, 29, 547–650 (1957), www.pmf.unizg.hr_download/repository/burbidge_RMP_29_547_1957.pdf.
321. O.J. Eggen, D. Lynden-Bell, & A.R. Sandage, *Evidence from the motions of old stars that the Galaxy collapsed*, Astrophysical Journal, 136, 748–766. [1962ApJ...136..748E].
322. A. Hewish & M. Ryle, *The Cambridge radio telescope*, Memoirs of the Royal Astronomical Society, 67, 97–105 (1955), [1955MmRAS..67...97H].
323. M. Ryle & P.A.G. Scheuer, *The spatial distribution and the nature of radio stars*, Proceedings of the Royal Society of London. Series A, Mathematical and Physical Sciences, 230, 448–462 (1955), [1955RSPSA.230..448R].
324. G. Westerhout, *The early history of radio astronomy*, Annals of the New York Academy of Sciences, 198, 211–218. [1972NYASA.198..211W]
325. E. Raimond & R. Genee, Astrophysics and Space Science Library No. 208, Kluwer, ISBN 0-792-34150-3 (1996), [1996ASSL..207.....R].
326. www.nrao.edu/archives/Sullivan/sullivan_transcript_seeger_1975.shtml.
327. R.H Frater, W.M. Goss, & H. Wendt, *Four Pillars of Radio Astronomy: Mills, Christiansen, Wild, Bracewell*, Springer, ISBN 978-3-319-65598-7 (2017).
328. M. Ryle & A.C. Neville, *A radio survey of the North Polar region with a 4.5 minute of arc pencil-beam system*, Monthly Notices of the Royal Astronomical Society, 125, 39–56 (1962), [1962MNRAS.125...39R].
329. P.A.G. Scheuer, *The development of aperture synthesis at Cambridge*, im: *The early years of radio astronomy – Reflections fifty years after Jansky's discovery*, W.T. Sullivan (ed.), Cambridge University Press, 249–266 (1984), [1984eyra.book..249S].
330. *Large Radio-Telescopes*, Proceedings of the OECD Symposium on Large Antennae for Radioastronomy held in Paris, 12–14 December, 1961. Organisation for Economic Co-operation and Development (1961), [1961lrt..conf......].
331. W. Baade & R. Minkowski, *Identification of the radio sources in Cassiopeia, Cygnus A, and Puppis A*, Astrophysical Journal, 119, 206–214 (1954), [1954ApJ...119..206B].
332. H.J. Habing, *Gijsbert van Herk (1907–1999)*, (incorrectly attributed to W.B. Burton), Bulletin of the American Astronomical Society, 32, 1689–1690 (2000). [2000BAAS...32.1689B], aas.org/obituaries/gijsbert-van-herk-1907-1999.
333. I.R. King, *The structure of star clusters. III. Some simple dynamical models*, Astronomical Journal, 71, 64–75 (1966) and references therein. [1966AJ.....71...64K].
334. E.R. Hill, *The component of the Galactic gravitational field perpendicular to the Galactic plane, K_z*, Bulletin of the Astronomical Institutes of the Netherlands, 15, 1–44 (1960). [1960BAN....15....1H].
335. A. Ollongren, *Three-dimensional Galactic stellar orbits*, Bulletin of the Astronomical Institutes of the Netherlands, 16, 241–296 (1962), [1962BAN....16..241O], also [1962AnLun..18....5O].
336. https://en.wikipedia.org/wiki/Confocal_conic_sections#/media/File:Ell-hyp-konfokal.svg.
337. G. Contopoulos, *A third integral of motion in a Galaxy*, Zeitschrift für Astrophysik, 49, 273–291 (1960), [1960ZA.....49..273C].
338. H.C. van de Hulst, *The third integral of motion for low-velocity stars*, Bulletin of the Astronomical Institutes of the Netherlands, 16, 235–239 (1962). [1962BAN....16..235V].
339. V.C. Rubin, *Structure and evolution of the Galactic System*, Physics Today, 13, 32–35 (1960).
340. D.C. Mortion, *The 1960 summer course at Nyenrode Castle*, in: the Symposium 'Remembering Adriaan Blaauw's 100th birthday' (2014), www.astro.rug.nl/~khan/blaauw100/presentations/D1-Morton.pdf.

Chapter 13. Galactic Structure and the Dwingeloo Telescope

341. Harry Mulisch, *De ontdekking van de hemel*, ISBN 978-9-0234-2822-0 (1992). English translation ISBN 0-6-708-5668-1.
342. www.quora.com/How-is-Carl-Sagans-famous-snowflake-quote-true.
343. Bill Bryson*A short history of nearly everything*, ISBN 0-767-90817-1 (2003).
344. Huib J. Zuidervaart. *Telescopes from Leiden Observatory and other collections. 1656–1859. A Descriptive Catalogue*, Leiden 2007, Museum Boerhaave, p.115. www.dwc.knaw.nl/pub/zuidervaart/2007_Telescopes_from_Leiden_Observatory.pdf.
345. https://commons.wikimedia.org/wiki/File:LeidenObservatory1961c.jpg.
346. R. Minkowski, *Problems of extragalactic spectroscopy*, Annales d'Astrophysique, 23, 385–396 (1960), [1960AnAp...23..385M].
347. R. Minkowski, *Internal dispersion of velocities in other galaxies*, in: IAU Symposium 15, G.C. McVittie (ed.), Macmillan Press, 112–118 (1962), [1962IAUS...15..112M].
348. P.C. van der Kruit, & K.C, Freeman, *Galaxy Disks*, Annual Review of Astronomy and Astrophysics, 49, 301–371 (2011), section 9, [2011ARA&A..49..301V].
349. G. Illingworth, & P.L. Schechter, *Velocity and velocity dispersion profiles in NGC 3115*, Astrophysical Journal, 481–496 (1982), [1982ApJ...256..481I].
350. V. Rubin, *An interesting voyage*, Annual Review of Astronomy and Astrophysics, 49, 1–28 (2011), [2011ARA&A..49....1R].
351. H.C. van de Hulst, *Observing the Galactic magnetic field*, Annual Review of Astronomy & Astrophysics, 5, 167–182 (1967), [1967ARA&A...5..167V]
352. H.C. van de Hulst, E. Raimond & H. van Woerden, *Rotation and density distribution of the Andromeda nebula derived from observations of the 21-cm line*, Bulletin of the Astronomical Institutes of the Netherlands, 14, 1–16 (1957), [1957BAN....14....1V].
353. M. Schmidt, *The distribution of mass in M31*, Bulletin of the Astronomical Institutes of the Netherlands, 14, 17–19 (1957), [1957BAN....14...17S].
354. A.J, Kalnajs, *Spiral structure viewed as a density wave*, Proceedings of the Astronomical Society of Australia, 2, 174–177 (1973), [1973PASAu...2..174K].
355. P.C. van der Kruit, *The peculiar spiral galaxy NGC 3310: II. The velocity field of the ionized gas*, Astronomy & Astrophysics, 49, 161–171 (1976), [1976A&A....49..161V].
356. W.B. Burton, *Preliminary discussion of 21-cm observations of the Sagittarius arm and the systematic motion of the gas near its edge*, Bulletin of the Astronomical Institutes of the Netherlands, 18, 247–255 (1966), [1966BAN....18..247B].
357. W.W. Shane & G.P. Bieger-Smith, *The galactic rotation curve derived from observations of neutral hydrogen*, Bulletin of the Astronomical Institutes of the Netherlands, 18, 263–292 (1966), [1966BAN....18..263S].
358. W.B. Burton, *Galactic structure derived from neutral hydrogen observations using kinematic models based on the density-wave theory*, Astronomy & Astrophysics, 10, 76–96 (1971), [1971A&A....10...76B].
359. W.B. Burton & W.W. Shane, *Neutral hydrogen in the Sagittarius and Scutum spiral arms*, in: IAU Symposium 38, W. Wilhelm Becker & G.I. Contopoulos (eds.), Reidel, 397–414 (1970), [1970IAUS...38..397B].
360. H. van Woerden, B.P. Wakker, U. Schwarz, & K. de Boer, *High Velocity Clouds*, Astrophysics and Space Science Library No. 312, Kluwer, ISBN 978-1-402-02578-5, [2004ASSL..312.....V].
361. G.P. Smith, *A peculiar feature at* xx, xx, Bulletin of the Astronomical Institutes of the Netherlands, 17, 203–208 (1963), [1963BAN....17..203S].
362. F.J. Lockman, R.A. Benjamin, A.J. Heroux, & G.I. Langston, *The Smith Cloud: A high-velocity cloud colliding with the Milky Way*, The Astrophysical Journal, 679, article L21, 2008, [2008ApJ...679L..21L].

363. *Toch nog een beetje botsen met de wolk van Gail Smith*, de Volkskrant, January 19, 2008, page 5.
364. W. Otterspeer, *The Bastion of Freedom: Leiden University in the present and past*, Leiden, ISBN 9-789-08728-0307 (2008).
365. D.M. Baneke, *De vette jaren: de Commissie-Casimir en het Nederlandse wetenschapsbeleid 1957–1970*, Studium, 5, 110–127 (2012). www.gewina-studium.nl/articles/abstract/10.18352/studium.8195/.
366. Centraal Bureau voor de Statistiek. www.cbs.nl/en-gb.

Chapter 14. Westerbork, Retirement and Beyond

367. *Fifty years of Westerbork radio Observatory: A continuing journey to discoveries and innovations*, R. Strom, A. van Ardenne & S. Torchinsky (eds.), ASTRON, ISBN 978-9-080-54349-0 (2018).
368. R.G. Strom, Journal of Astronomical History and Heritage, 21, 95–96 (2018), [2018JAHH...21...95R].
369. J.W.M. Baars & B.G. Hooghoudt, *The Synthesis Radio Telescope at Westerbork. General lay-out and mechanical aspects*, Astronomy and Astrophysics, 31, 323–331 (1974) [1974A&A....31..323B]
370. *Study Week on Nuclei of Galaxies*. D.J.K. O'Connell (ed.), Pontificiae Academiae Scientiarum Scripta Varia (1971), [1971swng.conf.....O].
371. P.C. van der Kruit, *Observations of core sources in Seyfert and normal galaxies with the Westerbork synthesis radio telescope at 1415 MHz*, Astronomy and Astrophysics, 15, 110–122 (1971). [1971A&A....15..110V].
372. From an undated magazine cutting of Paris Match Europe.
373. D.S. Mathewson, P.C. van der Kruit, & W.N. Brouw, *A high.resolution radio continuum survey of M51 and NGC 5195 at 1415 MHz*, Astronomy and Astrophysics, 17, 468–468 (1972), [1972A&A....17..468M].
374. G.K. Miley, G.C. Perola, P.C. van der Kruit, & H. van der Laan, *Active galaxies with radio trails in clusters*, Nature 237, 269–272 (1972). [1972Natur.237..269M].
375. A.S. Wilson, Y. Yang, & G. Cecil, *Chandra observations and the nature of the anomalous arms of NGC 4258 (M106)* Astrophysical Journal, 560, 689–697 (2001). [2001ApJ...560..689W].
376. arxiv.org/archive/astro-ph/Astrophysics.
377. P.C van der Kruit (2005), *Citation analysis and the NASA Astrophysics Data System (ADS)*, www.astro.rug.nl/~vdkruit/jea3/homepage/ads.pdf.
378. P. Kamphuis & P.C. van der Kruit, *Citations and impact of Dutch astronomy*, arxiv.org/abs/1011.5311 (2010).
379. H.J.T. Buist & P.C. van der Kruit, *Report on the citation analysis of the Netherlands School for Astronomy (NOVA)*, www.astro.rug.nl/~vdkruit/jea3/homepage/NOVAreportII.pdf (2016).

Chapter 15. The Last Horizon

380. M. Ryle, *The counts of radio sources*, Annual Review of Astronomy and Astrophysics, 6, 249–266 (1968). [1968ARA&A...6..249R].
381. Newspaper cutting from de 'Telegraaf', January 11, 1986, p. T21.
382. 'Het Parool', July 17, 1987, p. 9.
383. *Memory of the Netherlands*, Royal National Library, translation Frits Ham. www.geheugenvannederland.nl/en/geheugen/pages/collectie/Frederik+van+Eeden/De+water+lelie.

Appendices

384. www.dwc.knaw.nl/toegangen/digital-library-knaw/?pagetype=publist&search_author=PE00001201
385. https://ssd.jpl.nasa.gov/sbdb.cgi.
386. genealogy.math.ndsu.nodak.edu.; Kapteyn is number 112114 and Oort number 112244.
387. https://nl.wikipedia.org/wiki/Lijst_van_muurformules_in_Leiden.
388. https://commons.wikimedia.org/wiki/Category:Wall_formulas?uselang=nl#mw-sub categories.
389. 11fountains.nl/en/.

Index

Symbols

A

P. C. van der Kruit, *Jan Hendrik Oort*, Astrophysics and Space Science Library 459, https://doi.org/10.1007/978-3-030-17801-7

C

E

L

M

Q

R

MIX
Papier aus verantwortungsvollen Quellen
Paper from responsible sources
FSC® C105338

If you have any concerns about our products,
you can contact us on
ProductSafety@springernature.com

In case Publisher is established outside the EU,
the EU authorized representative is:
Springer Nature Customer Service Center GmbH
Europaplatz 3, 69115 Heidelberg, Germany

Printed by Libri Plureos GmbH
in Hamburg, Germany